Principles of

RADIOISOTOPE METHODOLOGY

Third Edition

Principles of RADIOISOTOPE METHODOLOGY

Third Edition

by

GRAFTON D. CHASE, Ph.D.
Professor of Chemistry
Philadelphia College of Pharmacy and Science

and

JOSEPH L. RABINOWITZ, Ph.D.
School of Dental Medicine
University of Pennsylvania
Philadelphia

Printed in the United States of America

Standard Book Number 8087-0308-0

Library of Congress Catalog Card Number 66-19903

Second Printing 1968
Third Printing 1969
Fourth Printing 1970

PREFACE TO THE THIRD EDITION

The job of remodelling a textbook is a neverending one as needs and conditions change and as new methods and procedures are developed. Time was when a textbook need only include topics up to twenty or thirty years prior to its publication. Progress has altered this situation. Today many topics twenty or thirty years old are obsolete or obsolescent. In addition, most fields of science have expanded into so many new areas that it is difficult to prepare a comprehensive text covering all these new facets of knowledge and, at the same time, to maintain a current coverage of the established ones.

In preparing the third edition an effort has been made to increase the usefulness of the book through the inclusion of new chapters and the expansion of existing ones. Many helpful comments received from those who use the book, from reviewers, and from the authors' associates have been embodied in a variety of improvements.

More than twenty new experimental procedures have been included. Among these are experiments in gas chromatography, absolute alpha counting, beta calibration techniques and tracer studies in biological systems. Radiation biology is the subject of a new chapter included to provide a more comprehensive coverage of the biological sciences. Chapter 15—Nuclear Medicine is also new with this edition and provides detailed instructions for thirteen diagnostic procedures utilizing radioisotopes in current use at the Philadelphia Veterans Administration Hospital. Discussion of the physical aspects of the atom, including atomic and nuclear structure, has been doubled in scope. The greater variety of experiments and topics discussed in this edition should provide more flexibility in the use of the book as a teaching text for students in diverse disciplines.

The authors take pleasure in acknowledging the help and encouragement of Dr. Arthur Osol, President of the Philadelphia College of Pharmacy and Science, Dr. Louis A. Reber, Chairman of the Department of Chemistry at the Philadelphia College of Pharmacy and Science, Dr. Arlyne T. Shockman, Chief of Radioisotope Service, Veterans Administration Hospital, Philadelphia, Dr. William H. Blahd, Chief of Radioisotope Service, Veterans Administration Hospital, Los Angeles, Dr. William H. Parsons, Professor of Mathematics and Physics, Massachusetts College of Pharmacy, and Dr. Jacob Nevyas, Professor of Chemistry, Pennsylvania College of Optometry. The authors also wish to acknowledge the assistance of Messrs. Benjamin Weiss, Clayton Delamater, Robert Garber, Terry L. Benney and Robert E. David, Mrs. Ellen Gilligan and Miss JoAnn L. Erwetowski in various phases of manuscript preparation.

PREFACE TO THE FIRST EDITION (Abstracted)

Only since the end of World War II have radioisotopes become available in sufficient quantities for use in general fields of research and in the control of industrial processes. The last few years in particular have witnessed a tremendous increase in the number and scope of the applications of radioisotopes, principally as a valuable tool in established fields of scientific investigation. Research workers in chemistry, biology, pharmacology, physiology, biochemistry and many other areas of science began to look into the possibilities of radioisotopes in their work and to seek training in the handling and use of radioisotopes and the instruments associated with such use. It was to supply this training that courses in *Radioisotope Methodology* were established at the Philadelphia College of Pharmacy and Science in 1953 and at the U.S. Veterans Administration Hospital in Philadelphia in 1954. These courses embodied not only didactic instruction but laboratory work with radioisotopes and measuring instruments as well. The students in these courses were originally provided with mimeographed notes on basic theory and procedures for carrying out a list of experiments. The success of these courses inspired the expansion of a portion of these notes into the present volume.

This book is essentially a textbook on methods of procedure. It is not meant as a complete treatise on the subject and is not written for the specialist in physics, but rather as an introductory guide for the technician, the advanced medical student and the researcher who may find the use of radioisotopes valuable in his work. The fundamental theory of each individual discipline when encountered, e.g., chemistry, biology, etc., is covered only sufficiently to satisfy the need for an understanding of the experiments involved. Each chapter contains a section on basic theory and a laboratory section with recommended experiments. Some are meant for the beginner in radioisotope techniques, while others are designed to present a challenge to the more advanced laboratory worker or outstanding student. The instructor should use his own judgment in the selection of experiments, choosing them according to the background and training of his students.

The graphs and drawings presented in this book have been simplified to permit a clearer understanding of the intended concept. All electrical, electronic and isotopic terminology, as well as all designations of radioisotope standards, have been made to conform with current practice in the United States. Also, the instruments described for measuring radiations are those currently available in the United States. It is to be expected that with changes and improvements in available instruments, some changes in instruction will become necessary, but that the principles presented will remain the same.

G.D.C.
J.L.R.

Philadelphia, July, 1966

CONTENTS

Chapter

Chapter

Chapter

Chapter

PART III – APPLICATIONS OF RADIOISOTOPES

Chapter

PART I

INTRODUCTION TO THE RADIOISOTOPE

CHAPTER 1
Atomic and Molecular Structure

LECTURE OUTLINE AND STUDY GUIDE

I. STRUCTURE OF MATTER
   A. Definitions
   B. Molecules and Atoms
   C. Nuclear Numerics—A, Z and N

II. PARTICLES
   A. Classification—leptons, mesons, nucleons and hyperons
   B. Properties of electrons, protons and neutrons

III. RADIATION FROM RADIOACTIVE NUCLEI
   A. Types of Radiation
   B. Properties of Radiation

IV. HISTORICAL SURVEY
   A. Radioactivity and Radiation
   B. Atomic Models—Thomson, Rutherford, Bohr and Schrödinger
   C. Investigations of the Nucleus—Nuclear composition, models and reactions

V. MECHANICS
   A. The Quantum
   B. Relativity

VI. THEORY OF THE NUCLEAR ATOM
   A. Defects of the Thomson Model
   B. Scattering Experiments of Geiger and Marsden
   C. Rutherford Interpretation of Scattering Experiments—Concentrated Positive Charge

VII. ATOMIC STRUCTURE
   A. Atomic Dimensions and Properties
   B. Bohr Model
      1. Assumptions of the theory
      2. Calculations of orbit radii
      3. Calculation of electron velocity
      4. Calculation of frequency of revolution

VIII. ATOMIC SPECTRA
   A. Energy Relationships of Orbital Electrons
   B. Reduced Mass
   C. Sommerfeld Hypothesis and Elliptic Orbits
   D. Magnetic Quantum Number
   E. Spin
   F. Pauli Exclusion Principle

IX. X-RAY SPECTRA
   A. Moseley's Law
   B. Screening Factor
   C. Nomenclature and Calculation of Characteristic Lines

X. PARTICLES AND WAVES
   A. Assumptions of de Broglie Equation
   B. Derivation of de Broglie Wavelength
   C. Applications of de Broglie Theory to the Atom

XI. WAVE MECHANICS
   A. Wave Motion
      1. Traveling waves
      2. Standing waves
      3. Boundary conditions
   B. Schrödinger Equation
      1. Wave function
      2. Eigenvalues
      3. Eigenfunctions

STRUCTURE OF MATTER

Nearly two thousand years ago Lucretius wrote a treatise on the nature of things. He called it "De Rerum Natura" and in it proposed that if any matter found in nature were subdivided into smaller and smaller particles, one would ultimately obtain a particle incapable of further subdivision. These indivisible particles he called corpuscles or atoms. We now know how close Lucretius came to the truth.

Matter is anything which occupies space and comprises all substances of which the universe is composed. The term *substance* signifies a type of homogeneous matter possessing a definite chemical composition (as opposed to a mixture or heterogeneous matter). The chemical elements and pure compounds are substances. Each is characterized by a specific melting point, boiling point and by other specific physical and chemical properties.

When matter is subdivided, as suggested by Lucretius, the smallest "particle" possessing all the original physical and chemical properties is the *molecule*. Molecules of elements or of compounds may be separated into components called *atoms*. Molecules of an element (e.g. O_2 or H_2) contain but a single type of atom whereas molecules of compounds (e.g. H_2O) are composed of different types of atoms. We have seen, too, that the atom is no longer considered indivisible but consists of a positively charged *nucleus*, in which is found almost the complete mass of the atom, and which in turn is surrounded by a swarm of *orbital electrons*. In a neutral atom the number of orbital electrons is equal to the positive charge on the nucleus. These electrons are responsible for the chemical properties associated with the atom. When the number or arrangement of these electrons changes, the chemical properties of the atom are changed.

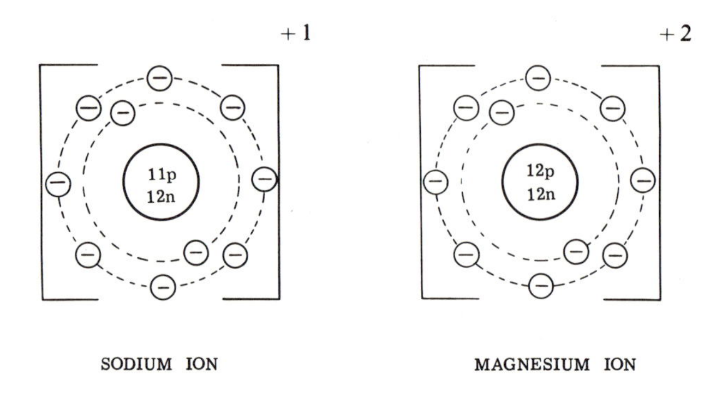

Figure 1.1—Sodium and magnesium ions are structurally similar except for the number of protons in the nucleus.

The identity of an atom is determined by the charge of the nucleus. It is this charge which distinguishes sodium from magnesium (see figure 1.1) or any of the other elements. This charge is equal to the number of protons in the nucleus and is known as the *atomic number* Z. The nucleus is also found to contain neutrons, the number of which is the *neutron number* N. The *mass number* A is equal to the sum of the protons and neutrons in the nucleus, thus

$$A = Z + N \tag{1.1}$$

This relationship was first reported by Heisenberg and Majorana in 1932.

PARTICLES

Particles have been divided into four groups according to their masses. In order of increasing rest mass the groups are:

1. *Photon*—The photon is the only member of this group. The symbol is γ. A photon has zero rest mass.

2. *Leptons*—A particle with a mass less than that of the proton. All leptons have a spin of ½.

   a. *Neutrinos*—Two distinct types of neutrinos are known, each with its own discrete antiparticle. They have been given the symbols ν_e and ν_μ.

   b. *Electrons*—The negatron e^- and its antiparticle, the positron e^+.

   c. *Muons*—These particles were once called mu mesons but a muon is not a meson and is no longer classified as such. The symbol is μ^+ or μ^- depending upon the charge.

3. *Mesons*—The mass of a meson is greater than that of a lepton but less than that of a nucleon. Mesons have a spin of zero. Symbols of the known mesons are π^+, π^0, π^-, K^+, K^-, K^0, K_1, K_2 and η.

4. *Baryons*—These are heavy particles; the lightest is the proton. Known baryon spins are all ½.

   a. *Nucleons*—Particles found within the atomic nucleus are called nucleons. Included are the proton p and the neutron n.

   b. *Hyperons*—All hyperons have a mass greater than the neutron. Symbols of the known hyperons are Λ, Σ^+, Σ^-, Ξ^-, Ξ^0 and Ω^-.

Three particles seem to be the primary building blocks of atoms. These are the electron, the proton and the neutron. The name *electron* was proposed by G. J. Stoney in 1891 for the unit of electricity. Six years later the value of the charge-to-mass ratio (e/m) for the electron was published almost simultaneously by Wiechert, Kaufmann and J. J. Thomson. J. S. Townsend made the first measurement of e, the electronic charge. The electronic charge was again measured in 1913 by R. A. Millikan, by means of his famous oil-drop experiment. The value of the charge is 1.592×10^{-19} coulombs or 4.802×10^{-10} esu. The electron mass m_0 is 9.106×10^{-28} gram.

The mass of the electron is often used as a unit of mass. Thus the mass of the proton is 1836 m_0 and the mass of the neutron is 1837 m_0. Similarly, the electronic charge is often used as a unit of charge. Thus if the charge of the electron is taken as -1, then the charge of the proton is $+1$. The neutron, as the name suggests, is neutral, and hence has a charge of zero.

Although it is possible to express the mass and charge of a particle by an exact numerical value, this is not true in the case of particle size. The approximate size of electrons, protons and neutrons is of the order of 10^{-13} cm but the exact dimensions, if such could be measured, are a function of the energy (or velocity) of the particle. Frequently the de Broglie wavelength (Cf. p. 21) is used as a measure of the size of a particle.

The name *proton* (Gr; protos—first) was suggested by Rutherford in 1920 for the elementary unit of positive charge observed in positive-ray experiments. The proton has long been recognized as being identical with the nucleus of the hydrogen atom.

In 1920 Rutherford postulated the existence of the *neutron*. Neutrons had been observed but not identified by W. Bothe and H. Becker in 1930 when certain light elements, especially beryllium, were bombarded with alpha particles. Final identification of the neutron was accomplished by Chadwick in 1932.

The *neutrino* is an unusual particle. Postulated by W. Pauli in 1930 and its existence assumed by E. Fermi in 1934 in the development of his theory of beta decay, it was not until 1956 that Reines and Cowan reported the results of their experiments in which they obtained direct evidence for the existence of the neutrino. It has zero charge and apparently zero rest mass. These properties account for its evasive nature and for the fact that its existence was not demonstrated by direct means until long after this strange particle has been postulated to maintain the conservation laws. The important role of the neutrino in beta decay will be discussed in the chapter on radioactive decay.

RADIATION FROM RADIOACTIVE NUCLEI

Radioactive nuclei may emit different types of radiation. Among these the more important are alpha, beta and gamma.

Alpha radiation is radiation consisting of a flow of alpha particles. These particles are identical with helium nuclei; that is, they are helium atoms without the two orbital electrons. Thus they are compound particles, consisting of two protons and two neutrons. As an alpha particle loses energy it attracts electrons and ultimately becomes a neutral helium atom.

Beta radiation is a flow of electrons and may be either of two types, negative or positive. There are two types of electrons, the negative electron or *negatron*, and the positive electron or *positron*. These two particles are identical except for charge, which is -1 for the negatron and $+1$ for the positron. One is the *antiparticle* of the other. When emitted from radioactive nuclei they are called beta particles and are represented by the symbols β^- and β^+, respectively. Otherwise, they may be represented by the symbols e^- and e^+.

Gamma radiation is electromagnetic radiation and, consequently, is basically different from alpha and beta radiation, which are particulate. Gamma rays are more penetrating than either alpha or beta. Like all electromagnetic radiation, gamma radiation travels with a velocity c of 3.0×10^{10} cm/sec and exhibits properties of photons or quanta.

HISTORICAL SURVEY:

Radioactivity was observed as early as 1867 when Niepce de Saint-Victor noticed a fogging of silver chloride emulsions by uranium salts. He reported that blackening of the emulsions occurred even when they were separated from the uranium salts by thin sheets of paper. He did not recognize the true cause of the blackening (radioactivity) but attributed the effect to luminescence phenomena.

When Henri Becquerel repeated some of these experiments in 1896, he too believed that fluorescence was associated with the fogging of his photographic plates. Becquerel did not realize either that he had observed the result of a process of spontaneous transmutation, akin to one of the long-sought goals of the alchemist, and was surprised and puzzled to learn that photographic plates were still darkened by uranium salts even when the salts had received no previous "activation" with light. It was not until 1898, after Marie Curie had studied the radiation emitted by uranium, that a truly new phenomenon was recognized. The fogging of the photographic plates was now known to be caused by some sort of penetrating radiation similar to the x-rays of Roentgen. Produced not only by uranium and thorium salts, radiation was also observed from two new elements, polonium and radium, which had just been discovered by Marie Curie.

By 1899 Rutherford had demonstrated that the radiation emanating from uranium salts and from a variety of other salts and minerals was of two types, which he called *alpha* and *beta*. In 1900 a third type of radiation was observed by P. Curie and Villard; they called it *gamma*. It was only natural that much curiosity should be aroused concerning the origin of this radiation. To appreciate the imagination and foresight required of Rutherford and Soddy—who in 1903 proposed the *theory of radioactive disintegration* to explain its origin—one must realize that it was not until 1908 that Rutherford and Geiger determined the charge of the alpha particle and not until 1909 that Rutherford and Royds identified the alpha particle as a positively charged helium nucleus. It should be further realized that the *nuclear theory*, as taught even at the high school level today, was unknown until 1911 when Rutherford demonstrated by alpha-particle-scattering experiments that the positive charge of an atom must be concentrated at one point (the nucleus) rather than uniformly distributed throughout the volume of the atom as had been suggested by J. J. Thomson.

Rutherford's nuclear theory was the break-through which made possible a rapid series of developments in both atomic and nuclear theory. For example, in 1912, J. J. Thomson, using positive-ray analysis (a forerunner of mass spectrometry) to measure particle mass, showed that two types of neon existed, with masses of 20 and 22. The following year these two forms of neon, differing only in mass, were separated by F. W. Aston by fractional diffusion through clay pipe stems, thus confirming the work of Thomson. F. Soddy proposed the name *isotope* (Gr; equal place) to denote a different mass of the same element.

By 1913 Niels Bohr had proposed his theory of atomic structure which applied the principles of Planck's quantum theory (1900) to Rutherford's planetary model of the atom. Proof of the Bohr theory was provided by spectral data, the utility of which is based on the principle that the energy associated with a particular line in the spectrum of an element is equal to the energy lost by an electron in passing from one specific orbit (energy level) to another in the Bohr model of the atom. Yet, the spectra which provided evidence for validity of the Bohr atom ultimately pointed to its shortcomings. Certain hyper-fine structures of the spectra cannot be explained by the Bohr theory.

Bohr had dealt with the atom from a particulate point of view, in which he assumed that particles called electrons revolve about a particle called the nucleus. In 1923, when Count Louis de Broglie suggested that the dualism of the wave and particle functions of radiation (waves versus photons) might apply to material particles as well, Schrödinger quickly applied the de Broglie theory to the atom. According to de Broglie, every particle in motion has associated with it a wavelength. Based upon the wave function, the *wave mechanical theory* proposed by Schrödinger in 1926 differs considerably from the rather rigid mechanical model of Bohr but presents the most acceptable atomic model known today.

As our knowledge of the extranuclear structure of the atom was extended by persistent investigation, information about the nucleus, too, was gradually forthcoming. Thus Rutherford's investigations led to the very important observation of *artificial transmutation* first performed in 1919. In the years that followed many notable experiments were performed, many discoveries made, and many concepts proposed. Especially noteworthy are the contributions of L. Meitner, W. Heisenberg, F. W. Aston, P. A. M. Dirac, H. Geiger, W. Müller, R. J. Van de Graaff, C. D. Anderson, J. D. Cockcroft, E. T. S. Walton, J. Chadwick, E. O. Lawrence, M. S. Livingston, I. Curie, M. F. Joliet and W. Pauli. These researches culminated in the discovery of fission, which led to the first atomic pile, constructed by E. Fermi in 1942 in Chicago. This development not only made possible production of the nuclear bomb, but also made practical large-scale use of radioisotopes in research, medicine, industry, and in numerous other fields. Over 1300 species of atoms (nuclides) are presently known, and these may have an infinite variety of applications. It can be safely said that almost everything made today has at some point in its development or production been benefited by the use of isotopes.

MECHANICS, THE QUANTUM AND RELATIVITY

The motion of objects in space and the energy changes associated with this motion are described by means of mechanics. While recognizing the granularity or particulate nature of matter (i.e., the existence of atoms and molecules) Newtonian mechanics held that energy changes are smooth and continuous and that mass is independent of velocity. Although accurate, to within the limits of experimental error, at moderate velocities and for large objects, Newtonian mechanics does not explain phenomena involving velocities approaching that of light nor phenomena observed in the realm of atomic and nuclear dimensions.

Relativistic corrections must be made for matter traveling at high velocities. Mass increases with velocity and approaches infinity as the velocity approaches c. The relativistic mass correction is:

$$m = \frac{m_0}{\sqrt{1 - \beta^2}} \tag{1.2}$$

where $\beta = v/c$.

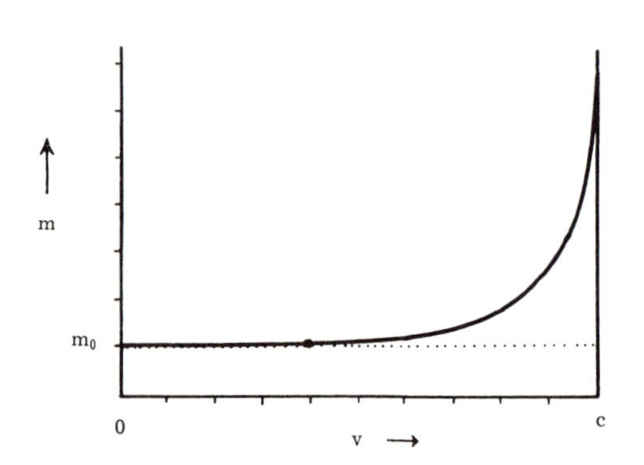

Figure 1.2—RELATIVISTIC MASS INCREASE

Mass and energy are interrelated by the Einstein equation, total energy E (ergs) being related to mass m (grams) by

$$E = mc^2 \tag{1.3}$$

The velocity of light c is here expressed in cm/sec. The potential energy U for a particle of atomic dimensions may be considered as the energy that would be released upon annihilation of the particle and which is therefore attributable to its rest mass. Thus

$$U = m_0c^2 \tag{1.4}$$

Since the total energy E is equal to the sum of the potential energy U and the kinetic energy T,

$$T = E - U = mc^2 - m_0c^2 \tag{1.5}$$

By use of equation (1.2) one obtains

$$T = m_0c^2 \left(\frac{1}{\sqrt{1 - \beta^2}} - 1\right) \tag{1.6}$$

When one considers phenomena in the microscopic world (at dimensions of about 10^{-8} cm and smaller) it again becomes necessary to modify Newtonian mechanics: the granularity of matter becomes a major factor and the quantum must be considered. Many atomic and nuclear processes can be explained only by means of quantum mechanics. The mode of radioactive decay, decay rate, nuclear fission, x-ray production and numerous other events are regulated by quantized states of motion.

For most observable electrical phenomena current appears to be continuous yet it is not. The smallest unit charge is that possessed by the electron. This charge e ($= 1.602 \times 10^{-19}$ coulombs), known as the electronic charge, is also called the quantum of charge, being the smallest possible quantity of electricity. Evidence to support the theory of the basic or unitary nature of electronic charge is provided by the fact that the smallest unit of positive charge, i.e., the charge of the proton and positron, is exactly equal to that of the electron though of opposite sign.

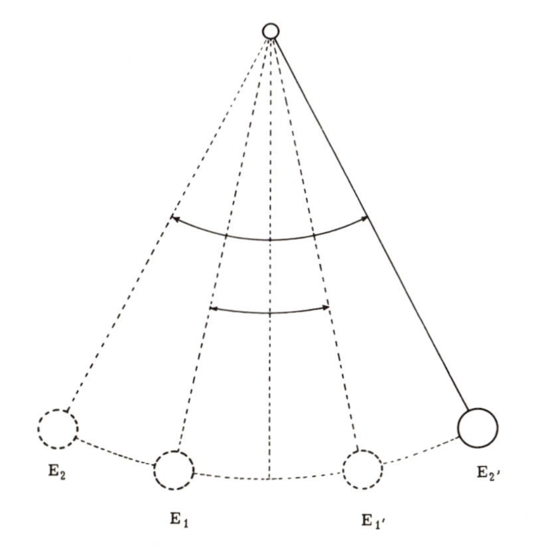

Figure 1.3—QUANTIZED OSCILLATOR. The oscillator is allowed only certain discrete energy states.

Energy is granular since it too is quantized. Because the wave motion of electromagnetic radiation is described by the action of a simple linear oscillator, such as a pendulum or the rotation of an electron in an atomic orbit, the energy of such a system is of special interest. In 1900 Max Planck assumed an oscillator cannot have any energy but only certain discrete energies given by

$$E = (n + \tfrac{1}{2}) h\nu \quad (1.7)$$

where n is a small whole number, h is a constant* and ν is the oscillator frequency. That is to say that the energy of the oscillator can only increase or decrease by an amount of energy $h\nu$ or by a whole multiple n of this basic energy package.

Thus if an oscillator jumps from quantum state n_2 to quantum state n_1 it will lose an amount of energy

$$\Delta E = E_2 - E_1 = n_2 h\nu - n_1 h\nu = \Delta n h\nu \quad (1.8)$$

Note that ΔE represents a change in the amplitude of oscillation but not in the frequency of oscillation.

Angular momentum p_ϕ (= mva) is also of particular interest in atomic structure theory. According to quantum theory p_ϕ can only increase in increments of $h/2\pi$. (Because the quantum of angular momentum $h/2\pi$ is used so frequently it is often represented by the symbol $\hbar$). Thus

$$p_\phi = mva = \frac{nh}{2\pi} = n\hbar \quad (1.9)$$

THEORY OF THE NUCLEAR ATOM

Sir J. J. Thomson (1) had suggested a theory of atomic structure to explain the scattering of charged particles passing through small thicknesses of matter. The atom as proposed by him was supposed to consist of a number of negatively charged particles scattered throughout a sphere consisting of an equal amount of positive electricity. The deflection of a negatively charge particle passing through the atom was thought to result from two causes: (a) Repulsion by the negative particles distributed throughout the atom and (b) Attraction of the positive electricity of the atom. In passing through the atom the particle was supposed to experience multiple encounters, each causing a small deflection. The total deflection on passing through the atom was taken as $\sqrt{m}\,\theta$ where m is the number of encounters and θ is the average deflection per encounter.

E. Rutherford (2) stated that the Thomson model does not allow a large deflection for an α particle unless it is assumed that the sphere of positive electricity is concentrated within a minute volume. The observations of Geiger and Marsden (3) on the scattering of α rays indicated that some of the α particles suffered a deflection of more than a right angle at a single encounter. They found that about 1 in 20,000 was turned

*Known as Planck's constant, $h = 6.625 \times 10^{-34}$ joule sec or 6.625×10^{-27} erg sec.

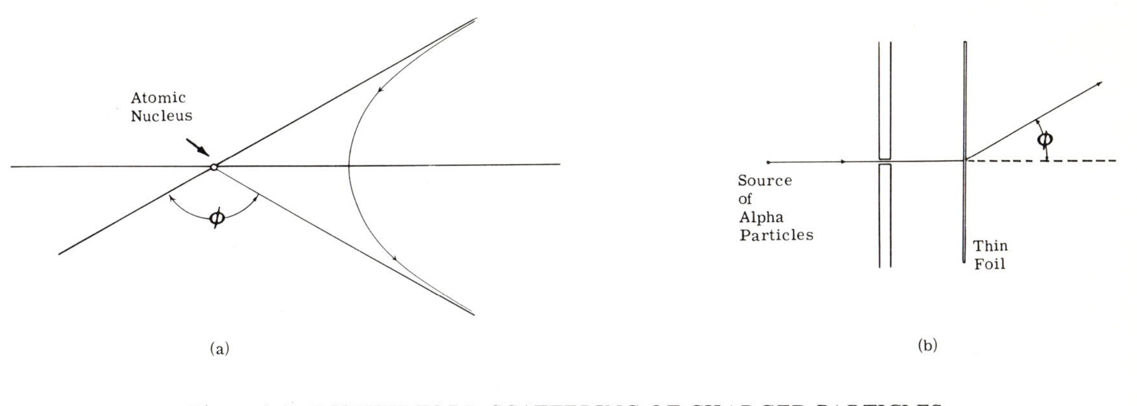

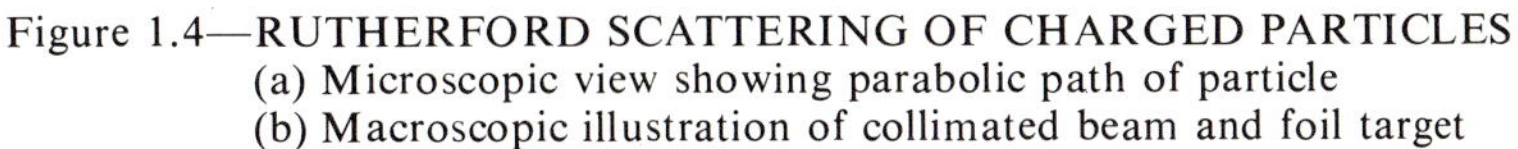
Figure 1.4—RUTHERFORD SCATTERING OF CHARGED PARTICLES
(a) Microscopic view showing parabolic path of particle
(b) Macroscopic illustration of collimated beam and foil target

through an angle of more than 90° in passing through a layer of gold foil 0.00004 cm thick. It was supposed by Rutherford that deflection through a large angle is due to a single encounter.

Rutherford (2) suggested that the atom consists of a central charge concentrated at a point and that the large single deflections of α and β particles are mainly due to their passage through the strong central field rather than to interactions with the surrounding sphere of negative charge. Thus if a particle of mass m, charge E and velocity v strikes a thin foil of thickness t, the fraction p of the particles which will be scattered through an angle greater than ϕ will be given by

$$p = \frac{\pi}{4} ntb^2 \cot^2\frac{\phi}{2} \tag{1.10}$$

where n is the number of atoms per unit volume of foil and $b = \frac{2ZeE}{mv^2}$

BOHR THEORY OF ATOMIC STRUCTURE

Experimental evidence has shown that the most acceptable visual model of the atom is that of a positively charged nucleus about which oscillate one or more electrons. The radius of an atom is approximately 10^{-8} centimeters. The nuclear radius is approximately 10^{-13} centimeters, about the same as the approximate radius of an electron. To appreciate these dimensions let a hydrogen atom be magnified until its diameter is one mile. The proton would then be comparable to a golf ball around which is revolving the electron, represented by a thin balloon about the same size as the golf ball. If the golf ball weighed 184 grams the balloon representing the electron would weigh 0.1 gram. This was the model suggested in 1911 by Rutherford's alpha-ray-scattering experiments and which was quickly developed by Niels Bohr into a useful atomic model capable of explaining the origin of atomic spectra.

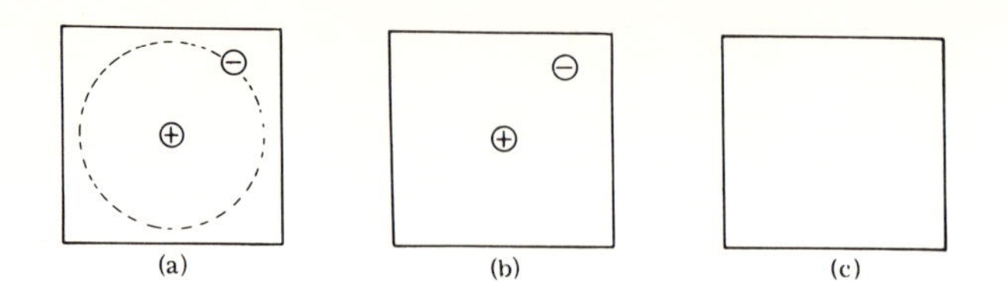

Figure 1.5—THE BOHR HYDROGEN ATOM. (a) Schematic representation showing the nucleus (a single proton), the orbital electron and the orbit of the electron. (b) Same atom but with orbit omitted to illustrate the atom more nearly as it would appear if one had the means to see it. (c) Same atom, drawn to same scale, but with size of the nucleus and electron reduced to the proper scale.

Bohr proposed his theory on the basis of three assumptions:

1. That the atom takes the form of Rutherford's model with almost all of the mass concentrated in a positively charged nucleus about which rotate negative electrons, each of which is attracted to the nucleus by coulombic force.

2. That Planck's quantum theory is obeyed so that energy can be radiated only in multiples of $h\nu$. That is, $E = h\nu$.

3. That an orbital electron does not radiate energy, in contradiction to the classical electromagnetic theory which predicts that an electron undergoing acceleration should radiate energy. An electron in a circular orbit, for example, has an acceleration

$$\frac{dv}{dt} = \frac{v^2}{a}$$

where v is the velocity of the electron and a is the radius of the orbit. If energy loss through radiation occurred, the electron would spiral into the nucleus and be annihilated.

Earnshaw's theorem postulates that a system of electrical charges cannot be in equilibrium unless an additional force of a different type is exerted. In the Bohr atom this additional force is a central force and is simply an application of Newton's second law of motion

$$F = m \cdot \frac{dv}{dt}$$

$$\frac{Ze \cdot e}{a^2} = m_0 \cdot \frac{v^2}{a} \tag{1.11}$$

where Ze is the nuclear charge, e the electronic charge, a the radius of the orbit, m_0 the mass* and v the velocity of the electron in its orbit. According to equation (1.11) any orbit, and thus any energy, should be possible. Spectral data have been used to show that this is not the case. Application of the quantum theory restricts the orbits to those where the angular momentum p_ϕ is equal to a multiple of $h/2\pi$. That is,

$$p_\phi = m_0 va = \frac{nh}{2\pi} \tag{1.12}$$

*Use of the rest mass m_0 rather than the relativistic mass m introduces only a minor error in the case of the lighter elements because the electron velocity is small compared to c.

where n is a small whole number. It is called the *principal quantum number*. By combining equations (1.11) and (1.12) it can now be shown that the velocity of the electron in any given orbit may be calculated from the values of known physical constants.

$$v = \frac{2\pi Ze^2}{nh} \tag{1.13}$$

Also, equation (1.12) shows that the radius of an orbit is given by

$$a = \frac{nh}{2\pi m_0 v} \tag{1.14}$$

Substitution of equation (1.13) for v in (1.14) gives an expression for the radius in terms of fundamental constants.

$$a = \frac{n^2h^2}{4\pi^2Ze^2m_0} \tag{1.15}$$

Although the frequency of revolution of an electron about the nucleus does not correspond to a spectral-line frequency, such information is of interest since it helps to provide a more complete understanding of atomic mechanisms. Frequency of revolution f can be calculated by dividing the velocity of the electron by the circumference of the orbit.

$$f = \frac{v}{2\pi a} = \frac{4\pi^2Z^2e^4m}{n^3h^3} \tag{1.16}$$

Table 1.1—THE HYDROGEN ATOM. Values of orbital radius, electron velocity and frequency of revolution for the first four quantum levels.

Principal Quantum Number n	Orbital Radius a (cm)	Electron Velocity v (cm/sec)	Frequency of Revolution f (rev/sec)
1	5.366×10^{-9}	2.16×10^{8}	6.38×10^{15}
2	2.146×10^{-8}	1.08×10^{8}	7.99×10^{14}
3	4.829×10^{-8}	7.18×10^{7}	2.37×10^{14}
4	8.586×10^{-8}	5.40×10^{7}	9.98×10^{13}

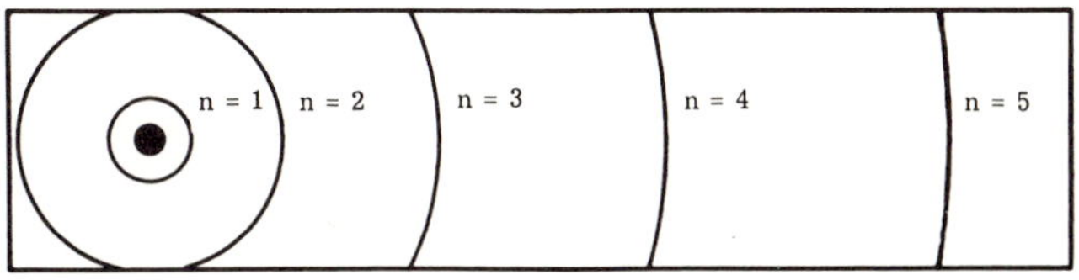

n = 1 n = 2 n = 3 n = 4 n = 5

Figure 1.6—BOHR ORBITS FOR HYDROGEN. Bohr's theory predicts that the radius, a, should be proportional to n^2. Thus, for instance, the ratio of the orbital radii for n = 3 and n = 5 should be 9:25. (For helium Z = 2. Thus each calculated value for radius would be reduced by 1/2.)

ATOMIC SPECTRA:

The potential energy of an electron in an orbit is expressed in terms of the energy required to remove an electron from an orbit of radius a to an infinite distance from the nucleus. Because the potential energy of an electron at rest at an infinite distance from the nucleus is arbitrarily assigned a value of zero it will be seen that the potential energy of an orbital electron is therefore negative and is given by

$$U = \int_{\infty}^{a} \frac{Ze^2}{a^2}\, da = -\frac{Ze^2}{a} \tag{1.17}$$

Kinetic energy $T = \frac{1}{2}mv^2$. Thus by use of equation (1.11) it is seen that the kinetic energy of an orbital electron is

$$T = \frac{Ze^2}{2a} \tag{1.18}$$

The total energy E of the orbital electron is the sum of the potential and kinetic energy and hence is equal to the sum of U and T as given by equations (1.17) and (1.18) respectively. Thus

$$E = -\frac{Ze^2}{a} + \frac{Ze^2}{2a} = -\frac{Ze^2}{2a} \tag{1.19}$$

Here the total energy is expressed as a function of the radius a. It is convenient to express energy in terms of fundamental constants only. This can be done by the proper substitution of equation (1.15) in equation (1.19). The result is

$$E = -\frac{2\pi^2 Z^2 e^4 m_0}{n^2 h^2} \tag{1.20}$$

When an electron passes from one orbit to another, represented by quantum numbers n_2 and n_1 respectively, the loss of energy $(E_2 - E_1)$ is emitted as electromagnetic radiation of energy $h\nu$. This is the result of the second assumption in the formulation of the Bohr atom and is expressed mathematically by

$$E_2 - E_1 = h\nu = \frac{2\pi^2 Z^2 e^4 m_0}{h^2}\left(\frac{1}{n_1^2} - \frac{1}{n_2^2}\right) \tag{1.21}$$

A general interrelationship of spectral lines was postulated by Ritz in 1908 who suggested that a new spectral line could be expected if its wave number was equal to the sum or difference of the wave numbers of lines already discovered. This postulate is known as the *Ritz combination principle*. The significance of a "difference" is illustrated by equation 1.21 by the terms within parentheses.

According to classical mechanics an excited atom should emit light with a frequency f according to equation 1.16 but it follows from equation 1.21 that quantum mechanics predicts a frequency ν given by

$$\nu = \frac{2\pi^2 Z^2 e^4 m_0}{h^3}\left(\frac{1}{n_1^2} - \frac{1}{n_2^2}\right) \tag{1.22}$$

The calculated frequencies are significantly different for small values of n but as n approaches infinity, ν approaches f. If $n_1 = n_2 - 1$, then

$$\nu = \frac{2\pi^2 Z^2 e^4 m_0}{h^3}\left(\frac{1}{(n-1)^2} - \frac{1}{n^2}\right) \tag{1.23}$$

$$= \frac{2\pi^2 Z^2 e^4 m_0}{h^3}\left(\frac{2n - 1}{(n-1)^2\ (n)^2}\right) \tag{1.24}$$

hence, the limiting value of ν as $n \to \infty$ is

$$\nu = \frac{4\pi^2 Z^2 e^4 m_0}{n^3 h^3} \tag{1.25}$$

which is identical to equation 1.16. That quantum mechanics reduces to classical mechanics for large values of n is known as the *correspondence principle*.

The validity of the Bohr atom can be checked by use of equation 1.21 but comparison of theory with spectroscopic data is performed more conveniently in terms of wave number $\bar{\nu}$ where

$$\bar{\nu} = \frac{1}{\lambda} = \frac{\nu}{c} \tag{1.26}$$

so that

$$\bar{\nu} = \frac{2\pi^2 Z^2 e^4 m_0}{ch^3}\left(\frac{1}{n_1^2} - \frac{1}{n_2^2}\right) \tag{1.27}$$

The Rydberg constant R is defined by

$$R \equiv \frac{2\pi^2 e^4 m_0}{ch^3} \tag{1.28}$$

and equation 1.27 becomes

$$\bar{\nu} = Z^2 R\left(\frac{1}{n_1^2} - \frac{1}{n_2^2}\right) \tag{1.29}$$

The theoretical value of R for hydrogen, calculated by substituting known values for e, Z, m_0, c and h into equation 1.28 is found to be 109,750 cm^{-1}. The value obtained from spectroscopic data is 109,677.58 cm^{-1}.

The Bohr theory can be refined by taking into account the motion of the nucleus. In the case of the simple hydrogen model, rotation of the electron about the proton will cause the proton to move, and both will rotate about a common center. This motion of the nucleus, not taken into account in the preceding derivation of the Bohr model, will alter the kinetic energy of the atom by a small amount. Correction for this effect is obtained by use of the *reduced mass* μ in place of the electronic mass m where

$$\mu = \frac{mM'}{m + M'} \tag{1.30}$$

M' is the mass of the nucleus. As M' approaches ∞, μ approaches m.

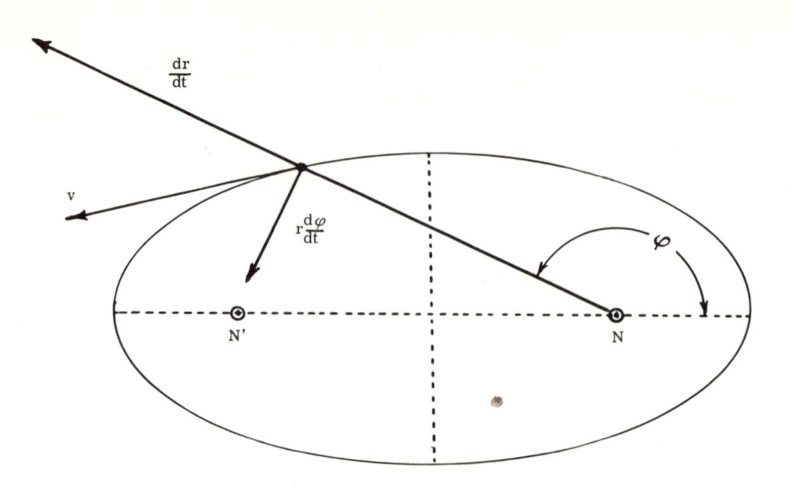

Figure 1.7—Elliptical orbit of an electron showing the components of radial and angular momentum

Although the Bohr planetary model of the atom agrees well with observed data for the hydrogen atom, it is somewhat less than satisfactory for the quantitative explanation of observed data for more complex atoms. In order to salvage the Bohr model, Sommerfeld advanced a hypothesis to account for the splitting of spectral lines to form doublets. Hitherto it had been assumed that the electron orbits are circular but, in general, periodic motion under the influence of a central force leads to elliptic orbits with the central body, e.g., the atomic nucleus, situated at a focus. The circle is, of course, a special case of an ellipse with major and minor axes the same length. The momentum of a particle moving in an ellipse can be resolved into two components, one along the radius vector and the other at right angles to it. (See figure 1.7.) These are the radial momentum

$$p_r = m\,dr/dt \tag{1.31}$$

and the angular momentum

$$p_\varphi = mr^2\,d\varphi/dt. \tag{1.32}$$

Following the treatment used by Bohr, it may be supposed that the angular momentum is quantized and can have only specific values given by

$$p_\varphi = kh/2\pi = (l+1)\,h/2\pi \tag{1.33}$$

where k is the azimuthal quantum number. The length of a semi-major axis a is found to be

$$a = \frac{n^2h^2}{4\pi^2me^2Z} \tag{1.34}$$

while the length of a semi-minor axis b is

$$b = \frac{knh^2}{4\pi^2 me^2 Z} \tag{1.35}$$

One result of this treatment is an expression for the ratio of the lengths of the axes.

$$\frac{a}{b} = \frac{n}{k} = \frac{n}{l+1} \tag{1.36}$$

The quantum number l, usually used instead of k, is defined by $l = k - 1$. The resulting orbits for the first three shells of hydrogen are illustrated in figure 1.8. For $k = 0$ the electron must oscillate through the nucleus. Although the physical model of Bohr and Sommerfeld would not allow such motion, the wave mechanical model does allow passage of the electron through the nucleus. On occasion the electron reacts with the nucleus; the event is called "electron capture."

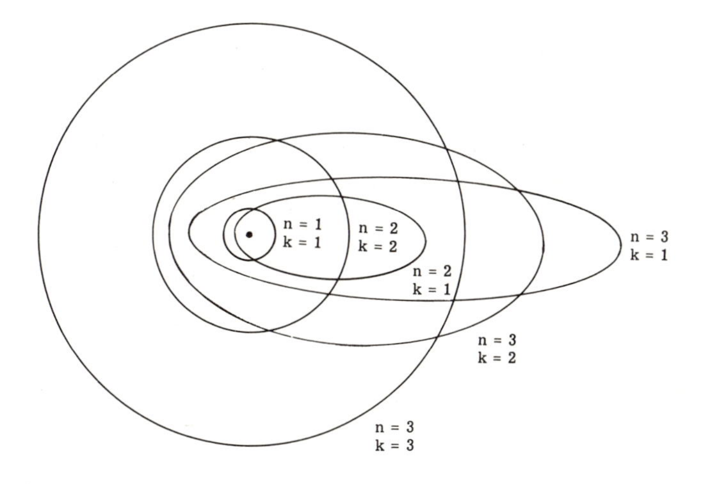

Figure 1.8—Orbits for the first three shells of the hydrogen atom

The azimuthal quantum number l defines the orbital angular momentum p the magnitude of which must be integral units of the Planck constant $\hbar$. Thus the orbital angular momentum is $l\hbar$. It is customary to represent the various orbit defined by the azimuthal quantum number by means of a letter.

For the following values of l:	0	1	2	3	4	5	6	7
The corresponding symbols are:	s	p	d	f	g	h	i	j

A level possessing a given orbital angular momentum l contains $2l + 1$ possible discrete states of space orientation, each represented by a specific value of the magnetic quantum number m. Thus for $l = 3$, m may have values of +3, +2, +1, 0, −1, −2 or −3.

Table 1.2—ELECTRONIC CONFIGURATION OF THE ATOM ACCORDING TO THE BOHR MODEL

Quantum Number				Distribution of Electrons	
n	l	m	s	in Subshells	in Shells
1	0	0	+½	(1s)	K
1	0	0	−½	2	2
2	0	0	+½	(2s)	L
2	0	0	−½	2	
2	1	−1	+½	(2p)	
2	1	−1	−½		
2	1	0	+½		8
2	1	0	−½	6	
2	1	1	+½		
2	1	1	−½		
3	0	0	+½	(3s)	M
3	0	0	−½	2	
3	1	−1	+½	(3p)	
3	1	−1	−½		
3	1	0	+½		
3	1	0	−½	6	
3	1	1	+½		
3	1	1	−½		18
3	2	−2	+½	(3d)*	
3	2	−2	−½		
3	2	−1	+½		
3	2	−1	−½		
3	2	0	+½		
3	2	0	−½		
3	2	1	+½	10	
3	2	1	−½		
3	2	2	+½		
3	2	2	−½		
4	0	0	+½	(4s)	N
4	0	0	−½	2	
4	1	−1	+½	(4p)	
4	1	−1	−½		
4	1	0	+½		

*With potassium (Z = 19) irregularities begin to appear in the sequence in which electrons are added. The energy of a 4s electron is less than that of a 3d electron. Consequently, the 4s orbits must be filled before the 3d orbits.

To account for the splitting of spectral lines energy differences must exist but the calculated total energy of an elliptic orbit, regardless of eccentricity, was the same as that of the circular orbit with which it is associated. Sommerfeld therefore considered the relativistic change in mass as the velocity of the electron in an elliptic orbit changes. This consideration provided an additional energy term to account for the dependence of energy on the azimuthal quantum number k (or $l = k - 1$).

In applying the planetary theory of atomic structure to complex atoms it was found necessary to put more than one electron in an orbit. Although the utilization of elliptic orbits reduced this problem, it was still necessary to put two electrons in a single orbit. Clues to the problem were forthcoming through studies of the fine structure of atomic spectra and by the work of Zeeman on the splitting of spectral lines in magnetic fields. Uhlenbeck and Goudsmit proposed the *theory of electron spin* to explain the splitting of spectral lines and it is now believed that electrons actually spin like a top. Since an electron can spin either parallel or antiparallel to the vector l—which represents the angular momentum of the electron in its orbit—the spin quantum number can have a value of either $+\frac{1}{2}$ or $-\frac{1}{2}$, the sign describing the direction of spin. The value of *spin momentum* is

$$p_s = \frac{sh}{2\pi} = s\hbar \qquad (1.37)$$

The *total angular momentum* for a single electron is equal to the vector sum of the orbital momentum p_φ and the spin momentum p_s, but since each is equal to the product of $\hbar$ and the respective quantum number, it is therefore also true that the total quantum number j is the vector sum of l and s. It is found that j only assumes a value which is either equal to $\frac{1}{2}$ or which differs from $\frac{1}{2}$ by a small whole number. A magnetic quantum number m describes the motion of the electron relative to the magnetic field of the atom.

The quantum numbers can be summarized as follows:

n = principal quantum number. Describes the major energy level of the electron.
Permissible values are n = 1 2 3 4 5 6
The corresponding symbols are K L M N O P

k = azimuthal quantum number. Usually replaced by l.

l = azimuthal quantum number where $l = k - 1$. May have values of 0, 1, ⋯, (n − 1), where n is the principal quantum number.
Permissible values are l = 0 1 2 3 4 5
The corresponding symbols are s p d f g h
The orbital angular momentum = $l\hbar$.

m = magnetic orbital quantum number. It represents the possible states of space orientation relative to a magnetic field and may have one of $(2l + 1)$ possible discrete values, namely, $m = l, (l - 1), \ldots, 1, 0, -1, \ldots, (-l + 1), -l$.

s = spin quantum number. Permissible values are $+\frac{1}{2}$ and $-\frac{1}{2}$.
The spin angular momentum = $s\hbar$.

j = total quantum number. It is the vector sum of l and s. ($j = l \pm s$)

Permissible values are j = 1/2 3/2 5/2 7/2 9/2 11/2 13/2
Calculated as follows:

l	s	j
0	1/2	1/2
1	+1/2	3/2
1	−1/2	1/2
2	+1/2	5/2
2	−1/2	3/2
3	+1/2	7/2
3	−1/2	5/2

The total angular momentum = $j\hbar$.

That no two electrons in the same atom behave in identical ways is now generally accepted as fact. Though having no theoretical basis, the *exclusion principle* enunciated by W. Pauli in 1925 states that it is impossible for any two electrons in the same atom to have their four quantum numbers identical. To date no exceptions to this statement have been found.

X-RAY SPECTRA

Although the Bohr model of the atom explained the spectra of a few light elements, it fails to account for the observed optical spectra for others. Nevertheless, because of its usefulness in explaining atomic processes it is still of value in discussing the origin of x-rays.

Moseley studied the x-ray spectra of about 38 elements. He observed two distinct series of spectra, later identified with the K and L quantum levels. In passing from one element to the next there was a regular shift in frequency. The frequency was related to the atomic number by a relationship now known as Moseley's law

$$\nu^{1/2} = K(Z-\sigma) \tag{1.38}$$

where K and σ are constants. The constant σ is now called the screening constant. It is a correction which must be applied to the nuclear charge Z to correct for the effect due to the repulsion of an electron by the other orbital electrons. Thus if equation 1.29 is corrected for screening and the velocity of light is introduced, then

$$\nu = \bar{\nu}c = (Z-\sigma)^2 Rc\left(\frac{1}{n_1^2} - \frac{1}{n_2^2}\right) \tag{1.39}$$

and if

$$K^2 = Rc\left(\frac{1}{n_1^2} - \frac{1}{n_2^2}\right) \tag{1.40}$$

then

$$\nu = (Z - \sigma)^2 K^2 \tag{1.45}$$

Taking the square root of equation 1.45 yields equation 1.38.

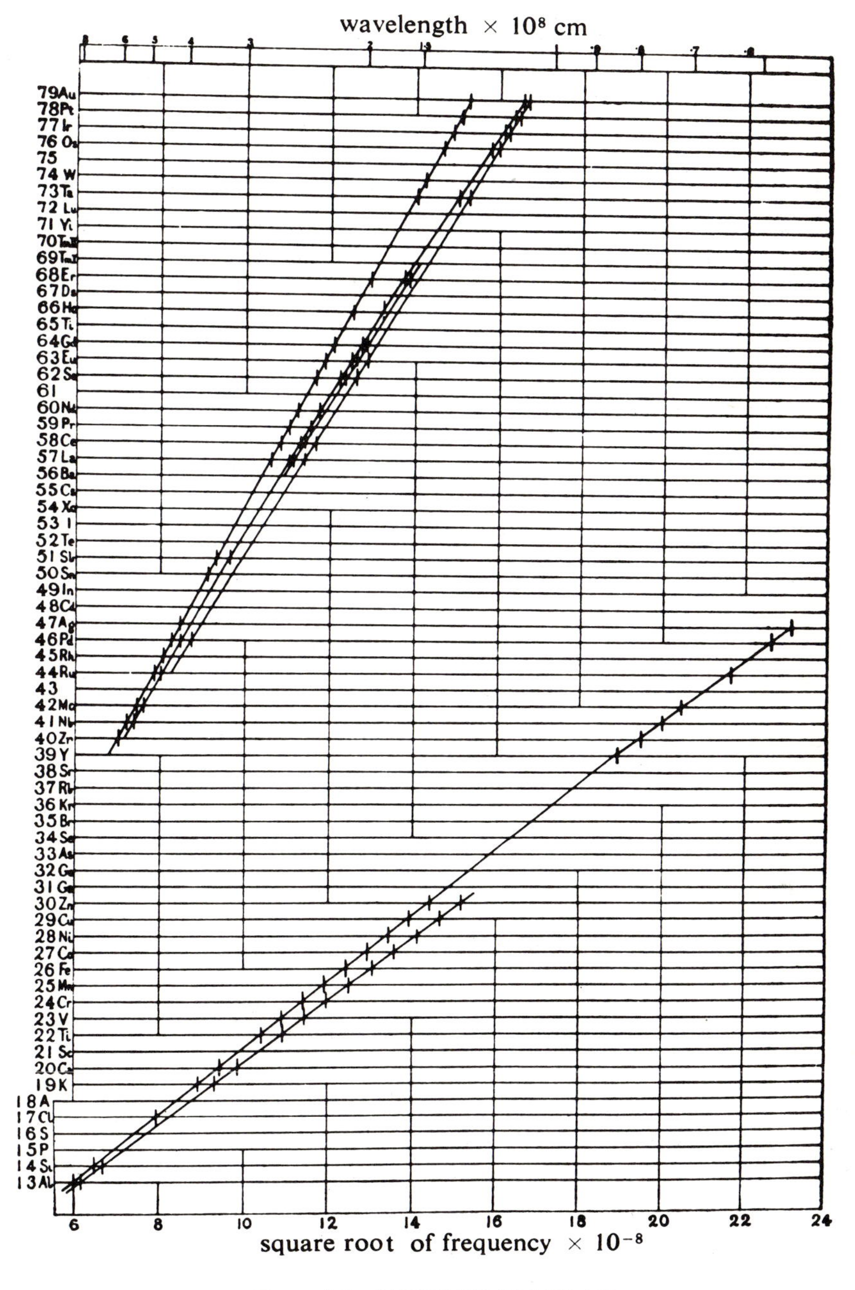

Figure 1.9—MOSELEY DIAGRAM
(From Moseley, *Phil. Mag.* 26, 1024 (1913).

According to Compton and Allison (4) the value of σ can be estimated as $\sigma = p + \frac{1}{2}(q-1)$ where q is the number of electrons in the final energy level of the radiating electron, and p is the number of electrons in levels closer to the nucleus.

Characteristic x-rays are identified according to the energy transitions producing them. The K series of x-rays arises from vacancies in the K shell. A vacancy in a K shell can be filled by an electron from the L or M shell which give rise to K_α and K_β lines respectively. Vacancies occurring in the L shell, filled by electrons from the M, N or O shells, give rise to the L series of x-ray lines. The energy level diagram in figure 1.5 shows the origin of the major lines. A table of values for these lines (in keV) has been published by Fine and Hendee (5). Wavelengths of the x-ray emission lines can be calculated with sufficient accuracy for most radioisotopic work by means of equation 1.21. K_α lines are calculated by setting $n_1 = 1$, $n_2 = 2$ and $\sigma = 1$. The L_α lines can be calculted by setting $n_1 = 2$, $n_2 = 3$ and $\sigma = 7.4$.

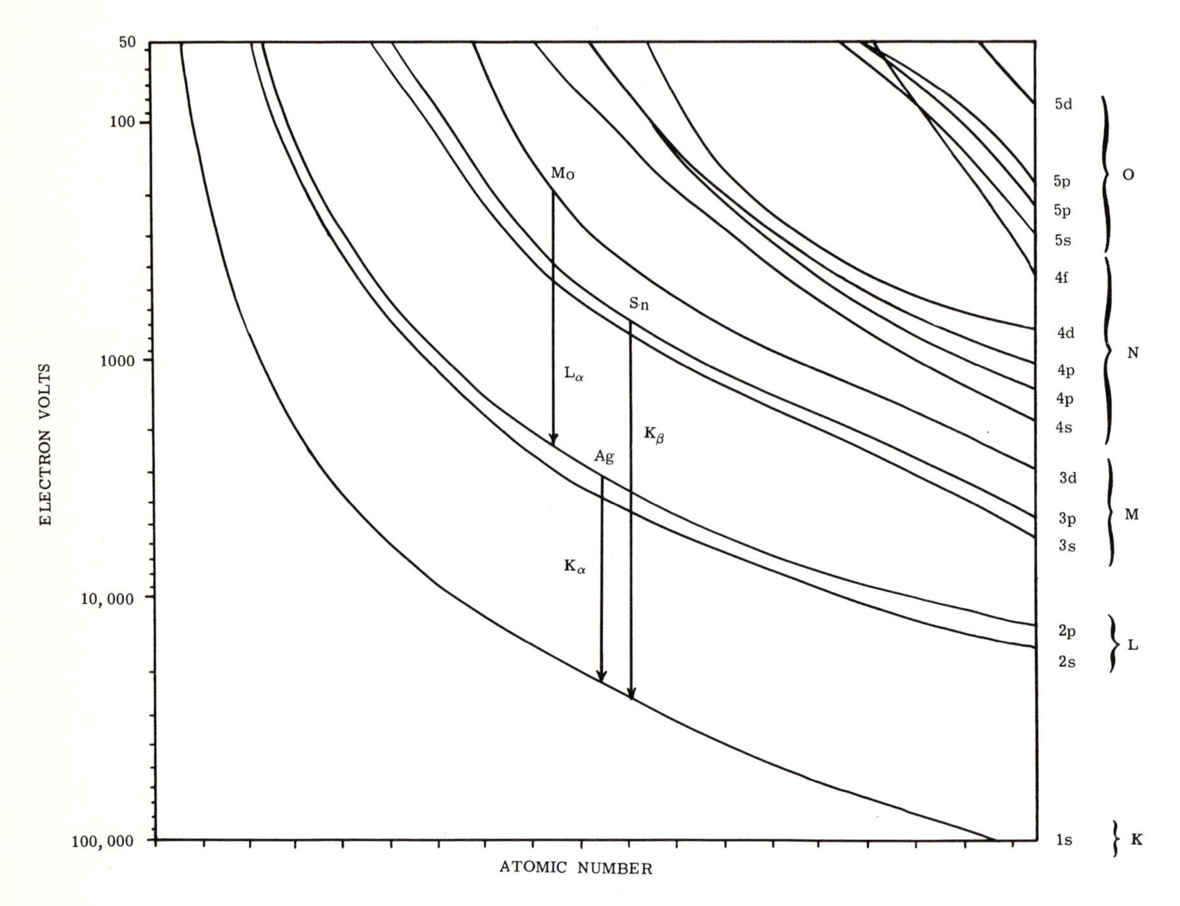

Figure 1.10—X-RAY ENERGY LEVEL DIAGRAM. Transitions illustrated are those resulting in the production of the $K_\alpha$line x-ray for silver, the K_β x-ray for tin and the L_α x-ray of molybdenum. Exact values of these energy levels illustrated here will be found in Moore (7).

PARTICLES AND WAVES

Having pondered the dualism of the nature of light, de Broglie suggested that if electromagnetic radiation, with its very definite wave characteristics, could also behave as though composed of particles (photons or quanta), then perhaps material particles such as protons and electrons could, in turn, have an associated wavelength. The relation between mass m and energy E had been shown by Einstein to be $E = mc^2$, where c is the velocity of light. Planck had shown the energy of a photon to be proportional to the frequency ν (equation 1.7). These two well known relationships may be equated

$$E = mc^2 = h\nu \tag{1.46}$$

Here, m is the mass equivalent of the photon. Further, the momentum p (equal to the product of mass and velocity) is given by

$$p = mc = \frac{h\nu}{c} = \frac{h}{\lambda} \tag{1.47}$$

(Wavelength $\lambda = c/\nu$). In this instance p is the momentum of a photon. de Broglie suggested that this equation for momentum may be extended to include a material particle. Thus if the velocity c is replaced by the particle velocity v, then the momentum p of the particle is given by

$$p = mv = \frac{h}{\lambda} \tag{1.48}$$

The wavelength λ, known as the de Broglie wavelength, is therefore

$$\lambda = \frac{h}{mv} = \frac{h}{p} \tag{1.49}$$

Thus de Broglie's equation defines the wavelength of the hypothetical matter-waves. An electron, for example, traveling with velocity v, will have associated with it a wavelength λ. This wavelength serves as a measure of the size of the particle.

The mass m is the relativistic mass. It is often more convenient to use the de Broglie equation in the form

$$\lambda = \frac{h}{m_0 v}\sqrt{1-\beta^2} \tag{1.50}$$

where m_0 is the rest mass of the particle and $\beta = v/c$.

It is of interest to note that if the circumference of a Bohr orbit ($2\pi a$) is divided by the de Broglie wavelength of an electron in that orbit, then the number of de Broglie wavelengths per orbit is found to be equal to the principle quantum number n. That is

$$2\pi a = n\lambda = nh/mv \tag{1.51}$$

from which the angular momentum p_ϕ of the orbital electron can readily be derived

$$p_\phi = m_0 va = nh/2\pi \tag{1.52}$$

Here it should be noted that equation (1.52) is identical to equation (1.9), one of the basic assumptions of the Bohr theory.

The wave nature of electrons has not only been verified experimentally by the Davisson-Germer experiment but if a wave motion is associated with orbital electrons, it follows of necessity that an orbit be an integral number of de Broglie wavelengths for reinforcement of the wave to occur. It was these vibrating electrons which led Schrödinger to his theory of atomic structure.

Example: What is the de Broglie wavelength of a 100 eV electron?
The kinetic energy T = ½ mv², or

$$v = \sqrt{\frac{2T}{m}} = \sqrt{\frac{2(100\,eV)(1.6 \times 10^{-12}\ erg/eV)}{9.1 \times 10^{-28}\ g}}$$

$$v = 5.9 \times 10^{8}\ cm/sec$$

The wavelength is then found from the de Broglie equation

$$\lambda = \frac{h}{mv} = \frac{6.6 \times 10^{-27}\ erg\ sec}{(9.1 \times 10^{-28}\ g)(5.9 \times 10^{8}\ cm/sec)}$$

$$\lambda = 1.22 \times 10^{-8}\ cm = 1.22\ A$$

Note: A relativistic correction was not necessary in this instance because the velocity of the electron is only about 20% of that for light.

Table 1.3—SELECTED DE BROGLIE WAVELENGTHS

Particle	Mass (g)	Velocity (cm/sec)	Wavelength (cm)
Electron	9.1×10^{-28}	1	7.3
Electron	9.1×10^{-28}	10	0.73
Electron	9.1×10^{-28}	100	0.073
1-volt electron	9.1×10^{-28}	5.9×10^{7}	1.22×10^{-7}
100-volt electron	9.1×10^{-28}	5.9×10^{8}	1.22×10^{-8}
100-volt alpha-particle	6.6×10^{-24}	6.9×10^{6}	1.44×10^{-10}
10-keV alpha particle	6.6×10^{-24}	6.9×10^{7}	1.44×10^{-11}
1-MeV alpha particle	6.6×10^{-24}	6.9×10^{8}	1.44×10^{-12}

WAVE MECHANICS

The wave-mechanical model proposed by Erwin Schrödinger in 1926 differs considerably from the planetary model of Bohr. Since de Broglie had shown that every particle in motion has an associated wave characteristic, it seemed reasonable to Schrödinger to develop a model based on the wave function itself. The result was a more acceptable model which explained accurately the same experimental data which had strained the Bohr theory. The Schrödinger atom is statistical since his theory predicts the probability of finding an electron at a given location. It allows electrons a greater degree of freedom in that they may occupy positions other than those defined by the

Bohr model. Although it is not feasible to present a rigorous development of the wave-mechanical model here, a few highlights should be considered.

In a beam of electrons the wave motion is not restricted; the wave is said to be a *traveling wave*. But if an electron is restricted to a circular path, as it is in the case of the atom, restrictions are also imposed on the wave motion. In order that reinforcement of the wave occur the orbital circumference must equal an integral number of wavelengths (see equation 1.55). When this condition is met a resonant state exists and a *standing wave* is produced. A similar condition is represented by vibrating strings. The vibrations of a violin string, resulting in the formation of standing waves, are restricted by certain boundary conditions; the ends of the string are held stationary so that a node must occur at each end.

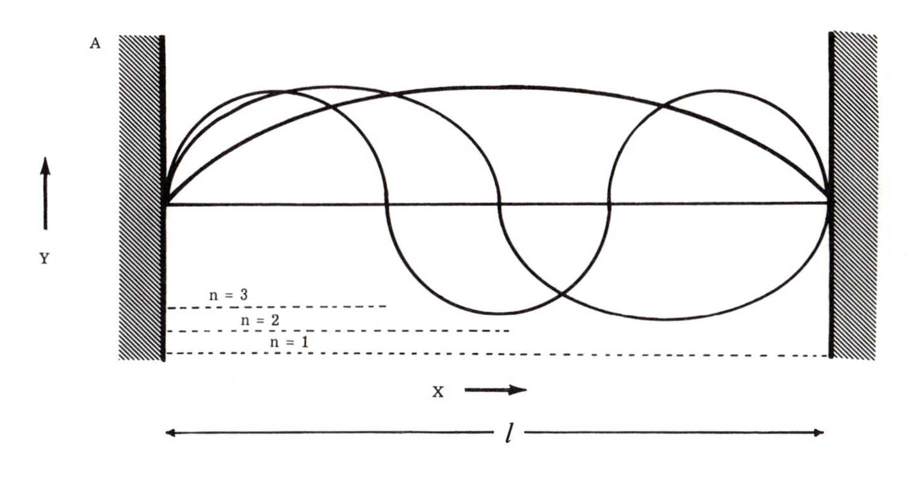

Figure 1.11—STANDING WAVES IN A VIBRATING STRING

For boundary conditions such as those described, the length of the string l is determined by

$$l = n\frac{\lambda}{2} \qquad n = 1, 2, 3, \ldots\ldots \tag{1.53}$$

or

$$\lambda = \frac{2l}{n} \tag{1.54}$$

It can therefore be seen, for a given value of l, that the wavelength is "quantized" by the requirement of equation (1.54). The equation for the motion of a string vibrating with amplitude A is

$$y_n = A \sin n\pi(x/l) \tag{1.55}$$

and by use of equation (1.54) one obtains

$$y_n = A \sin 2\pi(x/\lambda_n) \tag{1.56}$$

or, since x/t is equal to velocity,

$$y_n = A \sin 2\pi \nu t \qquad (1.57)$$

where ν is the frequency of vibration.

While the disturbance produced in a vibrating string is measured by y, the *wave function* Ψ measures the disturbance of matter waves. To illustrate the significance of the wave function let the string (see figure 1.11) be replaced by a particle which is only allowed to move along the x axis and which is also restricted by the same two walls. The associated de Broglie wavelength, given by $\lambda = h/p$ is also restricted by equation (1.54). The momentum of the particle is therefore quantized since

$$p = \frac{h}{\lambda} = \frac{nh}{2l} \qquad (1.58)$$

and, since $p = mv$ and $T = ½mv^2$,

$$p = \sqrt{2mT} \qquad (1.59)$$

By combining equations (1.58) and (1.59) the expression for the kinetic energy T of the confined particle is obtained.

$$T = \frac{n^2h^2}{8ml^2} \qquad n = 1, 2, 3, \ldots\ldots \qquad (1.60)$$

It should be noted especially that the particle cannot possess a continuously variable energy but only the discrete energies defined by equation (1.60). Confining the particle has resulted in the quantization of its energy.

In the above illustration, y is the amplitude function for one dimension, i.e., along the x-axis. If extended to three dimensions the amplitude factor is usually represented by Ψ. Thus, while the wave function for one dimension (see appendix I) can be expressed in the form

$$\frac{\partial^2 y}{\partial x^2} - \frac{1}{u^2}\frac{\partial^2 y}{\partial t^2} = 0 \qquad (1.61)$$

the wave function for three dimensions is

$$\frac{\partial^2 \Psi}{\partial x^2} + \frac{\partial^2 \Psi}{\partial y^2} + \frac{\partial^2 \Psi}{\partial z^2} - \frac{1}{u^2}\frac{\partial^2 \Psi}{\partial t^2} = 0 \qquad (1.62)$$

or simply

$$\nabla^2\Psi - \frac{1}{u^2}\frac{\partial^2 \Psi}{\partial t^2} = 0 \qquad (1.63)$$

Equation 1.63 is now divided into two terms, one containing only time and the other only space coordinates. This can be done by letting $\Psi = \psi\, f(t)$ where $f(t) = \exp \sqrt{2}\,\pi\bar{\nu}ut$.

It takes the form

$$\frac{\nabla^2\psi}{\psi} - \frac{1}{u^2\, f(t)} \frac{\partial^2 f(t)}{\partial t^2} = 0 \tag{1.64}$$

Solution of this equation then gives

$$\nabla^2\psi = -4\pi^2\nu^2\psi \tag{1.65}$$

By use of the relationships $\nu = 1/\lambda$, $\lambda = h/p$ and $p = \sqrt{2mT} = \sqrt{2m(E-U)}$ the Schrödinger equation is obtained

$$\nabla^2\psi + \frac{8\pi^2 m}{h^2}(E-U)\,\psi = 0 \tag{1.66}$$

Solutions to the Schrödinger equation will depend upon the particular boundary conditions, just as in the case of the vibrating string. The resulting allowed values of E are called *eigenvalues* and the corresponding wave functions ψ are called the *eigenfunctions.* (Eigen in German means proper, characteristic, specific or special.)

The significance of ψ was pointed out by Max Born who suggested that $\psi^2 dV$ represents the probability of the electron occupying the volume element dV. For example, solution of the Schrödinger equation in r (to determine the probability distribution of the electron as a function of its distance from the nucleus) yields the following relationships for the 1s and 2s electrons:

$$\psi_{1s} = \frac{1}{\sqrt{\pi}} \left(\frac{Z}{a_0}\right)^{3/2} e^{-Zr/a_0} \tag{1.67}$$

$$\psi_{2s} = \frac{1}{4\sqrt{2\pi}} \left(\frac{Z}{a_0}\right)^{3/2} \left(2 - \frac{Zr}{a_0}\right) e^{-Zr/a_0} \tag{1.68}$$

In these equations $a_0 = h^2/4\pi^2\mu e$ where h is the Planck constant, μ is the reduced mass and e is the electronic charge. The wave functions for the s electrons are spherically symmetrical and depend only on the distance r. Wave functions for p and d electrons are not spherically symmetrical.

The probability of observing an electron in a volume element dV at a distance r from the nucleus is illustrated by plotting ψ^2 versus r. This is shown in figure 1.12. It is seen that the greatest probability is that the electron will be very close to the nucleus. On the other hand, an orbit-like structure becomes apparent if the probability of the electron being located in a spherical shell of radius r and thickness dr is considered. The volume of a spherical shell is $4\pi r^2\, dr$*. Therefore

$$\psi^2 dV = \frac{Z^3}{\pi a_0^3} \left(e^{-2Zr/a_0}\right) (4\pi r^2\, dr) \tag{1.69}$$

*The volume of a sphere is expressed by $V = \frac{4}{3}\pi r^3$. Differentiation with respect to r gives $dV/dr = 4\pi r^2$. Thus $dV = 4\pi r^2\, dr$.

The probability of finding an electron in a spherical shell is proportional to $4\pi r^2\psi^2$. In figure 1.13 this quantity is plotted as a function of r for the 1s and the 2s electrons of the hydrogen atom. It will be noted that the greatest probability is found for a radius equal to that calculated according to the Bohr theory.

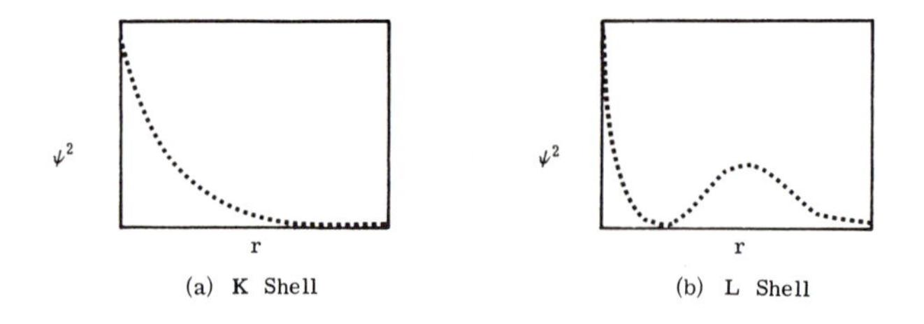

Figure 1.12—PROBABILITY OF OBSERVING AN "S" ELECTRON AT A DISTANCE r FROM THE NUCLEUS FOR THE HYDROGEN ATOM.

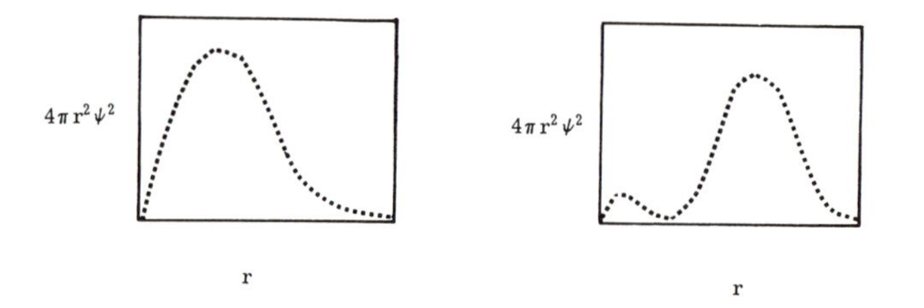

Figure 1.13—PROBABILITY OF OBSERVING AN "s" ELECTRON IN A SPHERICAL SHELL AT A DISTANCE r FROM THE NUCLEUS FOR THE HYDROGEN ATOM.

REFERENCES

1. Thomson, J. J., Camb. Lit. and Phil. Soc. XV, pt. 5 (1910).
2. Rutherford, E., "The Scattering of α and β Particles by Matter and the Structure of the Atom", The London, Edinburgh and Dublin Phil. Mag. and J. of Sci., 6th Series, 21, 669 (1911).
3. Geiger, H., and E. Marsden, Proc. Roy. Soc. lxxxii, 495 (1909).
4. Compton, A. H., and S. K. Allison, "X-rays in Theory and Experiment", Van Nostrand (1935). p. 37.
5. Fine, S., and C. F. Hendee, "X-ray Critical-Absorption and Emission Energies in kev", Nucleonics 13, No. 3, 36-37 (1955).
6. Garrett, A. B., "The Neutron Identified: Sir James Chadwick", J. Chem. Ed., 39, 638 (1962).
7. Moore, C. E., "Atomic Energy Levels", Circular of the National Bureau of Standards 467, Vol. I, 1949; Vol. II, 1952; Vol. III, 1958.
8. Kuhn, H. G., "Atomic Spectra", Academic Press, New York (1962).
9. Stafford, F. E., and J. H. Wortman, "Atomic Spectra, A Physical Chemistry Experiment", J. Chem. Ed., 39, 630-632 (1962).
10. Keller, R. N., "Energy Level Diagrams and Extranuclear Building of the Elements", J. Chem. Ed., 39, 289-293 (1962).
11. Garrett, A. B., "The Neutron Identified: Sir James Chadwick", J. Chem. Ed., 39, 638-639 (1962).
12. Physics Staff of University of Pittsburgh, "An Outline of Atomic Physics", Wiley (1937).
13. Semat, H., "Introduction to Atomic and Nuclear Physics", Rinehart (1959).
14. Romer, A., "The Restless Atom", Doubleday (1960).
15. Davis, H. M., "The Chemical Elements", Science Service (1959).

16. Ruderman, M. A., and A. H. Rosenfeld, "An Explanatory Statement on Elementary Particle Physics", Amer. Scientist 48, No. 2, 209 (June 1960).
17. da C. Andrade, E. N., "The Birth of the Nuclear Atom", Sci. Amer., p. 93 (Nov. 1956).
18. Darrow, K. K., "The Quantum Theory", Sci. Amer., p. 47 (March 1952).
19. Gamow, G., "The Exclusion Principle", Sci. Amer., p. 74 (July 1959).
20. Brown, G. I., "A Simple Guide to Modern Valance Theory", Longmans (1953).
21. Chew, G. F., M. Gell-Mann and A. H. Rosenfeld, "Strongly Interacting Particles", Sci. Amer., p. 74 (Feb. 1964).
22. Gell-Mann, M., and E. P. Rosenbaum, "Elementary Particles", Sci. Amer., p. 72 (July 1957).
23. Fowler, W. B. and N. P. Samios, "The Omega-Minus Experiment", Sci. Amer., p. 36 (Oct. 1964).
24. Hill, R. D., "Resonance Particles", Sci. Amer., p. 39 (Jan. 1963).
25. Feld, B. T., "High-Energy Nuclear Physics", Nucleonics, 34 (May 1955).
26. Snell, A. H., "A Survey of the Particles of Physics", Amer. Scientist 45, No. 1, 44 (Jan. 1957).
27. Noyes, H. P., "The Physical Description of Elementary Particles", Sci. Amer. 45, No. 5 (Dec. 1957).
28. Kuhn, H. G., "Atomic Spectra", Academic Press (1962).
29. Dawson, J. K., "Electronic Structure of the Heaviest Elements", Nucleonics 10, No. 9, 39-45 (1952).

CHAPTER 2
The Atomic Nucleus

LECTURE OUTLINE AND STUDY GUIDE

I. NUCLEAR COMPOSITION
  A. Theories on Nuclear Composition
  B. Nuclear Terms Related to Composition
    1. Nuclides
    2. Isotopes
    3. Isobars
    4. Isotones
  C. Conventions for Isotopic Symbols

II. NUCLEAR STRUCTURE
  A. Forces
    1. Gravitational
    2. Electromagnetic
    3. Nuclear
    4. Weak interactions
  B. Nuclear Models
    1. Potential well
    2. Shell
    3. Statistical or compound nucleus
    4. Liquid drop
    5. Individual particle
    6. Collective
    7. Cloudy crystal ball
    8. Close-packed spheron

III. NUCLEAR EQUATIONS
  A. Conventions for Writing Equations
  B. Balancing Equations

IV. NUCLEAR ENERGETICS
  A. Mass-Energy Relationship
  B. Units and Standard of Mass
  C. Binding Energy
    1. Total binding energy
    2. Binding energy per nucleon
  D. Calculation of Q
  E. Exergonic versus Endergonic Reactions

V. NUCLEAR STABILITY
  A. Neutron-to-Proton Ratio
  B. The Stability Diagonal
  C. Even-Odd Rules
  D. Binding Energy
    1. Mass defect and total binding energy
    2. Binding energy per nucleon
    3. Separation Energy
    4. Packing fraction
  E. Magic Numbers

VI. NUCLEAR REACTIONS
  A. Spontaneous Versus Induced Reactions
  B. Radioactive Decay
  C. Fission
  D. Fusion
  E. Reactor Reactions
  F. Cyclotron Reactions

VII. UNITS OF RADIOACTIVITY
  A. Radium as a Standard
  B. The Curie Unit

THEORIES ON NUCLEAR COMPOSITION

Although Rutherford had proposed the existence of the neutron as early as 1920, the nucleus was first thought to contain protons and electrons. Thus helium, for example, was believed to contain four protons and two electrons to account for a mass of four and a net charge of +2. In retrospect one can calculate that this is improbable, if not impossible. From wave mechanics we know that the size of a particle can be estimated in terms of its associated de Broglie wavelength. To exist within the nucleus (as small as 2×10^{-13} cm) a particle must have a de Broglie wavelength equal to or smaller than the diameter of the nucleus containing it. While a proton associated with a wavelength of 2×10^{-13} cm would possess an energy of about 25 MeV, an electron would possess energy of the order of 100 MeV and would emerge violently from the nucleus. Even for a wavelength of 10^{-12} cm the electron energy is about 20 MeV, the corresponding proton energy now being of the order of 1 MeV. Another aspect of the problem created by electrons within the nucleus is illustrated by the relationship between the hydrogen molecule ion and the deuteron. According to the proton-electron theory of nuclear structure both would contain two protons and one electron yet the deuteron is about a million times smaller. Extremely strong forces would be required to confine the electron to such a small volume and evidence for the existence of these forces should be found in the hydrogen spectrum.

Some scientists considered the possibility that alpha particles are the basic building blocks of nuclei while others condemned this hypothesis on the basis that one should not conclude that a particle exists within the nucleus just because it is emitted from the nucleus. The same precaution would also apply to beta particle emission; emission of a beta particle is not conclusive evidence that the electron was within the nucleus before its emission.

Finally the basic problem of nuclear composition was resolved with Chadwick's identification of the neutron in 1932 and the hypothesis proposed the same year by Werner Heisenberg that the neutron and proton are the basic building blocks of the nucleus. Thus the mass and charge of the helium nucleus could be explained by assuming the presence of two protons and two neutrons and the deuteron could be explained by assuming it to be composed of one proton and one neutron. Recent investigators suggest that the neutron and proton are actually two quantum states of the same basic particle, the nucleon, and that one is converted into the other by means of meson exchange or a transition involving the creation of an electron and a neutrino.

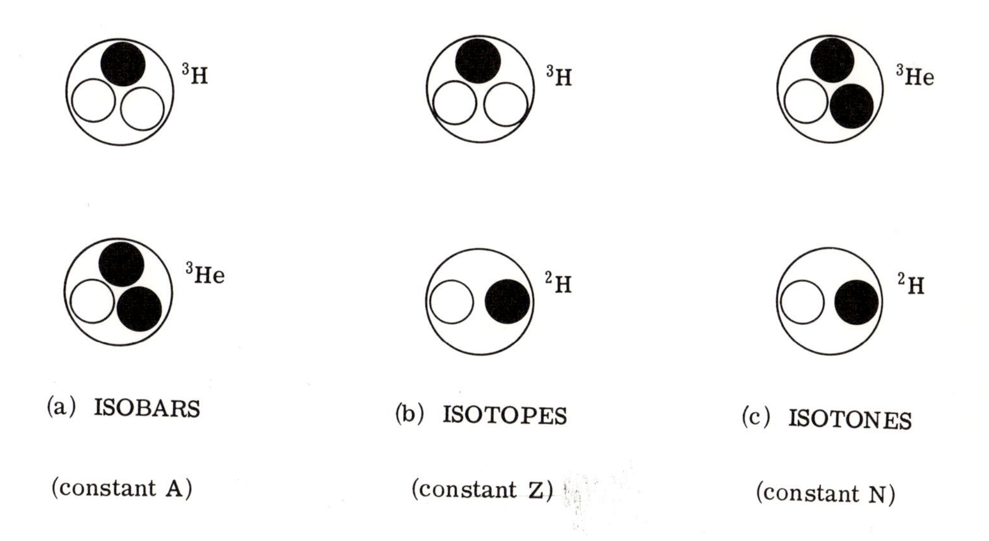

Figure 2.1—ISOBARS, ISOTOPES AND ISOTONES. The dark circles represent protons and the light circles represent neutrons. Only nuclei of the atoms are illustrated.

NUCLEAR TERMS

The term *nuclide* is used to signify any one of the more than 1300 known species of atoms characterized by the number of protons and neutrons in the nucleus. The number of protons is represented by the *atomic number* Z and the number of neutrons by the *neutron number* N. From equation (1.1) it is seen that the *mass number* A is the sum of Z and N.

Mass numbers range from A = 1 for the proton or neutron to about A = 258. Except for A = 5 and A = 8, nuclides are found in nature with mass numbers from A = 1 to A = 238. The others have been prepared artificially. Nuclides with the same mass number are called *isobars*. Thus hydrogen-3 (tritium) and helium-3 are isobars.

Isotopes are nuclides of the same Z but different N and consequently of different A. Deuterium and tritium are therefore isotopes of hydrogen because both contain the same number of protons in the nucleus.

Isotones are nuclides with the same N but different Z and consequently different A. Tritium and helium-4 are isotones because each possesses the same number (two) of neutrons.

CONVENTIONS FOR ISOTOPIC SYMBOLS

To identify nuclides the mass number is written as a superscript of the chemical symbol for the element. In the United States, in past years, the mass number has generally been written as a right superscript (e.g., Na^{24}) while in most other countries it has been written as a left superscript (i.e., ^{24}Na). The latter convention has the advantage of simplifying the writing of nuclidic ions (e.g., $^{24}Na^+$ rather than Na^{24+}) and is now generally preferred throughout the world.

The atomic number Z is expressed as a left subscript and the neutron number N as a right subscript (e.g., $^{24}_{11}Na_{13}$). However, since N is known from the difference N = A − Z and since Z can have but a single value for a particular element which is, of course, identified by the chemical symbol, both are considered superfluous when writing the symbol. The sodium isotopes are therefore completely identifiable by use of the symbols ^{20}Na, ^{21}Na, ^{22}Na, ^{23}Na, ^{24}Na, ^{25}Na and ^{26}Na.

NUCLEAR FORCES

When charged particles of like sign are packed tightly together, as they are in the nucleus, coulombic forces should cause them to fly apart. Because this does not occur we must conclude that an even more powerful force must exist to hold the nucleus together. A bold theory to explain this nuclear force, based on an analogy with electromagnetic radiation, was suggested by Yukawa (4) in 1935. He wrote:

> "At the present stage of the quantum theory little is known about the nature of interaction of elementary particles. Heisenberg considered the interaction of "Platzwechsel" between the neutron and the proton to be of importance to the nuclear structure.
>
> "Recently Fermi treated the problem of β-disintegration on the hypothesis of "neutrino." According to this theory, the neutron and the proton can interact by emitting and absorbing a pair of neutrino and electron. Unfortunately the interaction energy calculated on such assumption is much too small to account for the binding energies of neutrons and protons in the nucleus.

> "To remove this defect, it seems natural to modify the theory of Heisenberg and Fermi in the following way. The transition of a heavy particle from neutron state to proton state is not always accompanied by the emission of light particles, i.e., a neutrino and an electron, but the energy liberated by the transition is taken up sometimes by another heavy particle, which in turn will be transformed from proton state into neutron state. If the probability of occurrence of the latter process is much larger than that of the former, the interaction between the neutron and the proton will be much larger than in the case of Fermi, whereas the probability of emission of light particles is not affected essentially.
>
> "Now such interaction between the elementary particles can be described by means of a field of force, just as the interaction between the charged particles is described by the electromagnetic field. The above considerations show that the interactions of heavy particles with this field is much larger than that of light particles with it.
>
> "In the quantum theory this field should be accompanied by a new sort of quantum, just as the electromagnetic field is accompanied by the photon.
>
> "...... so that the quantum accompanying the field has the proper mass $m = \lambda h/c$.* Assuming $\lambda = 5 \times 10^{12} cm^{-1}$, [equivalent to a wavelength of 2×10^{13}cm, an estimate of nuclear size] we obtain for m a value 2×10^2 times as large as the electron mass. As such a quantum with large mass and positive or negative charge has never been found by the experiment, the above theory seems to be on a wrong line. We can show, however, that, in the ordinary nuclear transformation, such a quantum can not be emitted into outer space."

Mesons similar to the particles predicted by Yukawa, and having a mass of about 200 electron masses, were observed in 1937 among the particles produced by cosmic rays but experiments to determine the properties of these new particles revealed a serious discrepancy—the probability of their production was 10^{14} times the probability of their absorption by nuclei. To explain this problem Marshak suggested the existence of two kinds of mesons. In 1947, only a few weeks following Marshak's prediction, another particle with a mass of 276 electron masses was discovered by Powell in cosmic ray tracks obtained in the Bolivian Andes. The lighter particle is now called the muon while that discovered by Powell is called the pi meson. The pi meson, or pion as it is sometimes called, has the properties predicted by Yukawa for the quantum associated with proton-neutron interactions.

COMPARISON OF KNOWN FORCES

There are four known types of force or interaction. The force holding the universe together is *gravity*, the force holding atoms together (coulombic forces considered in the discussion of the Bohr model, van der Waals' forces, adhesive and cohesive forces) is *electromagnetic*, forces holding nuclei together are called *nuclear*, while *weak interactions* are those involved in beta decay and are the Fermi forces to which Yukawa referred in the excerpt above.

Gravitational force occurs between objects possessing mass. It is a weak force but extends over a great range. The force F between two objects of masses m_1 and m_2 and separated by a distance r is given by Newton's law of universal gravitation:

$$F = G \frac{m_1 m_2}{r} \tag{2.1}$$

where G is the universal gravitational constant (not the acceleration constant g). It can be seen that gravitational force follows the inverse square law. The work W required to separate masses from a distance r to infinity is:

*Here Yukawa uses the symbol λ for wave number; elsewhere in the text the symbol $\bar{\nu}$ is used. In addition h is used in place of $\hbar$. Adopting the symbols used in this text the equation becomes $m = \hbar\bar{\nu}/c = \hbar/\lambda c = h/2\pi\lambda c$.

$$W = \int_r^\infty G \frac{m_1 m_2}{r^2} dr = G \frac{m_1 m_2}{r} \tag{2.2}$$

while the potential energy $U = -Gm_1m_2/r$. Gravitation is a central force. The quantum acting in a gravitational field is called the graviton and is thought by some to be the neutrino.

Electrostatic force is stronger than gravitational force but, like gravitational force, is central and follows the inverse square law. The force between two charges Q_1 and Q_2 is given by Coulomb's law

$$F = k \frac{Q_1 Q_2}{r^2} \tag{2.3}$$

In the cgs system $k = 1$ for a vacuum while in the mks system $k = 1/4\pi\epsilon$ where ϵ is the permittivity. Magnetic forces also come into play if one of the charges is moving relative to the other. Magnetic attraction is a tensor force because it depends upon the alignment as well as the separation.

The quantum acting in an electromagnetic field is the photon. The mode of action of the quantum can be explained by means of the uncertainty principle

$$\Delta E \cdot \Delta t = h/2\pi \tag{2.4}$$

and since $\Delta E = \Delta mc^2$ it follows, in the case of the electron, for example, that there is an uncertainty in the rest mass depending upon the time

$$\Delta m \cdot \Delta t = \frac{h}{2\pi c^2} \tag{2.5}$$

The uncertainty in the rest-mass energy corresponding to this uncertainty in the rest mass is an amount $\Delta E = h\upsilon$. Hence

$$\Delta t = \frac{h}{2\pi \Delta E} = \frac{h}{2\pi h\upsilon} = \frac{1}{2\pi\upsilon} \tag{2.6}$$

and the distance traveled by the photon in time Δt is

$$d = c\Delta t = \frac{c}{2\pi\upsilon} = \frac{\lambda}{2\pi} \approx \frac{\lambda}{6} \tag{2.7}$$

During the time Δt the electron loans energy for the formation of the photon. The photon then travels from the electron a distance of up to one-sixth of a wavelength before it must retreat in order to repay its energy debt to the electron. A red photon may extend about 1000 Angstroms while a radio-frequency photon may extend as far as 50 meters. For charges close together an exchange of photons is thought to occur. This picture of energy exchange through photons accounts for coulombic forces and explains how these forces can act at a distance. These photons which are continually emitted and reabsorbed constitute the field about the electron.

Nuclear forces are extremely strong. Nuclear force is in part central like gravitational force and is part tensor like magnetic force. Yukawa (3) states,

> "The potential of force between the neutron and the proton should, however, not be of the Coulomb type, but decrease more rapidly with distance. It can be expressed, for example, by
>
> $$+ \text{ or } - g^2 \frac{e^{-\lambda r}}{r}$$
>
> where g is a constant [sometimes called nuclearity, mesonic charge or mesic charge] with the dimension of electric charge, i.e., $cm^{3/2}$ sec^{-1} $g^{1/2}$ and λ with the dimension cm^{-1}."

This nuclear force is always attractive, never repulsive. It is the same for neutrons and protons and equal to zero for electrons. The work required to separate two nucleons from a distance r to infinity is

$$W = k' \frac{g^2}{r} e^{-r/a} \tag{2.8}$$

In this relationship the term $e^{-r/a}$ gives nuclear forces their typical short-range properties. Because the ratio r/a must be dimensionless the units of a must equal those of r (length). The distance a can be evaluated as follows:

Assuming nuclear forces to involve the transfer of a π-meson from one nucleon to another, the energy change ΔE resulting from such a transfer is

$$\Delta E = m_\mu c^2 = 135 \text{ MeV} \tag{2.9}$$

where m_μ is the mass of the π-meson. By use of the uncertainty principle as expressed by equation 2.4 the time for the transfer is

$$\Delta t = \frac{h}{2\pi m_\mu c^2} = 4.6 \times 10^{-24} \text{ sec} \tag{2.10}$$

and the distance traveled by the meson in time Δt is

$$a = c\Delta t = \frac{h}{2\pi m_\mu c} = 1.4 \times 10^{-13} \text{ cm} \tag{2.11}$$

A meson, emitted and reabsorbed within the short period of 10^{-24} seconds, would be undetectable or virtual yet there is substantial evidence to suggest that each nucleon is surrounded by a cloud of pions. For example, if a proton-rich substance such as paraffin is bombarded with a beam of neutrons, the number of protons ejected in a direction parallel with the incident neutron beam and with energies essentially equal to the incident neutron energies cannot be explained by means of simple collision theory. But it is explained by assuming an exchange of pions (see figure 2.2.). Here a virtual positive pion associated with a proton is captured by a passing neutron. In the process the neutron becomes a proton and the proton becomes a neutron. It is believed that mesonic exchanges of this sort also occur within the nucleus creating a resonant condition among the nucleons. It will be recalled that a resonance is generally accepted as accounting for the unusual chemical stability observed for the benzene ring and it is

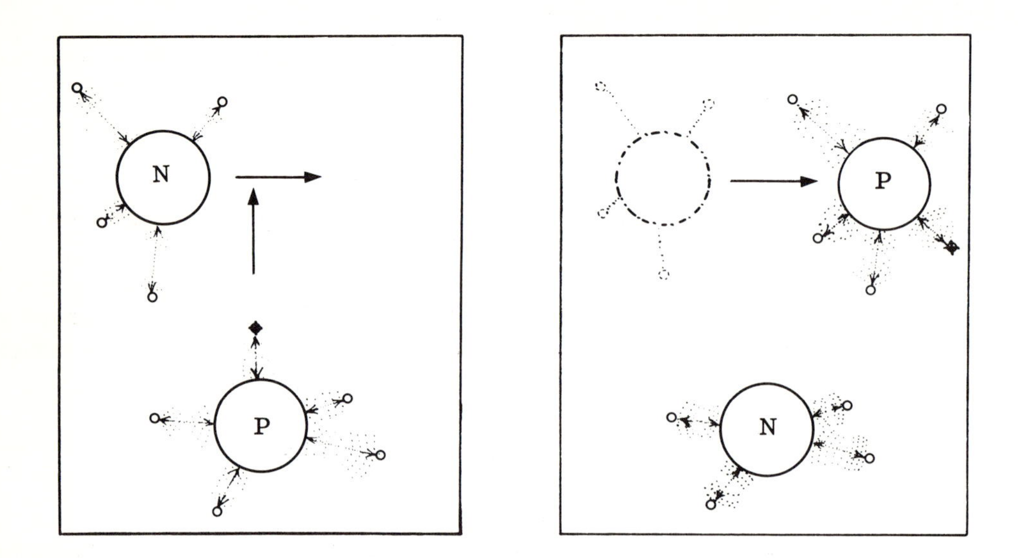

Figure 2.2—COLLISION OF A NEUTRON WITH A PROTON. In half of such collisions a positive pion is exchanged and the neutron and proton appear to change positions.

therefore only natural to associate resonance within the nucleus with stability also. Nuclear forces associated with this increased stability and which involve an exchange of mesons are called "exchange forces."

While Yukawa's mechanism accounts for the very strong attractive force required to hold the nucleus together, at still shorter distances nuclear forces become repulsive. If this were not so nuclei would collapse. A reversal from an attractive force to a repulsive force is possible only if a part of the nuclear force is non-central. Non-central forces are produced by spin. It will be recalled that neutrons and protons, as well as all other particles, are spinning like tops. Nature has simplified the situation through quantized energy states by allowing only two values of spin, i.e., + or − ½ unit. In terms of angular momentum spin can have values of $\pm \frac{1}{2}\hbar$. In the simplest situation the case of only two nucleons may be considered. Each is spinning with an angular momentum of + or − ½. If the spins are parallel the total *spin angular momentum* is 1; if the spins are anti-parallel the total spin angular momentum is 0. Spin produces a force field of the tensor type. Thus a non-central tensor force is produced by nucleons with parallel spins. If the spins are anti-parallel no tensor force exists; the only forces are central.

Another type of non-central force is *spin-orbit force* which results from an interaction between the spins of the nucleons and their orbits. Again the spins of the nucleons must be parallel but the spin angular momentum vectors of the nucleons may be parallel, anti-parallel or perpendicular to the orbital angular momentum vectors. Each of these three possible orientations is not only associated with a different magnitude of force, but the force may actually change from one of attraction to one of repulsion.

Recently an interesting hypothesis was proposed by Sternglass (35, 36). He suggests that nuclear forces arise from interactions between electron-positron pairs. Although it is

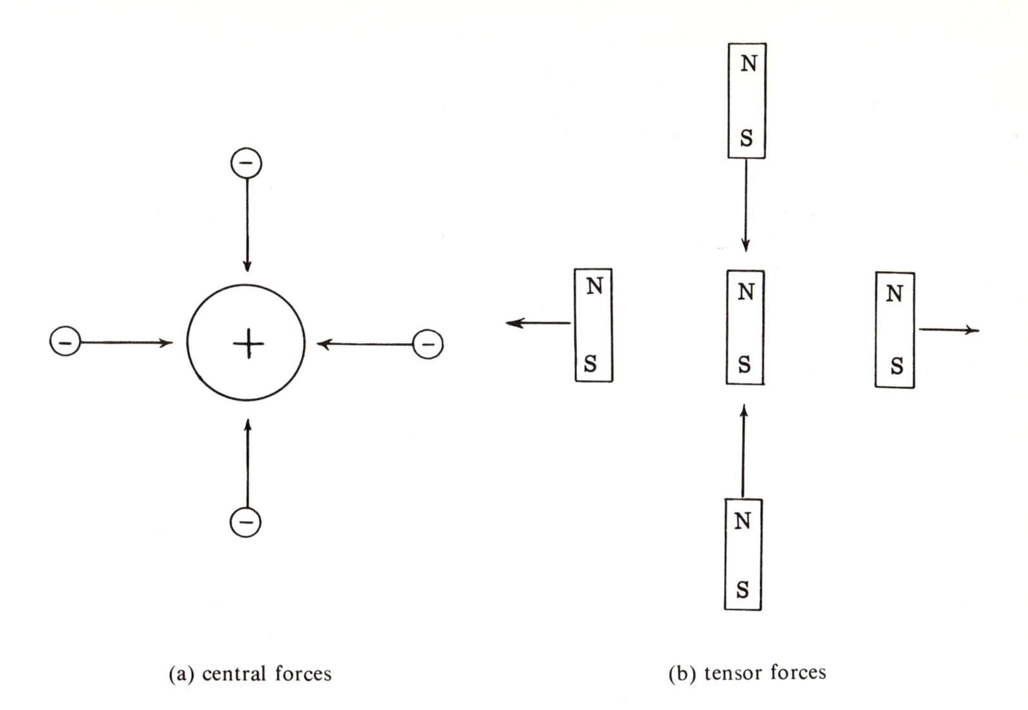

Figure 2.3—CENTRAL AND TENSOR FORCES. Electrostatic and gravitational forces are central since they are independent of orientation while magnetic forces are tensor.

generally accepted that these forces arise from interactions of π-mesons, Sternglass presents evidence that these mesons are, in reality, positron-negatron pairs spinning at relativistic velocities. Because the spin is quantized only certain conditions are allowed. Calculated states correspond to relativistic mass increases for the spinning pair which agree to within a few per cent of observed masses for π-mesons and several other heavy particles. Rotating at relativistic velocities the positron-negatron pairs produce strong magnetic fields which would interact with similar fields in adjacent mesons. The force between two rotating pairs (pions) depends on their orientation and are therefore tensor forces.

NUCLEAR MODELS

A good model must be capable of explaining not only the arrangement of the nucleons within the nucleus but all experimentally observed facts as well. Numerous nuclear models have been proposed, each successful in explaining the behavior of the nucleus for some situations but not for all. Sometimes the models are even in apparent contradiction with one another, yet each has served a useful purpose.

The first attempt to describe the nucleus was the *potential model* proposed by Wigner. Energy levels within a nucleus are represented by a potential energy well. In a simple nuclear transition a nucleon may be thought to change from a higher to a lower energy level by releasing the energy difference in the form of electromagnetic radiation. This process is analogous to the case of an orbital electron seeking a more stable configuration by release of an x-ray or visible photon. While this model has fallen into disuse, it was the forerunner of the *shell model* also similar in many respects to the Bohr atom. In the shell model it is supposed that nucleons occupy shells, each representing a specific energy level.

Experiments involving the capture of neutrons and protons by nuclei demonstrated the existence of very sharp resonant absorption peaks. From the uncertainty principle (see equation 2.4) it follows that if ΔE, the width of a peak, is very small, Δt must be comparatively great, greater by several orders of magnitude than the time required for the incident particle to traverse a nuclear diameter. It was supposed, therefore, that a compound nucleus is probably formed followed, after a finite time, by its decay. This model has been called the *Bohr model,* the *statistical model* or the *compound nucleus model.* This model, too, served as the forerunner of another, the *liquid drop model,* a modification suggested by Niels Bohr and John Wheeler. Nuclei are compared to a droplet of water or mercury with nucleons packed tightly together rather than spaced widely apart as in a solar system. This model was most successful in explaining the mechanism of fission.

For a time the liquid drop model was so successful that the shell model appeared doomed. Then, in 1949, Mayer, Jensen, Haxel and Suess independently observed that the facts could be made to fit a slight modification of the original shell model. It was found that the interaction between two nucleons is a function of their relative orientation and spin and that the resulting spin-orbit coupling forces were very strong. An individual nucleon was also assumed to move or interact relative to the nucleus as a whole rather than primarily with neighboring nucleons. This model is the *individual particle theory.* Its rationale may be seen by comparison to the periodic table of Mendeleef. The properties and characteristics of the nuclei reappear periodically. These characteristics include (1) stability—related to the so-called "magic numbers", (2) low neutron cross sections, (3) lack of spin or magnetic moment and (4) uniformity of shape as defined by the quadrupole moment*.

Still other models have been advanced, among which are the "collective model" of Aage Bohr and Ben Mottelson and the "cloudy-crystal-ball model" of Feshbach, Porter and Weisskopf. The most recent contribution is the "close-packed-spheron theory" of Pauling.

THE POTENTIAL MODEL

Just as visible light and x-rays are emitted by atoms when orbital electrons move to lower energy states, gamma rays are emitted by nuclei as the nucleons seek the lowest energy or ground state. Where the visible and x-ray spectra provide valuable information concerning atomic structure, gamma rays, in turn, provide information with respect to nuclear structure. Nuclei are frequently represented as a *potential energy well* in which the nucleons are assigned energy levels.

* *Quadrupole moment*—The shape of a nucleus is described by its electric quadrupole moment. For nuclei which are perfect spheres the quadrupole moment is zero. If the nucleus has a cigar shape the quadrupole moment is positive; if disc shaped it is negative. The numerical value of quadrupole moment is a measure of the extent of asphericity. The observed quadrupole moments could be explained by the liquid drop model but not by the early shell model.

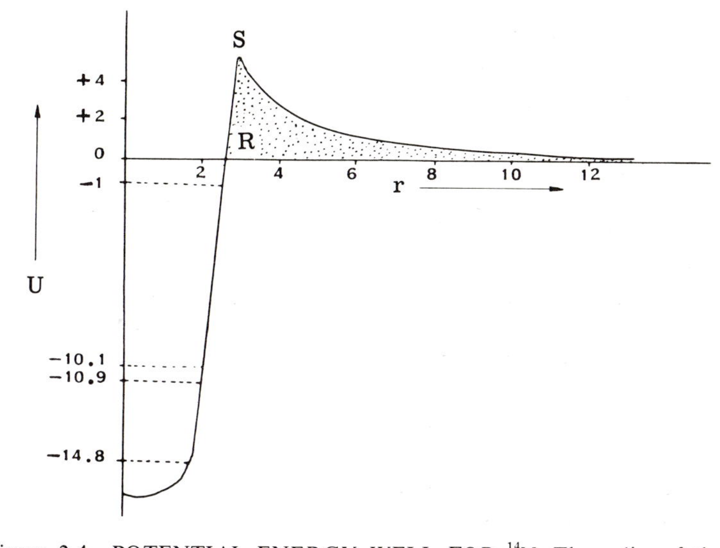

Figure 2.4—POTENTIAL ENERGY WELL FOR ^{14}N. The radius of the nucleus is depicted by the width of the well. R is the nuclear radius. U is the potential energy. For a particle approaching from the right, the curve gives the energy required of the particle to approach the nucleus to a distance r. If the particle passes S and drops into the well it has been captured. It will then fall to one of the horizontal dotted lines representing energy levels within the nucleus. For a particle to escape from the nucleus it must have an energy S to get over the barrier.

THE LIQUID-DROP MODEL

When a high energy particle collides with a nucleus a delay is observed before it, or other resulting radiation, is released by the nucleus. If the nucleus had an open structure, like that expected if it were similar to a miniature solar system, the particle should pass through without delay. Experimentally, however, it has been shown that this does not occur. In nuclear spectroscopy a controlled beam of particles is caused to impinge on a target. Particles of certain specific energies are selectively absorbed and others possessing certain specific values of energy are emitted. This selective or resonant absorption and emission results in the production of very sharp spectral peaks representing very narrow bands of energy. It follows from the uncertainty principle that $\Delta E \cdot \Delta t = \hbar$ (see equation 2.4) or that the energy width Γ is related to the decay time τ by

$$\tau = \hbar \Gamma^{-1} \tag{2.12}$$

Since the range of energies Γ is so narrow it follows that τ must be relatively great. From known values of particle velocity and nuclear diameter one can readily calculate the time required for the particle to traverse a nuclear diameter. It is found that this value is many orders of magnitude smaller than τ. One would therefore conclude that the incident particle remains in the nucleus for a finite time. Although little is known about the nature of such a compound nucleus it is presumed that the particle undergoes many collisions with the nucleons since its mean free path would be quite small compared to a nuclear diameter. During this time the energies of the various particles within the nucleus are statistically redistributed. Ultimately one of them accumulates sufficient energy to escape from the nucleus, the range of energies, i.e., band width, for a particular process being inversely proportional to the decay time of the compound nucleus as defined by equation 2.12. This theory has been called the *Bohr model*, the *statistical model* or the *compound-nucleus model*.

In 1939 Niels Bohr and John Wheeler recognized a certain analogy between the compound-nucleus model and a droplet of liquid. Thus a modification of the compound-nucleus model became known as the *liquid-drop model*. Many nuclear properties could be explained quite accurately by means of this new model.

Since nucleons, whether protons or neutrons, are of nearly the same size and since the nucleons do indeed seem to be packed tightly together, the volume V of a nucleus is nearly proportional to $A^{1/3}$. Thus the radius R of a nucleus is given approximately by

$$R \simeq 1.2 \times 10^{-13} A^{1/3} \text{ cm} \quad (2.13)*$$

Assuming a nearly spherical shape, the volume is then given by

$$V = 4/3\pi R^3 \simeq 3.1\pi A \times 10^{-39} \text{ cm}^3 \quad (2.14)$$

From the value 1.66×10^{-24} g for the mass of the amu the density of the nucleus is found to be approximately 1.7×10^{14} g/cm³. Thus one cm³ of nuclear material would weigh about 200 million tons.

An analogy with a water droplet can also be made with respect to surface tension. A given particle experiences attraction from other particles about it. For an interior particle these forces will, on the average, be equal in all directions with the result that the net force will be approximately zero. But for a particle on the surface that is not so because it will experience a net force toward the center of the droplet. The result is a phenomenon known as surface tension. For a water droplet the magnitude of this force is about 70 dynes per cm while for the nucleus it is about 9×10^{19} dynes per cm.

The compact structure suggested by the liquid drop model also explains the interaction of neutrons with nuclei upon collision more accurately than the earlier shell model. In addition, it provides a mechanism for nuclear fission and provides satisfactory agreement between calculated and observed values of nuclear binding energies.

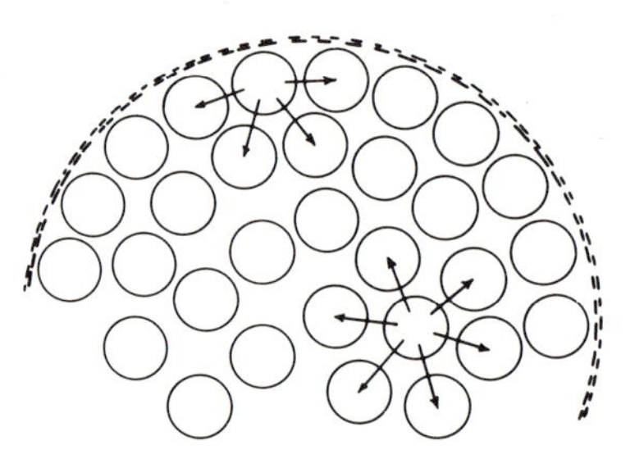

Figure 2.5—BOHR'S LIQUID DROP MODEL. Surface tension is explained by attraction of interior particles for those on the surface of the droplet.

*Reported values for the coefficient range from about 1.1×10^{-13} to 1.45×10^{-13}. The Fermi ($=10^{-13}$cm) is often used to express nuclear radii.

THE SHELL MODEL

Even when the nucleus was thought to contain protons and electrons a shell theory was considered as a possible explanation of nuclear structure. Following the discovery of the neutron, consideration was again given to the possibility of the nuclear particles being arranged in shells. In view of the success of the Bohr theory of the atom it was plausible to assume that the nucleus might possess a similar type of structure.

Even-Odd Rules—Evidence of a shell structure is given by the tendency of nucleons to pair. Of the stable nuclei the greatest number have an even number of both protons and neutrons which suggests that stability is achieved when a shell is filled. Only four stable nuclides have an odd number of both protons and neutrons. A summary of stable nuclei is given in table 2.1.

Table 2.1

Z	N	Number of Stable Nuclei
even	even	164
even	odd	55
odd	even	50
odd	odd	4

It would also seem pertinent to point out here a somewhat similar situation concerning the radioactive nuclides. It will be noted, from the chart of the nuclides, how consistently radioisotopes with an even Z and/or an even N have a greater half-life than that of an adjacent isotope or isotone.

Magic Numbers — The observation of so-called magic numbers of neutrons and protons which tend to produce a particularly stable configuration lends further impetus to the shell theory. Although reference was made to such an observation by Elsasser in 1933, it was not until 1948 that the relationship was rediscovered and explained by Maria Goeppert-Mayer and Edward Teller. The magic numbers now known are 2, 8, 20, 28, 50, 82 and 126. It is observed, for example, that helium nuclei (alpha particles) with two neutrons and two protons are exceptionally stable. Oxygen with eight neutrons and eight protons, calcium with twenty of each nucleon and lead with 82 protons and 126 neutrons show similar evidence of stability. A list of nuclides corresponding to the magic numbers is found in table 2.2.

Magic numbers can be identified from experimental data in various ways. Several of the more obvious are based on a consideration of:

1. *Natural abundance of the nuclides* —^{4}He constitutes almost 100% of natural helium. All of its isotopes, isotones and isobars are either less abundant, less stable or non-existent. The same statement can be made of ^{16}O and ^{40}Ca.

2. *Relative number of stable isotopes or isotones*—Tin (Z = 50) has 10 stable isotopes, more than any other nuclide; adjacent elements indium (Z = 49) and antimony (Z = 51) have only two stable nuclides each. Other relationships can be discovered by consulting table 2.2 and the chart of the nuclides.

3. *Decay energies* —When alpha or beta decay energies are plotted as a function of Z or N, discontinuities are observed in otherwise smooth curves. These discontinuities occur at values of Z and N corresponding to magic numbers.

Table 2.2—NUCLIDES CORRESPONDING TO MAGIC NUMBERS

2	8	20	28	50	82	126
			Neutrons			
^{4}He	^{15}N	^{36}S	^{48}Ca	^{86}Kr	^{136}Xe	^{208}Pb
	^{16}O	^{37}Cl	^{50}Ti	^{87}Rb	^{138}Ba	^{209}Bi
		^{38}Ar	^{51}V	^{88}Sr	^{139}La	
		^{39}K	^{52}Cr	^{89}Y	^{140}Ce	
		^{40}Ca	^{54}Fe	^{90}Zr	^{141}Pr	
				^{92}Mo	^{142}Nd	
					^{144}Sm	
			Protons			
^{4}He	^{16}O	^{40}Ca	^{58}Ni	^{112}Sn	^{204}Pb	
	^{17}O	^{42}Ca	^{60}Ni	^{114}Sn	^{206}Pb	
	^{18}O	^{43}Ca	^{61}Ni	^{115}Sn	^{207}Pb	
		^{44}Ca	^{62}Ni	^{116}Sn	^{208}Pb	
		^{46}Ca	^{64}Ni	^{117}Sn		
		^{48}Ca		^{118}Sn		
				^{119}Sn		
				^{120}Sn		
				^{122}Sn		
				^{124}Sn		

Individual-Particle Theory—In view of the preceding evidence for certain periodic properties of nuclei, a periodicity strikingly similar in nature to that observed by Mendeleef for the elements, interest was renewed in searching for a nuclear model analogous to the atomic model of Bohr. That the facts could be made to fit a slight modification of the original shell model was independently observed by Maria Goeppert-Mayer at the University of Chicago and J. D. H. Jensen at Heidelberg. This new model was to become known as the *individual-particle theory.* Success of the individual-particle theory lay in the recognition of the importance of spin-orbit coupling.

An individual nucleon was assumed to move or interact relative to the nucleus as a whole rather than primarily with neighboring nucleons. For simplicity this assumption is generally made in the case of the atom where the interaction of a particular electron is assumed to be entirely with the nucleus, the other orbital electrons being ignored because of their insignificant mass. To apply this same concept to the nucleus, where all particles are of approximately equal mass and where they are tightly packed together, seems unreasonable yet it has met with considerable success.

In chapter 1 it was seen that orbital angular momentum is quantized and is therefore an integral multiple of the Planck constant $\hbar$. That is, $p_\varphi = l\hbar$. Further, a given quantum level l contains $2l + 1$ discrete states of space orientation. These states were identified by a magnetic quantum number m, where m is an integer and $-l \leqslant m \leqslant l$. For a given quantum level l, all values of m represent the same energy. For each m there are two possible spin orientations, the spin angular momentum p_φ having the values + or $- \frac{1}{2}\hbar$. The Pauli exclusion principle does not allow more than one electron to have the same set of quantum numbers. Since there are $2l + 1$ states of m, each containing up to two electrons, then each energy level l can be occupied by up to $2(2l + 1)$ electrons. One might refer to this model as the individual orbit model of the atom.

In the individual particle theory the same basic approach which has been so successful with the atom has been applied to the nucleus. One objection which could be raised immediately is that the two situations are, in reality, so different.

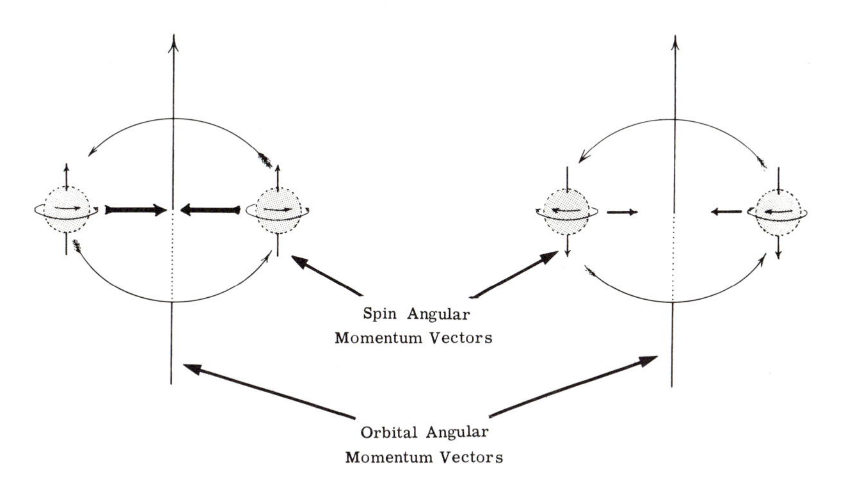

Figure 2.6—SPIN-ORBIT COUPLING. If two particles spin in the same direction as their orbit, the spin-orbit force is strong; when they spin in opposite directions the force is weak.

In the nucleus, for example, it would be expected that one nucleon would collide with another long before completing one orbit. However, such an event is not likely because the Pauli exclusion principle forbids it. In the nucleus, however, there are two kinds of particles—neutrons and protons. Here an application of the Pauli exclusion principle means that a given quantum level l cannot contain more than $2(2l + 1)$ particles of a kind or a total of $4(2l + 1)$ nucleons.

Spin-orbit coupling, mentioned briefly in regard to the atom, becomes especially important in the case of the nucleus. The Russell-Saunders l-s coupling, characteristic of the atom, occurs predominately in light nuclei while l-l coupling becomes more important with medium weight and heavy nuclei. In table 2.3 nucleons are listed according to their quantum numbers without regard for the energy associated with a level and without consideration for spin-orbit coupling. A comparison can be made with a similar table (table 1.2) for the atom.

One of the principal difficulties with the original shell theory was its inability to account for the magic numbers beyond 20. The old model is represented by the first three

Table 2.3—SHELL MODEL OF THE NUCLEUS. Nucleons are listed according to their quantum number. (See Table 2.1 for comparison.)

Quantum Number			Distribution of Nucleons		
l Orbital Angular Momentum	s Spin	j Total Angular Momentum	2j + 1 Nucleons of One Kind in a Subsubshell	$2(2l + 1)$ Nucleons of One Kind in a Subshell	Subshell
0	±1/2	1/2	2	2	1s
1	+1/2	3/2	4		
	−1/2	1/2	2	6	1p
2	+1/2	5/2	6		
	−1/2	3/2	4	10	1d
3	+1/2	7/2	8		
	−1/2	5/2	6	14	1f
4	+1/2	9/2	10		
	−1/2	7/2	8	18	1g
5	+1/2	11/2	12		
	−1/2	9/2	10	22	1h
6	+1/2	13/2	14		
	−1/2	11/2	12	26	1i
0	±1/2	1/2	2	2	2s
1	+1/2	3/2	4		
	−1/2	1/2	2	6	2p
2	+1/2	5/2	6		
	−1/2	3/2	4	10	1d
3	+1/2	7/2	8		
	−1/2	5/2	6	14	2f
4	+1/2	9/2	10		
	−1/2	7/2	8	18	2g
0	±1/2	1/2	2	2	3s
1	+1/2	3/2	4		
	−1/2	1/2	2	6	3p
2	+1/2	5/2	6		
	−1/2	3/2	4	10	3d
0	±1/2	1/2	2	2	4s

Table 2.4—SHELL MODEL OF THE NUCLEUS. Nucleons listed according to energy. Magic numbers correspond to $\Sigma\,(2j + 1)$.

Old Shell Model			New Model—Result of Spin-Orbit Coupling			
Total to a Shell	Number per Subshell	Subshell	Splitting of Energy Levels	j	2j + 1	Σ (2j + 1)
2	2	1s		1/2	2	2
				3/2	4	
8	6	1p		1/2	2	8
				5/2	6	
	10	1d		1/2	2	
20	2	2s		3/2	4	20
				7/2	8	28
	14	1f				
				3/2	4	
40	6	2p		5/2	6	
				1/2	2	
				9/2	10	50
	18	1g		5/2	6	
	10	2d		7/2	8	
				3/2	4	
70	2	3s		1/2	2	
				11/2	12	82
	22	1h				
				9/2	10	
	14	2f		7/2	8	
				5/2	6	
				3/2	4	
112	6	3p		1/2	2	
				13/2	14	126
	26	1i				

columns of table 2.4. The order in which the shells are listed is, in part, determined by parity; all levels within a shell must have the same parity. That is, they must contain either only even or only odd values of l. With such a sequence the resulting series of numbers—2, 8, 20, 40, 70 and 112—does not agree with the observed series for the magic numbers.

Each nucleon is known to rotate just as an orbital electron rotates or spins. The spin of a nucleon is quantized and may have a value of + or − $\frac{1}{2}\hbar$ depending upon whether the spin is parallel or antiparallel to the orbital angular momentum l. The vector sum of these rotations imparts a *total angular momentum* j to the nucleus as a whole, thus $j = l + \frac{1}{2}$ or $j = l - \frac{1}{2}$. The term "spin" is also used to mean total angular momentum and should not be confused with "spin of a nucleon". Spin causes a nucleus to behave as a tiny magnet.

Nuclear spin can be measured experimentally and should agree with the theoretical value arrived at by use of a particular nuclear model.

Finally, the result of introducing spin-orbit coupling into the shell theory of the nucleus can be seen by considering the right side of table 2.4. The sequence for listing the values for spin (i.e., j) must, for the most part, be determined by experiment. There is no question, however, concerning the correctness of the order in which they are listed since ample experimental proof, especially through nuclear spin measurements, has been obtained to support it.

CLOSE-PACKED-SPHERON THEORY

A theory which is basically an extension of the shell theory but which is compatible with both it and the liquid-drop model was proposed by Pauling in late 1965. This new model explains, for example, the asymmetry with which heavy nuclei fission. In addition, while the shell model explains why certain numbers can be magic, it does not point out why about 20 others corresponding to the filling of subsubshells are not. The new model does.

Pauling assumes that in nuclei the nucleons form small clusters called spherons. These spherons are generally helions (two protons and two neutrons), tritions (a proton

Table 2.5—NUCLEON CONFIGURATIONS FOR THE MAGIC NUMBERS
From Pauling (66).

Magic Number	Mantle	Core or Outer Core	Inner Core
2	$1s^2$		
8	$1s^2 1p^6$		
20	$2s^2 1p^6 1d^{10}$	$1s^2$	
50	$2s^2 2p^6 1d^{10} 1f^{14} (1g9/2)^{10}$	$1s^2 1p^6$	
82	$3s^2 2p^6 2d^{10} 1f^{14} 1g^{18} (1h\,11/2)^{12}$	$2s^2 1p^6 1d^{10}$	$1s^2$
126	$3s^2 3p^6 2d^{10} 2f^{14} 1g^{18} 1h^{22} (1i\,13/2)^{14}$	$2s^2 2p^6 1d^{10} 1f^{14}$	$1s^2 1p^6$

and two neutrons) and a dineutron (two neutrons). The close-packed-spheron theory differs from the liquid drop model in having spherons rather than nucleons as the basic building block.

Maximum stability is achieved when each spheron ligates about itself the maximum number of neighbor spherons in a closest-packed structure. Nuclei may consist of one, two or three layers of spherons depending upon the number of spherons in their structure. These layers are called the *mantle*, the *outer core* and the *inner core*. The nucleon configurations for the magic numbers are found in table 2.5.

In the close-packed spheron theory the magic numbers represent the completion of a shell (K, L, M, ... with $2n^2$ neutrons or protons), or the completion of a shell by a core layer and the completion of a shell and a $j = l + \frac{1}{2}$ subsubshell.

NUCLEAR EQUATIONS

The conventions used for writing nuclear equations may be illustrated by the first observed nuclear reaction reported by Rutherford in 1919. He bombarded nitrogen with alpha particles and observed the formation of protons. These observations are represented by the reaction

$$^{14}_{7}N + ^{4}_{2}He \longrightarrow ^{17}_{8}O + ^{1}_{1}H \qquad (2.15)$$

As a chemical equation must balance, so must a nuclear equation. It will be noted that the sum of the mass numbers of the reactants $(14 + 4 = 18)$ is equal to the sum of the mass numbers of the products $(17 + 1 = 18)$. Likewise, the sum of the atomic numbers of the reactants $(7 + 2 = 9)$ is equal to the sum of the atomic numbers of the products $(8 + 1 = 9)$.

Often an abbreviated equation is used in place of the formal notation illustrated by equation 2.15. Because the atomic numbers are defined by the chemical symbol, Z is sometimes omitted. Thus equation 2.15 becomes

$$^{14}N + ^{4}He \longrightarrow ^{17}O + ^{1}H \qquad (2.16)$$

By convention this notation may be shortened still further. ^{4}He is an alpha particle and is replaced by the symbol α. Similarly, the hydrogen nucleus, being a proton, is represented by p. The equation is then written

$$^{14}N\ (\alpha, p)\ ^{17}O \qquad (2.17)$$

The *target nucleus*, in this case ^{14}N, should precede the parenthesis, while the bombarding particle and the particle or photon produced, are placed within the parentheses.

NUCLEAR ENERGETICS

Each nuclide has an *atomic mass* or *isotopic mass* M usually expressed in atomic mass units (amu) and a *mass number* A. The isotopic mass is very important in nuclear energetics because energy and mass are interrelated. This relationship is given by the Einstein equation,

$$E = mc^2 \qquad (2.18)$$

and the energy equivalent of a change in mass Δm is

$$\Delta E = \Delta mc^2 \qquad (2.19)$$

Such changes in mass occur not only when matter is annihilated, as by a collision of antiparticles, but also when protons and neutrons are changed from one configuration to another. Thus the sum of the masses of two deuterons is not equal to the mass of an alpha particle, even though the total number of each type of nucleon remains constant. This results because of the difference in energy with which the nucleons are bound in a deuteron and in an alpha particle.

In 1961 the International Union of Chemistry and Physics adopted ^{12}C as the standard of mass and the unit of mass was defined as 1/12th the mass of the ^{12}C atom. The absolute value of the mass unit is

$$1 \text{ amu} = 1.660420 \times 10^{-24} \text{g} = 931.48 \text{ MeV} \qquad (2.20)$$

By means of mass spectrometry the mass of the other atoms can be compared to the mass of the ^{12}C atom with an accuracy better than one part in 10^5. The reported values for M are generally those for the neutral atom including the orbital electrons. The mass of a nucleus, stripped of all orbital electrons is generally represented by the symbol M′. To obtain nuclear mass M′ from atomic mass M, electron masses must be subtracted and the sum of the electron binding energies must be added

$$M' = M - Zm_0 + Z\bar{B}_e \qquad (2.21)$$

Here $\bar{B}_e$ is the average binding energy per electron. Usually the binding energy term is very small and can be neglected.

Although masses of the nuclides are nearly integral multiples of one amu (to within 0.1 amu) the exact masses are less than the sum of the masses of the component nucleons. This difference in mass is equivalent to the binding energy of a nucleus. A calculation of the nuclear binding energy B_{He} for the helium nucleus will serve as an illustration. The helium nucleus contains two protons and two neutrons. The mass change ΔM during formation of the helium nucleus from the individual nucleons would then be

$$\begin{aligned} \Delta M &= 2M_H + 2M_n - M_{He} \qquad (2.22) \\ &= 2(1.007825) + 2(1.008665) - 4.002604 \\ &= 0.030376 \text{ amu} \end{aligned}$$

Note: M_H is used rather than M_p to account for the mass of the electrons in the neutral atom.

Mass defect is the difference between the mass of an atom and the sum of the free masses of its constituents. In the above example in which two neutrons and two hydrogen atoms combine to form a helium atom the mass defect is 0.030376 amu as calculated by use of equation 2.22.

The *binding energy* of nuclei may be compared with the binding energy which holds atoms together to form molecules with the difference that the binding energy of the nucleus is over 10^6 times greater. Mass defect represents the total binding energy of a nucleus and in view of the equivalence of mass and energy through equation 2.18 they may be considered to be equivalent.

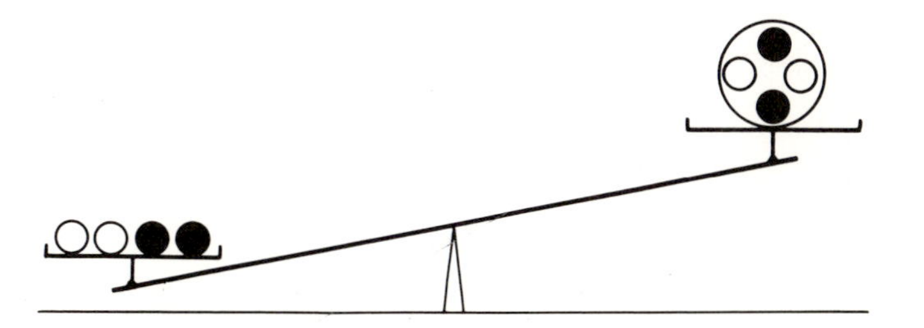

Figure 2.7—MASS DEFECT AND BINDING ENERGY. Energy is released when two neutrons and two protons combine to form a helium nucleus. The result is a loss of mass called the mass defect.

If the mass change in amu, ΔM, is converted to grams, Δm, then by use of equation 2.18 the energy is obtained in ergs. To obtain the energy in electron volts the conversion factor

$$1 \text{ amu} = 931.48 \tag{2.23}$$

is useful. The binding energy of the helium nucleus is

$$B_{He} = 931.48\,(0.030376) = 28.29 \text{ MeV} \tag{2.24}$$

The *average binding energy per nucleon* is B/A. Therefore, the value for the ^4He nucleus is 28.29/4 = 7.07 MeV. B/A is sometimes called the *binding fraction.*

When a nuclear reaction occurs an amount of energy Q is either released or absorbed. The binding energy B is a special case of Q. If energy is released, Q is positive and the reaction is said to be *exergonic* or *exoergic.* If energy is absorbed (i.e., external energy must be supplied to initiate and to sustain the reaction) Q is negative and the reaction is *endergonic* or *endoergic.* Q can be calculated if the isotopic masses of all reactants and products are known. For the reaction given by equation (2.15),

$$\begin{aligned} Q &= 931\,[M_{^{14}N} + M_{^{4}He} - (M_{^{17}O} + M_{^{1}H})] \\ &= 931\,[14.00307 + 4.00260 - (16.99913 + 1.00782)] \\ &= -1.19 \text{ MeV (endoergic)} \end{aligned} \tag{2.25}$$

The value −1.19 MeV is the energy per atom consumed in the reaction by conversion of energy into mass, since, in this example, the sum of the masses of the products is greater than the sum of the masses of the reactants. The source of this energy is the kinetic energy of the alpha particle and is equivalent to 2.70×10^{10} calories per gram atom. By comparison with chemical reactions the energy involved in nuclear reactions is generally greater by a factor of about 10^6.

NUCLEAR STABILITY

Certain combinations of protons and neutrons produce stable nuclei whereas others do not. A critical examination of the parameters resulting in stability or in instability reveals certain useful relationships even though some of these relationships are entirely or partially empirical.

The *neutron-proton ratio* is equal to or very nearly equal to unity for the light nuclides. Here nuclear stability is achieved when the number of neutrons is approximately equal to the number of protons. For deuterium, helium-4, lithium-6, boron-10, carbon-12, oxygen-16 and others, the ratio is exactly 1.0. Above calcium-40 ($Z = 20$, $N = 20$) no stable nuclide exists with equal numbers of neutrons and protons for as Z increases, the repulsive coulombic forces increase at a greater rate than the attractive nuclear forces. The addition of extra neutrons is necessary to increase the average distance between protons within the nucleus thereby reducing the coulombic force according to the inverse-square law. For heavy nuclei the n/p ratio is 1.50 or greater. For stable mercury-200 the ratio is 1.50, while for lead-208, the heaviest stable nuclide, the ratio is 1.53.

If a nuclide is too rich in neutrons it will be unstable. Stability may be achieved by conversion of a neutron into a proton within the nucleus

$$n \longrightarrow p + e^- + \upsilon \quad (2.26)$$

On the other hand, if a nucleus is too rich in protons it is also unstable. Stability is now achieved through an increase in the n/p ratio. This can be accomplished by

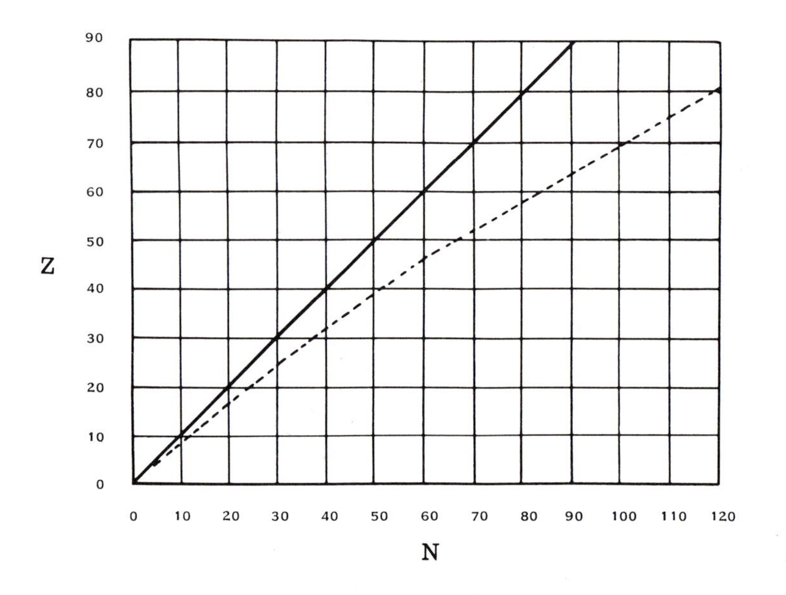

Figure 2.8—NEUTRON-PROTON RATIO. The solid diagonal line represents a neutron-proton ratio of one. The curved dashed line indicates the neutron-proton ratios of the stable nuclides.

conversion of a proton into a neutron within the nucleus. Two possible reactions are

$$p \longrightarrow n + e^{+} + \nu \quad (2.27)$$

$$p + e^{-} \longrightarrow n + \nu \quad (2.28)$$

The first reaction (2.27) results in the emission of a positron. The second reaction (2.28) involves electron capture by the nucleus. In both cases a neutrino is emitted to conserve energy and momentum.

The *stability diagonal,* indicated on some charts of the nuclides, is a line drawn through the most stable nuclides. Because most elements have several stable isotopes, the location of the line is often a compromise but it nevertheless serves as a very useful guide. The farther a nuclide falls from the stability diagonal the less stable it is likely to be. The experimental stability curve follows closely the theoretical relation (68):

$$Z = \frac{A}{2 + 0.0146\,A^{2/3}} \quad (2.29)$$

Nuclear stability is a function of the pairing of nucleons as predicted by the shell model of the nucleus. The effect of pairing is summarized by the *even-odd rules:* (1) stable nuclei of even Z are more numerous than stable nuclei of odd Z, (2) stable nuclei of even N are more numerous than stable nuclei of odd N, (3) stable nuclei of even A are more numerous than stable nuclei of odd A, and (4) nearly all stable nuclei of even A have even Z; the exceptions are ^{2}H, ^{6}Li, ^{10}B and ^{14}N. A summary of these data is presented in table 2.1.

The *packing fraction* f is defined by

$$f = \frac{M - A}{A} \quad (2.30)$$

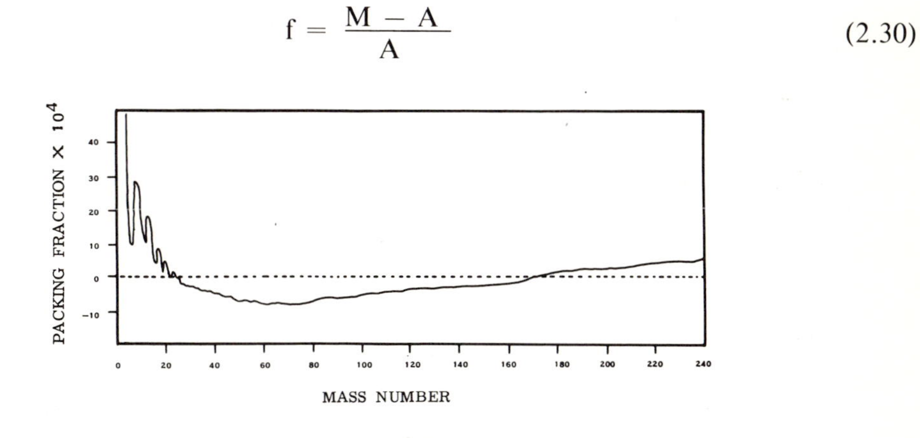

Figure 2.9—PACKING FRACTION CURVE

For convenience, values of f are generally multiplied by 10^4. If the packing fraction is plotted as a function of A, a minimum is observed from A = 50 to A = 60. Here one finds the most stable nuclides, such as iron and nickel. Where f increases to large values, for the light elements fusion becomes a possibility and for the heavy elements fission becomes a possibility.

One of the most satisfactory ways of measuring nuclear stability is by means of *separation energies* S. The separation energy is the energy required to separate a single nucleon (proton or neutron) or an aggregate of nucleons (deuteron, alpha particle, etc.) from a specific nucleus. The masses of the parent and daughter nuclei and of the particles involved must be known. Separation energies provide the means to predict if alpha decay, for example, is possible. The case of polonium-210, known to decay by alpha emission, is cited as an example.

$$\begin{array}{cccc} {}^{210}\text{Po} & \longrightarrow & {}^{206}\text{Pb} & + \quad {}^{4}\text{He} \\ 210.0495 & & 206.0386 & \quad 4.00388 \end{array} \qquad (2.31)$$

$$S = 931\,(M_{^{206}Pb} + M_{^{4}He} - M_{^{210}Po}) \qquad (2.32)$$

$$S = 931\,(206.0386 + 4.00388 - 210.0495)$$

$$S = -6.5 \text{ MeV}$$

A negative value of S implies that decay by alpha emission is energetically possible (since S = −Q and the decay of polonium-210 is therefore exoergic) while a positive value would imply that the reaction is endoergic and hence energetically impossible without the addition of external energy.

NUCLEAR REACTIONS

Nuclear reactions may be either spontaneous or induced. Spontaneous reactions occur only when nuclear instability exists. Such nuclides are said to be radioactive. Certain unstable heavy nuclides also undergo spontaneous fission. All spontaneous reactions are exoergic.

Radioactive decay by negatron emission occurs when the n/p ratio is too high. Occasionally, when the n/p ratio is extremely high, as it would be in the case of a fission product, a neutron may be emitted but this process is not found with the commonly used isotopes. If the isotope has an excess of protons, decay may occur either by positron emission or by electron capture, depending upon the available energy. Radioactive decay processes are discussed in detail in chapter 6.

Induced reactions occur when nuclei are struck by high volocity particles. The kinetic energy of the particle supplies the energy required to initiate the reaction. Such reactions may be endoergic or exoergic. Induced reactions may involve a neutral particle (a neutron) or a charged particle (proton, deuteron, etc.). Where neutron reactions are involved, a source of neutrons such as a nuclear reactor is required. Neutrons cannot be accelerated in cyclotrons or similar particle accelerators and consequently cannot be produced by such machines except as the product of a reaction caused by the bombardment of a suitable target with a charged particle, e.g., ^{3}H (d,n)^{4}He. Reactions occurring in a pile include fission, neutron capture or activation and transmutation.

Fission is an exergonic process in which a heavy nucleus is split into approximately equal pieces. Fission may be initiated when a heavy nucleus is struck by and absorbs a neutron. The activated nucleus pulsates rapidly like a drop of liquid and splits into two, and occasionally into three, pieces. The nuclides produced most frequently by fission lie near atomic number 42 (molybdenum) and atomic number 56 (barium) but may range from $Z = 30$ to $Z = 64$. It is thus seen that fissioned nuclei rarely split into two equal pieces. Over 40 modes of splitting are known which produce more than 80 direct products and over two hundred product nuclides, including those formed by subsequent decay.

Fission of some nuclei (^{233}U, ^{235}U, ^{239}Pu, ^{241}Am, etc.) may be caused by either fast or slow neutrons while other more stable nuclei (^{232}Th, ^{238}U, etc.) require fast neutrons to initiate fission. In other instances fission has been produced by bombardment with protons, deuterons or gamma rays. The greater the stability of the target nucleus, the greater the energy of the bombarding particle must be in order to initiate fission. A typical example is the fission of a ^{235}U nucleus.

$$^{235}_{92}U + ^{1}_{0}n \longrightarrow ^{236}_{92}U \longrightarrow ^{90}_{36}Kr + ^{144}_{56}Ba + 2\,^{1}_{0}n$$

The fission products $^{90}_{36}Kr$ and $^{144}_{56}Ba$ have an excessively high n/p ratio and consequently have very short half lives. Some of these highly radioactive products may even emit neutrons but most decay by negatron emission. For the cited cases the following fission product chains would be anticipated. (See equation 2.26.)

$$^{90}_{36}Kr \longrightarrow ^{90}_{37}Rb \longrightarrow ^{90}_{38}Sr \longrightarrow ^{90}_{39}Y \longrightarrow ^{90}_{40}Zr$$

$$^{144}_{56}Ba \longrightarrow ^{144}_{57}La \longrightarrow ^{144}_{58}Ce \longrightarrow ^{144}_{59}Pr \longrightarrow ^{144}_{60}Nd$$

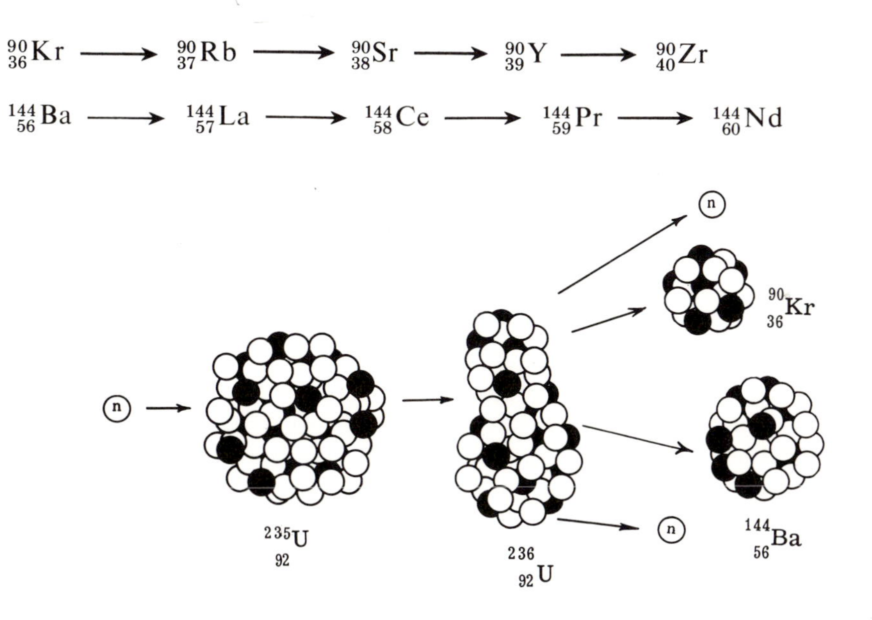

Figure 2.10—FISSION OF ^{235}U

Fusion is the joining of light nuclei to form a heavier nucleus. In order to bring about the fusion of nuclei the components must be brought together against very great forces of electrostatic repulsion. This is accomplished by the acceleration of the particles to very high velocities. By the kinetic theory these velocities represent temperatures of millions of degrees. Only then will the particles collide with sufficient energy to fuse. Fusion reactions are exoergic and release large amounts of energy. If a sufficient amount of this energy is retained by the system the process of fusion is sustained. Such a sustained fusion reaction accounts for the vast amount of energy produced by the sun. Typical fusion reactions which may be useful for the production of power are

$$^{2}H + ^{2}H \longrightarrow ^{3}He + ^{1}n + 3.25\ MeV$$

$$^{2}H + ^{2}H \longrightarrow ^{3}H + ^{1}p + 4.0\ MeV$$

$$^{2}H + ^{3}H \longrightarrow ^{4}He + ^{1}n + 17.6\ MeV$$

$$^{2}H + ^{3}He \longrightarrow ^{4}He + ^{1}p + 18.3\ MeV$$

If reactions of this type can be harnessed there will be no shortage of fuel for power for many years, for although there is only one ^{2}H atom for every 6700 atoms of ^{1}H in nature, it has been calculated that the deuterium in a gallon of water will provide approximately 150 times as much energy as is produced by a gallon of oil.

Most commercial isotopes are produced in nuclear reactors. Several methods of production currently employed illustrate *neutron induced reactions.*

1. *Fission*—Fragments obtained from the fission of uranium atoms yield radioactive nuclides with atomic number in the range from 30 to 64. These nuclides may be separated, concentrated and purified. Often reactor products contain several isotopes of a particular element. For the preparation of high purity materials one of the following processes is generally used.

2. *Activation* (by neutron capture)—The *(n,γ) Process* is the most common method for producing isotopes and involves the capture of a neutron with the subsequent emission of a gamma ray. Since there is no change in the number of protons, the chemical identity of the target remains the same, and the result is a mixture of radioactive and stable isotopes.

$$^{197}Au\,(n, \gamma)\,^{198}Au$$

Occasionally a radioactive product may decay to a daughter which is also radioactive. The daughter nuclide can then be separated from the target nuclide which results in a product of high specific activity, i.e., a product possessing a large number of radioactive atoms relative to the number of stable atoms. The production of ^{77}As illustrates the technique.

$$^{76}Ge\,(n, \gamma)\,^{77}Ge$$

$$^{77}Ge \xrightarrow[(11h)]{\beta^-} {}^{77}As$$

3. *Transmutation* is a process in which one element is converted into another. This type of reaction is illustrated by the production of phosphorus-32 from sulfur.

$$^{32}S\,(n, p)\,^{32}P$$

(n,p) reactions require high energy neutrons in order to eject the proton from the nucleus. The advantage of such a production method lies in the ease with which the radioactive

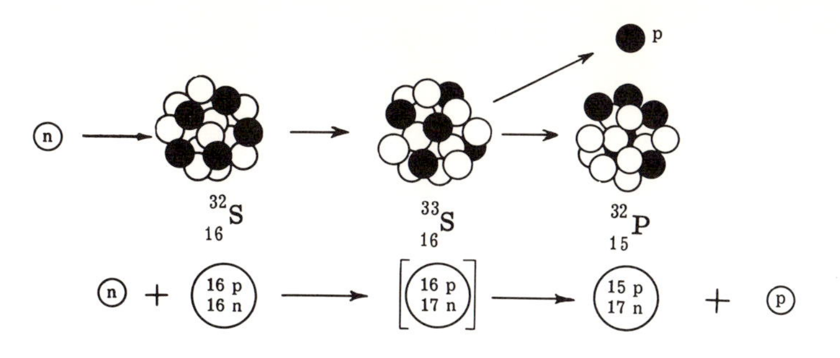

Figure 2.11—TRANSMUTATION OF SULFUR TO PHOSPHORUS. The sulfur-33 produced as an intermediate in this reaction is in an excited state and immediately decays to phosphorus-32.

product, in this case ^{32}P, can be separated from the target ^{32}S by chemical means. Argon-37 is produced by transmutation from calcium-40. In this case an alpha particle is ejected by the incident neutron.

$$^{40}\text{Ca}(\text{n}, \alpha)\,^{37}\text{Ar}$$

Cyclotron reactions—By means of cyclotrons and other particle accelerators, charged particles may be accelerated to a sufficiently high energy to react with target nuclei. Sodium-22 is prepared by deuteron bombardment of a magnesium target. The reaction is

$$^{24}\text{Mg}(\text{d}, \alpha)\,^{22}\text{Na}$$

Because the operation of cyclotrons and similar devices depends upon magnetic and electrostatic interactions with the accelerated particles, only charged particles such as deuterons, protons, etc. may be used.

UNITS OF RADIOACTIVITY

Because radium was the first radioactive element of importance and utility, it was only natural that it should become the unit of radioactivity. Thus the *curie* was originally defined in terms of the disintegration rate associated with one gram of radium. An accurate measurement of the rate was found to be extremely difficult. The value 3.7×10^{10} disintegrating atoms per second per gram of radium was adopted as the value of the curie. Though originally applicable only to radium, the curie unit is now applied to all radioactive nuclides. Thus one curie of sulfur-35 means that amount of sulfur-35 required to provide 3.7×10^{10} disintegrating atoms per second. It is often more convenient to express activities in terms of megacuries, millicuries or microcuries.

1 megacurie (MCi) $= 3.7 \times 10^{16}$ dps $= 10^{6}$ curies
1 curie (Ci) $= 3.7 \times 10^{10}$ dps
1 millicurie (mCi) $= 3.7 \times 10^{7}$ dps $= 10^{-3}$ curie
1 microcurie (μCi) $= 3.7 \times 10^{4}$ dps $= 10^{-6}$ curie

REFERENCES

1. Jensen, J. H. D., "The History of the Theory of Structure of the Atomic Nucleus", Science 147, 1419-1423 (1965).
2. Kramer, A. W., "Nuclear Energy—What It Is and How It Acts", Atomics (1961-1965). A series of articles which has been running continuously in Atomics since July 1961. Now available as reprints in four volumes at $1.00 per volume from Editor, Atomics, Technical Publishing Company, 308 East James St., Barrington, Ill.
3. Yukawa, Hideki, "On the Interaction of Elementary Particles I", Proceedings of the Physico-Mathematical Society of Japan (3) 17, 48 (1935). Complete reprint appears in reference 4 below.
4. Beyer, R. T. (ed.), "Foundations of Nuclear Physics", Dover (1949).
5. Fermi, Enrico, "Nuclear Physics", Univ. Chicago Press (1950).
6. Eisenbud, Leonard and Eugene P. Wigner, "Nuclear Structure", Princeton Univ. Press (1958).
7. de Broglie, Louis, "Physics and Microphysics", Harper (1960).
8. Hoffman, Banesh, "The Strange Story of the Quantum", Dover (1959).
9. Hughes, Donald J., "The Neutron Story", Doubleday (1959).
10. Bethe, H. A., "Elementary Nuclear Theory", Wiley (1947).
11. Physics Staff Univ. of Pittsburgh, "An Outline of Atomic Physics", Wiley (1937).
12. Semat, Henry, "Introduction to Atomic and Nuclear Physics", Rinehart (1959).
13. Evans, Robley D., "The Atomic Nucleus", McGraw-Hill (1955).
14. Friedlander, Gerhart, and Joseph W. Kennedy, "Nuclear and Radiochemistry", Wiley (1964).
15. Bragg, Sir William, "Concerning the Nature of Things", Dover (1925).
16. Romer, Alfred, "The Restless Atom", Doubleday (1960).
17. Catalog and Price List, "Radioisotopes Special Materials and Services", Oak Ridge National Laboratory (1960).
18. Hahn, Otto, "The Discovery of Fission", Sci. Amer., p. 76 (Feb. 1958).
19. Ruderman, M. A. and A. H. Rosenfeld, "An Explanatory Statement on Elementary Particle Physics", Amer. Scientist 48, No. 2, 209 (June 1960).
20. Beard, David Breed, "The Atomic Nucleus", Amer. Scientist 45, No. 4, 333 (Sept. 1957).
21. Peierls, R. E., "The Atomic Nucleus", Sci. Amer., p. 75 (Jan. 1959).
22. Hofstadter, Robert, "The Atomic Nucleus", Sci. Amer., p. 55 (July 1956).
23. Weisskopf, Victor F. and E. P. Rosenbaum, "A Model of the Nucleus", Sci. Amer., p. 84 (Dec. 1955).
24. Mayer, Maria G., "The Structure of the Nucleus", Sci. Amer., p. 22 (March 1951).
25. Lauritsen, T., "Energy Levels of Light Nuclei", Ann. Rev. Nuclear Sci. 1, 67 (1952).
26. Nier, A. O., "Mass and Relative Abundance of Isotopes", Ann. Rev. Nuclear Sci. 1, 137 (1952).
27. Villars, F., "The Collective Model of Nuclei", Ann. Rev. Nuclear Sci. 7, 185 (1957).
28. Fishbach, Herman, "The Optical Model and Its Justification", Ann. Rev. Nuclear Sci. 8, 49 (1958).
29. Post, Richard F., "Fusion Power", Sci. Amer., p. 73 (Dec. 1957).
30. Cowling, T. G., "Nuclear Reactions in Stars", Amer. Scientist 49, No. 2, 182 (June 1961).
31. Millikan, Robert A., "Electrons (+ and —), Protons, Photons, Neutrons, Mesotrons, and Cosmic Rays", Univ. Chicago Press (Rev. 1947).
32. Semat, Henry and Harvey E. White, "Atomic Age Physics", Rinehart (1959).
33. Brown, G. I., "A Simple Guide to Modern Valency Theory", Longmans (1953).
34. Marshak, Robert E., "The Nuclear Force", Sci. Amer., p. 99 (March 1960).
35. Sternglass, E. J., "Relativistic Electron-Pair System and the Structure of Neutral Mesons", Phys. Rev., 123, 391 (1961).
36. Anon., "Theory Proposed to Explain Nuclear Forces", Chem. and Eng. News, p. 40 (Feb. 11, 1963).
37. Bromley, D. A., "Structure of Nuclei Probed by New Van de Graaff", Nucleonic 21, No. 5 48-51, 54, 55 (May, 1963).
38. Flowers, B. H., "The Structure of the Nucleus", J. Chem. Ed. 37, 610-615 (1960).
39. Marshak, Robert E., "Pions", Sci. Amer., p. 84 (Jan. 1957).
40. Marshak, Robert E., "The Multiplicity of Particles", Sci. Amer., p. 23 (Jan. 1952).
41. Mayer, Maria G., "The Shell Model", Science 145, 999-1006 (1964).
42. Dunn, T. M., D. S. McClure and R. G. Pearson, "Some Aspects of Crystal Field Theory", Harper and Row (1964).
43. Rabi, I. I., "The Atomic Nucleus, A New World to Conquer", Science 108, 673-675 (1948).
44. Anon., "Atomic Weight Scales Unified?", Chem. and Eng. News, p. 76 (Sept. 8, 1958).
45. Anon., "A Tale of Atomic Weights", Chem. and Eng. News, p. 92 (Jan. 26, 1959).
46. Anon., "IUPAC Revises Atomic Weight Values", Chem. and Eng. News, p. 42 (Nov. 20, 1961).
47. Labbauf, Abbas, "The Carbon-12 Scale of Atomic Masses", J. Chem. Ed. 39, 282-286 (1962).
48. Present, R. D., "The Liquid-Drop Model for Nuclear Fission", Nucleonics 3, 25 (Sept., 1948).
49. Lauritsen, T., W. A. Fowler and C. C. Lauritsen, "Energy Levels of Light Nuclei", Nucleonics 2, 18 (April 1948).
50. Milner, C. J., "Model of Nuclear Structure", Nucleonics 4, 56 (Jan. 1949).
51. Brightsen, R. A., "Nuclear Abundances and Closed Shells in Nuclei", Nucleonics 6, 14-24 (April 1950).

52. Pollard, E., "Measurement of Nuclear Energy Change Values", Nucleonics 2, 1 (April 1948).
53. Green, A. E. S., "Nomogram for Estimating Nuclear Reaction Energies", Nucleonics 13, 34-37 (Feb. 1955).
54. Purkayastha, B. C., "Fission of Atomic Nuclei—I", Nucleonics 3, 2 (Nov. 1948). II—Nucleonics 3, 2 (Dec. 1948).
55. Inthoff, W., "Fission-Neutron Reaction Cross Sections", Nucleonics 13, 67 (Nov. 1955).
56. Katcoff, S., "Fission Product Yields from U, Th and Pu", Nucleonics 16, 78-85 (April 1958).
57. Safford, G. J., and W. W. Havens, Jr., "Fission Parameters for U^{235}", Nucleonics 17, 134 (Nov. 1959).
58. Hughs, D. J., "New 'World Average' Thermal Cross Sections", Nucleonics 17, 132 (Nov. 1959).
59. Rochlin, R. S., "Fission-Neutron Cross Sections for Threshold Reactions", Nucleonics 17, 54 (Jan. 1959).
60. Katcoff, S., "Fission-Product Yield from Neutron-Induced Fission", Nucleonics 18, 201 (Nov. 1960).
61. Anon., "Fusion", Nucleonics 16 (Sept. 1958).
62. Bishop, Amasa S., "Controlled Fusion", Nucleonics 15, 128 (Sept. 1957).
63. Hofstadter, R., "Structure of Nuclei and Nucleons", Science 136, 1013 (1962).
64. Zucker, A. and D. A. Bromley, "Nuclear Physics: A Status Report", Science 149, 1197 (1965).
65. Anon., "Pauling Suggests Nuclear Model that Could Help Relate Ideas on Particle Behavior", Chem. and Eng. News, p. 23 (Oct. 18, 1965).
66. Pauling, L., "The Close-Packed-Spheron Theory and Nuclear Fission", Science 150, 297 (1965).
67. Leachman, R. B., "Nuclear Fission", Sci. Amer., p. 49 (Aug. 1965).
68. Rainwater, L. J., and C. S. Wu, "Principles of Nuclear Particle Detection", Nucleonics 1, 13-23 (Sept. 1947).

PART II

EXPERIMENTAL TECHNIQUES

CHAPTER 3
Introduction to Radiation Measurement

LECTURE OUTLINE AND STUDY GUIDE

I. PROCEDURES FOR USE OF RADIO-ISOTOPES
   - A. The Radiochemical Laboratory
     1. Laboratory organization
     2. Special techniques and precautions
   - B. Regulations Governing Radioisotope Usage
     1. A.E.C. regulations
     2. General laboratory regulations

II. PREPARATION OF SAMPLE FOR COUNTING (*Experiment* 3.1)
   - A. Preparation of Counting Card
   - B. Preparation of a Planchet Mount
   - C. Use of Micropipette
     1. Calibration
     2. Technique for rinsing
   - D. Procedure for Drying a Sample of Radioactive Material
   - E. Technique for Pipetting Large Samples

III. CONSTRUCTION AND OPERATION OF GEIGER TUBES (*Experiment* 3.2)
   - A. Construction
   - B. Mechanism of Detection of Radiation
     1. Direct production of ions by charged particles
     2. Production of electrons by gamma rays
   - C. Collection of Ions Produced in a Gas
   - D. Auxiliary Counting Equipment (scalers)

IV. CHARACTERISTICS OF GEIGER TUBES (*Experiment* 3.2)
   - A. Factors Tending to Perpetuate Discharge
   - B. Factors Tending to Limit Discharge
   - C. Quenching
     1. External quenching
     2. Self quenching
   - D. Resolving Time
   - E. Characteristic Counting Rate Curve
     1. Starting potential
     2. Threshold
     3. Plateau and plateau slope
     4. Discharge region
     5. Operating voltage
     6. Changes in characteristic curve with aging of tube
   - F. Memory

THE RADIOCHEMICAL LABORATORY:

Radiochemical laboratories are commonly divided into two areas: (1) the work area or laboratory proper; and (2) the separate room or space in which quantitative measurements of radioactivity are made, this being commonly designated as the counter room or area. It is especially important that the counter room be kept free from radioactive contamination so that we may maintain not only a low background of radioactivity for the instruments but also healthful working conditions for ourselves.

All preparative work for radiochemical experiments must be performed in the work area—never in the counter room. Moreover, all such work should be restricted to designated individual working areas. Such individual areas may be defined by a large flat metal tray lined with an absorbent paper. Special paper is available for this purpose. Such paper is constructed so that the upper surface is highly absorbent while the under layer consists of a hard, water-resistant paper which retards liquids from soaking through to the table surface. Normally, all work should be performed within the tray so that contamination from accidental spills may be confined within this area. As a safeguard against contamination of the table tops within the work area, these too should be covered with the absorbent paper. By this means we may limit not only the extent of the contamination in case of a spill but we may also facilitate the disposal of the spilled material.

Pipettes and equipment which have become contaminated through normal use should be placed on a small tissue within the confines of the pan or in a designated location. Solid wastes should be disposed of in the "radioactive" waste can and liquid wastes should be poured into designated containers.

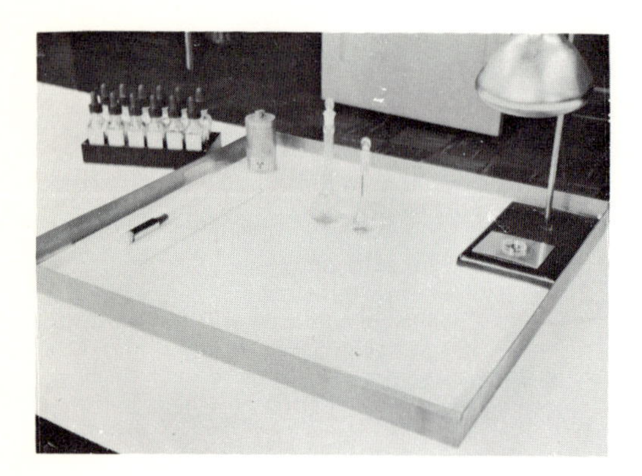

Figure 3.1—Work area defined by tray.

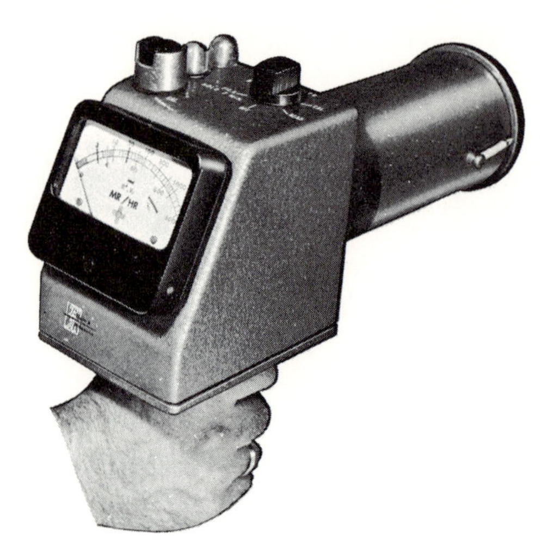

Figure 3.2—Single Scale Logarithmic "Gun Type" Survey Meter (Model 414). (*Baird-Atomic, Inc.*)

Normally, rubber gloves should be worn in the laboratory. At the conclusion of the work, the hands are washed with the *gloves still on.* The gloves are then checked for contamination. If found free of contamination, the gloves are then removed and the hands themselves are washed.

Rubber gloves, worn by students while working in the laboratory proper, should not be worn in the counting rooms. Indeed, cooperation is especially essential in this respect so that the counting rooms can be kept free from even trace contamination.

At the conclusion of the working period, the hands, feet and clothing should be monitored. A beta-gamma survey meter is suitable for this purpose. However, when one has been working with alpha emitters or weak beta emitters, a laboratory monitor employing a special detector for alpha and weak beta rays should be used. Any contamination should be reported at once to the instructor. The working area and the table tops should also be monitored for the detection of spills which may have occurred unobserved.

REGULATIONS GOVERNING RADIOISOTOPE USAGE:

The maximum permissible weekly dose of radiation was set at 300 milliroentgens (2). However, the accumulated dosage limits have been revised downward (3,6). Although there is no concordant conclusion as to the level constituting a safe maximum dose (4,5), it is generally believed that all radiation is harmful no matter how small the dose.

Exposure limits to radiation recommended by the National Committee on Radiation Protection and Measurements are published by the National Bureau of Standards. On April 23, 1959, the National Bureau of Standards announced that the recommendations published in NBS Handbook 59 (issued September 24, 1954 and amended April 15, 1958) have again been revised and have been published in a new NBS Handbook 69. On May 4, 1959, the Atomic Energy Commission distributed proposed amendments to the regulations currently in effect to bring the regulations into accord with the latest recommendations of the NCRPM (7). These were adopted and new regulations (8), published in November, 1960, became effective January 1, 1961. Limits on whole body exposure dose were reduced from 3.75 to 1.25 rems per calendar quarter for individuals 18 or more years of age. Limits for minors under 18 are 10% of the limits for adults. Limits for accumulated dose were set at 5 rems per year above age 18. The complete set of "Rules and Regulations," Title 10, Part 20—"Standards for Protection Against Radiation" may be obtained from the Atomic Energy Commission (10).

It is the policy of most laboratories to urge their employees not to expose themselves to any more radiation than is absolutely necessary. As a means of personal check on radiation exposure, each worker should wear a pocket ionization chamber or a film badge.

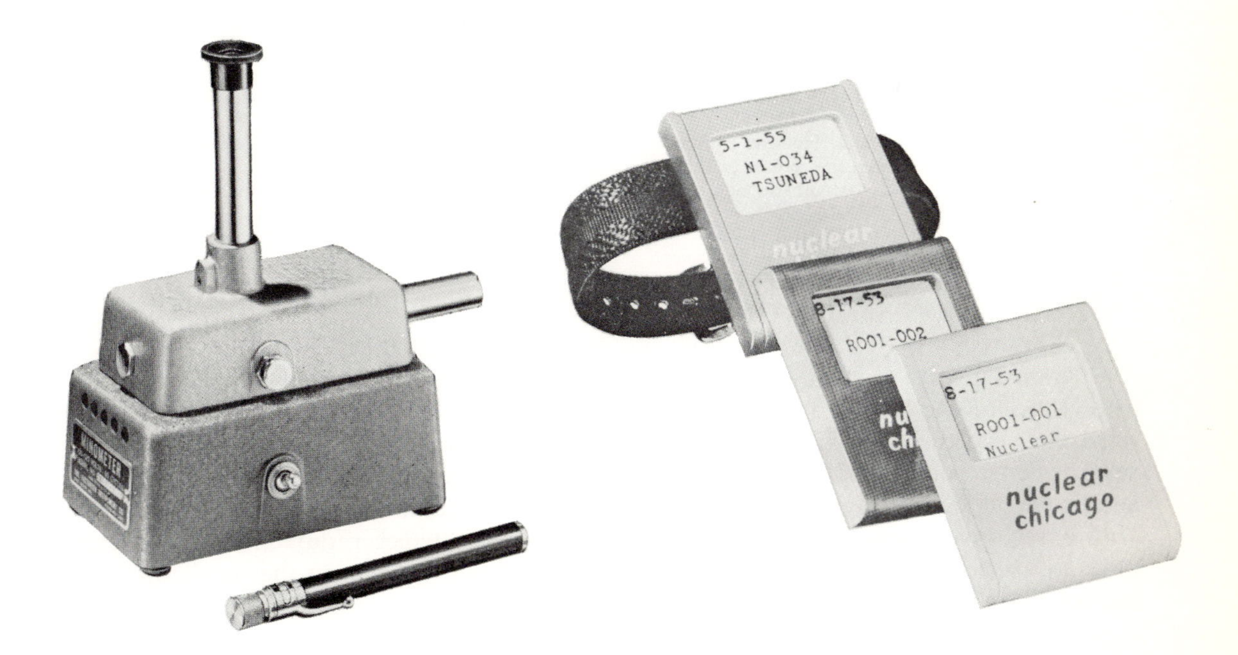

Figure 3.3—Pocket Dosimeter and Charger (*The Victoreen Instrument Company*)

Figure 3.4—Film Badges. (*Nuclear-Chicago Corp.*)

The Atomic Energy Commission requires that certain written records be maintained in radioisotope laboratories (8). Where student cooperation is requested in this respect, each student should see that the necessary records are properly completed before he leaves the laboratory.

The following laboratory regulations have been adapted from a list of those in use at the Oak Ridge Institute of Nuclear Studies and are in accord with standard safety requirements in effect at A.E.C. laboratories (17).

LABORATORY REGULATIONS:

1. Eating, drinking, smoking and the use of cosmetics in the laboratory are not permitted.
2. Pipetting or the performance of any similar operation should not be done by mouth suction.
3. Before a worker leaves the laboratories, the hands should be washed first, then checked with a beta-gamma survey meter. Contamination remaining after thorough washing should be reported.
4. If, in the course of work, personal contamination is suspected, a survey with a suitable instrument should be made immediately. This should be followed by the required cleansing and a further survey. Routine precautionary surveys should be made at intervals.
5. No person should work with active materials if he has any breaks in the skin on the hands unless he wears rubber gloves. All such breaks should be reported to the instructor in charge before work begins.
6. No person should work in the laboratories without wearing a pocket ionization chamber. If available, two pocket ionization chambers (dosimeters) should be worn.
7. Active liquid wastes should be poured into the labeled containers provided. They should *never* be poured into a standard drain.
8. Active solid wastes and contaminated materials should be placed in trash cans labeled "contaminated" or in designated containers.
9. Good housekeeping is encouraged at all times. Spillage should be prevented, but in the event of such an accident the following procedure should be followed:
   a. The liquid should be blotted up. (Wear rubber gloves.)
   b. All disposable materials contaminated by the spill and the cleaning process should be placed in a "contaminated" trash can.
   c. The area of the spill and the type of activity (e.g., ^{131}I) should be clearly marked.
10. No apparatus should be washed in the public water-sewage system if it contains any activity appreciably above background when measured with a counter type survey meter.
11. In general, active materials and contaminated materials are to be retained within the radioisotope laboratory and at specific points within the laboratory.
12. All wounds, spills and other emergencies should be reported to the instructor immediately.
13. Before leaving the laboratory, be sure all written records have been completed.

INTRODUCTION TO INSTRUMENTATION:

Almost every procedure involving the use of radioactive substances requires a knowledge of the basic equipment used for radiation detection and measurement. Perhaps the best known device for the detection of radiation is the Geiger counter. This term to most people implies a device such as the beta-gamma survey meter—a relatively small, portable instrument such as is used by prospectors. This instrument is provided with a sensitive probe on the end of a cord. It will detect radioactive substances by causing a series of clicks to be heard in the earphones or by causing the needle of a small meter to be deflected.

Instruments operating on this same basic principle are used in the laboratory but are normally somewhat more refined than the portable meter mentioned above. Such equipment employs a detecting device such as a Geiger-Müller tube and a recording device such as a scaler or ratemeter.

The *Geiger-Müller tube* is a sensing device or energy converter since it converts the energy of the ionizing radiation into electrical impulses. A large number of different types of radiation sensing elements have been developed. In addition to the Geiger Müller detector they include the proportional counter, the ionization chamber and the scintillation counter, as well as others discussed in Chapters 8 and 9. Each of these detectors differs in its mode of operation and in its sensitivity to a particular type of radiation. They are all similar in that they all convert the energy of radiation into electrical energy.

The electrical energy of the sensing element is fed into an electronic recorder. The recorder, known as a *scaler*, is nothing more than an adding machine which keeps a record of the total number of impulses emitted by the Geiger tube or other detector. In addition to their function as adding machines, most scalers also contain a high voltage supply necessary for the operation of the detector and some provision for a timer.

If one is not interested in the total number of pulses received but rather in the rate at which they are being received, a *ratemeter* may be used to good advantage. In this case the output of the detector (possibly a Geiger tube) is fed into the ratemeter which is similar in function to the speedometer in an automobile. If desired, the output of the ratemeter can be made to drive the pen of a recorder which will provide a continuous record of the radioactivity as a function of time.

Embodied in this chapter are the fundamental procedures necessary for the operation of basic counting equipment as well as simple techniques of sample preparation.

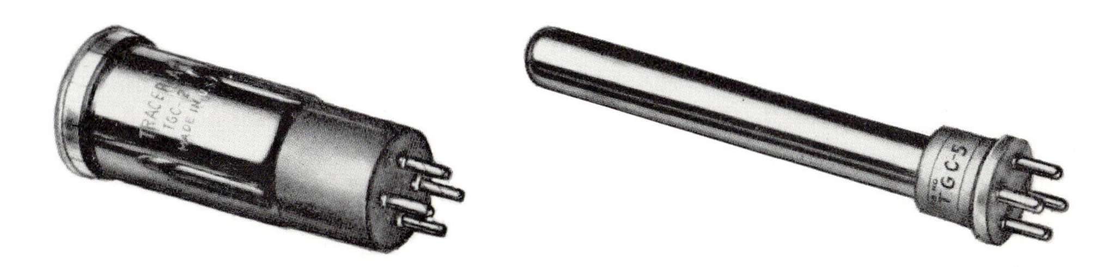

Figure 3.5—Geiger-Müller Tubes. (*Tracerlab, Inc.*)

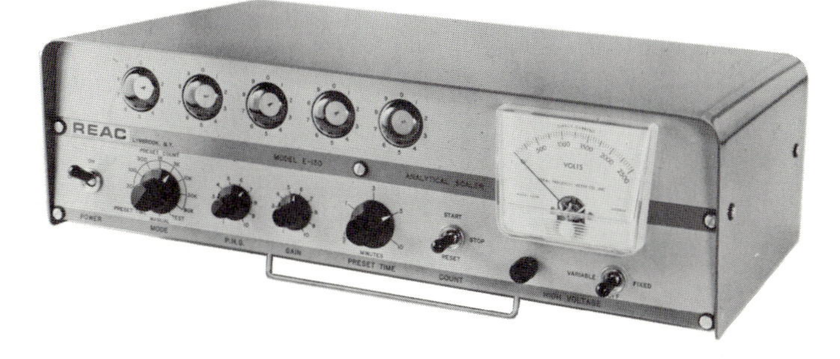

Figure 3.6—SCALER (*Courtesy of Radiation Equipment and Accessories Corp.*)

EXPERIMENT 3.1 SAMPLE PREPARATION TECHNIQUES

OBJECTIVES:

To practice various methods for preparing and mounting a sample for radioactive measurements.

To obtain familiarity with micro-measuring techniques and with remote handling techniques.

THEORY:

Before a measurement of radioactivity can be made, the sample must be suitably prepared for counting. Numerous techniques of sample preparation have had to be developed for not only must one consider the physical state of the sample, its chemical properties and the properties of the radiation emitted, but also the construction and characteristics of available detectors. This discussion and experiment are therefore limited to certain simple basic procedures while the more advanced and special techniques of sample preparation have been deferred to Chapters 8, 9 and 10.

Card mounts have been employed for many years and, although not used so extensively today, still serve a useful function. This type of mount is illustrated in figure 3.7. An aluminum, cardboard or plastic card is used, in the center of which is cut a 1″ diameter hole. A piece of thin plastic film is fastened to the card by strips of cellophane tape as illustrated by the dotted lines in the diagram. The card is then placed flat on the table with the plastic film on the bottom. The sample is measured onto the film in the center of the hole and dried. If desired a similar film can be fastened to the opposite side of the card as a cover with the sample protectively sandwiched between the two plastic films. Care must be taken that the radioactive liquid does not touch and contaminate the aluminum or cardboard card.

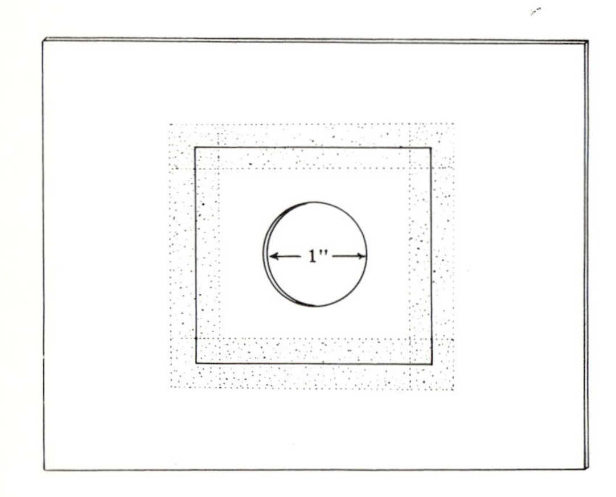

Figure 3.7—CARD MOUNT.

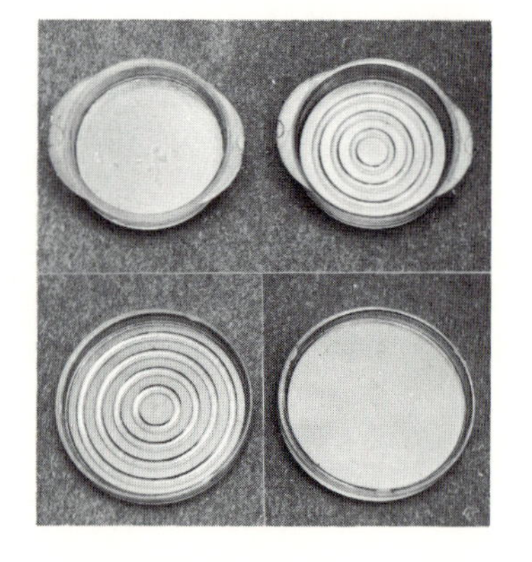

Figure 3.8—PLANCHETS (*Plastics Engineering*)

Planchet mounts are very convenient, reproducible and relatively inexpensive. A planchet is a little cup into which the sample can be measured. Planchets are made of glass, plastic and a variety of metals and come in various shapes and sizes. The sample is measured into the center of the planchet and dried. It is then ready for assay. Solids such as $BaCO_3$ may be powdered and suspended in acetone. An aliquot of the suspension is then pipetted onto a planchet and dried.

Micropipettes are available in sizes ranging from 1 microliter (0.001 ml also called a lambda) to 500 microliters. These pipettes are made "to contain." Hence they should be rinsed and the rinse liquid added to the measured sample.

To facilitate rinsing these pipettes without contaminating a large volume of liquid it is convenient to place a drop or two of rinse liquid on a small piece of cellophane. After delivery of the sample the pipette is filled from one of the drops. Unused portions of rinse liquid are dried and the cellophane discarded in the radioactive waste can.

Mouth suction should never be used in the Radiochemical Laboratory. A variety of devices are available for applying suction to a micropipette. A small (number 00 or smaller) one-hole rubber stopper is inserted on the end of a 1-ml syringe. When the upper end of the pipette is inserted in the opposite end of the stopper, filling of the pipette can be controlled easily with one hand after a little practice. An excellent pipetting device, having a screw type control, is manufactured by Clay-Adams. Larger pipettes, from 2 ml to 100 ml, are easily manipulated with the help of a pipetting device such as the Propipetter. These pipetting devices are illustrated in figures 3.10 and 3.11.

To wash pipettes, insert the upper end into a small bore tubing connected to a vacuum line or aspirator. A trap consisting of a side arm flask serves to retain any radioactivity washed out of the pipette. Wash liquids are drawn through the soiled pipette, a suggested sequence being—water containing detergent, distilled water, alcohol, ether, and finally air until dry.

Drying of samples is efficiently accomplished with an infrared lamp held by a clamp and ringstand above the sample. There is a temptation to hasten drying by having the lamp too close to the sample. Too rapid drying will be indicated by warped planchets where plastic is used. A hair dryer used to deliver a blast of hot air is often very helpful.

A more even and reproducible distribution of the sample may be obtained by the use of a rotating platform beneath the infrared lamp to turn the sample continuously during the drying operation. Such platforms are available commercially or can be made without too much difficulty.

Often it will be found that the sample will not wet the planchet. Better reproducibility is obtained if the sample wets the planchet and spreads evenly over a given surface. A dropper bottle containing 0.1% Duponal or other detergent should be at hand. One drop of detergent solution added to the sample will often produce the desired results.

Loss of sample due to volatilization can be avoided by the addition of a drop of a specific reagent to fix the active element. Iodine is a typical example. A drop or two of the following reagent will retard the volatilization of iodine:

REAGENT FOR IODINE

Sodium hydroxide	0.8	grams
Sodium thiosulfate	0.5	grams
Potassium iodide	0.03	grams
Water, to make	1000	ml

Remote pipetting techniques are necessary for measuring highly radioactive solutions. Devices for this purpose are commercially available. Such devices permit the operator to work at distances of one foot or more from the radioactive substance and allow the use of massive lead shielding for personnel protection.

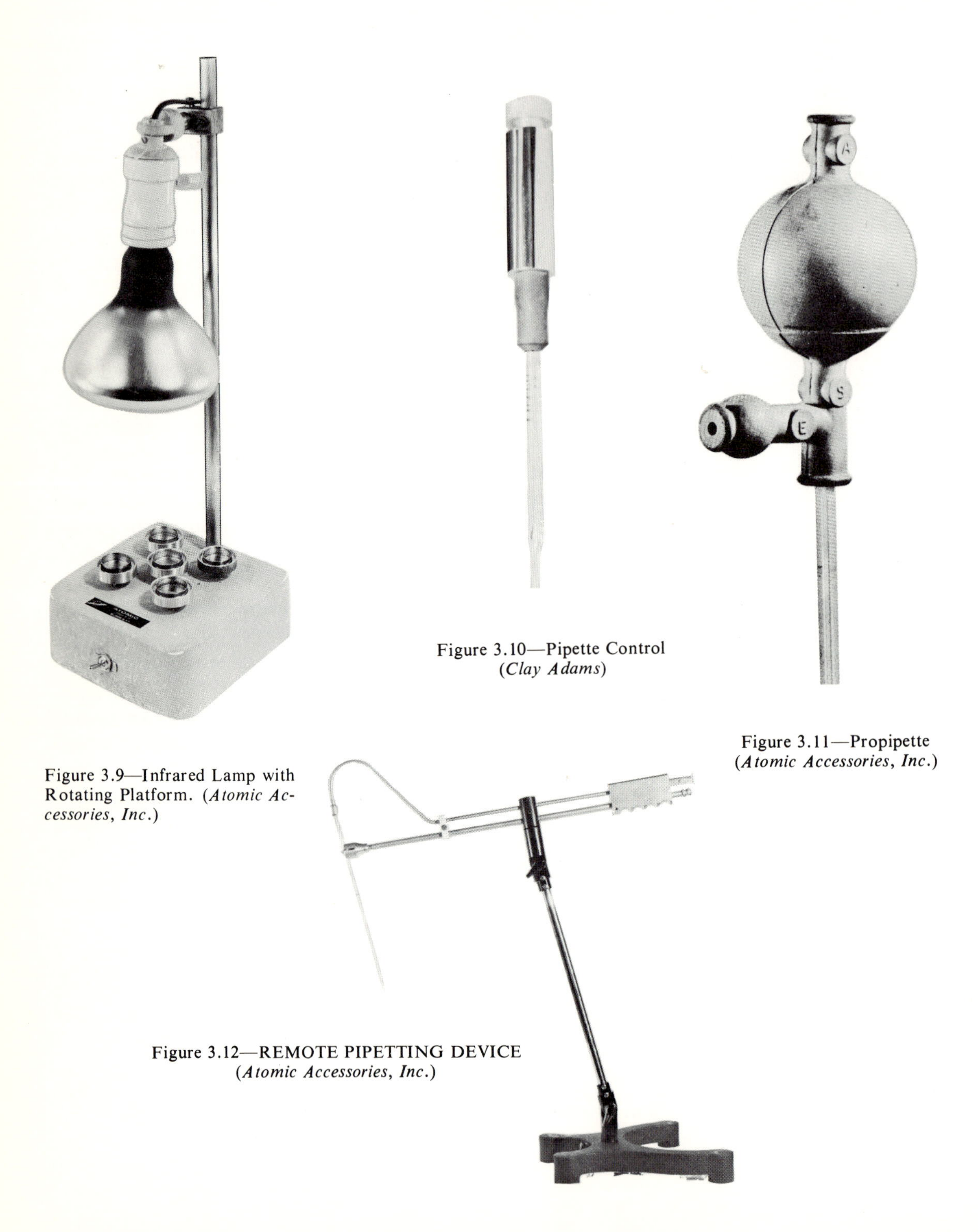

Figure 3.9—Infrared Lamp with Rotating Platform. (*Atomic Accessories, Inc.*)

Figure 3.10—Pipette Control (*Clay Adams*)

Figure 3.11—Propipette (*Atomic Accessories, Inc.*)

Figure 3.12—REMOTE PIPETTING DEVICE (*Atomic Accessories, Inc.*)

APPARATUS:

Solution of radioactive sample*; rubber gloves; micropipettes; mounting cards; plastic film; cellophane tape; planchets; remote pipeter; lead shield with mirror; infrared lamp.

PROCEDURE:

Always wear rubber gloves when handling radioactive substances. Upon completion of your work wash your hands with the *gloves on* to decontaminate them, dry and check for residual contamination on the gloves with a survey meter. If contamination is present wash hands again. In particular, check for contamination between the fingers. When gloves are free of contamination remove them and wash the hands.

Do not wear rubber gloves outside the radiochemical laboratory and do not wear them in the counting room. Refer to the section "The Radiochemical Laboratory," page 62 for laboratory suggestions and rules.

1. Practice the use of micropipetting techniques with water using a 20 or 25 microliter (lambda) pipette.

2. When proficiency in the use of micropipettes has been attained, obtain a sample mounting card. Cover the bottom of the hole in the card with a piece of plastic film, holding the film in place with strips of cellophane tape. Be sure that the film is taut and free of wrinkles. Measure an amount of radioactive solution**, calculated to provide 1,000 to 5,000 counts per minute, on the film in the center of the hole being careful not to allow the solution to contact the card. *NEVER* pipette by mouth, always use a syringe. Micropipettes are calibrated to "contain" rather than to "deliver." Therefore, rinse twice with distilled water, placing the rinse water on the taut plastic film also. (For a source of rinse water place two or three drops of distilled water on a small piece of cellophane. Do not dip pipette into a beaker of water.) Dry the sample carefully under the heating lamp and cover with a piece of plastic film, holding the plastic film in place with strips of cellophane tape. Label properly to identify the sample.

3. Laboratory counters may have different types of sample holders. It may therefore be found advantageous to prepare a similar sample in a planchet. Place the same volume of the radioactive solution in the center of the planchet and dry under the heat lamp as before. Label the planchet.

4. Practice the use of remote pipetting techniques with water behind a lead shield. Attempt to attain some degree of proficiency in pipetting and other operations while observing your work in the mirror only.

*If a long-lived isotope such as Ra-DEF or ^{90}Sr is used, the samples prepared in this experiment may be used later as reference sources. In the selection of the isotope used, this advantage must be weighed against the disadvantage of long term contamination of the laboratory in the event of a spill. Short-lived isotopes such as ^{131}I or ^{32}P may therefore be more desirable.

**Strength of the solution required to provide 1,000-5,000 counts per minute will depend upon source type, counting geometry, etc., but as an estimate, 25 to 50 microliters of a solution containing 1 microcurie of activity per ml should be satisfactory.

EXPERIMENT 3.2 GEIGER-MÜLLER COUNTERS

OBJECTIVES:

To measure the characteristics of a Geiger-Müller counter.
To determine the plateau and operating voltage for a G-M counter.
To measure the slope of the plateau of a G-M characteristic curve.

THEORY:

A Geiger-Müller counter is a device used for the detection and measurement of radiation. Basically, it consists of a pair of electrodes surrounded by a gas especially selected for the ease with which it can be ionized. When radiation ionizes the gas, the ions so produced travel to the electrodes between which is maintained a high electrical potential. The motion of the ions to the electrodes constitutes an electric current which is detected and recorded by a scaler. Thus, each particle or ray of radiation passing through the Geiger-Müller tube causes a short pulse of current to flow, the number of such pulses being a measure of the intensity of the radiation.

Geiger-Müller counters are supplied in a variety of forms. A typical "end-window" type of tube is depicted in figure 3.5. This tube is so named because it has a thin window at one end through which the radiation passes. It consists of a metal or glass cylindrical envelope the inside of which has been coated with a conducting material. The wall of the tube constitutes the negative electrode known as the cathode. In the center, concentrically aligned, is a fine wire which serves as the anode. It is charged positively, often to approximately 1200 volts, with respect to the cathode.

The space between the electrodes is filled with a gas, helium or argon usually being used. The window prevents the escape of the gas to the atmosphere, yet is sufficiently thin so that it does not prevent the passage of radiation into the tube to any appreciable extent, especially the passage of beta particles for which this type of tube is most useful.

A beta particle entering the counter produces a number of ion pairs consisting of electrons and positively charged ions of the gas filling the tube. Under the influence of the electrical field between the electrodes, the electrons travel to the center wire or anode. In the process, however, they themselves acquire enough energy that their collisions with gas molecules result in the formation of still other ion pairs. This process continues with the resulting formation of an avalanche of electrons and positive ions. The positive ions

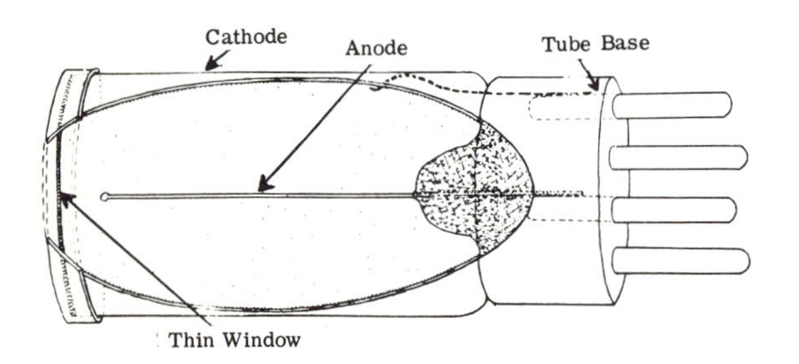

Figure 3.13—Cut-away of Geiger-Müller Counter

travel toward the outer envelope, the cathode, also causing additional ion pairs to be produced. Thus, once ionization is initiated, the tube would continue to discharge continuously unless turned off or *quenched* by some other process.

One way to quench a tube in order to restore it to its original quiescent state is to remove the high voltage momentarily. This can be done electronically and is called *external quenching*. Another method more commonly used today is to employ *internal quenching* by mixing a small quantity of a polyatomic gas with the counter gas to absorb some of the energy of the electrons and positive ions after an ionizing event. In the process the polyatomic molecule is decomposed. If a substance like alcohol or butane is used as a quenching agent, the tube is said to be *"organic-quenched."* Such a tube has a useful life of about 10^8 counts because the molecules are decomposed irreversibly. *"Halogen-quenched"* tubes utilize chlorine, bromine and their compounds, as quenching agents. These tubes have a much longer life because the atoms normally recombine.

The net behavior of a Geiger-Müller tube during an ionizing event is the result of two opposing groups of factors, those tending to perpetuate discharge and those tending to limit discharge as discussed in the preceding paragraphs.

Figure 3.14 illustrates a typical characteristic curve for a Geiger-Müller tube. If a sample is placed beneath a tube and the voltage which is impressed on the tube slowly increased, a voltage will be reached at which the G-M tube just begins to perceive a few counts as indicated by the scaler. This is the *starting potential.* Now as the voltage is increased very slightly a very rapid increase in the counting rate is observed. This voltage is known as the *threshold.* Beyond the threshold, further increases in the voltage over quite a range will produce little effect on the counting rate. This region, known as the *plateau,* should have a slope of less than 3% for good tubes. Within the plateau region the proper operating voltage is selected. The *operating voltage* should be selected relatively close to the threshold voltage (within the lower 25% of the plateau) to help preserve the life of the tube. Also, the operating voltage should be selected at a point where the

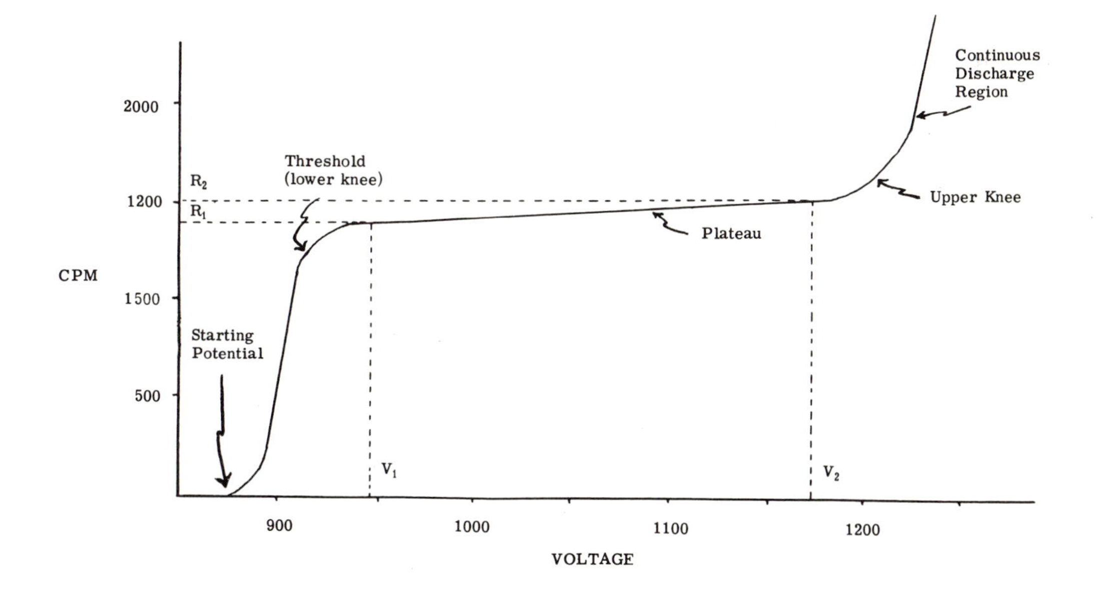

Figure 3.14—Characteristic Curve for Geiger-Müller Tube

plateau shows minimum slope. If the voltage is indiscriminately increased beyond the plateau region, the region of *continuous discharge* is reached and the tube may be seriously damaged.

The *plateau slope* is of value since it serves as a figure of merit for a counter tube. The slope $\Delta R/\Delta V$ defined using the notation of figure 3.14 as simply $(R_2 - R_1)/(V_2 - V_1)$ is meaningless because the merit of the tube could be "improved" through the simple expedient of reducing the sample activity. The *normalized plateau slope* is used as the figure of merit. It is calculated as the percentage change in counting rate R divided by the percentage change in applied voltage V using the threshold values as the base.

$$\frac{100\,(R_2 - R_1)/R_1}{100\,(V_2 - V_1)/V_1} = \frac{(R_2 - R_1)\,V_1}{(V_2 - V_1)\,R_1} \tag{3.1}$$

Very often the slope is expressed as the per cent increase in counting rate per volt (or per 100 volts) increase in the applied voltage. Thus the *relative plateau slope* is expressed as

$$\frac{100\,(R_2 - R_1)/R_1}{V_2 - V_1} = \text{per cent per volt} \tag{3.2}$$

or

$$\frac{100\,(R_2 - R_1)/R_1}{V_2 - V_1} \times 100 = \text{per cent per 100 volts} \tag{3.3}$$

A good tube should have a slope of less than 10% per 100 volts. Often the slope is as little as 3% per 100 volts.

After a Geiger-Müller tube has been exposed to a high intensity gamma source the background of the tube may be abnormally high for some time after the source has been removed. The reason for this *"memory"* is not entirely understood but it is probably due to the formation of activated molecules of gas within the tube.

Once ionization of a Geiger tube is initiated, the tube becomes insensitive for a short interval of time. This interval is called the *resolving time.* It represents the time during which two or more ionizing particles striking the sensitive portion of the tube will be counted as a single particle. As a result of this phenomenon, the number of counts recorded will be less than the actual number of particles passing into the tube. The difference between the true and observed count is known as the coincidence loss. Resolving time and coincidence loss are considered in detail in Experiment 5.2.

Geiger tubes are not equally sensitive to alpha, beta and gamma radiations. This is explained by considering both the properties of the radiation and the properties of the Geiger tube. To initiate discharge of the tube, the radiation must first reach the sensitive volume. Alpha particles, being the least penetrating, may be absorbed by the window unless it is very thin. Beta particles are more penetrating and gamma rays are very penetrating. Thus thicker windows can be tolerated with the latter two types of radiation.

Considering only the radiation which reaches the sensitive volume of the tube and assuming zero coincidence loss, it is found that the efficiency of the Geiger tube is essentially 100% for alpha particles, nearly 100% for beta particles but only 1 or 2% for gamma radiation. Ironically, the property of gamma rays which enables them to penetrate the window of the tube so efficiently and thus to enter the sensitive volume of the tube, i.e., their ability to pass through relatively thick layers of matter without interacting with the matter, now works against them for most of them now pass through the sensitive volume of gas also without interacting. Thus, the low efficiency for gamma rays is explained by the need of the gamma ray to interact either with a gas atom or with an atom in the wall of the tube to dislodge an electron which will in turn produce ionization resulting in the discharge of the tube. The processes by which gamma rays interact with matter are discussed in Chapter 5.

Geiger tubes are available in a variety of forms, the most common being the *end-window* variety and the *side-window* variety. End-windows consist of a thin film of mica, mylar or other suitable substance especially selected to retain the gas in the tube while at the same time offering minimum interference to the passage of radiation. Obviously, this window is extremely delicate and must not be touched with the fingers. Broken windows frequently account for Geiger tube casualties in the laboratory. Side-windows are normally about ten times thicker than end-windows, the window itself serving to support the end of the tube. Side window tubes are useless for the detection of alpha particles but are quite useful for the detection of energetic betas. On the other hand side-window tubes are more sensitive to gamma radiation than are end-window tubes presenting the same cross section to the radiation. The thicker window increases the probability of interaction of the gamma rays to eject an ionizing electron. Certain Geiger tubes especially designed to be gamma sensitive have bismuth coated cathodes and sometimes steel windows. For these tubes the gamma efficiency is often as high as 6 or 8%.

GENERAL DIRECTIONS FOR OPERATION OF A SCALER:

Although each make and model of scaler will differ in certain details, they are basically similar in operation. The student may need to make minor changes in the following directions to adapt them to the specific instrument being used.

1. Make sure that the MASTER SWITCH and HIGHVOLTAGE SWITCH are both in the OFF position and that the HIGH VOLTAGE ADJUST is in the extreme counter-clockwise position.
2. Plug the line cord into the 110 volt, 60 cycle outlet and check to ensure that the Geiger tube is connected to the G-M input jack. If ready-made cables are not available for connecting the Geiger tube to the scaler, the anode should be connected to the positive terminal of the high-voltage supply while the cathode is connected to the negative terminal (chassis ground). Pin connections for most three- and four-pin tubes are shown in figure 3.15. If an external timer is required, check to be sure that it is plugged into the proper outlet on the back of the scaler.
3. Turn on the MASTER SWITCH and allow about one minute for warm-up. It is good practice *not* to turn the MASTER SWITCH off unless the instrument will be used only once per week or less. A pilot lamp should indicate that power is being applied to the instrument.
4. Place a sample near the detector. Turn the COUNT SWITCH to the COUNT position and turn the HIGH VOLTAGE SWITCH to the ON position. Gradually increase the high voltage until the scaler begins to indicate a count. This is the threshold voltage for the tube.
5. Adjust the high voltage to the proper operating voltage for the tube. If this is not known a plateau curve should be plotted—see "Procedure" for this experiment.
6. With the high voltage properly adjusted, turn the COUNT SWITCH to STOP and depress the RESET SWITCH. The scaler is now ready for operation.

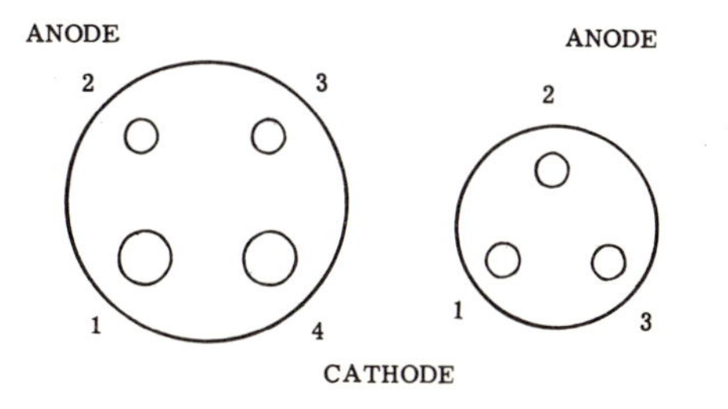

Figure 3.15—Pin Connections for most three and four pin Geiger-Müller tubes. (Illustration shows bottom of tube base or bottom side of socket.)

CAUTION—If the high voltage is incorrectly set too high, the tube will go into continuous discharge and be ruined. Should this occur, as evidenced, for example, by the scaler jamming with all interpolation lights on, *turn the high voltage down immediately.* Placing the COUNT SWITCH in STOP position does *not* remove high voltage from the tube and will not prevent its being damaged.

Continued Operation—If a scaler or other electronic instrument is to be used more often than once a week the main power to the instrument should not be removed at any time during this period of use except for repair. Instrument failures occur most frequently at the time they are turned on as a result of the surge of current through cold filaments and abnormal voltages momentarily applied to various components before the instrument has a chance to warm up. The result is more frequent failure of tubes and other components, more down-time and increased maintenance problems. Place the equipment in "stand-by" operation. For some scalers this is equivalent to the condition existing at the end of step 3 above. Scalers having a "60 cycle" or "test" position should remain *counting* in this position but *without* high voltage applied to the Geiger tube.

APPARATUS:

End-window Geiger-Müller counter; scaler with high voltage power supply; a prepared mount of Ra-DEF or other beta emitter.

PROCEDURE:

To plot a characteristic curve for a Geiger-Müller tube, set the high voltage control to its lowest (counterclockwise) position and insert a prepared sample into the sample holder. Turn on the high voltage switch and turn the "count-stop" switch to the count position. Increase the high voltage slowly until the counter begins to operate. Observe and record the counting rate at various voltages until the counting rate begins to rise above the plateau value. It is suggested that readings be obtained at increments of 50 volts. A total count of at least 5,000 should be accumulated at each voltage setting. It will not be necessary to correct for background for this experiment.

In order to avoid discharging and ruining the tube, do not allow the counting rate to rise more than 10 to 15 per cent above the plateau value.

DATA:

Tube type(end-window, side-window)____________________
(halogen, organic quench)____________________
Tube designation or number____________________
Operating voltage marked on tube____________________volts
Scaler used____________________
Source of radioactivity____________________

VOLTAGE	TOTAL COUNTS	COUNT TIME	COUNTS PER MINUTE

CALCULATIONS:

Plot a graph as illustrated in figure 3.14 showing the counting rate as the ordinate and the voltage on the tube as the abscissa. Linear graph paper should be used.

Indicate the best operating voltage for the tube.__________volts. How does this voltage compare with that suggested by the manufacturer?______________volts.

Calculate the normalized plateau slope, the slope as per cent per volt and the slope as per cent per 100 volts.

Normalized slope (Equation 3.1) ________
(Equation 3.2) ________ % per volt
(Equation 3.3) ________ % per 100 volts

NOTE: Although the preferred method for determining the operating voltage is that given above, it will often be found satisfactory to operate the tube at approximately 75 to 100 volts above the threshold.

Suggestions for further work:

Repeat the experiment using other types of Geiger-Müller tubes.

Measure the effect of source type on the plateau slope.

Determine the effect of temperature on the plateau slope.

REFERENCES:

The Radiochemical Laboratory and Health Physics

1. "Outsmarting the Atomic Hazard", Chemical and Engineering News, p. 652 (Feb. 14, 1955).
2. Federal Register, Title 10, Chapter 1—AEC, Part 20 (Jan. 29, 1957).
3. "A-Worker Radiation Dose Limits Cut", Chemical and Engineering News, p. 6328 (Dec. 24, 1956).
4. "Fallout and Radiation Hazard—Experts Disagree", Chemical and Engineering News, p. 16-19 (June 24, 1957).
5. "Radiation Picture Still Blurred", Chemical and Engineering News, p. 36-37 (Aug. 18, 1958).
6. "AEC Adopts New Exposure Limits", Chemical and Engineering News, p. 36 (Dec. 30, 1957).
7. "Somatic Radiation Dose for the General Population", by The Report of the Ad Hoc Committee of the National Committee on Radiation Protection and Measurements of 6 May 1959, Science, 131, 482 (Feb. 19, 1960).
8. Federal Register, Title 10, Chapter 1—AEC, Part 20 (Nov. 17, 1960).
9. Federal Register, Title 10, Chapter 1—AEC, Part 20 (Dec. 30, 1960).
10. Rules and Regulations, Title 10, Chapter 1, Code of Federal Regulations, U.S. AEC, Supt. of Documents, U.S. Gov't. Printing Office, Washington, D.C. (Jan. 1965).
11. Boursnell, J. C., "Safety Techniques for Radioactive Tracers", Cambridge University PRESS (1958).
12. Braestrup, C. B., and H. O. Wyckoff, "Radiation Protection", Thomas (1958).

13. "Safe Handling of Radioisotopes", International Atomic Energy Agency, International Publications, Inc. (1958).
14. Lane, W., R. Fuller, L. Graham and J. Macklin, "Contamination and Decontamination of Laboratory Bench-Top Materials", Nucleonics, 11, 49 (Aug. 1953).
15. Talboys, A. P., and S. S. Copp, "Contamination and Decontamination of Rubber Gloves", Nucleonics, 11, 60 (Sept. 1953).
16. Anon., "Chicago Radiochemical Plant is Closed as a Potential Radiation Hazard", Chem. and Eng. News, 39, 21 (May 22, 1961).
17. Anon., "General Rules and Procedures Concerning Radioactive Hazards", Nucleonics 1, 60-67 (Dec. 1947).
18. Blake, Martin I., "Radiation Hazards", Amer. Prof. Pharm. 24, 447-452 (1958).
19. Anon., "Retention Values for Laboratory Surface Coatings", Nucleonics 21, No. 2, 75 (Feb. 1963).
20. Morgan, Karl Z., "Permissible Exposure to Ionizing Radiation", Science 139, 565-571 (1963).
21. Clark, H. M., "Criteria for Experiments in Radiochemistry", J. Chem. Ed. 40, 618-626.
22. Rinaker, R. E., "Personnel Protection with Pocket Ionization Chambers", Nucleonics 2, 78 (Jan. 1948).
23. Henriques, F. C., and A. D. Schreiber, "Administration and Operation of a Radiochemical Laboratory", Nucleonics 2, 1 (March 1948).
24. Cohn, W. E., "Toxicity of Inhaled or Ingested Radioactive Products", Nucleonics 3, 21 (July 1948).
25. Lapp, R. E., and H. L. Andrews, "Health Physics", Nucleonics 3, 60 (Sept. 1948).
26. Solomon, A. K., and C. A. Foster, "A Hood for Work with Radioactive Isotopes", Anal. Chem., 21, 304-6 (1949).
27. Tompkins, P. C., O. M. Bizzell and C. D. Watson, "Practical Aspects of Surface Decontamination", Nucleonics 7, 42-54, 87 (Aug. 1950).
28. "Guide for the Selection of Equipment for Radioactivity Laboratories", Nucleonics 7 (Nov. 1950). A series of articles:
   a. Fields, P., and D. C. Stewart, "Laboratory Equipment. $<$ 1 mc, α-emitter", p. R-5 - R-8.
   b. Manov, G. G., "Laboratory Equipment. $<$ 1 mc, β-emitter", p. R-9 - R11.
   c. Stang, L. G., "Laboratory Equipment. 1 mc - 1c, γ-emitter", p. R-12 - R-17.
   d. Hawkins, M. B., "Laboratory Equipment. 1 mc - 1c, β-emitter", p. R-18 - R-19.
   e. Turner, L. D., "Laboratory Equipment. $>$ 1 c, Physical", p. R-20 - R-22.
   f. Gifford, J. G., "Laboratory Equipment. $>$ 1 c, Chemical", p. R-22 - R-24.
29. Eaton, S. E., and R. J. Bowen, "Decontaminable Surfaces for Milicurie-Level Laboratories", Nucleonics 8, No. 5, 27-37 (1951).
30. Brinkerhoff, J., C. A. Ziegler, R. Bersin and D. J. Chleck, "Continuous Air Monitor for Tritium", Nucleonics 17, No. 2, 76-81 (Feb. 1959).
31. Anon., "Pocket Dosimeters? . . . Film Badges . . . Or Both?", Nucleonics 17, No. 5, 116 (May 1959).

Basic Manipulations and Sample Preparation

32. Steyermark, Al, et al., "Report on Recommended Specifications for Microchemical Apparatus. Volumetric Glassware. Microliter Pipets", Anal. Chem. 30, 1702-1703 (1958).
33. Anderson, H. H., "Automatically Adjusting Micropipets and Micropycnometers", Anal. Chem., 24, 579-583 (1952).
34. Smith, R. E., and J. F. Bronson, "An Improved Radioactivity Measuring Cup", Science 107, 603 (1948).

Geiger-Müller Counters and Accessory Equipment

35. Brown, S. C., "Theory and Operation of Geiger-Müller Counters". I, "The Discharge Mechanisms", Nucleonics, 2, 10-22 (June 1948). II, "Counters for Specific Purposes", Nucleonics, 3, 50-64 (August 1948). III, "The Circuits", Nucleonics, 3, 46-61 (October 1948).
36. Korff, Serge A., "Counters", Scientific American, p. 40 (July 1950).
37. Clark, L. B., Sr., "Recent Developments in the Production of Halogen-Quenched Geiger-Müller Counting Tubes", Rev. Sci. Instr., 24, 641-643 (1953).
38. "New Forms of Geiger-Müller Counters", Nucleonics, 13, 17 (September 1955).
39. Price, W. J., "Nuclear Radiation Detection", McGraw-Hill (1958).
40. Van Duuren, K., A. J. M. Jasper and J. Hermsen, "G-M Counters", Nucleonics, 17, 86 (June 1959).
41. Healea, M., "Bibliography. Geiger and Proportional Counters", Nucleonics 1, 68-75 (Dec. 1947).
42. Bousquet, A. G., "Counting Rate Meters versus Scalers", Nucleonics 4, 67-76 (Feb. 1949).
43. Brown, S. C., "Counter Fillings", Nucleonics 4, 139-141 (May 1949).
44. Curtiss, L. F., "The Geiger-Müller Counter", National Bureau of Standards Circular 490 (1950). U.S. Gov't. Print. Off.
45. Newell, R. R., "Binary vs. Decade Scalers", Nucleonics 10, No. 2, 82-83 (1952).
46. Streit, G. H., and W. R. Kennedy, "Automatic Tester for G-M Tubes", Nucleonics 10, No. 6, 61-62 (1952).
47. Porter, W. C., "Extending the Efficiency Range of G-M Counters", Nucleonics 11, No. 3, 32 (1953).
48. Singer, L., and W. D. Armstrong, "Liquid-Sample Geiger Counter", Nucleonics 11, No. 8, 55 (1953).
49. Reddie, J. S., and W. C. Roesch, "Two Better Ways to Determine Geiger-Müller Tube Age", Nucleonics 14, No. 7, 30-32 (July 1956).
50. Armstrong, F. E., "Simple, Inexpensive Device Checks G-M Tube Operation", Nucleonics 14, No. 8, 79-80 (Aug. 1956).
51. Bukstein, E., "Basic Decade Counters", Radio and TV News, p. 34-36 (Aug. 1958).
52. Kawin, B., and F. V. Huston, "Digital or Analog Methods for Radioisotope Measurement?", Nucleonics 22, No. 7, 86-92 (July 1964).

CHAPTER 4
Indeterminate Errors in Measurement (Statistics)

LECTURE OUTLINE AND STUDY GUIDE

I. INTRODUCTION TO EXPERIMENTAL MEASUREMENTS
  A. Basic Terms
    1. Variable
    2. Population
    3. Average or mean
      a. Sample mean
      b. True mean
    4. Accuracy
      a. Bias
      b. Precision
  B. Sources of Variation
    1. Errors in measurement
      a. Determinate errors (See chapter 5)
      b. Indeterminate errors
    2. Real difference between populations
  C. Calculations involving errors
  D. Probability
    1. Addition law
    2. Multiplication law
  E. Distribution of data
    1. Binomial distribution
    2. Poisson distribution
    3. Gaussian distribution

II. GAUSSIAN OR NORMAL CURVE (*Experiment* 4.1)
  A. Parameters of Normal Distribution
    1. Arithmetic mean
    2. Standard deviation
  B. Cumulative Normal Distribution
    1. Relative error (parameter of the error)
    2. Summation probability
      a. "One-tailed" vs "two-tailed" probability
      b. "Greater-than" vs "less-than" probability
  C. Confidence Limits
    1. Level of significance
    2. Risk
  D. "Errors" Used to Express Confidence
    1. Probable error
    2. Standard deviation
    3. Nine-tenths error
    4. Ninety-five hundredths error
    5. Ninety-nine hundredths error
  E. Fit of Data to Normal Curve
    1. Sample standard deviation vs standard deviation
    2. Application of "errors" to determine fit

III. APPLICATIONS OF STATISTICS
  A. Comparison of Samples (*Experiment* 4.2)
    1. Counting rate errors
    2. Detection of abnormal deviations
    3. "t" test of significance
  B. Efficient Distribution of Counting Time
    1. Effect of background on counting statistics
    2. Minimizing the error
  C. Chi-Square Test (*Experiment* 4.4)
    1. Fit of data to Poisson curve
    2. Evaluation of behavior of counting equipment
  D. Chauvenet's Criterion (*Experiment* 4.5)
    1. Probability of occurrence of deviations
    2. Rejection of suspected data

INTRODUCTION:

If duplicate determinations of activity are made on the same radioactive sample, one immediately following the other (with all necessary precautions having been taken to preserve exactly the same geometry) it will be found with very rare exceptions that the two values will differ from each other.

Had we the ability to isolate a single radioactive atom and to watch it until it decayed, there would be no method available to us to predict at exactly which instant decay of the atom would take place. If, on the other hand, we consider a large number of atoms, it becomes possible for us to predict how many atoms will decay within a certain period of time. This situation is comparable to that confronting the life insurance companies. When they insure a single life, they have no means of predicting exactly how long that individual will live. If, however, they consider a large number of lives, it then becomes possible for them to predict how many of the insured will die during a particular period of time. It can be seen that the larger the number of atoms or the greater the number of people considered in such a calculation, the more accurate will be the prediction percentage-wise.

It is the purpose of this chapter to illustrate the random decay of atoms and to investigate methods by which problems associated with random decay can be overcome.

SOURCES OF VARIATION:

The purpose of a laboratory procedure is to measure a characteristic of an object, a system or an event. If the value of the measured characteristic is found to fluctuate it is called a *variable*. If a group of objects, items or events have a particular value of a variable in common they constitute a *population*.

There are at least three basic objectives for conducting an experiment:

1. to characterize a particular variable for a given population; e.g., to determine the branching ratio for the decay of a particular nuclide.

2. to determine what relationship may exist between two variables; e.g., to determine the relationship between radioactivity and time for a particular sample, i.e., to determine its half-life.

3. to determine whether an observed difference is a real difference between two populations or is a chance variation within a single population (see experiment 4.2).

In each of the cases cited it is necessary to control or to account for the source of variation. Recognition of factors leading to a variation in measured quantities is therefore required. The cause of a variation may be: (1) *errors in measurement* or (2) *variation representing a true difference between two populations.*

ERRORS IN MEASUREMENT:

If one is to make exact measurements, it is necessary to eliminate all errors insofar as possible. *Error* may be defined as the deviation of an observed value from the absolute or true value. There are two types of errors:

1. *Determinate* (constant and systemic errors)—Included as determinate errors are those caused by malfunction of equipment, incomplete precipitation in gravi-

metric procedures, effects caused by temperature changes, backscattering and self absorption of radiation. Errors of this type can be eliminated through careful planning and control of an experiment. (Determinate errors are considered in chapter 5.)

2. *Indeterminate* (random or accidental and observational errors)—Indeterminate errors are those which are beyond the control of the experimenter. In replicate measurements of the mass of an object, all systemic errors having been carefully eliminated, the data will be scattered about an average value or mean. In the measurement of radioactivity, a scatter of data is caused by the random nature of radioactive decay.

INDETERMINATE ERRORS IN RADIOACTIVITY MEASUREMENT

Radioactive decay is a random process. Not only is there a consistent change in the activity of a specific sample due to the half-life of the nuclide, but there is also a fluctuation in the decay rate of a particular sample from one instant to the next due to the random nature of radioactive decay.

If during a given interval of time we observe the decay of n_1 atoms and during a second equal interval of time the decay of n_2 atoms, the probability that $n_1 = n_2$ is very small. If we were to make N replicate determinations of the activity n of the sample under similar conditions of geometry and free from constant or systemic error, we might obtain data similar to those shown in table 4.1.

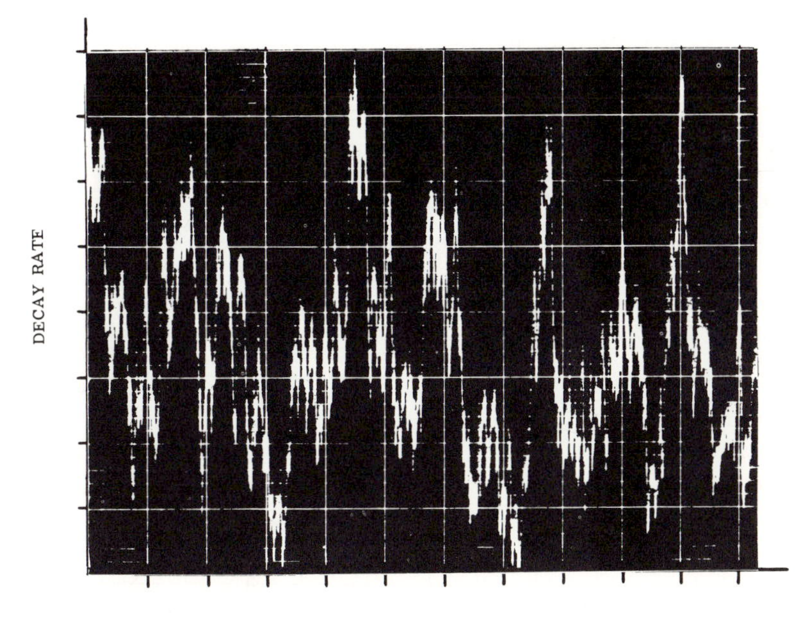

Figure 4.1—RANDOM DECAY OF A NUCLIDE. The variation of decay rate has been recorded as a function of time.

THE AVERAGE OR MEAN

In the measurement of radioactivity, the best value for the average is the arithmetic mean. The *sample mean* $\bar{n}$ is calculated by adding the number of counts per observation (the n's) and dividing this sum by the total number of observations N. The sample mean is expressed by

$$\bar{n} = \frac{1}{N} \sum_{i=1}^{N} n_i \tag{4.1}$$

Using the data of table 4.1, we obtain $\bar{n} = 980/10 = 98$.

The theoretical average for an infinite number of measurements is the actual or *true mean* μ, sometimes called the universe or population mean. Using the data of table 4.1, we would assume that a mean of 98 counts represents a reasonable estimate of μ. However, if we were to repeat the experiment, collecting a duplicate set of 10 observations, the value obtained for $\bar{n}$ would most likely be some value differing from 98 by some small amount. If it were, say, 96 then we would be correct in assuming that the value of $\bar{n} = 97$ (based on the average of all twenty determinations) more nearly represents the true mean or average count of the sample. Thus $\bar{n}$ approaches μ as N becomes very large (assuming there are no determinate errors to bias the results).

(NOTE: μ is the value we would really like to know. If it could be measured directly, statistical considerations would become unnecessary. Unfortunately this is not possible. It is necessary to use n or $\bar{n}$ as an estimate.)

Table 4.1—TYPICAL COUNTING DATA

Observation i	Number of Counts n	Sample Deviation $n-\bar{n}$	Square of Deviation $(n-\bar{n})^2$
1	106	+8	64
2	103	+5	25
3	94	−4	16
4	87	−11	121
5	118	+20	400
6	100	+2	4
7	96	−2	4
8	82	−16	256
9	86	−12	144
10	108	+10	100
	980	0	1134

ACCURACY

For a measurement to be accurate, it must be both precise and unbiased. *Precision* is a measure of agreement among individual measurements. Precision is influenced primarily by indeterminate errors such as random decay and may be measured in terms of the deviation of individual samples from the sample mean (i.e., $n - \bar{n}$). *Bias* is influenced primarily by determinate errors which introduce a constant error into the data. Bias may be measured in terms of the deviation of the sample mean from the true mean (i.e., $\bar{n} - \mu$). These terms are illustrated in figure 4.2(b).

Figure 4.2(a) — ON AN AVERAGE THE DUCK WAS DEAD. A hunter fired both barrels of a shotgun at a duck. The first hit two feet in front, the second hit two feet behind. On an average the duck was dead. What the hunter really wanted was meat on the table. In duck hunting one wants to keep trying until a single shot hits the mark. But in estimating the activity of a radioactive source the best estimate is usually the average. (*Adapted from "Remington's Pharmaceutical Sciences" with permission of the Philadelphia College of Pharmacy and Science.*)

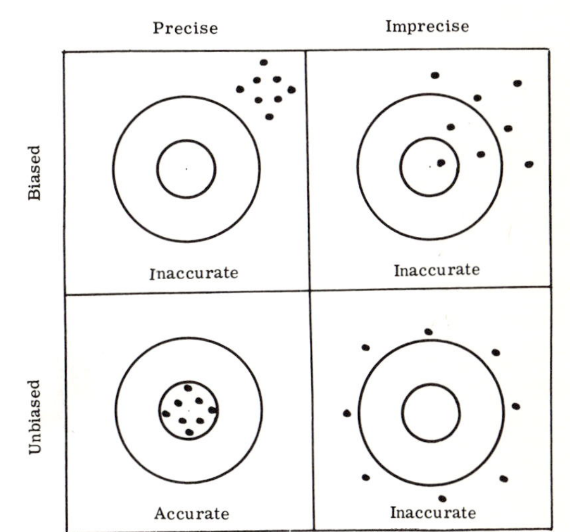

Figure 4.2(b) — DIAGRAM ILLUSTRATING BIAS, PRECISION AND ACCURACY. (*From "Remington's Pharmaceutical Sciences" through the courtesy of the Philadelphia College of Pharmacy and Science and Eli Lilly and Company.*)

CALCULATIONS WITH VALUES INCLUDING ERRORS

When normally simple mathematical operations such as addition, subtraction, multiplication and division are conducted with numbers which include an error, it is necessary to handle the errors in a separate and distinct manner. These operations are illustrated for the quantities A and B which include the errors a and b respectively.

Addition: $$(A \pm a) + (B \pm b) = (A + B) \pm \sqrt{a^2 + b^2} \quad (4.2)$$

Subtraction: $$(A \pm a) - (B \pm b) = (A - B) \pm \sqrt{a^2 + b^2} \quad (4.3)$$

Multiplication: $$(A \pm a)(B \pm b) = (AB)\left(1 \pm \sqrt{a^2/A^2 + b^2/B^2}\right) \quad (4.4)$$

Division: $$(A \pm a)/(B \pm b) = (A/B)\left(1 \pm \sqrt{a^2/A^2 + b^2/B^2}\right) \quad (4.5)$$

PROBABILITY

Probability p is the chance that a particular phenomenon will occur and is represented by a scale of values from zero to unity. The probability of occurrence of an absolute certainty is represented by $p = 1$, and of an absolute impossibility by $p = 0$. When a coin is flipped, it is obvious that there is a 50-50 chance that "heads" will show. A 50-50 chance is represented by $p = 0.5$.

Addition Law—If an event can occur in one of several ways, the sum of the probabilities for each way is equal to the probability of the event occurring at all. Returning to the coin as an example, we know the probability of obtaining either "heads" or "tails" is an absolute certainty. The probability of obtaining "heads" is $p = 0.5$ and of obtaining "tails" is $p = 0.5$. Thus, the probability that the coin will land one way or the other is $p = 0.5 + 0.5 = 1.0$.

Multiplication Law—What is the probability of flipping "heads" twice in a row? For the first flip, $p = 0.5$ and, since the results of the second flip in no way depend upon the results of the first, $p = 0.5$ for the second flip too. The probability that heads will occur twice in a row is then $p = 0.5 \times 0.5 = 0.25$. This is illustrated by the following table.

	Possible Combinations			
	1	2	3	4
First flip Second flip	Heads Heads	Heads Tails	Tails Heads	Tails Tails

Since there is an equal probability of occurrence for each of the four possible combinations, the chance of obtaining the first combination of "heads" twice is, therefore, one in four or 25%. This corresponds to $p = 0.25$.

BINOMIAL DISTRIBUTION

In the preceding paragraph it was seen that two "heads" could be expected once in four tries, that two "tails" could be expected with the same frequency but that a "head-tail' combination is expected twice in four tries. These facts are expressed by

$$1\ H^2 + 2\ HT + 1\ T^2 \tag{4.6}$$

which is an expansion of the term $(H + T)^2$.

If this process were extended to include a third flip of the coin there would be eight combinations as follows:

	Combinations							
	1	2	3	4	5	6	7	8
First flip	H	H	H	T	T	T	H	T
Second flip	H	H	T	H	T	H	T	T
Third flip	H	T	H	H	H	T	T	T

Each of the eight combinations would have an equal probability of occurrence.

These data can be expressed by

$$1\,H^3 + 3H^2T + 3\,HT^2 + 1\,T^3 \tag{4.7}$$

which is an expansion of $(H + T)^3$. It can be demonstrated that the general expression for n flips (i.e., n determinants) is given by the general equation for a binomial:

$$(H + T)^n = H^n + nH^{n-1}T + \frac{n\,(n-1)}{1\cdot 2}H^{n-2}T^2 + \frac{n(n-1)(n-2)}{1\cdot 2\cdot 3}H^{n-3}T^3 + \frac{n(n-1)(n-2)(n-3)}{1\cdot 2\cdot 3\cdot 4}H^{n-4}T^4 \ldots nHT^{n-1} + T^n \tag{4.8}$$

Data dealing with the occurrence or non-occurrence of events generally follow a binomial distribution. Chi-square test is considered in experiment 4.4 where it is used to test radiation counting equipment for proper function.

POISSON DISTRIBUTION

Radioactive decay is a perfectly random process. If replicate measurements of radioactivity were made using the same radioactive source with constant sample geometry, a plot of these data would yield a curve illustrated by the dotted line in figure 4.3. This is known as a Poisson distribution and is given by the expression

$$P_n = \frac{\mu^n e^{-\mu}}{n!} \tag{4.9}$$

Let the Poisson equation be illustrated by assuming a true average count μ of 20. The probability of obtaining a count of 18 is

$$P_{18} = \frac{20^{18} e^{-20}}{18 \cdot 17 \cdot 16 \cdot 15 \text{-----} \cdot 1} = 0.0844$$

Thus, the probability of obtaining a count of 18 is 0.0844. (This means 0.0844 in 1 because the probability of obtaining a count of some value is 1. This could also be read as 0.844 in 10 or 8.44 in 100 or 844 in 10,000).

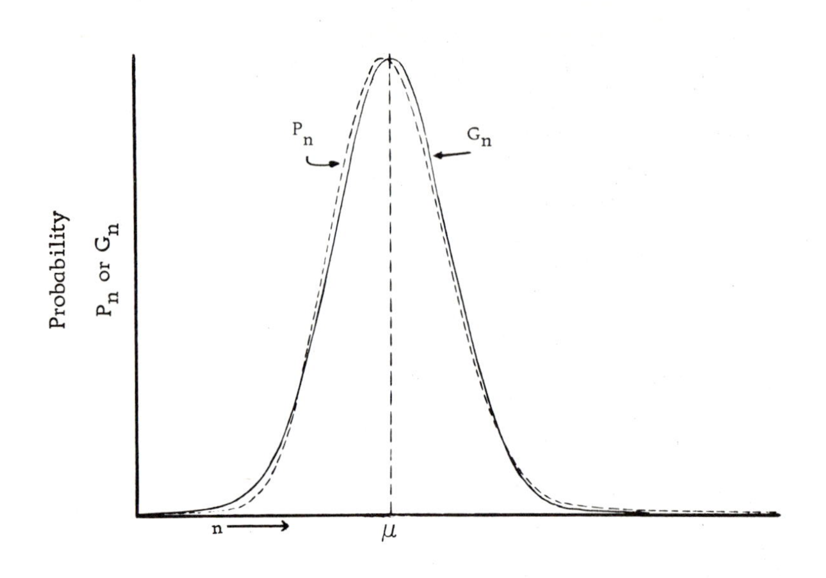

Figure 4.3—Comparison of the Poisson and Gaussian Distributions

GAUSSIAN DISTRIBUTION

Although the Poisson distribution describes the random decay of radioactive atoms, it is often cumbersome to use. For this reason the Gaussian probability—almost as accurate as the Poisson in the most widely used range—is usually used. It is given by the equation

$$G_n = \frac{1}{\sqrt{2\pi\mu}} e^{-(n - \mu)^2/2\mu} \qquad (4.10)$$

If again it is assumed $\mu = 20$, then the probability of obtaining a count of 18 is:

$$G_{18} = \frac{1}{\sqrt{2\pi \cdot 20}} e^{-(18 - 20)^2/2 \cdot 20}$$

$$G_{18} = 0.0807$$

There is a difference between these two curves (figure 4.3); the Gaussian curve is symmetrical whereas the Poisson curve is not. The Poisson curve passes through the origin.

In the problem given above, where the true mean count was taken as 20, it was seen that the probability ($G_n = G_{18}$) of obtaining a count of 18 was 0.0807. Because the normal curve is symmetrical, $G_{22} = 0.0807$ also. This is because the error is 2 for both G_{18} and G_{22}. If we were to add all probabilities from $G_{-\infty}$ to $G_{+\infty}$ the sum would be unity.

EXPERIMENT 4.1 THE GAUSSIAN OR NORMAL CURVE

OBJECTIVES:

To investigate the statistics of radioactive measurements.
To determine the fit of observed data to a gaussian or normal curve.

THEORY:

The Gaussian or normal distribution, defined by equation 4.10 in terms of the error $(n-\mu)$ and the mean μ, describes the distribution of perfectly random events. The Gaussian distribution may also be defined by means of the parameters: μ, the *mean,* which defines the center of the distribution and σ, the *standard deviation,* which defines the spread or dispersion of the data. The standard deviation is a unit of measure and is equal to the square root of the mean.

$$\sigma = \sqrt{\mu} \quad \text{or} \quad \mu = \sigma^2 \tag{4.11}$$

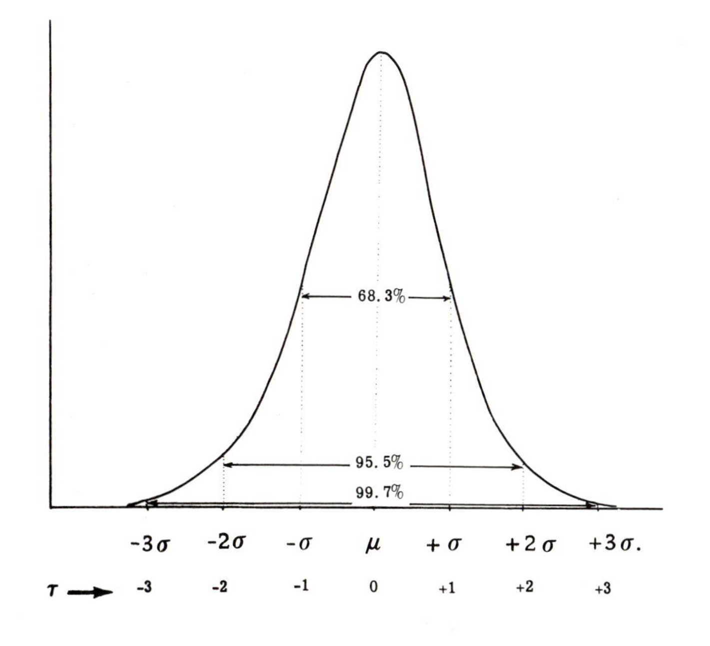

Figure 4.4—THE NORMAL DISTRIBUTION CURVE. Percentages indicate that 68.3% of the observations will be in error by less than $\pm\sigma$, while 95.5% of the observations will be in error by less than $\pm 2\sigma$.

Equation 4.10 can therefore be expressed as

$$G_n = \frac{1}{\sigma\sqrt{2\pi}} e^{-(n-\sigma^2)^2/2\sigma^2} \tag{4.12}$$

The Cumulative Normal Frequency Distribution (Summation Probability) — Rather than calculate the probability of obtaining a specific count, it is usually more useful and informative to know the magnitude of the error to be expected; i.e., the probability of the error being less than or more than a particular value.

For convenience, the error is measured in units of σ, the standard deviation. The *relative error* τ, a parameter of the error, is equal to the number of standard deviations in the error.

$$n - \mu = \tau\sigma \quad \text{or} \quad \tau = \frac{n-\mu}{\sigma} \tag{4.13}$$

Thus, the probability G_τ of the relative error lying between τ and $\tau + d\tau$.

$$G_\tau d\tau = \frac{1}{\sqrt{2\pi}} e^{-\tau^2/2} d\tau \tag{4.14}$$

Also, because there is unit probability that the relative error lies between $-\infty$ and $+\infty$,

$$\int_{-\infty}^{+\infty} G_\tau d\tau = 1 \tag{4.15}$$

and the summation probability G_s is given by

$$G_s = \int_{-\tau}^{+\tau} (2\pi)^{-1/2} e^{-\tau^2/2} d\tau \tag{4.16}$$

or, since the distribution is symmetrical, by

$$G_s = 2\int_0^{\tau} (2\pi)^{-1/2} e^{-\tau^2/2} d\tau \tag{4.17}$$

In both cases the result will be the probability that the relative error will be equal to or *less* than $\pm\tau$. To calculate the probability of the relative error being equal to or *greater* than $\pm\tau$, the limits of integration should be taken from τ to ∞. Tables of probability for values of τ of both types (i.e., "less than" and "greater than") are available. Both are equally useful, and the choice lies with the individual.

(NOTE: If p is the probability of occurrence of an event and q is the probability that an event will not occur, then $p + q = 1$. On this basis "greater than" and "less than" tables are related.)

An inspection of the curve in figure 4.4 gives the impression of a curve with two tails. This term is indeed used to describe the curves. Thus, a "two-tailed" probability is based upon the entire area between the limits of $\pm\tau$. A "one-tailed" probability is based upon the area either to the left *or* to the right of the mean μ *but not both;* that is, only $+\tau$ *or* $-\tau$. For a "one-tailed" calculation, the coefficient 2 should be dropped from equation 4.17.

Data for a "two-tailed... more than" probability for various values of τ are given in table 4.2. These data have been plotted to provide the curve in figure 4.4. Data for a "one-tailed... less than" probability are found in table 4.3.

In the example previously considered (page 82), the probability of observing a result in error by 2 or more from the true average count of 20 is determined by first solving for τ by the use of equations 4.11 and 4.13. It is found that $\tau = 2/\sqrt{20} = 0.447$. By the use of table 4.2, it is found through interpolation of the data given, that the probability p = 0.65 (for τ = 0.447) which means there are 65 chances in 100 of observing a result in error by 2 or more from the true average count of 20.

Table 4.2—CUMULATIVE NORMAL FREQUENCY DISTRIBUTION

τ	p	τ	p
0.0	1.000	2.5	0.0124
0.1	0.920	2.6	0.0093
0.2	0.841	2.7	0.0069
0.3	0.764	2.8	0.0051
0.4	0.689	2.9	0.0037
0.5	0.617	3.0	0.00270
0.6	0.548	3.1	0.00194
0.7	0.483	3.2	0.00136
0.8	0.423	3.3	0.00096
0.9	0.368	3.4	0.00068
1.0	0.317	3.5	0.00046
1.1	0.272	3.6	0.00032
1.2	0.230	3.7	0.00022
1.3	0.194	3.8	0.00014
1.4	0.162	3.9	0.00010
1.5	0.134	4.0	0.0000634
1.6	0.110	4.1	0.0000414
1.7	0.090	4.2	0.0000266
1.8	0.072	4.3	0.0000170
1.9	0.060	4.4	0.0000108
2.0	0.046	4.5	0.0000068
2.1	0.036	4.6	0.0000042
2.2	0.028	4.7	0.0000026
2.3	0.022	4.8	0.0000016
2.4	0.016	4.9	0.0000010

Table 4.3—AREA OF A NORMAL DISTRIBUTION CURVE BETWEEN THE ORDINATE OF THE MEAN μ AND THE INDICATED VALUE OF τ.
(Values of p for a "one-tailed," "less-than" Cumulative Normal Frequency Distribution)*

τ	00	01	02	03	04	05	06	07	08	09
0.0	00000	00399	00798	01197	01595	01994	02392	02790	03188	03586
0.1	03983	04380	04776	05172	05567	05962	06356	06749	07142	07535
0.2	07926	08317	08706	09095	09483	09871	10257	10642	11026	11409
0.3	11791	12172	12552	12930	13307	13686	14058	14431	14803	15173
0.4	15542	15910	16276	16640	17003	17364	17724	18082	18439	18793
0.5	19146	19497	19847	20194	20540	20884	21226	21566	21904	22240
0.6	22575	22907	23237	23565	23891	24215	24537	24857	25175	25490
0.7	25804	26115	26424	26730	27035	27337	27637	27935	28230	28524
0.8	28814	29103	29389	29673	29955	30234	30511	30785	31057	31327
0.9	31594	31859	32121	32381	32639	32894	33147	33398	33646	33891
1.0	34134	34375	34614	34850	35083	35314	35543	35769	35993	36214
1.1	36433	36650	36864	37076	37286	37493	37698	37900	38100	38298
1.2	38493	38686	38877	39065	39251	39435	39617	39796	39973	40147
1.3	40320	40490	40658	40824	40988	41149	41309	41466	41621	41774
1.4	41924	42073	42220	42364	42507	42647	42786	42922	43056	43189
1.5	43319	43448	43574	43699	43822	43943	44062	44179	44295	44408
1.6	44520	44630	44738	44845	44950	45053	45154	45254	45352	45449
1.7	45543	45637	45728	45818	45907	45994	46080	46164	46246	46327
1.8	46407	46485	46562	46638	46712	46784	46856	46926	46995	47062
1.9	47128	47193	47257	47320	47381	47441	47500	47558	47615	47670
2.0	47725	47778	47831	47882	47932	47982	48030	48077	48124	48169
2.1	48214	48257	48300	48341	48382	48422	48461	48500	48537	48574
2.2	48610	48645	48679	48713	48745	48778	48809	48840	48870	48899
2.3	48928	48956	48983	49010	49036	49061	49086	49111	49134	49158
2.4	49180	49202	49224	49245	49266	49286	49305	49324	49343	49361
2.5	49377	49396	49413	49430	49446	49461	49477	49492	49506	49520
2.6	49534	49547	49560	49573	49585	49598	49609	49621	49632	49643
2.7	49653	49664	49674	49683	49693	49702	49711	49720	49728	49736
2.8	49744	49752	49760	49767	49774	49781	49788	49795	49801	49807
2.9	49813	49819	49825	49831	49836	49841	49846	49851	49856	49861
3.0	49865	49869	49874	49878	49882	49886	49889	49893	49897	49900
3.1	49903	49906	49910	49913	49916	49918	49921	49924	49926	49929
3.2	49931	49934	49936	49938	49940	49942	49944	49946	49948	49950
3.3	49952	49953	49955	49957	49958	49960	49961	49962	49964	49965
3.4	49966	49968	49969	49970	49971	49972	49973	49974	49975	49976
3.5	49977	49978	49978	49979	49980	49981	49981	49982	49983	49983
3.6	49984	49985	49985	49986	49986	49987	49987	49988	49988	49989
3.7	49989	49990	49990	49990	49991	49991	49992	49992	49992	49992
3.8	49993	49993	49993	49994	49994	49994	49994	49995	49995	49995
3.9	49995	49995	49996	49996	49996	49996	49996	49996	49997	49997
4.0	49997	49997	49997	49997	49997	49997	49998	49998	49998	49998

(All values of p should be read as a decimal; e.g., for $\tau = 1.36$ read p = 0.41309)

*To obtain values for a "two-tailed", "less than" distribution, multiply values in this table by 2.

Confidence Limits — It will be recalled that we would like to measure the true mean count μ but that in practice it is the observed count n which is measured, and n differs from μ by an error of $(n-\mu)$. If A represents the numerical value of n, then the count should be reported as $(A \pm a)$ where a is the error in A. Yet we cannot be sure of the absolute limits of "a". The best we can do is to indicate the degree of confidence that the error "a" as given is not exceeded.

With reference to table 4.2, which lists the probability of observing a counting error greater than a certain defined value, we would speak of a *level of significance* p, a *confidence level* of $(1-p)$, a *confidence interval* of $100(1-p)\%$ or of a *risk* of $100p\%$. Thus, if we choose a high level of significance, say, $p = 0.05$ (corresponding to $\tau = 1.96$), then the confidence level is 0.95 or 95% with a 5% risk that the observed error will be greater than "a".

For example, let it be supposed that a total of 10,000 counts is recorded ($A = 10{,}000$). Assuming a normal distribution, what numerical value can be assigned the error "a" with a 95% confidence? The error $(n-\mu) = \tau\sigma$ and $\sigma = \sqrt{\mu}$. Therefore, $\sigma = 100$ and $n - \mu = (1.96)(100) = 196$. Thus, we can say, with 95% confidence, that the observed count is $(A \pm a) = 10{,}000 \pm 196$ counts (for $p = 0.05$) and that the error in the measurement ($\pm$ 196) represents a relative error of 1.96% or less. That is, in only 5% of replicate observations will the data be in error by more than 1.96%.

The basic confidence level is that defined by the *standard deviation,* when $\tau = 1.00$ and $p = 0.3173$. Known as the *standard error* or *one-sigma level* ($n - \mu = \tau\sigma = 1.0\sigma$), there is a 31.73% chance that this level will be exceeded or a 68.27% chance that it will not be exceeded.

The confidence of data may also be reported in terms of one of the following error values:

The *probable error* $P = 0.6745\sqrt{\mu}$, and there is a 50-50 chance that any specific error will be larger than the probable error. The probable error can also be calculated from the standard deviation by the formula, $P = 0.6745\sigma$.

The *nine-tenths error* is so named because there are nine chances out of ten that the error in a specific determination will be smaller than this value. It is commonly used in reporting radioactivity measurements, but results possessing deviations with a probability as high as this are not very significant.

The *ninety-five hundredths error* is so called because there is a 95% chance that the error in a specific determination will be smaller than this value. Data falling within this level of chance variation are considered significant. It is sometimes called the 0.05 level of significance.

The *ninety-nine hundredths error* signifies a confidence that the errors of 99% of the determinations will be less than this value. This degree of confidence is highly significant.

Fit of Data to a Normal Curve—A true normal distribution can be defined by the parameters μ and σ as previously described. It is hoped that laboratory data will take the form of a normal curve, but perfect conformity cannot be expected since data cannot be obtained under ideal (perfect) conditions but rather only under existing practical conditions subject to errors introduced by the equipment, operator and environment. Thus neither μ nor σ can be known with exactitude. In practice, the sample count n or the sample mean $\bar{n}$ is used as an *estimate* of μ, and the *sample standard deviation* s is not equivalent to σ. As laboratory conditions approach ideal conditions and as n becomes

Table 4.4—"ERRORS" USED TO DEFINE CONFIDENCE LEVELS

p = probability of observing an error larger than $n - \mu = \tau\sqrt{\mu}$		
Name of Error	τ	p
Probable	0.6745	0.5000
Standard Deviation (Standard error)	1.0000	0.3173
Nine-tenths	1.6449	0.1000
Ninety-five Hundredths	1.9600	0.0500
Two Sigma	2.0000	0.0455
Ninety-nine Hundredths	2.5758	0.0100
Three Sigma	3.0000	0.0027

very large, s approaches σ. Consider methods used to determine s.

The best measure of s is given by

$$s = \sqrt{\frac{1}{N-1}\Sigma(n-\bar{n})^2} \tag{4.18}$$

Using the data in table 4.1, we find $\Sigma(n-\bar{n})^2 = 1134$ and $N = 10$. Therefore

$$s = \sqrt{\frac{1}{10-1}(1134)} = \sqrt{126} = 11.2$$

The quantity $N-1$ is the number of *degrees of freedom** which is one less than the number of observations.

The distribution of data from table 4.1 is therefore best described by the parameters $\bar{n} = 98$ and $s = 11.2$. The deviation $n-\bar{n}$ for 31.73% of the observations should exceed 11.2. We find that three out of ten or 30% of the observations recorded in table 4.1 meet this requirement. This does not mean, however, that the data correspond to the Gaussian curve defined by μ and σ. The best estimate of σ is given by

$$\sigma \approx \sqrt{n} \quad \text{or} \quad \sigma \approx \sqrt{\bar{n}} \tag{4.19}$$

Here $\sigma \approx \sqrt{98} = 9.9$. It is seen that five out of ten or 50% of the observations fall outside of this range. Thus the spread of data given in table 4.1 is greater than that of a true Gaussian distribution. Such a distribution may be described as *platykurtic.* If the spread of data had not been so great as that expected for a normal distribution, the spread would be described as being *leptokurtic.*

* The number of degrees of freedom DF is one less than the number of observations N. For example, let $10 be divided among three people. If the first receives $5 and the second $3, then the third must get $2. While there was freedom to determine the amounts received by the first two, there was no choice in the case of the third. Hence $DF = N-1$.

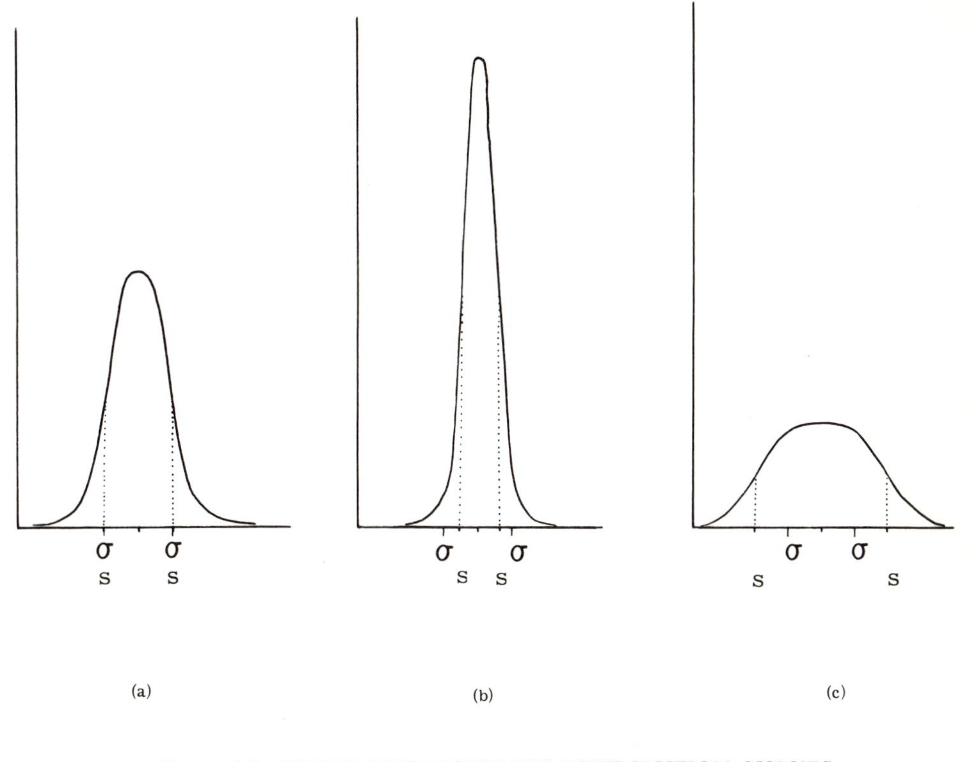

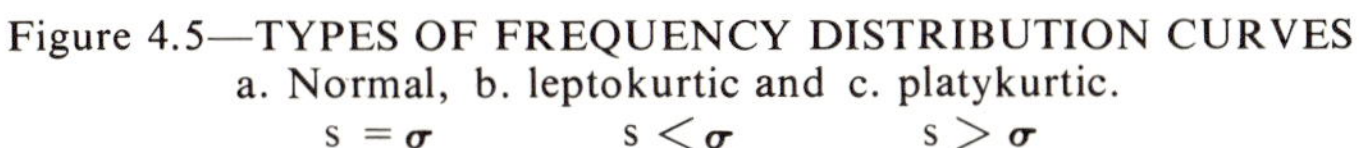

Figure 4.5—TYPES OF FREQUENCY DISTRIBUTION CURVES
a. Normal, b. leptokurtic and c. platykurtic.
$s = \sigma$ $s < \sigma$ $s > \sigma$

The *standard deviation of the mean* s_m is defined by

$$s_m = \frac{s}{\sqrt{N}} \tag{4.20}$$

For the data given in table 4.1, $s_m = 11.2/\sqrt{10} = 3.57 \approx 4$. One could then say that the sample used to obtain the data represented in table 4.1 had an activity of 98 ± 4 cpm where the error 4 is the *sample standard error*.

PROCEDURE:

Place a radioactive sample on a suitable shelf so that 1,000 to 2,000 cpm are obtained and make twenty 1-minute observations. The observations must be made in a continuous series without long interruptions—and the source, counter and other conditions must not be disturbed throughout.

Record the results of the twenty observations. These will be used to determine the validity of the Gaussian distribution.

DATA:

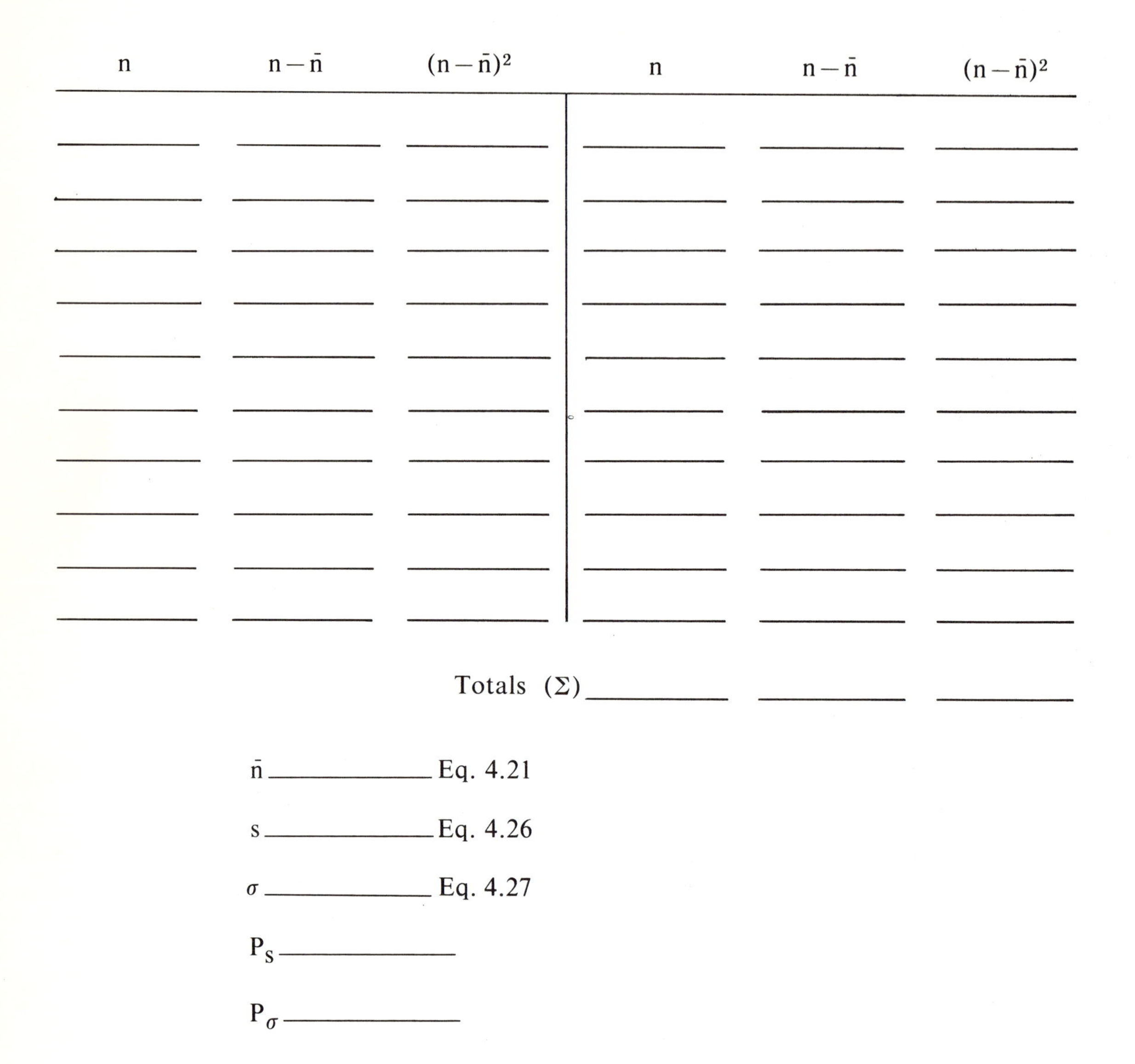

n	$n - \bar{n}$	$(n - \bar{n})^2$	n	$n - \bar{n}$	$(n - \bar{n})^2$
______	______	______	______	______	______
______	______	______	______	______	______
______	______	______	______	______	______
______	______	______	______	______	______
______	______	______	______	______	______
______	______	______	______	______	______
______	______	______	______	______	______
______	______	______	______	______	______
______	______	______	______	______	______
______	______	______	______	______	______
		Totals (Σ)	______	______	______

$\bar{n}$ ____________ Eq. 4.21

s ____________ Eq. 4.26

σ ____________ Eq. 4.27

P_s ____________

P_σ ____________

Refer to values of $n - \bar{n}$ above. Note the percentage of times deviation $(n - \bar{n})$ is greater than each of the following:

s ________% Eq. 4.26 (Theoretical value 31.7%)

σ ________% Eq. 4.27 (Theoretical value 31.7% if distribution is Gaussian)

P_s ________% (Theoretical value 50.0%)

P_σ ________% (Theoretical value 50.0% if distribution is Gaussian)

CALCULATIONS:

1. The total count of each observation n, uncorrected for background, dead time, etc., may be used. Compute the "errors" by calculation of the following quantities in order. Results are recorded above.

a. The average count $\bar{n}$, that is, the sums of all the counts divided by the number of observations N

$$\bar{n} = \frac{\Sigma n}{N} \tag{4.21}$$

b. The sample deviation of each count from the average

$$n - \bar{n} \tag{4.22}$$

c. The square of the deviations of each count

$$(n - \bar{n})^2 \tag{4.23}$$

d. The sum of the square of sample deviations

$$\Sigma\,(n - \bar{n})^2 \tag{4.24}$$

e. The sample variance

$$s^2 = \frac{1}{N-1}\,\Sigma\,(n - \bar{n})^2 \tag{4.25}$$

f. The sample standard deviation, s

$$s = \sqrt{\frac{1}{N-1}\,\Sigma\,(n - \bar{n})^2} \tag{4.26}$$

g. The algebraic sum of the deviations $\Sigma\,(n - \bar{n})$ should be equal to zero. This serves as a mathematical check.

h. The standard deviation σ, where $\bar{n}$ is used as an estimate of μ

$$\sigma \approx \sqrt{\bar{n}} \tag{4.27}$$

i. The probable error P

$$P = 0.6745\,\sigma \tag{4.28}$$

j. The probable % error $= 100\ P/\bar{n}$

2. Show that the observations follow a Gaussian distribution by making the following statistical tests:

a. Show that the standard deviation calculated using equation 4.26 is nearly equal to that obtained using equation 4.27. (Too great a difference is suggestive of a determinate error. Application of the chi-square test, experiment 4.4, will verify this point.)

b. Count the number of times that the deviation $n - \bar{n}$ is greater than the standard deviation and show that this occurs in approximately a third (31.7%) of the observations.

The value of s calculated from equation 4.26 should give a better fit than that obtained from equation 4.27, this value having been calculated with consideration of the individual values of n. Calculation of σ from equation 4.27 assumes $\bar{n} = \mu$ and that the distribution is truly Gaussian.

c. Also show that the number of times the deviation is greater than the probable error is approximately 50%.

Table 4.5
STATISTICAL SIGNIFICANCE OF A GIVEN TOTAL NUMBER OF COUNTS

Total Count (μ)	Standard Deviation[1] (σ)	Probable Error[2] (P)	Probable % Error[3] (100 P/μ)	Two-sigma Confidence[4]
30	5.47	3.68	12.27%	36.47%
40	6.32	4.26	10.65	31.60
50	7.07	4.77	9.54	28.28
55	7.42	5.00	9.09	26.98
60	7.75	5.22	8.70	25.83
70	8.37	5.64	8.06	23.91
80	8.94	6.03	7.53	22.35
90	9.49	6.40	7.10	20.87
100	10.00	6.74	6.74	20.00
200	14.14	9.53	4.77	14.14
300	17.32	11.67	3.89	10.88
400	20.00	13.48	3.46	10.00
500	22.36	15.07	3.01	8.94
600	24.49	16.51	2.75	8.16
700	26.46	17.83	2.55	7.56
800	28.28	19.06	2.38	7.07
900	30.30	20.22	2.25	6.67
1,000	31.62	21.31	2.13	6.32
1,500	38.72	26.10	1.74	5.16
2,000	44.72	30.14	1.51	4.47
2,500	50.00	33.70	1.35	4.00
3,000	54.77	36.90	1.23	3.65
3,500	59.16	39.87	1.15	3.38
4,000	63.24	42.60	1.06	3.16
4,500	67.08	45.20	1.00	2.98
5,000	70.71	47.60	0.95	2.83
5,500	74.16	49.98	0.91	2.70
6,000	77.45	52.20	0.87	2.58
6,500	80.62	54.30	0.83	2.48
7,000	83.66	56.40	0.81	2.39
7,500	86.60	58.40	0.78	2.31
8,000	89.44	60.30	0.75	2.24
8,500	92.20	62.10	0.73	2.17
9,000	94.87	63.90	0.71	2.11
9,500	97.47	65.70	0.69	2.05
10,000	100.00	67.40	0.67	2.00
20,000	141.42	95.32	0.48	1.41
30,000	173.20	116.74	0.39	1.15

1. $\sigma = \sqrt{\mu}$
2. $P = 0.6745\ \sigma$
3. 50% confidence level. For a total count μ the per cent error listed will not be exceeded in 50% of the observations.
4. Two-sigma or 96% confidence level. For a total count μ, the per cent error listed will not be exceeded in 96% of the observations.

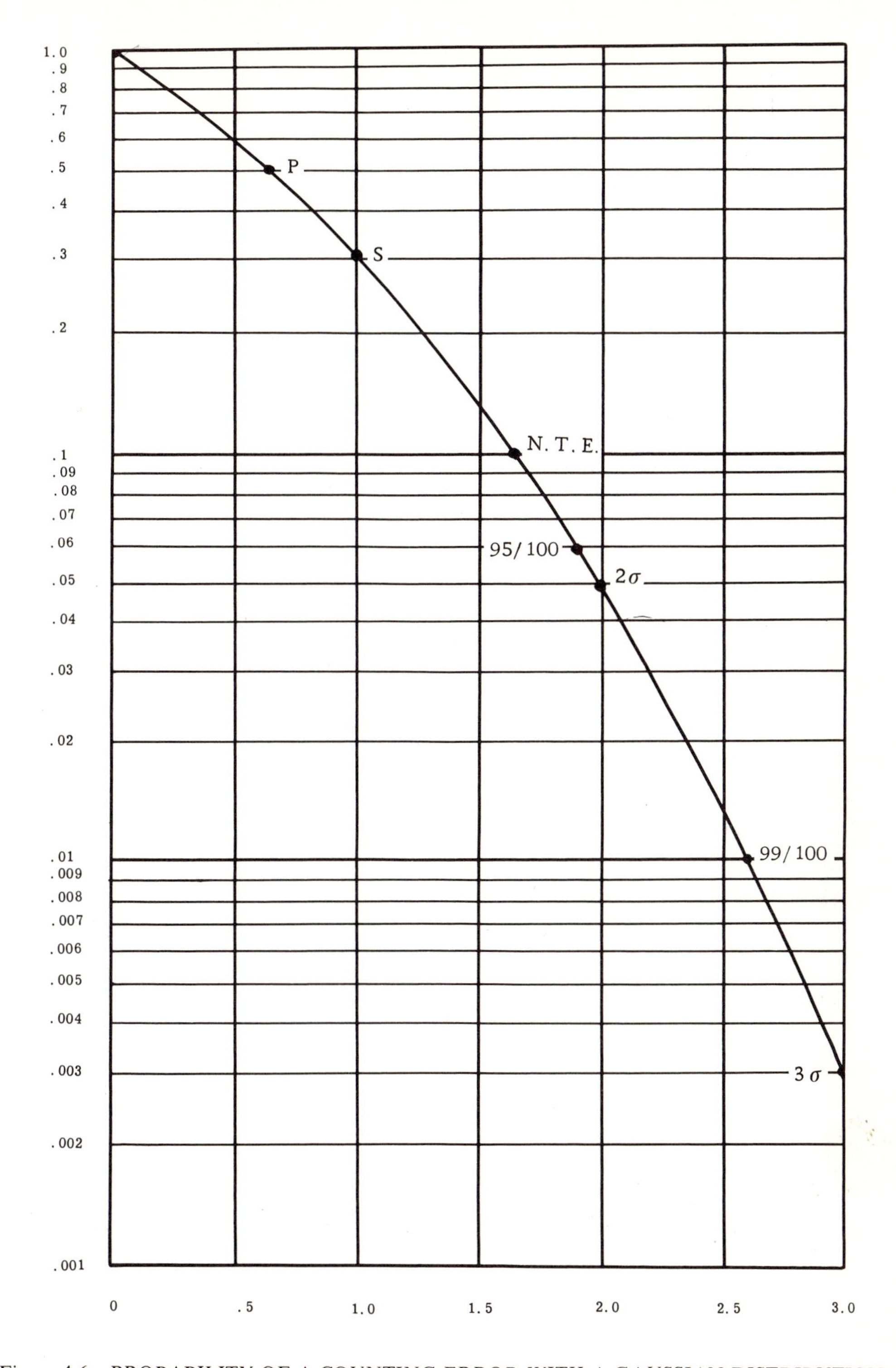

Figure 4.6—PROBABILITY OF A COUNTING ERROR WITH A GAUSSIAN DISTRIBUTION. (Probability of observing a counting error greater than $n - \mu$ $(= \tau\sqrt{\mu})$ versus τ. A "two-tailed" probability, constructed from the data in table 4.2)

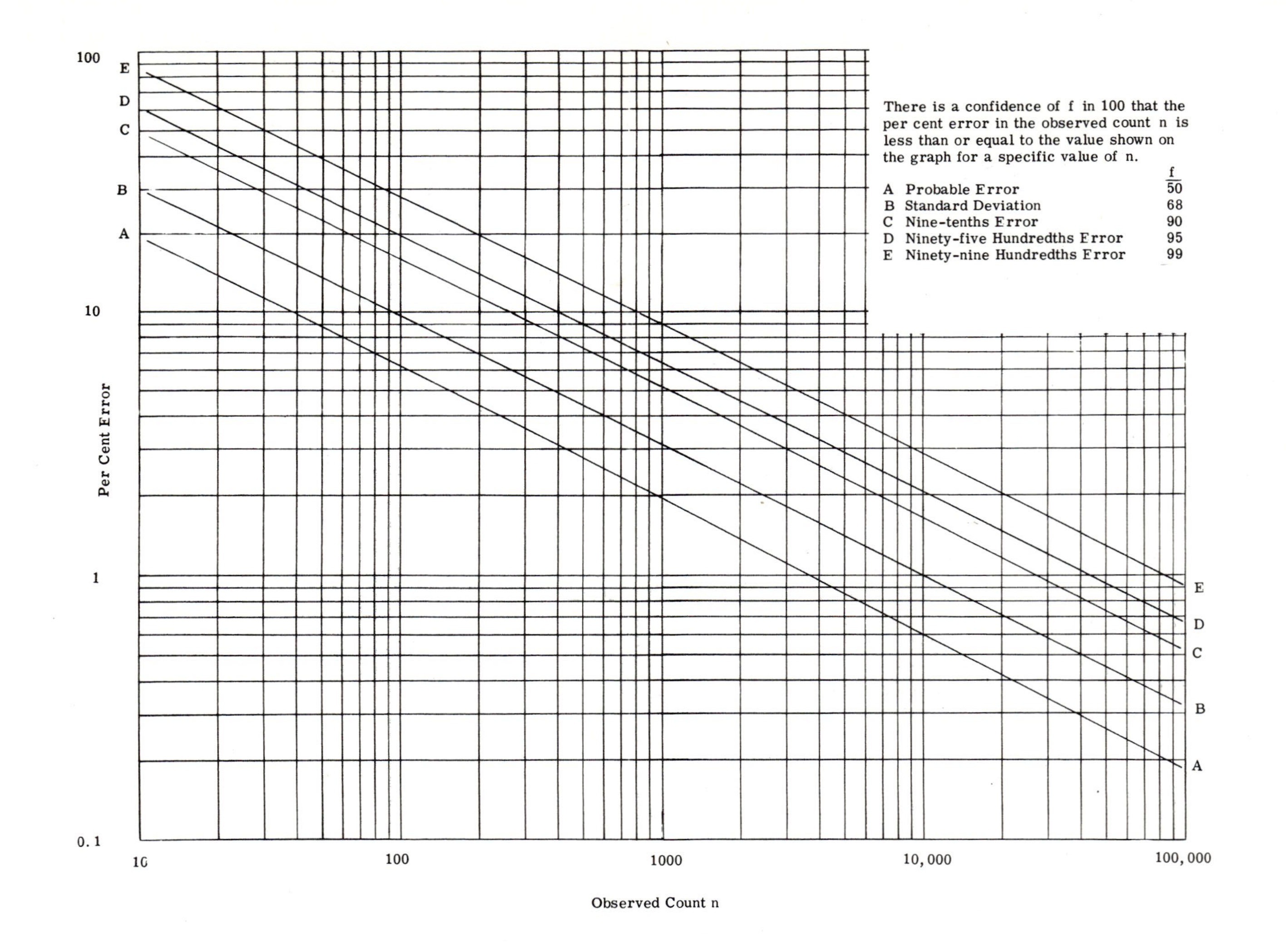

Figure 4.7—SIGNIFICANCE OF COUNTING ERRORS (Confidence Limits)

EXPERIMENT 4.2 COMPARISON OF SAMPLES

OBJECTIVES:

To apply statistics to the analysis of counting rate errors.

To illustrate a method to determine if a variation represents a true difference between two populations. (The t test of significance.)

THEORY:

When two samples are compared, it must be determined whether or not the difference between the observed activities is "significant." A "significant" difference means that the measured difference of the activities may be real and not just a difference caused by the randomness of radioactive decay.

The test applied here is known as *the t test of significance** where τ' stands for a relative error and is defined by

$$\tau' = \frac{\bar{n} - \mu}{s_m} = \frac{\bar{n} - \mu}{s/\sqrt{N}} \tag{4.29}$$

Equation 4.29 should be compared with equation 4.13. If τ is the relative error then τ' may be considered the sample relative error. If there are no determinate errors $\tau' \longrightarrow \tau$ as $N \longrightarrow \infty$.

The activity r of a sample is expressed in terms of the number of counts n per unit of time t:

$$r = n/t \tag{4.30}$$

The error $n - \mu$ in an observation is related by

$$n - \mu = \tau\sigma \tag{4.31}$$

In practice it is convenient to use $n - \bar{n}$ as an approximation of $n - \mu$ and $\sqrt{n}$ as an approximation of σ. Thus,

$$n - \bar{n} = \tau\sqrt{n} \tag{4.32}$$

The error in the determination of rate is, therefore,

$$\frac{n}{t} - \frac{\bar{n}}{t} = r - \bar{r} = \frac{\tau\sqrt{rt}}{t} = \tau\sqrt{r/t} \tag{4.33}$$

*Since t is used in this text to denote time and τ is used for the relative error, τ' is used for the t test of significance.

In a statistical comparison of duplicate samples, one must consider the error of the difference between the results of two observations. It is seen from equation 4.3 that this is measured as the square root of the sums of the squares of the individual errors. Expressed in terms of variance, the following expression is derived in standard texts on statistics.* The sample variance of difference $s^2{}_D$ is equal to the variances of each sample observation:

$$s_D^2 = s_{m_1}^2 + s_{m_2}^2 = \frac{s_1^2}{N_1} + \frac{s_2^2}{N_2} \tag{4.34}$$

Thus, the *sample standard deviation of difference* s_D for $N_1 = N_2 = 1$ is

$$s_D = \sqrt{s_1^2 + s_2^2} \tag{4.35}$$

and the number of standard deviations (i.e., standard errors) in the difference is τ, expressed by

$$\tau = \frac{\text{difference}}{\text{standard error of the difference}} = \frac{r_1 - r_2}{s_D} = \frac{r_1 - r_2}{\sqrt{s_1^2 + s_2^2}} \tag{4.36}$$

Examination of equations 4.33 and 4.36 will show that $\sqrt{r/t}$ may be used as an estimate of s. Therefore, $s^2 = r/t$ and

$$\boxed{\tau = \frac{|\, r_1 - r_2 \,|}{\sqrt{r_1/t_1 + r_2/t_2}}} \tag{4.37}$$

The point in question is this—If the activity of samples is determined, what is the probability that the difference between the observed values is due to chance variation within a single population? This probability p is determined by first calculating τ by the use of equation 4.37. Then the probability p is found in table 4.2.

Example—Two radioactive samples were counted. The first sample was counted for two minutes and found to average 1104 cpm. The second sample was counted for three minutes and found to average 1068 cpm. Does this difference in counting rate represent a true difference between populations or is it a chance variation of a single population caused by the nature of random decay? The calculation follows:

$$\tau = \frac{1104 - 1068}{\sqrt{\frac{1104}{2} + \frac{1068}{3}}} = \frac{36}{30} = 1.2$$

The value 0.230 for p is found opposite the value 1.2 for τ in table 4.2. This means that there is a 23.0% chance of observing a difference of 36 counts or more in duplicate observations attributable to chance variation within a single population. Conversely, there is a probability of 0.77 that this variation represents a true difference between two populations.

*Bennett, C. A., and N. L. Franklin "Statistical Analysis in Chemistry and the Chemical Industry, (Wiley), page 50.

In the following experiments duplicate samples will be prepared. They will therefore represent the same population. If application of the τ' test of significance to the observed activities suggests that the variation represents a true difference between two populations it would also suggest that the technique of the student needs refinement.

EQUIPMENT AND REAGENTS:

Microliter pipettes; 10-ml beakers; semi-micro stirring rods; any suitable radioactive solution containing about one microcurie of activity per milliliter (e.g., ^{32}P, ^{131}I); planchets; infrared lamp; end-window or side-window Geiger tube with associated counting equipment; microfiltration assembly for 1″ filter paper; one-inch circles of filter paper; 10% nitric acid; 1% silver nitrate solution; stock iodide solution prepared as follows:

Iodide Stock Solution

^{131}I as iodide - about 0.1 microcurie per ml
Sodium iodide - about 20 mg/ml

PROCEDURE AND DATA:

1. PART A—COMPARISON OF EVAPORATED SAMPLES

Measure exactly the same volume of a radioactive solution with a micropipette into two identical planchets. Since the experiment is a check on technique as well as statistics, this procedure should be conducted with extreme care. The identity of the radionuclide used is not important, provided its radiation is capable of detection with an end-window Geiger tube. It is suggested that a volume of about 25 microliters be used and that the activity in this volume be sufficient to provide between one and five thousand cpm (e.g., a solution containing about 1 μCi of ^{32}P per ml has been used). Taking all necessary precautions to assure accuracy in preparing the duplicate samples, dry under an infrared lamp and count. (When the counting rate of two samples is very nearly the same, the coincidence correction can be omitted when making only a relative comparison.)

Count each sample for 0.5-, 1.0- and 5.0-minute intervals.

Sample	Total Counts n	Time t	Activity cpm r	Relative Error τ	Probability p
1	________	0.5 min.	________		
2	________	0.5 ″	________	________	________
1	________	1.0 min.	________		
2	________	1.0 ″	________	________	________
1	________	5.0 min.	________		
2	________	5.0 ″	________	________	________

2. PART B – COMPARISON OF PRECIPITATED SAMPLES

Prepare two identical samples of silver radioiodide. Exactly 500 microliters of Iodide Stock Solution are added to about 10 ml of water contained in each of two small beakers. Silver iodide is then precipitated by adding 1% $AgNO_3$ solution dropwise until precipitation is complete, the solutions having been acidulated by addition of 1 ml of 10% nitric acid. Quantitatively collect the precipitates by filtration as described in experiment 10.1. The precipitates are then dried by washing with 10 ml each of alcohol and ether, drawing the wash solvents through the filter with an aspirator. Mount the papers and allow to dry. Count for 0.5, 1.0 and 5.0 minute intervals.

Sample	Total Counts n	Time t	Activity cpm r	Relative Error τ	Probability p
1	______	0.5 min.	______		
2	______	0.5 "	______	______	______
1	______	1.0 min.	______		
2	______	1.0 "	______	______	______
1	______	5.0 min.	______		
2	______	5.0 "	______	______	______

CALCULATIONS:

The count n has been measured for each of three periods of time (0.5, 1.0 and 5.0 minutes) for each of duplicate pipetted samples and for each of duplicate precipitated samples. For each case calculate the activity r in counts per minute, the relative error τ and the probability p that the deviation of duplicate samples might be expected to equal or exceed the observed value.

Suggestions for further work:

Having mastered good technique in measuring 25-lambda quantities of liquid and larger, repeat part A of the procedure using 10-lambda, 5-lambda and 1-lambda quantities of radioactive solution and check on the reproducibility of your results.

EXPERIMENT 4.3 EFFICIENT DISTRIBUTION OF COUNTING TIME

OBJECTIVES:

To determine the effect of background on the counting statistics.

To calculate the most efficient distribution of counting time when the background count is large relative to the total count.

THEORY:

Radioactivity measurements cannot be made without consideration of the background. In practice the total count (sample plus background) is recorded by the counter. The background count must be measured by a separate operation and then subtracted from the total count to give the net activity of the sample.

Statistical fluctuations which pertain to the sample itself also pertain to the background. Thus, equation 4.33 from experiment 4.2 applies to the background as well as to the sample, itself:

$$r_b - \bar{r}_b = \tau \sqrt{r_b/t_b} \tag{4.38}$$

where, $r_b - \bar{r}_b$ is the background error, τ is the parameter of the error, r_b is the background activity and t_b is the time during which the background was measured.

Since the presence of the background introduces additional fluctuations into our measurements, it is necessary that we determine just what error is introduced by it. The calculation of the sample activity r_s involves the error of the difference of two quantities, $r_t - r_b = r_s$, the total activity r_t and the background activity r_b. This is analogous to the situation presented in experiment 4.2, and equation 4.35 from experiment 4.2 applies here as well.

$$s_D = \sqrt{s_t^2 + s_b^2} \tag{4.39}$$

and

$$s_D = \sqrt{r_t/t_t + r_b/t_b} \tag{4.40}$$

where s_D is the sample standard error of the difference.

Example—A sample was counted for 5 minutes and averaged 50 counts per minute. The background was counted for 4 minutes and averaged 24 counts per minute. What is the probable error? (i.e., the probable error of the difference P_D). From table 4.4 we find $\tau = 0.6745$. Therefore

$$P_D = 0.6745\, s_D = 0.6745 \sqrt{r_t/t_t + r_b/t_b}$$
$$= 0.6745 \sqrt{50/5 + 24/4} = 0.6745 \sqrt{10 + 6} = 2.7$$

Thus, the sample activity r_s is 26.0 ± 2.7 cpm (P)

Minimizing the error. When the activity of a sample is low compared to the background, it is advantageous to calculate the most efficient distribution of counting time between the sample activity and the background activity in order to minimize the error. It can be shown that for a particular time distribution, the counting error will be at a minimum.

Error in the observed activity of the sample is measured by the standard sample deviation of the difference s_D or by the sample variance s_D^2. Conditions resulting in minimum sample error can be calculated by setting the derivative of equation 4.40 equal to zero, or the derivative of the variance equal to zero.

The following symbols have been used:

t_t = time used to count the total activity r_t.
t_b = time used to count the background activity r_b.
$r_s = r_t - r_b$ = net sample activity. (Note: t_s is meaningless)
$t = t_t + t_b$ = total time to measure both total and background activities.

For a given observation r_t, r_b, r_s and t are not variable and hence are treated as constants in the differentiation.

$$\frac{d(s_D^2)}{dt_t} = 0 = \frac{d}{dt_t}\left(\frac{r_t}{t_t} + \frac{r_b}{t - t_t}\right) = \frac{0 - r_t}{t_t^2} + \frac{0 - r_b(0-1)}{(t - t_t)^2}$$

$$0 = -\frac{r_t}{t_t^2} + \frac{r_b}{(t - t_t)^2} = -\frac{r_t}{t_t^2} + \frac{r_b}{(t_b)^2}$$

From which:

$$\boxed{t_t/t_b = \sqrt{r_t/r_b}} \qquad (4.41)$$

For ease in computation, the ratios indicated in equation 4.41 have been plotted in figure 4.8.

Example—Suppose a measured activity, uncorrected for background, is 4,000 cpm and the background is 40 cpm. Then r_t = 4,000 cpm and r_t/r_b = 4,000/40 = 100. Referring to the graph, we see this corresponds to t_t/t_b = 10, which means that for the most efficient distribution of counting time with minimum counting error the sample should be counted for a period of time ten times longer than the time used to measure the background.

PROCEDURE AND DATA:

1. Select a sample having an activity of about 500 cpm. Make four observations of activity: a 1-minute, a 3-minute, a 5-minute and a 10-minute measurement.
2. Remove the sample and make a one-minute background count.

	Counts n_t	Time t_t	Activity cpm r_t	Sample Activity $r_s \pm P$	Total Time t
1.	________	1.0 min.	________	________	2.0 min.
2.	________	3.0 min.	________	________	4.0 min.
3.	________	5.0 min.	________	________	6.0 min.
4.	________	10. min.	________	________	11.0 min.

Background (r_b) ________________ (1.0 minute)

CALCULATIONS:

1. For each observation calculate the activity and include the probable error. How do you account for the decrease in probable error P with an increase in counting time on the sample?

2. Calculate which of the above four observations gave the most efficient distribution of counting time. Observation No.________

3. Assume that you have a large number of measurements to make and can allot only 6 minutes total time to the determination of sample activity and background. All samples have activities of the same order of magnitude as the one used to collect the above data. Calculate the most efficient distribution of the 6 minutes between measurement of background and measurement of sample activity.

Background (t_b) ______minutes

Sample (t_t) ______minutes

Total time (t) __6.0__minutes

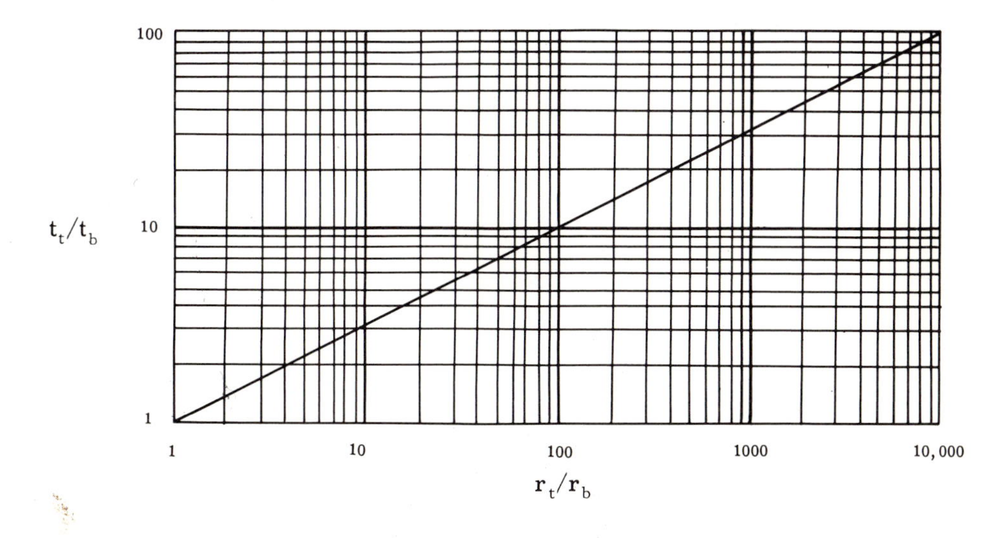

Figure 4.8—MOST EFFICIENT DISTRIBUTION OF COUNTING TIME.

EXPERIMENT 4.4 CHI-SQUARE TEST

OBJECTIVES:

To evaluate counter behavior statistically by means of the "chi-square test."

THEORY:

In experiment 4.1 proper operation of the counting equipment was assumed and the "goodness of fit" of data—accumulated in a series of identical measurements—to a Poisson curve was determined. The validity of the Poisson probability for describing radioactive decay has been demonstrated. Thus, if the procedure for the analysis of data used in experiment 4.1 is reversed, then on the premise that the Poisson distribution is valid, the equipment can be checked for proper operation by determining the goodness of fit of the data to a Poisson curve.

One method for the detection of non-statistical behavior was presented in experiment 4.2. This method was useful for comparing the results of duplicate analysis but is hardly adequate as a test for the measuring equipment.

The chi-square test, developed by Max Pearson, is a more rigorous test, the results of which may be based upon a large number of observations. It, too, is a test of the distribution of data. The derivation of this test is beyond the scope of this book. The discussion will, therefore, be restricted to its use and application.

For the Poisson probability, chi-square is defined as:

$$\chi^2_N = \frac{(n_1-\bar{n})^2 + (n_2-\bar{n})^2 + \ldots\ldots\ldots(n_N-\bar{n})^2}{\bar{n}} = \frac{1}{\bar{n}}\sum_{i=1}^{N}(n_i-\bar{n})^2 \tag{4.42}$$

Chi-square may also be defined in terms of the activities

$$\chi^2_N = \frac{(r_1-\bar{r})^2 + (r_2-\bar{r})^2 + \ldots\ldots\ldots(r_N-\bar{r})^2}{\bar{r}/t} = \frac{t}{\bar{r}}\sum_{i=1}^{N}(r_i-\bar{r})^2 \tag{4.43}$$

where $r = n/t$

The distribution of chi-square is known. Values of chi-square are given in table 4.6. It is seen that chi-square depends on the number of counting observations but not explicitly on the counting time. Referring to table 4.6, we can see, for example, for a series of 5 observations that there is a 50% chance of the value of chi-square being larger than 3.357 and an equal probability of its being smaller; that the probability of chi-square being as large as 7.779 is only 10% and that there is only a 1% chance of its being as large as 13.277. (Note: 5 observations corresponds to 4 degrees of freedom.)

Table 4.6—TABLE OF CHI-SQUARE

Degrees of Freedom* (N−1)	There is a probability of						
	0.99	0.95	0.90	0.50	0.10	0.05	0.01
	that the calculated value of Chi-Square will be equal to or greater than						
2	0.020	0.103	0.211	1.386	4.605	5.991	9.210
3	0.115	0.352	0.584	2.366	6.251	7.815	11.345
4	0.297	0.711	1.064	3.357	7.779	9.488	13.277
5	0.554	1.145	1.610	4.351	9.236	11.070	15.086
6	0.872	1.635	2.204	5.348	10.645	12.592	16.812
7	1.239	2.167	2.833	6.346	12.017	14.067	18.475
8	1.646	2.733	3.490	7.344	13.362	15.507	20.090
9	2.088	3.325	4.168	8.343	14.684	16.919	21.666
10	2.558	3.940	4.865	9.342	15.987	18.307	23.209
11	3.053	4.575	5.578	10.341	17.275	19.675	24.725
12	2.571	5.226	6.304	11.340	18.549	21.026	26.217
13	4.107	5.892	7.042	12.340	19.812	22.362	27.688
14	4.660	6.571	7.790	13.339	21.064	23.685	29.141
15	5.229	7.261	8.547	14.339	22.307	24.996	30.578
16	5.812	7.962	9.312	15.338	23.542	2 .296	32.000
17	6.408	8.672	10.085	16.338	24.769	27.587	33.409
18	7.015	9.390	10.865	17.338	25.989	28.869	34.805
19	7.633	10.117	11.651	18.338	27.204	30.144	36.191
20	8.260	10.851	12.443	19.337	28.412	31.410	37.566
21	8.897	11.591	13.240	20.337	29.615	32.671	38.932
22	9.542	12.338	14.041	21.337	30.813	33.924	40.289
23	10.196	13.091	14.848	22.337	32.007	35.172	41.638
24	10.856	13.848	15.659	23.337	33.196	36.415	42.980
25	11.534	14.611	16.473	24.337	34.382	37.382	44.314
26	12.198	15.379	17.292	25.336	35.563	38.885	45.642
27	12.879	16.151	18.114	26.336	36.741	40.113	46.963
28	13.565	16.928	18.939	27.336	37.916	41.337	48.278
29	14.256	17.708	19.768	28.336	39.087	42.557	49.588

* The number of degrees of freedom is usually one less than the number of observations N.

A probability of 0.1 or larger is considered reasonable. On the other hand, high values of probability, 0.9 or greater, indicate that the spread of data is not so great as predicted by the Poisson distribution. (A probability of 0.9 for 5 determinations means there is a 90% chance of chi-square being larger than 1.064 and only a 10% chance of its being smaller).

Example:

Refer to the sample data given in table 4.1.

$$\chi^2_{10} = \frac{1134}{98} = 11.5$$

From table 4.6 it is seen that the probability of obtaining a value of 11.5 for chi-square is greater than 0.1 and less than 0.9 since 11.5 lies between 4.168 and 14.684 (for 9 degrees of freedom). Thus, the distribution of data is reasonably in accord with Poisson and the instrument is operating properly.

PROCEDURE AND CALCULATIONS:

Using the data accumulated in experiment 4.1, calculate the value of chi-square. Assuming the Poisson distribution to hold, compare the value obtained for chi-square to the value given in table 4.6 and determine if the instrument is working properly.

EXPERIMENT 4.5 CHAUVENET'S CRITERION

OBJECTIVES:

To apply the Chauvenet criterion for the rejection of data not representative of a Poisson distribution.

THEORY:

Occasionally, when successive analyses are made on the same sample, the results of one or more measurements will be at considerable variance with the mean. If a large number of observations are made, a single incorrect result averaged with the others will introduce only a small error; but, if the total number of observations is small, the error introduced by an erroneous observation will create a large error.

Obviously, one must use some criterion in deciding if a result should be included or discarded because of a variance greater than that expected by the Poisson probability. Chauvenet's criterion states that an observation should be discarded if the probability of its occurrence is equal to or less than 1/(2N), where N is the number of observations.

For simplicity, these limits are given in table 4.7 in terms of the ratio of the deviation to the standard deviation, i.e., $(n-\bar{n})/s$, for various numbers of observations N.

Example:

Refer to the sample data in table 4.1.

Determine if the count of 118 counts per minute should be discarded from the data assuming that all counts recorded were 1 minute observations.

$$n - \bar{n} = 118 - 98 = 20 \text{ cpm}$$
$$s = \sqrt{\bar{n}} = \sqrt{98} = 9.9 \text{ cpm (see equation 4.27)}$$
$$(n - \bar{n})/s = 20/9.9 = 2.02 = \tau$$
$$N = 10$$

From table 4.7 the value of $(n-\bar{n})/s$ for 10 observations should not exceed 1.96. Thus, the result of 118 counts per minute should be discarded because the calculated value of τ is 2.02.

A direct approach based upon principles discussed earlier in this chapter may be considered. We have seen from equation 4.32 that $(n-\bar{n}) = \tau\sqrt{n}$. Where replicate measurements have been made, $\bar{n}$ is used for n and

$$\tau = \frac{n-\bar{n}}{\sqrt{\bar{n}}} = \frac{n-\bar{n}}{s} \tag{4.44}$$

Table 4.7*

N	$(n-\bar{n})/s = \tau$	N	$(n-\bar{n})/s = \tau$
2	1.15	30	2.40
3	1.38	35	2.45
4	1.54	40	2.50
5	1.65	50	2.58
6	1.73	75	2.71
7	1.80	100	2.81
8	1.86	200	3.02
9	1.91	250	3.09
10	1.96	300	3.14
12	2.04	400	3.23
15	2.13	500	3.29
20	2.24	1000	3.48
25	2.33		

*For the number of observations N indicated in column one, the value of $(n-\bar{n})/s$ must not exceed that given in the second column.

By reference to table 4.2 or to figure 4.5, the probability corresponding to the calculated value of τ may be found. If this probability is equal to or less than 1/2N, the observation should be rejected.

Table 4.8 — PROBABILITY OF OCCURRENCE OF DEVIATIONS

Ratio of Deviation to probable error (n − n̄)/P	Probable occurrence (%)	Odds against (to 1)
1.0	50.00	1.00
1.1	45.81	1.118
1.2	41.83	1.39
1.3	38.06	1.63
1.4	34.50	1.90
1.5	31.17	2.21
1.6	28.05	2.57
1.7	25.15	2.98
1.8	22.47	3.45
1.9	20.00	4.00
2.0	17.73	4.64
2.1	15.67	5.38
2.2	13.78	6.25
2.3	12.08	7.28
2.4	10.55	8.48
2.5	9.18	9.90
2.6	7.95	11.58
2.7	6.86	13.58
2.8	5.89	15.96
2.9	5.05	18.82
3.0	4.30	22.24
3.1	3.65	26.37
3.2	3.09	31.36
3.3	2.60	37.42
3.4	2.18	44.80
3.5	1.82	53.82
3.6	1.52	64.89
3.7	1.26	78.53
3.8	1.04	95.38
3.9	0.853	116.3
4.0	0.698	142.3
4.1	0.569	174.9
4.2	0.461	215.8
4.3	0.373	267.2
4.4	0.300	332.4
4.5	0.240	415.0
4.6	0.192	520.4
4.7	0.152	655.3
4.8	0.121	828.3
4.9	0.0950	1,052.
5.0	0.0745	1,341.
6.0	0.0052	19,300.
7.0	0.00023	4.27×10^5
8.0	0.0000068	1.47×10^7

Ratio of deviation to standard deviation (n − n̄)/s	Probable occurrence (%)	Odds against (to 1)
0.6745	50.00	1.00
0.7	48.39	1.07
0.8	42.37	1.36
0.9	36.81	1.72
1.0	31.73	2.15
1.1	27.13	2.67
1.2	23.01	3.35
1.3	19.36	4.17
1.4	16.15	5.19
1.5	13.36	6.48
1.6	10.96	8.12
1.7	8.91	10.22
1.8	7.19	12.92
1.9	5.74	16.41
2.0	4.55	20.98
2.1	3.57	26.99
2.2	2.78	34.96
2.3	2.14	45.62
2.4	1.64	60.00
2.5	1.24	79.52
2.6	0.932	106.3
2.7	0.693	143.2
2.8	0.511	194.7
2.9	0.373	267.0
3.0	0.270	369.4
3.1	0.194	515.7
3.2	0.137	726.7
3.3	0.0967	1,033.
3.4	0.0674	1,483.
3.5	0.0465	2,149.
3.6	0.0318	3,142.
3.7	0.0216	4,637.
3.8	0.0145	6,915.
3.9	0.00962	10,390.
4.0	0.00634	15,770.
5.0	5.7×10^{-5}	1.7×10^6
6.0	2.0×10^{-7}	5.0×10^8
7.0	2.6×10^{-10}	3.9×10^{11}

Example:

Using once again the sample data from table 4.1, we find:

$$\frac{118-98}{\sqrt{98}} = \frac{20}{9.9} = 2.02$$

From table 4.2 it is seen that $p = 0.045$ for $\tau = 2.02$. From Chauvenet's criterion this probability of occurrence should not be less than $1/2N$. For $N = 10$, the value of $1/2N$ is 0.05. But 0.045 is less than 0.05, so the result should be rejected.

If data are expressed as rates rather than as counts, we obtain an applicable relationship by dividing equation 4.40 by t:

$$\frac{n}{t} - \frac{\bar{n}}{t} = r - \bar{r} = \frac{\tau\sqrt{\bar{r}t}}{t} = \tau\sqrt{r/t} \qquad (4.45)$$

and

$$\tau = \frac{r - \bar{r}}{\sqrt{r/t}} \qquad (4.46)$$

PROCEDURE AND CALCULATIONS:

Examine the experimental data obtained in experiment 4.1. Determine which, if any, of the observations should be rejected on the basis of Chauvenet's criterion.

REFERENCES

1. Kuyper, A. C., "The Statistics of Radioactivity Measurement", J. Chem. Ed., 36, 128-132 (1959).
2. Bennett, C. A. and Franklin, N. L., "Statistical Analysis", John Wiley and Sons (1954).
3. Jarrett, A. A., "Statistical Methods Used in the Measurement of Radioactivity", AECU-262 (Mon P-126) Oak Ridge National Laboratory.
4. Friedlander, G., and J. W. Kennedy, "Nuclear and Radiochemistry", John Wiley and Sons (1956).
5. Snedecor, G. W., "Statistical Methods Applied to Experiments in Agriculture and Biology", 5th ed., Iowa State College Press (1956).
6. Moroney, M. J., "Facts from Figures", Penguin Books.
7. Bauer, E. L., "A Statistical Manual for Chemists", Academic Press (1960).
8. Rainwater, L. J. and C. S. Wu, "Applications of Probability Theory to Nuclear Particle Detection", Nucleonics 1, 60-64 (Oct. 1947).
9. Loevinger, R., and M. Berman, "Efficiency Criteria in Radioactivity Counting", Nucleonics 9, 26-29 (July 1951).
10. Elmore, W. C., "Statistics of Counting", Nucleonics 6, 26-34 (Jan. 1950).
11. Thomas, A., "How to Compare Counters", Nucleonics 6, 50-53 (Feb. 1950).
12. Browning, W. E., Jr., "Charts for Determining Optimum Distribution of Counting Times", Nucleonics 9, No. 3, 63-67 (1951).
13. Liebhafsky, H. A., H. G. Pfeiffer and E. W. Balis, "Statistical Operating Rule for Analytical Chemists", Anal. Chem. 23, 1531-1534 (1951).
14. Dean, R. B., and W. J. Dixon, "Simplified Statistics for Small Numbers of Observations", Anal. Chem., 23, 636-638 (1951).
15. Scheffe, H., "Theoretical Backgrounds of the Statistical Methods—Some Basic Concepts of Probability and Statistics", Ind. and Eng. Chem. 43, 1292-1294 (1951).

16. Mosteller, F., "Theoretical Backgrounds of the Statistical Methods—Underlying Probability Model Used in Making a Statistical Inference", Ind. and Eng. Chem., 43, 1295-1297 (1951).
17. Bennett, C. A., "Application of Tests for Randomness", Ind. and Eng. Chem. 43, 2063-2067 (1951).
18. Brownless, K. A., "Correlation Methods Applied to Production Process Data", Ind. and Eng. Chem. 43, 2068-2071 (1951).
19. Hader, R. J., and W. J. Youden, "Experimental Statistics", Anal. Chem. 24, 120-124 (1952).
20. Randolph, L. K., "Use of the Range in the Statistical Evaluation of a Biological Assay", J.A.Ph.A., Sci. Ed., 41, 438-440 (1952).
21. Davidson, W. C., "Nomogram for Counting Time", Nucleonics 11, No. 9, 62 (1953).
22. Riedel, O., "Statistical Purity in Nuclear Counting", Nucleonics 12, No. 6, 64-67 (June 1954).
23. Lark, P. D., "Application of Statistical Analysis to Analytical Data", Anal. Chem. 26, 1712-1715 (1954).
24. Linnig, F. J., J. Mandel and J. M. Peterson, "A Plan for Studying the Accuracy and Precision of an Analytical Procedure", Anal. Chem. 26, 1102-1110 (1954).
25. Horton, W. S., "Location of Symmetric Peaks by a Simple Least Squares Method", Anal. Chem. 27, 1190-1191 (1955).
26. Zilversmit, D. B., and A. L. Orvis, "Calculating Average Counting Rates", Nucleonics 14, No. 7, 75 (July 1956).
27. Anon., "How to Save Time When Counting Weak Samples", Nucleonics 15, No. 11, 118-120 (Nov. 1957).
28. Jaffey, A. H., "Statistical Tests for Counting", Nucleonics 18, No. 11, 180-184 (Nov. 1960).
29. Anon., "Is My Nuclear Instrumentation Putting Out Valid Data?", Picker Scintillator IV, No. 4 (May 20, 1960).
30. Herberg, R. J., "Counting Statistics for Liquid Scintillation Counting", Anal. Chem. 33, 1308 (1961).
31. Reynolds, S. A., "Choosing Optimum Counting Methods and Radiotracers", Nucleonics 22, 104-105 (Aug. 1964).
32. Daugherty, K. E., and R. J. Robinson, "A Statistical Analysis of Results", J. Chem. Ed. 41, 51 (1964).
33. Kac, M., "Probability", Sci. Amer., p. 92-108 (Sept. 1964).
34. Brucer, M., "Statistical Lying", Medical Science, p. 66-73 (Nov. 1964).
35. Little, J. M., "An Introduction to the Experimental Method", Burgess Pub. Co. (1961).
36. Batson, H. C., "An Introduction to Statistics in the Medical Sciences", Burgess Pub. Co., Minneapolis (1961).

CHAPTER 5
Determinate Errors in Radioactivity Measurement

LECTURE OUTLINE AND STUDY GUIDE

I. CORRECTIONS FOR THE INSTRUMENT AND ITS ENVIRONMENT
   A. Background (*Experiment* 5.1)
   B. Resolving time (*Experiment* 5.2)
   C. Variation of Instrument Efficiency (*Experiment* 5.3)

II. GEOMETRY CORRECTIONS
   A. General Equation for Simple Beta Emitter
   B. Physical Geometry Factor (G) (*Experiment* 5.4)
   C. Forescattering in Air (f_A)
   D. Backscattering (f_B) (*Experiment* 5.5)
   E. Scattering by Walls and Structure (f_H) (*Experiment* 5.5)
   F. Self Absorption and Self Scattering in Source (f_S) (*Experiment* 5.6)
   G. Absorption by Air, Window and Sample Cover (f_W)

III. ERRORS ORIGINATING FROM SAMPLE GEOMETRY
   A. Effect of Point Source Displacement on Counting Rate
      1. Vertical displacement (*Experiment* 5.4)
      2. Horizontal displacement
   B. Extended Thin Sources
   C. Self Absorption and Self Scattering in Source (*Experiment* 5.6)
      1. Constant volume
      2. Constant total activity
      3. Constant specific activity

IV. ERRORS ARISING FROM SOLUTION PREPARATION (*Experiment* 5.7)
   A. Meaning of "Carrier Free" Solutions
   B. Adsorption on Walls of Container
   C. Use of Carriers

CAUSE OF VARIATION:

In chapter 4 the cause of variation in measured values of a characteristic was shown to be (1) errors in measurement or (2) variation representing a true difference between two populations. Errors in measurement are of two types: (a) *indeterminate errors* which were the subject of chapter 4 and (b) *determinate errors* which constitute the subject matter of the present chapter.

Determinate errors are those errors which are controllable by the experimenter. He should therefore be aware of the types of determinate errors which are likely to occur so he can take the necessary precautions to prevent them.

RADIOACTIVITY MEASUREMENT:

The measurement of radiation may be performed with one of two basic objectives in mind, namely, (1) to determine the absolute disintegration rate of a sample or (2) to make only a relative comparison of the activity of one sample with respect to the activity of another. When relative comparisons are made, a knowledge of the absolute disintegration rate need not be known.

Consider first the determination of the absolute disintegration rate A ($=dN/dt$). Let us suppose the radioactivity of a sample has been measured with a Geiger tube and scaler. The sample was placed, say, on the second shelf of the sample holder, the scaler was properly adjusted and used and the count indicated by the mechanical register and interpolation lights of the scaler has been recorded. What, now, is the relationship of the observed activity r, which has been recorded (in counts per second), to the true activity A (disintegrations per second), i.e., the absolute disintegration rate, dN/dt?

First, the observed activity or counting rate must be corrected for *coincidence loss* (see experiments 3.2 and 5.2) to correct for the resolving time or time during which the tube is not sensitive to radiation. This correction results in a true counting rate slightly higher than the observed counting rate.

Secondly, the *background* must be subtracted (see experiment 5.1). Not all of the recorded counts are caused by the sample. Some are produced by cosmic radiation, natural radioactivity in the building, etc. All of these constitute the background and must be subtracted from the gross count rate. Application of the coincidence correction and the background correction to the observed count r gives the corrected net count rate R.

The corrected net count rate of the sample R (expressed, for example, in counts per second) may now be related to the absolute beta activity A of the sample (expressed in disintegrations per seconds) by applying a series of geometry corrections. A general equation for a simple beta emitter, assuming 100% efficiency for the detector, and assuming that one beta particle is emitted by each disintegrating atom, is

$$A = \frac{R}{G\, f_A\, f_B\, f_H\, f_S\, f_W} \tag{5.1}$$

G is the *physical geometry factor.* It is the ratio of the number of beta particles emitted in a direction included by the solid angle formed by the sample and the window of the tube to the total number of beta particles emitted by the sample. G includes, by its very nature, a correction for the choice of shelf. It is discussed in detail in experiment 5.4.

Forescattering in air (f_A) is caused by elastic collisions of the beta particles with molecules of oxygen and nitrogen as they travel toward the Geiger tube. Beta particles directed initially toward the window of the tube are sometimes deflected by these collisions so that the deflected betas do not strike the tube and are not counted. On the other hand, some betas, directed initially so that their path would miss the tube, are deflected into the tube window by similar collisions with air molecules. In practice, the forescattering factor, f_A, is difficult to evaluate. It is measured more easily in combination with f_W, the factor for absorption by the air, the window of the Geiger tube and the sample cover if used (see experiments 7.1 and 7.2). f_W is a measure of the loss of beta particles through inelastic collisions. The inelastic collision of a beta particle with an atom causes an electron to be ejected from the atom, a positively charged ion thus being formed. The quantitative evaluations of f_A and f_W are deferred to chapter 7.

Not only are beta particles forescattered by the air, but they are also *backscattered* by the sample support or planchet and *sidescattered* by the housing supporting the Geiger tube and sample holder. Backscattered beta particles actually undergo a change in direction of approximately 180 degrees. On passing into the planchet, they undergo collisions with the atoms composing the planchet. If the planchet is composed of a high atomic number element, the effect is very large, causing as much as a 60 or 70% increase in the observed count. A detailed study of backscattering is made in experiment 5.5 wherein the backscatter factor f_B is evaluated.

Sidescattering effects may be just as pronounced as backscattering effects. With properly designed sample holders, the sidescattering factor f_H can be made very nearly equal to unity. Thus, for example, if a card mount such as that illustrated in figure 3.7 is used to support a sample, it can be assumed that the larger the card the farther the housing and card support will be removed from the sample and the closer f_H will approach unity. Where lead shielding is used to reduce background, it is important not to have the shielding too close to the sample and window of the Geiger tube if scattering phenomena are to be minimized.

Finally, one must consider absorption of the radiation by the sample itself. This is termed *self absorption* or *self scattering.* The correction factor applied is f_S. The problems associated with self absorption are outlined in experiment 5.6.

When it is desired to make only a relative comparison of the activities of two or more samples, especially if of the same radioactive species, the procedure is considerably simplified. Fortunately the majority of radioactive measurements fall into this category. Where relative results only are desired, it is necessary to observe one basic precaution. That is, one must reproduce faithfully the exact geometry and counting conditions for all samples, both with respect to the equipment used and to the sample itself.

EXPERIMENT 5.1 BACKGROUND

OBJECTIVES:

To observe the factors contributing to background.

To measure the background count for various G-M tubes.

To observe the effect of massive shielding on the background count.

THEORY:

Even when no source of radiation exists in the vicinity of a Geiger-Müller counter, it will be observed that a small amount of radiation is being detected. This is known as background radiation, and the activity observed on the scaler is known as the background count. Organic-quenched end-window tubes show a normal background of 50 to 70 counts per minute, while halogen-quenched tubes show a background of 30 to 40 counts per minute.

Background may result from natural sources or from artificially produced radioactive sources. These may be classified as follows:

A. *Natural sources*
   1. Cosmic rays
   2. Natural radioactivity of the surroundings, e.g.,
      Naturally radioactive substances in lead, carbon-14 in wood and carbon containing compounds, carbon-14 and potassium-40 in the human body.
   3. Chemicals on the stockroom shelf—radium, uranium, thorium and potassium salts.

B. *Artificial radioactivity*
   1. Wrist watches—although the substances in the luminous paint may be naturally radioactive.
   2. Radioactive samples stored in a nearby room.
   3. Contamination of the counting equipment.
   4. Fallout—about 1% of the total background.
   5. Memory of G-M tubes following exposure to high level gamma radiation.

Background radiation will produce an error in measurements of radioactivity unless the background count is determined and subtracted from the total activity. The difference between the total activity and the background is the net activity of the sample.

Background may change during the course of an experiment and should always be determined both at the beginning and at the end of the working period. If the time of an experiment is prolonged, it is judicious to make an observation of background periodically, or even hourly. If many samples are being measured, the background should be checked repeatedly as a precaution against contamination of the equipment from spillage.

Several methods are available for the reduction of background. These are:

A. *Massive shielding*—a thick wall of lead or of iron around the G-M tube can be used to reduce the background to as little as 15 per cent of that obtained with an unshielded tube. Two inches of lead, for example, will reduce the background to about one fourth the unshielded value.

B. *Anticoincidence shield*—this consists of surrounding the G-M tubes with an umbrella or shield of other G-M tubes so connected electronically that if both the working tube and a shielding tube detect a count simultaneously, that count is rejected. Such rejection normally occurs when the source of the radiation lies external to the sample.

C. *Pulse height discrimination*—this method which is not very useful with Geiger tubes, consists of sorting the pulses produced by a detector according to size and accepting only those pulses of a size which would be produced by the sample.

A relationship* for evaluating counting procedures and equipment to determine the optimum conditions of counting efficiency and background count is given by the ratio

$$\frac{R_s^2}{R_b} \tag{5.2}$$

where R_b = background count rate
R_s = $R_t - R_b$ = net sample count rate
R_t = total count rate (sample plus background)

*Annual Reviews of Nuclear Science, *6*, 303 (1956).

Using the same radioactive source for all comparative measurements, the larger the ratio of R_s^2/R_b the better the counting conditions. This relationship is sometimes called the "figure of merit." Refer to experiment 4.3 for a statistical analysis of background and sample activities.

APPARATUS:

Halogen and organic-quenched Geiger-Müeller counters; side-window G-M counters, bismuth-coated cathode or other special types of G-M tubes; scaler with high voltage power supply; a large lead shield.

PROCEDURE:

Remove all sources of activity from the working area. Using a Geiger-Müeller tube and scaler adjust the high voltage to the proper operating voltage. Determine the background count on the unshielded tube. At least a three minute count should be made for each background determination and preferably a 5- or 10-minute count.

Turn off the high voltage and place the G-M tube inside a massive shield. Readjust the voltage to the proper operating point and redetermine the background.

Repeat this procedure using other G-M tubes of different types.

CAUTION: Note that different tubes have different operating potentials! If the proper operating potential is not marked on the tube, determine the approximate operating potential by placing a prepared source in the sample holder, slowly increasing the high voltage until the tube begins to count. Be sure the "count-stop" switch is in the count position. Then increase the potential about 75 or 100 volts above the threshold potential. Remove the source and proceed with the experiment. Do this for each new tube used.

DATA:

Tube Type	Manufacturer or Tube Designation	Operating Potential	Unshielded Background cpm	Background with Shield cpm
End-window (organic)				
End-window (halogen)				
Side-window				

EXPERIMENT 5.2 RESOLVING TIME

OBJECTIVES:

To illustrate methods by which the resolving time of a Geiger-Müller counter may be determined.

To determine the resolving time of a Geiger-Müller counter.

THEORY:

When radiation produces ionization in a G-M tube, the electrons travel to the anode very quickly, but the positive ions are much slower in action and require between 100 and 400 microseconds to reach the cathode. During the time it takes the positive ions to get to the cathode, the tube is insensitive. Several terms related to this event have been defined (5). *Deadtime* is the time interval after a pulse has occurred during which the counter is insensitive to further ionizing events. *Resolving time* is the minimum time interval by which two pulses must be separated to be detected as separate pulses.

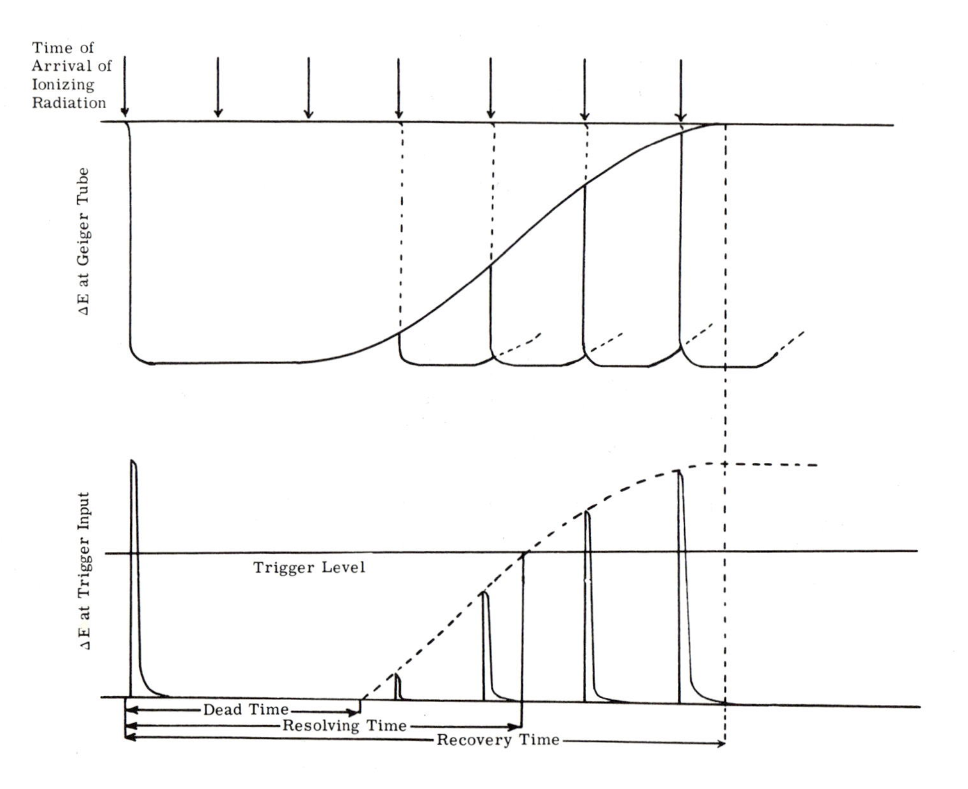

Figure 5.1—PULSE PRODUCTION BY IONIZING RADIATION.

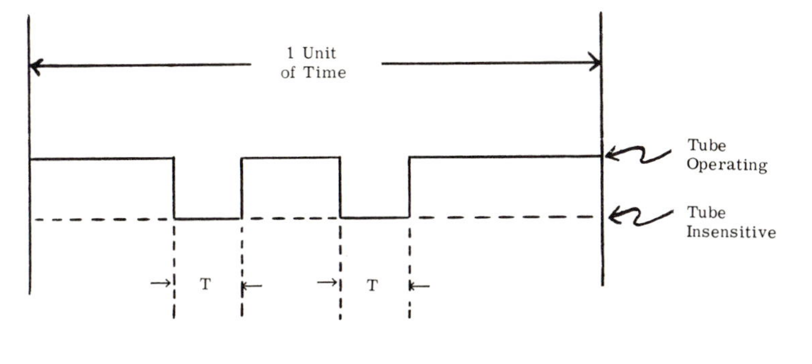

Figure 5.2—DIAGRAMATIC REPRESENTATION
OF RESOLVING TIME

Recovery time is the time interval that must elapse after a pulse has occurred before a full-size pulse can again occur. The phenomenon of a second or even third ionizing ray striking the tube during the resolving time is known as coincidence and the correction applied is known as the *coincidence correction.*

As a result of coincidence, the observed count is always less than the true count. An approximate correction for resolving time (based on an assumed resolving time of about 300 μs) can be made by adding 0.5% per 1000 cpm to the observed counting rate. Thus, if the observed count is 3000 cpm the correction to be added would be $3 \times 0.5\%$ or 1.5% of 3000. In this case the approximate true count would be 3045 cpm.

In a more accurate analysis of resolving time, one must consider the ratio of the total time to the time during which the tube is operative. Referring to figure 5.2, if during a period of one unit of time several ionizing events occur, the tube will not be sensitive for time = 1 (i.e., unit time), but only for a time equal to $1 - RT$, and the ratio of true count rate to observed count rate (R/r) is given by

$$\frac{R}{r} = \frac{1}{1-RT} \tag{5.3}$$

where R = true count rate (=N/t)
r = observed count rate (=n/t)
T = resolving time

When r is plotted as a function of R (figure 5.3), it is seen that r and R are nearly equal at very low count rates; but as R is increased further, the observed count rate begins to lag because of the increase in the coincidence loss. Ultimately, when very active samples are measured, the observed count rate is seen to decrease, finally decreasing to zero. Although one may not often encounter such active samples, it is very important to be cognizant of this behavior since one might be led to believe a sample is safe to handle whereas in reality it is dangerously radioactive.

Because the mathematics of equation 5.3 is especially complex, it is generally assumed that $(1-rT) \approx (1-RT)$ at the lower count rates ordinarily encountered in the laboratory. It can be seen from figure 5.3 that this is a valid assumption if not extended to high counting rates. Thus,

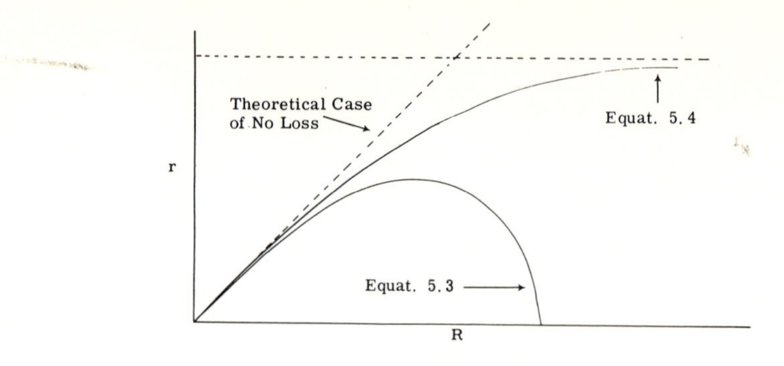

Figure 5.3—EFFECT OF COUNT RATE ON COINCIDENCE LOSS

$$\frac{R}{r} = \frac{1}{1 - rT} \tag{5.4}$$

and if resolving time is known, the true count can be calculated by

$$R = \frac{r}{1 - rT} \tag{5.5}$$

Resolving time can be measured by the *method of paired sources*. The activities of two sources are measured individually (r_1 and r_2) and then together (r_{12}). A convenient form for the sources is shown in figure 5.4 and consists of a mounting card (see figure 2.3) without a hole, cut lengthwise into two sections. A small quantity of radioactive material is placed on each card. The quantities should be approximately but need not be exactly equal and should each give a count rate of about 10,000 cpm.

One might expect that $r_1 + r_2 = r_{12}$ but because of resolving time, this is not so. It is true, however, that

$$R_1 + R_2 = R_{12} + R_b \tag{5.6}$$

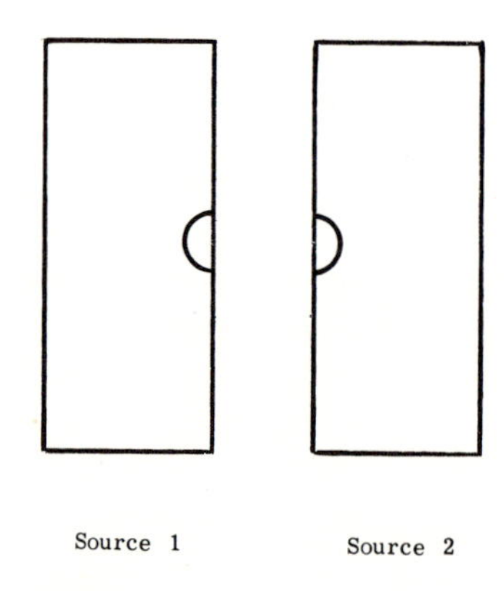

Figure 5.4—PAIRED SOURCES

where R_b is the background count and must be added to R_{12} since two background counts are included in the sum of R_1 and R_2, whereas R_{12} includes only one background. It follows then, from equations (3.5) and (3.6) that

$$\frac{r_1}{1 - r_1T} + \frac{r_2}{1 - r_2T} = \frac{r_{12}}{1 - r_{12}T} + \frac{r_b}{1 - r_bT} \tag{5.7}$$

Although all quantities except T can be measured, the form of this relationship is awkward for the calculation of T, and several approximate relationships have been proposed. For example, if background is considered negligible the fourth term may be omitted. Multiplying by the least common denominator, expanding and neglecting terms in T^2, the following relationship is easily derived:

$$T = \frac{r_1 + r_2 - r_{12}}{2r_1r_2} = \frac{\Delta}{2r_1r_2} \tag{5.8}$$

Preuss (3) describes paired sources suitable for conducting the above procedure and gives a modified form of equation 5.8.

$$T = \frac{\Delta}{2\,r_1r_2} \cdot \frac{1}{8\,r_{12}\,2r_b} \cdot \frac{\Delta\,r_{12}}{r_1r_2} \tag{5.9}$$

However, this equation appears to be in error and is not recommended.

Another relationship which has been proposed for the calculation of resolving time follows:

$$T = \frac{2(r_1 + r_2 - r_{12})}{(r_1 + r_2)\,r_{12}} \tag{5.10}$$

Still another relationship is found in the U.S.P. XV monograph on coincidence correction:

$$T = T_1\,[1 + \frac{T_1}{2}(r_{12} - 3r_b)] \tag{5.11}$$

in which T is the resolving time and

$$T_1 = \frac{r_1 + r_2 - r_{12} - r_b}{2(r_1 - r_b)\,(r_2 - r_b)}$$

r_1, r_2, r_b = activities of samples 1 and 2 and the background respectively, and

r_{12} = activity of the combined samples.

After the value of resolving time has been determined, observed sample activities can be corrected for coincidence loss by means of equation 5.5.

APPARATUS:

Geiger tube in sample holder, scaler, paired sources and blank cards.

PROCEDURE:

Prepare two radioactive sources on identical mounts each having approximately the same activity in the range of 10 thousand counts per minute. Carefully determine the activity of the sources individually and combined, being sure to retain identical geometry and backscattering effects.

Measure the values of r_1, r_2, r_{12} and r_b as follows. Note the sequence in which the measurements are made. This sequence is used to permit the smallest number of changes possible, thereby allowing the possibility of a more reproducible geometry.

1. Place source one (S-1) in position with blank two (B-2) next to it as illustrated in figure 5.5 (a). Count for five minutes. Calculate the activity in counts per minute. This is r_1.

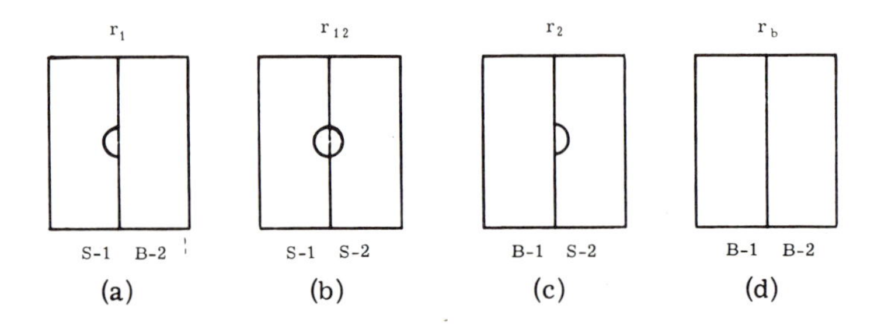

Figure 5.5—MEASUREMENT OF RESOLVING TIME BY THE METHOD OF PAIRED SOURCES.

2. Carefully remove blank two (B-2) and replace it with source two (S-2) as shown in figure 3.4 (b). Count for five minutes. Calculate the activity in counts per minute. This is r_{12}.

3. Carefully remove source one (S-1) and replace it with blank one (B-1) as illustrated in figure 3.4(c). Count for five minutes. Calculate the activity in counts per minute. This is r_2.

4. Carefully remove source two (S-2) and replace it with blank two (B-2) as shown in figure 3.4 (d). Count for five minutes. Calculate the activity in counts per minute. This is r_b.

NOTE: In view of the term involving a difference, represented by Δ in equations 5.8 and 5.9, a high degree of accuracy is essential. If time permits, it is recommended that the procedure be repeated and that average values be used in the calculations. As pointed out by Preuss (5), reproducible geometry is essential. Using the average values of repetitive observations will aid in reducing geometry errors.

DATA:

	Observed Activities (cpm)			
Observation	r_1	r_2	r_{12}	r_b
1				
2				
3				
Average				

CALCULATIONS:

1. Calculate the resolving time by means of each of the equations given and tabulate below:

	Resolving Time		
Equation	Minutes	Seconds	Microseconds
5.8			
5.9			
5.10			
5.11			
Best value			

2. Correct each of the observed activities (r_1, r_2, r_{12} and r_b) for coincidence to obtain R_1, R_2, R_{12} and R_b respectively. Add r_1 and r_2 and compare with r_{12}. Similarly add R_1 and R_2 and compare the sum with the value of R_{12} and the sum of R_{12} and R_b.

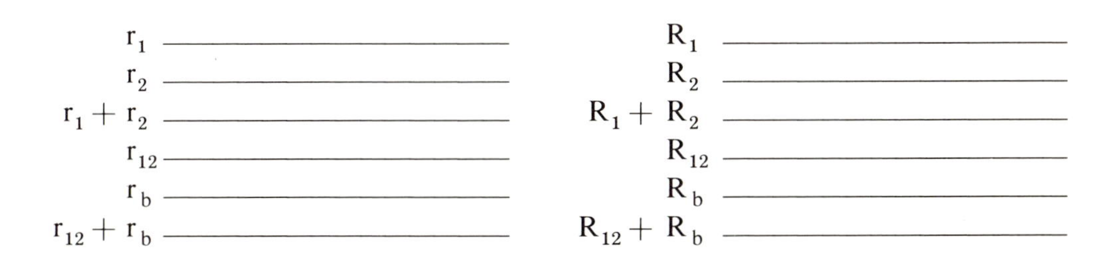

If the calculations are correct,

$$r_1 + r_2 > r_{12} + r_b$$

but

$$R_1 + R_2 = R_{12} + R_b$$

EXPERIMENT 5.3 INSTRUMENT EFFICIENCY

OBJECTIVES:

To observe and measure changes in instrument efficiency.
To illustrate a method for correcting data for changes in instrument efficiency.
To study the causes for changes in instrument efficiency.

THEORY:

The *activity* of a sample is a measure of the number of radioactive atoms decaying per unit of time. Exact *absolute activities* of samples are required only in certain types of work. For most purposes it is necessary to know only *relative activities* from which can be calculated the percentage of the radioactive substance. To illustrate this point, the following examples are given:

Case 1.—Let it be supposed that an animal is given an oral dose of radioactive substance. Suppose the activity of the total dose given was 20,000 cpm. If a sample of blood is then removed from the animal and found to give an activity of 200 cpm per ml, one can calculate:

$$\text{\% of dose/ml of blood} = \frac{200}{20{,}000} \times 100 = 1\%$$

Case 2.—Let it now be supposed that a second person checks these results using another instrument. If this instrument is only 50% as efficient as the one used in case 1, the results obtained will be 10,000 cpm on the total dose and 100 cpm on one ml of blood. Thus

$$\text{\% of dose/ml of blood} = \frac{100}{10{,}000} \times 100 = 1\%$$

The result for case 2 is identical with that for case 1. Thus, it is seen that only *relative activities* were required for making the calculation. Instrument efficiency introduced no error so long as the efficiency was constant for a particular set of measurements.

Not only will the efficiency of one instrument differ from that of another, but the efficiency of the same instrument may and does vary from day to day or even from hour to hour. A change in the efficiency of an instrument during the course of a series of measurements can introduce serious errors into the calculations.

Case 3.—Suppose, in the illustrations previously cited, that the dose is given on one day and the blood sample collected the next day. The operator measures the activity of the total dose as 20,000 cpm but he does not measure the blood activity until the next day. He assumes the instrument is functioning exactly as it has been the previous day, but it is actually less efficient. He thus measures the blood activity to be 180 cpm per ml and calculates:

$$\text{\% of dose/ml of blood} = \frac{180}{20{,}000} \times 100 = 0.9\%$$

In this case an error has been introduced through a change in instrument efficiency.

Changes in the efficiency of a counter may be brought about in a variety of ways. Changes in room temperature and atmospheric pressure will alter the efficiency of the

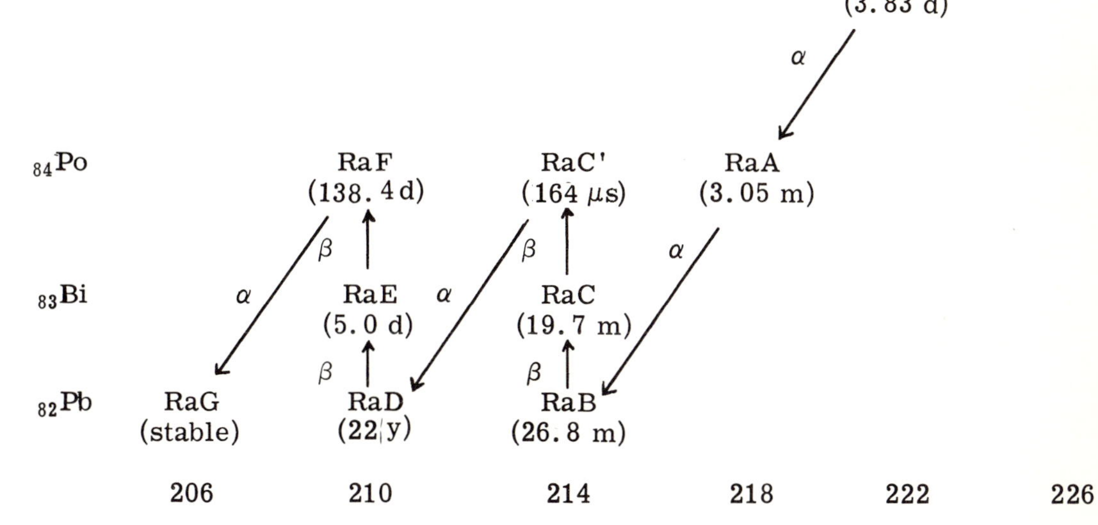

FIGURE 5.6—DECAY SCHEME FOR RADIUM-226.

Geiger counter. Changes in line voltage will affect the electronics as well as the Gieger tube efficiency. Errors in resetting the high voltage can also change the efficiency. Aging of the G-M tube results in an alteration of the tube characteristics—shortened plateau, steeper plateau and lower threshold. All of these changes are difficult to control, so that in practice it is easier and more accurate to determine the *relative efficiency* of the instrument and to normalize the efficiency. Normalization of data should be limited to changes of less than 10%. If the efficiency changes by more than 10%, one should look for trouble of a more serious nature.

Prepared mounts of radioactive substances having long half-lives are used as standards for the measurement of relative efficiencies. One of the most commonly used standards is Ra-DEF. Card and planchet mounts of Ra-DEF when prepared in experiment 3.1 can be used in this and in furture experiments for this purpose.

Radium decomposes to produce radon, a gas, which is collected in small glass tubes known as radon "seeds." These are used in the treatment of disease. But the radon quickly decays to radium-A, then to radium-B, radium-C, radium-C′ and finally to radium-D. In time, essestially all elements preceding Ra-D in the series decay, and only Ra-D remains in equilibrium with Ra-E and Ra-F. This Ra-DEF equilibrium mixture provides an excellent source for a standard. It provides an energetic beta particle as well as an alpha particle and decreases in activity about 3% per year. If only the energetic Ra-E beta is desired, the alpha and weak beta particles can be eliminated with 7 mg/cm^2 of absorber.

Case 4.—Again consider the situation presented in case 3 where the activity of the total dose was measured on one day and the blood activity on the following day, but include this time a Ra-DEF standard:

1st day

Activity of total dose	20,000 cpm
Activity of Ra-DEF standard	6,000 cpm

2nd day

Activity of blood	180 cpm/ml
Activity of Ra-DEF standard	5,400 cpm

Although the activity of the Ra-DEF standard will decrease about 3% per year it is assumed that the true activity of the standard has not changed from the first day to the second. It is seen, however, that the activity observed for the standard is not the same on each of the two days. This is assumed to be caused by a change in instrument efficiency.

Normalize all activities to the conditions prevailing on the first day. The instrument efficiency on the second day is calculated and found to be 0.9 that of the first day.

$$5{,}400/6{,}000 = 0.90$$

Hence, the true activity of the blood on the second day is found by dividing the observed activity by the efficiency.

$$180/0.90 = 200 \text{ cpm/ml}$$

and

$$\% \text{ of dose/ml of blood} = \frac{200}{20{,}000} \times 100 = 1\%$$

APPARATUS:

Geiger tube; scaler; Ra-DEF or other long-lived standard source such as Strontium-90 or U_3O_8.

PROCEDURE:

1. Adjust the high voltage to the proper operating potential and measure the background count. Place the standard radioactive source in the same holder, recording the shelf and other information required to reproduce the exact geometrical arrangement at a later date. Measure the activity of the standard. At least a 5 minute count should be made of both the background and the standard. Calculate the activities in counts per minute and subtract the background count from the gross standard count to obtain the net activity of the sample.

Turn the high voltage adjust to the extreme counterclockwise position and turn off the high voltage switch.

2. Repeat procedure 1 adjusting the high voltage to as nearly the same position as possible. Measure the background and the activity of the standard, reproducing the geometry as accurately as possible. Compare the results with those obtained in procedure 1.

3. Repeat procedure 1 on each of several subsequent days.

DATA:

Geiger tube used ______________________

Scaler ______________________

High voltage setting ______________________

Sample holder and shelf ______________________

Date	Hour	Background	Standard	Net Count	Efficiency
					1.0 (ref.)

CALCULATIONS:

At the time of the first observation, the efficiency of the instrument is arbitrarily given a value of unity. Determine the efficiency of the instrument at the time each of the subsequent observations were made.

Suggestions for further work:

Line voltage dependence of the counter can be determined by connecting the scaler to the 110 volt 60 cycle main through an autotransformer. Change the voltage applied to the scaler from 100 volts to 120 volts in 5 volt increments. For each setting of the autotransformer, readjust the high voltage to the G-M tube to the original value and measure the activity of a suitable source. Plot voltage vs. the activity.

Most scalers contain electronic regulation to compensate for line voltage changes. If good compensation has been achieved, the activity as measured above will not change appreciably.

CAUTION: At low line voltages, the electronic regulation of the high voltage supply may lose control thus allowing the high voltage to increase to an excessive value, causing damage to the Geiger tube.

Uranium Oxide, U_3O_8, is also a very useful reference standard and may be used in place of the Ra-DEF standard suggested in this experiment. Uranium oxide can be purchased in pure form from chemical suppliers or can be prepared from uranyl nitrate by precipitation with ammonium hydroxide and subsequent ignition. Uranium oxide is also useful as an absolute standard. One mg of pure U_3O_8 undergoes 724 disintegrations per minute.

To prepare a mount of uranium oxide, place the U_3O_8 on the center of a mounting card. Weigh the oxide if an absolute standard is desired, otherwise estimate about 50 mg. Place a drop of collodion, thinned with an equal volume of ether, on the uranium oxide and work into a paste with a fine wire or applicator stick. If an absolute standard is being prepared, there must be no loss of material on the stirrer, and the mount must be very thin so self absorption will be negligible. Allow to dry. If the mount is to be used as a beta reference, cover the sample with a thin aluminum foil to remove the alpha particles and the weak betas.

EXPERIMENT 5.4 THE GEOMETRY FACTOR
(Shelf Ratio and the Inverse Square Law)

OBJECTIVES:

To study the effect of vertical displacement of a point source.
To measure the shelf ratios of a sample holder.

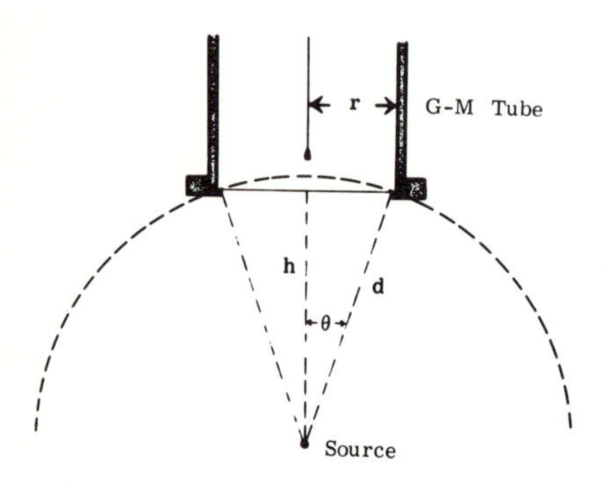

Figure 5.7—PHYSICAL GEOMETRY FACTOR, G

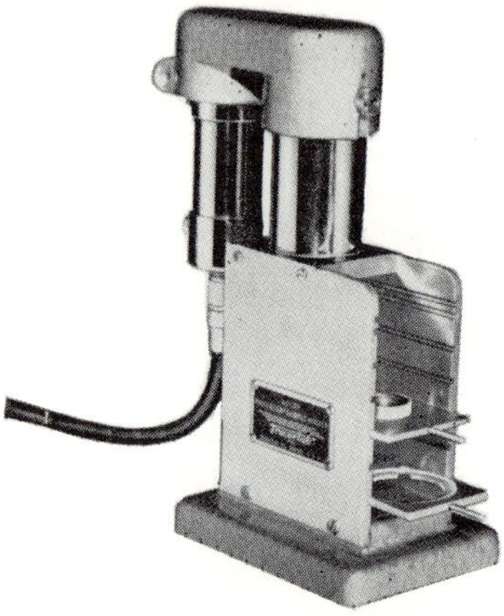

Figure 5.8—TUBE HOLDER AND SAMPLE SUPPORT (*Tracerlab, Inc.*)

THEORY:

Radiation is emitted from a radioactive source equally in all directions. The fraction G of the radiation which enters the window of the Geiger tube is a function of the radius r of the tube and the distance h from the point source of radiation to the window of the tube. G, called the *geometry factor,* is equal to the ratio of the area S′ on the sphere enclosed by the window to the area of the whole sphere.

$$G = \frac{S'}{4\pi d^2} \tag{5.12}$$

Since the area enclosed by the window is given by $S' = 2\pi d^2(1 - \cos\theta)$

$$G = \frac{2\pi d^2(1 - \cos\theta)}{4\pi d^2} = \frac{1 - \cos\theta}{2} \tag{5.13}$$

Expressing the cosine in terms of h and r

$$G = \frac{1}{2}\left(1 - \frac{h}{\sqrt{h^2 + r^2}}\right) \quad (5.14)$$

The solid angle ω subtended by the window of the tube is

$$\omega = \frac{S'}{d^2} \quad (5.15)$$

The area of the whole sphere $S = 4\pi d^2$ and the solid angle subtended by a whole sphere is $4\pi d^2/d^2 = 4\pi$ steradians. If a counter is designed to detect radiation given off in all directions from a radioactive source, it is known as a 4π counter. If the counter detects radiation emitted in one hemisphere only, it is known as a 2π counter. The relationship between G and ω is

$$G = \omega/4\pi \quad (5.16)$$

If the source is not placed too close to the window of the tube, it can be assumed that the area of the window is approximately equal to the area S′ on the sphere enclosed by the window and that h is approximately equal to d. Thus,

$$G \simeq \frac{\pi r^2}{4\pi d^2} = \frac{r^2}{4d^2} \simeq \frac{r^2}{4h^2} \quad (5.17)$$

The presence of d^2 (or h^2) in the denominator of equation 5.17 indicates that the *inverse square law* applies and that the observed activity should be proportional to the reciprocal of the square of the distance of the source from the tube assuming no absorption of radiation occurs. Although the inverse square law may be considered universally applicable to all radiation, pronounced absorption effects are encountered with both alpha and beta radiation.

It is sometimes convenient to take advantage of this change in activity with distance by moving less active samples toward the counter and samples which are too active away from the tube in order to obtain a suitable counting rate. It is for this purpose that sample holders are made with adjustable shelves.

It becomes necessary, however, when comparing the activities of two samples not measured on the same shelf, to know the *shelf ratio* so that the activities can be normalized. The shelf ratio is the ratio of the sample activity measured on a given shelf to the activity of the same sample measured on the reference shelf. Any shelf may be used as the reference although it is best to use the shelf most often used, possibly the second from the top of the sample support.

APPARATUS:

G-M tube and sample holder; scaler; beta sources (e.g., ^{32}P, ^{90}Sr, ^{14}C, Ra-DEF); gamma source (e.g., ^{226}Ra, ^{60}Co); side-window G-M tube with beta shield; meter stick or calibrated mounting board.

PROCEDURE:

PART A—DETERMINATION OF SHELF RATIOS

1. Using a beta source, determine the observed count rate r with the sample placed first on one shelf and then on the next in a sample support. The sample used should not give an activity in excess of 10,000 cpm on any shelf. Background is measured at the beginning and at the end of the experiment. The average value of background is used for the calculation of the corrected activity R.

2. For comparison, shelf ratios are calculated for at least two species of nuclide. Values for the maximum beta energies are in the Nuclides Chart. Values of shelf ratio should be examined to determine if there is a difference from one nuclide to another, and if so, the trend with respect to the beta energy. Results should be explained in a qualitative manner.

DATA:

Background: Initial ________ cpm Final ________ cpm Average ________ cpm

	Source ________ E_β ________			Source ________ E_β ________		
Shelf	Observed Activity r (cpm)	Corrected Activity R (cpm)	Shelf Ratio	Observed Activity r (cpm)	Corrected Activity R (cpm)	Shelf Ratio
1						
2			1.0			1.0
3						
4						
5						
6						

PART B—ENERGY DEPENDENCE OF β-COUNTER EFFICIENCY

1. Measure the activity of each of several calibrated beta sources on each shelf of the sample holder. The sources used may be commercial sources such as those illustrated in figure 11.5. They should be of about equal activity and should represent a wide range of beta energies.

2. Adjust all observed activities r for background and resolving time to obtain the corrected activity R.

3. From the absolute activity A of each standard source, calculate the counter efficiency ϵ on each shelf using the relationship $\epsilon = R/A$. (The values given for A may have to be corrected for decay.) Values of efficiency may be multiplied by 100 to convert to a percentage form.

DATA:

Shelf	Source___ E_β___MeV: A___dpm			Source___ E_β___MeV: A___dpm			Source___ E_β___MeV: A___dpm		
	Observed Activity r	Corrected Activity R	Efficiency $\epsilon = R/A$	Observed Activity r	Corrected Activity R	Efficiency $\epsilon = R/A$	Observed Activity r	Corrected Activity R	Efficiency $\epsilon = R/A$
1									
2									
3									
4									
5									
6									

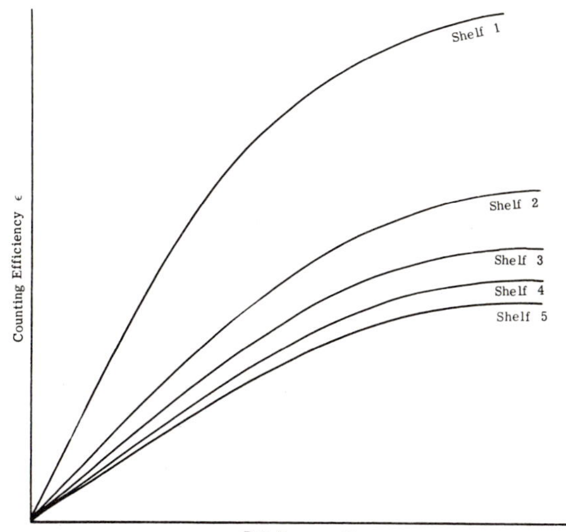

Figure 5.9—THE ENERGY DEPENDENCE OF β-COUNTER EFFICIENCY.

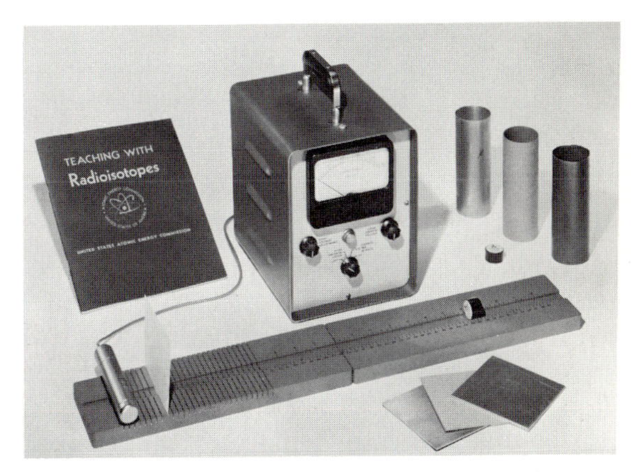

Figure 5.10—APPARATUS FOR INVESTIGATING THE INVERSE-SQUARE LAW. (*Courtesy Radiation Equipment and Accessories Corporation.*)

PART C—THE INVERSE-SQUARE LAW

1. With a device such as that illustrated in figure 5.10, or by means of a meter stick to measure the source to detector distance, the activity of a gamma source is measured at different measured distances from the detector. A side-window G-M tube equipped with an aluminum shield to remove beta particles should be used so that only gamma rays are counted.

2. On linear graph paper, the activity corrected for background is plotted versus $1/d^2$.

3. The experiment is repeated with the aluminum shield removed from the side-window G-M tube. Again data are plotted as a function of $1/d^2$. Agreement with the inverse square law is in this case compared with the agreement obtained using pure gamma radiation. Any differences observed should be explained, with special attention given to the properties of the radiation. The theoretical consideration will be less complicated if a pure beta-emitting nuclide is used here. Will the energy of the beta particle have any influence on the results?

EXPERIMENT 5.5 SCATTERING OF BETA RADIATION

OBJECTIVES:

To demonstrate the backscattering and sidescattering effects of beta radiation and its importance in analytical measurements.

To show the dependence of backscattering of beta radiation on (1) the atomic number of the scattering material, (2) the thickness of the backing material, (3) the energy of the radiation and (4) the physical geometry.

THEORY:

In making measurements of sample activities, consideration must be given to the manner in which the sample is mounted. With reference to backscattering, particular consideration must be given to the composition of the backing material (the planchet) and its thickness.

Radiation is emitted from a source in all directions. That which is emitted in a direction within the solid angle subtended by the window of the G-M tube will be counted. In addition, that radiation which is emitted in a direction other than toward the tube but which strikes matter and is deflected or scattered in such a way that it enters the counter will also be recorded. Radiation which passes away from the counter into the material composing the mount supporting the sample and is subsequently scattered toward the tube is said to be *backscattered.* The process is known as *backscattering.* Backscattering of beta particles is illustrated in figure 5.11.

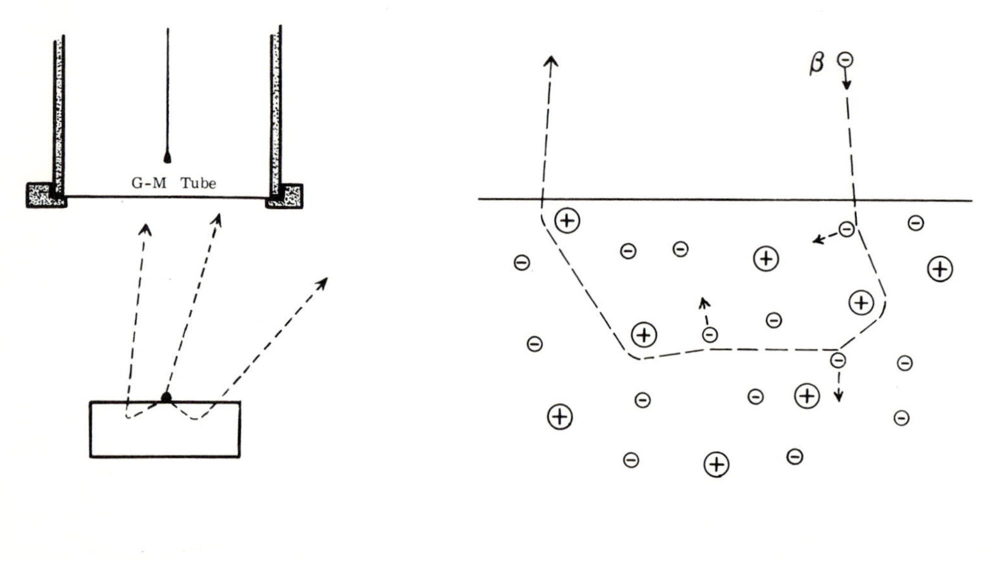

Figure 5.11

Figure 5.12

The manner in which backscattering occurs is depicted diagramatically in figure 5.12. The beta particle entering the dense matter of the backing material undergoes a series of *collisions*. A collision does not occur in the lay sense of the term but refers to any interaction, coulombic or otherwise, between the beta particle and the nuclei and electrons in the backing material. A collision may be *elastic* or *inelastic,* but in either case the result is not only a change in direction but usually a decrease in the energy of the beta particle as well.

Per cent backscattering is given by the relationship

$$\%B = \frac{r - R}{R} \times 100 \qquad (5.18)$$

where r = activity measured with a backing
R = activity measured with no backing or negligible backing (no backscattering)

the *backscattering factor* f_B is defined by

$$f_B = \frac{r}{R} = \frac{\%B}{100} + 1 \qquad (5.19)$$

The observed per cent backscattering (or backscattering factor) is dependent upon numerous factors. Among those of major importance are:

1. Atomic number of the backing material
2. Thickness of the backing material
3. Air and window absorption
4. Energy distribution of the incident beta particles
5. Distance from the source (or backscattering material) to the window of the tube.

Of these, the fourth and fifth items are largely dependent upon the third. Thus, if all air could be removed from the experimental system and the G-M tube could be operated without a window, only the first two parameters listed above would be of real consequence. Although such experimental conditions would be difficult to achieve, data corresponding to such an experimental arrangement can be estimated by the use of calibrated absorbers with subsequent extrapolation of the data to zero total absorber (see experiment 7.2).

Effect of Atomic Number of Backing Material — It has been demonstrated that backscattering of beta particles is strictly proportional to the atomic number Z within each period of the Periodic Table, but abrupt changes in slope occur at atomic numbers 2, 10, 18, 36 and 54 corresponding to the rare gases.

Effect of Thickness of Backing Material—If the backing material is infinitely thin, so as not to exist at all, backscattering will be zero. As the thickness of the backing material is increased, backscattering will also increase until a limiting value known as *saturation backscattering* is attained. Beyond this point additional increases in the thickness of the backing material will produce no increase in backscattering. This theoretical limit is reached when the backing material has a thickness equal to one-half the range

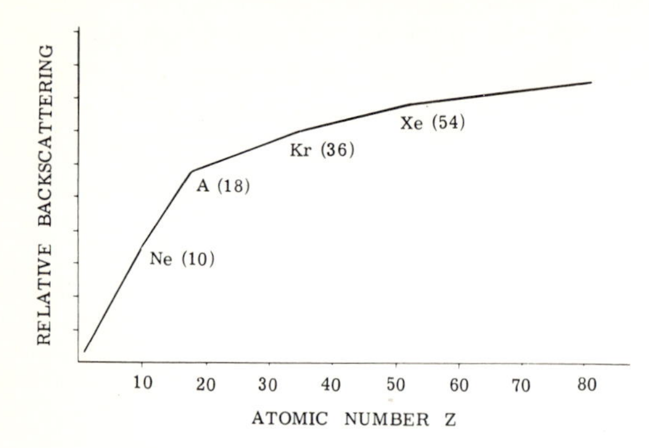

Figure 5.13—EFFECT OF ATOMIC NUMBER OF BACKING MATERIAL ON BACKSCATTERING OF BETA PARTICLES

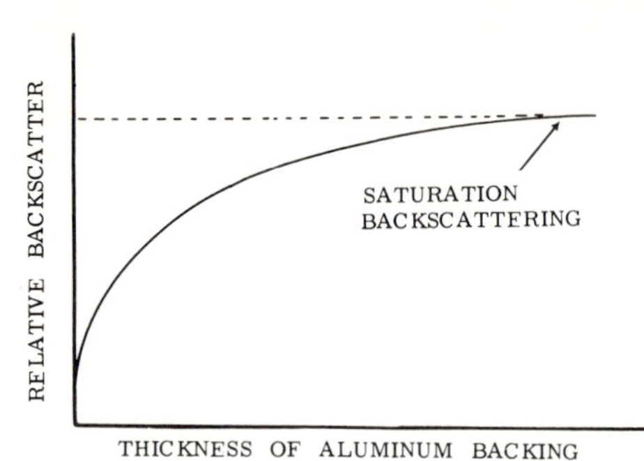

Figure 5.14—EFFECT OF THICKNESS OF BACKING MATERIAL ON BACK-SCATTERING OF BETA PARTICLES

of the beta particle. In practice, the value is closer to 0.3 of the range. The linear thickness of a particular material required to produce saturation backscattering is approximately proportional to the reciprocal of the density of the material.

Effect of Air and Window Absorption—Air and window absorption account for the differences observed in backscattering when two radioactive sources with different beta energies are compared under similar conditions of geometry or when measurements are made using the same radioactive source at different distances (different shelves) from the tube window.

The effect of beta energy differences is illustrated in figure 5.15, where typical data for carbon-14 and strontium-90/yttrium-90 are represented by the dotted lines. If both the carbon and strontium backscatter data are corrected for air and window absorption by extrapolation to zero total absorber, the solid curve in figure 5.15 is obtained for both. The uncorrected data for carbon-14 fall below those for strontium-90/yttrium-90 because the carbon-14 beta energies are initially weaker and are, therefore, more strongly absorbed after being backscattered. Consequently, a smaller number reach the detector.

The effect of the source-to-window distance is likewise a function of beta energies and of air and window absorption, but the trends which may exist in this case are further complicated by accompanying changes in sidescattering when a standard sample support is used.

Sidescattering—When beta particles are deflected by the sample support or housing into the radiation detector, they are said to be sidescattered, and the increase in count rate from this cause is measured by the sidescattering factor f_H. With wide sample supports, in which the walls of the housing are well removed from the direct pathway from the source to the window, f_H may approach unity. With narrow sample supports, f_H will be somewhat greater.

A reduction in f_H is also effected through the use of low Z materials in the construction of the sample support. For example, the inner surface of a lead castle may be lined with aluminum, sample cards may be prepared from sheets of aluminum or plastic and the card support is usually constructed of plastic for this purpose.

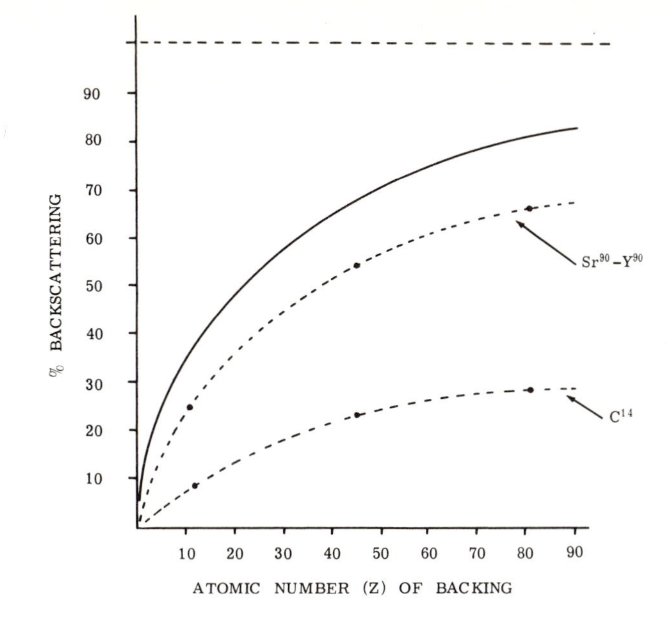

Figure 5.15—EFFECT OF ABSORPTION OF RADIATION ON BACKSCATTERING. When air and window absorption corrections are applied, all data fall on the solid curve. (Breaks in curves at Z's for rare gases not shown.)

APPARATUS:

Geiger tube and sample holder; scaler; card mount; solution of a beta source (e.g., ^{32}P, Ra-DEF); thin plastic film; approximately one-inch square samples of backing materials (glass, aluminum, iron, nickel, copper, zinc, silver, lead); cellophane tape; side-window Geiger tube; beta source, sealed or in a planchet; cylinders of cardboard, aluminum and lead.

PROCEDURE:

PART A—BACKSCATTERING

Place an appropriate amount of a beta source (e.g., ^{32}P or Ra-DEF solution) on a piece of thin plastic film fastened across the hole in an aluminum or cardboard card. Dry under the infrared lamp and determine the counting rate of the sample with this thin or negligible backing. The activity, when measured at approximately 2 cm from the tube window should be between 1,000 and 5,000 cpm.

Now fasten one of the additional backing materials under the sample as shown in figure 5.16, holding it in place with cellophane tape. For accurate results be sure the backing material is held snugly against the plastic mount. Determine the activity of the sample.

Repeat using all other available backing materials.

Measure the background and subtract this value from all readings.

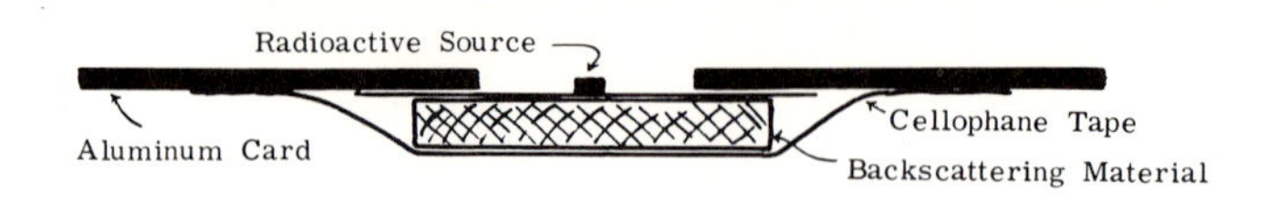

Figure 5.16—METHOD FOR MEASUREMENT OF BACKSCATTERING.

DATA:

Background ____________ cpm Instrument used ____________

Source used ____________ Beta energy (max.) ____________ MeV

Backing Material	Atomic Number	Observed Activity	Corrected Activity	% Back-scatter	Backscatter factor, f_B
None	0			0.0%	1.000
Al	13				
Fe	26				
Ni	28				
Cu	29				
Zn	30				
Ag	47				
Pb	82				

CALCULATIONS:

Calculate the per cent backscattering using equation (5.18) and the backscattering factor using equation (5.19) for each of the backing materials. In making these calculations, consider R to be the activity, corrected for background, obtained with the sample on the thin plastic film only.

On linear graph paper plot the per cent backscattering as the ordinate against the atomic number as the abscissa.

PART B—SIDESCATTERING

Support a side-window tube in a vertical position so the end of the tube is about 5 cm above the table top. If the tube is shielded, remove the shield. See figure 5.17. Place a beta source directly beneath the end of the tube and record the activity measured with a scaler or rate meter. Carefully raise the geiger tube and place one of the cylinders —made of cardboard, aluminum, lead or other material—over the source as illustrated. Replace the geiger tube to exactly the same position and record the measured activity. Measure and record the activity using each of the cylinders available.

Is there any relationship between the density of the material out of which the cylinder is composed and the increase in sidescattering?

DATA:

Composition of Cylinder	Observed Activity	Per cent increase in activity
No cylinder		0.0%
Cardboard		
Aluminum		
Lead		

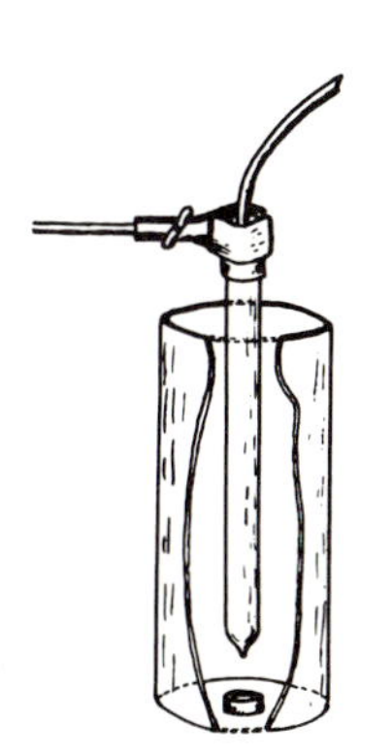

Figure 5.17—DEMONSTRATION OF BETA PARTICLE SIDESCATTERING

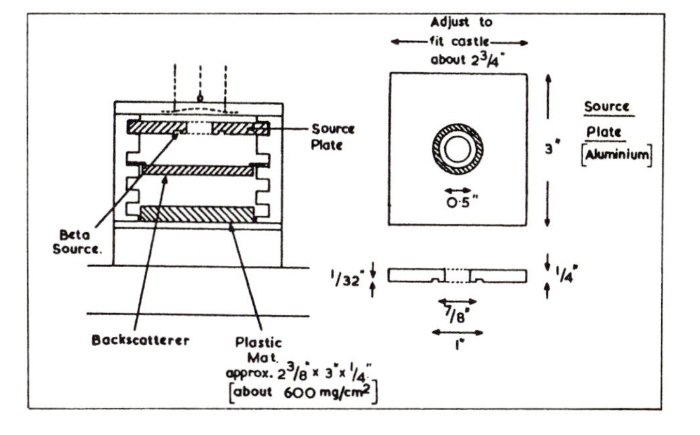

Figure 5.18—APPARATUS TO DETERMINE THE EFFECT OF THICKNESS OF THE BACKING MATERIAL ON PER CENT BACKSCATTER. (*From R. A. Faires, Isotope School, A.E.R.E. Harwell.*)

Suggestion for further work:

1. Determine the effect of thickness of the backing material on the per cent backscattering. The apparatus illustrated in figure 5.18 will provide a convenient method of measurement. The radioactive source is deposited in a circular groove so radiation does not enter the tube directly. Calibrated aluminum absorbers are used for the backscatters. A plastic mat is placed in the bottom of the chamber to reduce backscatter from the floor. The per cent backscatter is plotted as a function of the backscatterer thickness.

2. Determine the effect of beta energy on the per cent backscattering. This can be done by repeating Part A using another beta emitter in place of the original source.

3. To show the effect of physical geometry on backscattering, Part A is repeated exactly as described but using a different shelf.

EXPERIMENT 5.6 ABSORPTION OF RADIATION BY SAMPLE

OBJECTIVES:

To study the counting errors originating from the sample geometry.

To observe the effects of self absorption and self scattering. In particular, to determine the effect of changes on the measured activity of (1) concentration at constant volume, (2) at constant total activity and (3) volume at constant specific activity.

THEORY:

An important consideration in the measurement of activity of a sample is the *sample geometry*. One aspect of sample geometry is the absorption of radiation by the solvent or by the sample itself. This is especially important in the case of weak beta emitters, but even with high energy particles consideration must be made of absorption. Although gamma radiation is also scattered and absorbed, gamma radiation is so penetrating that the effect of self absorption is generally negligible.

LIQUID SAMPLES

In this experiment three specific examples of sample preparation techniques are considered, namely, constant volume, constant total activity and constant specific activity.

Constant volume—In figure 5.19 are depicted two samples of equal volume. Sample 2 contains twice as many radioactive atoms as sample 1. The sketch of sample 1 shows three radioactive atoms. The beta particles emitted by two of the atoms emerge from the sample and enter the G-M tube. The beta particle from the third atom is absorbed by the sample and never reaches the G-M tube. Thus, one-third of the betas is absorbed.

Sample 2, with twice the activity of sample 1, also absorbs one-third of the beta particles, in this case 2 out of 6. It is evident that the percentage of emitted beta particles

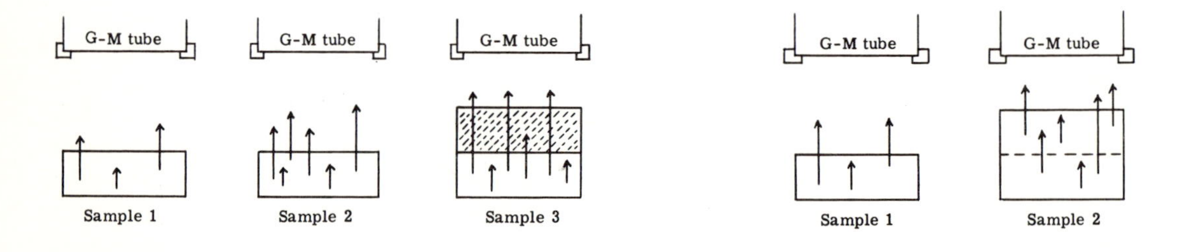

Figure 5.19—ABSORPTION FOR CASE OF CONSTANT VOLUME (Samples 1 and 2) AND CONSTANT TOTAL ACTIVITY (Samples 2 and 3)

Figure 5.20—ABSORPTION FOR CASE OF CONSTANT SPECIFIC ACTIVITY.

registered by the G-M tube remains the same. Thus, the activity recorded by the G-M tube is proportional to the concentration of radioactive atoms in the sample.

Constant total activity—In this case we wish to compare two samples, each containing the same number of radioactive atoms but differing in thickness or volume. Consider samples 2 and 3 in figure 5.19. Each shows six radioactive atoms, but because of the increased volume (shaded area) of sample 3, and hence increased absorber thickness, fewer beta particles leave the sample. In the illustration only 3 betas emerge from sample 3 as contrasted to 4 from sample 2.

In the case of constant total activity, it can be seen that the measured activity is roughly inversely proportional to the volume and theoretically approaches zero at infinite dilution.

Constant Specific Activity—This term refers to samples possessing the same activity per unit of weight or per unit of volume. Consider the two samples illustrated in figure 5.20. In sample 1, two of three beta particles emitted leave the sample. Sample 2 consists of double the volume of the same solution. From the upper half of this sample, two of three beta particles emerge (just as in the case of sample 1) but from the lower half, only one beta emerges instead of two, the second being absorbed by the upper half of the sample.

Now it can be seen that doubling the volume of sample does not result in a doubling of the observed activity, even though the total number of radioactive atoms is doubled.

When measurements of radioactivity are made on liquid samples or on samples in solution, the problems presented by absorption of the radiation in the solution can be minimized through the simple expedient of making all measurements of activity on equal volumes of solution (Constant Volume). When this is done, the observed activity r is proportional to the total activity in the aliquot measured. See figure 5.21a.

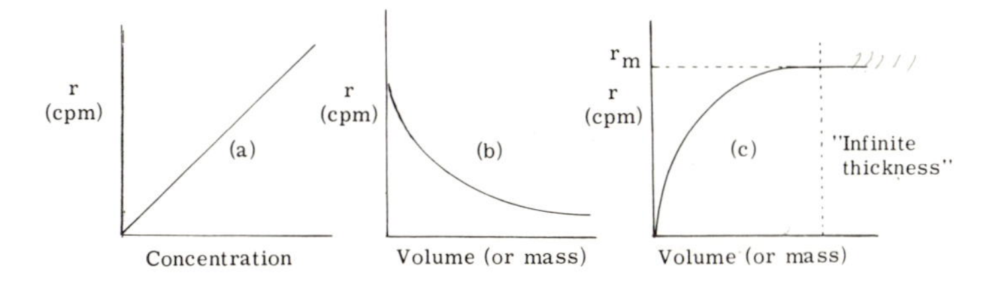

Figure 5.21—OBSERVED-ACTIVITY RELATIONSHIPS FOR THE CASES OF (a) Constant Volume (or Constant Mass), (b) Constant Total Activity and (c) Constant Specific Activity.

SOLID SAMPLES

Dilution to Constant Mass—The technique of diluting liquid samples to constant volume can be applied to solid samples as well. Solid samples are diluted to constant mass. A sufficient quantity of inactive material, usually of the same chemical composition as the radioactive sample, is added to increase the total mass to a constant value. The entire sample is then mixed well and the activity measured.

Several disadvantages to the use of this method exist, however. First, dilution of the sample with inert material alters its specific activity and in this or other ways may interfere with subsequent tests or analyses which may be desired. Secondly, any inert material which is added to the sample will reduce its activity. Longer counting times may be necessary to compensate for the resultant decrease in count rate. Because of this second point, radio-assays for carbon-14, sulfur-35 or other isotopes emitting short-range radiation may actually be rendered useless.

There are several other approaches to the general problem of self absorption in solids. Three specific cases may be cited: (1) the use of an infinitely thin sample, (2) the use of an infinitely thick sample, and (3) the use of a sample of intermediate thickness.

Infinitely Thin Samples—Solid samples are generally prepared (1) by evaporation of a solution to dryness or (2) by collection of a precipitate on a filter followed by drying of the precipitate. When samples are prepared by evaporation of an aliquot of a solution to dryness, it is frequently possible to assume that the mass of the residue is negligible and that it will have no measurable effect on the counting rate due to self absorption. Such a sample may be said to be "infinitely thin," and for such samples self absorption need not be considered. The assumption of infinite thinness is certainly valid, for example, for carrier-free solutions of isotopes of comparatively short half-life or for solutions containing but a trace of carrier or impurities.

When samples are prepared by precipitation, the weight of precipitate is generally appreciable. For such samples one cannot assume negligible self-absorption losses, and one of the following procedures must be used.

Ponderable Samples—For a source of ponderable mass, a self-absorption correction must be applied. The self-absorption coefficient f_S can be evaluated from a theoretical consideration of absorption by assuming an exponential absorption of a collimated beam of beta particles (36, 37). The expression is

$$f_S = \frac{r}{R} = \frac{1-e^{-ux}}{ux} \tag{5.20}$$

where
r = observed count rate, cpm
R = true activity of the sample, cpm
e = base of natural logarithms
u = absorption coefficient, cm^2/mg
x = sample thickness, mg/cm^2

When this equation is plotted, a curve of the form shown in figure 5.21c is obtained. A value of 0.28 for u (the absorption of carbon-14 betas in $BaCO_3$) is in good agreement with experimental values. The count rate r is seen to approach a limit as the quantity of sample is increased. Beyond a certain value known as "infinite thickness," no further increase in count rate is observed.

Infinitely Thick Samples—If a sample consists of a large quantity of a precipitate, or of a large quantity of a residue obtained upon the evaporation of a solution, the sample may possess a mass greater than infinite thickness. When infinitely thick samples are counted, the observed activity, corrected for background and coincidence loss, will be proportional to the specific activity of the sample. Thus, relative comparisons of total activity can be obtained directly, and if the absolute activity of any one sample is known, the absolute activity of other samples of the same chemical and physical composition can be calculated. This method is especially useful, for example, for the assay of carbon-14 recovered in copious quantity as $BaCO_3$.

Samples of Intermediate Thickness—When one does not have the good fortune to have an infinitely thin or an infinitely thick sample, it becomes necessary to normalize the data to an arbitrary standard mass or thickness of sample. A general procedure which may be used follows: (^{14}C as $BaCO_3$ is used as the example.)

1. Using measured aliquots of a known stock solution of radioactive sodium carbonate, a series of samples of barium carbonate of increasing mass are precipitated by the addition of excess barium chloride solution. The precipitates are collected on standard filters, washed and dried. The samples are weighed and the observed activity r of each sample is determined. Note that the true specific activity of each sample should be the same.

2. From this data a plot of activity r versus sample mass m is prepared (see figure 5.21c). The purpose of this graph is to determine the thickness of an "infinitely thick" sample thereby defining the working range required. Infinitely thick samples may be treated as described in the paragraph above.

3. The observed or "apparent" specific activity S_a (expressed in cpm/mg) is calculated for each sample. $S_a = r/m$ where r = observed count rate and m = mass of sample.

4. A value of 100% specific activity is assigned to the sample selected as "normal". The percentage of normal activity for all other specific activities is then calculated. These percentages are plotted as a function of the sample mass m. (see figure 3.20).

5. Unknown samples of ^{14}C are oxidized to CO_2 which in turn is converted to Na_2CO_3 by absorption in NaOH solution. Samples are then precipitated as $BaCO_3$ as described above, filtered and the precipitates washed and dried. Unknown $BaCO_3$ samples are weighed and the activities are measured.

6. The activities of all unknown samples are normalized for comparison.
   a. From the weight of unknown, the per cent of standard activity is read from the graph (figure 5.22).

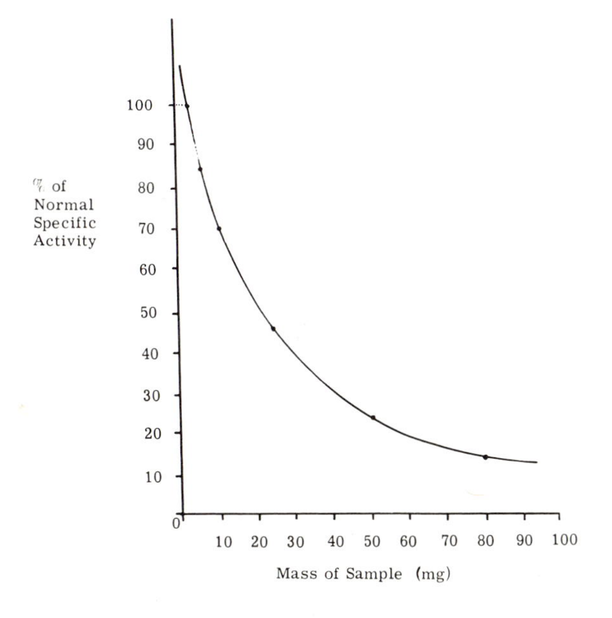

Figure 5.22—CALIBRATION CURVE FOR SELF ABSORPTION.

b. Specific activities S_a are corrected to 100% to obtain the normalized specific activity S_n (i.e., the activity the sample would have if its mass were reduced until equal to that of the norm by removal of non-radioactive $BaCO_3$).

$$S_n = S_a \, [100\%/(\% \text{ of normal})]$$

c. Normalized total activities r_n are calculated.

$$r_n = mS_n \text{ or } r_n = r[100\%/(\% \text{ of normal})]$$

Sample Data

Sample No.	m Mass of Sample (mg)	r Observed Activity (cpm)	S_a Apparent Spec. Act. (cpm/mg)	Per Cent of Normal	S_n Normalized Spec. Act. (cpm/mg)	r_n Normalized Total Act. (cpm)
Known Calibration Samples						
1	3.0	1030	343	100	343	1030
2	8.5	2470	291	85	343	2920
3	16.7	3960	237	69	343	5730
4	29.5	4720	160	46.6	343	10120
5	54.3	4850	89	26.0	343	18660
6	81.5	4875	60	17.5	343	27900
Typical Unknown Samples						
A	20.0	1500	75	61	123	2460
B	60.0	600	10	23.5	42.5	2550
C	40.0	600	15	36.0	41.7	1666

MATERIALS REQUIRED:

End-window Geiger tube, sample support and scaler; planchets (7 dram plastic vials one inch in diameter and cut off at a height of one inch hold 10 ml of solution; Technical Associates Model planchets will hold 5 ml of solution); Pipettes with manual pipetting device; stock radioactive solution (10% uranyl nitrate solution or a solution containing between 0.01 and 0.1 μCi per ml of one of the following: ^{143}Pr, ^{185}W, ^{35}S or ^{45}Ca).

PROCEDURE AND DATA:

For the sake of convenience in manipulation and to reduce the time required for performance of the experiment, the principles of self absorption are demonstrated by means of liquid samples only. To minimize the possibility of contamination of sample holders and shields through spillage, the bottom of the holder and other critical surfaces should be covered with absorbent paper. Samples should be transported to the counting room in a small tray also lined with absorbent paper.

Quantities of solutions specified in the procedure are for 10-ml planchets. For planchets of other sizes (even two milliliter planchets can be used) volumes must be adjusted proportionately.

All measurements are to be made on liquid samples. Do not dry these samples.

Constant Volume—Prepare separate samples in each of six planchets. To the volume of stock radioactive solution listed in the second column add the volume of water listed in the third column. Take the background count as equal to the activity of the planchet containing the 10 ml of water. Stir each solution well with a thin glass rod. (DO *NOT* DRY)

Relative Concentration	Stock Solution ml	Water ml	Observed Activity cpm	Corrected Activity cpm
00	0.00	10.00		0
20	2.00	8.00		
40	4.00	6.00		
60	6.00	4.00		
80	8.00	2.00		
100	10.00	0.00		

Constant Total Activity—Place 2.0 ml of stock radioactive solution in a planchet and measure the activity.(DO *NOT* DRY) Add water in 2.0 ml increments and determine the activity of each dilution. Stir the solution well with a thin rod after each addition of water.

Total Volume of Soln. (ml)	Total added water (ml)	Observed Activity cpm	Corrected Activity cpm
2.0	None		
4.0	2.0		
6.0	4.0		
8.0	6.0		
10.0	8.0		

Constant Specific Activity—Measure the activity of an empty planchet. This is the blank (background). Place 2.0 ml of stock radioactive solution in the planchet and measure the activity. Add 2.0 ml increments of this *same solution* (not water) and measure the activity after each addition.

Total Volume of Solution	Observed Activity cpm	Corrected Activity cpm
0.00 (Blank)		0
2.0		
4.0		
6.0		
8.0		
10.0		

CALCULATIONS:

Using linear graph paper, plot a graph for each of the following three cases:

1. Constant volume: concentration vs activity (counts per minute corrected for background.)
2. Constant total activity:—Total volume of solution vs activity.
3. Constant specific activity:—Total volume of solution vs activity.

Always plot the *dependent* variable as the ordinate and the *independent* variable as the abscissa. Remember to plot the blank, where applicable, as a point at the origin (i.e. zero concentration and zero activity).

Determine which plots are linear. Consider each situation from a theoretical point of view. Is there agreement between theory and experiment?

What is the practical application of this experiment?

NOTE: In the case of "constant total activity" and "constant specific activity" the total volume of solution changed from one planchet to another. Consequently the distance from the sample to the detector will vary. If the samples are sufficiently distant from the window of the tube the effect will be negligible but if they are close the data will be distorted.

EXPERIMENT 5.7 PHENOMENA OF "CARRIER-FREE" SOLUTIONS

OBJECTIVES:

To observe some of the properties of "carrier free" solutions.

To learn how to avoid certain errors in handling "carrier free" solutions.

THEORY:

The term *carrier free* (abbreviated C.F.) signifies a solution in which all atoms of a particular element are radioactive. For example, in a solution of sodium radiophosphate, $Na_2H^{32}PO_4$, C.F., there would exist no stable phosphorus. Thus, if the solution possessed, for example, a total activity of 1 microcurie, the total amount of phosphorus in the solution would amount to only 3.5×10^{-6} micrograms or 1.5×10^{-5} micrograms of $Na_2HP^*O_4$. (An asterisk is often used to denote a radioactive atom, thus $P^* = {}^{32}P$.)

In such dilute solutions it is not unusual for all or some of the solute (and hence the radioactivity) to be absorbed on the walls of the container or on the surface of a precipitate. There are several ways in which the problem can be overcome.

Siliconed glassware has been used successfully with carrier free solutions. In this case, adsorption of solute is prevented by keeping the solution from wetting the walls of the container. Volumetric glassware so treated with silicones should be recalibrated.

Presaturation of the walls of the container with the same chemical form of the radioactive element retards adsorption. If, for example, it is desired to dilute a carrier free solution of $Na_2HP^*O_4$ in a volumetric flask, rinse out the volumetric flask first with a 1% solution of Na_2HPO_4. It is best to fill the flask and let it stand for several hours. Remove the sodium phosphate solution and rinse out the flask until washings contain no trace of phosphate. The walls of the flask have now adsorbed phosphate to the point of saturation. The flask can not be used for C.F. solutions of the same chemical compound.

Carrier Solutions can be used for dilution if there is no need for the solution to be carrier free. Carrier solutions must be prepared especially for each radioactive element and even for each chemical form of each element. Several typical carrier solutions are:

For phosphate (PO_4^{-3})
0.001 *M* phosphoric acid.

For iodide (I^-)

Sodium hydroxide	0.8	gram
Sodium thiosulfate	0.4	gram
Potassium iodide	0.02	gram
Water, to make	1000	ml

For cobalt (Co^{++})
0.001 *M* $CoCl_2$ solution

PROCEDURE FOR COATING GLASSWARE WITH SILICONE

Several types of silicone products (38-40) are available for treatment of glass surfaces including the methylpolysiloxanes† and the partially hydrolyzed methylchlorosiloxanes††. Desicote‡ is a solution of a partially hydrolyzed methylchlorosiloxane type compound.

To apply a silicone coating, the glassware must first be thoroughly cleaned and *dried.* The apparatus to be siliconed is best coated by filling with the silicone solution. Be sure the apparatus is absolutely dry. Equipment too large to fill with the solution can be coated by turning in various positions and by swirling until the entire inner surface is wet with the silicone solution. The solution is then poured back into the stock bottle and the excess allowed to drain from the siliconed container. For this procedure a 1 to 2% solution of the Dow Corning fluids or a 5 to 10% solution of the General Electric Dri-Film in carbon tetrachloride or similar solvent should be used. Desicote is used without further dilution.

The solvent may be removed by air drying thereby leaving a thin film of the silicone on the surface of the glass, but baking in an oven is recommended to produce a more durable coating. In the case of the methylchlorosiloxanes, dry in an oven at 100°C. for 15 to 30 minutes. The methylpolysiloxane fluids must be cured at 300°C. for one hour or for 2 hours at 275°C. The chlorinated compounds react with traces of water remaining adsorbed on the surface of the glass and do not require this severe curing process.

† Dow Corning 200 Fluids, Dow Corning Corporation, Midland, Michigan.
††Dri-Film, General Electric Silicone Products Department, Waterford, N.Y.
‡ Beckman Instruments Company.

APPARATUS:

Flasks (50 to 150 ml, each with a mouth wide enough to admit a side-window Geiger tube); Micropipettes; carrier free solution† of sodium phosphate ^{32}P; 1 ml pipettes; Geiger tubes and scaler; silicone solution; 1% Na_2HPO_4 solution; 0.001 M H_3PO_4; precipitated chalk ($CaCO_3$); 1% silver nitrate solution; 1% sodium chloride solution; hydrochloric acid, dilute.

PROCEDURE

PART A—ADSORPTION ON WALLS OF CONTAINER

1. Clean four flasks very thoroughly with chromic acid cleaning solution and then rinse them thoroughly with distilled water. For the sake of uniformity, dry all four flasks in an inverted position in an oven.
2. Fill the first flask with silicone solution. Be sure the flask is absolutely dry. Rotate the flask so that all of the interior surface is wetted with the solution and then pour the solution back into the stock bottle. Allow the flask to drain well and dry it in the oven.
3. Do not treat the second flask.
4. Fill the third flask with a 1% solution of Na_2HPO_4 and allow it to stand for a few minutes. Pour the solution back into the stock bottle and rinse the flask thoroughly with distilled water.
5. Do not treat the fourth flask.
6. Make dilutions of a ^{32}P solution. To each flask add distilled water or carrier (as indicated on the chart below) until it is half full. To each flask add 1 microcurie of carrier-free phosphorus-32 solution and fill to the mark with the appropriate diluting fluid. Mix well.
7. Measure 1 ml of each solution into a separate planchet. *Do not dry* these solutions, but measure the activities of the four solutions immediately.
8. Hold the solutions in their flasks for a period of at least 24 hours to allow adsorption on the walls of the containers to take place. Flasks can be sealed by pressing a piece of aluminum foil over the mouth of each. Again measure 1 ml portions of each solution into separate planchets and measure the activities.

Flask	Treatment	Diluting Solution	Activity of 1 ml of Solution		Activity of Empty Flask
			Immediately	Next Day	
1	Siliconed	Water			
2	None	Water			
3	1% Na_2HPO_4	Water			
4	None	0.001 *M* H_3PO_4			

‡ Phosphotope (Squibb) contains dibasic sodium phosphate as a carrier, to provide a total phosphorus content of less than 0.5 mg per ml. Consequently this or other radioactive phosphate solution containing carrier will not yield the anticipated results in this experiment.

9. Transfer the solutions to the liquid waste container. Thoroughly rinse each flask with small portions of distilled water. Drain well and measure the relative activity adsorbed on the wall of each flask by use of a side-window tube as illustrated in figure 5.23.

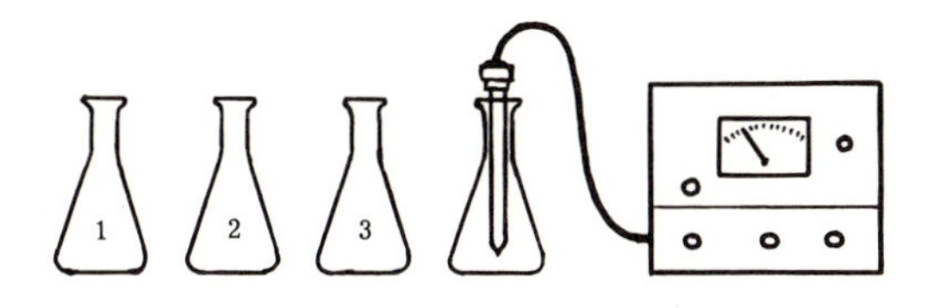

Figure 5.23—DETERMINATION OF ADSORPTION OF "CARRIER-FREE" ISOTOPE ON WALLS OF CONTAINER. (*Courtesy Baird-Atomic, Inc.*)

CALCULATIONS:

1. Determine the percentage change in the strength of each solution over the 24 hour period.

2. List the procedures in the order of effectiveness for preserving the original strength of the solution.

PART B—ADSORPTION ON PRECIPITATES

1. Prepare 10 ml of carrier-free phosphate solution containing about 0.1 microcurie of ^{32}P.

2. To 2 ml of this solution in a 3-ml test tube, add about 10 mg of calcium carbonate. Stopper and shake. Centrifuge, transfer 1.0 ml of the supernatant to a planchet and measure the activity. Compare with the activity of 1.0 ml of the original solution in a similar planchet.

3. To 2 ml of ^{32}P solution (from step 1) in a 3-ml test tube, add 3 drops of 1% silver nitrate solution. Mix and add an excess of hydrochloric acid to precipitate the silver as the chloride. Mix well and centrifuge. Transfer 1.0 ml of the supernatant to a planchet and measure the activity.

4. Repeat step 3 but use a 1% sodium chloride solution in place of the hydrochloric acid. Compare the activity of 1.0 ml of this supernatant with that obtained in step 3. Does the pH have any effect on the adsorption of trace amounts of phosphate?

REFERENCES

Resolving Time (Dead Time)

1. Rainwater, L. J., and C. S. Wu, "Applications of Probability to Nuclear Particle Detection", Nucleonics 2, 42 (Jan. 1948).
2. den Hartog, H., "Speed of Operation of Geiger-Müller Counters", Nucleonics 5, 33-47 (Sept. 1949).
3. Preuss, L. E., "Constant Geometry for Dead-Time Determination", Nucleonics 10, No. 2, 62-63 (1952).
4. Porter, W. C., "Extending the Efficiency Range of G-M Counters", Nucleonics 11, No. 3, 32-35 (1053).
5. Damon, P. E., and P. N. Winters, "Resolution Losses in Counters and Trigger Circuits", Nucleonics 12, No. 12, 36-39 (Dec. 1954).
6. Covell, D. F., M. M. Sandomire and M. S. Eichen, "Automatic Compensation of Dead Time in Pulse Analysis Equipment", Anal. Chem. 32, 1086 (1960).
7. Stearns, R. L., and J. F. Mucci, "Using the Decay of Ba^{137} to Determine Resolving Time in G-M Counting", J. Chem. Ed. 38, 29-30 (1961).
8. Scott, J. H., "Slide Rule for Making Dead Time Correction", Nucleonics 19, No. 9, 90-92 (Sept. 1961).

Geometry Factor

9. Cook, G. B., J. F. Duncan and M. A. Hewitt, "Geometrical Efficiency of End-Window G-M Counters", Nucleonics 8, 24-27 (Jan. 1951).
10. Amith, A., and W. W. Meinke, "Radial Response of Typical End Window G-M Tubes", Nucleonics 11, No. 5, 60-61 (1953).
11. Kalmon, B., "Experimental Method for Determination of Counting Geometry", Nucleonics 11, No. 7, 56 (1953).

Window Absorption

12. Chang, C., and C. S. Cook, "Relative Transmission of Beta Particles Through G-M Counter Windows", Nucleonics 10, No. 4, 24-27 (1952).
13. Upson, U. L., "Evaluating Beta Energy Dependence in End-Window G-M Tubes", Nucleonics 11, No. 12, 49-54 (1953).

Scattering of Radiation

14. Christian, D., W. W. Dunning and D. S. Martin, Jr., "Backscattering of Beta Rays in Windowless G-M Counters", Nucleonics 10, No. 5, 41-43 (1952).
15. Hine, G. J., and R. C. McCall, "Gamma-Ray Backscattering", Nucleonics 12, No. 4, 27-30 (April 1954).
16. Müller, R. H., "Interaction of Beta Particles with Matter", Anal. Chem. 29, 969 (1957).
17. Müller, D. C., "Interaction of Beta Particles with Organic Compounds", Anal. Chem. 29, 975 (1957).
18. Gray, P. R., D. H. Clarey and W. H. Beamer, "Interaction of Beta Particles with Matter—Analysis of Hydrocarbons by Beta-Ray Backscattering", Anal. Chem. 31, 2065-2068 (1959).
19. Johns, D. H., and O. R. Alexander, "Magnetic Domains Complicate Backscatter Technique", Nucleonics 19, No. 9, 94-96 (Sept. 1961).
20. Cowing, R. F., and E. DeAmicis, "Variations in Total Counts of P^{32} and I^{131} for Dishes of Different Atomic Number", Science, 108, 187 (1948).

Self Absorption

21. Yankwich, P. E., and J. W. Weigl, "The Relation of Backscattering to Self-Absorption in Routine Beta-Ray Measurements", Science 107, 651 (1948).
22. Yankwich, P. E., "Loss of Radioactivity from Barium Carbonate Samples", Science 107, 681 (1948).
23. Faul, H. and G. R. Sullivan, "Density Correction in Beta-Ray Assaying of Rock and Mineral Samples", Nucleonics 4, 53 (Jan. 1949).
24. Wick, A. N., H. N. Barnet and N. Ackerman, "Self Absorption Curves of C^{14}—Labeled Barium Carbonate, Glucose and Fatty Acids—Influence of Physical Form and Sample", Anal. Chem. 21, 1511-1513 (1949).
25. Aten, A. H. W., Jr., "Corrections for Beta Particle Self-Absorption", Nucleonics 6, 68-74 (Jan. 1950).
26. Rossi, H. H., and R. H. Ellis, Jr., "Distributed Beta Sources in Uniformly Absorbing Media", Part I—Nucleonics 7, 18-25 (July 1950); Part II—Nucleonics 7, 19-25 (Aug. 1950).
27. Schweitzer, G. K., and B. R. Stein, "Measuring Solid Samples of Low-Energy Beta Emitters", Nucleonics 7, 65-72 (Sept. 1950).
28. Comar, C. L., et al, "Use of Calcium-45 in Biological Studies", Nucleonics 8, No. 3, 19-31 (1951).
29. Dixon, W. R., "Self-Absorption Correction for Large Gamma-Ray Sources", Nucleonics 8, No. 4, 68-72 (1951).
30. Rose, G., and E. W. Emery, "Effects of Solution Composition in a G-M Counter for Liquid Samples", Nucleonics 9, No. 1, 5-12 (1951).
31. Nervik, W. E., and P. C. Stevenson, "Self-Scattering and Self-Absorption of Betas by Moderately Thick Samples", Nucleonics 10, No. 3, 18-22 (1952).

32. Field, E. L., "Estimation of Self Absorption within a Homogeneous Cylindrical Source", Nucleonics 11, No. 9, 66-68 (1953).
33. Gora, E. K., and F. C. Hickey, "Self-Absorption Correction in Carbon-14 Counting", Anal. Chem. 26, 1158-1161 (1954).
34. Katz, J., "Self-Absorption Correction for Carbon-14 Assay", Science 131, 1886 (1960).
35. Massini, P., "Self-Absorption Correction for Isotopes Emitting Weak Beat Rays", Science 133, 877 (1961).
36. Miller, W. W., and T. D. Price, "Research with Carbon-14", Nucleonics 1, No. 3, 4-22 (Nov. 1947).
37. Libby, W. F.,
Ind. Eng. Chem. 19, 2 (1947).

Behavior of Carrier-Free Solutions

38. Hershenson, H. M., and L. B. Rogers, "Errors in Volumetric Analysis Arising from Adsorption", Anal. Chem. , 219-220 (19).
39. Bingenheimer, L. E., Jr., "Silicone Coating of Laboratory Glassware", J.A.Ph.A., Pr. Ed. 17, 797-799 (1956).
40. McKeever, R. M., and W. B. Swafford, "Magnesia Magma in Silicone Coated Bottles", Amer. J. Pharm., 142-144 (April 1961).
41. Gilbert, P. T., Jr., "
Science 114, 637 (1951).
42. Bouner, N. A., and M. Kahn, "Some Aspects of the Behavior of Carrier-Free Tracers—I", Nucleonics 8, 46-59 (Feb. 1951). —II, Nucleonics 8, 40-61 (March 1951).
43. Gevantman, L. H., and J. F. Pestaner, "Counting Losses in I^-, I_2 on Paint and Metal Surfaces", Nucleonics 14, No. 11, 109-111 (Nov. 1956).
44. Preiss, I. L., and R. W. Fink, "Carrier-Free Solution Storage in Glass", Nucleonics 15, No. 10, 108 (Oct. 1957).

CHAPTER 6
Radioactive Decay

LECTURE OUTLINE AND STUDY GUIDE

I. MODES OF RADIOACTIVE DECAY
   A. Alpha Decay
   B. Beta Decay
      1. Radioisotopes having a high Neutron-proton ratio
         a. Negatron emission
         b. Energetics of beta decay
      2. Radioisotopes having a low neutron-proton ratio
         a. Positron emission
            Energetics of positron emission
            Annihilation radiation
            Positronium
         b. Electron capture
   C. Atomic De-excitation
      1. Fluorescence
      2. Auger effect
   D. Nuclear De-excitation
      1. Nuclear isomerism
      2. Gamma-ray emission
      3. Internal conversion

II. DECAY SCHEMES
   A. Types
      1. Simple
      2. Complex
   B. Application of Decay Schemes
      1. Radiation standards
      2. Dosage calculations
      3. Absolute counting

III. BASIC RATE EQUATION FOR RADIOACTIVE DECAY (*Experiment* 6.1)
   A. Derivation
   B. Radioactive Constants
      1. Disintegration constant
      2. Half-life, $t_{1/2}$
      3. Average life, τ
   C. Graphic Representation

IV. DETERMINATION OF RADIOACTIVE CONSTANTS (*Experiment* 6.1)
   A. Graphical Method from Semi-log Plots of Decay Curves
   B. Analytical Methods
      1. Exponential form of equation
      2. Logarithmic form of equation
   C. Method for Long-Lived Radioisotopes (*Experiment* 6.4)

V. MIXTURES OF INDEPENDENTLY DECAYING ACTIVITIES (*Experiment* 6.2)
   A. Equation
   B. Resolution into Component Curves

VI. RADIOACTIVE EQUILIBRIA (Experiment 6.3)
   A. General Equation
   B. Secular Equilibrium
   C. Transient Equilibrium
   D. Case of No Equilibrium
   E. Case of Many Successive Decays—Bateman Solution
   F. Branching Decay (*Experiment* 6.4)

MODES OF RADIOACTIVE DECAY:

In chapter 2 it was seen that nuclei are composed of combinations of nucleons (protons and neutrons) and that certain combinations of these nucleons (i.e., certain nuclides) possess a high degree of stability while others are relatively unstable.

It will also be recalled from chapter 2 that the more important factors related to nuclear stability include: (1) the neutron-to-proton ratio, the most stable ratio being that indicated by the stability diagonal, (2) pairing of nucleons as indicated by the even-odd rules and (3) the binding energy which, in turn, is related to the mass defect and the packing fraction. Thus, one radioactive nuclide may have a high proton-to-neutron ratio while for another the ratio may be low. A radioactive nuclide may have an odd number of protons and an even number of neutrons while for another the reverse may be true. Because radioactive nuclides do differ one from the other in so many respects, it is not surprising, then, that various modes of decay are possible, depending upon the nature of the nuclide and the type of instability.

Unstable nuclei are said to be radioactive because they emit radiation as they undergo spontaneous decay. This radiation is emitted either from the nucleus itself or as a result of alterations in the configuration of the orbital electrons about it. The nature of this radiation is a function of the mode of decay of the particular nuclear species. Types of radiation resulting from radioactive decay include: (1) alpha particles, (2) positrons, (3) negatrons and (4) electromagnetic radiation.

DECAY BY ALPHA PARTICLE EMISSION:

The alpha particle consists of two protons and two neutrons and is, therefore, a relatively large particle. In order that a nucleus be capable of releasing so large a particle, the nucleus must be relatively large itself. Experimental facts support this concept since it has been found, with few exceptions, that nuclei which decay by alpha emission have an atomic number of 83 or more. A typical example is radon-222.

$$^{222}_{86}Rn_{136} \longrightarrow {}^{218}_{84}Po_{134} + \alpha + 5.58 \text{ MeV} \tag{6.1}$$

In this reaction the Z and N of the nucleus both decrease by 2 while the mass number A decreases by 4. Alpha decay can be expressed by a generalized equation:

$$^{A}_{Z}X_{N} \longrightarrow {}^{A-4}_{Z-2}Y_{N-2} + {}^{4}_{2}He_{2} + Q \tag{6.2}$$

Although the total energy change for the decay of radon-222 is 5.58 MeV, the energy of the alpha particle is only 5.48 MeV. The difference of 0.10 MeV is the energy of recoil imparted to the newly formed polonium nucleus. If the validity of non-relativistic mechanics is assumed, this recoil energy T_n can be calculated readily from consideration of the laws of conservation of energy and conservation of momentum. The total energy of the transition Q is equal to the sum of the kinetic energies of the recoiling nucleus and the alpha particle

$$Q = T_n + T_\alpha = \tfrac{1}{2}mv^2 + \tfrac{1}{2}m_\alpha v_\alpha^2 \tag{6.3}$$

and the momentum of the recoiling nucleus is equal to the momentum of the alpha particle

$$mv = m_\alpha v_\alpha \tag{6.4}$$

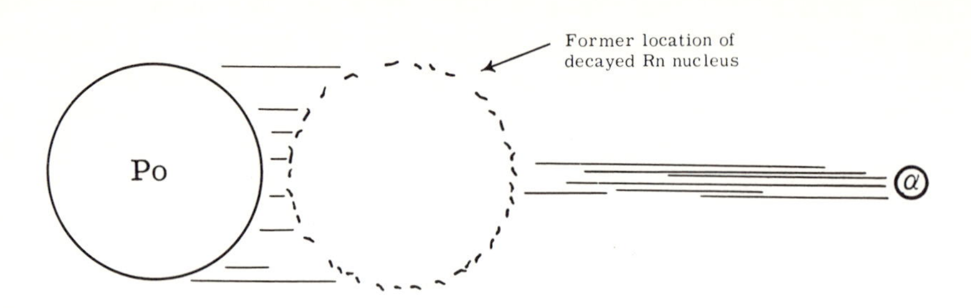

Figure 6.1—ALPHA DECAY. A radon-222 nucleus has decayed into a polonium-218 nucleus and an alpha particle. The Po nucleus recoils when the alpha particle is ejected.

By squaring equation (6.4) and rearranging terms one obtains

$$mv^2 = \frac{m_\alpha^2 v_\alpha^2}{m} \qquad (6.5)$$

Substitution of equation (6.5) into equation (6.3) and rearranging terms yields

$$Q = \tfrac{1}{2} m_\alpha v_\alpha^2 \left(\frac{m_\alpha}{m} + 1\right) \qquad (6.6)$$

Thus

$$Q = T_\alpha \left(\frac{m_\alpha}{m} + 1\right) \qquad (6.7)$$

or

$$T_\alpha = \frac{mQ}{m_\alpha + m} \qquad (6.8)$$

and

$$T_n = Q - T_\alpha = Q\left(1 - \frac{m}{m_\alpha + m}\right) \qquad (6.9)$$

In these relationships the kinetic energy can be expressed in MeV and the ratio of masses can be replaced by the nuclidic masses. Mass numbers can be used as estimates of the nuclidic masses.

$$T_\alpha = \frac{MQ}{M_\alpha + M} \approx \frac{AQ}{A_\alpha + A} \qquad (6.10)$$

$$T_n = Q\left(1 - \frac{M}{M_\alpha + M}\right) \approx Q\left(1 - \frac{A}{A_\alpha + A}\right) \qquad (6.11)$$

For the polonium-218 mentioned above, the calculated recoil energy is 0.101 MeV while the alpha particle energy is 5.48 MeV.

Alpha particles from a particular nuclear transformation are monoenergetic. In the decay of thorium-228, for example, it can be seen that at least five specific values

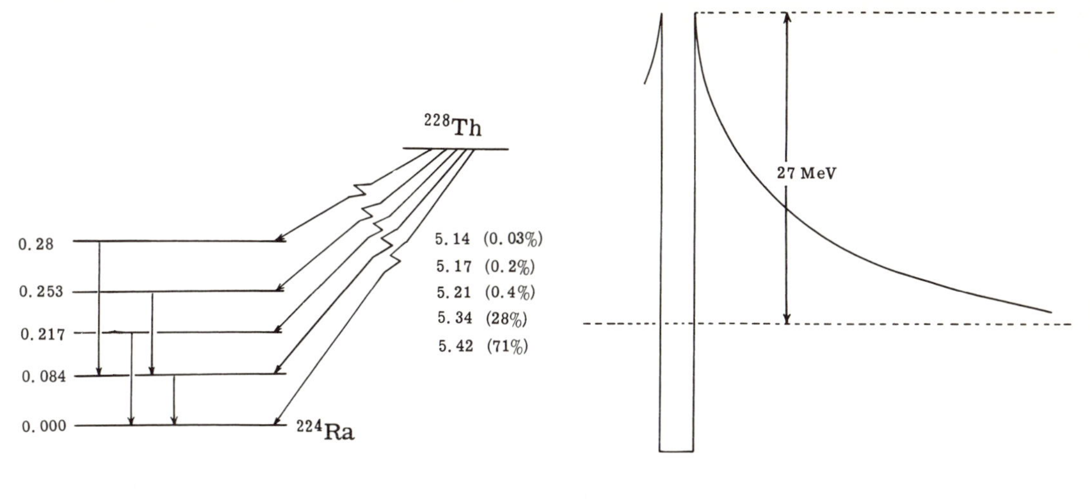

Figure 6.2—ENERGY LEVEL DIAGRAM FOR THE ^{228}Th-^{224}Ra TRANSITION.

Figure 6.3—POTENTIAL ENERGY WELL FOR THE ALPHA DECAY OF ^{228}Th.

of alpha particle energy are possible, one corresponding to each decay pathway. While 71% of the thorium-228 nuclei decay directly to the ground state of radium-224 by emission of a 5.42 MeV alpha particle, 28% of the nuclei emit a 5.34 MeV alpha particle in decaying to an excited state of radium-224. The difference of 0.084 MeV is ultimately released as electromagnetic radiation.

Thermodynamically the energy change accompanying a reaction in one direction is equal in magnitude, but opposite in sign, to that associated with the same reaction in the opposite direction. Accordingly, the height of the coulombic barrier restraining an alpha particle within a nucleus is equal to that of the barrier of repulsion that an alpha particle would have to overcome in order to enter the nucleus. This energy is equal to $2Ze^2/R$ where Z is the atomic number of the daughter nucleus and R is its radius. For radium-224, $Z = 88$ and R is approximately 9.4×10^{-13} cm. The height of the barrier is then calculated to be

$$E = \frac{2Ze^2}{R} = 4.3 \times 10^{-5} \text{ ergs} = 27 \text{ MeV} \tag{6.12}$$

Because the energy accompanying the reverse reaction is the same, the alpha particle emitted from thorium-228 should have an energy of about 27 MeV. But, this is not the case! The energy associated with the alpha decay of thorium-228 is only about 6 MeV. This apparent discrepancy was explained by G. Gamow (1) and by R. W. Gurney and E. U. Condon (2) in 1928. While classical mechanics requires that the energy be 27 MeV as calculated above, wave mechanics accounts for a small probability that the alpha particle be found beyond the barrier. This is a quantum effect which would be observed on a larger scale but for the very small value of the Planck constant. To illustrate, let $h = 1$. You are crossing a very high mountain in your automobile. Suddenly you are on the other side of the mountain without having actually crossed over it. This is what happens to the alpha particle. The phenomenon is known as the *tunnel effect*.

Although an inverse relationship between α-particle energy and half-life had been observed, it was first formulated quantitatively by H. Geiger and J. M. Nuttall (3) in 1911. This expression, called the Geiger-Nuttall rule, relates the decay constant λ with alpha particle range *R:*

$$\log \lambda = a + b \log R \tag{6.13}$$

The slope b is a constant but the intercept a assumes a different value for each of the radioactive series. Because the range and energy are related in an almost linear fashion it is possible to equate λ and E by means of a similar expression:

$$\log \lambda = a' + b' \log E \tag{6.14}$$

BETA DECAY:

There are three types of beta decay: (1) Negatron emission (β^-), (2) Positron emission (β^+), and (3) Electron capture (EC).

When a nucleus has an excess of neutrons relative to its more stable isobars (i.e., it is proton deficient), greater stability will be achieved by conversion of a neutron into a proton. This process is called *negatron emission* or *negatron decay*

$$n \longrightarrow p + {}_{-1}e^0 + \bar{\nu} \tag{6.15}$$

When a nucleus has an excess of protons relative to its more stable isobars (i.e., it is neutron deficient), greater nuclear stability is achieved by conversion of a proton into a neutron. This conversion can be accomplished by either *positron emission (positron decay)*

$$p \longrightarrow n + {}_{+1}e^0 + \nu \tag{6.16}$$

or by *electron capture*

$$p + {}_{-1}e^0 \longrightarrow n + \nu \tag{6.17}$$

DECAY BY NEGATRON EMISSION:

Negatron emission resulting from β^- decay occurs if the neutron-proton ratio is too great. It is illustrated by the particle reaction for negatron decay (equation 6.15). This reaction can also be written in the form

$$n \longrightarrow p + \beta^- + \bar{\nu} \tag{6.18}$$

since ${}_{-1}e^0$ and β^- are equivalent. Although a free neutron will undergo spontaneous decay with a half life of about 12 minutes, the situation is considerably different if the neutron is bound within a nucleus. The difference is manifest, for example, through the wide range of half-lives observed for radioisotopes which decay by this process. Half lives from a few microseconds to over 10^{10} years have been observed.

Negatron emission is illustrated by the decay of phosphorus-32.

$$^{32}_{15}P_{17} \longrightarrow ^{32}_{16}S_{16} + {}_{-1}e^0 + \bar{\nu} \quad (6.19)$$

In the case of this particular reaction, it could be predicted, from a consideration of the even-odd rules, that ^{32}P would be unstable, having an odd number of both protons and neutrons, while ^{32}S might be expected to possess a high degree of stability. However, a single criterion such as the even-odd rule cannot be used to predict the decay scheme. For example, in the decay of ^{45}Ca

$$^{45}_{20}Ca_{25} \longrightarrow ^{45}_{21}Sc_{24} + \beta^- + \bar{\nu} + 0.25\,\text{MeV} \quad (6.20)$$

it is seen that the proton number has changed from even to odd while the neutron number has changed from odd to even, one tending to cancel the effect of the other. Reference to the "Chart of the Nuclides" will show, however, that scandium-45 is located somewhat closer to the stability diagonal than is calcium-45.

In negatron decay the mass number A remains constant while Z increases by 1 and N decreases by 1. Negatron decay can be expressed by a generalized equation:

$$^{A}_{Z}X_{N} \longrightarrow {}_{Z+1}^{\;\;A}Y_{N-1} + \beta^- + \bar{\nu} \quad (6.21)$$

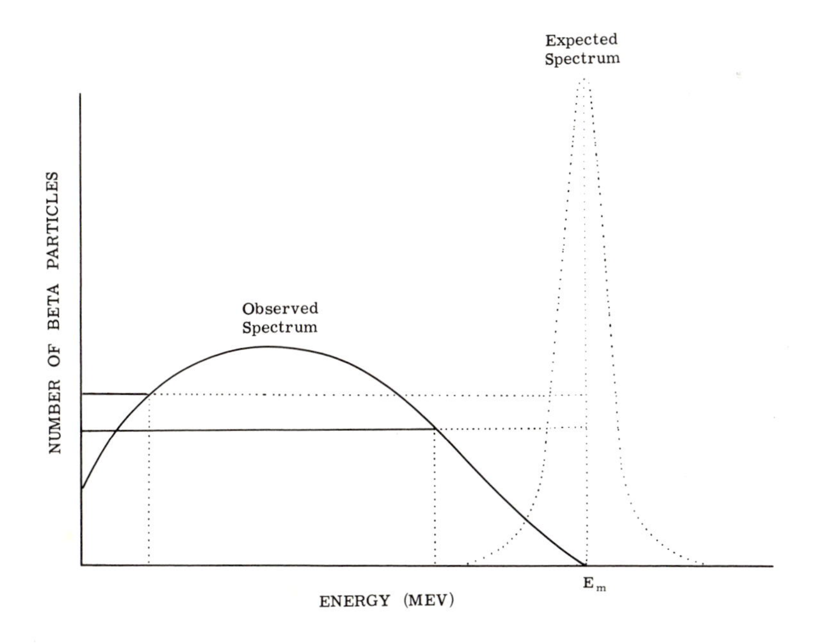

Figure 6.4—BETA SPECTRUM. The energy of the beta particle is shown by the horizontal solid line. The balance of the energy (horizontal dotted line) is that of the neutrino.

Unlike alpha particles which are monoenergetic from a given source, beta particles are emitted with a range of energies lying between zero MeV and the maximum energy E_m for a particular species of isotope. This can be shown by means of a magnetic field which can be used to spread out beta radiation from a particular source into a continuous spectrum (figure 6.4). It has not been easy to explain this fact. Considering the law of conservation of energy, one would expect all beta particles from a given source to have the same kinetic energy. For example, at the moment ^{32}P decays to ^{32}S the energy of the nucleus changes from energy level E_1 to energy level E_2, a change ΔE of 1.71 MeV. Some of this energy is associated with the emitted beta particle, but the disposition of the balance of the energy posed an enigma. The suggestion was made by Pauli and developed by Fermi (1934) that the emission of a beta particle is accompanied by the simultaneous emission of another particle, the *neutrino*. Recent experiments have demonstrated the existence of the neutrino. Its mass is essentially zero and its charge is zero, but it does possess momentum and energy.

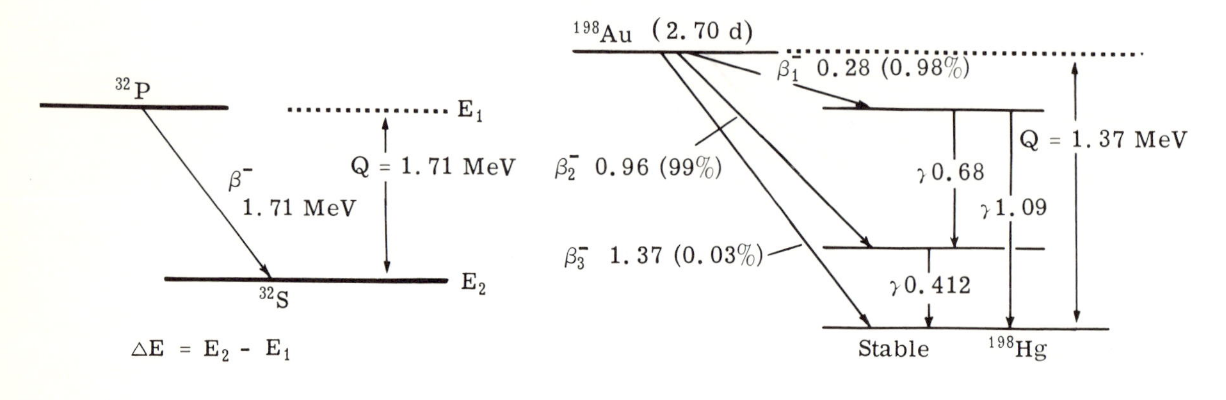

Figure 6.5—ENERGY LEVEL DIAGRAM FOR THE DECAY OF ^{32}P.

Figure 6.6—BETA DECAY FOLLOWED BY GAMMA-RAY EMISSION. Competitive decay modes for ^{198}Au are shown in addition to the mode discussed in the text.

In many instances of alpha and beta decay, decay does not occur with a direct transition to the ground (stable) state of the daughter but often results first in the formation of an excited state of the daughter. The case of gold-198 is an example. In about 99% of the decay processes, a 0.96 MeV (E_m) beta particle is emitted with the formation of an excited state of ^{198}Hg. Return of the nucleons of ^{198}Hg to the ground state is associated with the emission of a gamma ray with an energy (0.412 MeV) corresponding to a specific transformation of the nucleus from one energy state to another. Gamma rays resulting from nuclear transformations are monoenergetic.

The decay of gold-198 is shown in figure 6.6. Vertical distances on a decay scheme represent energies and are drawn to scale.

Energetics of β^- Decay — If the mass-energy conservation law is applied to β^- decay where a nucleus of mass $_ZM'$ decays to a nucleus of mass $_{Z+1}M'$ one obtains the relation

$$_ZM' = {}_{Z+1}M' + m_o + m_\nu + T_\beta + T_\nu + T_{M'} + T_\gamma \qquad (6.22)$$

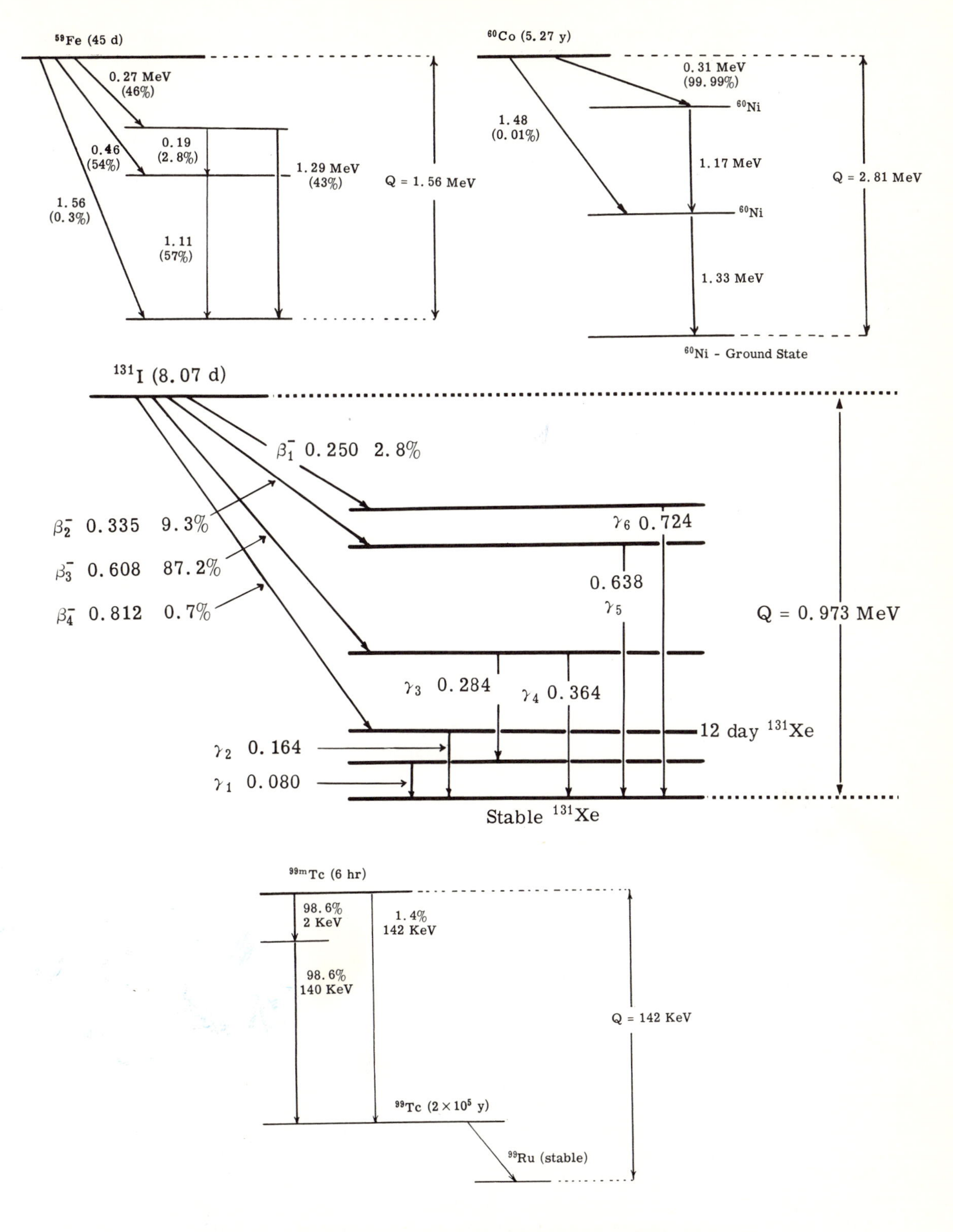

Figure 6.7—DECAY SCHEMES FOR COMMONLY USED ISOTOPES WHICH DECAY BY NEGATRON EMISSION.

Since the mass of the neutrino m_ν is presumably zero and the recoil energy of the nucleus $T_{M'}$ may be considered negligible, then

$$_Z M' = {}_{Z+1}M' + m_0 + T_m + T_\gamma \tag{6.23}$$

where $T_m = T_\beta + T_\nu$ and is equal to the maximum value for the kinetic energy of the beta particle.

If Z electron masses are added to each side of equation (6.23) then the relationship can be expressed in terms of atomic mass M rather than the nuclear mass M′. Thus

$$_Z M = {}_{Z+1}M + T_m + T_\gamma \tag{6.24}$$

The decay energy Q_{β^-} is then given by

$$Q_{\beta^-} = {}_Z M - {}_{Z+1}M = T_m + T_\gamma \tag{6.25}$$

DECAY BY POSITRON EMISSION:

Positron emission, a type of beta decay, occurs if the neutron-proton ratio is not great enough for stability of the nucleus. This process is represented by the particle reaction of equation (6.16). This reaction can also be written in the form

$$p \longrightarrow n + \beta^+ + \nu \tag{6.26}$$

since $_{+1}e^0$ and β^+ are equivalent.

Spontaneous decay of free protons does not occur. Proton decay is energetically impossible since the sum of the neutron and positron masses exceeds the proton mass.

$$\begin{aligned} M_n &= 1.0086654 \text{ amu} = 939.550 \text{ MeV} \\ M_e &= \underline{0.0005486} \qquad\;\; = \underline{\;0.511\;} \\[1em] M_n + M_e &= 1.0092140 \text{ amu} = 940.061 \text{ MeV} \\ M_p &= \underline{1.0072766} \qquad\;\; = \underline{938.256} \\[1em] \Delta M &= 0.0019384 \text{ amu} = \;\;1.805 \text{ MeV} \end{aligned}$$

Because this reaction is endoergic, 1.805 MeV of energy must be supplied from the surroundings of the proton in order to convert it into a neutron and a positron. (It has been assumed that the neutrino rest mass is zero and has therefore not been considered in the calculation.) While this energy is not likely to be supplied to a free proton, if the proton is contained within a nucleus the necessary energy may be supplied by the other nucleons providing that the atomic mass difference between the parent and daughter nuclides is greater than $2m_0c^2$ as required in equation (6.30).

Positron decay is illustrated by the case of zinc-65:

$$^{65}_{30}Zn_{35} \longrightarrow ^{65}_{29}Cu_{36} + \beta^{+} + \nu \tag{6.27}$$

In positron decay the mass number A does not change since a decrease in Z of 1 is just offset by an increase of N by 1. Positron decay can be represented by a generalized equation

$$^{A}_{Z}X_{N} \longrightarrow {}_{Z-1}^{A}Y_{N+1} + \beta^{+} + \nu \tag{6.28}$$

Energetics of β^{+} *Decay*—Application of the mass-energy law to β^{+} decay yields the general relationship

$$_{Z}M' = {}_{Z-1}M' + m_0 + T_m + T_{\gamma} \tag{6.29}$$

Addition of Z electrons to each side of the equation yields

$$_{Z}M = {}_{Z-1}M + 2m_0 + T_m + T_{\gamma} \tag{6.30}$$

and the decay energy $Q_{\beta^{+}}$ is therefore given by

$$Q_{\beta^{+}} = {}_{Z}M - {}_{Z-1}M = 2m_0 + T_m + T_{\gamma} \tag{6.31}$$

Annihilation Radiation—All positrons emitted during β^{+} decay are ultimately annihilated as the result of an interaction with a negative electron. The positron is the antiparticle of the negatron and will therefore experience a very short life in an environment so rich in negative electrons.

The kinetic energy of a positron is reduced primarily through inelastic collisions. These collisions are with atoms and molecules composing the matter through which it passes and result in their excitation or ionization. Because the cross-section for energy loss in this way is so much greater than the cross section for annihilation, most positrons will lose most of their kinetic energy before undergoing annihilation. Having lost nearly all of its kinetic energy a positron may then react by:

1. immediate annihilation in a collision with a negatron.
2. capture by an atom or molecule followed by annihilation.
3. interaction with a negatron to form positronium followed by annihilation.

In each instance the fate of the positron is annihilation. In the process the mass of two electrons is converted into electromagnetic energy.

$$\Delta E = 2\,m_0c^2 = 1.022 \text{ MeV} \tag{6.32}$$

The total energy of the reaction is generally released as two photons, each carrying away 50% of the energy. Thus the annihilation radiation most frequently observed has an energy of 1.022/2 or 0.511 MeV.

The decay scheme for ^{65}Zn is shown in figure 6.8. Only 1.5% of ^{65}Zn nuclei decay by positron emission. The remaining 98.5% decay by electron capture. The gamma-ray spectrum of ^{65}Zn is shown in figure 6.9. The 0.511 MeV annihilation radiation and the 1.11 MeV gamma ray are easily seen. The relative heights of these two peaks can be explained by considering the branching ratio for ^{65}Zn decay and the energy dependence of the detector efficiency.

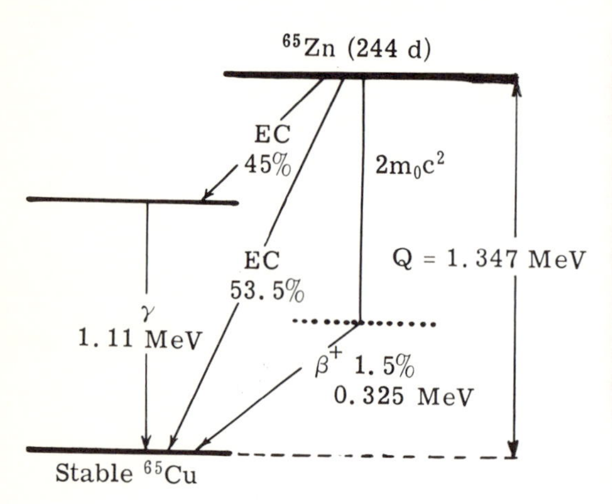

Figure 6.8—DECAY SCHEME OF ^{65}Zn. The vertical line labeled $2m_0c^2$ represents the amount of energy required to create an electron pair.

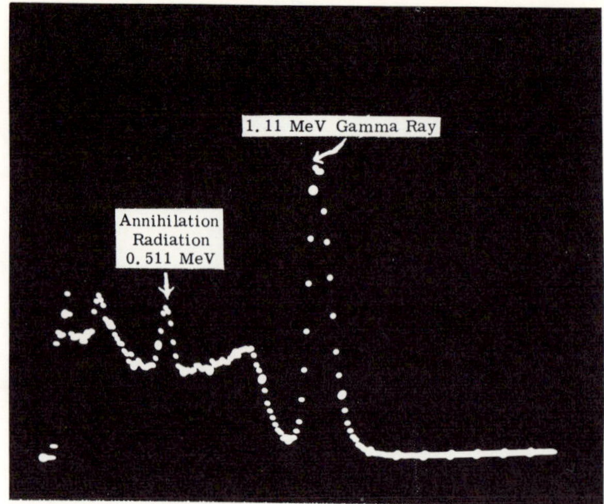

Figure 6.9—GAMMA-RAY SPECTRUM OF ^{65}Zn. Photons representing annihilation radiation are responsible for the peak at 0.511 MeV.

Positronium—If one could replace the proton of hydrogen (i.e., ^1H) with a positron the resulting atom would be positronium, but because the positron and negatron have the same mass, one does not revolve about the other. Instead, they revolve about a common center of mass. Considered as a nuclide one would assign $Z = 0$, $A = 0$ and $N = 0$. Its atomic mass would be $M = 2m_0$. The first evidence for the existence of positronium was presented by Deutsch (6) in 1951.

The binding energy of positronium is only 6.8 eV, less than the ionizing potential of most molecules. However, this means that the combined kinetic energy of the negatron and positron must be 6.8 eV for positronium formation to occur. If the kinetic energy is reduced to less than this value the positron is annihilated in a free collision with an electron (see method #1 above).

The ground state of positronium is defined by the quantum numbers $n = 1$ and $l = 0$. This corresponds to an S state. This ground state is split into two levels, a singlet state ^1S and a triplet state ^3S corresponding to parapositronium and orthopositronium respectively. The triplet state is metastable with an energy only 0.013 eV above the singlet state.

Parapositronium (singlet state, ^1S) results when the spins are antiparallel. The half-life of this form is about 10^{-10} seconds. Parapositronium decays into two gamma rays, each possessing an energy of 0.511 MeV. They are emitted in opposite directions as required by the law of conservation of momentum.

Orthopositronium (triplet state, ^3S) results when the spins are parallel. The half-life of the triplet state is 1.4×10^{-7} seconds. Orthopositronium decays into three photons. Since both linear and angular momentum must be conserved the emission of these three photons is planar and they are emitted at angles of 120° to each other, each carrying away one-third of the energy or 0.341 MeV each.

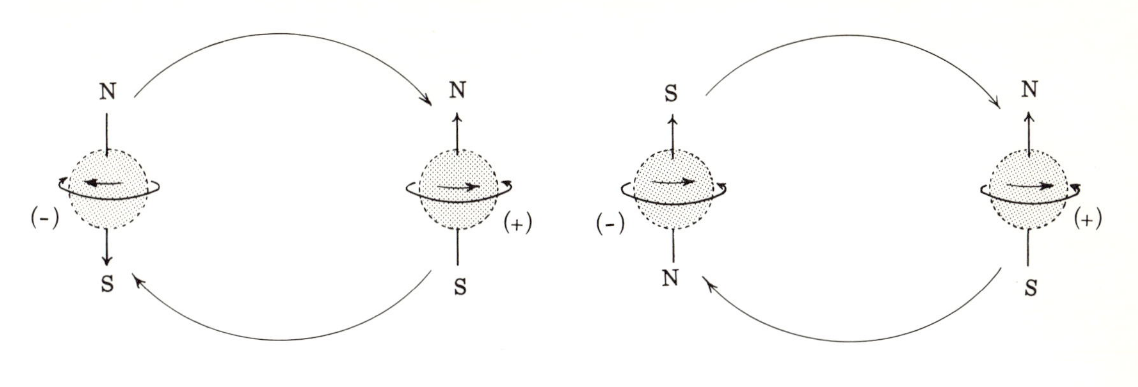

Parapositronium
Singlet state ^1S
Spins are antiparallel
Magnet poles are parallel
Disintegrates into 2 photons
Half-life $= 10^{-10}$ seconds

Orthopositronium
Triplet state ^3S
Spins are parallel
Magnet poles are antiparallel
Disintegrates into 3 photons
Half-life $= 1.4 \times 10^{-7}$ seconds

Figure 6.10—POSITRONIUM.

ELECTRON CAPTURE (EC):

Electron capture is a nuclear transition in which a nucleus captures one of its own orbital electrons and emits a neutrino. Because a K-electron is involved in about 90% of such events, this process is often called "K-capture", but for many EC transitions about 10-12% of electrons captured are L-electrons and about 1% are M-electrons. The reaction for electron capture can be represented by

$$e^- + p \longrightarrow n + \nu \tag{6.33}$$

Thus in electron capture decay Z decreases by 1, N increases by 1 and A remains the same. Electron capture is expressed by the generalized equation

$$^{A}_{Z}X_{N} + {}_{-1}e^{0} \longrightarrow {}^{\;\;A}_{Z-1}Y_{N+1} + \nu \tag{6.34}$$

Electron capture occurs when a nucleus contains an excess of protons relative to its more stable isobars, and while electron capture occurs in every case of β^+ decay as a competitive process, where sufficient energy is unavailable for β^+ decay, electron capture is the only process for converting protons into neutrons. Electron capture and β^+ decay as competitive processes is illustrated by the decay of ^{65}Zn (figure 6.8). Electron capture as the only allowed process is illustrated by the decay of ^{55}Fe (figure 6.11).

Mass-energy relationships—Electron capture energetics are expressed by

$${}_{Z}M' + m_o = {}_{Z-1}M' + m_\nu + T_{M'} + T_\nu + T_\gamma \tag{6.35}$$

Since the neutrino mass m_ν is presumably zero and the recoil energy of the nucleus $T_{M'}$ can usually be neglected, the expression can be simplified to

$$_ZM' + m_o = {}_{Z-1}M' + T_\nu + T_\gamma \tag{6.36}$$

If $Z-1$ electrons are added to each side of equation 6.36 the mass-energy relationship will be expressed in terms of atomic masses

$$_ZM = {}_{Z-1}M + T_\nu + T_\gamma \tag{6.37}$$

and the transition energy Q_{EC} is approximated by

$$Q_{EC} = {}_ZM - {}_{Z-1}M = T_\nu + T_\gamma \tag{6.38}$$

Extranuclear De-excitation—With the loss of an electron in the K, L or M shell, a vacancy exists which is filled by electrons from higher energy levels. This process is accompanied by the emission of x-rays characteristic of the daughter nuclide. For example, the decay of ^{55}Fe results in the emission of the characteristic x-ray of manganese. It is by means of these x-rays that a nuclide decaying by electron capture can be detected.

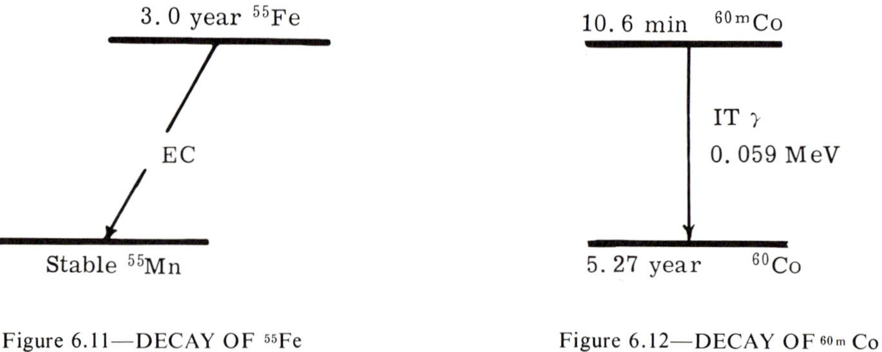

Figure 6.11—DECAY OF ^{55}Fe BY ELECTRON CAPTURE.

Figure 6.12—DECAY OF ^{60m}Co BY ISOMERIC TRANSITION.

Fluorescence and the *Auger Effect* are competing processes for the de-excitation of atoms. *Fluorescence* is the emission of electromagnetic radiation as the result of electronic excitation of an atom by absorption of energy from either corpuscular or electromagnetic radiation. *The fluorescent yield* is the probability that an electronically excited atom will emit electromagnetic radiation rather than Auger electrons in reestablishing the ground state. The energy of the fluorescent radiation usually corresponds to that of the x-ray region of the spectrum.

The *Auger effect* is the de-excitation of an atom by electron emission. An orbital electron excited by a fluorescent photon (usually an x-ray) by an *internal photoelectric effect* is called an *Auger electron* and is ejected from the atom with an energy equal to the difference between the energy of the fluorescent photon and the binding energy of the electron B_e, thus

$$T_a = h\nu - B_e \quad (6.39)$$

The *Auger yield* is the probability of de-excitation by emission of an Auger electron rather than by fluorescence. The *Auger coefficient* is equal to the ratio of the number of Auger electrons to the number of fluorescent photons.

ISOMERIC TRANSITION (IT):

Nuclear Isomerism—Nuclides having the same atomic number Z and the same mass number A but which exist in different quantum states are called *nuclear isomers.* This situation may be compared with that of an excited atom and the processes involved in the extranuclear de-excitation of atoms. Here, nucleons may be considered as occupying excited quantum states; upon the release of energy the nucleons occupy quantum states representing lesser values of potential energy. The *ground state* is the quantum condition representing minimal potential energy. Sometimes a nucleus will exist in a higher energy level *(metastable state)* for a measurable length of time. A small letter m following the mass number is used to indicate the metastable state. For example, when cobalt is irradiated with neutrons ^{60m}Co is produced. The decay of ^{60m}Co to ^{60}Co is illustrated in figure 6.12. While the transition of ^{60m}Co has a half life of 10.6 minutes, the half-life of most transitions is of the order of a microsecond or less. The transition of one nuclear isomer to another is called *isomeric transition.*

Nuclear de-excitation—Decay of a nucleus by either an α- or a β-decay process may result in the formation of the ground state of the daughter nucleus (e.g., the β^- decay of ^{32}P, figure 6.5) but more frequently the result is the formation of an excited state of the daughter nucleus (e.g., ^{198}Au, figure 6.6, ^{59}Fe and ^{60}Co, figure 6.7). De-excitation of a nucleus involves the release of energy through an isomeric transition. This may be accomplished by means of one of the following processes:

1. Gamma-ray emission
2. Internal conversion
3. Pair production

While the first two processes are relatively common, pair production occurs less frequently since it requires a minimum transition energy of 1.02 MeV for the formation of a negatron-positron pair. In those instances when both the initial and final states of a nucleus have zero angular momentum, gamma rays cannot be emitted and only conversion electrons or pairs can be produced.

Internal conversion—De-excitation of a nucleus by emission of an orbital electron through a coupling between the excited nucleus and the electron is known as internal conversion. The ejected electron, called a *conversion electron,* is usually from the K, L or M shell and has an energy equal to the transition energy minus the binding energy of the electron. Conversion electrons therefore have certain discrete values of energy and appear as line spectra. If beta decay precedes internal conversion the line spectra will be seen superimposed on the continuous beta-ray spectrum (see figure 6.13).

The *conversion coefficient (conversion factor* or *conversion ratio)* is the ratio of the number of conversion electrons N_e to the number of quanta N_γ emitted for a given mode of de-excitation.

$$\alpha = \frac{N_e}{N_\gamma} \quad (6.40)$$

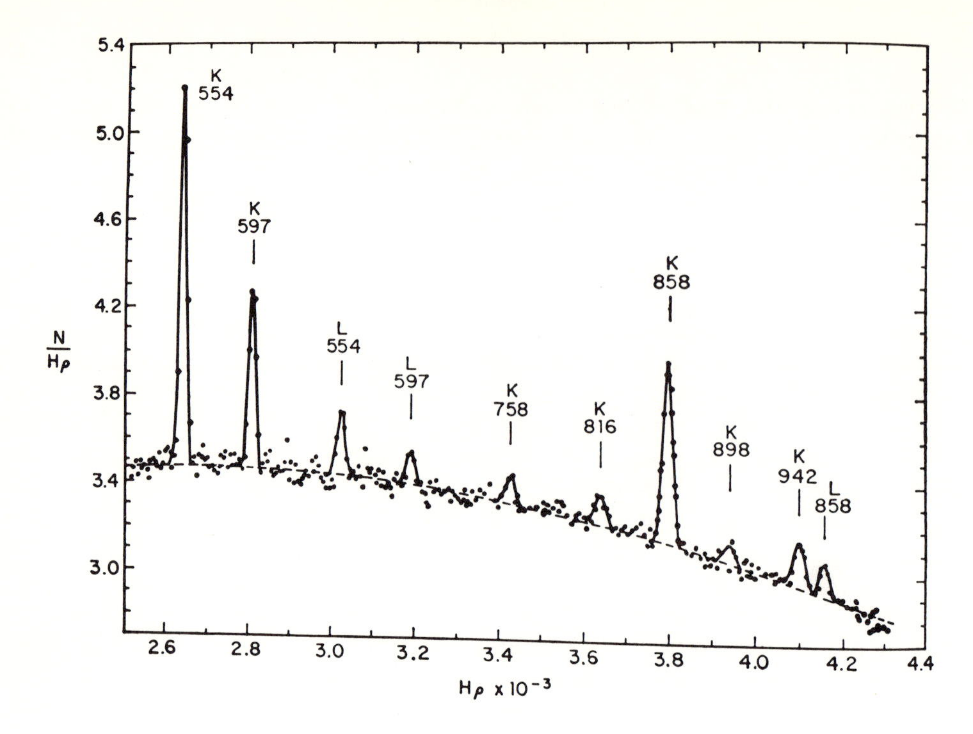

Figure 6.13—BETA SPECTRUM SHOWING CONVERSION ELECTRON LINES. [*From Physical Reviews* 116, 143 (1959) *with permission of American Institute of Physics, Inc., and M. E. Bunker, et al.*]

Partial internal conversion coefficients α_K, α_L, and α_M represent the fraction of internal conversion electrons arising from the K, L and M shells respectively. The total conversion coefficient is equal to the sum of the partial conversion coefficients. That is $\alpha = \alpha_K + \alpha_L + \alpha_M$.

DECAY SCHEMES:

We have seen in the preceding paragraphs that radioactive decay can occur by any one of, or a combination of, several processes. For many nuclides the decay scheme can be considered simple because radioactive decay is by a single type of decay process. The electron capture decay of ^{55}Fe is a typical example, as well as the many "pure beta emitters"*, so-called because no gamma ray is emitted by the nucleus following beta decay. Pure beta emitters include many commonly used nuclides—^{3}H, ^{14}C, ^{32}P, ^{35}S, ^{45}Ca, ^{90}Sr and ^{90}Y. All atoms of each of these particular nuclides decay by the same route.

The decay of some nuclei is complex because decay occurs by more than a single route. This is true for ^{198}Au (figure 5.4) and for ^{65}Zn (figure 5.5). Fortunately the

*The term "pure beta emitter" is misleading if one considers the gross sample because the interaction of beta particles with matter (see chapter 7), including the radioactive sample itself, will result in the formation of some electromagnetic radiation as well.

probability of decay by a particular route for a given nuclide is constant. Thus, for ^{65}Zn the probability of decay by beta emission is 0.015, and hence it is possible to predict that, on the average, a beta particle will be produced by 1.5% of the disintegrations. Iodine – 131 is an example of a nuclide having an especially complex decay scheme (see figure 6.7).

During the decay of ^{131}I, beta particles representing four distinct energy transitions are emitted. The most important of these transitions is that which occurs in 87.2% of the disintegrations and results in a maximum beta energy $E_m = 0.608$ MeV. The most prominent gamma ray energy is 0.364 MeV but at least five other gamma ray energies appear in the spectrum.

The unit of radioactive decay, the curie, is defined as 3.7×10^{10} disintegrating atoms per second. But nuclear detection equipment does not detect disintegrating atoms directly. Rather, detection equipment is used to measure the number of beta particles, gamma rays, or other radiation emitted by the nuclide. In addition, detectors are especially sensitive to only one type of radiation. Thus, if one is to relate the number of gamma rays detected to the number of disintegrating atoms, one must not only consider the geometry, instrument efficiency and the other criteria of radiation measurement, but the decay scheme as well.

Consider the situation where the same quantity (i.e., the same number of microcuries) of three different nuclides is measured by means of a gamma sensitive detector (figure 6.14). Without some knowledge of the decay schemes, the experimenter would be led to believe that there was no activity in the first sample (^{32}P) and that the activity of the second sample (^{59}Fe) was only half that of the third (^{60}Co) when in reality the same amount of each nuclide was present. It can be seen that while each will emit the same number of beta particles, the number of gamma rays emitted by each will be considerably different.

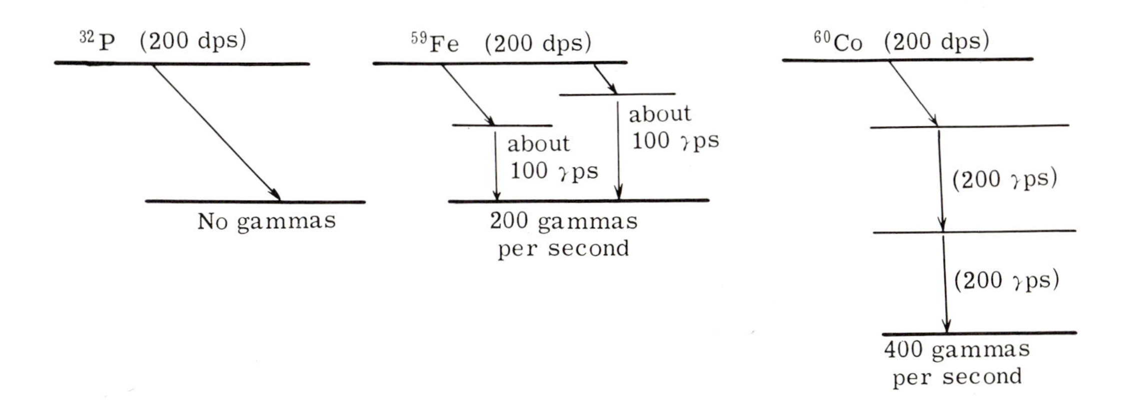

Figure 6.14—EFFECT OF DECAY MODE ON MEASURED RADIATION. If each nuclide has an activity of 200 dps, each will emit 200 beta particles per second, but the ^{32}P will emit no gamma rays while ^{59}Fe will emit 200 per second and ^{60}Co will emit 400 per second.

If the curie quantity of a radioisotope is to be related to an observed activity, the importance of the decay scheme becomes apparent in the calibration of radioactive standards, in absolute counting and in dosage calculation.

RATE OF RADIOACTIVE DECAY:

There is no way in which we can influence the decay of radioactive atoms. We can neither retard their rate of decay by refrigeration or any other means nor accelerate their rate of decay by the use of heat, pressure, etc. Thus, the amount of radioactivity or the number of radioactive atoms which we have on hand at any time is undergoing a consistent, continuous change.

Fortunately this change in the number of radioactive atoms follows a very orderly process. If we know the number of atoms on hand and their decay constant, we can calculate exactly how many atoms will remain at any future time. It is the purpose of the experiments in this chapter to illustrate how the necessary calculations can be made.

EXPERIMENT 6.1 HALF-LIFE

OBJECTIVES:

To illustrate a method for determining the half-life of a radioactive material.

To demonstrate the method of relativistic correction for instrument efficiency (see experiment 5.3).

DISCUSSION:

The decay of all species of radioactive atoms is a first order reaction in that the rate of decay is proportional to the number of radioactive atoms present. This statement is expressed by the equation

$$-dN/dt = \lambda N \tag{6.41}$$

where dN/dt is the disintegration rate of the radioactive atoms, λ is the *decay constant* and N is the number of radioactive atoms present at time t. The minus sign is indicative of a decrease in the total number of atoms with time. The decay rate, $-dN/dt$, is an average value and is subject to fluctuation in accordance with Poisson statistics. Multiplication of equation 6.41 by dt/N gives

$$-dN/N = \lambda dt \tag{6.42}$$

Integration of equation 6.42 between the limits of N_1 and N_2, and t_1 and t_2, yields the working equation

$$\ln(N_1/N_2) = \lambda(t_2 - t_1) \tag{6.43}$$

where N_1 is the number of atoms present at time t_1, and N_2 is the number of atoms present at time t_2.

It is often more convenient to use the equation obtained by integrating between the limits of N_0 and N, and t_0 and t where $t_0 = 0$, thus

$$-\int_{N_0}^{N} \frac{dN}{N} = \int_{t_0}^{t} \lambda\, dt \qquad (6.44)$$

and

$$\boxed{\ln \frac{N_0}{N} = \lambda t} \qquad (6.45)$$

where N_0 = original number of atoms present
N = number of atoms remaining at time t.

Expressed in exponential form

$$\boxed{N = N_0 e^{-\lambda t}} \qquad (6.46)$$

where e is the base of natural logarithms. The term $e^{-\lambda t}$ is called the *decay factor.* Values for the decay factor are found in Appendix F where x is equal to the product λt. Calculations of decay factors are also made with ease by use of the log scales of a slide rule.* The unit of the decay constant λ depends upon the unit of time. For example, if t is in hours, λ is in hrs^{-1} and if t is in years then λ must be in yrs^{-1}.

The absolute activity $A (= -dN/dt)$ is proportional to N, the number of radioactive atoms present. It is also true that the relative activity R will be proportional to the number of radioactive atoms present in the sample if constant geometry is maintained for all measurements. Therefore, equation 6.46 can be modified to give

$$\frac{N}{N_0} = \frac{A}{A_0} = \frac{R}{R_0} = e^{-\lambda t} \qquad (6.47)$$

and it follows that

$$A = A_0 e^{-\lambda t} \quad \text{and} \quad R = R_0 e^{-\lambda t} \qquad (6.48)$$

Applying the same proportionality to equation 6.45 gives

$$\ln \frac{N_0}{N} = \ln \frac{A_0}{A} = \ln \frac{R_0}{R} = \lambda t \qquad (6.49)$$

The units of A may be curies, millicuries, disintegrations per second (dps) or disintegrations per minute (dpm) while the units of R may be counts per second (cps) or counts per minute (cpm).

* Set the hairline at 0.500 on the LLOO scale. Locate the half-life on the B scale; set this point under the hairline also. Now read the decay factor on the LLOO scale opposite elapsed time on the B scale. (Also see references 18-22.)

If the number of radioactive atoms remaining at any given instant is plotted against time, a curve similar to that illustrated in figure 6.15 will be obtained. If the logarithm of the number of radioactive atoms is plotted as a function of time a straight line similar to that illustrated in figure 6.16 results.

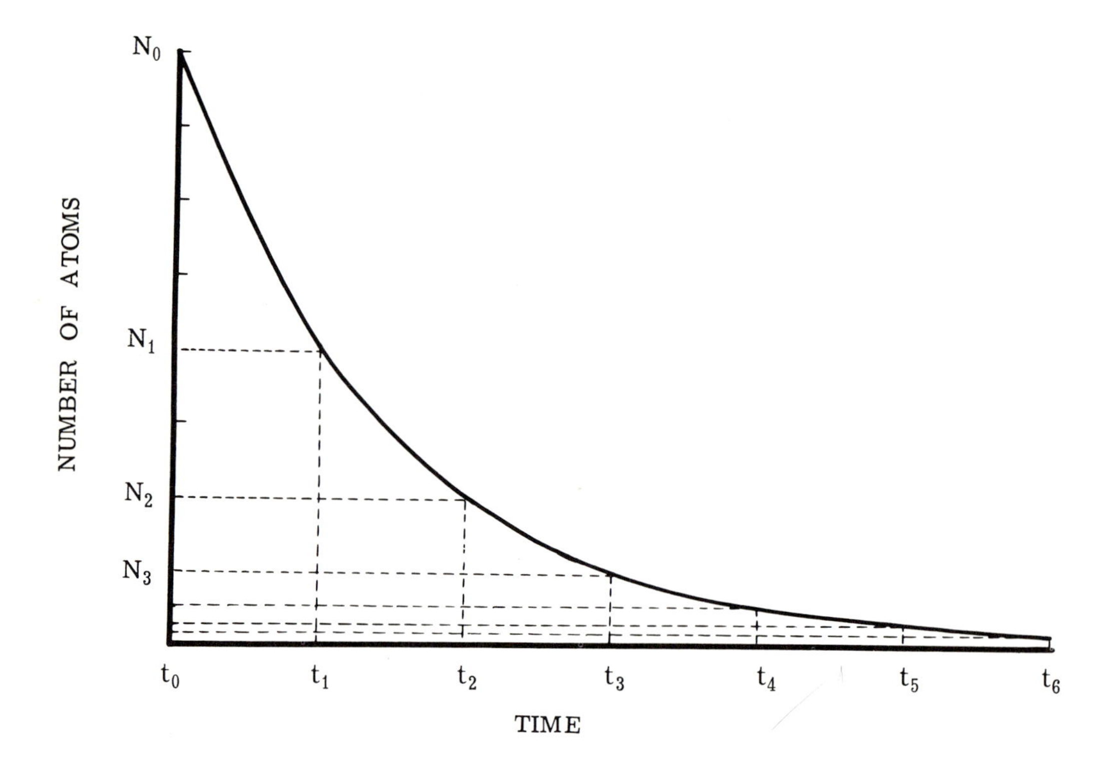

Figure 6.15—GENERALIZED DECAY CURVE.

Half-life—The time required for the disintegration of one-half the atoms in a radioactive sample is called the half-life. In figure 6.15 it can be seen that at time t_0 there are N_0 atoms present, and at time t_1 only N_1 atoms remain. But $N_1 = N_0/2$; therefore, $t_1 - t_0 = t_{1/2}$, the half-life. It is also seen that $N_2 = N_1/2$ and that the half-life is also equal to $t_2 - t_1$. Substituting these values into equation we obtain

$$\ln \frac{N_1}{N_1/2} = \lambda (t_2 - t_1) = \lambda t_{1/2}$$

$$\ln 2 = \lambda t_{1/2} = 2.30 \log 2$$

$$\boxed{t_{1/2} = \frac{0.693}{\lambda}} \qquad (6.50)$$

If either the half-life or the decay constant for a particular radioactive species is known, the other value can be calculated by use of equation 6.50.

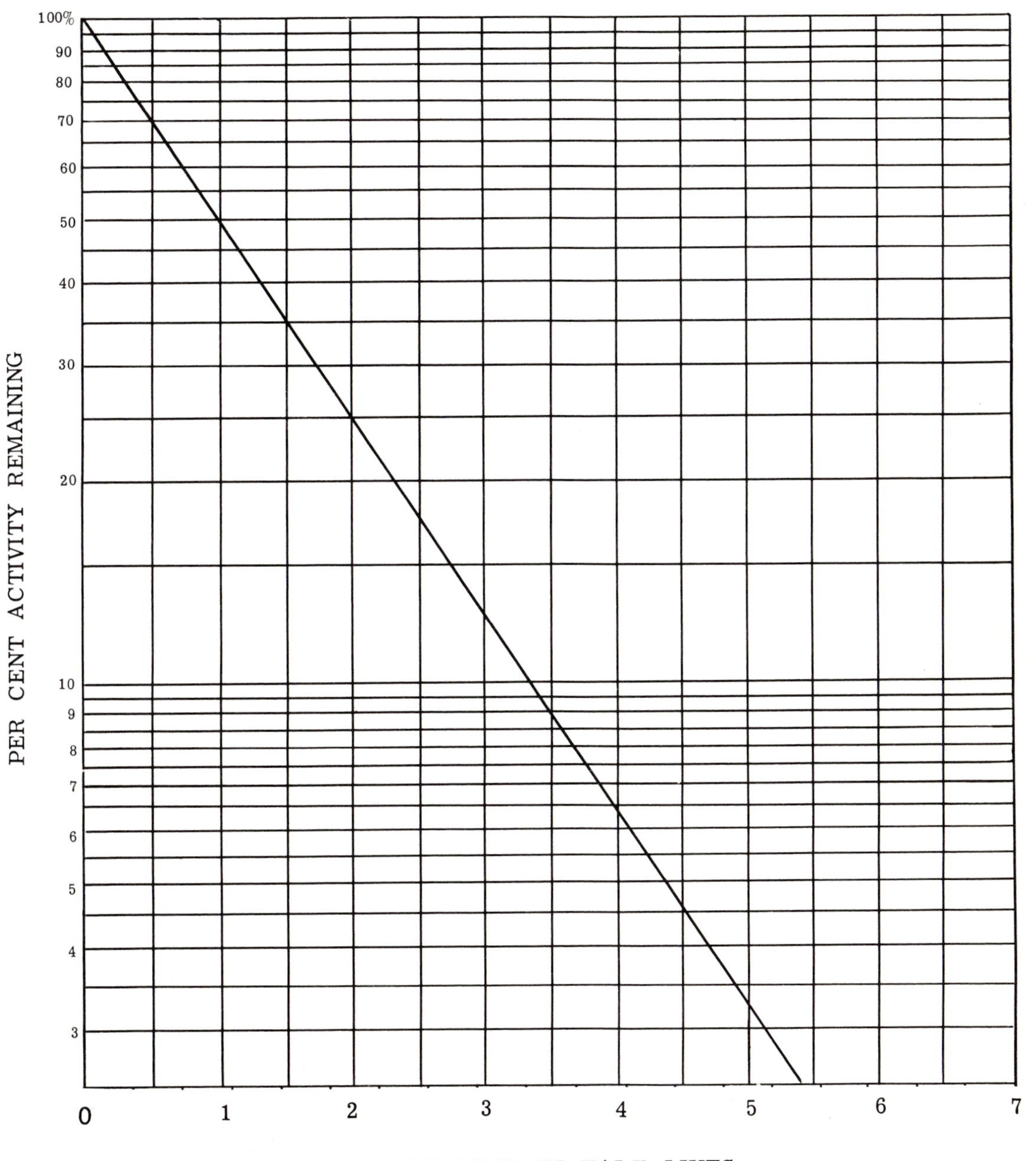

Figure 6.16—GENERALIZED DECAY CURVE. Compute the elapsed time in half lives by dividing the elapsed time by the half-life of the isotope. The per cent activity remaining in the sample can then be determined from the curve.

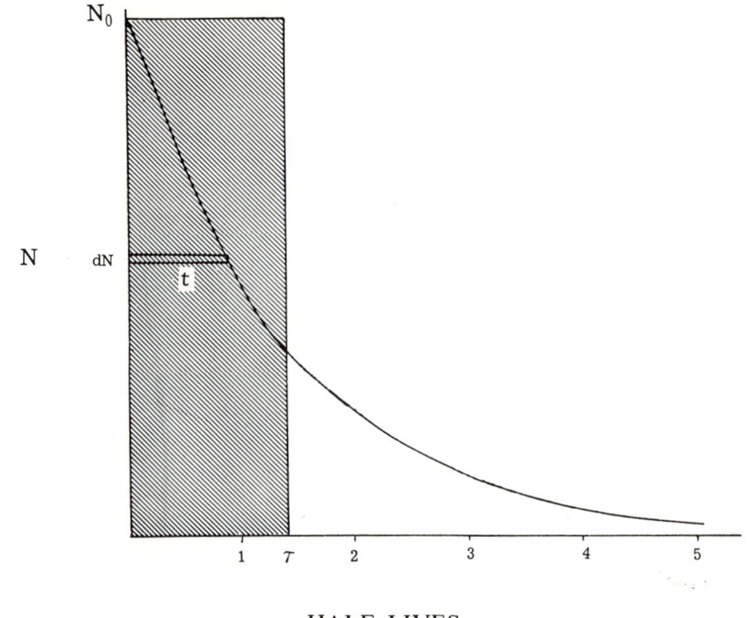

Figure 6.17—AVERAGE LIFE. If all radioactive atoms had the same lifetime, none would decay until $t = \tau$ at which time all would decay simultaneously. The areas under the two curves are equal. $\tau = 1.41\ t_{\frac{1}{2}}$

Average life—If from a given time t_0 we were to measure, individually, the times $t_1, t_2, \ldots . t_i$ required for each atom to decay, then the sum of these times divided by the number of atoms present at time t_0 would give the *mean life* or *average life* τ.

$$\tau = -\frac{1}{N_0}\int_{N_0}^{0} t\ dN \tag{6.51}$$

and since $-dN = \lambda N\,dt$ and $N = N_0 e^{-\lambda t}$ (6.52)

$$\tau = \frac{1}{N_0}\int_0^{\infty} \lambda\ t\ N_0\ e^{-\lambda t}\,dt = \lambda\int_0^{\infty} t e^{-\lambda t}\ dt \tag{6.53}$$

which upon integration gives

$$\boxed{\tau = \frac{1}{\lambda} = \frac{t_{1/2}}{0.693} = 1.44\ t_{1/2}} \tag{6.54}$$

Example 1. A sample of $Na_2H^{32}PO_4$ had an activity of 1,000 cpm at "zero time." Calculate the activity one should anticipate 10 days later. The half-life of ^{32}P is 14.3 days.

$$R = R_0 e^{-\lambda t} = 1{,}000\ e^{-0.693(10)/14.3} = 606 \text{ cpm}$$

Example 2. How many atoms of ^{32}P are required to produce 1 mc of activity?

$$A = -dN/dt = \lambda N$$

The activity A = 1 mc = 3.7×10^7 dps (i.e., 3.7×10^7 atoms sec^{-1}) and $\lambda = 0.693/t_{1/2}$. Here the half-life must be expressed in seconds, thus

$$N = (3.7 \times 10^7 \text{ atoms sec}^{-1})(1.235 \times 10^6 \text{ sec}) / 0.693$$
$$= 6.58 \times 10^{13} \text{ atoms}$$

Measurement of Half-Life—The method of choice for measuring the half-life of a radioactive nuclide will depend upon the magnitude of the half-life to be measured. For long half-lived nuclides one might calculate the half-life from measured values of specific activity and the isotopic abundance, the latter being determined by means of mass spectrometry. For the measurement of a very short half-life a coincidence technique might be used. Where the half-life lies within a range of from about a minute to about a year the most obvious approach is a direct observation of the decay of the activity of the sample.

When observations of decay must be made over a period of a few hours or more changes in instrument efficiency must be considered. To compensate for changes in instrument efficiency activity measurements on the sample are compared with the activity of a long-lived standard. Use of a long-lived standard eliminates the need for maintenance of instrument precision for more than about an hour at a time. The use of such standards is discussed in experiment 5.3.

Calculations for Short-Lived Nuclides—When a very short half-life is measured, especially when the activity is also low, it is necessary to count the samples for a time comparable to the half-life. The decrease in activity occuring during the time required for a single measurement may introduce a significant error if neglected. One solution to the problem, based on a series equation, was suggested by Schular (26). Another approach, proposed by Wagner (27), is more direct and is relatively simple to apply to practice. Wagner's method follows:

By combining equations 6.41 and 6.46 one obtains

$$-dN/dt = \lambda N_0 e^{-\lambda t} \tag{6.55}$$

Let n_1 = the number of decay events observed during the first of several equal time intervals. Then

$$n_1 = \lambda N_0 \int_0^{t_1} e^{-\lambda t} dt = N_0(1 - e^{-\lambda t_1}) \tag{6.56}$$

and for the next successive equal period

$$n_2 = \lambda N_0 \int_{t_2}^{t_3} e^{-\lambda t} dt = N_0(e^{-\lambda t_2} - e^{-\lambda t_3})$$
$$= N_0 e^{-\lambda t_2}(1 - e^{-\lambda(t_3 - t_2)}) \tag{6.57}$$

Since all measurements are made for equal periods of time, $t_3 - t_2 = t_1$ and

$$n_2 = N_0 e^{-\lambda t_2}(1 - e^{-\lambda t_1}) \tag{6.58}$$

Equation 6.56 is divided by equation 6.58 to obtain

$$n_1 / n_2 = e^{\lambda t_2} \tag{6.59}$$

The half-life is then expressed by

$$t_{1/2} = \frac{0.693\ t_2}{\ln (n_1/n_2)} \tag{6.60}$$

Converting to the form of an equation for a straight line one obtains

$$\ln n_2 = \ln n_1 - \frac{0.693}{t_{1/2}} t_2 \tag{6.61}$$

The slope is $-0.693/t_{1/2}$ (i.e., $-\lambda$). Converted to base 10 logarithms for convenience and generalized to permit checking the half-life by any number of observations instead of only two yields

$$\boxed{\log n_i = \log n_1 \quad -\lambda t_i/2.30} \tag{6.62}$$

Here the slope is $-\lambda/2.30$

APPARATUS AND REAGENTS:

Solution of a radioactive sample with "unknown" half-life (e.g., ^{32}P, ^{131}I); long-lived radioactive reference source (e.g., Ra-DEF, ^{90}Sr); Geiger tube and scaler; planchets; micropipettes.

PROCEDURE:

1. A quantity of the solution of sample with "unknown" half-life is transferred to a planchet and dried by means of an infrared lamp. The activity of this sample should be 5,000-10,000 cpm.
2. The following measurements are made:
   a. Background
   b. Sample activity
   c. Standard activity

All measurements should be made using reproducible geometry. The time devoted to each measurement should be consistent with observed activities, statistical accuracy and the half-life of the sample.

3. Step 2 is repeated as frequently as necessary to provide ample data, well distributed in time, to permit an accurate construction of the decay curve.

DATA:

		Activity of Standard		Activity of Sample		
Elapsed Time	Background cpm	Observed cpm	Corrected cpm	Observed cpm	Corrected cpm	Normalized cpm

CALCULATIONS:

1. All observed activities should be corrected for resolving time and background count. If the resolving time is not known, 0.5% per 1,000 cpm can be used as an estimate of the concidence correction if a Geiger tube was used for measuring the activities.

2. The sample activity must be corrected for changes in instrument efficiency by making use of the standard activity to normalize all data. A relative efficiency of unity is assumed for zero time. Either the observed value of activity is adjusted for changes in efficiency to obtain the normalized value or, alternatively, one can simply calculate the ratio between the sample and standard activities and plot the log of the ratio vs. time in step 3.

3. Using semi-log paper the normalized sample activities (or sample/standard ratios) are plotted as a function of time. From equation 6.43 it can be seen that

$$\ln (R_1 / R_2) = \frac{\lambda}{2.30} (t_2 - t_1)$$

and that the function in log R is a straight line. Thus the best *straight* line should be drawn through the experimental points.

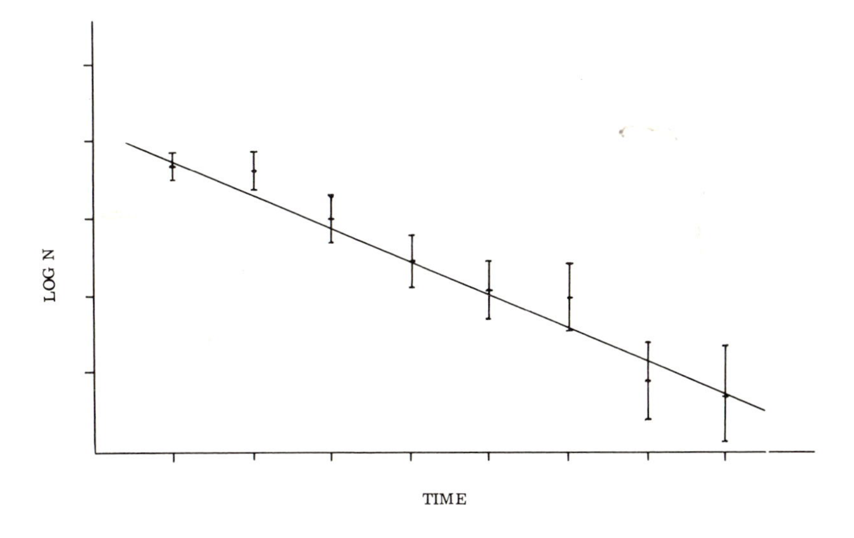

Figure 6.18—SEMI-LOG PLOT OF DECAY CURVE. Vertical I-shaped lines indicate the standard error of the measurement.

4. From the slope, $\lambda/2.30$, the value of λ is determined from which the half-life can then be calculated by use of equation 6.50. The measured values of the constants are then compared with those reported in the literature.

5. For each of the activities plotted on the decay curve, calculate the standard deviation. Through each point plotted on the decay curve, draw a short vertical line. On this vertical line mark off, with small horizontal lines, distances corresponding to the calculated standard deviation (see figure 6.18). In how many cases is the observed error greater than the standard error?

	Measured Value	Literature Value
Decay constant		
Half-life		

Suggestions for further work:

If a neutron source is available, isotopes with half-lives of a few seconds (e.g., 24 second silver) to an hour (54 minute indium) or several hours (2.6 hour manganese or 2.3 hour dysprosium) can be prepared for half-life measurements. It is interesting to follow the decay of indium with a ratemeter and recorder. These isotopes offer the advantage that only a single class period is required for completion of the experiment. A standard reference source for efficiency correction need not be used.

Other short-lived nuclides are available by use of chemical separation techniques to separate them from their longer-lived parents:

Yttrium-90—Salutsky and Kirby (29) describe a method for the preparation of carrier-free yttrium-90 solutions. Yttrium-90 remains in solution when the strontium-90 parent is precipitated as strontium nitrate in 80% nitric acid. Yttrium-90 emits an energetic beta particle and has a half-life of 64 hours.

Protoactinium-234—If a very short half-life is desired, 1.18 minute protoactinium-234 is readily separated from uranyl nitrate by ion exchange. According to the method of Braunstein and Young (31), 100 ml of a 10% uranyl nitrate solution is passed through an ion exchange column containing Amberlite IR-120 (a cation exchange resin). The column is washed consecutively with 100 ml of water, 200 ml of 0.3 *N* H_2SO_4 and 100 ml of water. Uranyl ion is eluted leaving thorium-234 on the column. Then by use of 5% HCl, ^{234}Pa is eluted from the column while the parent ^{234}Th remains behind on the column. As ingrowth of the protoactinium will occur in a very short time, the column can be "milked" about every five minutes to obtain a fresh quantity of ^{234}Pa.

Bismuth-210—This nuclide is separated from lead-210 by deposition on a nickel foil. In the technique described by Smith and Wood (32) the radiolead is added to 0.5 *N* HCl. Most of the lead collects on the bottom of the container as lead chloride. When a nickel foil is immersed in a small quantity of the supernatant, ^{210}Bi plates out as the nickel displaces it.

EXPERIMENT 6.2 MIXTURE OF INDEPENDENTLY DECAYING ACTIVITIES

OBJECTIVES:

To study the change in activity of a sample consisting of two independently decaying radioisotopes.

To illustrate a method for resolving the decay curve for the mixture into two simple decay curves.

THEORY:

If a sample consists of a mixture of independently decaying radioisotopes (such as phosphorus-32 and sodium-24), the presence of one will have no effect on the decay rate of the other, and the total activity of the sample will merely be the sum of the individual activities. Thus, the total activity A of the mixture is given by the equation

$$A = A_1 + A_2 + \ldots\ldots = -dN_1/dt - dN_2/dt - \ldots\ldots \quad (6.63)$$

$$A = \lambda_1 N_1 + \lambda_2 N_2 + \ldots\ldots \quad (6.64)$$

and since $N = N^0 e^{-\lambda t}$

it follows that $A = \lambda_1 N_1^0 e^{-\lambda_1 t} + \lambda_2 N_2^0 e^{-\lambda_2 t} + \ldots\ldots \quad (6.65)$

For a simple source consisting of a single species of nuclide, the logarithm of the activity plotted against time results in a straight line plot. This is illustrated by the dotted line (1) of figure 6.19.

A similar plot for the decay of a single species of nuclide but of longer half-life than depicted by curve (1) is illustrated by curve (2) of figure 6.19.

If these two nuclides are mixed together, they will still decay as they did separately, but the observed activity of the mixture (the sum of the two activities) is given by the solid curved line in figure 6.19.

If constant geometry is maintained, the observed activity R will be proportional to the absolute activity A and the observed activity of a mixture of independently decaying nuclides will be

$$R = R_1^0 e^{-\lambda_1 t} + R_2^0 e^{-\lambda_2 t} + \ldots\ldots\ldots \quad (6.66)$$

where R_1^0 and R_2^0 are the zero time activities of the two components measured independently. When only two components are present, Perkel (36) suggests, for $\lambda_1 > \lambda_2$ that equation 6.66 be multiplied through by $e^{\lambda_2 t}$ to give

$$R\, e^{\lambda_2 t} = R_1^0 e^{(\lambda_2 - \lambda_1)t} + R_2^0 \quad (6.67)$$

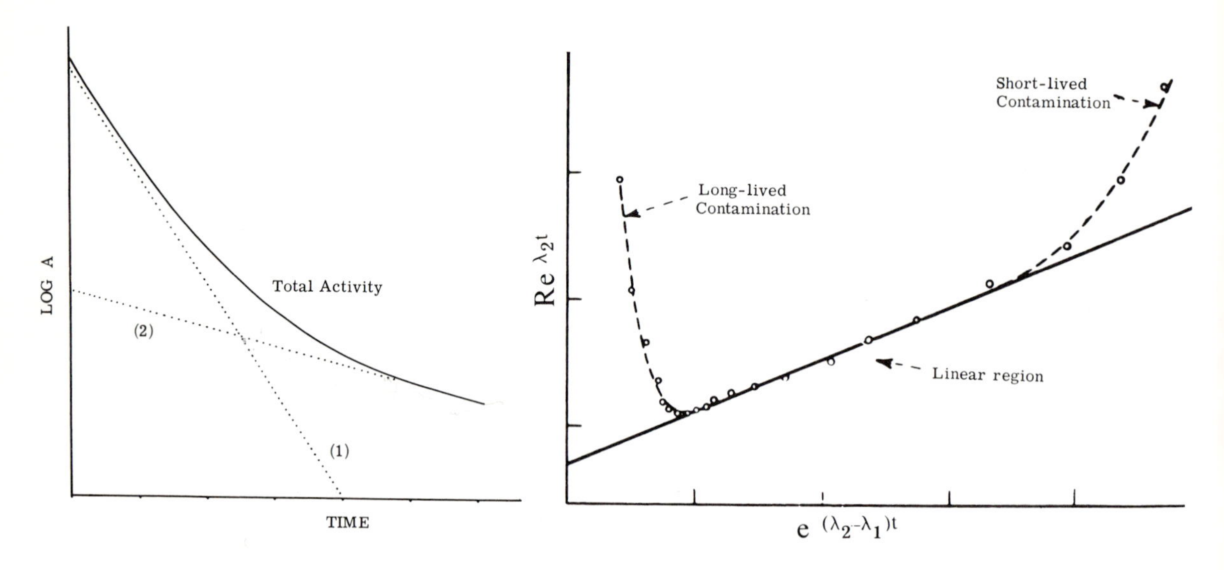

Figure 6.19—DECAY OF A MIXTURE OF INDEPENDENTLY DECAYING RADIOISOTOPES.

Figure 6.20—TYPICAL TWO-COMPONENT LINEAR PLOT. A plot according to Perkel (36) showing the effect of contamination. (*From Nucleonics* 15, *No.* 6, *June* 1957.)

which is of the form $y = mx + b$ where $y \equiv Re^{\lambda_2 t}$ and $x \equiv e^{(\lambda_2 - \lambda_1)t}$. When $Re^{\lambda_2 t}$ is plotted vs $e^{(\lambda_2 - \lambda_1)t}$ on linear graph paper a straight line should be obtained with a y-intercept equal to R_2^0. A typical plot is shown in figure 6.20.

APPARATUS AND REAGENTS:

Solutions of two radioactive elements, preferably with half lives differing by at least a factor of four or five, e.g. ^{198}Au ($t_{1/2}$ = 2.7 days) and ^{32}P ($t_{1/2}$ = 14.2 days); radioactive reference standard (Ra-DEF); planchets, micropipettes. ^{24}Na and ^{131}I also provide a suitable combination. Selection will depend upon the availability of the particular isotopes.

If a neutron source is available, an activated source may be used. The authors have used prepared samples for their classes consisting of mixtures of two elements and activated in a neutron howitzer. For a short-lived mixture silver and indium mixtures have been used. For a longer lived combination indium and samarium serve well.

PROCEDURE:

1. Into a planchet measure a sufficient quantity of one of the radioactive solutions (e.g., ^{198}Au) to give an activity of about 5,000 cpm. To the same planchet add a sufficient quantity of the second radioactive solution (e.g. ^{32}P) to increase the activity by approximately 5,000 cpm, thus giving a total activity of about 10,000 cpm.

If a volatile substance such as ^{131}I is used, be sure to "fix" the sample in the planchet to prevent mechanical loss through evaporation or sublimation. Dry the combined sample under a heating lamp.

2. The following measurements are made:
   a. background
   b. sample activity
   c. activity of standard source, if used

All activities should be measured using reproducible geometry. The time allocated for each measurement should be consistent with the sample activity, statistical accuracy and the half-lives of the nuclides being measured.

3. Step 2 should be repeated as frequently as necessary to permit an accurate analysis of the data.

DATA:

		Activity of Standard		Activity of Sample		
Elapsed Time	Background cpm	Observed cpm	Corrected cpm	Observed cpm	Corrected cpm	Normalized cpm

CALCULATIONS:

1. Correct all readings for resolving time. If resolving time is not known, add 0.5% per 1,000 cpm.

2. Subtract background from both standard reference and sample readings. This should be done after the coincidence correction has been made.

3. Correct the sample activity for instrument efficiency by using the activity of the standard sample to normalize to efficiency at "zero" time.

4. Using semi-log paper, plot the net corrected activity of the sample vs time. Observations must be made over a sufficient length of time to allow considerable decay of the short-lived component. The lower right hand portion of the curve will then be linear.

5. Extrapolate the linear, lower, right-hand portion of the graph to the left until it intercepts the ordinate (zero time). This is dotted line (2) in figure 6.19 and represents the contribution of the long-lived isotope to the total activity.

6. Subtract the activity of the long-lived component from the total activity to obtain the activity of the short-lived component. Do this at five or six points along the time axis thus providing a number of points which are connected by the best straight line to give line (1) in figure 6.18.

7. Analyze the data according to the method of Perkel.

EXPERIMENT 6.3 RADIOACTIVE EQUILIBRIA

OBJECTIVES:

To demonstrate various types of radioactive equilibria.

To plot a growth decay curve illustrating the formation of a radioactive daughter element and the establishment of a state of secular equilibrium between the parent and daughter nuclides.

THEORY:

The decay of many a radioactive nucleus results in the formation of a daughter isotope which is itself radioactive. The radioactive daughter in turn undergoes decay, thus becoming a parent, to produce a new daughter atom. A general equation for a series of decays of this type is

$$A \xrightarrow{\lambda_1} B \xrightarrow{\lambda_2} C$$

where A decays to B which in turn undergoes decay to form C. λ_1 and λ_2 are the decay constants for the transitions of A to B and of B to C, respectively.

Let N_1 and N_2 represent the number of atoms of A and B, respectively, present at any time t, and let N_1^0 and N_2^0 be the number of atoms of A and B present at zero time.
Then

$$N_1 = N_1^0 e^{-\lambda_1 t} \quad (6.68)$$

$$dN_1/dt = -\lambda_1 N_1 \quad (6.69)$$

$$dN_2/dt = \lambda_1 N_1 - \lambda_2 N_2 \quad (6.70)$$

Substitution of equation 6.68 into 6.70 then gives

$$dN_2/dt = \lambda_1 N_1^0 e^{-\lambda_1 t} - \lambda_2 N_2 \quad (6.71)$$

The solution of this linear differential equation for N_2 as a function of time is

$$N_2 = \frac{\lambda_1}{\lambda_2 - \lambda_1} N_1^0 (e^{-\lambda_1 t} - e^{-\lambda_2 t}) + N_2^0 e^{-\lambda_2 t} \quad (6.72)$$

For the special case where only atoms of A are present initially, i.e., $N_2^0 = 0$ at $t = 0$, the last term of equation 6.72 is equal to zero and

$$N_2 = \frac{\lambda_1}{\lambda_2 - \lambda_1} N_1^0 \left(e^{-\lambda_1 t} - e^{-\lambda_2 t}\right) \tag{6.73}$$

Secular equilibrium is a limiting case of radioactive equilibrium in which the half-life of the parent is many times greater than the half-life of the daughter. That is, $\lambda_1 <<< \lambda_2$. The difference between the half-lives of the parent and daughter is usually a factor of 10^4 or better, so that the activity of the parent shows no appreciable change during many half-life periods of the daughter. The decay of radium -226 to radon -222 is a typical example.

$$^{226}Ra \xrightarrow[(t_{1/2} = 1620 \text{ years})]{\lambda_1 = 1.3 \times 10^{-11} \sec^{-1}} {}^{222}Rn \xrightarrow[(t_{1/2} = 3.8 \text{ days})]{\lambda_2 = 2.1 \times 10^{-6} \sec^{-1}} {}^{218}Po$$

For this secular equilibrium between radium and radon, equation 6.73 can be simplified still further because λ_1 is negligible compared to λ_2. Also, after a period of time t equal to many half-lives of radon, the product of t and λ_2 becomes very great and $e^{-\lambda_2 t}$ approaches zero. Hence,

$$N_2 = \frac{\lambda_1}{\lambda_2} N_1^0 e^{-\lambda_1 t} \tag{6.74}$$

and substitution of equation 6.68 into equation 6.74 gives

$$N_1\lambda_1 = N_2\lambda_2 \tag{6.75}$$

Equation 6.75 illustrates the fact that the relative numbers of atoms of parent and daughter are inversely proportional to their decay constants. Another important concept of secular equilibrium is that the daughter has an apparent half-life equal to that of the parent. This is especially important, for example, in the case of radium-DEF where the

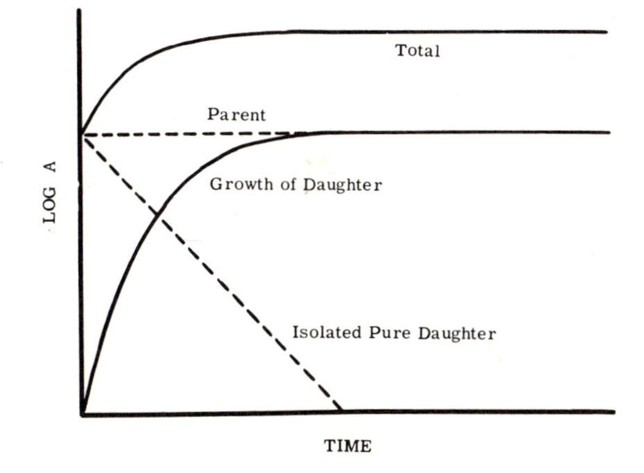

Figure 6.21—SECULAR EQUILIBRIUM.

energetic betas of Ra-E are used for standardization, yet the short-lived Ra-E has an apparent half-life of 22 years, the same as the parent Ra-D.

$$\text{Ra-D} \xrightarrow[0.025 \text{ MeV beta}]{t_{1/2} = 22 \text{ y}} \text{Ra-E} \xrightarrow[1.17 \text{ MeV beta}]{t_{1/2} = 5.0 \text{ d}} \text{Ra-F}$$

Radioactive fallout has attached considerable importance to strontium-90. Actually more damage is attributable to the daughter yttrium -90 which produces much more energetic beta particles

$$^{90}\text{Sr} \xrightarrow[0.61 \text{ MeV beta}]{t_{1/2} = 28 \text{ y}} {}^{90}\text{Y} \xrightarrow[2.18 \text{ MeV beta}]{t_{1/2} = 64 \text{ h}} {}^{90}\text{Zr (stable)}$$

Transient equilibrium is similar to secular equilibrium in that the half-life of the parent is greater than the half-life of the daughter but differs from secular equilibrium in that the half-lives differ only by a small factor (about 10×) rather than a large factor (10^4 or greater). That is, $\lambda_1 < \lambda_2$.

As t becomes very large, $e^{-\lambda_2 t}$ becomes negligible compared to $e^{-\lambda_1 t}$ and the term $e^{-\lambda_2 t}$ approaches zero. Accordingly, equation 6.73 simplifies to

$$N_2 = \frac{\lambda_1}{\lambda_2 - \lambda_1} N_1^0 e^{-\lambda_1 t} \tag{6.76}$$

and substitution of equation 6.68 into equation 6.76 gives

$$N_1\lambda_1 = N_2(\lambda_2 - \lambda_1) \tag{6.77}$$

As in the case of secular equilibrium, when equilibrium is established, both parent and daughter activities decrease at equal rates, the rate of decrease being dependent on the half-life of the parent. However, because the half-life of the parent is much shorter in the case of transient equilibrium, there is a pronounced decrease in the total activity with time. This is illustrated in figure 6.22.

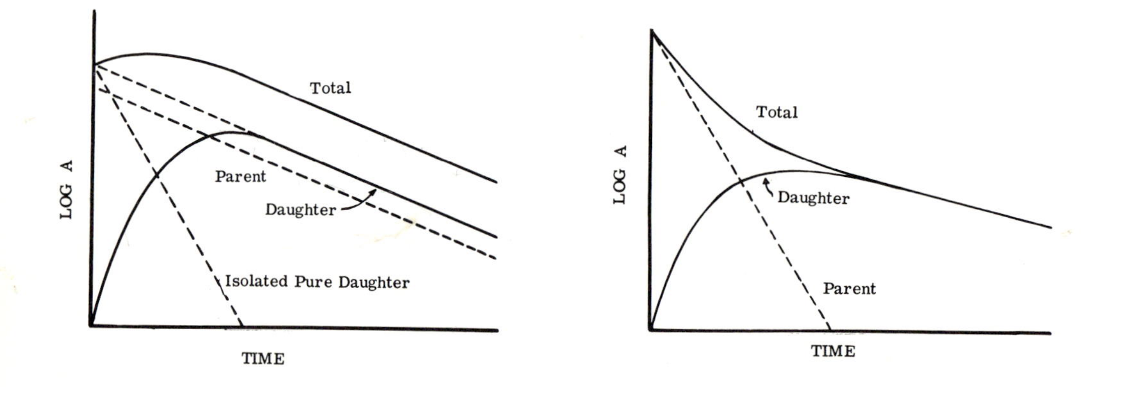

Figure 6.22—TRANSIENT EQUILIBRIUM.

Figure 6.23—NO EQUILIBRIUM.

No equilibrium is established if the half-life of the parent is less than that of the daughter. This is illustrated in figure 6.23. If one starts with pure parent, the activity of the daughter will increase, pass through a maximum and then decrease. At the maximum of the daughter curve $dN_2/dt = 0$. The time required to reach this maximum can be calculated by differentiation of equation 6.72, setting the differential equal to zero and solving for t.

The case of many successive decays has been considered by H. Bateman (42, 43) who has developed a general solution to the problem. The case of many successive decays is represented by any radioactive series in which radioactive daughters are produced.

$$A \xrightarrow{\lambda_1} B \xrightarrow{\lambda_2} C \xrightarrow{\lambda_3} D \xrightarrow{\lambda_4} E \xrightarrow{\lambda_5} F \xrightarrow{\lambda_6}$$

Thus, starting with pure A, the total number of atoms N_n of any daughter at time t can be calculated.

$$N_n = C_1 e^{-\lambda_1 t} + C_2 e^{-\lambda_2 t} + \ldots\ldots + C_n e^{-\lambda_n t} \tag{6.78}$$

where

$$C_1 = \frac{\lambda_1 \lambda_2 \ldots\ldots\ldots\ldots \lambda_{n-1}}{(\lambda_2 - \lambda_1)(\lambda_3 - \lambda_1)\ldots\ldots(\lambda_n - \lambda_1)} N_1^0$$

and

$$C_2 = \frac{\lambda_1 \lambda_2 \ldots\ldots\ldots\ldots \lambda_{n-1}}{(\lambda_1 - \lambda_2)(\lambda_3 - \lambda_2)\ldots\ldots(\lambda_n - \lambda_2)} N_1^0$$

In the use of the Bateman solution, a term must be included for each member of the radioactive series.

Branching decay is often encountered in radioactive decay schemes. It is illustrated by the general case of A decaying to both B and C.

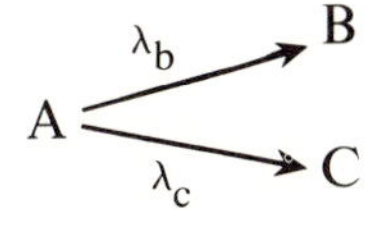

The rate equation for the disappearance of the parent A is

$$-dN_a/dt = N_a(\lambda_b + \lambda_c) \tag{6.79}$$

Thus the disappearance of A is first order. The number of atoms of A remaining at time t is given by

$$N_a = N_a^0 e^{-(\lambda_b + \lambda_c)t} \tag{6.80}$$

and the ratio of the activities is given by

$$\frac{A_a}{A_a^o} = \frac{R_a}{R_a^o} = e^{-(\lambda_b + \lambda_c)t} \tag{6.81}$$

where A_a and R_a are the absolute and relative activities at time t while A_a^0 and R_a^0 are the absolute and relative activities, respectively, at zero time. The rate equation for the formation of the daughter B is

$$dN_b/dt = \lambda_b N_a = \lambda_b N_a^0 e^{-(\lambda_b + \lambda_c)t} \tag{6.82}$$

Integration gives

$$N_b = \frac{-\lambda_b}{\lambda_b + \lambda_c} N_a^0 e^{-(\lambda_b + \lambda_c)t} + \text{constant} \tag{6.83}$$

If $N_b = 0$ at $t = 0$, then constant $= \lambda_b N_a^0/(\lambda_b + \lambda_c)$ and

$$N_b = \frac{\lambda_b}{\lambda_b + \lambda_c} N_a^0 \left[1 - e^{-(\lambda_b + \lambda_c)t}\right] \tag{6.84}$$

For daughter C

$$N_c = \frac{\lambda_c}{\lambda_b + \lambda_c} N_a^0 \left[1 - e^{-(\lambda_b + \lambda_c)t}\right] \tag{6.85}$$

Both daughters B and C have the same apparent half-life although they are formed from the parent with different decay rates.

The half-life for the decay of A is

$$t_{1/2} = 0.693/(\lambda_b + \lambda_c) \tag{6.86}$$

and is related to the *partial half-lives* by

$$\boxed{\frac{1}{t_{1/2}} = \frac{1}{t_{1/2b}} + \frac{1}{t_{1/2c}}} \tag{6.87}$$

where $t_{1/2b}$ is the partial half-life for the decay of A to B, and $t_{1/2c}$ is the partial half-life for the decay of A to C.

A comparison of branching decay to a parallel resistive electrical circuit, with resistance being the counterpart of partial half-life, may be helpful in understanding the mathematics of this type of decay. Branching decay is discussed further in experiment 6.4 where ^{40}K is considered as a typical example.

APPARATUS AND REAGENTS:

Part A—Ra-DEF solution (0.1 to 0.5 μCi/ml); bismuth carrier solution (5% bismuth nitrate); lead carrier solution (5% lead nitrate or lead acetate); 10% sodium hydroxide; 10% sulfuric acid; 3-ml test tubes or centrifuge tubes; semi-micro centrifuge.

Part B—Solution of ^{137}Cs (0.1 to 0.5 μCi/ml); barium carrier solution (5% barium nitrate).

PROCEDURE:

PART A—EQUILIBRIA OF THE SYSTEM ^{210}Pb-^{210}Bi.

The decay of lead-210 provides an interesting system for study

$$\underset{(\text{RaD})}{^{210}\text{Pb}} \xrightarrow[\beta\ 0.061\ \text{MeV}]{22\ \text{y}} \underset{(\text{RaE})}{^{210}\text{Bi}} \xrightarrow[\beta\ 1.17\ \text{MeV}]{5.0\ \text{d}} \underset{(\text{RaF})}{^{210}\text{Po}} \xrightarrow[\alpha\ 5.30\ \text{MeV}]{139\ \text{d}} \underset{(\text{RaG})}{^{206}\text{Pb}}$$

In this experiment ^{210}Pb is separated from ^{210}Bi by means of a precipitation technique. There being no suitable holdback carrier for polonium, ^{210}Po will coprecipitate. Either or both of the following schemes may be used. If both are used an opportunity is provided for a comparison of the two chemical procedures.

Scheme I—Precipitation of lead as $PbSO_4$.

$$\left.\begin{matrix} Pb^{+2} \\ Bi^{+3} \\ (Po) \end{matrix}\right\} \xrightarrow{H_2SO_4} \frac{PbSO_4\ (Po)}{Bi^{+3}} \quad \text{[I.a.]}$$

$$Bi^{+3} \xrightarrow{NH_4OH} \underline{BiONO_3} \quad \text{[I.b.]}$$

To an aliquot containing about 0.1 μCi of Ra-DEF solution in a 3-ml test tube is added a drop or two of bismuth carrier solution and enough 10% H_2SO_4 solution to produce a total volume of about 1 ml. The solution should be clear. If it is not it may mean that the solution is insufficiently acid to prevent hydrolysis of the bismuth salt. On the other hand, if the solution is too acid, lead will form the bisulfate which is soluble. To the clear solution add a drop or two of lead carrier solution. Mix well and promote the formation of the lead sulfate precipitate by scraping the wall of the tube with a stirring rod. Centrifuge and transfer the supernatant to another 3-ml test tube. Wash the precipitate with 10% H_2SO_4 to assure removal of the bismuth, then wash with water to remove the acid. The precipitate is transferred to a planchet. This transfer is accomplished easily by suspending the precipitate in a small quantity of water or acetone, the suspension is transferred by means of a dropper and the liquid is evaporated. This is sample I.a.

The supernatant containing the bismuth is treated with ammonium hydroxide to precipitate the basic bismuth salt. This precipitate is washed with water and transferred to a planchet by suspension in a small volume of water or acetone. This is sample I.b.

An aliquot of Ra-DEF solution, equal in volume to that used in the above procedure, is transferred to a planchet and evaporated to dryness. This is sample I.c. and will serve as a reference.

Scheme II—Precipitation of bismuth as the basic salt.

$$\left.\begin{array}{l} Pb^{+2} \\ Bi^{+3} \\ (Po) \end{array}\right\} \xrightarrow{NaOH} \begin{array}{l} \left.PbO_2^{-2}\right\} \xrightarrow{H_2SO_4} \underline{PbSO_4} \quad [II.a.] \\ \underline{BiONO_3\,(Po)} \quad [II.b.] \end{array}$$

To about 0.5 ml of water in a 3-ml test tube is added an aliquot of Ra-DEF solution, equal in volume to that used for Scheme I of the procedure, and about 2 drops of lead carrier solution. Sodium hydroxide solution is added with constant stirring until any precipitate formed redissolves indicating lead has been converted to the plumbite form. A slight excess of sodium hydroxide should be added followed by about two drops of bismuth carrier solution. The solution is centrifuged and the supernatant containing the lead is removed from the precipitate of the basic bismuth salt. Wash the precipitate with water and transfer to a planchet. This precipitate, containing the ^{210}Bi is II.b.

To the supernatant containing the ^{210}Pb as plumbite add sulfuric acid until the solution gives an acid reaction. Place the test tube in a beaker of boiling water and digest for a few minutes. Stir frequently to promote the precipitation of lead sulfate. Centrifuge and discard the supernatant, wash the precipitate with a few volumes of water and transfer it to a planchet. This is sample II.a.

An aliquot of Ra-DEF solution, equal in volume to that used in the above precedure, is transferred to a planchet and evaporated to dryness. This is sample II.c. and will serve as a reference. If sample I.c. was prepared in the preceding section from the same volume of solution I.c. and II.c. will be identical and one can be omitted.

Treatment of the samples—The activity of each sample is measured with an end-window Geiger tube. A thin absorber (about 10 mg/cm^2) should be placed between the sample and the tube to remove weak beta radiation from the ^{210}Pb and to absorb the polonium alpha particles. With this arrangement only the energetic beta particles of ^{210}Bi should be detected.

If a good separation has been effected the activities of samples I.a. and II.a. should be very low. As ingrowth of ^{210}Bi occurs in the ^{210}Pb samples the activity will be observed to increase.

The activity of each sample should be measured daily, if possible, for about two weeks. Activity is plotted as a function of time on linear graph paper. If no loss was incurred in the chemical separation the sum of the observed activities of samples a and b should equal that of c.

PART B—EQUILIBRIA OF THE SYSTEM ^{137}Cs-^{137m}Ba.

Cesium-137 decays with a 30-year half-life to barium-137m. The latter is a metastable nuclear isomer which decays to the ground state with a 2.6 minute half-life. This experiment, illustrating the separation of these two nuclides by precipitation of barium as the sulfate, is based on the procedure of Choppin and Nealy (51).

Saturation Value—An aliquot of the ^{137}Cs stock solution calculated to yield an observed activity of 2×10^4 cpm is added to 2 ml of 1*N* H_2SO_4 in a test tube which fits into a well counter. The activity of this solution is measured. This activity represents the saturation value, that is, the activity expected when ingrowth of ^{137m}Ba is complete.

Decay of ^{137m}*Ba*—The solution used to measure the saturation value is warmed in a water bath and 0.5 ml of barium carrier solution is added. The precipitate is digested about ½ minute and then centrifuged for ½ minute. The supernatant is then transferred to another test tube and the tube containing the barium sulfate is placed in the well counter. Readings of the scaler are recorded every 15 seconds without stopping the counter for 6 to 7 minutes. The difference between consecutive 15-second readings is plotted as a function of time. If the logarithm of the difference is plotted vs time a straight line should result.

Ingrowth of ^{137m}Ba—Approximately 10 minutes after the first precipitation another 0.5 ml of barium carrier solution is added to the warm supernatant. The mixture is digested and centrifuged and the supernatant is transferred to another tube as before. Place the tube containing the ^{137}Cs supernatant in the well and take a reading of the scaler every 15 seconds as before. These activities are plotted on the same graph as the decay data above.

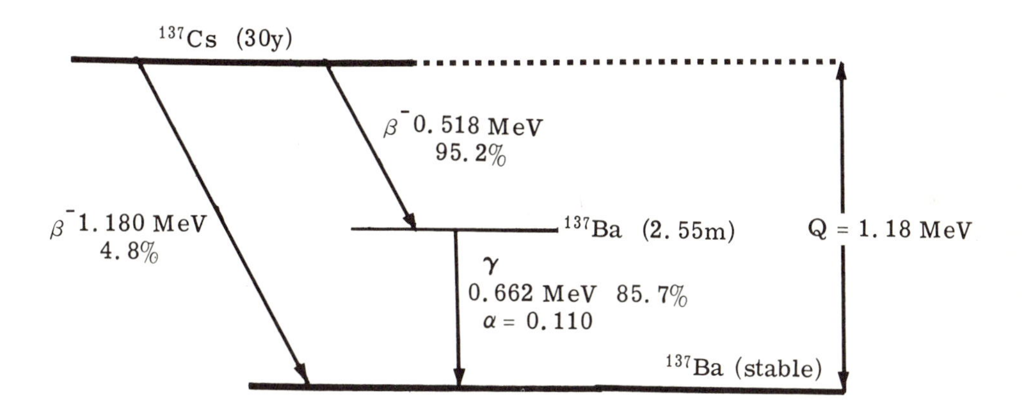

Figure 6.24—DECAY SCHEME FOR CESIUM-137
[After Merritt and Taylor (58)].

EXPERIMENT 6.4 HALF-LIFE OF LONG-LIVED ISOTOPES

OBJECTIVES:

To demonstrate the determination of the half-life of an isotope having a half-life too great to be measured by the decay method (experiment 6.1).

To determine the half-life of potassium-40.

To illustrate a case of branching decay and the relationship of the decay constants.

DISCUSSION:

Naturally occurring radioactive isotopes must have a half-life nearly equal to or greater than the age of the earth (about 3×10^9 years). One of the most important of these is ^{40}K. The element potassium as found in nature, with an atomic weight of 39.102, has three isotopes, ^{39}K (93.10%), ^{40}K (0.012%), and ^{41}K (6.88%) as determined by mass spectroscopy. Hence, only 0.012% of the potassium atoms are radioactive.

Two modes of decay have been observed for ^{40}K. It can decay by beta emission to ^{40}Ca or by electron capture (K-capture) to ^{40}Ar. Both of these daughter atoms are stable.

The gamma emission (and hence the decay route by K-capture) is found to be equal to 13% of the beta emission. The total decay constant is the sum of the partial decay constants of the two processes, electron capture and beta emission:

$$\lambda = \lambda_c + \lambda_\beta$$

but, since $$\lambda_c = 0.13\,\lambda_\beta$$

then $$\lambda = 1.13\,\lambda_\beta$$

and, since $$t_{1/2}\,\lambda = 0.693 = t_{1/2\beta}\,\lambda_\beta$$

then $$t_{1/2} = t_{1/2\beta}/1.13$$

where $t_{1/2}$ is the *true half-life* for ^{40}K and $t_{1/2\beta}$ is the *partial half-life* for β decay.

The decay constant λ_β must be calculated from the relationship

$$dN_\beta/dt = -\lambda_\beta N$$

where N is the number of radioactive atoms present in the sample and dN_β/dt is the number of atoms decaying per unit of time by beta decay.

The number of radioactive atoms N present in the sample can be calculated from the weight of the sample used. The value of dN_β/dt, the number of atoms decaying per unit of time by beta decay, can be determined by the use of a calibrated beta counter.

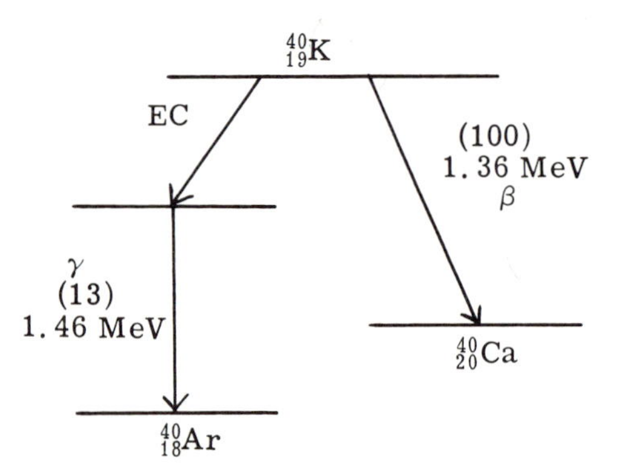

Figure 6.25—BRANCHING DECAY OF K^{40}.

PROCEDURE:

Place exactly 2 ml of a 2.0 molar potassium chloride solution in a planchet and dry under the infrared lamp. Measure the activity of the sample in the windowless flow counter. At least a 10 minute count should be made of both background and sample. (Preferably a 30 minute count of each, or better, a preset count of 1000 or more).

DATA:

Sample activity (uncorrected) ________________ cpm
Background ________________ cpm
Sample activity (corrected) ________________ cpm
Efficiency of counter (measured) ________________ %

CALCULATIONS:

If the efficiency of the counter has not been measured, an approximate efficiency of 60% may be assumed. The approximate value of 60% is based on 50% efficiency for a 2π counter plus 10% for backscattering. For measurements of high accuracy, self-absorption should be considered but may be ignored in this experiment. A method for the absolute assay of beta radioactivity of naturally radioactive potassium is presented by Suttle and Libby (58).

Decay rate ($\lambda_\beta N$) = ________________ dpm
________________ dis per year

Number of atoms of potassium per 2 ml of 2 *M* KCl ____________
N = Number of atoms of ^{40}K per 2 ml of 2 *M* KCl ____________

λ_β = (dis/yr)/N = ____________ $year^{-1}$
$t_{1/2\ \beta} = 0.693/\lambda_\beta$ = ____________ years
$t_{1/2} = t_{1/2\beta}/1.13$ = ____________ years
$t_{1/2}$ (literature) = ____________ years

REFERENCES

Modes of Decay

1. Gamow, G., "Zur Quantentheorie des Atomkernes", Z. Physik 51, 204 (1928).
2. Gurney, R. W., and E. U. Condon, "Quantum Mechanics and Radioactive Disintegration", Nature 122, 439 (1928).
3. Geiger, H., and J. M. Nuttall, "The Ranges of the α-Particles from Various Radioactive Substances and a Relation Between Range and Period of Transformation", Phil. Mag. 22, 613 (1911); 23, 439 (1912).
4. Perlman, I., "Alpha Radioactivity and the Stability of Heavy Nuclei", Nucleonics 7, 3-18, 25 (Aug. 1950).
5. Ore, A., and J. L. Powell, "Three Photon Annihilation of an Electron-Positron Pair", Phys. Rev. 75, 1696-1699 (1949).
6. Deutsch, M., "Evidence for the Formation of Positronium in Gases", Phys. Rev. 82, 455-456 (1951).
7. Deutsch, M., "Three-quantum Decay of Positronium", Phys. Rev. 83, 866-867 (1951).
8. Pond, T. A., "The Formation of Triplet Positronium in Gases", Phys. Rev. 85, 489 (1952).
9. DeBenedetti, S., and R. T. Siegel, "The Three-photon Annihilation of Positrons and Electrons", Phys. Rev. 94, 955-959 (1954).
10. DeBenedetti, S., and H. C. Corben, "Positronium", Ann. Rev. Nucl. Sci. 4, 191-218 (1954).
11. Corben, H. C., and S. DeBenedetti, "The Ultimate Atom", Sci. Amer. 191, 88-92 (Dec. 1954).
12. DeBenedetti, S., "New Atoms (Positronium and Mesonic Atoms)", Nuovo Cimento 4, 1209-1270 (1956).
13. Hughs, V. W., "Positronium Formation in Gases", J. Appl. Phys. 28, 16-22 (1957).
14. Wiles, D. R., "Principles of β-Decay Theory", Nucleonics 11, No. 11, 32-35 (1953).
15. Mandeville, C. E., and M. V. Scherb, "Nuclear Disintegration Schemes and the Coincidence Method", Nucleonics 3, 2 (Oct. 1948).
16. Edwards, R. R., and T. H. Davies, "Chemical Effects of Nuclear Transformations", Nucleonics 2, 44 (June 1948).
17. Rose, M. E., and J. L. Jackson, "The Ratio of L_1 to K Capture", Phys. Rev. 76, 1540 (1949).

Calculation of Decay

18. Estabrook, G. M., "Factors for Converting Decay Constants", Nucleonics 18, No. 11, 209 (Nov. 1960).
19. Ellis, R. H., Jr., "Log-Log Scale Becomes Decay Table", Nucleonics 14, No. 6, 86 (June 1956).
20. Rose, M. E., "Internal Conversion Coefficients", Interscience (1958).
21. Knauss, H. P., "Slide Rule Calculations of Radioactive Decay", Science 107, 324 (1948).
22. Miller, W. B. Jr., "Nomogram for Estimating Decay of I^{131} and P^{32}", Nucleonics 9, No. 4, 58-59 (1951).
23. Aitken, P. B., "Ruler for Drawing Radioactive Decay Curves", Nucleonics 10, No. 6, 64 (1952).

Measurement of Half-Life

24. Anders, O. U., and W. W. Meinke, "P^{32} Half-Life Determination", Nucleonics 15, No. 12, 68-70 (Dec. 1957).
25. Locket, E. E. and R. H. Thomas, "The Half-Lives of Several Radioisotopes", Nucleonics 11, No. 3, 14 (1953).
26. Kuroda, P. K., and Y. Yokoyama, "Determination of Short-Lived Decay Products of Radon in Natural Waters", Anal. Chem. 26, 1509-11 (1954).
27. Schuler, R. H., "Short-Lived Radioactivity: Correction for Long Counting Periods", Nucleonics 10, No. 11, 96-97 (1952).
28. Wagner, P. T., "Calculating Half-Life for Low Activities of Short-Lived Nuclides", Nucleonics 13, No. 4, 54-56 (April 1955).
29. Stoenner, R. W., O. A. Schaeffer and S. Katcoff, "Half-Lives of Argon-37, Argon-39, and Argon-42", Science 148, 1325 (1965).
30. Salutsky, M. L., and H. W. Kirby, "Preparation and Half Life of Carrier-Free Yttrium-90", Anal. Chem. 27, 567-569 (1955).
31. Gorbics, S. G., W. E. Kunz and A. E. Nash, "New Values for Half-Lives of Cs^{137} and Co^{60} Nuclides", Nucleonics 21, No. 1, 63 (Jan. 1963).
32. Braunstein, J., and R. H. Young, "Measurement of the Half Life of $UX_2(Pa^{234})$ with a Dipping Counter", J. Chem. Ed., 38, No. 1, 31 (1961).
33. Smith, W. T., and J. H. Wood, "A Half-Life Experiment for General Chemistry Students", J. Chem. Ed. 36, 492 (1959).
34. Booth, A. H., "Measurement of the Half-Life of UX_2", J. Chem. Ed. (March 1951).

Resolution of Decay Data

35. Monk, R. G., A. Mercer and T. Downham, "Evaluation of Radioactive Decay Data", Anal. Chem. 35, 178 (1963).
36. Perkel, D. H., "Resolving Complex Decay Curves", Nucleonics 15, No. 6, 103-106 (June 1957).
37. Freiling, E. C., and L. R. Bunney, "Resolution of Two-Component Decay Data", Nucleonics 14, No. 9, 112 (Sept. 1956).
38. Van Liew, H. D., "Semilogarithmic Plots of Data Which Reflect a Continuum of Exponential Processes", Science 138, 682 (1962).

Radioactive Equilibria

39. Fresco, J., E. Jetter and J. Harley, "Radiometric Properties of the Thorium Series", Nucleonics 10, No. 8, 60-64 (1952).
40. Stehn, J. R., and E. F. Clancy, "Nomogram for Radioisotope Buildup and Decay", Nucleonics 13, No. 4, 27 (April 1955).
41. Sawle, D. R., "Continuous Alpha Monitor", Nucleonics 14, No. 8, 90-96 (Aug. 1956).
42. Kirby, H. W., "Decay and Growth Tables for the Naturally Occurring Radioactive Series", Anal. Chem. 26, 1063-71 (1954). (N.B.—Corrections appear on page 1513)
43. Bateman, H., "
Cambridge Phil. Soc. Proc. 15, 423-7 (1910).
44. Flanagan, F. J., and F. E. Senftle, "Tables for Evaluating Bateman Equation Coefficients for Radioactivity Calculations", Anal. Chem. 26, 1595-1600 (1954).
45. Marcus, Y., "Parent Activity from Daughter Growth", Nucleonics 19, No. 3, 76 (March 1961).
46. Geiger, K. W., J. S. Merritt and J. G. V. Taylor, "New Branching Ratio for Kr^{85}", Nucleonics 19, No. 1, 97-101 (Jan. 1961).
47. Hayes, R. L., and W. R. Butler, Jr., "Growth and Decay of Radionuclides—A Demonstration", J. Chem. Ed. 37, 590-592 (1960).
48. Semmelrogge, W. F., and F. Sicilio, "A Study of Secular Equilibrium Using Ce^{144}-Pr^{144}", J. Chem. Ed. 42, 427 (1965).
49. Aronson, A. L., and P. B. Hammond, "Faster Analysis for Pb^{210} in Biological Specimens", Nucleonics 22, No. 2, 90, 92 (Feb. 1964).
50. Parekh, P. P., and M. Sankar Das, "Genetically Related Radionuclides—Growth-Decay Studies of ThB-ThC", J. Chem. Ed. 40, 354-355 (1963).
51. Kirby, H. W., and D. A. Kremer, "Simplified Procedure for Computing the Growth of Radioactive Decay Products", Anal. Chem. 27, 298-299 (1955).
52. Choppin, G. R., and C. L. Nealy, "Demonstration of a Parent-Daughter Radioactive Equilibrium Using ^{137}Cs- ^{137m}Ba", J. Chem. Ed. 41, 599 (1964).
53. Stevens, W. H., "Kr^{85} Branching Ratio", Chem. Eng. News, p. 5 (Jan. 2, 1961).

Methods for Long-Lived Isotopes

54. Cork, J. M., "Nuclear Energy Levels for Some Long-Lived Isotopes", Nucleonics 7, 24-33 (Nov. 1950).
55. Rowlands, S., "Methods of Measuring Very Long and Very Short Half-Lives", Nucleonics 3, 2 (Sept. 1948).
56. Lockett, E. E., and R. H. Thomas, "Half-Lives of Cobalt-60, Thallium-204, and Europium-152", Nucleonics 14, No. 11, 127 (Nov. 1956).
57. Cali, J. P., and L. F. Lowe, "Direct Determination of the Half-Lives of Nine Nuclides", Nucleonics 17, No. 10, 86-88 (Oct. 1959).
58. Merritt, J. S., and J. G. V. Taylor, "Decay of Cesium-137 Determined by Absolute Counting Methods". Anal. Chem. 37, 351-354 (1965).
59. Suttle, A. D., and W. F. Libby, "Absolute Assay of Beta Radioactivity in Thick Solids—Application to Naturally Radioactive Potassium", Anal. Chem. 27, 921-927 (1955).

CHAPTER 7
Properties of Radiation and Interaction with Matter

LECTURE OUTLINE AND STUDY GUIDE

I. INTRODUCTION
   A. Properties of Radiation
   B. Interaction of Radiation with Matter
      1. Types of collisions
      2. Collision probability

II. ALPHA PARTICLES (*Experiment* 7.1)
   A. Structure
   B. Source
   C. Energy (monoenergetic nature)
   D. Velocity
   E. Range and Range-Energy Relationship
      1. Mean range
      2. Extrapolated range
      3. Cut-off distance
      4. Straggling
         a. Random variations in energy loss along path
         b. Energy loss per ion pair formed
      5. Stopping power
      6. Range-density relationship
   F. Nature of Interaction of Alpha Particles with Matter
      1. Ionization
      2. Specific ionization
      3. Delta tracks or delta rays
      4. Scattering of alpha particles

III. BETA PARTICLES
   A. Properties of Beta Particles
   B. Relativistic Variation of Mass
   C. Beta Spectra
      1. Shape
      2. Maximum energy
      3. Average energy
      4. Neutrino theory
   D. Interaction of Beta Particles with Matter
      1. Interactions with nuclei
         a. Elastic—Rutherford scattering
         b. Inelastic—Bremsstrahlung production (*Experiment* 7.2, Part 1)
      2. Interactions with orbital electrons
         a. Ionization and excitation
         b. Production of characteristic x-rays
   E. Deflection by Magnetic Field (*Experiment* 7.2, Part 2)
   F. Absorption of Beta Particles by Matter (*Experiment* 7.2, Part 3)
      1. Mass absorption coefficient
      2. Linear absorption coefficient
   G. Range-Energy Relationships (*Experiment* 7.3)
      1. Measurement of range
      2. Feather analysis
      3. Glendenin range-energy equations

IV. GAMMA QUANTA
   A. Properties
      1. Electromagnetic spectrum
      2. Energy relationship
   B. Sources
   C. Interactions of Gamma Photons with Matter
      1. Rayleigh scattering
      2. Photodisintegration
      3. Nuclear resonance scattering
      4. Bragg scattering
      5. Photoelectric effect
      6. Compton scattering (*Experiment* 7.4)
      7. Pair production
   D. Absorption of Gamma Radiation (*Experiment* 7.5)
      1. Dependence of absorption coefficients on nature of absorbing medium
         a. Linear coefficient
         b. Mass coefficient
         c. Half-thickness or half-value-layer
      2. Dependence of absorption coefficients on gamma ray energy
         a. Photoelectric effect
         b. Compton effect
         c. Pair production
         d. Total

3. Dependence of absorption on geometry
   a. Total, true and scatter absorption coefficients
   b. Narrow-beam or "good" geometry
   c. Broad-beam or "poor" geometry
   d. Buildup

V. PROPERTIES OF NEUTRONS (*Experiment* 7.6)
   A. Sources
      1. Nuclear bombardment
         a. Using a radioactive source (e.g., alpha-beryllium sources)
         b. Using a particle accelerator (neutron generators)
      2. Fission
   B. Classification of Neutrons by Energy
   C. Types of Interactions of Neutrons
      1. Elastic scattering
      2. Inelastic resonance scattering
      3. Capture reactions
         a. Activation
         b. Transmutation
         c. Spallation
   D. Cross Section
      1. Total cross section
      2. Partial cross sections
         a. Absorption cross section
         b. Scattering cross section
   E. Beam Calculations
      1. Macroscopic absorption cross section
      2. Microscopic absorption cross section

INTRODUCTION:

The importance of understanding the properties of radiation lies, first, in the fact that the operation of every detecting device known for any type of radiation depends upon one or more of the particular properties of the radiation being measured. Secondly, —a knowledge of the mechanism of the energy loss of particles as they pass through matter is necessary for the proper design of measuring instruments and for the intelligent interpretation of the data obtained. All interpretation of data is based on an understanding of the geometry effects which must invariably be considered. Thirdly,—the selection of proper shielding for the safe manipulation of radioactive substances relies on the nature of the radiation and its ability to penetrate matter. The harmful effects of radiation on tissue are highly dependent on the ability of the radiation to ionize matter, as well as on the energy of the incident radiation. These factors alone are more than sufficient to justify a very thorough investigation into the basic properties of radiation.

In radiochemical laboratories alpha, beta and gamma radiations are encountered most frequently, although the number of laboratories which have access to neutron sources is increasing. Positrons and the heavy nuclei which constitute fission products are also encountered in some laboratories. However, the scope of this chapter will be limited to the four most common types, namely, alpha, beta and gamma radiations, and neutrons.

PROPERTIES OF RADIATION:

Radiation may consist of a stream of particles (alpha or beta particles or neutrons) or of electromagnetic radiation propagated through space and matter from a radioactive source. Particles may be charged (alpha and beta) or uncharged (neutrons). Electromagnetic radiation may possess a relatively small amount of energy (x-rays) or a large amount of energy (gamma rays) per photon. These differences in the nature of radiation create important differences in their modes of interaction with matter. The principal characteristics of radiation are summarized in table 7.1. Specific and more detailed information is presented in each experiment.

Table 7.1—PROPERTIES OF RADIATION

RADIATION	CHARGE	APPROXIMATE ENERGY RANGE	APPROXIMATE RANGE		PRIMARY SOURCE OF RADIATION	APPROXIMATE RELATIVE SPECIFIC IONIZATION
			AIR	WATER		
Particles						
Alpha	+2	3-9 MeV	2-8 cm	20-40 μ	Heavy nuclei	2,500
Beta—β^-	−1	0-3 MeV	0-10 m	0-1 mm	Nuclei with high n/p ratio	100
β^+	+1	0-3 MeV	0-10 m	0-1 mm	Nuclei with low n/p ratio	100
Neutrons	0	0-10 MeV	0-100 m	0-1 m	Nuclear bombardment	0.1
Electromagnetic						
X-rays	None	eV-100KeV	mm-10 m	μ-cm	Orbital electron transitions	10
Gamma rays	None	10KeV-10MeV	cm-100 m	mm-10 cm	Nuclear transitions	1

TYPES OF COLLISIONS OF RADIATION WITH MATTER:

Unless the space through which radiation passes is completely devoid of matter (i.e., is a vacuum), interactions resulting from collisions of radiation with matter will occur. Such interactions may result in attenuation (absorption), scattering (change in direction), ionization and excitation of atoms or the conversion of one type of radiation into another (photoelectrons from photons). These phenomena occur upon *collision* of the radiation with nuclei and orbital electrons, with "free" particles or with the field surrounding a particle.

A collision may involve a physical impact of one particle with another, but the term includes any interaction of nuclei, particles or photons in which an interchange of momentum, energy, charge or other quality occurs. For example, charged particles will interact through their coulombic fields upon close approach to each other. As a result, their paths are deflected and a change occurs in the momentum and energy of one or both without direct physical contact.

Collisions are said to be *elastic* if there is no change in the internal energy or total kinetic energy of the colliding particles or quanta. On the other hand, an *inelastic* collision results in a change in the internal energy of one or more of the colliding systems (e.g., ionization of an atom) as well as in the total kinetic energy of the systems. In still other collisions, complete absorption of the incident radiation may occur.

Observations of the interactions of radiation with matter may be *macroscopic* or *microscopic*. For example, if the gross intensity of radiation is measured before and after its passage through a particular body of matter, such as a sheet of lead, the attenuation or absorption of the radiation in the material can be calculated. Here a macroscopic observation has been made. Macroscopic measurements are based upon the mean result of many individual collisions usually involving various types of interactions.

Much is to be gained, however, from a consideration of collisions at the atomic or subatomic level, too. From this microscopic point of view, attention is centered on an individual collision or single type of collision and the process by which it occurs. The modes by which various particles or photons can interact with matter are summarized in table 7.2.

Table 7.2—INTERACTIONS OF RADIATION WITH MATTER

INCIDENT RADIATION	IN COLLISION WITH	TYPE OF COLLISON		
		ELASTIC	INELASTIC	COMPLETE ABSORPTION
Alpha	Nucleus	Rutherford scattering	Bremsstrahlung production (negligible)	Transmutation (First observed nuclear reaction)
	Orbital Electrons	(negligible)	Ionization & Excitation (principal interaction)	(none)
Electrons	Nucleus	Rutherford scattering (principal cause of backscattering)	Bremsstrahlung production (Especially important with high Z absorbers)	Electron capture (K-capture)
	Orbital Electrons	Causes some scattering	Ionization & Excitation (characteristic x-ray production)	Annihilation (if incident electron is a positron)
Neutrons	Nucleus	Moderation or slowing down of neutrons	Resonance scattering (Excitation of nucleus)	Activation and other nuclear reactions (Capture)
	Orbital Electrons	(negligible)	(negligible)	(none)
Photons	Nucleus	Thomson Scattering (negligible)	Nuclear resonance Mossbauer effect	Photodisintegration
	Orbital Electrons	Rayleigh scattering (coherent) Thompson Scattering	Compton scattering* (partial absorption of photon energy)	Photoelectric effect
	Field	Delbruck scattering (negligible)	(negligible)	Pair production (predominant with high energy photons)

*A Compton collision with a free electron is elastic

COLLISION KINETICS:

Newton's second law of motion relates the force exerted by a body to the time rate of change in its momentum, p

$$F = dp/dt \quad \text{or} \quad dp = F\,dt \tag{7.1}$$

During a collision there is a change in the momentum of a particle Δp where

$$\Delta p = p_2 - p_1 = \int_{p_1}^{p_2} dp = \int_{t_1}^{t_2} F\,dt \tag{7.2}$$

If two particles of mass m_a and m_b collide, the forces F_a and F_b exerted by these particles must be of equal magnitude and opposite in direction. This is a result of *Newton's third law of motion.* Thus,

$$\Delta p_a = \int F_a\,dt = -\Delta p_b = -\int F_b\,dt \tag{7.3}$$

since at every instant F_a must equal F_b. From equation 7.3 the *law of conservation of momentum* follows—if the resultant external force acting on a system is zero (or constant), the resultant vector sum of the momentum for the system components must be zero (or constant). This is simply expressed by

$$\Delta p = \Delta p_a + \Delta p_b = 0 \tag{7.4}$$

Consider a system of two particles, an incident particle of mass m_a and velocity v which collides elastically with a particle of mass m_b initially at rest. The incident particle is deflected through an angle θ and continues in the new direction with a velocity v_a. The struck particle acquires a velocity v_b at an angle θ with the original direction of motion of the incident particle. The center of mass of the system m_c moves with a velocity v_c in the direction of v and neither m_c nor v_c is altered by the collision.

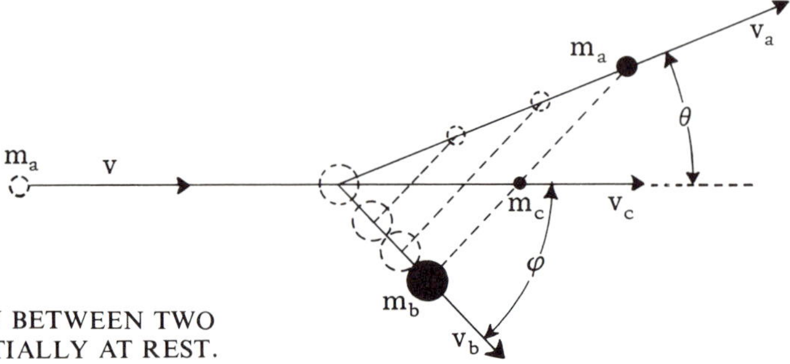

Figure 7.1a—AN ELASTIC COLLISION BETWEEN TWO PARTICLES, ONE OF WHICH IS INITIALLY AT REST.

Momentum is equal to the product of mass and velocity. Applying the law of conservation of momentum one obtains

$$m_a v = (m_a + m_b) v_c$$

If both sides of the equation are divided by the sum of the masses an expression for the velocity of the center of mass is obtained

$$v_c = v \frac{m_a}{m_a + m_b} \tag{7.5}$$

Conservation of momentum in the direction of the x-axis yields

$$m_a v = m_a v_a \cos\theta + m_b v_b \cos\varphi \tag{7.6}$$

and in the direction of the y-axis

$$m_a v_a \sin\theta = m_b v_b \sin\varphi \tag{7.7}$$

Application of the law of energy conservation yields the relation

$$\tfrac{1}{2} m_a v^2 = \tfrac{1}{2} m_a v_a{}^2 + \tfrac{1}{2} m_b v_b{}^2 \tag{7.8}$$

The conservation equations, equations 7.6, 7.7 and 7.8, can be solved simultaneously for any desired term.

COLLISION PROBABILITY:

The probability of occurrence of a collision is measured by the *cross section.* The *geometric cross section* of a particle is πR^2 where R is the radius of the particle. The *reaction cross section* σ may be thought of as the apparent cross sectional area of the bombarded particle as seen by the incident particle. Thus, for a given interaction, σ may be greater or less than πR^2 and indeed may even be equal to zero if a particular reation does not take place.

The unit of cross section is the *barn* b (= $10^{-24} cm^2$) a measure of area. The value of cross section depends not only on the properties of the matter through which the radiation passes, but also on the properties of the radiation itself.

Some of the properties of matter which influence cross section are the mass number A, atomic number Z, neutron number N, density ρ of the material and the charge and mass of the target particles. Cross section depends upon the nature of the incident radiation. For incident particles it is necessary to consider the mass, charge, velocity (energy) and size (deBroglie wavelength). In the case of electromagnetic radiation, cross section is a function of the energy (wavelength) of the incident radiation.

CERENKOV RADIATION:

The bluish-white light emitted from transparent substances in the vicinity of strong radioactive sources is called Cerenkov radiation. Mallet investigated the phenomenon in 1926 and found the spectrum of the light to be continuous. He also noted that the effect was different than fluorescence and other luminescent phenomena. Apparently nothing further was done until the years 1934 to 1938 when Cerenkov investigated the phenomenon extensively. In 1937 Frank and Tamm proposed a theory to explain the radiation and in 1940 Ginsberg proposed an explanation in terms of quantum mechanics.

The velocity of light in a vacuum is about 3.0×10^{10} cm/sec but the velocity of light c′ in a medium of index of refraction η is $c' = c/\eta$. For water, $\eta = 1.33$ and the velocity of light in water is therefore $c' = 2.26 \times 10^{10}$ cm/sec. When a particle travels in excess of the velocity c′ in a medium it loses energy in the form of Cerenkov radiation.

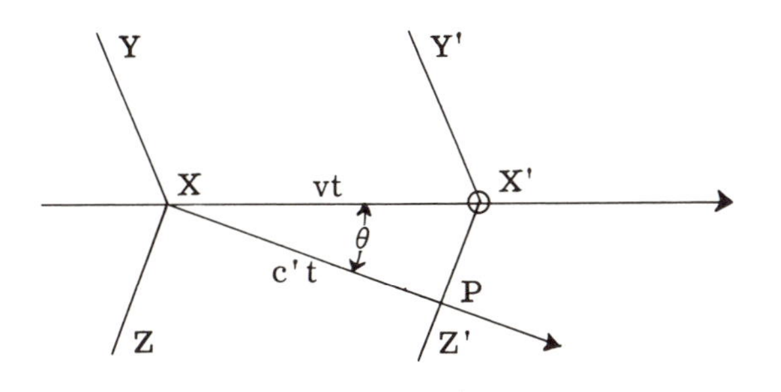

Figure 7.1b—CERENKOV RADIATION. For a particle traveling in the direction XX′ with a velocity v, energy, is released as visible light at an angle θ from the path of the particle and with a velocity c′.

Let the particle illustrated in figure 7.1b travel in a direction from X to X′ with a velocity v. Let it be assumed that the distance XX′ is so short that there is no appreciable decrease in v. The distance traveled by the particle in time t is $XX' = vt$. As it travels along this path waves emanate from it with reinforcement of the waves taking place along YXZ. Thus the light is not propagated in all directions but only in a direction perpendicular to the wave front. While the particle travels a distance $XX' = vt$ the light travels a distance $XP = c't$. Therefore $\cos\theta = c'/v$ and since $\cos\theta < 1$ then $c' < v$.

It will be recalled that $c' = c/\eta$ and that $\beta = v/c$ or $v = \beta c$. Therefore,

$$\cos\theta = \frac{c'}{v} = \frac{c}{\eta} \times \frac{1}{\beta c}$$

or

$$\boxed{\cos\theta = \frac{1}{\beta\eta}} \qquad (7.9)$$

Equation 7.9 is called the Cerenkov relation.

EXPERIMENT 7.1 PROPERTIES AND RANGE OF ALPHA PARTICLES

OBJECTIVES:

To study the basic properties of alpha particles.

To observe the nature of the interaction of alpha particles with matter and to measure their range.

THEORY:

The alpha particle is a helium nucleus. It is composed of two protons and two neutrons, and thus carries a charge of plus two. Support for this structure was presented in 1903 by Ramsay and Soddy, who observed that helium is constantly being produced from radium and that radioactive minerals invariable contain helium. Further support for the relationship between the alpha particle and the helium atom was advanced in 1909 by Rutherford and Royds. They placed radon in a thin-walled tube A (figure 7.2), which was enclosed in a second tube B. A spectral tube C was attached to the top of B. A vacuum was produced in tubes B and C, and after several days the gas collected in B was compressed into C by raising the mercury column. Helium was detected in C by spectroscopic examination of the gas discharge. Thus, it can be concluded that when a fast moving alpha particle comes to rest it takes up two electrons to become a helium atom.

Alpha particles emitted from a particular species of nucleus are monoenergetic. The energy of most alpha particles falls in the range between 3 and 8 MeV. By comparison, chemical reactions have energies of only 5 to 30 eV per molecule.

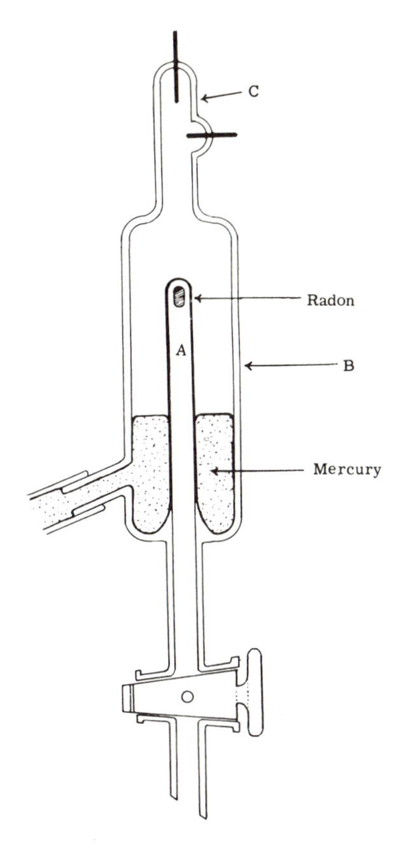

Figure 7.2–APPARATUS USED TO IDENTIFY HELIUM FORMATION FROM ALPHA PARTICLES.

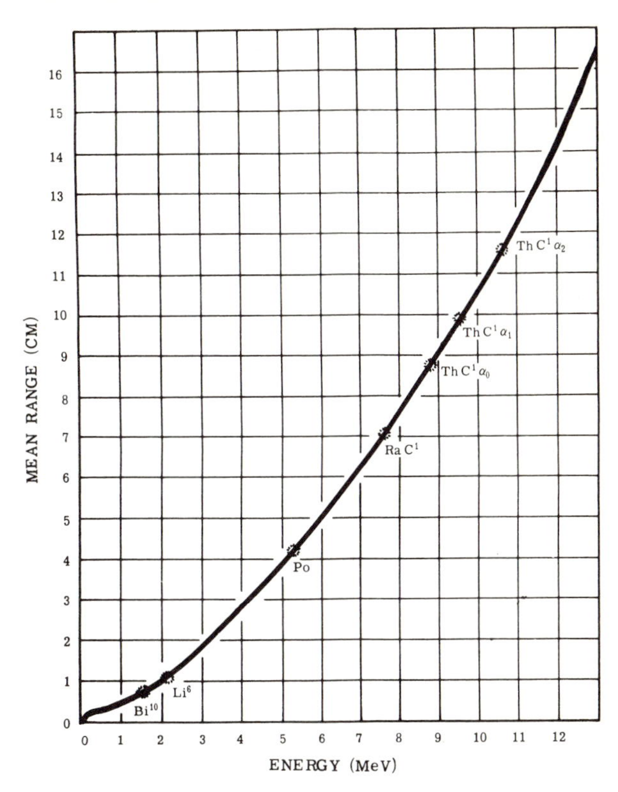

Figure 7.3—RANGE-ENERGY RELATIONSHIP FOR ALPHA PARTICLES IN AIR.

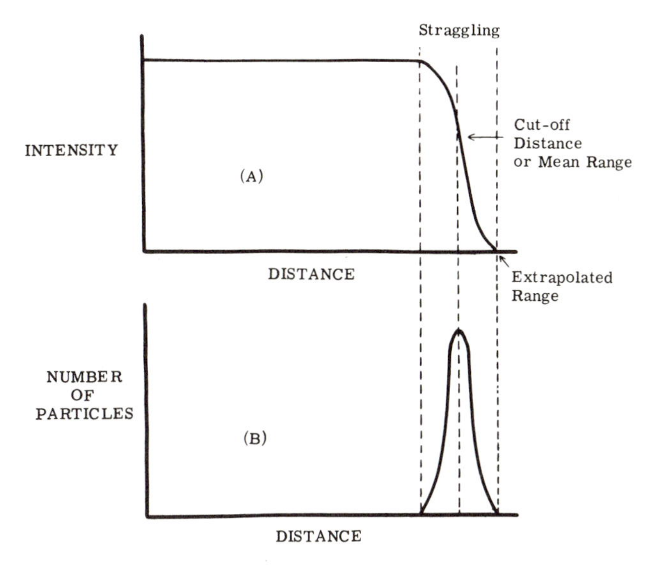

Figure 7.4

The velocity of the alpha particle upon emission from the nucleus depends upon the radioelement from which it originates, but is of the order of 1.4 to 2.0 × 10^9 cm per second. In 1910, Geiger showed that the initial velocity v_0 in cm per second is related to the range in air R in cm by the equation

$$v_0^3 = a R \qquad (7.10)$$

where the constant a has the value of about 1.0 × 10^{27}.

The range of an alpha particle in air is roughly approximated by the relation

$$R \simeq E \qquad (7.11)$$

where R is the range in air in cm and E is the energy in MeV. Thus, the range for alphas is approximately 3 to 8 cm for energies from 3 to 8 MeV, respectively. This relationship is illustrated in figure 7.3. For monoenergetic alpha particles from a single source, one would expect all of the alpha particles to exhibit the same range. This is nearly true. If the intensity of a beam of collimated monoenergetic alpha particles is measured as a function of distance from the source, a graph is obtained similar to that in figure 7.4 (A). There is no sharp *cut-off distance,* but the alpha particles *straggle* about the *mean range,* the average value for the range. The graph shown in figure 7.4 (B) is the differential of (A) and follows a Gaussian distribution.

Stopping power is a measure of the energy loss of alpha particles in a substance compared to the energy loss in air as the standard. The *relative atomic stopping power* of a substance is the ratio of the energy lost by alpha particles per atom of the substance encountered to the energy lost by alpha particles of similar energy per atom of air encountered. The *relative electronic stopping power* of a substance is the ratio of the energies lost per *electron,* air again being the reference substance. The *linear relative stopping power* S is given by the expression

$$S = \frac{R_a}{R_s} \qquad (7.12)$$

R_a and R_s representing the ranges in air and in the substance, respectively. Since the range R is approximately inversely proportional to the density ρ of the material through which the particles are passing, that is

$$R_a \rho_a = R_s \rho_s \qquad (7.13)$$

then we may write

$$S = \frac{\rho_s}{\rho_a} \qquad (7.14)$$

For aluminum, the approximate theoretical stopping power S is calculated from the densities as follows:-

$$S_{Al} = \frac{\rho_{Al}}{\rho_{air}} = \frac{2.74}{1.29 \times 10^{-3}} = 2.1 \times 10^3$$

This means that the stopping power of aluminum is 2100 times greater than that of air, and hence that the range in aluminum should be 1/2100 times as great as the range in air. For a 5 MeV alpha particle, the range in air can be estimated (figure 7.3) to be 4 cm. The range in aluminum is therefore $4/2100 = 1.9 \times 10^{-3}$ cm or 19 microns.

The interaction of alpha particles with matter is a random process. Because of their great mass, the path of alpha particles through a gas describes an almost straight line. By virtue of their double positive charge, alpha particles are powerful ionizers. Consequently, a very heavy track is produced in a cloud chamber. The energy required to strip an electron from an atom of gas is between 25 and 40 eV. Most values lie between 30 and 35 eV. The value frequently used for air is 32.5 eV. The number of ion pairs (I. P.) produced by an alpha particle of energy E_α is readily calculated from the expression

$$\text{I. P.} = \frac{E_\alpha}{32.5} \tag{7.15}$$

Occasionally, an electron is torn from its atomic orbit by the alpha particle and hurled through space with such energy that the electron also becomes capable of producing ion pairs. This results in the formation of *delta rays,* as illustrated in figure 7.5.

Figure 7.5—DELTA RAYS (tracks).

The *specific ionization* is defined as the number of ion pairs produced per unit length of path (per mm) of the ionizing radiation. Specific ionization is dependent upon the residual energy of the particles (see figure 7.6). As the alpha particle continuously loses energy along its path through ionization processes, the specific ionization likewise changes continuously. As the alpha particle loses energy, its velocity decreases. At lesser velocities the alpha particle spends more time in the sphere of influence of the atoms it passes—this process is referred to as a collision—and hence the alpha particle has a greater opportunity to attract an electron, thereby dislodging it from its orbit.

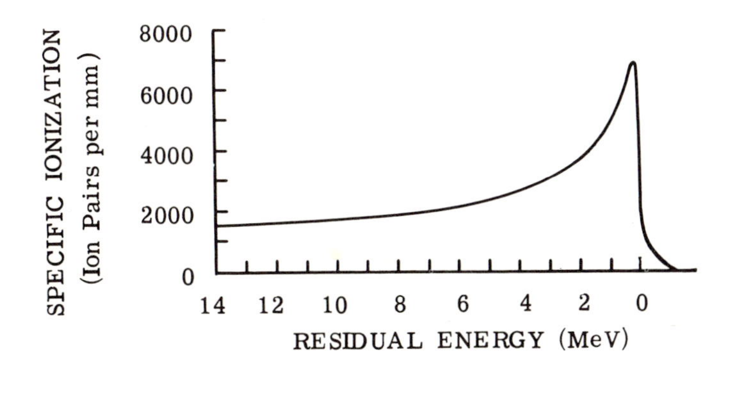

Figure 7.6—BRAGG CURVE FOR ALPHA PARTICLES.

APPARATUS:

A—Source of alpha particles, proportional counter with amplifier or Geiger counter, scaler and sample holder.

B—In addition to above, 2-liter side-arm vacuum flask; vacuum pump, mechanical or water aspirator; pressure gauge or manometer and barometer.

PROCEDURE A—(DISTANCE VARIABLE)

Although the exact arrangement of the sample and counter will depend upon available equipment, one suggested arrangement is that shown in figure 7.7. The counter, with a thin end window of known thickness, is mounted above. If the clamp is loosened, the tube can be slid up or down to vary the distance to the source.

If a mixed alpha source (Ra-DEF) is used, the detector should be a proportional counter adjusted to operate on the alpha plateau. (For details on the operation of the proportional counter refer to experiment 8.3). The proportional counter should be insensitive to beta and gamma radiation when properly adjusted. If the source is a pure alpha emitter such as ^{210}Po, a thin, end-window Geiger tube can be used.

A ^{210}Po source can be prepared very easily from a solution of Ra-DEF containing a small amount of lead carrier to retard absorption of Ra-D. Simply place a piece of silver in the solution and agitate for about ten minutes. Rinse and dry. For this purpose a blank silver disc such as used in experiment 11.4 works very well.

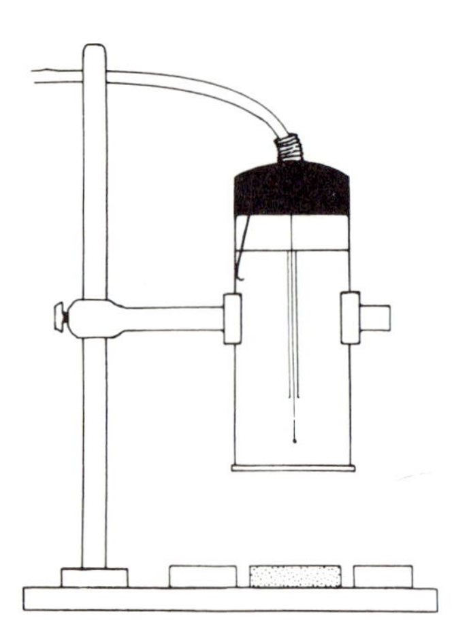

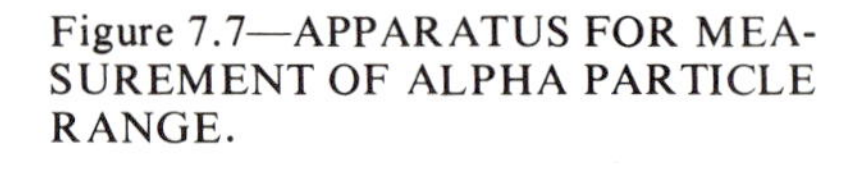

Figure 7.7—APPARATUS FOR MEASUREMENT OF ALPHA PARTICLE RANGE.

Measure the activity of the sample as a function of the distance from the window to the source. Do not use any absorbers. Because the alpha particle source is not collimated, it is necessary to correct for the inverse-square law if a point source is used. Multiply each activity by the square of the distance and plot Rd^2 versus d, where R is the activity in counts per minute and d is the distance in cm. The result should be similar to figure 7.4(A). If a thin, extended source is used, the inverse square law does not apply unless the distance from the source to the detector is greater than the diameter of the source.

Distance d	Activity R cpm	Rd^2 (use this column for point source only)	Air Absorption mg/cm^2	Total Absorber Thickness mg/cm^2

Assuming the density of air to be 1.2 mg/cm^3, calculate the air absorption in mg/cm^2. Add the window thickness in mg/cm^2 to obtain the total absorber thickness. Because of straggling, the range is computed as that distance which gives a 50% reduction in activity. This is the range in mg/cm^2.

Convert this to range in air, including the air equivalent of the counter window.

Repeat the above measurements, using a thin, calibrated aluminum or mica absorber (1.0 mg/cm^2 or less) between the source and the counter. Again calculate the range in air, making the necessary corrections for the air equivalent of the aluminum absorber. Compare this result with the result obtained without the added absorber.

Compute the energy of the alpha radiation using figure 7.3.

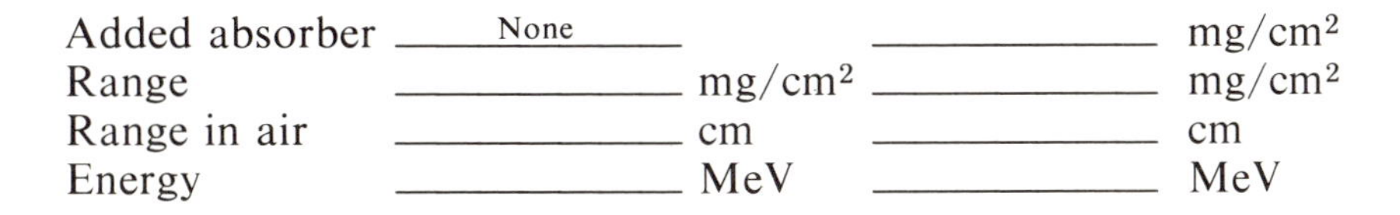

Added absorber	None		______	mg/cm^2
Range	______	mg/cm^2	______	mg/cm^2
Range in air	______	cm	______	cm
Energy	______	MeV	______	MeV

PROCEDURE B—(PRESSURE VARIABLE)

An alpha source is suspended beneath a Geiger tube by means of a small bracket. A one-hole rubber stopper which fits the neck of a 2-liter side-arm vacuum flask is carefully cut through on one side to the hole to permit placing the cable of the Geiger tube in the hole. The apparatus is assembled as illustrated in figure 7.8. The stopper must be inserted firmly into the neck of the flask so leakage will not occur where the cut was made. The distance from the alpha source to the window of the tube should be slightly greater than the range of alpha particles in air at atmospheric pressure. If the tube window is 1-inch or less in diameter no trouble should be encountered with breakage when a vacuum is applied.

The side arm of the flask is connected to a source of vacuum. This may be a mechanical vacuum pump or a water aspirator. Provision should also be made to measure the pressure in the flask and to regulate the pressure by bleeding air into the flask.

The Geiger tube is connected to a scaler and the alpha activity is measured as a function of pressure in the flask. Note is made of the pressure at which the change in activity is greatest. Since the density of the air is proportional to pressure, the range of the alpha particles at atmospheric pressure is calculated by the use of equation 7.13 modified as follows:

$$\frac{R_s}{R_a} = \frac{\rho_a}{\rho_s} = \frac{P_a}{P_s}$$

The air equivalent for the tube window must be added to the measured range to obtain the total range.

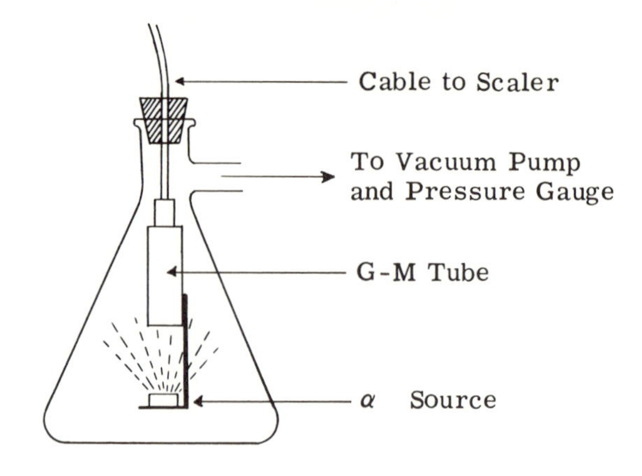

Figure 7.8—VACUUM APPARATUS FOR THE MEASUREMENT OF ALPHA PARTICLE RANGE.

EXPERIMENT 7.2 PROPERTIES OF BETA PARTICLES

OBJECTIVES:

To study the basic properties of beta particles.

To observe the nature of the interaction of beta particles with matter and to measure the mass absorption coefficient for a given isotope.

THEORY:

Experimental evidence shows the beta particle to be identical with the electron. It has a rest mass m_0 of 9.1×10^{-28} grams and a charge Q of 1.6×10^{-19} coulombs.* Thus, we conclude that the principal distinction between an electron and a beta particle is the source or origin. An electron emitted from a nucleus is called a beta particle.

The *velocity* of a beta particle is dependent on its energy. Velocities range from zero continuously up to about 2.9×10^{10} cm/sec, or nearly the velocity of light. Because of their much smaller mass, beta particles must travel at much greater velocities in order to possess an amount of energy equal to that of alpha particles of a given velocity. Classically, the energy of the beta particle is given by the expression

$$E_\beta = \tfrac{1}{2} mv^2 \qquad (7.16)$$

*The electronic mass is 9.1×10^{-28}g or 5.5×10^{-4} amu. The electronic charge is 1.6×10^{-19} coulombs in the mks system. In the cgs system the electronic charge is 1.6×10^{-20} emu or 4.8×10^{-10} esu.

Here, E is the energy in ergs, m the mass in grams, and v the velocity in cm/sec. This equation is quite useful for small values of v, but at higher velocities the effective mass and so also the charge-to-mass ratio e/m, depend upon the velocity of the beta particle. The following relativistic correction, as proposed by Einstein, is required

$$m = \frac{m_0}{\sqrt{1 - \beta^2}} \tag{7.17}$$

Here, $\beta = v/c$, m is the effective mass, m_0 the rest mass of the electron, v the velocity of the particle, and c the velocity of light. The magnitude of the mass variation with velocity is given in the following table.

Table 7.3—RELATIVISTIC MASS INCREASE FOR ELECTRONS

Energy MeV	$\beta = v/c$	m/m_0
0.000	0.00	1.0
0.046	0.40	1.1
0.128	0.60	1.3
0.34	0.80	1.7
0.66	0.90	2.3
1.12	0.95	3.2
3.11	0.99	7.0

Because high energy electrons are relativistic, total energy $W = mc^2$. Since the potential energy $U = m_0c^2$, then the kinetic energy $T = W - U$ is given by

$$T = (m - m_0)\, c^2 \tag{7.18}$$

Substitution of equation 7.17 in equation 7.18 then gives an expression for the beta particle energy in terms of the rest mass and the velocity.

$$E_\beta = 6.24 \times 10^5 T = 6.24 \times 10^5 m_0c^2 \left(\frac{1}{\sqrt{1 - \beta^2}} - 1 \right) \tag{7.19}$$

where T is expressed in ergs and E_β in MeV.

Beta spectra—It is known that beta rays have a continuous spectrum (see chapter 6) and that a part of the decay energy is carried away by neutrinos. The shapes of these spectra differ among nuclides, the shape being a function of the type of nuclear transition. (Beta transitions are referred to as "allowed," "forbidden," etc. to indicate nuclear conditions related to stability). Ra-E (^{210}Bi) which is frequently used as a standard, decays by a "forbidden" transition and hence has a spectrum especially rich in low energy beta rays (see figure 7.9. The *average energy* of the beta particles lies between one-fourth and one-third of E_m. In some instances sharp peaks are found in the beta spectra. These are caused by beta emitters which produce *conversion electrons.* ^{137}Cs is a typical example.

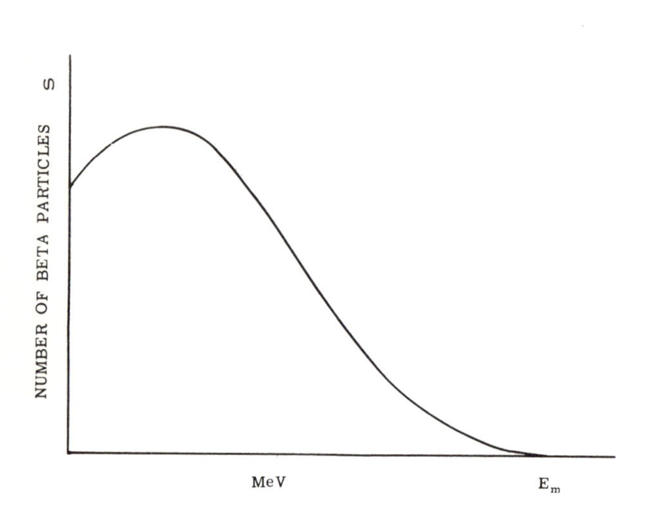

Figure 7.9—BETA SPECTRUM FOR Ra-E (^{210}Bi).

Figure 7.10—RUTHERFORD SCATTERING OF A BETA PARTICLE (elastic).

Beta-particle interactions with nuclei—A collision of a beta particle with an atomic nucleus involves a coulombic interaction in which the electron is sharply deflected in its path. If this interaction is elastic, the process is called *Rutherford scattering,* and the energy of the emergent beta particle is essentially equal to that prior to the collision. This can be deduced from the fact that the total kinetic energy for the colliding systems has not changed (the collision was elastic) and since the atom is at least several thousand times heavier than the electron, the recoil energy of the atom will be negligible. Rutherford scattering is primarily responsible for back-scattering of beta particles.

When electrons are slowed down in the coulombic field of an atomic nucleus electromagnetic radiation called *bremsstrahlung* is produced. (See figure 7.11). This radiation is characteristic of the target and of the beta particle energy, but is a continuous band and usually amounts to about 1% of the total radiation. The percentage of bremsstrahlung production increases with the atomic number of the absorbing material. Hence, for shielding for protection against beta radiation, it is customary to use a material of low atomic number, such as plastics. Bremsstrahlung are produced by the inelastic interaction of a beta particle with the nucleus. The total kinetic energy of the colliding systems is less by an amount equal to the energy radiated as bremsstrahlung.

Beta-ray interactions with orbital electrons—Particles possessing like charges repel each other. The coulombic repulsion between a beta particle and one of the orbital elec-

Figure 7.11—BREMSSTRAHLUNG PRODUCTION (inelastic collision).

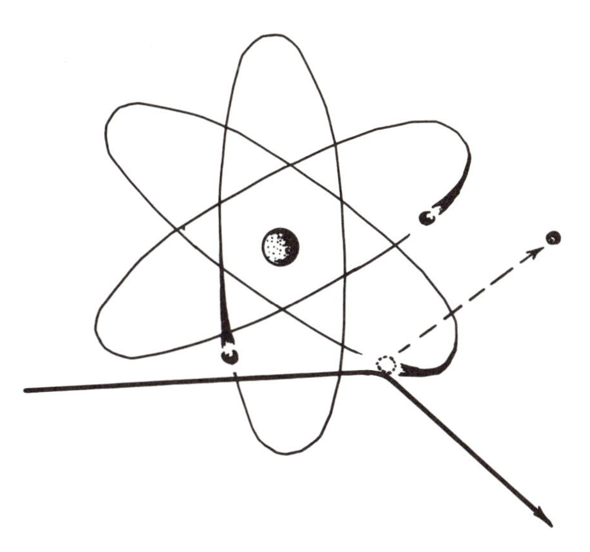

Figure 7.12—IONIZATION (inelastic collision).

trons in the substance being traversed by a beta ray may be sufficient to expel the electron completely from its atom. The atom then becomes a positively charged ion. After this ionization process, the final energy of the electron E_f is less than the initial energy E_i by an amount equal to the sum of the binding energy ϕ of the ejected electron and its kinetic energy. That is,

$$E_f = E_i - (\phi + \tfrac{1}{2}mv^2) \tag{7.20}$$

In collisions involving the expulsion of K, L or M electrons from an atom, *characteristic x-rays* are produced as electrons fall back into the ground state.

$$E_{photon} = h\nu = E_{excited} - E_{ground} \tag{7.21}$$

Deflection of Beta Particles by a Magnetic Field—Charged particles interacting with a magnetic field are deflected at right angles to the lines of force. In figure 7.13 a negative beta particle traveling away from the observer is deflected to the left while a positron is equally deflected to the right. A heavier, positively charged alpha particle is also deflected to the right but to a lesser extent. The paths of the particles are defined by

$$Hev = \frac{mv^2}{r} \tag{7.22}$$

where H is the field strength in gauss, e the charge of the particle (1.6×10^{-20} emu for the electron), m the mass of the particle (grams), r the radius of curvature (cm), and v the velocity of the particle (cm/sec).

Absorption of beta particles in matter—The absorption of beta particles by matter, from a macroscopic point of view, is a function of the distance traveled by the particles

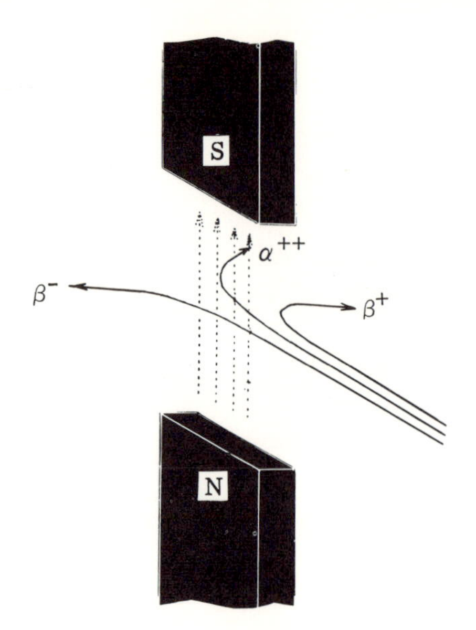

Figure 7.13—DEFLECTION OF CHARGED PARTICLES BY A MAGNETIC FIELD.

through the absorbing material and the density of the material. The product of these two variables, the density-distance has the units grams cm^{-2} or often, for convenience, mg cm^{-2}. This value is used to express the absorber thickness.

Let a beam of beta particles of intensity I_0 impinge upon an absorber (see figure 7.15). Some of these beta particles will be absorbed so the intensity of the emerging beam is I. It has been observed that the absorption of beta particles is approximately an exponential function of the density ρ of the absorber and the distance X through the absorber. Hence, the absorption of beta particles is approximately analogous to the absorption of light, and the mathematical relationship for beta absorption takes the same form as the familiar Beer-Lambert Law.

$$\ln \frac{I_0}{I} = u' X \rho \qquad (7.23)$$

Since X is measured in cm and the density in mg/cm^3, the units of the product $X\rho$ are mg/cm^2. Consequently u′, the *mass absorption coefficient**, has the units cm^2/mg.

*1. The letter u′ is used rather than μ' for the beta mass absorption coefficient because the coefficient, even from theory, is not a true constant for beta absorption. This is also illustrated in figure 7.16.

2. If the Lambert law only is considered, rather than the combined Beer-Lambert law, ρ drops out of equation 6.13 and the coefficient u then becomes the *linear absorption coefficient* with units of cm^{-1}.

Figure 7.14—CALIBRATED ABSORBER SET (*Atomic Accessories, Inc.*)

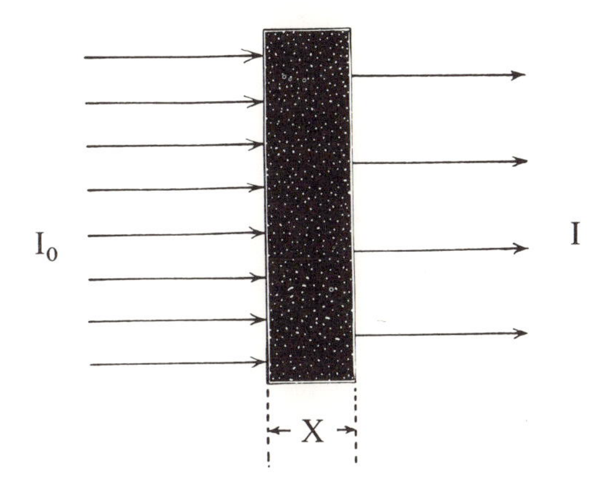

Figure 7.15—ABSORPTION OF BETA PARTICLES. The density of the absorber is ρ. The distance through the absorber is X. The absorber "thickness" $x = \rho X$.

Because the observed activity R is proportional to intensity I, equation 7.23 can be written

$$\ln \frac{R_0}{R} = u' X \rho \qquad (7.24)$$

where R_0 is the observed activity without absorber and R the observed activity with absorber. For convenience let the absorber thickness $x = X\rho$. Then

$$u' = \frac{1}{x} \ln \frac{R_0}{R} \qquad (7.25)$$

If the logarithm of the activity is plotted against the thickness of absorber placed between the sample and the detector, a curve similar to the solid line in figure 7.16 will be obtained. A straight line is not obtained but rather one which is slightly curved. In upper left part, however, the curve may be assumed to be linear and the slope can be determined.

It is important to note that the observed activity is a function of the position of the absorber. As the absorber is moved to a position closer to the source, the activity will be seen to increase. This increase in activity is caused by scattering of the beta particles from the absorber. The effect is shown by the dotted line of figure 7.16. The absorber should therefore be located close to the window of the detector as illustrated in figure 7.17.

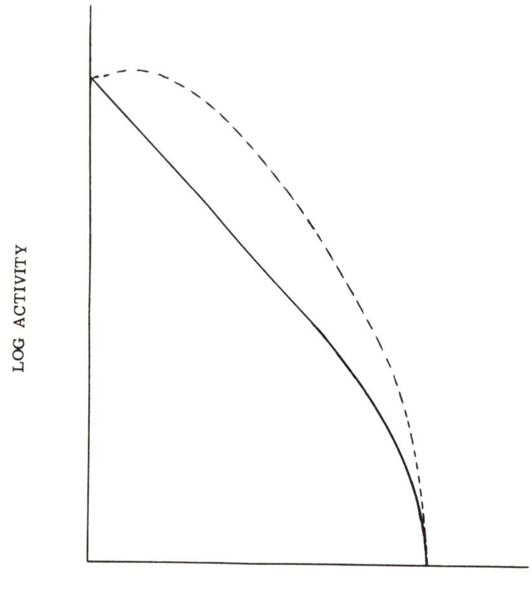

Figure 7.16 — BETA-RAY ABSORPTION CURVE. An ideal curve is shown by the solid line. The dotted line shows the effect of scattering from the absorber when the absorber is close to the source.

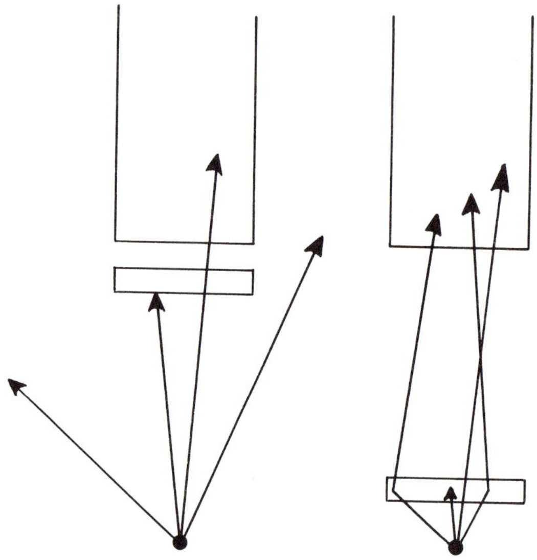

Figure 7.17—INFLUENCE OF ABSORBER LOCATION IN MEASUREMENTS OF BETA RAY ABSORPTION. (a) if good geometry is used beta-particle scattering is minimized. (b) if the absorber is placed directly on or near the source, beta particles are scattered into the tube resulting in the dotted curve of figure 7.16.

APPARATUS AND REAGENTS:

Part A—Geiger tube and scaler; scintillation or other gamma-efficient detector; thin lead sheets; ^{32}P source, preferably but not necessarily in a special plastic rod mount; ^{60}Co source.

Part B—Geiger tube and scaler or ratemeter; magnetron magnet; ^{32}P or ^{90}Sr source in special plastic rod mount.

Part C—Geiger tube and scaler; tube and sample holder; set of aluminum absorbers; beta sources.

PROCEDURE:

PART A—DEMONSTRATION OF BREMSSTRAHLUNG PRODUCTION*

1. Prepare a ^{32}P source with sufficient activity to provide about 2,000 cpm at a distance of one foot from an end-window Geiger tube. For safety in handling this source, deposit the radioactive solution in a well made by drilling a ⅛″ hole about ¼″ deep in the end of a 1″ plastic rod. Dry under an infrared lamp.

2. Observe the activity with the source held as in step 1 of figure 7.18. The activity should be several thousand counts per minute.

3. Cover the source with a 1⁄16″ thick sheet of lead and notice the decrease in observed activity even when the source, still covered with the thin lead sheet, is moved close to the window of the G-M tube (step 2). Note: The G-M tube has an efficiency of nearly 100% for beta particles but an efficiency of only 1 or 2% for electromagnetic radiation.

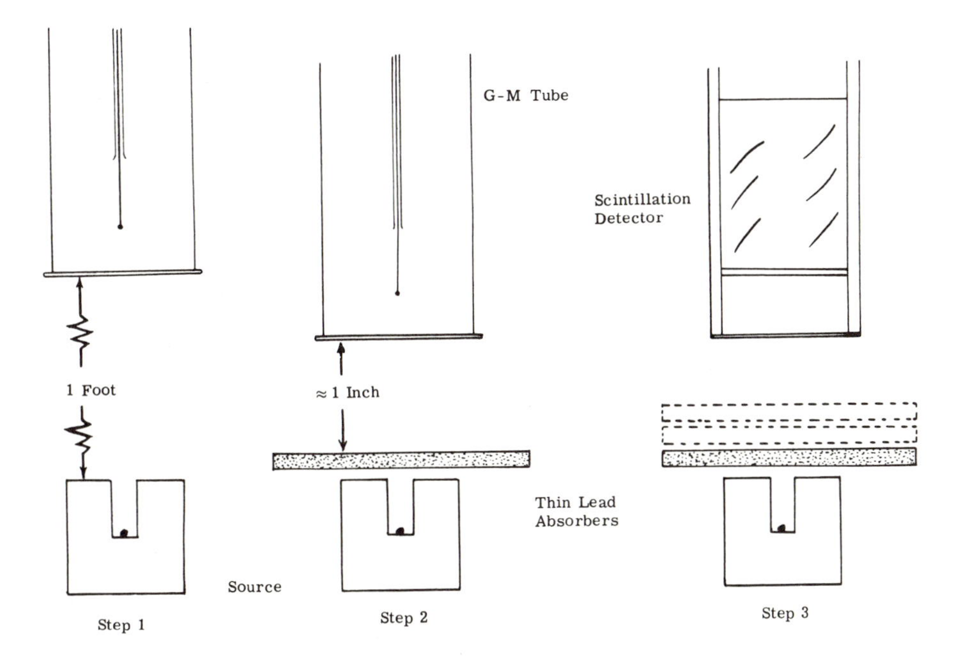

Figure 7.18—DEMONSTRATION OF BREMSSTRAHLUNG PRODUCTION.

4. Replace the G-M tube with a scintillation detector or other detector with a high efficiency for electromagnetic radiation. If a scintillation detector is used, the high voltage should be increased to just below the noise level in order that it be sensitive to low energy electromagnetic radiation. The increase in activity observed with the scintillation counter over that observed with the G-M counter is caused by bremsstrahlung.

* Chase, G. D., "Radioisotopes in the Educational Laboratory", Symposium on Nuclear Education, January 15, 1960, sponsored by Baird Atomic, Inc.

5. When additional sheets of lead are placed between the source and the scintillation detector, the activity decreases rapidly, indicating the "soft" nature of the bremsstrahlung. For comparison, this step should be repeated with a ^{60}Co source, a "hard" gamma emitter.

6. With a single sheet of lead between the source and the scintillation detector, observe the activity. Now measure the activity with a ⅛″ sheet of plastic, first between the lead and the detector and then between the lead and the source. How do you account for your observations?

7. If a gamma-ray spectrometer is available, the bremsstrahlung spectrum may be plotted. Place a thin sheet of lead between the source and the crystal detector with the lead snug against the source. (See experiment 7.6.) Note especially the shape of the spectrum and the predominance of low energy radiation.

Note: Success of this experiment depends upon the proper adjustment of the high voltage for the scintillation detector in steps 4 to 6. If it is too low, the low energy bremsstrahlung will not be detected. If it is too high, results will be masked by noise.

PART B—DEFLECTION OF BETA PARTICLES BY A MAGNETIC FIELD*

1. Either ^{32}P or ^{90}Sr can be used as a source of beta particles, but ^{90}Sr has the advantage of a long half-life which eliminates the need to prepare the source more than once. The beta source is carefully placed at the bottom of a ⅛″ hole drilled to a depth of 1″ to 1½″ in the end of a 1″ to 1¼″ plastic rod. Since the beta particles do not have sufficient energy to penetrate the plastic walls of this source-mount, the results is an "electron gun" with a thin stream of beta particles emerging from the hole.

2. The apparatus is assembled as shown in figure 7.19. It is suggested that a glass-walled Geiger tube be used. Metal-walled tubes will function properly, but care is required to prevent breakage of the window should the tube be drawn violently against the magnet.

3. When the magnet (a "surplus" magnetron magnet with a field strength of about 1500 gauss) is placed in position, beta particles leaving the end of the tube are deflected by the magnet and strike the Geiger counter. The ratemeter now shows a measureable de-

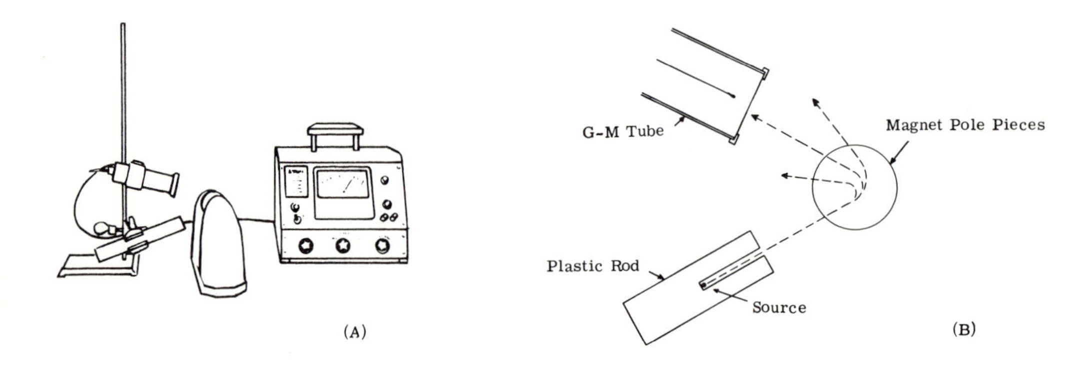

Figure 7.19—DEFLECTION OF BETA PARTICLES BY A MAGNETIC FIELD
(Courtesy Baird-Atomic, Inc.)

*See footnote page 205.

flection and with suitable adjustment can be made to read near full-scale. When the magnet is removed, the activity registered by the ratemeter should decrease nearly to zero. Placing the magnet in position, but with poles reversed, again produces essentially no activity above background.

PART C—BETA PARTICLE ABSORPTION

The absorption of radiation in a given medium can be used to advantage as an aid to the identification of a radioactive isotope. This is illustrated by the following paragraphs taken from the United States Pharmacopeia XVI.

"The absorption coefficient (u') using aluminum absorbers of graduated thickness, expressed in mg/cm², is commonly determined to characterize the beta radiation emitted by a radioactive isotope as by the following procedure."

(Note: The following procedure from the U.S.P. XVI is specifically for ^{131}I. It must be modified, as noted, when applied to other nuclides. Further, the method described here is not satisfactory for isotopes which emit beta particles of low energy such as ^{14}C and ^{35}S.)

"Place the radioactive substance, suitably mounted for counting, under a Geiger-Müller counter. Make activity determinations individually and successively, using at least six different thicknesses of aluminum chosen from the range of 10 to 60 mg/cm² and a single absorber thicker than 250 mg/cm². Obtain the net beta activity as the various absorbers used by subtracting the activity found with the 250 mg/cm² or greater absorber.* Plot the logarithm of the net beta-activity as a function of the total absorber thickness. The total absorber, thickness is the thickness of the aluminum in mg/cm² plus the thickness of the Geiger-Müller window (as stated by the manufacturer), plus the air equivalent thickness (the cm distance of sample from the Geiger-Müller window $\times$ 1.205) all expressed in mg/cm². A straight line results.

"Choose two of the absorber thicknesses 20 mg/cm² or more apart and which fall on the plot, and calculate the mass absorption coefficient by the equation

$$u' = \frac{1}{x_2 - x_1} \ln \frac{R_1}{R_2} \tag{7.26}$$

where x_1 and x_2 = total absorber thickness greater that 10 mg/cm² and differing by a least 20 mg/cm², x_2 being the thicker absorber
and R_1 and R_2 = net beta activity at x_1 and x_2 absorbers respectively."

"For characterization of the isotope, the absorption coefficient should be within plus or minus 5 per cent of that found for a sample of the same isotope of known purity when both are determined by identical procedures."

"Activity at zero total absorber may be determined by plotting a curve identical to the one described for determination of the absorption coefficient and extrapolating the straight line plot to zero absorber, taking into consideration the mg/cm² thickness of the sample coverings, the air, and the Geiger-Müller window."

When beta-ray absorption is used for the identification of ^{32}P, notice should be made of the modification made in the U.S.P. XVI procedure.

"Use at least six different thicknesses of aluminum chosen from the range 50 mg/cm² to 140 mg/cm² in the preparation of the curve. Calculate the mass absorption coefficient using two absorber thicknesses, x_1 and x_2 which fall on the curve, where x_1 and x_2 represent total absorber thicknesses greater than 50 mg/cm² and differ by at least 20 mg/cm², x_2 being the thicker absorber."

* Absorber thickness for beta emitting isotopes should be great enough to just absorb all beta particles. The value of 250 mg/cm² is for ^{131}I. For other isotopes the value of maximum range can be determined from E_m for the isotope by the use of the range-energy relationships shown in figure 7.24.

It is suggested that each student be given a specimen of a known radioisotope, e.g., iodine-131, as well as several unknown specimens, one of which is of the same species as the known. Calculate the mass absorption coefficient of each sample and identify the unknown which compares with the known.

DATA:

Known Source________________

Total Absorber mg/cm^2	Total Counts	Total Time	Observed Activity r	Corrected Activity R

Unknown Source # ________________

Total Absorber mg/cm^2	Total Counts	Total Time	Observed Activity r	Corrected Activity R

CALCULATIONS:

Plot the data on semi-log paper. Calculate the slope u′ for each curve and report below.

	u′
Known Source	________
Unknown Source # ______	________
Unknown Source # ______	________

EXPERIMENT 7.3 BETA-PARTICLE RANGE AND MAXIMUM ENERGY (Feather Analysis)

OBJECTIVES:

To study the Feather method for measuring the maximum range of beta particles.

To apply various range-energy relationships for the determination of the maximum beta energy.

THEORY:

Experimentally it has been found that beta particles from a radioactive source give a range distribution which closely resembles an exponential curve (see experiment 7.2). That the curve should be even approximately exponential is somewhat coincidental. The effect has been explained as the combined result of two phenomena: (1) An electron passing through matter undergoes many collisions, both elastic and inelastic, which leads to much straggling; and (2) The energy spectrum of beta particles emitted from a nucleus is not monoenergetic.

It might well be added that the first phenomenon—collisions of electrons with matter—produces the observed straggling of electrons by scattering, wherein the path of the electron is deflected, and also by absorption of the energy of the electron, with the result that its velocity is eventually decreased to zero.

Referring to figure 7.16, one might surmise that the maximum range of the beta particles from a given source could be measured easily by incrementally increasing the absorber thickness until the counting rate just drops to zero. Certanly this would appear to be the least complicated method, but it is, unfortunately, not a practical routine method. It is in the lower right hand portion of the curve (figure 7.16) that the greatest uncertainty is incurred. This is due to the very low measured activities. The reduction of statistical errors when using very thick absorbers would involve impractically long counting times.

Feather* describes a comparison method for the routine determination of maximum beta range, wherein the range of the unknown is compared to the range of beta particles from a standard isotope which has been calibrated previously by a more tedious method, such as the one described above. Feather used Radium-DEF as the standard, the energetic beta particle emanating from Radium-E constituting the usable radiation. E_m for Ra-E betas is 1.17 MeV.

We now know that Feather made an unfortunate selection for his reference standard which, in the light of knowledge at the time his method was published, appeared to be a wise choice. Ra-E (^{210}Bi) decays by a "forbidden" transition. The beta-ray spectrum of Ra-E is therefore skewed toward the low energy region (see figure 7.9). For this reason the absorption curve for Ra-E will differ somewhat from those produced by nuclides with a higher percentage of higher energy beta particles. It is also known that the value for the maximum range of the Ra-E beta particles is about 505 mg/cm^2. However, good agreement has been obtained using Feather's value of 476 and the experiment has been based upon this figure. If desired, other standards may be used, e.g., ^{204}Tl. The appropriate value for the maximum range of the new standard must then be used in place of the value 476 mg/cm^2 for Ra-E.

MATERIALS REQUIRED:

Ra-DEF source; phosphorus-32 source; calibrated absorbers.

* Feather, N., Proc. Cambridge Phil. Soc. *34*, 599 (1938)

PROCEDURE:

In this experiment the beta range of an unknown isotope (phosphorous-32, iodine-131 or other suitable beta emitter) will be compared to the beta range for Radium-E by the Feather method. From the data the value of the maximum beta range R_m for the unknown isotope can be calculated. Then, by means of an equation or graph, the maximum beta energy E_m will be evaluated from R_m.

1. *The "Unknown"*

Prepare a sample of the radioactive unknown on a thin plastic mount or in a planchet which will give approximately 10,000 cpm on the second shelf of the sample holder. Measure the activity of the sample with the thinnest absorber on the first shelf. (The thinnest absorber has zero thickness and consists of nothing but a plastic ring with no aluminum foil. It is used to create constant scattering for all readings). Then insert each absorber in the top shelf in sequence, beginning with the thin absorbers, and take a count with each one. Where practical, at least 5,000 total counts should be registered; preferably 10,000. The sample should not be moved throughout these measurements. Continue the series of observations using increasingly thicker absorbers until there is no appreciable further decrease in activity.

"Unknown" Sample

ADDED ABSORBER THICKNESS MG/CM²	TOTAL ABSORBER THICKNESS MG/CM²	TOTAL COUNTS	ACTIVITY cpm	ACTIVITY CORRECTED FOR BACKGROUND	ACTIVITY CORRECTED FOR BACKGROUND AND COINCIDENCE

Added absorber refers to the indicated thickness of the absorber placed on the first shelf. To obtain *total absorber,* the absorption of the window, the absorption due to air between the sample and the window, and the thickness of any cover, if used, which was placed over the sample for protection, must be added to the "added absorber" thickness.

A. Distance from source to window ____________ cm
B. Air density ____1.2____ mg/cm³
C. Air absorption (A × B) ____________ mg/cm²
D. Window density (marked on tube) ____________ mg/cm²
E. Sample cover (if used) ____________ mg/cm²
F. Absorber correction (C+D+E) ____________ mg/cm²

The *total absorber* is then equal to the added absorber + F.

The *background correction* to be applied must include not only the natural background from the surroundings and from cosmic radiation, but also that due to a gamma component—e.g., when using ^{131}I or ^{60}Co as the unknown—and also the contribution caused by bremsstrahlung. These situations are illustrated in figure 7.20. Curve "A" depicts the typical curve anticipated from a pure beta emitter. Even at large values of absorber thickness, the "tail" does not decrease appreciably below a certain value. Curve "B" illustrates the increased background anticipated in the case of a gamma emitting isotope.

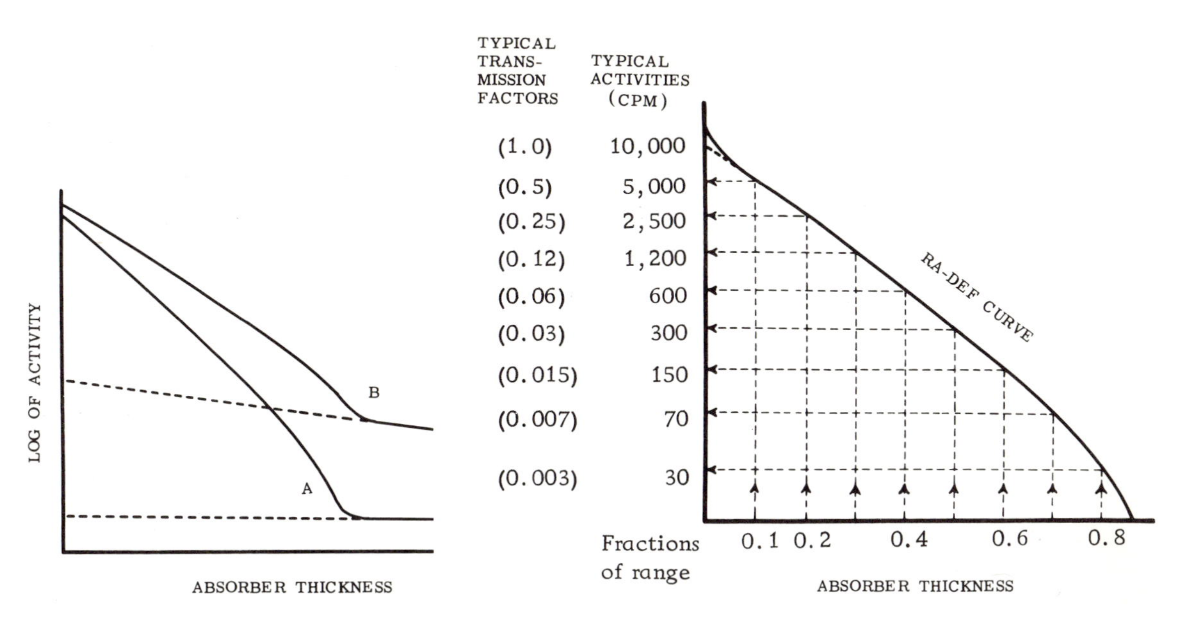

Figure 7.20 — TYPICAL BETA-PARTICLE ABSORPTION CURVES.

Figure 7.21—FEATHER ANALYSIS. Calculation of transmission factors.

Radium-DEF Sample

ADDED ABSORBER THICKNESS MG/CM²	TOTAL ABSORBER THICKNESS MG/CM²	TOTAL COUNTS	ACTIVITY cpm	ACTIVITY CORRECTED FOR BACKGROUND	ACTIVITY CORRECTED FOR BACKGROUND AND COINCIDENCE

Calculate the sample activities corrected for background and for dead time. Record on the data sheet.

2. *Radium-DEF Standard*

Replace the "unknown" sample with a radium-DEF standard which also has an activity of about 10,000 cpm on the second shelf. Repeat the procedure outlined in paragraph 1, using the standard in place of the unknown.

Range of Ra-DEF in aluminum = 476 mg/cm^2.

CALCULATION OF RANGE:

1. Plot the results of the Ra-DEF standard on semi-log graph paper. The corrected activity (cpm) should be plotted as the ordinate, and the total absorber thickness (mg/cm^2) as the abscissa.

2. Extrapolate this curve to zero total absorber thickness and, in the reverse direction, data should be plotted to as near 476 mg/cm^2 as possible with accuracy. Extrapolation to zero total absorber thickness corrects for the absorption due to air and the

window of the G-M tube and the cellophane, if used, covering the sample. Extrapolate with consideration of the overall shape of the curve. Ignore any sharp upward deflection of the curve near zero absorber thickness which is caused by the presence of weak beta components (from Ra-D). The value 476 mg/cm² is the maximum range.

3. Divide the abscissa into fractions of 1/10 of the range; i.e., at intervals of 47.6 mg/cm², beginning at zero mg/cm². (See figure 7.21).

4. Project lines from each interval of 0.1 of the range on the abscissa up to the curve and then to the ordinate, and mark off activities on the ordinate as read from the curve. Typical values of activity are indicated in figure 7.21. From these activities, at the points of intersection of the projections with the ordinate, will be calculated the transmission factors.

5. Assign to the total corrected activity (cpm) at zero total absorber thickness a *transmission factor* of unity. Assign a relative value to other ordinate intercepts. The transmission factors represent the fractions of the total number of beta particles penetrating the absorbers through 0.1, 0.2, 0.3, etc., of the total range.

6. Plot the results for the "unknown" isotope on semi-log paper in the identical manner used for Ra-DEF. Extrapolate the graph to zero total absorber thickness. Insert the activity of the unknown at zero total absorber thickness in the first space of the last column in the chart below.

7. Multiply the activity of the unknown at zero total absorber thickness by each transmission factor, respectively, and record the resulting activities in the above chart.

Tenths of Range	Ra-DEF Activity cpm	Transmission Factor	Unknown Activity at zero total absorber thickness multiplied by transmission factor
0		1.0	
1			
2			
3			
4			
5			
6			
7			
8			
9			
10			

8. Along the ordinate of the unknown curve, lay off points corresponding to these calculated activities as indicated by the arrows in figure.

9. From these points project lines to the curve and then to the abscissa. These points (on the abscissa) correspond to absorber thicknesses representing 0.1, 0.2, 0.3, etc., of the range. Record results in column B in the chart below.

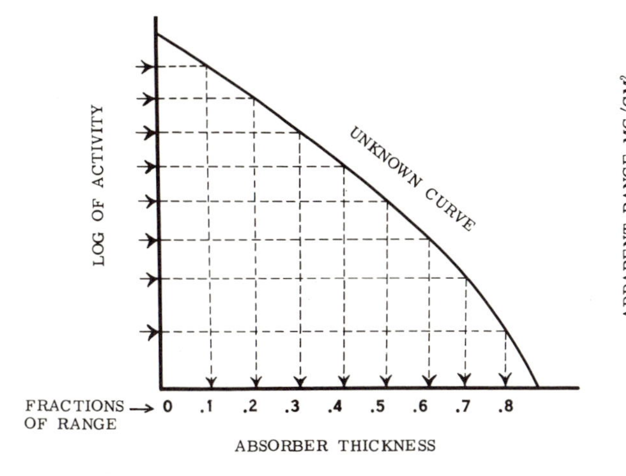

Figure 7.22—FEATHER ANALYSIS. Evaluation on tenths of range of unknown isotope.

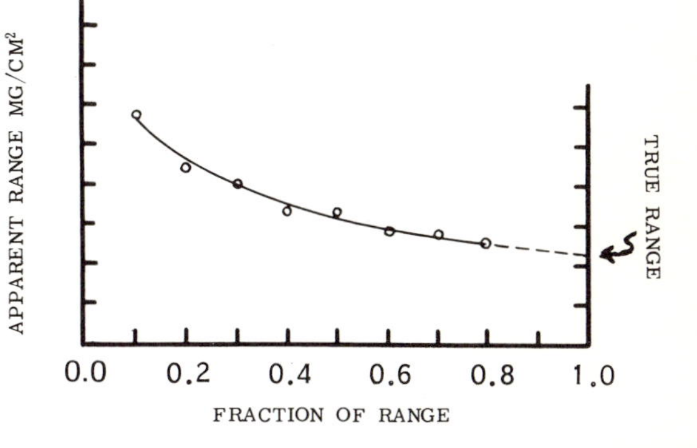

Figure 7.23—EXTRAPOLATION TO TRUE RANGE.

A	B	
Fraction of range	Unknown absorber thickness mg/cm²	Apparent Range (B/A) mg/cm²
0.1		
0.2		
0.3		
0.4		
0.5		
0.6		
0.7		
0.8		
0.9		
1.0		

10. The *apparent range* is calculated by dividing the absorber thickness by the fraction of the range.

11. The *true range* is calculated by plotting values for the apparent range versus fraction of the range. The resulting curve is then extrapolated to 10/10ths of the range. This extrapolated value is the true range.

CALCULATION OF ENERGY:

Many attempts have been made to formulate an empirical relation between the range of beta particles and their energy. One of the most satisfactory is that of Glendenin, which applies to energies greater than 0.8 MeV and to ranges greater than 300 mg/cm^2.

$$R_m = 542 E_m - 133$$

where R_m is the maximum range in mg/cm^2 and E_m is the maximum energy of the beta particles in MeV.

To relate ranges and energies below values where the Glendenin equation is not valid, use the graph (figure 7.24) compiled from the data of Glendenin (19), Varden (29), Madgwick (25) and Marshall and Ward (27).

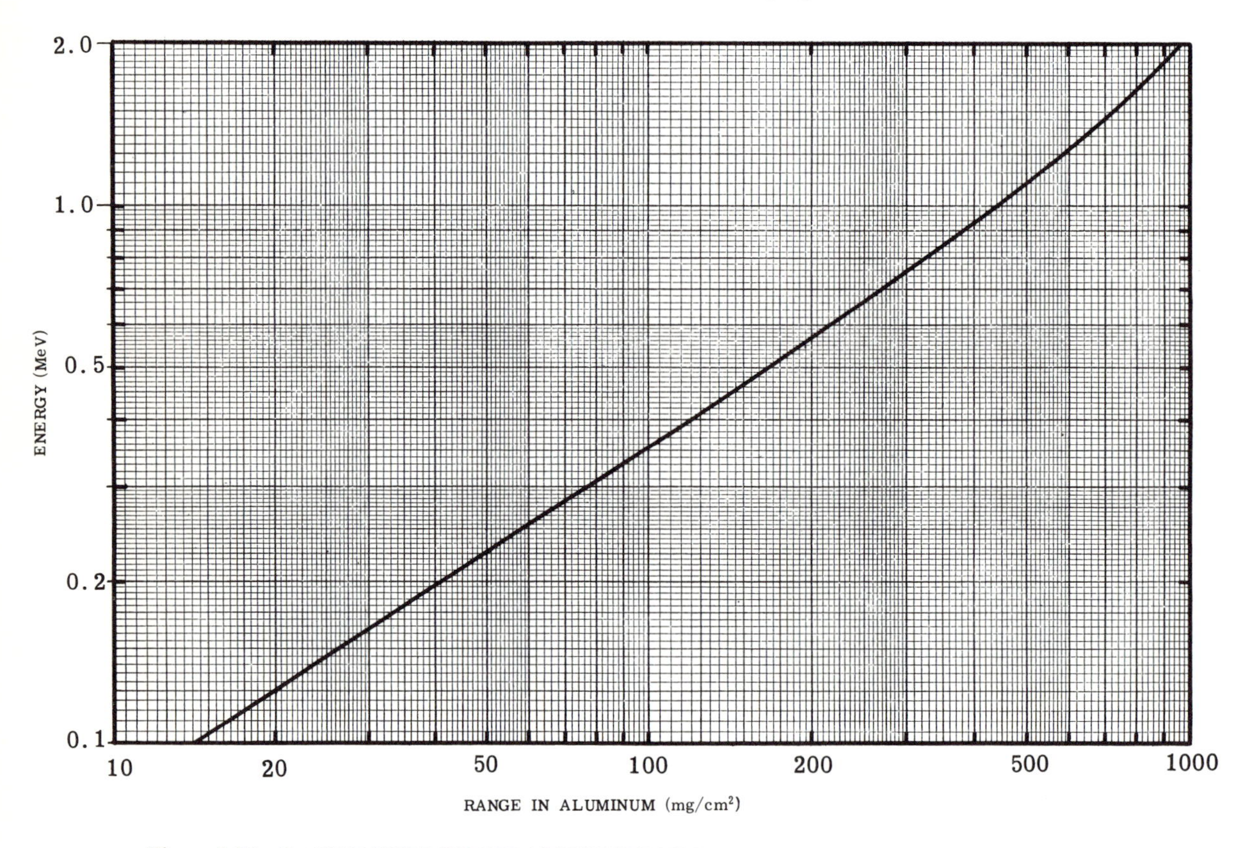

Figure 7.24—RANGE-ENERGY RELATIONSHIP FOR BETA PARTICLES IN ALUMINUM.

REPORT:

1. Calculate the maximum range of the beta particles of the unknown by plotting the three graphs indicated above. R_m ____________ mg/cm²
2. Calculate the maximum energy of these particles from their maximum range, using the equation of Glendenin. E_m ____________ MeV
3. Report the identity of the "unknown" sample____________

EXPERIMENT 7.4 PROPERTIES OF GAMMA RAYS AND THEIR INTERACTION WITH MATTER

OBJECTIVES:

To study the nature and source of gamma rays.

To observe the modes of interaction of gamma rays with matter.

To demonstrate the scattering of gamma rays by matter, and to note the effect of scattering on quantitative measurements of gamma emitters.

THEORY:

Gamma radiation, unlike alpha and beta radiation, is an electromagnetic wave. Gamma radiation is thus radiated as photons or quanta of energy which travel with the velocity of light, $c = 3.0 \times 10^{10}$ cm/sec. Gamma radiation differs from x-rays, visible light, radio waves, etc., only in wavelength λ or frequency ν, as illustrated in the diagram of the electromagnetic spectrum (figure 7.25). Wavelength and frequency are related to the velocity of light by the equation,

$$\lambda = \frac{c}{\nu} \tag{7.27}$$

The energy of a photon can be calculated by use of the relationship

$$E = h\nu \tag{7.28}$$

where E is the energy in ergs, h is Planck's constant (6.624×10^{-27} erg sec) and ν is the frequency in vibrations per second. Another useful relationship can be derived relating wavelength with the energy, as follows:

$$\lambda_{cm} = \frac{1.24 \times 10^{-10}}{E_{MeV}} \tag{7.29}$$

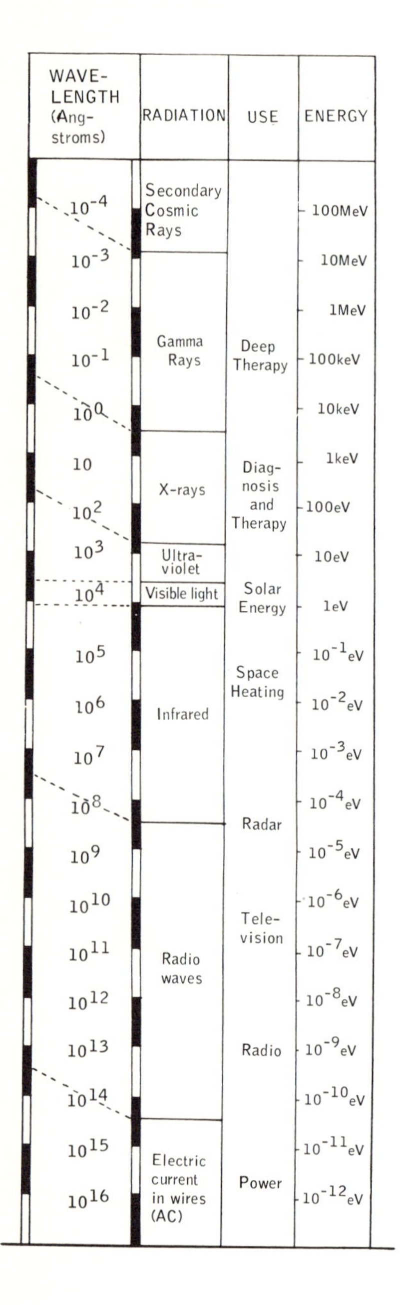

Figure 7.25—ELECTROMAGNETIC SPECTRUM.

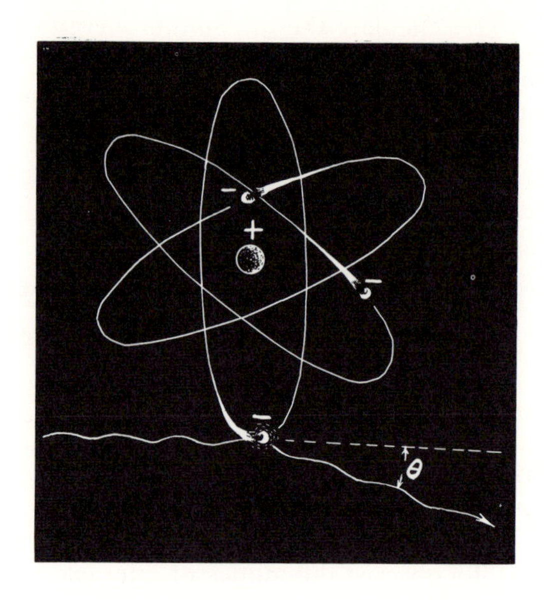

Figure 7.26—THOMSON SCATTERING.

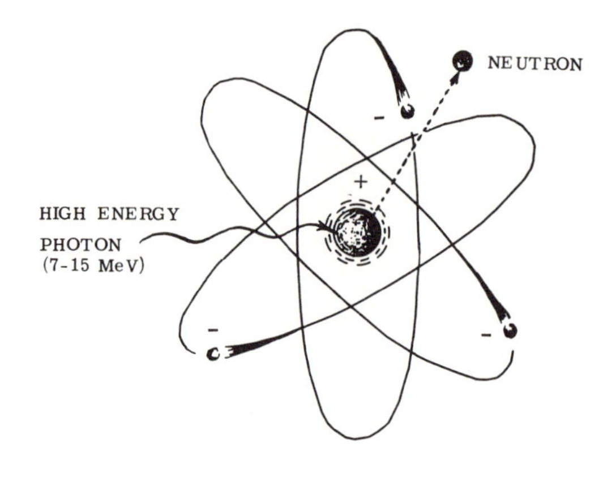

Figure 7.27—PHOTODISINTEGRATION.

A significant difference between x-rays and gamma rays is their source. The source of x-rays is extranuclear, that of gamma rays intranuclear. The process by which x-rays are produced is identical to that producing bremsstrahlung (see page 200). In an x-ray tube, high velocity electrons are generated and allowed to impinge on a target. The electrons undergo inelastic collisions with target nuclei which result in the deceleration of the electrons and the emission of electromagnetic radiation. Inelastic collisions with orbital electrons cause ionization and the production of characteristic x-rays.

Gamma rays are produced by nuclear energy transitions. For example, when cobalt-60 decays by beta emission, nickel-60 is produced. The process not only involves the emission of a beta particle, but of two gamma quanta as well. Electromagnetic radiation is also emitted during decay by electron capture. These topics are discussed in chapter 6.

When a photon interacts with matter, the collision may occur with a nucleus, an electron or with the field about the nucleus. This collision may be elastic, inelastic or may result in the complete absorption of the photon. These interactions are summarized in table 7.2. Interactions of some importance are Rayleigh scattering, photodisintegration, nuclear resonance scattering, Bragg scattering and Thomson scattering while the photoelectric effect, the Compton effect and pair production are of major importance in the utilization of radioisotopes.

Rayleigh scattering—If the interaction between a photon and an orbital electron is insufficient to produce ionization or excitation of the atom, the collision is elastic and the energy of the photon is the same after the collision as it was before. Rayleigh scattering rarely results in scattering of the photon through more than a few degrees.

Thomson scattering—It was first thought that x-rays should be reflected from a mirror the same as visible electromagnetic radiation. Soon after their discovery it was observed that they are scattered rather than reflected. This scattering was explained by J. J. Thomson as an interaction of x-rays with orbital electrons. The electrons are presumed to absorb the electromagnetic radiation, oscillate in an excited state and re-radiate the x-radiation in random directions.

Photodisintegration—Here the interaction involves a collision of a high energy photon with a nucleus. The photon is completely absorbed in the process, and a neutron, proton or alpha particle is ejected from the excited nucleus. The following are typical examples of photodisintegration:

$$^{2}_{1}D + \gamma \longrightarrow {}^{1}_{1}H + {}^{1}_{0}n$$
$$^{9}_{4}Be + \gamma \longrightarrow {}^{8}_{4}Be + {}^{1}_{0}n$$

Nuclear resonance scattering—If the vibrational frequency of a nucleus is equal to that of an incident photon, absorption of the photon may occur. The photon is then re-emitted from the excited nucleus. In 1961, R. L. Mössbauer received the Nobel prize for his resonance experiments with the iron isotope ^{57}Fe excited by radiation from the decay of ^{57}Co. Detection of this Mössbauer effect is accomplished by means of a Doppler effect by moving the ^{57}Fe target with respect to the ^{57}Co source to compensate for the energy of recoil of the ^{57}Fe nucleus. Mössbauer's work is of considerable importance to physics.

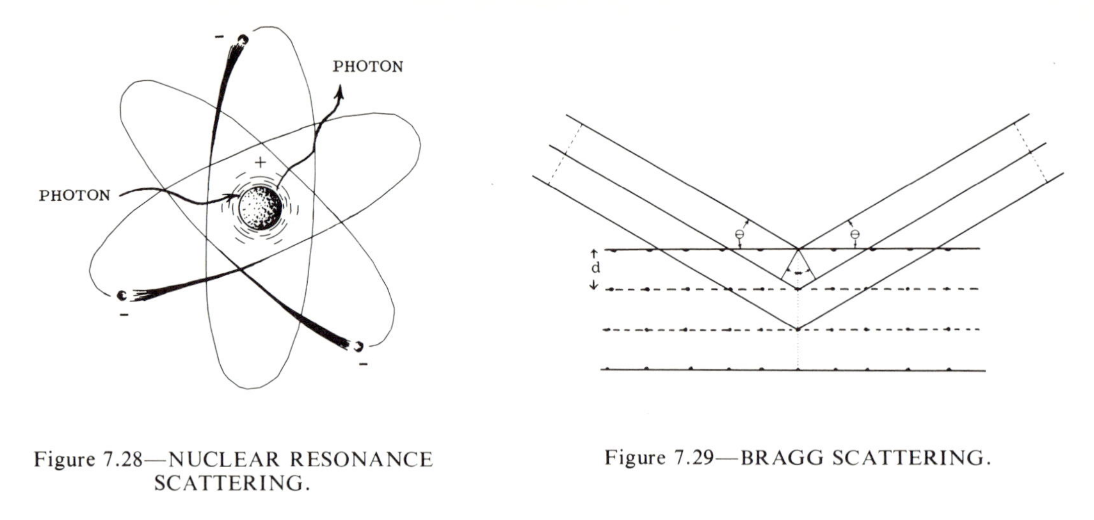

Figure 7.28—NUCLEAR RESONANCE SCATTERING.

Figure 7.29—BRAGG SCATTERING.

Bragg scattering occurs with gamma rays in the same manner that the Bragg scattering of x-rays (x-ray diffraction) takes place at a crystal face. It is an elastic coherent type of scattering. It is said to be coherent because there are definite phase relationships between the incident and scattered waves.

$$n\lambda = 2d \sin \theta \qquad (7.30)$$

If the angle θ at which the radiation strikes the face of the crystal is related to the wavelength λ of the incident radiation and the distance d between planes of atoms within the crystal by the Bragg equation (7.30) such that n is a small, whole number, reinforcement of the intensity of radiation will occur. This phenomenon, incidentally, affords us one of the most accurate procedures by which the wavelength may be determined.

The *Photoelectric effect* occurs principally when the photon energy is low. The inelastic collision of the photon with the orbital electron results in the complete ejection of the electron and the production of an ion pair. The kinetic energy of the ejected electron is given by the equation,

$$\tfrac{1}{2}mv^2 = h\nu - \varphi \qquad (7.31)$$

where φ is the work function or binding energy of the electron. This means that the total energy imparted to the electron by the photon is equal to that required to remove it an infinite distance from the nucleus φ plus the kinetic energy of the electron $\tfrac{1}{2}mv^2$. In this process, a K-electron is usually involved. The photoelectric effect is most pronounced if the atomic number Z of the absorbing material is high.

The *Compton effect* or Compton scattering is especially important for gamma rays of medium energy (0.5 to 1.0 MeV). It involves a collision between a photon and an electron in which a part of the energy of the photon is imparted to the electron. The photon emerges from the collision in a new direction and with reduced energy. Considering that both energy and momentum must be conserved in the collision, we can derive the following relationship:

$$\lambda' - \lambda = (h/mc)(1 - \cos \phi) \qquad (7.32)$$

Thus, the change in wavelength of the photon is related to the scattering angle by equation 7.32. The term h/mc is called the Compton wavelength λ_0 and has the value 2.43×10^{-10}cm.

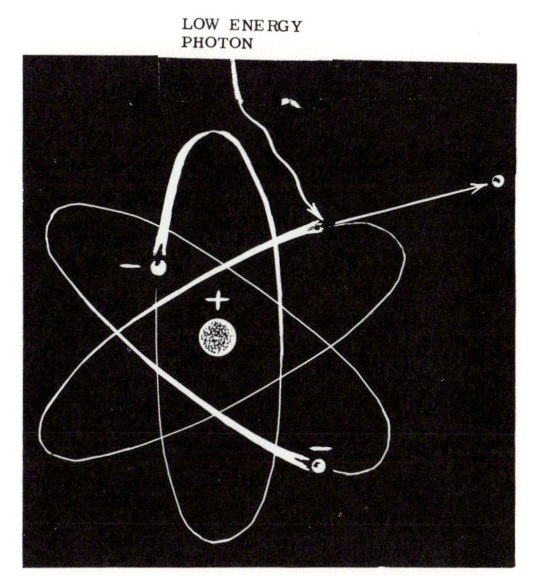

Figure 7.30—PHOTOELECTRIC EFFECT.

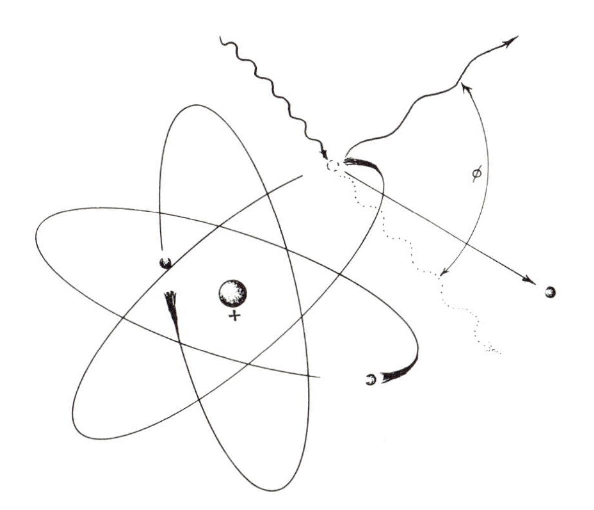

Figure 7.31—COMPTON SCATTERING.

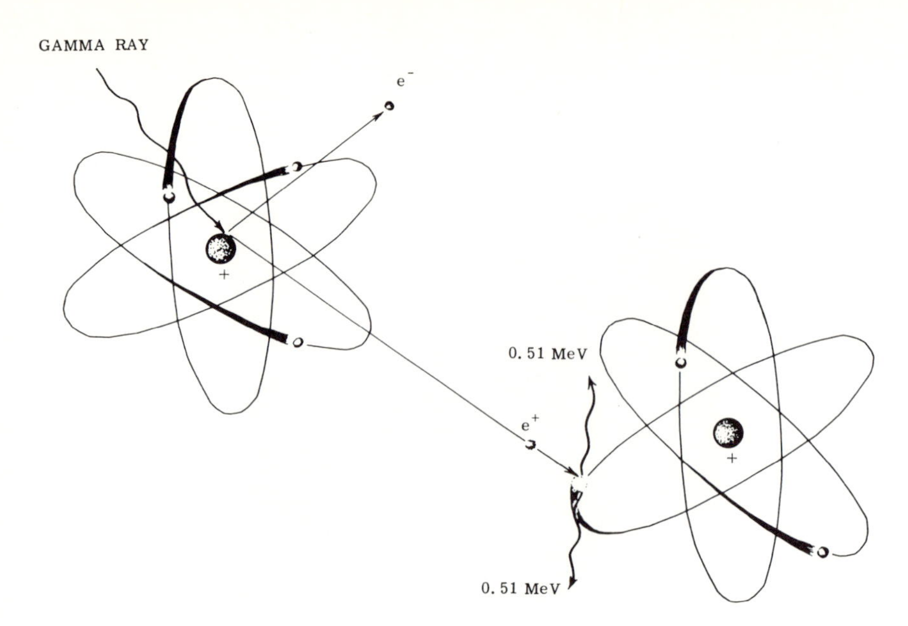

Figure 7.32—PAIR PRODUCTION.

Pair production is involved only with gamma rays having energies greater than 1.02 MeV. The energy of the gamma ray is converted into an electron and a positron in the region of a strong electromagnetic field such as that surrounding the nucleus. Photon energy in excess of 1.02 MeV appears as the kinetic energies of the electron and positron produced.

$$E = h\nu = 2m_0c^2 + T_{e-} + T_{e+} \tag{7.33}$$

where $2\ m_0c^2 = 1.02$ MeV and represents the energy required to form the pair of particles according to the Einstein equation, $E = mc^2$. T_{e-} and T_{e+} are the kinetic energies of the electron and positron, respectively.

PROCEDURE:

PART A – DEMONSTRATION OF COMPTON SCATTERING

Arrange the apparatus as shown in figure 7.33. A gamma source such as cobalt-60 is placed in position as shown, with about 4 inches of lead between it and the scintillation detector.

See experiment 9.1 for further information on scintillation detectors. An empty beaker is placed in front of the detector so that the source, the beaker and the detector form a right angle. Measure the activity. The high voltage should be adjusted to just below the noise level to provide high sensitivity of the scintillation detector to the lower energy scattered radiation.

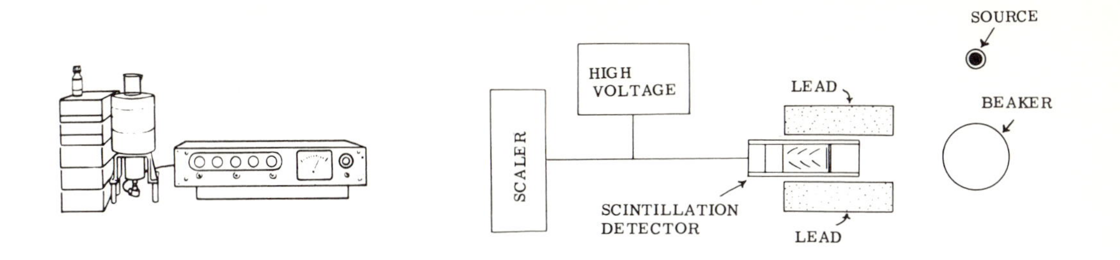

Figure 7.33—DEMONSTRATION OF COMPTON SCATTERING.

Now fill the beaker with water. Do not disturb the position of the source, detector or shielding. Again record the activity.

Activity (empty beaker)__________________cpm
Activity (beaker filled with water)____________cpm

How do you account for the change in activity?

PART B—EFFECT OF COMPTON SCATTERING ON QUANTITATIVE RESULTS

This experiment demonstrates radiation scattering problems associated with the measurement of iodine uptake by the thyroid. The scattering problems encountered here are also encountered in uptake measurements. The test tube containing ^{131}I solution represents the thyroid gland and the beaker represents the neck.

Arrange the apparatus as shown in figure 7.34. The scintillation detector should be surrounded with a lead shield or by lead bricks, as in the illustration, to reduce background. The high voltage is adjusted so that noise pulses are not detected by the scaler. Make the following observations.

1. Remove the source from the room and measure the background.

2. Place the source inside the empty beaker in position 1, close to the wall of the beaker and midway between the bottom and the top of the beaker. The source should

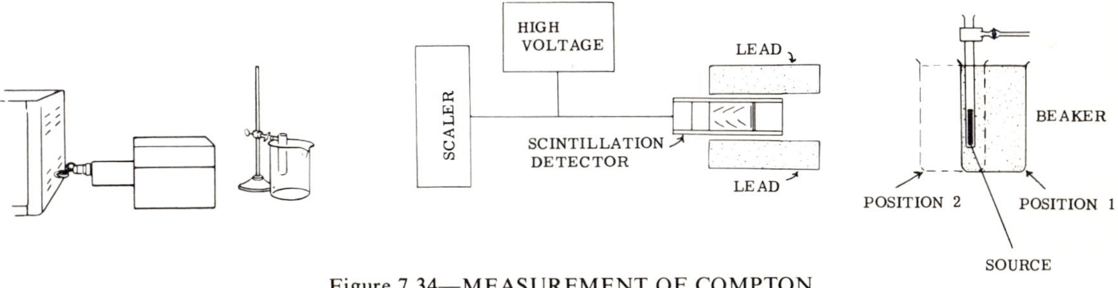

Figure 7.34—MEASUREMENT OF COMPTON SCATTERING.

also be in line with the detector. Measure the activity. The source used may be cobalt-60. It should be sealed in a water tight tube or glass vial.

3. Fill the beaker with water. Be careful not to disturb the position of the source and be sure that the source is covered with water. Again measure the activity.

4. Remove the beaker without disturbing the source. Empty out the water and replace the beaker in position 2. Measure the activity.

5. Fill the beaker again with water and measure the activity as before.

6. Repeat steps 1 through 5 using a source of iodine-131 in place of cobalt-60.

Background ______________ cpm

Source	Beaker	Beaker in Position	Activity		% change (Note if + or −)
			Observed	Corrected	
^{60}Co	empty	1			
^{60}Co	full	1			
^{131}I	empty	1			
^{131}I	full	1			
^{60}Co	empty	2			
^{60}Co	full	2			
^{131}I	empty	2			
^{131}I	full	2			

Explain the observed changes in activity in terms of Compton scattering, absorption and energy of the radiation. How can the ^{131}I content of a thyroid be measured with accuracy? Can all scattering errors be eliminated in such a measurement?

PART C—SPECTRUM AND ENERGY OF SCATTERED GAMMA RADIATION

The apparatus required to plot a gamma spectrum (figure 7.35) consists of a scintillation detector, a high voltage source, amplifier, pulse height analyzer, counting rate meter, and a recorder. Provision must be made for driving the base line synchronously with the recorder. The source must have a fairly high activity for satisfactory results when the spectrum of the scattered radiation is to be plotted. A suitable source consists of 0.1 to 1.0 millicurie of cobalt-60 housed in a well of lead to collimate the beam. A scattering material, such as a beaker of water, is placed at the apex of the angle formed with the source and the detector. As much lead as is convenient should be placed in the direct path between the source and the detector to reduce background.

With the cobalt-60 source (or other gamma source) in position, but with the scattering material removed, record the activity (cpm) as a function of gamma energy (MeV). This is the background. Now place the scattering material in position and again record activity versus gamma energy. Typical curves are shown in figure 7.36.

$$\lambda = \frac{c}{\nu} \text{ and } E_\gamma = h\nu$$

Therefore,

$$\lambda = \frac{hc}{E_\gamma}$$

Substitution into equation 7.32 gives

$$\frac{hc}{E'} - \frac{hc}{E_\gamma} = \frac{h}{mc} \quad (1 - \cos \phi)$$

or

$$\frac{mc^2}{E'} - \frac{mc^2}{E_\gamma} = 1 - \cos \phi$$

but

$$mc^2 = 0.51 \text{ MeV}$$

and

$$\frac{0.51}{E'} - \frac{0.51}{E_\gamma} = 1 - \cos \phi \tag{7.34}$$

Here, E_γ is the energy of the incident radiation and E' is the energy of the Compton scattered radiation.

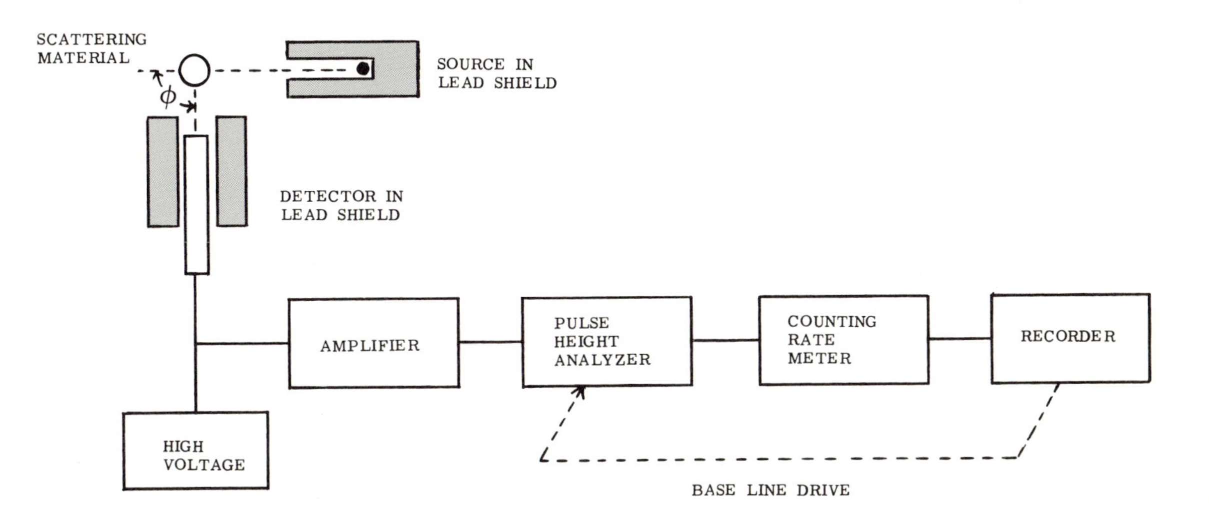

Figure 7.35—MEASUREMENT OF SPECTRAL DISTRIBUTION OF COMPTON SCATTERED RADIATION.

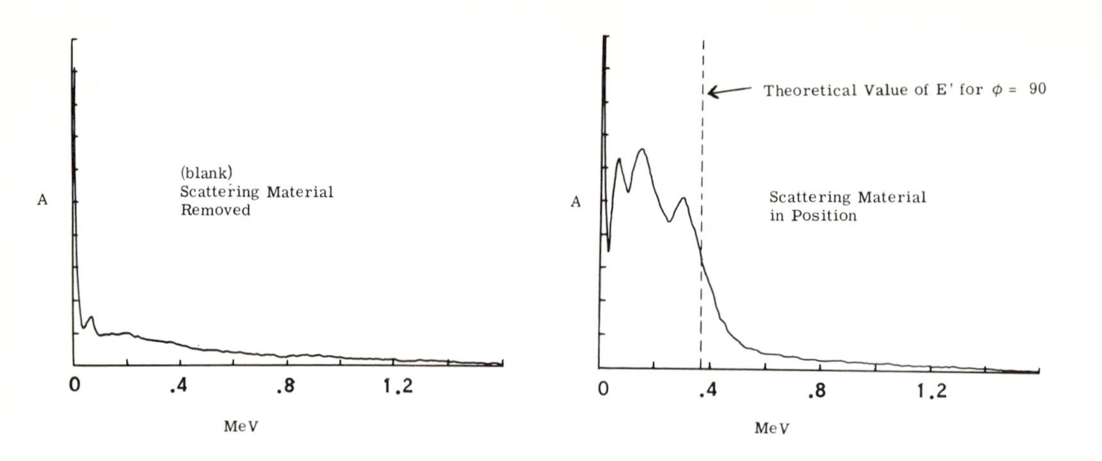

Figure 7.36—SPECTRAL DISTRIBUTION OF COMPTON SCATTERED RADIATION.

EXPERIMENT 7.5 ABSORPTION AND ENERGY OF GAMMA PHOTONS

OBJECTIVES:

To study the effect of the nature of the absorbing material on the absorption of gamma radiation.

To measure the linear and mass absorption coefficients for a given set of experimental conditions.

To estimate the energy of gamma emission from a given source.

THEORY:

As gamma radiation passes through matter, it undergoes absorption by interacting with atoms of the absorbing material, principally by the *photoelectric effect*, the *Compton effect*, and by *pair production*. The result is a decrease in the intensity of the radiation with the distance traversed through the absorbing material. The decrease in the energy of an incident beam of gamma radiation is exponential in form, as expressed by Lambert's law:

$$I = I_0 e^{-\mu X} \tag{7.35}$$

or

$$\ln (I/I_0) = -\mu X \tag{7.36}$$

Here, I_0 is the intensity of the incident beam of photons, I is the intensity after traversing a distance X through the substance and μ is the *linear absorption coefficient.* This relationship is depicted graphically in figure 7.38. It should be observed, from equation 7.35 or equation 7.36 that, theoretically, complete absorption of the radiation never occurs.

A useful concept regarding gamma absorption is the *Half-Value-Layer* (HVL) or the *half-thickness* $X_{1/2}$ which is defined as the distance of travel through an absorber required to decrease the intensity of a beam of gamma rays to one-half its initial value. Thus, after a ray has passed through a half-thickness of absorber, the intensity of the beam I is equal to $\frac{1}{2} I_0$. Rearranging equation 7.36 and substituting $\frac{1}{2} I_0$ for its equal, I, one obtains

$$\ln(I_0/I) = \mu X$$
$$2.303 \log (I_0/\tfrac{1}{2}I_0) = 2.303 \log 2 = \mu X_{1/2}$$
$$X_{1/2} = \text{HVL} = 0.693/\mu \qquad (7.37)$$

The significance of half-thickness with respect to the beam intensity is illustrated in figure 7.37.

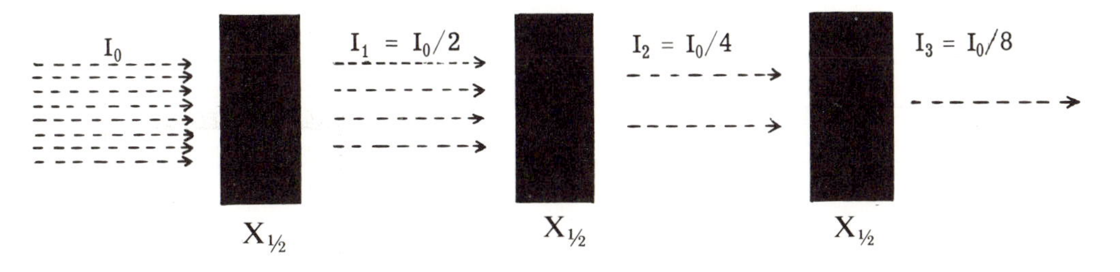

Figure 7.37—ATTENUATION OF GAMMA RADIATION. The significance of half-value-layer or half-thickness is illustrated.

Half-value-layers (values of half-thickness) for aluminum and lead are given in figures 7.40 and 7.41, respectively, in terms of $x_{1/2}$, where $x_{1/2} = \rho X_{1/2}$.

The *linear absorption coefficient* μ, the value of which depends upon the nature of the absorbing material, has the units cm^{-1}. The *mass absorption coefficient* μ' is defined as μ/ρ and has the units cm^2g^{-1}, where ρ is the density of absorbing material. If ln I is plotted versus absorber thickness in *centimeters,* the slope of the curve gives $-\mu$, the *linear absorption coefficient.* If ln I is plotted versus absorber thickness in grams per square centimeter (g/cm^2), the slope is equal to $-\mu/\rho$, the *mass absorption coefficient.*

It has been shown that when the log of the intensity is plotted versus the absorber thickness, an essentially straight line is obtained, indicating a linear relationship and the constancy of the coefficient μ. That this should be so seems somewhat miraculous in view of the fact that μ is the sum of no less than three other coefficients showing dependency on the gamma ray energy; namely, the atomic number, the mass the density of the absorbing medium, as well as other factors. Thus,

$$\mu = \tau + \sigma + K \qquad (7.38)$$

where τ = photoelectric absorption coefficient,
σ = Compton scattering coefficient,
and K = pair production coefficient

These three processes are discussed in experiment 7.4. Contributions from Rayleigh scattering, Bragg scattering, photodisintegration and nuclear resonance scattering are negligible.

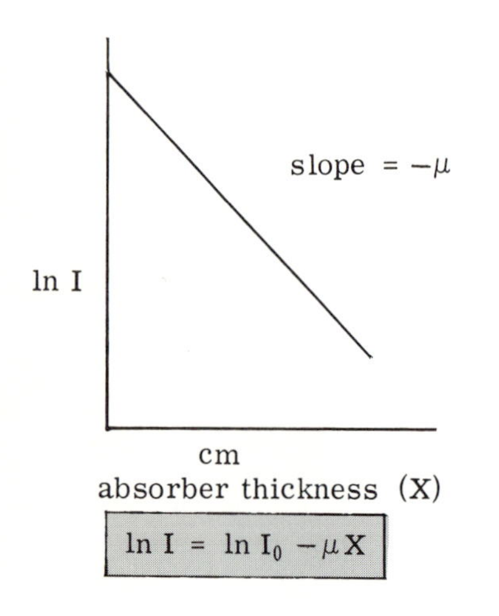

Figure 7.38—LINEAR ABSORPTION.

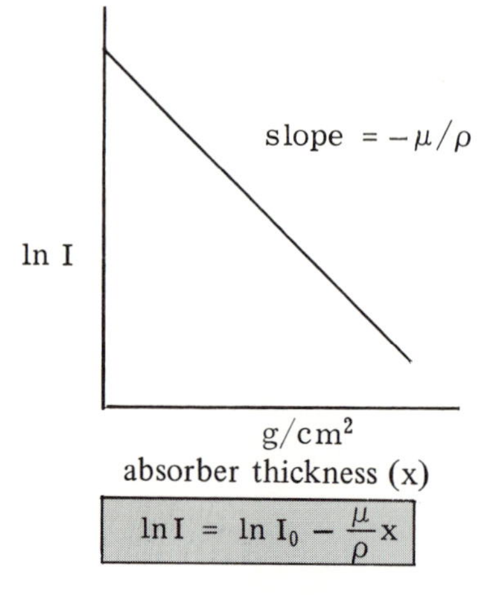

Figure 7.39—MASS ABSORPTION.

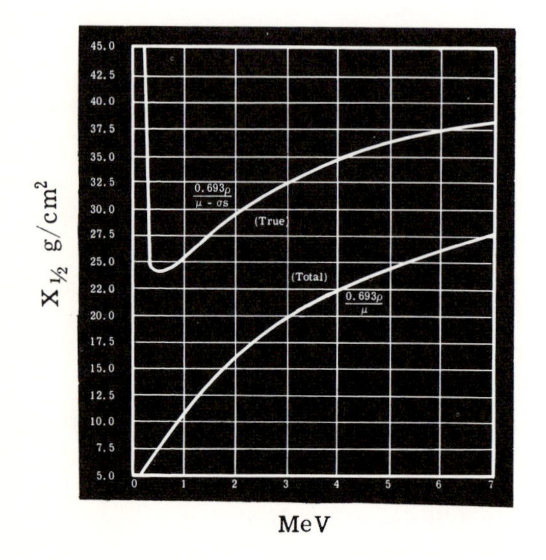

Figure 7.40—HALF-VALUE-LAYERS FOR ALUMINUM ABSORBERS.

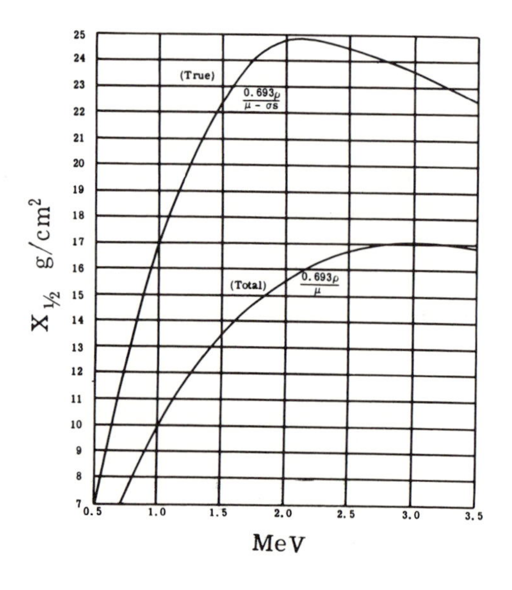

Figure 7.41—HALF-VALUE-LAYERS FOR LEAD ABSORBERS.

The photoelectric absorption coefficient is a function of the density ρ of the absorbing medium, the atomic number Z and the mass number A of the absorbing material, as well as of the wavelength λ and hence of the energy of the radiation. The coefficient is given by the expression,

$$\tau = 0.0089\, \rho\, (Z^{4.1}/A)\, \lambda^{n} \tag{7.39}$$

where n = 3 for the elements N, C and O,
and n = 2.85 for other elements below Fe.

It is evident that photoelectric absorption is most pronounced for low energy gamma radiations (less than 0.5 MeV) and that it increases rapidly with an increase in the atomic number of the absorber. The relationship between τ' and the gamma energy is shown in figure 7.42. The mass photoelectric absorption coefficient τ' is related to the linear photoelectric coefficient by $\tau' = \tau/\rho$.

The Compton scattering coefficient consists of two components

$$\sigma = \sigma_a + \sigma_s \tag{7.40}$$

σ_a is the scattering coefficient dependent on the loss of energy to electrons in collisions. σ_s is the scattering coefficient dependent on the loss of energy due to the scattering of photons out of the beam.

The relationships between the various absorption coefficients and the energies of the gamma radiations are illustrated in figure 7.42.

If the energy of a quantum is greater than 2×0.51 MeV, an electron-positron pair can be formed. The pair formation coefficient K can be evaluated from the relationship,

$$K = a\, NZ^2(E - 1.02) \tag{7.41}$$

where a is a constant, N the Avogadro number, Z the atomic number, and E the photon energy in MeV.

Geometry—the absorption coefficient μ, discussed in the preceding paragraphs, is more correctly referred to as the *total absorption coefficient.* This is the coefficient describing the decrease in beam intensity due both to *true absorption* and to the *scatter of radiation.* Thus,

Total absorption coefficient = True absorption coefficient + scatter absorption coefficient. This same statement is expressed in symbols by

$$\begin{array}{ccccc} \mu & = & (\tau + \sigma_a + K) & + & \sigma_s \\ \text{Total} & = & \text{True} & + & \text{scatter} \end{array} \tag{7.42}$$

If the total absorption coefficient is to be measured, it is necessary to establish "good geometry" so conditions required for narrow-beam attenuation are achieved. Conditions considered good are illustrated in figure 7.43 (c). Gamma rays from the source have been collimated so that only a narrow beam strikes the absorber. Compton scattered radiation produced in the absorber is then prevented from reaching the detector by additional shielding. Under such conditions, experimental values of the mass absorption

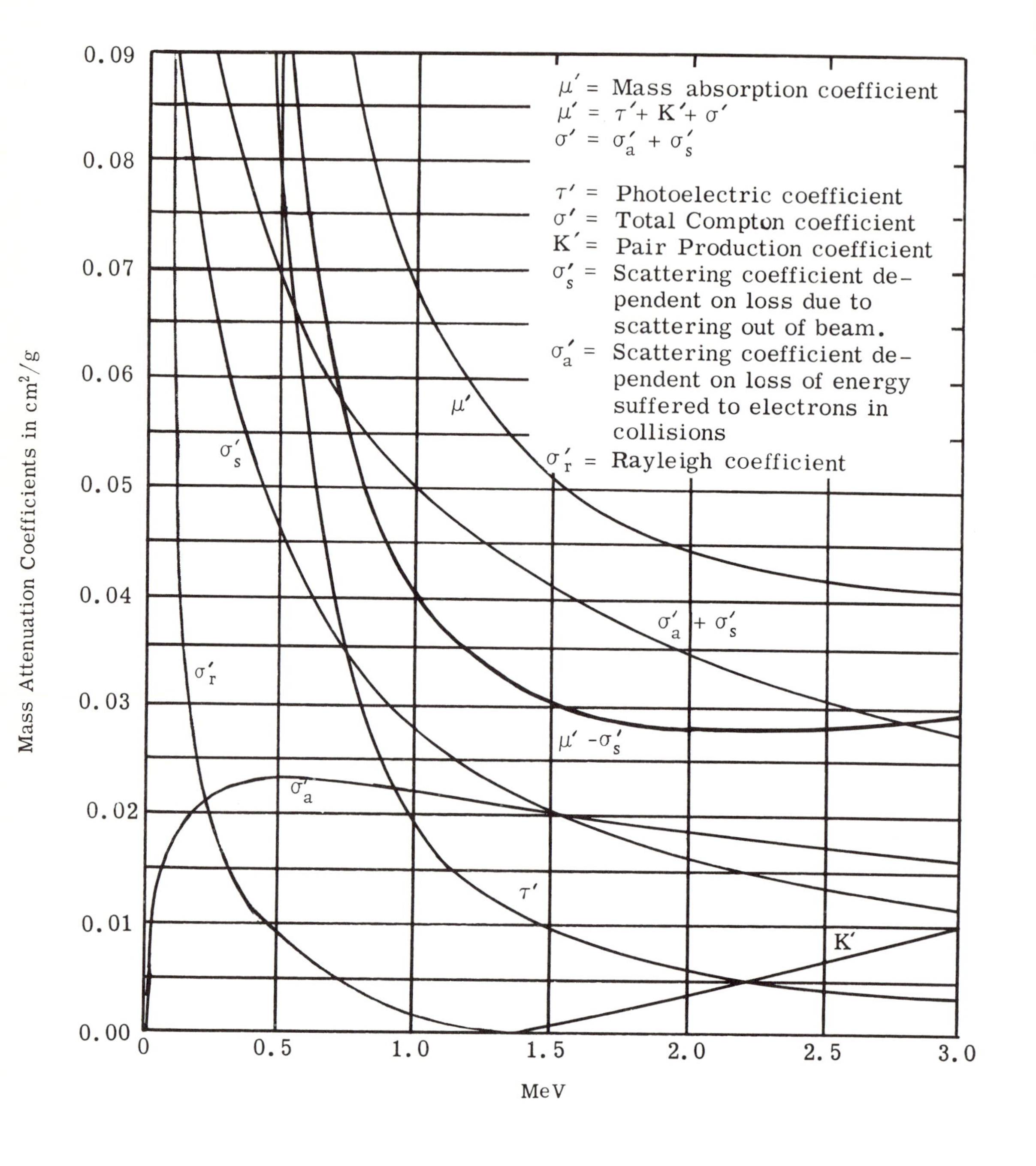

Figure 7.42—MASS ATTENUATION COEFFICIENTS FOR LEAD.

coefficient will approach those given in figure 7.42 for μ', and the values for the HVL will approach those given in figures 7.40 (Total) and 7.41 (Total).

When the geometry is "poor", the attenuation is called "broad-beam". Such conditions are shown in figure 7.43 (a) where Compton scattered radiation from the absorber as well as from the shield strikes the detector. Measured activities will be higher under poor conditions than under good conditions since the attenuation is less. The worst possible condition which can exist is that where the experimental value for mass absorption coefficient is equal to $\mu' - \sigma'_s$ as shown in figure 7.42 or where the values for the HVL approach those given by the curve marked "True' of figures 7.40 or 7.41.

Although good geometry is used to measure μ or μ', sometimes used for the identification of nuclides, it is important to consider scattering conditions prevailing when calculating the thickness of shielding required for safe handling or storage of gamma emitting isotopes. When in doubt, it is good practice to use $\mu' - \sigma'_s$ for the calculation because shielding almost always creates poor geometry.

Buildup—Narrow-beam geometry will yield the exponential curves shown in figures 7.38 and 7.39. Broad-beam geometry will cause a definite increase in activity. This departure from the narrow-beam curve is measured by the buildup factor B which is defined simply as the ratio of the observed activity to the activity expected from total absorption (i.e., good geometry).

$$B = \frac{\text{Rate observed using absorber of thickness x}}{\text{Rate calculated for total absorption using same absorber}}$$

Buildup can be estimated by

$$B = 1 + \mu X \qquad \text{or} \qquad B = 1 + \mu' x \tag{7.43}$$

where μX and $\mu' x$ are equal to the absorber thickness in mean free paths.

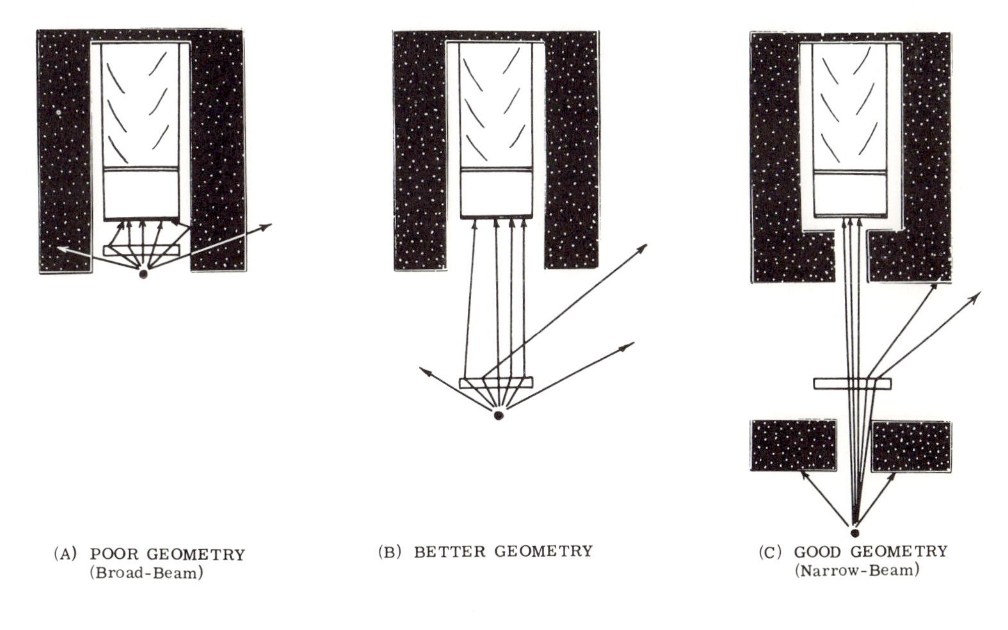

(A) POOR GEOMETRY (Broad-Beam)

(B) BETTER GEOMETRY

(C) GOOD GEOMETRY (Narrow-Beam)

Figure 7.43—BROAD-BEAM AND NARROW-BEAM GEOMETRY

APPARATUS:

Geiger or scintillation counting assembly; gamma-ray sources (e.g., ^{60}Co, ^{131}I); calibrated lead absorbers or sheets of lead and micrometer caliper; aluminum absorbers; ⅜″ plastic sheet.

PROCEDURE:

In this experiment the total absorption coefficient is to be determined. Various arrangements of equipment can be used, the choice depending upon what is available.

Use of G-M counter—If a Geiger counter is used as a detector, the counting arrangement shown in figure 7.44 is suggested. Directly above, and close to the source is placed a sheet of ⅜″ plastic to absorb the beta particles, thereby decreasing the production of bremsstrahlung. Gamma rays from the source will pass quite freely through the plastic. Lead absorbers of graduated thicknesses are then placed directly on top of the plastic. Finally, an aluminum absorber (400 mg/cm^2 to 600 mg/cm^2) is placed as close to the tube window as possible and allowed to remain there for all measurements. Its purpose is to filter out at least a part of the low energy scattered radiation. The source strength should be sufficient to give an activity of several thousand counts per minute with no lead absorber.

Use of scintillation detector—If a scintillation detector is available, it should be used in preference to the G-M tube shown in figure 7.44. All other conditions, including the use of the aluminum absorber and sheet of plastic, should remain unaltered. The major advantage in the use of a scintillation detector is its high sensitivity to gamma radiation

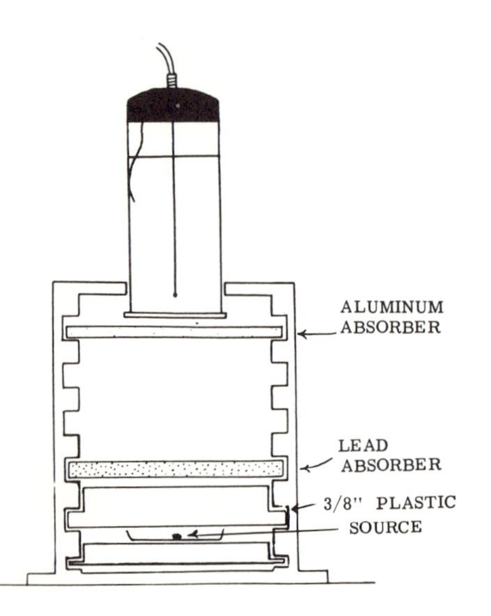

Figure 7.44—MEASUREMENT OF GAMMA-RAY ABSORPTION USING A GEIGER TUBE

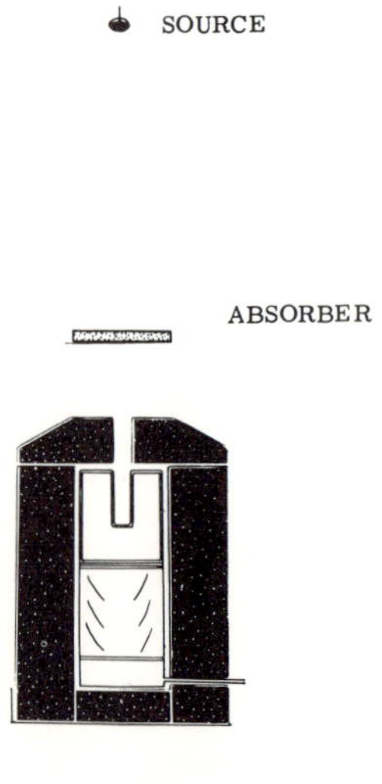

SOURCE

ABSORBER

Figure 7.45—USE OF A WELL-COUNTER TO MEASURE GAMMA-RAY ATTENUATION.

thus allowing the use of a much weaker radioactive source. Furthermore, the high voltage applied to the scintillation detector can be carefully increased until the counter just begins to record the primary radiation from the source (see experiment 9.1). It will then be insensitive to low energy scattered radiation. A more accurate measure of μ is thereby possible.

In a variation of this counting arrangement, a well-counter may be used (see figure 7.45). A ring stand is used to support the source and absorbers. The clamps are ring-clamps on which is placed a piece of cardboard.

Lead Absorber Number	Thickness X (cm)	Thickness x (g/cm^2)
1		
2		
3		
4		
5		
6		
7		
8		
9		
10		

Calibration of lead absorbers—Precalibrated lead absorbers may, of course, be used but these are unnecessary since a few pieces of sheet lead and a micrometer caliper will serve well. The thickness X (in cm) of each sheet of lead is measured with calipers and recorded. If the thickness x (in g/cm^2) is desired, it is calculated by use of the density since $x = X\rho$. The calibrated sheets of lead are used *in combination* to attain the desired thickness.

Measure the activity of the sample beginning with no lead absorber. Then proceed through the set of absorbers, singly or in combination, until the activity has decreased to 50% of the original value. Plot the resulting data on semi-log paper. Calculate the values of:

1. the half-value-layer
2. the linear absorption coefficient
3. the mass absorption coefficient
4. the gamma energy

From the known value of gamma energy and the use of figure 7.41 or 7.42, calculate the value of the buildup factor B.

Calibrated lead absorbers can be used, or uncalibrated lead discs or sheets can be calibrated by means of a micrometer.

	Lead Absorbers		
Absorber Thickness (cm)	Absorber Thickness (g/cm^2)	Observed Activity	Activity corrected for Background

Half-value-layer $X_{1/2}$ ________________ cm
Linear absorption coefficient μ ________________ cm^{-1}
Mass absorption coefficient μ' ________________ cm^2/g
Gamma energy (observed) ________________ MeV
Gamma energy (literature) ________________ MeV
Source ________________

Repeat this procedure using aluminum absorbers. Refer to figure 7.40 for the gamma ray energy. Calculate also the half-value-layer, the linear absorption coefficient and the mass absorption coefficient. Compare these values with those obtained with the use of lead absorbers and explain the results.

Use only the thicker aluminum absorbers and use them in combinations to obtain greater absorber thicknesses.

	Aluminum Absorbers		
Absorber Thickness (cm)	Absorber Thickness (g/cm^2)	Observed Activity	Activity corrected for Background

Half-value-layer $X_{1/2}$ ________________ cm
Linear absorption coefficient μ ________________ cm^{-1}
Mass absorption coefficient μ' ________________ cm^2/g
Gamma energy (observed) ________________ MeV
Gamma energy (literature) ________________ MeV
Source ________________

The linear absorption coefficients in cm^{-1} for certain materials at various energies are given below. Construct a similar table of $X_{1/2}$.

E	Pb		Fe		Al		H_2O	
MeV	μ	$X_{1/2}$	μ	$X_{1/2}$	μ	$X_{1/2}$	μ	$X_{1/2}$
0.2	5.0		1.06		0.33		0.14	
0.5	1.7		0.63		0.23		0.090	
1.0	0.77		0.44		0.16		0.067	
1.5	0.57		0.40		0.14		0.057	
2.0	0.51		0.33		0.12		0.048	

Construct a graph relating gamma-ray energy to HVL for each of the four absorbing materials listed in the above table.

EXPERIMENT 7.6 PROPERTIES OF NEUTRONS

OBJECTIVES:

To investigate the properties of neutrons and the nature of their interactions with matter.

To measure the relative scattering cross section and the total cross section of thermal neutrons for various substances.

THEORY:

The neutron had been postulated many years before Chadwick demonstrated its existence in 1932. Its mass is 1.68×10^{-24} grams (1.00898 amu), just slightly greater than that of the proton. It is an unstable particle with a half life of about 12 minutes and decays into a proton, a negatron and a neutrino. Because the neutron carries no charge, it is not influenced by magnetic and electrostatic fields. For this reason it is very difficult to control. It can be deflected only by means of collisions with other particles. Lack of a charge also accounts to a large extent for its great penetrating power. It is stopped only by very thick barriers. On the other hand, if neutrons are moving slowly, they are effectively absorbed by boron and cadmium.

Neutron Sources—Neutrons are produced by two general processes: (1) nuclear bombardment and (2) fission in a reactor. Most smaller laboratories rely on sources of the first type although the number of reactors in use has increased markedly in recent years.

So-called portable sources employing nuclear bombardment as a source of neutrons are of two types: (1) those employing a radioactive source and (2) those employing high voltage acceleration of charged particles.

Devices which use a radioactive source are called neutron howitzers, or, if a layer of uranium surrounds the neutron source—to provide neutron amplification through fission—the device is called a subcritical reactor. There are no common radioisotopes which emit neutrons directly, but there are several nuclear reactions by which neutrons are produced indirectly. Of particular note are neutron sources consisting of finely divided beryllium mixed with an alpha emitter in pellet form. Neutrons are produced by the (α, n) reaction of beryllium.

$$^{9}_{4}Be + ^{4}_{2}He \longrightarrow ^{1}_{0}n + ^{12}_{6}C + 5.76\ MeV \tag{7.44}$$

Other typical neutron producing reactions are

$$^{3}H\ (\alpha, n)\,^{6}Li \qquad \text{and} \qquad ^{9}Be\ (\gamma, n)\,^{8}Be$$

For each of these reactions, a radioactive source of the bombarding particle or photon is required. Other elements, including boron and lithium, will emit neutrons when subjected to alpha particle bombardment, but none with the efficiency of beryllium. Alpha sources commonly used for this purpose include radium-226, radium-DEF, polonium-210, thorium-228, plutonium-239 and actinium-227. Characteristics of these sources are shown in table 7.4.

Table 7.4—CHARACTERISTICS OF ALPHA-BERYLLIUM NEUTRON SOURCES

Source	Half-life	Yield (n/sec/curie)	Neutron Flux at 1 Meter ($n/cm^2/sec$)	Gamma Output mr-h-m/Ci
^{226}Ra:Be	1,620 years	1.2×10^7	95	850
Ra-DEF:Be	21 years	2.6×10^6	20.7	22
^{210}Po:Be	138 days	2.6×10^6	20.7	0.1
^{228}Th:Be	1.9 years	1.8×10^7	143	575
^{239}Pu:Be	24,360 years	2.2×10^6	17.5	3.7
^{227}Ac:Be	22 years	1.8×10^7	143	146

Plutonium-beryllium sources generally represent the best compromise of half-life, neutron yield, gamma flux and cost. Hertz (67) reports that a number of these sources were fabricated prior to 1956 at the Los Alamos Scientific Laboratory by the method of Tate and Coffinberry (82) in which stoichiometric quantities of plutonium and beryllium are heated together in a beryllium oxide crucible. The reaction between the elements is exothermic and results in the formation of the compound $PuBe_{13}$. The method of Richmond and Wells (79), developed at the Mound Laboratory, Miamisburg, Ohio, and used there until about 1960, involves placing a weighed pellet of plutonium into a weighed cup of beryllium which, in turn, is sealed in a tantalum can. When heated to about the melting point of beryllium the reaction occurs as in the Tate and Coffinberry procedure to form $PuBe_{13}$. Following decontamination of the tantalum can the source is enclosed in a stainless steel jacket as a further precaution.

A *neutron howitzer* consists of a neutron source surrounded by paraffin or water in a container of plastic, aluminum or stainless steel. Several ports or holes provide access to the neutron source as well as locations for sample irradiation and the means to produce a neutron beam. Paraffin and water are rich in protons and are therefore very effective neutron moderators. That is, through collisions with protons, fast neutrons are slowed down and become thermal neutrons.

Neutron generators for the production of neutrons are also available at a moderate price. In most of these accelerators, high energy deuterons are produced. As these deuterons strike a target of ^{3}H-zirconium hydride ($Zr^{3}H_4$), neutrons are produced by the reaction

$$^{3}H(d, n)^{4}He$$

One such unit is shown in figure 7.47. The neutron generator shown can generate up to 10^8 neutrons per second. This flux is approximately 10 to 50 times greater than that available from a neutron howitzer.

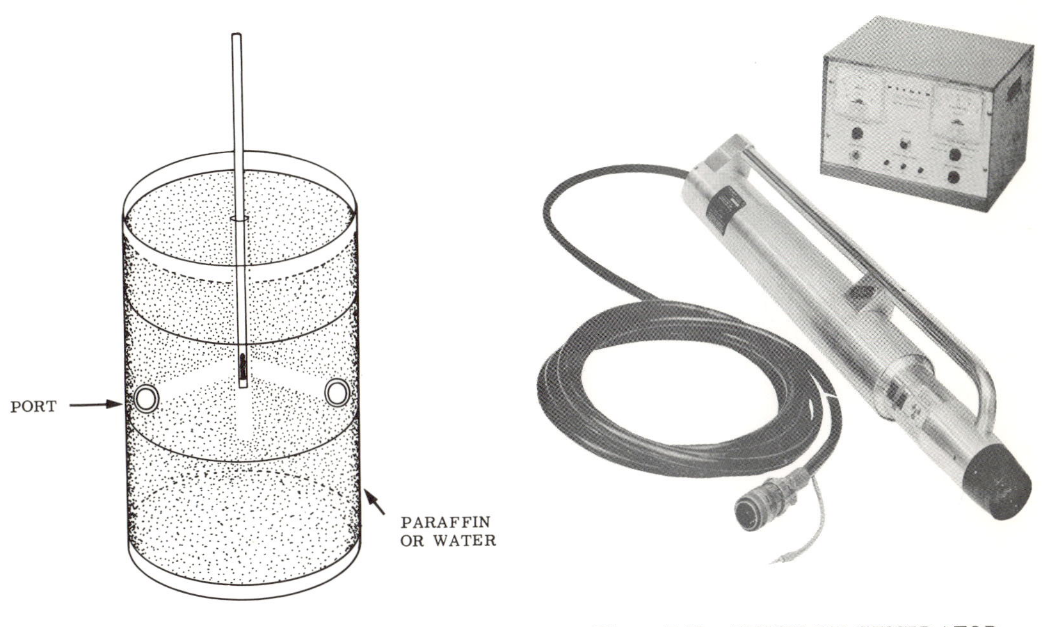

Figure 7.47—NEUTRON GENERATOR
(*Picker X-Ray Corp.*)

Figure 7.46—NEUTRON HOWITZER.

Classification of neutrons—Neutrons are classified according to their velocity. These classifications are summarized in table 7.5. *Thermal* neutrons are so named because their average energy is equal to the average kinetic energy of molecules at room temperature. The energy of these neutrons has a Maxwellian distribution as predicted by the kinetic theory. These slow thermal neutrons are formed by the slowing down of fast neutrons by a process involving numerous collisions with nuclei.

Interactions of neutrons—Because the neutron is electrically neutral, there is no barrier to prevent access of even slow neutrons to the atomic nucleus. Neutrons may interact with nuclei by colliding elastically or inelastically or by being captured by the nucleus.

Table 7.5—CLASSIFICATION OF NEUTRONS

NAME	ENERGY	PRINCIPAL INTERACTION	COMMENTS
Thermal	0.025 eV (at 22° C)	Capture	Mean velocity = 2200 m/sec Maxwellian distribution of energies
Intermediate (Resonance Neutrons)	0.5 ev - 10 keV	Elastic scattering and capture	Absorption follows 1/v law and shows absorption peaks
Fast	10 keV - 20 MeV	Elastic scattering	Most neutrons are ejected from nuclei with these energies
Relativistic	>20 MeV	Spallation	Relativistic mass much greater than rest mass

Elastic scattering—Elastic scattering is primarily responsible for the moderation (slowing down) of neutrons. In an elastic collision the total kinetic energy and the total momentum of the neutron and the nucleus with which it collides remain constant. That is, no energy is lost as electromagnetic radiation. Elements most often used as moderators are hydrogen and carbon. Collisions of neutrons with hydrogen are primarily elastic. Hydrogen is an especially efficient moderator because its mass is nearly equal to that of the neutron. (Consider the "moderation" of a billiard ball by (a) striking a small ball the size of a pea, (b) striking a large ball the size of a bowling ball and (c) striking another billiard ball. Assuming perfectly elastic collisions (c) would result in the greatest degree of moderation.) Paraffin, water and other substances rich in hydrogen are usually used.

The diffraction of neutrons by crystals involves a special type of elastic scattering known as *coherent scattering*. Coherent scattering requires the cooperative participation of the neutrons as waves. The Bragg scattering of photons (see figure 7.29) is an example of elastic coherent scattering. The same Bragg equation, $n\lambda = 2d \sin \theta$, applies. The wavelength λ is the de Broglie wavelength of the neutron.

Inelastic resonance scattering—Scattering which results in a loss in the total energy of the colliding systems is called inelastic. In a reaction of the type $(n, n'\gamma)$ where n is the incident neutron and n' is the slower neutron released from the nucleus—which may or may not be the same neutron as n—the energy difference is emitted as a photon. Moderation of neutrons by carbon is primarily by this process.

Capture reactions—For most nuclei the capture of a neutron results in an increase in energy of about 8 MeV plus the kinetic energy of the neutron. Hence, the compound nucleus formed upon neutron capture is in a high energy state. Stability is attained by emission of a particle or photon. The type of capture reaction depends upon the energy of the incident neutron.

1. *Slow neutron reactions* are the most common. The energy increase of the nucleus is only about 8 MeV and is generally insufficient to eject a particle. Hence, the reactions are generally of the (n, γ) type know as *activation.* The atomic number Z remains unchanged but A increases by one unit. Thus,

$$^{115}_{49}\mathrm{In} + ^{1}_{0}\mathrm{n} \longrightarrow ^{116}_{49}\mathrm{In} + \gamma \tag{7.45}$$

2. *Intermediate energy neutrons*—Capture may result in activation as above, but the compound nucleus produced may also have sufficient energy to overcome the binding energy and eject a particle (proton, deuteron or alpha particle). Reactions of this kind result in the formation of a new element and are called *transmutation.* A reaction of the (n, p) type is

$$^{14}_{7}N + ^{1}_{0}n \longrightarrow ^{14}_{6}C + ^{1}_{1}H \qquad (7.46)$$

A reaction of the (n, α) type is

$$^{6}_{3}Li + ^{1}_{0}n \longrightarrow ^{3}_{1}H + ^{4}_{2}He \qquad (7.47)$$

3. *Fast neutrons*—Kinetic energies up to 10 MeV contribute up to about 18 MeV to a nucleus. The binding energy of a nucleon is only about 8 MeV. Hence, two particles may be ejected in, for example, a reaction of the (n, 2n) type.

$$^{238}_{92}U + ^{1}_{0}n \longrightarrow ^{237}_{92}U + ^{1}_{0}n + ^{1}_{0}n \qquad (7.48)$$

4. *Relativistic neutrons*—Greater numbers of nucleons can be ejected by such high energy neutrons. Reactions such as (n, 3n) and (n, 2np) are possible.

Nuclear cross section—The probability of occurrence of a reaction is measured by its cross section σ. The unit of cross section is the barn b (= 10^{-24}cm^2). The *total cross section* σ_t is equal to the sum of the *absorption cross section* σ_a (also called the reaction cross section σ_r) and the *scattering cross section* σ_s. Thus,

$$\sigma_t = \sigma_a + \sigma_s \qquad (7.49)$$

Each of these *partial cross sections* can, in turn, be subdivided into components corresponding to the various absorption and scattering processes. There is, for example, a capture cross section σ_c and an activation cross section σ_{act}.

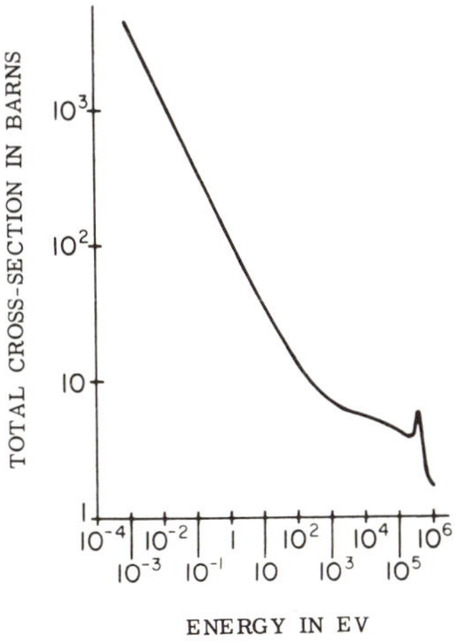

Figure 7.48 — TOTAL NEUTRON CROSS SECTION OF BORON (*Courtesy Picker X-Ray Corp.*)

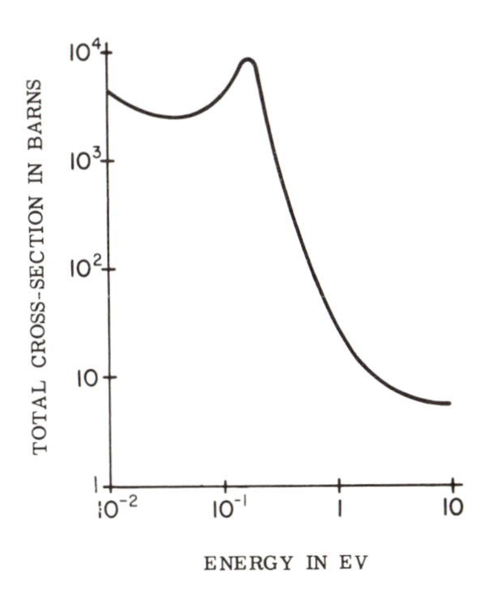

Figure 7.49 — TOTAL NEUTRON CROSS SECTION OF CADMIUM (*Courtesy Picker X-Ray Corp.*)

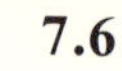

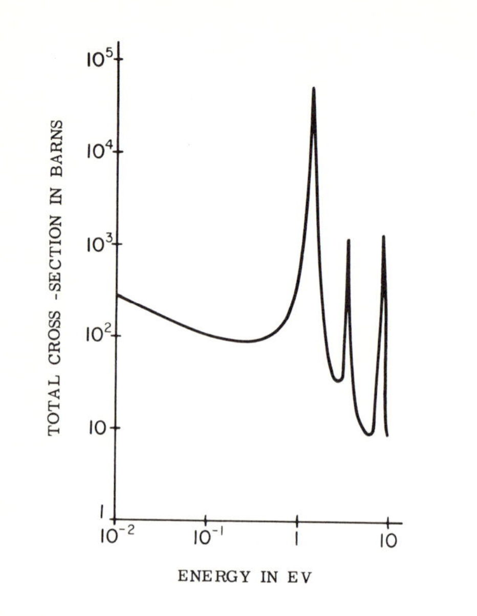

Figure 7.50 — TOTAL NEUTRON CROSS SECTION OF INDIUM (*Courtesy Picker X-Ray Corp.*)

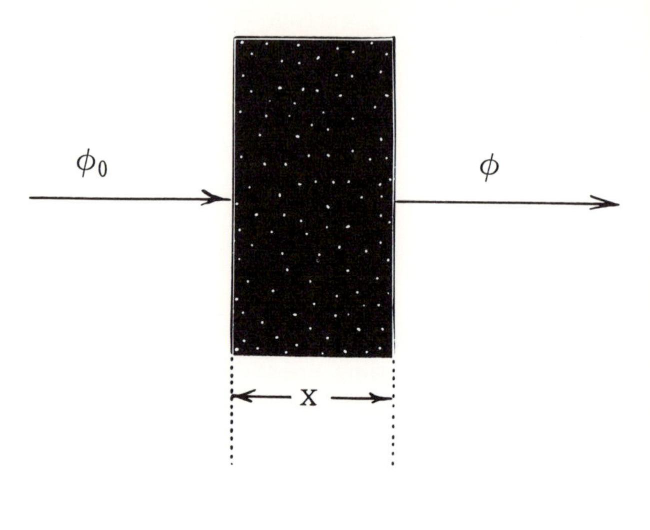

Figure 7.51—MACROSCOPIC ABSORPTION OF NEUTRONS.

Each mode of interaction is in competition with all other modes. Thus, the fraction of the collisions resulting in a particular type of reaction will depend upon the relative cross sections for the reactions involved.

Values of cross section are not constant but are a function of the neutron energy. The absorption cross section, for example, is inversely proportional to the velocity of the neutron for intermediate neutron energies. This is known as the 1/v region. The spectrum of boron (see figure 7.48) illustrates this energy dependency. In addition, absorption peaks may appear at certain energies. These peaks are caused by "resonance" when the incident neutron energy is equal to an energy level of the target nucleus. Cadmium (see figure 7.49) shows a resonance peak at 0.18 eV. Indium has several resonance peaks (see figure 7.50), the main peak exceeding 30,000 barns at 1.46 eV. In general, absorption cross sections decrease with an increase in neutron energy and for fast neutrons do not exceed the physical cross section πR^2.

Beam Calculations—Let a beam of neutrons of flux density ϕ_0 (n/cm²/sec) impinge upon an absorber of thickness X. Then the change in flux as a function of absorber thickness is

$$-\frac{d\phi}{dX} = \Sigma\,\phi \tag{7.50}$$

where Σ (cm^{-1}) is the *macroscopic absorption cross section* and is related to the *microscopic cross section* σ. Thus,

$$\Sigma = N\sigma \tag{7.51}$$

where N is the number of target atoms per cubic centimeter of target material. Integration of equation 7.50 gives

$$\ln \frac{\phi_0}{\phi} = \Sigma X \tag{7.52}$$

and

$$\phi = \phi_0 e^{-\Sigma X} \tag{7.53}$$

or

$$\phi = \phi_0 e^{-N\sigma X} \tag{7.54}$$

Thus, the number of events N′ per second (i.e., collisions of neutrons with nuclei per second) per unit area of absorber is

$$N' = \phi_0 - \phi = \phi_0 \left(1 - e^{-N\sigma X} \right) \tag{7.55}$$

Cross section is sometimes measured by the transmission method. In this method, the emergent flux density ϕ is measured for various values of absorber thickness X. Since from equation 7.54 it can be seen that

$$\ln \phi = -N\sigma X + \ln \phi_0 \tag{7.56}$$

which is the equation for a straight line, σ can be evaluated by plotting $\ln \phi$ versus X. The slope of the resulting line is $-N\sigma$ with an intercept at $\ln \phi_0$.

Integration between the limits of ϕ_1 and ϕ_2, and X_1 and X_2 followed by solution for σ gives

$$\sigma = \frac{\ln \phi_1 - \ln \phi_2}{N (X_2 - X_1)} \tag{7.57}$$

If the efficiency ϵ of the neutron detection system is held constant, then

$$\epsilon = \frac{R_1}{\phi_1} = \frac{R_2}{\phi_2} \tag{7.58}$$

and

$$\sigma = \frac{\ln R_1 - \ln R_2}{N (X_2 - X_1)} \tag{7.59}$$

APPARATUS:

Neutron howitzer or other equivalent neutron source; BF_3 neutron detector with preamplifier and scaler; neutron shields for collimation (e.g. sheets of cadmium or Marinelli beakers filled with borax); sheets of cadmium, aluminum, lead and paraffin.

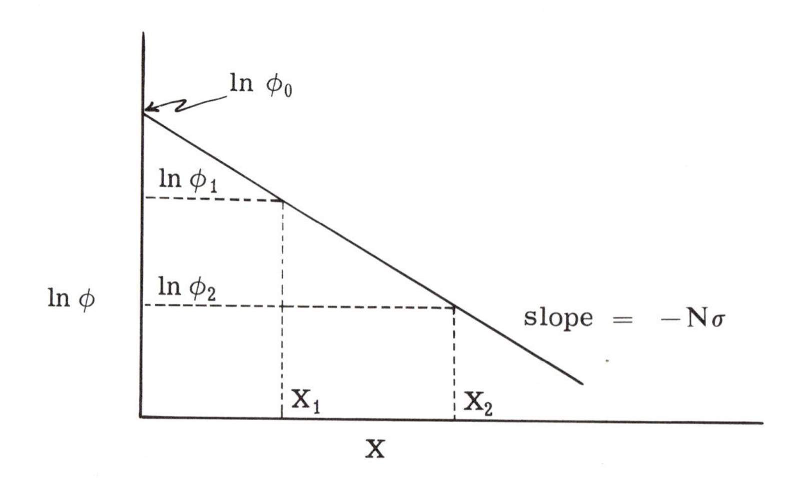

Figure 7.52—EVALUATION OF TOTAL CROSS SECTION BY TRANSMISSION METHOD.

PROCEDURE:

PART A—TOTAL CROSS SECTION BY TRANSMISSION METHOD

A BF_3 counter is mounted in front of the port of a neutron howitzer. Background, from neutrons emerging through the walls of the howitzer, is reduced by shielding and collimation of the neutrons. This may be done with sheets of cadmium, with borax sandwiched between sheets of plastic or by means of an arrangement of Marinelli beakers filled with borax and stoppered with glass wool or melted paraffin. The latter arrangement is illustrated in figure 7.53.

Sheets of various materials of known, graduated thicknesses are now placed between the detector and the port. Sheets of aluminum, lead and cadmium may be used for absorbers. The activity is measured for each thickness and plotted on semi-log paper to obtain a graph of the type shown in figure 7.52. The microscopic cross sections are calculated by the use of equation 7.50. Compare with the literature values for thermal neutron cross sections.

Compare lead vs. aluminum as a shielding material for (1) gamma radiation and (2) neutrons.

PART B—BACKSCATTERING OF NEUTRONS

A BF_3 counter is mounted in front of the port of a neutron howitzer and the activity is measured. Blocks of paraffin are placed behind the counter (i.e., with the counter between the paraffin and the port). Note the change in activity. Increase the thickness of paraffin by using several blocks together until the total thickness is at least 5 cm. Plot the activity as a function of the paraffin thickness.

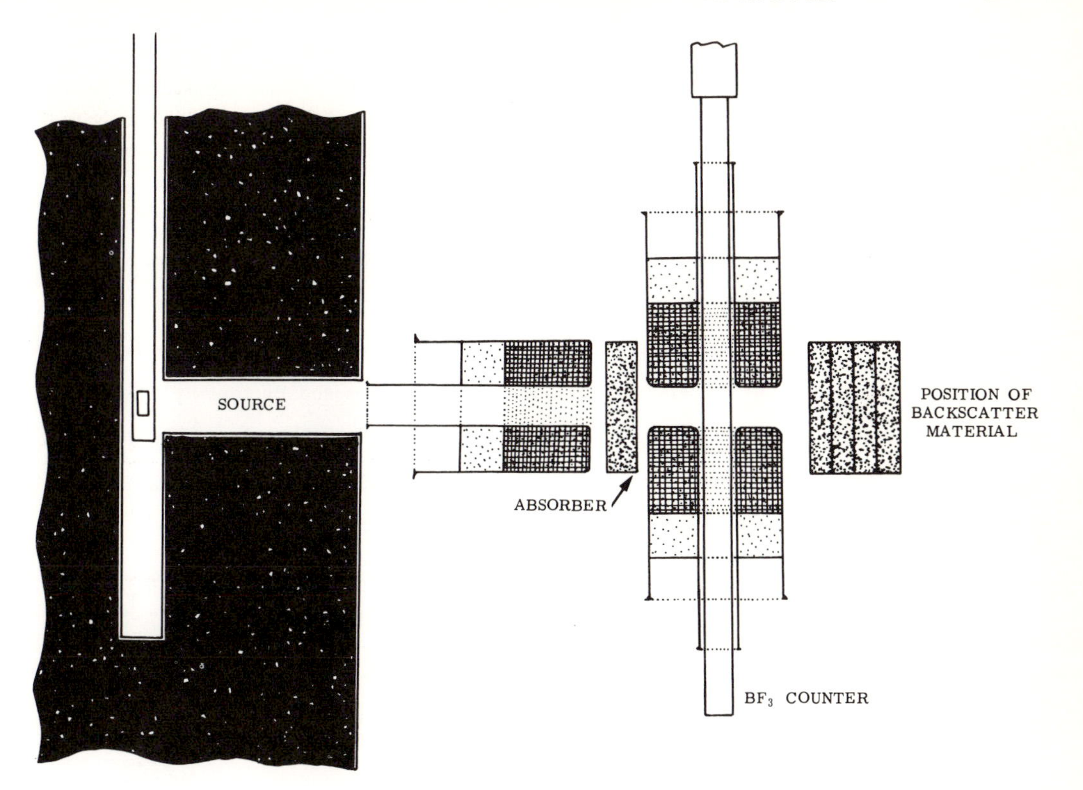

Figure 7.53—COLLIMATION OF NEUTRONS FOR ABSORPTION AND BACKSCATTERING MEASUREMENTS.

Repeat step 1 using sheets of other substances in place of paraffin.

Place a thin sheet of cadmium in front of the port of the howitzer to absorb the slow neutrons. Repeat step 1 with the paraffin. Is the percentage increase in activity, with increasing thickness of paraffin, greater or less than observed in step 1? Explain the results.

REFERENCES

Properties of Radiation

1. De Benedetti, S., "The Mössbauer Effect", Sci. Amer., p. 73 (April 1960).
2. Fluharty, R. G., "Interaction of Isotopic Radiation with Matter, I.", Nucleonics 2, 28-40 (May 1948).
3. Fluharty, R. G., "Interaction of Isotopic Radiation with Matter, II.", Nucleonics 3, 46-56 (July 1948).
4. Green, M. H., "Absorption of Monoenergetic X- and γ-Rays", Nucleonics 17, No. 10, 77 (Oct. 1959).
5. Morgan, G. W., "Some Practical Considerations in Radiation Shielding", Isotopes Div., Circ. B-4, USAEC (1948).
6. Shull, C. G., and E. O. Wollan, "X-Ray, Electron, and Neutron Diffraction", Science 108, 69-75 (1948).
7. Wertheim, G. K., "The Mössbauer Effect: A Tool for Science", Nucleonics 19, No. 1, 52-57 (Jan. 1961).

Alpha

8. Bragg, W. H., "Ionization of Various Gases by α-Particles", Phil. Mag. 13, 333-357 (1907).
9. Gold, R., "Beryllium-Hazard Detection Using Polonium-210 Alphas", Nucleonics 15, No. 11, 114-118 (Nov. 1957).
10. Marcley, R. G., "Apparatus for Measuring the Rutherford Scattering of Alpha Particles by Thin Metal Foils", Amer. J. Phys. 29, 349-354 (1961).

Beta

11. Brownell, G. L., "Interaction of Phosphorus-32 Beta Rays with Matter", Nucleonics 10, No. 6, 30-35 (1952).
12. Cerenkov, P. A., "Visible Radiation Produced by Electrons Moving in a Medium with Velocities Exceeding that of Light", Phys. Rev. 52, 499 (1938).
13. Cook, C. S., "Experimental Techniques in Beta-Ray Spectroscopy-II", Nucleonics 12, No. 2, 43-47 (Feb. 1954).
14. Duncan, J. F., and F. G. Thomas, "Three Beta-Absorption Methods . . . How do they Compare?", Nucleonics 15, No. 10, 82-85 (Oct. 1957).
15. Fano, U., "Gamma-Ray Attenuation. I, Basic Processes", Nucleonics 11, 8-12 (Aug. 1953).
16. Fano, U., "Gamma-Ray Attenuation. II, Analysis of Penetration", Nucleonics 11, 55 (Sept. 1953).
17. Feather, N., "Further Possibilities for the Absorption Method of Investigating the Primary -particles from Radioactive Substances", Proc. Cambridge Phil. Soc. 34, 599 (1938).
18. Fine, S., and C. F. Hendee, "X-Ray Critical-Absorption and Emission Energies in KEV", Nucleonics 13, No. 3, 36-37 (March 1955).
19. Glendenin, L. E., "Determination of the Energy of -particles and Photons by Absorption", Nucleonics 2, 26 (Jan. 1948).
20. Grodstein, G. W., "X-ray Attenuation Coefficients from 10 kev to 100 Mev", Nat. Bur. Stand. Circ. 583 (April 1957).
21. Harley, J. H., and N. Hallden, "Analyzing Beta Absorption Graphically to Identify Emitters", Nucleonics 13, No. 1, 32-35 (Jan. 1955).
22. Hayward, E., and J. H. Hubbell, "The Backscattering of Co^{60} Gamma Rays from Infinite Media", J. Appl. Phys. 25, 506 (1954).
23. Hine, G. J., "Secondary Electron Emission and Effective Atomic Numbers", Nucleonics 10, No. 1, 9-15 (1952).
24. MacGregor, M. H., "X-Ray Production with Linear Accelerators", Nucleonics 17, No. 2, 104-105 (Feb. 1959).
25. Madgwick, N., "The Absorption and Reduction in Velocity of β-Rays on Their Passage through Matter", Proc. Cambridge Phil. Soc. 23, 970 (1927).
26. Marshall, J. H., "How to Figure Shapes of Beta-Ray Spectra", Nucleonics 13, No. 8, 34-38 (Aug. 1955).
27. Marshall, J. H., and B. Ward, "Absorption Curves and Ranges for Homogeneous β-Rays", Can. J. Research A15, 39 (1937).
28. Reiffel, L., "Beta-Ray-Excited Low-Energy X-Ray Sources", Nucleonics 13, No. 3, 22-24 (March 1955).
29. Varden, A., and N. Madgwick, "Absorption of Homogeneous β-Rays", Phil. Mag. 29, 725 (1915).

Gamma

30. Anon., "Safe Source for Gamma Absorption Measurements", Nucleonics 17, No. 4, 149 (April 1959).
31. Bennett, G. A., "Nomogram for Calculating Shielding for Co^{60}", Nucleonics 8, No. 4, 55-58 (1951).
32. Bernstein, W., and R. H. Schuler, "Low Energy Scattered Radiation Inside a Cylindrical Co^{60} Source", Nucleonics 13, No. 11, 110-112 (Nov. 1955).
33. Berry, P. F., "Gamma-Ray Attenuation Coefficients", Nucleonics 19, No. 6, 62 (June 1961).
34. Brucer, M., "Marble as a Radiation Shield", Nucleonics 13, No. 1, 65-66 (Jan. 1955).
35. Chappell, D. G., "Gamma-Ray Attenuation", Nucleonics 14, No. 1, 40-41 (Jan. 1956).
36. Chappell, D. G., "Gamma Attenuation with Buildup in Lead and Iron", Nucleonics 15, No. 1, 52-53 (Jan. 1957).
37. Chappell, D. G., "Gamma-Ray Attenuation with Buildup in Water", Nucleonics 16, No. 7, 80 (July 1958).
38. Chappell, D. G., "Gamma-Ray Scattering from Thin Scatterers", Nucleonics 14, No. 7, 36-37 (July 1956).
39. Chasanov, M. G., and M. Shatzkes, "Attenuation of γ-Rays from an Infinite Plane", Nucleonics 16, No. 6, 63 (June 1958).
40. Cook, J. M., "Gamma Rays and Internal Conversion", Nucleonics 4, 24-34 (Jan. 1949).
41. Crasemann, B., and H. Easterday, "Locating Compton Edges and Backscatter Peaks in Scintillation Spectra", Nucleonics 14, No. 6, 63 (June 1956).
42. Fano, U., "Gamma-Ray Attenuation. Part I—Basic Processes", Nucleonics 11, No. 8, 8-12 (1953).
43. Fano, U., "Gamma-Ray Attenuation. Part II—Analysis of Penetration", Nucleonics 11, No. 9, 55-61 (1953).
44. Howland, P. R., N. E. Scofield and R. A. Taylor, "Fast Compton Scintillation Spectrometer", Nucleonics 14, No. 6, 50-53 (June 1956).
45. Johns, H. E., J. E. Till, and D. V. Cormack, "Electron Energy Distributions Produced by Gamma Rays", Nucleonics 12, No. 10, 40-46 (Oct. 1954).
46. Klahr, C. N., "Scattering Shields for Space Power", Nucleonics 19, 4, 110-112 (April 1961).
47. Larson, H. V., I. T. Myers, and W. C. Roesch, "Wide-Beam Fluorescent X-Ray Source", Nucleonics 13, No. 11, 100-102 (Nov. 1955).
48. Lindner, J. W., "Shielding-Glass Buildup Factors", Nucleonics 16, No. 10, 77 (Oct. 1958).
49. McCallum, "Preparation and Use of Neutron Sources", Nucleonics 5, 11-21 (1949).
50. Mittelman, P. S., and R. T. Liedtke, "Gamma Rays from Thermal Neutron Capture", Nucleonics 13, No. 5, 50-51 (May 1955).

51. Moteff, J., "Tenth-Value Thicknesses for Gamma-Ray Absorption", Nucleonics 13, No. 7, 24 (July 1955).
52. Steigelmann, W. H., Jr., "Gamma-Ray Shielding Design Curves", Nucleonics 19, No. 11, 148-150 (Nov. 1961).
53. Stewart, "Neutron Spectrum and Absolute Yield of a Plutonium: Beryllium Neutron Source", Phys. Rev. 98, 740 (1955).
54. Untermyer, S., et al., "Portable Thalium X-Ray Unit", Nucleonics 12, No. 5, 35-37 (May 1954).
55. Van Dilla, M. A., and G. J. Hine, "Gamma-Ray-Diffusion Experiments in Water", Nucleonics 10, No. 7, 54-58 (1952).
56. West, R., "Low Energy Gamma-Ray Sources", Nucleonics 11, No. 2, 20 (1953).
57. Whyte, G. N., "Density Effect in γ-Ray Measurements", Nucleonics 12, No. 2, 18-21 (Feb. 1954).
58. Wyard, S. J., "Radioactive-Source Corrections for Bremsstrahlung and Scatter", Nucleonics 13, No. 7, 44-45 (July 1955).
59. Wycoff, et al., "Broad and Narrow Beam Attenuation of 500-1400 Kv X-Rays in Lead and Concrete", Nucleonics 3, 62-70 (Nov. 1948).

Neutrons

60. Anon., "Table of Neutron Sources", Nucleonics 18, No. 12, 66-67 (Dec. 1960).
61. Atomic Energy of Canada Limited, "Neutron Sources and Their Characteristics", AECL Tech. Bull. NS-1, Commercial Products Div., Ottawa.
62. Baucom, H. H., Jr., "Nuclear Data for Reactor Studies", Nucleonics 18, No. 11, 198-200 (Nov. 1960).
63. Burrill, E. A., and M. H. MacGregor, "Using Accelerator Neutrons", Nucleonics 18, No. 12, 65-68 (Dec. 1960).
64. Dayton, I. E., and W. G. Pettus, "Effective Cadmium Cutoff Energy", Nucleonics 15, No. 12, 86-88 (Dec. 1957).
65. Heintzé, L. R., "Thermal Neutron Diffusion Lengths for Water, Lucite, Furfural, and an Expanded Plastic", Nucleonics 14, No. 5, 108-115 (May 1956).
66. Hennelly, E. J., "Intense Sb-Be Sources Make 10^{10} Neutrons/Sec.", Nucleonics 19, No. 3, 124-125 (March 1961).
67. Hertz, M. R., "Inspection and Recanning Program of PuBe Neutron Sources," MLM-1188, TID-4500, Jan. 7, 1964.
68. Kline, D. E., and F. J. Remick, "Gamma-Dose Enhancement from Neutron Capture in Cd.", Nucleonics 16, No. 3, 97-101 (March 1958).
69. Kruger, P., "Low-Level Neutron Dosimetry with Sensitive Foil Methods", Nucleonics 17, No. 6, 116-130 (June 1959).
70. Marshak, R. E., H. Brooks, and H. Hurwitz, Jr., "Introduction to the Theory of Diffusion and Slowing Down of Neutrons", Part I—Nucleonics 4, 10-22 (May 1949); Part II—Nucleonics 4, 43-49 (June 1949); Part III—Nucleonics 5, 53 (July 1949); Part IV—Nucleonics 5, 59-68 (Aug. 1949).
71. Martin, D. H., "Correction Factors for Cd-Covered-Foil Measurements", Nucleonics 13, No. 3, 52-53 (March 1955).
72. McCallum, K. J., "Preparation and Use of Neutron Sources", Nucleonics 5, 11-21 (July 1949).
73. National Bureau of Standards, "Measurement of Neutron Flux and Spectra for Physical and Biological Applications", Handbook 72, U.S. Dept. Comm. (July 1960).
74. National Bureau of Standards, "Measurement of Absorbed Dose of Neutrons and of Mixtures of Neutrons and Gamma Rays", Handbook 75, U.S. Dept. Comm. (Feb. 1961).
75. National Bureau of Standards, "Protection Against Neutron Radiation Up to 30 Million Electron Volts", Handbook 63, U.S. Dept. Comm. (Nov. 1957).
76. Nisle, R. G., "Neutron-Absorption Alignment Chart", Nucleonics 18, No. 3, 86-87 (March 1960).
77. Picker X-Ray Corporation, "Neutrons", Picker Scintillator, Special No. 3 (Nov. 1961).
78. Rausa, G. J., "Gamma Dose Rate from a Po-Be Source", Nucleonics 12, No. 2, 62 (Feb. 1954).
79. Richmond, J. L., and C. E. Wells, U.S. Patent 3,073,768 (Jan. 15, 1963).
80. Russel, J. L., Jr., "Relative Worth of Control Materials", Nucleonics 18, No. 12, 88-92 (Dec. 1960).
81. Schlumberger and Kamen, "Neutrons from Small Tubes", Nucleonics 18, No. 12, 69-76 (Dec. 1960).
82. Tate, R. E., and A. S. Coffinberry, "Plutonium-Beryllium Neutron Sources, Their Fabrication and Neutron Yield", Proc. Sec. U.N. Int. Conf. on Peaceful Uses of Atomic Energy, XIV, 427 (1958).
83. Tittle, C. W., "Slow-Neutron Detection by Foils, Part I", Nucleonics 8, No. 6, 5-9 (1951); Part II—Nucleonics 9, No. 1, 60-67 (1951).
84. Trowbelzkoy, E., and H. Goldstein, "Gamma Rays from Thermal-Neutron Capture", Nucleonics 18, No. 11, 171-173 (Nov. 1960).
85. Udy, Shaw and Boulger, "Properties of Beryllium", Nucleonics 11, No. 5, 52 (1953).
86. U.S. Atomic Energy Commission, "Atomic Energy and the Physical Sciences", USAEC, Supt. Doc. (Jan. 1950).
87. Van Wye, R. F., and J. G. Beckerley, "Calculation of Yields in Multiple Neutron Capture Processes", Nucleonics 9, No. 4, 17-21 (1951).
88. Wallace, P. R., "Neutron Distributions in Elementary Diffusion Theory, Part I", Nucleonics 4, 30-55 (Feb. 1949); Part II—Nucleonics 4, 48 (March 1949).
89. Wollan, E. O., and C. G. Shull, "Neutron Diffraction and Associated Studies, Part I", Nucleonics 3, 8 (July 1948); Part II—Nucleonics 3, 17 (Aug. 1948).
90. Zweifel, P. F., "Neutron Self-Shielding", Nucleonics 18, No. 11, 174-175 (Nov. 1960).

CHAPTER 8
Radiation Detection Based on Ion Collection

LECTURE OUTLINE AND STUDY GUIDE

I. METHODS OF ION COLLECTION

II. CHARACTERISTIC VOLTAGE-CURRENT CURVE
- A. Region of Simple Ionization
- B. Region of Gas Amplification
  1. Proportional region
  2. Limited proportional region (Semi-portional
  3. Geiger region
  4. Discharge region
- C. Voltage Gradients and Electrode Shapes

III. IONIZATION CHAMBERS
- A. Electroscopes (*Experiment* 8.1)
  1. Methods of use
     - a. Rate-of-drift method
     - b. Constant-deflection method
  2. Types of electroscopes
     - a. Gold-leaf electroscope
     - b. Landsverk
     - c. Lauritsen
     - d. Pocket chambers (dosimeters)
- B. Electrometers (*Experiment* 8.2)
  1. Vacuum-tube type
     - a. Mode of operation
       - (1) Integrating
       - (2) Differentiating
     - b. Hanford type meter or "Cutie-pie"
  2. Vibrating-reed electrometer
  3. Methods of use
     - a. High-resistance-leak method
     - b. Rate-of-charge method
  4. Source geometry
     - a. Internal sources
     - b. External sources

IV. PROPORTIONAL COUNTERS (*Experiment* 8.3)
- A. Characteristic Counting Rate-Voltage Curve
- B. Filling Gases
- C. Flow Counters and Filled Counters
- D. Coincidence Time

V. LIMITED PROPORTIONAL REGION

VI. GEIGER COUNTERS (*Experiments* 3.2 *and* 8.4)
- A. Characteristic Counting Rate-Voltage Curve
- B. Quenching Methods
  1. External quenching
     - a. Resistance method
     - b. Electronic methods
  2. Self quenching
     - a. Polyatomic gases (Organics, halogens)
     - b. Mechanisms
- C. Coincidence Time
- D. Memory

VII. SPECIAL OR MODIFIED COUNTERS
- A. Windowless-flow Counters (*Experiment* 8.4)
  1. Geiger
  2. Proportional
- B. Neutron Detectors (*Experiment* 8.5)
  1. Proportional
  2. Ion-chamber type
- C. Bernstein-Ballentine Counters

INTRODUCTION:

The purpose of this chapter is to acquaint the student with the types of instruments used for the detection and measurement of nuclear radiations, to elucidate the basic modes of operation of these instruments, and to show how these modes of operation are interrelated.

Instrumentation for the detection and measurement of radiation requires two basic elements—a sensing element and an indicating element. Sensing elements convert the energy of the radiation into electrical energy. The electrical energy may then pass directly to the indicating element—a scaler, meter, or recorder—or it may first pass through certain intermediate electronic circuits for the purpose of amplification or analysis.

Sensing elements for the detection of radiation depend upon the formation of ions for their operation. They can be divided into two general categories—(1) those which depend upon the collection of these ions; and (2) those which do not depend upon ion collection. Ionization chambers, proportional counters and Geiger counters require the collection and measurement of the ions produced while photographic methods, cloud chambers, and scintillation counters do not.

METHODS DEPENDING UPON ION COLLECTION:

The counting circuit shown in figure 8.1, consists simply of two parallel plates or electrodes across which a potential can be impressed by means of a battery. The current flowing in the circuit is read on a sensitive meter. The behavior of this circuit to radiation passing between the plates is perhaps most easily understood by an illustration which is somewhat artificial.

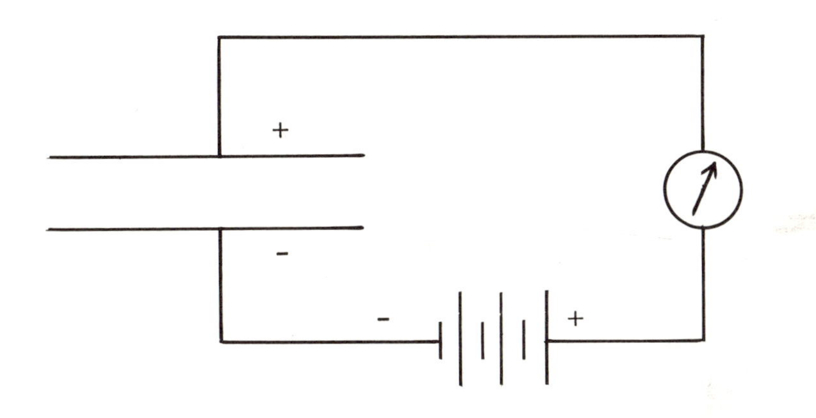

Figure 8.1—SIMPLIFIED CIRCUIT OF IONIZATION DETECTOR.

Radiation from a constant source is caused to impinge on the volume between the plates as the voltage across the plates is increased. A gas—air, helium, or argon—occupies the space between the electrodes. Each time an electron or alpha particle passes between the electrodes, the gas is ionized and a small current passes from one plate to the other. These small surges of current passing through the circuit cause the meter to deflect. The extent of the meter deflection, for a given set of electrodes and for a specific gas, depends upon two factors, i.e., the amount of ionization produced by the particle and the potential impressed across the plates by the battery. These relationships are illustrated in figure 8.2. The current is seen to increase with voltage, but not in a linear manner as one might anticipate from consideration of Ohm's law. The ionized gas in the detector does not behave as a simple passive resistive element.

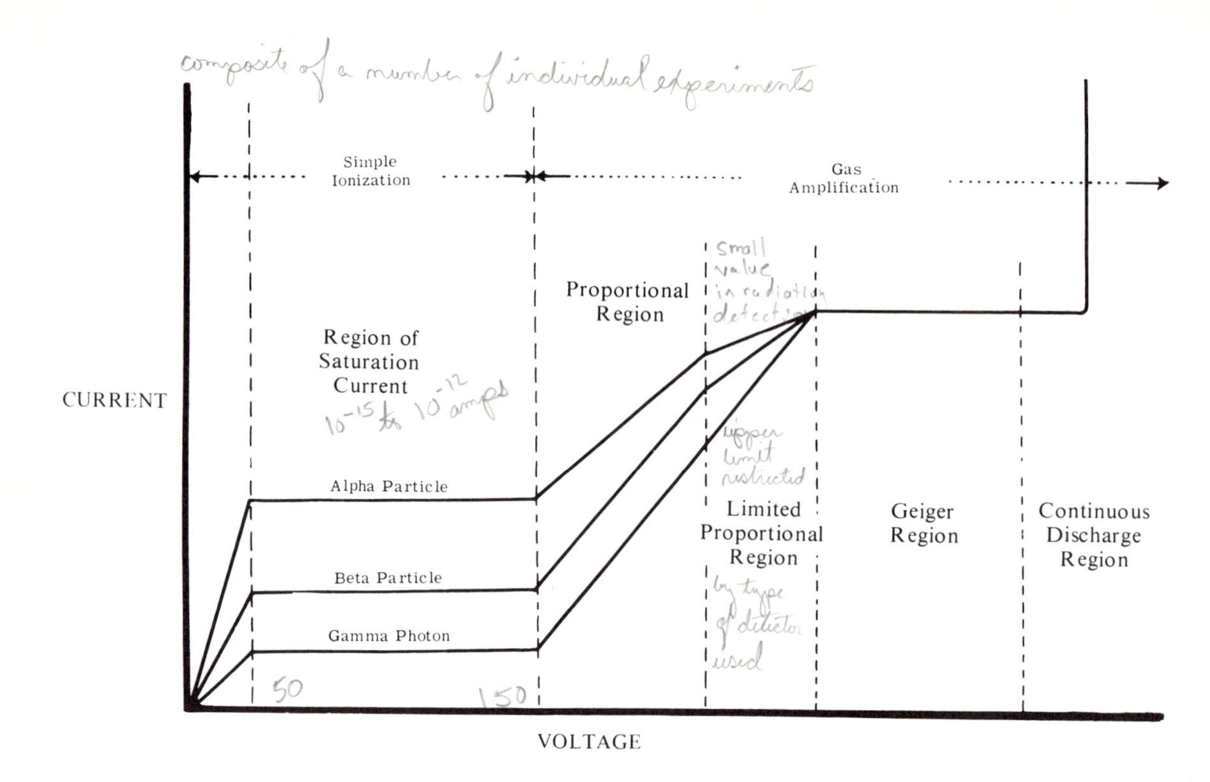

Figure 8.2—COMPOSITE CHARACTERISTIC CURVE FOR IONIZATION DETECTORS

REGION OF SIMPLE IONIZATION:

Collision of an alpha particle or beta particle with an atom of gas causes the gas to ionize. The electron and positive ion produced in this process are called an *ion pair*. The specific ionization for alpha particles is roughly 1000 times greater than that for beta particles, with the result that alpha particles produce about 1000 times as many ion pairs per unit of distance traveled through the gas in the counter as will a beta particle of average energy. If no potential is applied to the electrodes, recombination of these electrons and positive ions occurs, with the result that no current flows in the circuit. If, however, a small potential is applied, some of the electrons will be attracted to the anode (positive) and some of the positively charged ions will be attracted to the cathode (negative). Those electrons and ions not reaching the electrodes will recombine to form neutral atoms. As the potential difference is increased, the number of electrons and ions reaching the electrodes, before having a chance to recombine, will increase proportionately. If ionization is initiated by an alpha particle, a larger current will flow as compared to that initiated by a beta particle because of the much greater number of ion pairs produced.

As one continues to increase the potential applied to the electrodes, a voltage will be reached at which all ion pairs produced by the ionizing radiation will reach the electrodes and essentially no recombination of electrons and ions will take place. Saturation is now said to have occurred and the current flowing at this point is called the *saturation current*. Moderate increases in potential create no further increase in current, as all the ions and electrons produced are reaching the electrodes and the magnitude of the current is regulated by the number of primary ion pairs produced by the incident radiation. It is within this region—the region of the saturation current—that a simple ionization chamber is normally operated. The voltage applied under these conditions is usually in the order of 50 to 150 volts and the current flow in the order of 10^{-15} to 10^{-12} amperes.

PROPORTIONAL REGION:

A continued increase in potential above a "moderate" amount will again bring about an increase in current flow in the proportional region. This occurs in spite of the fact that all primary ion pairs produced by the incident radiation are being collected at the electrodes. This further increase in current is explained by the formation of *secondary ion pairs*. Approximately 30 to 35 electron volts is required to create a primary ion pair. Under the influence of higher potentials, some of the primary electrons attain energies in excess of these values and hence, should one of these electrons collide with an atom en route to the electrode, this atom, too, will become ionized. In this way, the number of ion pairs available for current conduction is increased. This process is called *gas amplification*. Gas amplification may vary from unity, near the region of simple ionization, to as much as 10^6 near the upper extreme of the proportional region. As gas amplification increases, the process of acceleration in the potential field, with subsequent further ion pair production, must occur many times. This region is known as the *proportional region* because the number of electrons and positive ions reaching the electrodes, and hence the magnitude of the current flow, is proportional to the number of primary ion pairs produced by the incident radiation. Thus, if an alpha particle produces 1000 times as many primary ion pairs as a beta particle, the magnitude of the current, amplified perhaps 10^5 times by gas amplification, but amplified equally in both cases, will still be 1000 times greater for an alpha particle than for a beta particle. This is a significant fact, enabling the experimenter to differentiate between alpha and beta radiation with the proportional counter.

LIMITED PROPORTIONAL REGION:

Gas amplification increases proportionately with the applied potential, but reaches a limit at the upper range of the proportional region. This limit is regulated by the total number of ion pairs which can be produced under a given set of conditions and is governed by the physical dimensions of the counter and the number of gas atoms present. If, for example, the gas amplification is 10^6, then an incident particle of radiation producing, say, 10^2 primary ion pairs would be responsible for the formation of a total of 10^8 ion pairs. On the other hand, a more powerful ionizing incident particle producing 10^5 primary ion pairs would be expected to yield a total of 10^{11} ion pairs as a result of gas amplification. If the aforementioned factors place an upper limit of 10^{10} on the total number of ion pairs which can possibly be produced, it is seen that although gas amplification for the first case was 10^6, it is only 10^5 for the second and the counter is not operating in a truly proportional manner. Hence, the expression *limited proportional region* has been used to describe this type of counter behavior. The limited proportional region has found little utility in the practical measurement of radiation.

GEIGER REGION:

If the applied voltage is still further increased, then all radiation, regardless of the number of primary ion pairs produced, will yield the same current flow. This is true because, in all cases, gas amplification will have reached its maximum value. The flood of ions produced has been referred to as an "avalanche" and is called a *Townsend avalanche* in honor of the researcher who first called attention to the phenomenon. In the Geiger region it is no longer possible to differentiate between alpha and beta radiation, as was the case in the proportional region. All ionization, regardless of its specific nature, will produce similar end results in the Geiger counter. This disadvantage is partially outweighed by the somewhat lesser complexity and cost of the equipment as compared to proportional counters. The reader is referred to chapter 2 for details of Geiger counter construction and operation.

REGION OF CONTINUOUS DISCHARGE:

At the upper end of the Geiger region, a potential is reached at which one begins to get *multiple discharges*, the rate of *multiple discharge* increasing rapidly with a relatively slight increase in potential. In the regions of multiple discharge or *continuous discharge*, either extremely high and, of course, false counting rates are observed, or the registered count rate may abruptly decrease. Under these conditions the Geiger tube may suffer permanent damage.

CONSTRUCTION DETAILS:

From the preceding discussion of the various operating regions of counters, one might obtain the notion that a single detector, operating under a single set of conditions, could be used to demonstrate each of these operating regions. Such is not the case. It is necessary to magnify the abscissa, a section at a time, using carefully selected conditions of physical dimensions of electrodes, gas composition and potential. In this way, the characteristics of each region can be studied. In reality, therefore, figure 8.2 represents a composite of a number of individual experiments.

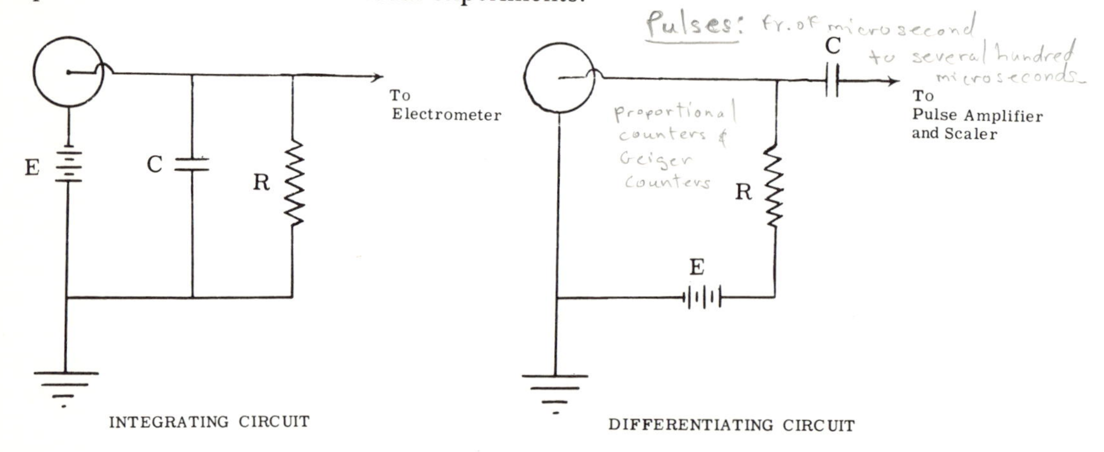

Figure 8.3—SIMPLIFIED BASIC CIRCUITS

A second important basic concept concerns the method of current measurement. Current flows through the detector in surges or pulses. Each particle of ionizing radiation causes a pulse of current to flow, the pulse lasting from a fraction of a microsecond to several hundred microseconds. If an *integrating circuit* is used, these pulses tend to be smoothed out and the recording device will indicate the average direct current flow through the chamber. A *differentiating circuit* produces the opposite effect. In this case, the pulses are shaped by the employment of the proper circuit components and no D.C. component of the current is permitted to reach the recording device. The recording device may be a counter or scaler which tallies the total number of discrete pulses originating in the ionization chamber, thereby indicating the number of ionizing particles entering the detector. Ionization chambers may be operated as either differentiating or integrating instruments. Proportional counters and Geiger counters are generally operated as differentiating types.

SOLID STATE DETECTORS:

Solid state detectors were first investigated about 1945. Recently they have received a great deal of attention and have become increasingly more important as radiation detectors as the advantages they offer are utilized. Among the advantages are (1) their high density with resulting great stopping power, (2) the good geometrical resolution resulting from their small physical size, (3) a short resolving time which permits high-speed counting and (4) a high efficiency for the conversion of incident energy into ions which gives excellent energy resolution.

There are several types of solid state detectors. Among them are (1) the crystal detectors, (2) the surface junction detectors and (3) the lithium-ion-drift detectors.

The basic mode of operation of conduction-type solid state detectors is similar to the operation of a simple ion chamber. In the ion chamber an incident particle passing through the gas produces ion pairs. In a solid detector an incident particle produces electrons and "holes" in the crystal structure. When a potential difference is applied between opposite surfaces of the crystal, the electrons migrate to the positive surface while the holes may or may not migrate, depending upon the nature of the crystal.

EXPERIMENT 8.1 IONIZATION CHAMBERS (ELECTROSCOPES)

OBJECTIVES:

To review the types of ionization chambers and the principle of their operation.

To demonstrate the calibration of a pocket chamber, a Landsverk electroscope and a Lauritsen electroscope.

To illustrate the use of these instruments for the measurement of radiation and/or radioactivity.

THEORY:

Technically, all counters whose operation involves gas ionization, including proportional counters and Geiger counters, are ionization counters. In this discussion, the term "ionization chamber" refers to those instruments operating in the region of simple ionization without gas amplification. (See figure 8.2). Ionization chambers are normally operated in the region of the saturation current, this being the region of least sensitivity to voltage changes and the region where the instrument response would be expected to be most nearly proportional to the ionization produced in the chamber.

Ionizing events which occur in an ionization chamber may be measured by means of either of two basic methods. In the first method, the ionization chamber behaves electrically as a simple capacitor. The chamber is charged by the momentary application of a potential across it. Exposure of the chamber to radiation causes the charge to be lost. The rate at which the charge is lost is a measure of the radiation intensity and the total loss of charge is a measure of the total radiation exposure. Instruments which operate in this manner are included in this experiment. They are all *integrating* instruments since they measure the accumulated ionizing energy.

In the second method, the ionization chamber behaves electrically as a current regulating device. The amount of current which flows through the chamber is proportional to the radiation intensity. Sensitive electrometers are used to detect and measure this current. Instruments which operate in this manner, e.g., chambers which are used in conjunction with a vibrating reed electrometer, are discussed in experiment 8.2. They may be either of the *integrating* or *differentiating* type. These terms, also, are explained in experiment 8.2.

The equivalent circuit of a simple condenser-type electroscope is shown in figure 8.4. It consists simply of a capacitor which can be charged either by momentary connection to a battery, as shown, or electrostatically by friction. A gold leaf or fiber attached to one plate of the condenser serves to indicate the relative electrical charge remaining on the capacitor. On refined instruments, such as the pocket chamber (dosimeter), the Landsverk electroscope or the Lauritsen electroscope, the position of the fiber is observed on an illuminated scale by means of a microscope.

Once charged, an electroscope (acting as a condenser) will retain the charge indefinitely unless a pathway is provided for its discharge. Discharge, exclusive of leakage

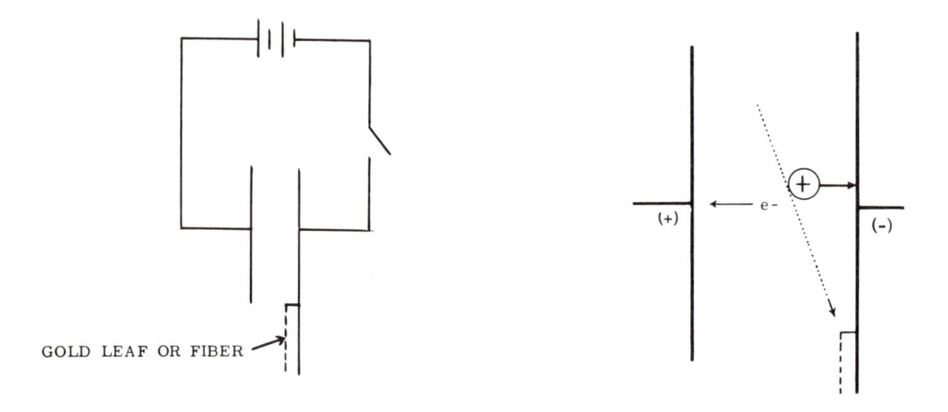

Figure 8.4

Figure 8.5

through the supporting dielectric, occurs as illustrated in figure 8.5. Ions, produced by ionizing radiation, migrate to oppositely charged electrodes, neutralizing the charge. As the charge is neutralized, the repelling electrostatic forces which hold the fiber away from its support (see figure 8.4) are reduced. As a result, the fiber moves toward the support. The rate at which it moves (drift) is proportional to the radiation intensity. Thus, the rate of drift may be used as a measure of the amount of radiation.

The classical *gold-leaf electroscope* is the simplest of ionization chambers. In this instance the surroundings of the electroscope may be considered to constitute the second plate of the condenser, for it is to (or from) the surroundings that the charge is conducted upon exposure of the instrument to ionizing radiation. In the case of the pocket chamber, the Landsverk electroscope and the Lauritsen electroscope, the body of the chamber is one plate of a condenser while a centrally located wire or probe is the other.

THE POCKET CHAMBER (POCKET DOSIMETER):

The pocket chamber has been designed specifically for use as a radiation protection device. Its appearance is generally similar to that of a fountain pen (see figure 3.3) so that it can be carried conveniently in a pocket. These chambers register the total accumulated dose. Scales are calibrated directly in roentgens per hour or milliroentgens per hour.

A charger, supplied as a separate unit, contains a battery or rectifier power supply. The protective cap, if any, is first removed from the end of the chamber, and the chamber then inserted into the charger. Some chargers have a momentary contact switch (charge button) which must be depressed while the potential control knob is adjusted to set the image of the fiber at zero. Upon replacing the protective cap the chamber is now ready for use.

The scale in some instances may be an integral part of the chamber. This type offers the advantage of being able to check the accumulated dose while "on the job." In other instruments the scale is in the charger. This type offers the advantage of economy. The basic mode of operation is the same for both types. For instruments of the latter type, which must be again inserted into the charger to be read, the charge button must not be depressed until after the reading is recorded.

PROCEDURE:

1. Charge a pocket chamber, setting the fiber image accurately at the zero mark.
2. Place a standard gamma source, e.g. radium, at a convenient measured distance d from the chamber. Calculate the exposure dose rate I_0 in roentgens per hour, at distance d from the source by use of the relation,

$$I_0 = \frac{\Gamma C}{d^2} \tag{8.1}$$

where Γ is the dose rate per hour at 1 cm, C is the number of millicuries of activity of the standard, and d is the distance in centimeters. Values of Γ for a number of radioisotopes in roentgens per millicurie per hour at 1 cm are given below. Γ is called the *specific gamma-ray constant* or *specific gamma-ray output*.

Table 8.1—VALUES OF SPECIFIC GAMMA-RAY CONSTANT
(Γ is given in cm^2 r mCi^{-1} hr^{-1})

Isotope	Γ	Isotope	Γ
Arsenic-76	2.4	Manganese-52	19.2**
Cesium-137	3.5	Manganese-54	4.5
Cobalt-58	5.6**	Mercury-203	1.3
Cobalt-60	13.2	Radium-226 (0.5 mm)	8.4*
Copper-64	1.1	Radium-226 (1.0 mm)	7.8
Gold-198	2.5	Rubidium-86	0.5
Iodine-130	12.3	Sodium-22	12.2
Iodine-131	2.3	Sodium-24	19.3
Iridium-192	5.1	Thulium-170	0.04
Iron-59	6.5	Zinc-65	3.0**

*The value for radium is that observed for a source encapsulated in 0.5mm of platinum or the equivalent, the usual form in which radium sources are supplied.

**X-ray emission, occurring as a result of electron capture, does not contribute appreciably to Γ at one meter but must not be ignored at a distance of a few cm.

The value of Γ for all pure beta emitting isotopes is zero. This includes ^{14}C, ^{32}P, ^{35}S, ^{3}H, ^{45}Ca, ^{90}Sr and others.

EXAMPLE: For 100 micrograms of radium (100 microcuries), calculate the dose rate at 10 cm.

$I_0 = (1000)(8.4)(0.1)/10^2$

$I_0 = 8.4$ mr/hr at 10 cm from 100 μg radium

3. After a period of time, calculated to cause the dosimeter to give a reading of approximately 50 per cent of full scale, remove the radioactive source. Read the pocket chamber and compare with the calculated value.

THE LANDSVERK ELECTROSCOPE (Analysis Unit):

The unit illustrated in figure 8.6 consists of a cylindrical, hollow base three inches in diameter and three inches high. The lower two-thirds of this cylinder serve as the wall of an ionization chamber, which is separated from the upper one-third by a thin plate. In the center of this plate, which forms the upper end of the chamber, there is a one-half inch hole containing a polystyrene insulator through which extends the collecting electrode.

In the upper one-third of the 3″ × 3″ cylindrical base are housed the quartz fiber electroscope and a friction charger or a transistor power supply which replaces the friction charger in more recent models. A microscope is mounted centrally on top of the unit. Observation by means of the microscope discloses the image of the quartz fiber focused on a scale divided into 100 equal divisions.

Figure 8.6—LANDSVERK ANALYSIS UNIT

OPERATING INSTRUCTIONS:

Friction Charged Model

1. Charge the electroscope by turning the charging knob *slowly* in the clockwise direction. When the knob is turned clockwise, a slipping clutch causes a contact arm to remain in contact with the electroscope, which thus becomes charged. If the knob is turned too rapidly, the fiber image may move across the reticle so rapidly that it is not seen. The fiber image may then travel beyond the zero point of the reticle and be "lost." It takes only about 3/4 of a turn after contact is made to charge the electroscope fully.

2. When the fiber image has been brought up to or a little past the zero mark of the reticle, reverse the charging knob about one quarter of a turn. Counter-clockwise rotation breaks contact between the charger and electroscope and causes the charger to become grounded. There is then no charge remaining on the charger which might affect the reading of the electroscope by induction.

3. With a stopwatch or timer, determine the rate of drift of the fiber in divisions per minute across the reticle. This is the background rate of drift.

4. Place the radioactive sample within, below or to the side of the ionization chamber, depending upon the type of measurement to be made. Repeat steps 1, 2 and 3 to establish the rate of drift of the fiber for the radioactive sample plus the background rate of drift.

5. Subtract the background rate of drift from the total rate to get the net rate of drift from the activity of the sample alone.

Transistor Power Supply Model

1. With the charge button depressed, adjust the fiber image to the zero mark of the reticle by means of the potentiometer knob. Depressing the charge button turns on the power supply and makes contact between it and the electroscope. The potentiometer functions as a voltage divider to select the proper potential for charging the instrument.

2. Release the charge button. As the charge button is released, the power supply turns off and contact is broken with the electroscope.

3. Follow steps 3, 4 and 5 above.

PROCEDURE:

PART A—BETA (OR ALPHA) MEASUREMENTS

1. Measure the background rate of drift for the electroscope. Be certain all sources of radiation have been removed from the immediate area.

2. Prepare a series of beta sources. For phosphorus-32 a range from 10^{-2} to 1.0 microcurie of activity, prepared in plastic planchets or other suitable mount, will provide a reasonable range of drift rates. These samples need not contain a specific amount of activity, but the activity of each sample, in microcuries, should be known accurately. Using dilutions of a shipment standardized by the supplier will meet this requirement.

3. Measure the rate of drift for each source in the electroscope and record the data. For beta (or alpha) measurements the source is centered on the base of the instrument. Thus, the source is within the chamber. One-inch planchets fit conveniently into the depression in the base.

4. Construct a calibration curve by plotting "sample activity" vs "rate of drift less background." Draw the best straight line through these points.

5. Calculate the sensitivity of the electroscope in divisions per minute per microcurie.
________div/min/μCi.

6. Measure the rate of drift for a known carbon-14 sample and calculate the sensitivity. To which isotope is the electroscope more sensitive, ^{32}P or ^{14}C? Explain.

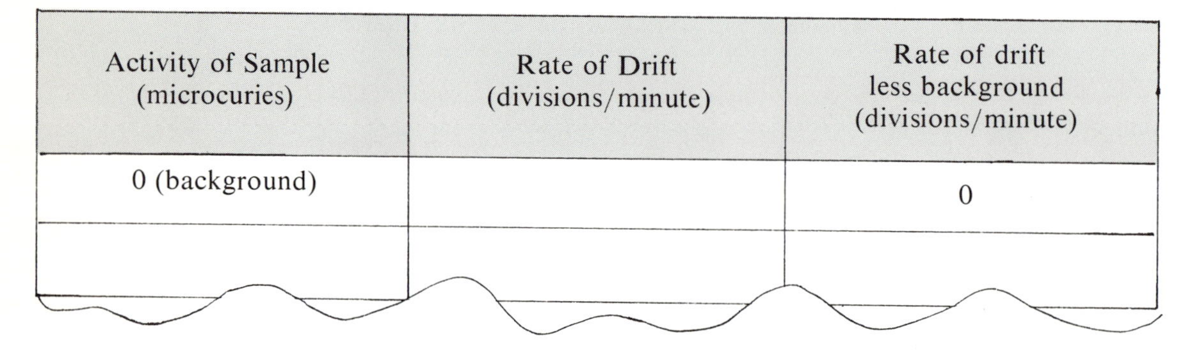

Activity of Sample (microcuries)	Rate of Drift (divisions/minute)	Rate of drift less background (divisions/minute)
0 (background)		0

PART B—GAMMA MEASUREMENTS

1. Measure the background rate of drift for the electroscope, all radioactive samples having been removed an appropriate distance from the instrument.

2. Check the electroscope for linearity. This is done as follows: First charge the instrument; then, place a gamma source near the electroscope and record the time required for the fiber to drift from 0 to 10, from 10 to 20, etc., accumulating a total of ten readings to 100 scale divisions. All drift times should be equal.

3. Place a standard gamma source, e.g., radium, at a convenient measured distance d from the *center* of the chamber. (Do *not* put the source inside the chamber as was done with beta emitting isotopes.) Determine the rate of drift as before in divisions per minute.

4. Calculate the dose rate (milliroentgens/hour) at distance d from the standard source. (See step 2 under the procedure for the pocket chamber on page 251 for calculation and values of Γ.)

Range	Time of Start	Time of Finish	Time for Range
0-10			
10-20			
20-30			
30-40			
40-50			
50-60			
60-70			
70-80			
80-90			
90-100			

5. Calculate the sensitivity S of the electroscope in divisions per minute per milliroentgen per hour (div/min/mr/hr). Thus,

$$S = \frac{\text{rate of drift}}{\text{dose rate}} = \frac{\text{div/min}}{\text{mr/hr}} = \text{div/min/mr/hr}$$

6. Replace the standard source with an "unknown" gamma-emitting isotope, placing it at a measured distance from the electroscope, and determine the rate of drift. Using the sensitivity S for the electroscope, calculate the dose rate (mr/hr).

7. Using the appropriate value of Γ, calculate the number of millicuries of activity in the "unknown" sample.

THE LAURITSEN ELECTROSCOPE:

Operating on the same basic principles, the Lauritsen electroscope is similar to the Landsverk unit with respect to mode of operation. The chamber is a cylindrical can 2-1/4 inches in diameter and 3 inches long, mounted on the end of a microscope. Lauritsen electroscopes are charged by means of batteries. The position of the gold-plated quartz fiber is observed on an illuminated scale by means of the microscope.

Unlike the Landsverk, in which a relatively sturdy probe extends into the sensitive volume of the chamber (so the fiber is at no time exposed), in the Lauritsen instrument the fiber itself is within the sensitive volume. To overcome the problem of disturbing the fiber with air currents, all measurements are made with the radioactive source external to the chamber. Three types of interchangeable aluminum cans (chambers) are supplied:

1. Lacquer window chamber for alpha counting. The lacquer window, supported on a wire screen, is extremely thin and *must not be touched.*

2. Aluminum window chamber for beta counting. The aluminum window, too, is extremely thin and must be handled with care.

3. Windowless chamber for gamma counting.

Removal of the cans should be avoided as much as possible to prevent repeated tangling of the fiber with accompanying hazard of breaking the gold coat.

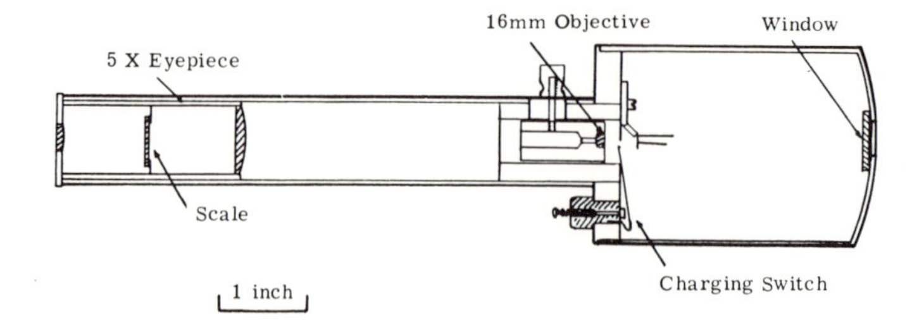

Figure 8.7—THE LAURITSEN ELECTROSCOPE
(*Courtesy Fred C. Henson Company*)

OPERATING INSTRUCTIONS (*Model* 5):

1. Place left hand middle finger on panel switch button and grasp potentiometer knob between thumb and forefinger of the same hand. Rotate knob full counter-clockwise.

2. Grasp the upright section of the gray instrument base between the thumb and middle finger of the right hand and place the forefinger on the red knob.

3. While pressing down on the panel switch button, look through the eyepiece. Now, press the red knob as far as it will go toward the can with the forefinger of the right hand. Slowly rotate the potentiometer knob clockwise until the fiber is on the left side of the scale. Remove right hand from instrument and left thumb and forefinger from the potentiometer knob, check fiber position, then remove left hand from instrument. The electroscope is now charged and ready for use. Whenever a fiber position reading is desired, simply press panel switch button and look through eyepiece.

Instrument drift (in addition to normal background), caused by soaking of charge off the fiber unit into the insulator, will generally continue for several hours after the initial charging of the instrument. This phenomenon is true of all electroscopes, including the pocket chamber and the Landsverk electroscope. For maximum precision, it is desirable to charge these units an hour or more before use, in order to allow an equilibrium to become established.

PROCEDURE:

PART A—CALIBRATION OF ELECTROSCOPE

Each instrument is calibrated by the manufacturer* and is supplied with a calibration certificate. Their standard condition for sensitivity is: the number of divisions discharge caused by 1 millicurie of radium placed 1 meter from the position of the fiber in the instrument for a period of 1 minute (with plain aluminum chamber). They use, as a radium source, a platinum-iridium needle containing radium which has been calibrated by the Bureau of Standards to have a gamma radiation equivalent to 5.51 milligrams of radium. It is used at such a distance as to be equivalent to 1 milligram at 1 meter.

*Fred C. Henson Co., 3311 E. Colorado St., Pasadena, California.

Background should be considered. It usually amounts to 1.5 to 5 divisions per hour, depending on the sensitivity of the fiber. Thus, as an example, if an electroscope had a sensitivity of 2.95 div/min/mCi at one meter and a background of 3.5 divisions per hour, the true sensitivity would be 2.95 div/min/mCi at one meter minus the background which is $\frac{3.5 \text{ div/hr}}{60}$. In this instance the background could be neglected.

Calibrate the Lauritsen electroscope against a standard gamma source employing the method described above. If a radium source is not available, other gamma sources can be used if a suitable correction is made for the differences in the values of specific gamma-ray constant Γ. (See part 2 of the procedure for pocket chambers.) Compare with the value given by the manufacturer. The procedure illustrated in figure 8.11 can be used.

PART B—MEASUREMENT OF UNKNOWN SOURCE

Measure the millicurie strength of an "unknown" gamma source. Follow one of the basic procedures previously described.

EXPERIMENT 8.2 IONIZATION CHAMBERS AND ELECTROMETERS

OBJECTIVES:

To review the fundamentals of radiation measurement employing ionization chambers and vacuum tube electrometers.

To illustrate the use of the vibrating-reed electrometer for measuring ionization chamber currents, to measure the characteristics of this apparatus and to demonstrate some of its applications.

THEORY:

In experiment 8.1 the function of the ionization chamber was that of an electrical condenser. It was charged by momentary contact with a battery or other potential source. The rate of discharge, caused by ionizing radiation and measured by means of a fiber electroscope in the role of a high impedance voltmeter, was a measure of the radiation dose rate.

The battery or potential source becomes a permanent circuit component of the equipment described in this experiment. In some cases the battery is connected in series with the ionization chamber and a condenser (capacitor); the rate of charge of the capacitor is a measure of the current flow through the ion chamber. In other cases, the battery, ion chamber and a resistor are series connected; voltages developed across the resistor are a function of the current flow.

It will be remembered that radiation is discontinuous. Thus, the currents produced in the ionization chamber are likewise discontinuous. The associated electronic circuitry can be used to *differentiate* these discontinuities of the current and thus indicate the number of ionizing events which occur in the ion chamber, or the circuitry can be used to *integrate* the current. Integrating circuits indicate the average current flow during a given interval of time and thus measure dose rate rather than decay rate.

A *differentiating circuit*, simplified to its essential elements, is shown in figure 8.8. The battery B applies a potential across the ionization chamber through resistor R. When ionizing radiation passes through the chamber, current flows. The instantaneous current

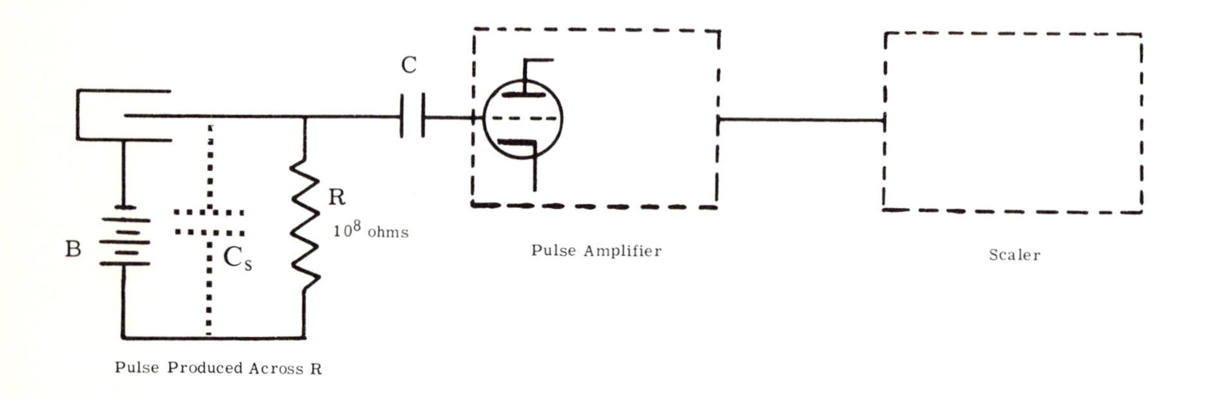

Figure 8.8—IONIZATION CHAMBER AND ASSOCIATED DIFFERENTIATING CIRCUIT

flowing through R produces an instantaneous potential, $e = iR$. The capacitance C_s, which is very small and may consist only of stray circuit capacity, is also charged momentarily but is quickly discharged through R. (Differentiating circuits are characterized by a small RC time constant.) As a result, a pulse (a rapid change in potential) is produced each time an incident ray passes through the ionization chamber. The pulse height is equal to the change in potential dE given by

$$dE = dQ/C_s$$

where dQ is the change in the charge (in coulombs) of the capacitor C_s. The limit of the change dQ is given by the quantity of current flowing through the ion chamber as a result of a particular ionizing event (or of simultaneous ionizing events). This small electrical pulse is amplified and finally counted by means of a scaler. Ionization chambers operated in this way offer no substantial advantage over proportional counters, however, and are seldom used with differentiating circuits.

The *integrating circuit* is more useful. Here, the current produced by ionizing radiation in the chamber is used to charge a condenser or to produce a potential across a resistor. These two modifications of the integrating circuit, used in conjunction with an electrometer, serve as the basis for two basic methods for measurement, known as the *rate-of-charge method* and the *high-resistance-leak method*.

Figure 8.9 illustrates the *rate-of-charge circuit*, reduced to its bare essentials for simplicity. When the shorting switch S_W is opened, ionizing radiation results in a flow of current through the ionization chamber charging capacitance C. The potential developed across C is given by

$$E = Q/C \tag{8.2}$$

where Q is the charge in coulombs and C the capacity in farads. The voltage E is measured by means of a vacuum tube electrometer. The change in E in time is a measure of the rate of production of ion pairs in the ionization chamber which, in turn, is a function of the dose rate. When the meter M reaches full-scale deflection, the shorting switch is momentarily depressed and the charging process is repeated.

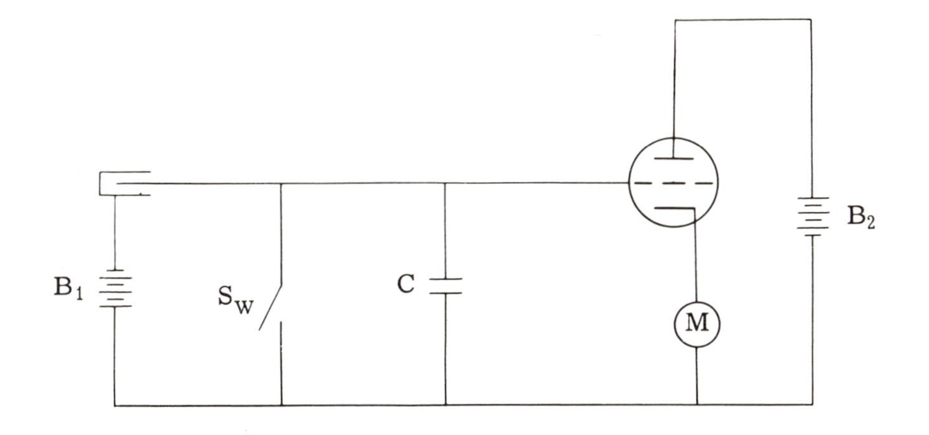

Figure 8.9—RATE-OF-CHARGE CIRCUIT

In the *high-resistance-leak circuit* (see figure 8.10), a high resistance R is placed in parallel with the capacitor. As a charge builds up on the capacitor, it leaks off through the resistor. An equilibrium potential E is attained in a period of time dependent upon the RC time constant of the resistor-capacitor combination. Potential E is indicated by the electrometer. At equilibrium, the current flow through R is equal to the current flow through the ionization chamber. Hence, if the value of R is known, and since E can be measured, the value of I can be calculated. A typical current will have a value in the order of 10^{-13} amperes. This current, flowing through the resistor, will develop a potential across the resistor in accord with Ohm's law:

$$E = IR \tag{8.3}$$

If R has a value of 10^{11} ohms, then a current of 10^{-13} amperes will produce a voltage of $10^{-13} \times 10^{11} = 10^{-2}$ volts across R. This voltage will be impressed on the grid of an electrometer tube which, in the circuit shown, is connected as a simplified cathode follower. Amplification occurs in the electrometer tube and the change in voltage is indicated on the meter. Following calibration, the current through the resistor, and hence through the ionization chamber can be calculated.

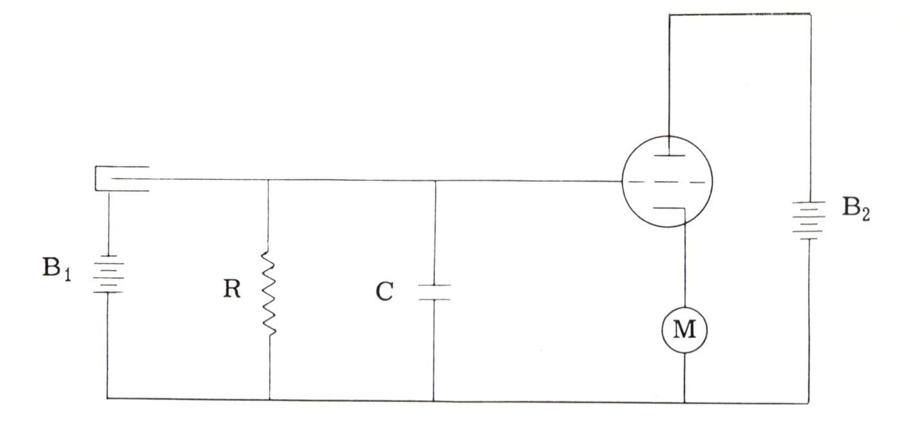

Figure 8.10—HIGH-RESISTANCE-LEAK CIRCUIT

SIMPLE VACUUM-TUBE ELECTROMETERS:

One of the best known ionization chamber instruments employing a simple vacuum-tube electrometer for voltage measurement is the survey meter commonly called a "cutie-pie." This instrument utilizes a high-resistance-leak circuit. The ionization chamber is cylindrical, the base of which is attached to the housing for batteries and electronic components. A pistol-grip handle is provided for convenience to allow one-hand operation. Use of this instrument is usually restricted to radiation protection. Thus, accurate calibration at frequent intervals is essential.

Calibrated radium or cobalt-60 sources are generally used for the periodic calibration of instruments. These calibrated sources are now available from numerous suppliers. Because such sources possess a rather high level of activity (usually from 1 to 10 millicuries), special care and technique in handling must be used.

PROCEDURE:

A physical arrangement for handling the source and survey meter with safety is illustrated in figure 8.11.

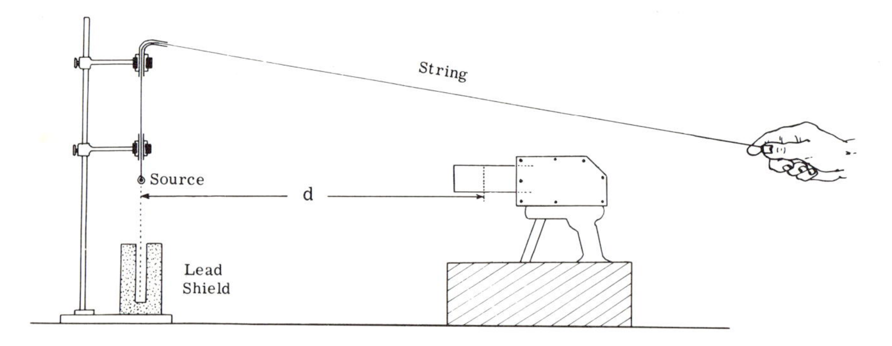

Figure 8.11—CALIBRATION OF A "CUTIE-PIE".

The strength of the source being known from the manufacturers' calibration certificate, the distance d from the source to the instrument necessary to give the desired dose rate is calculated from the relation,

$$d = \sqrt{\frac{\Gamma C}{r/hr}} \tag{8.4}$$

where C is the source strength in millicuries and the dose rate is given in milliroentgens per hour. Values of the specific gamma-ray constant (specific gamma-ray output) Γ are found from experiment 8.1, table 8.1. It should be remembered that encapsulation of a radium source introduces a certain degree of filtration of the radiation and hence influences the effective value of Γ. The measured values of Γ for a radium source with filtration are:

0.5 mm Pt-Ir*	8.4
1.0 mm Pt-Ir	7.8

2. The survey meter is now aimed at the point which the source will occupy when it is raised out of its shield. The meter is placed a distance d from this point. The distance should be measured to the center of the ionization chamber and not to the window of the chamber.

3. Allow the instrument to "warm up", check the zero setting and switch to the proper range.

4. While remaining well behind the instrument, yet in such position that the meter can be observed, raise the source out of the lead storage container by pulling the string.

5. Observe the meter reading and compare with the calculated value. If they do not agree, the calibration potentiometer of the instrument should be adjusted. Recheck the calibration.

6. Each range of the instrument should be calibrated. Depending upon the instrument ranges and upon the source strength, however, calibration may not be possible on the higher ranges of some instruments because of the error inherent in the measurement of small source-to-instrument distances.

7. When the calibration is completed, the source should be properly returned to storage.

VIBRATING-REED ELECTROMETERS:

The vibrating-reed electrometer is especially useful for the measurement of small currents, as low as 10^{-17} amperes, such as those produced by ionization chambers. The operation of this instrument is shown in simplified form in figure 8.12. Here the *high-resistance-leak circuit* is illustrated.

*An alloy containing 80% platinum and 20% iridium is generally used in the manufacture of radium needles.

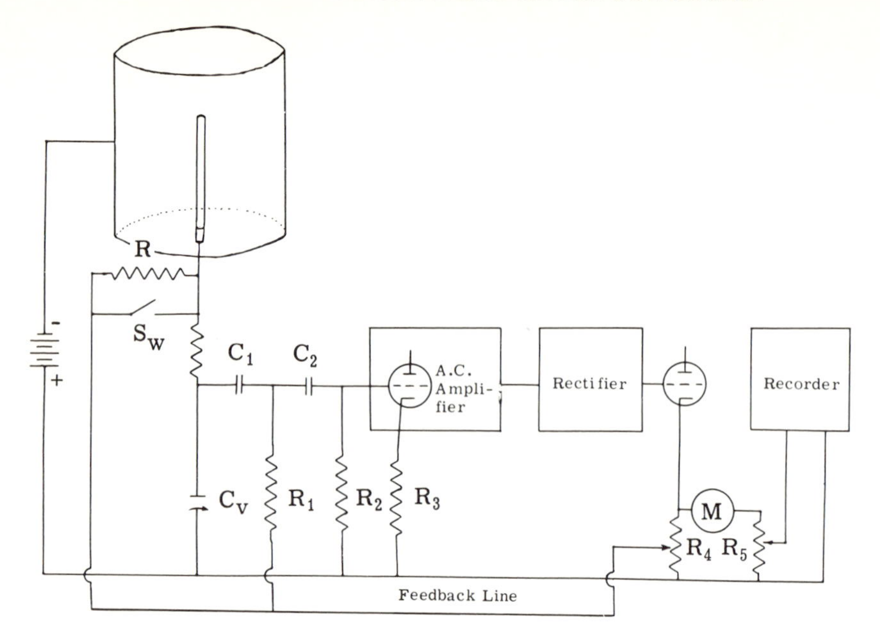

Figure 8.12—SIMPLIFIED SCHEMATIC OF CARY IONIZATION CHAMBER AND VIBRATING-REED ELECTROMETER. The high-resistance-leak circuit is shown.

Figure 8.13—CARY VIBRATING-REED ELECTROMETER. The ionization chamber (figure 8.17) attaches to top of preamplifier in place of dust cap shown here. (*Courtesy Applied Physics Corp.*)

Small currents flowing through the ionization chamber charge the capacitor C_V. This charge leaks off through resistance R. Values of R from 10^7 to 10^{12} ohms are commonly used and must be handled with great care. The potential developed across R is then measured by means of the vibrating-reed electrometer.

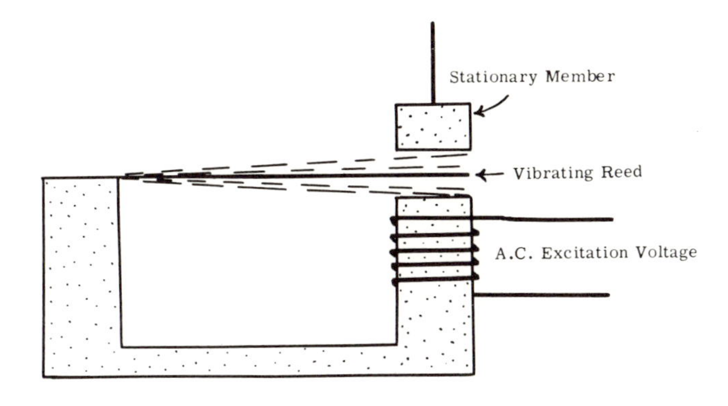

Figure 8.14—DIAGRAM OF THE VIBRATING REED, C .

Direct current amplification presents rather serious problems. Some form of direct coupling from one stage of amplification to the next is usually required, and thus, the problem of supplying the proper potential to each tube element is presented. Other serious problems in D.C. amplification include stability and zero drift. These problems are far less troublesome in A.C. amplification. In the vibrating-reed electrometer, D.C. potentials are converted to A.C. potentials by means of a vibrating-reed capacitor C_V before amplification.

Capacitor C_V consists of a stationary member and a vibrating member. The vibrating member (or reed) is made to oscillate at a fixed frequency, usually between 200 and 500 cycles per second. The D.C. potential across C_V is thus modulated. The amplitude of the A.C. signal produced is a function of the charge on the capacitor and is given by

$$dE = Q/dC \tag{8.5}$$

where dE is the change in potential (the A.C. component) produced by a change in capacitance dC (caused by vibration of the reed) for a given charge Q (a function of the potential across R). The A.C. potential thus produced passes through the coupling capacitors C_1 and C_2 to the A.C. amplifier. The amplified signal is then rectified and indicated on a meter M. A recorder may also be connected to the electrometer as shown. Stability is increased by feeding a portion of the output signal back into the input. The zero setting of the meter is adjusted by means of R_4 with the switch S_W closed.

The sensitivity of the vibrating-reed electrometer is illustrated by the following calculation. Assume the ionization chamber has been charged with 0.001μc of $^{14}CO_2$ and that resistor R is 10^{12} ohms. The average beta energy for carbon-14 is 0.045 MeV, 35 eV are required to produce an ion pair in air, and an ampere is equal to 6.28 × 10^{18} electronic charges per second. Thus,

Ion pairs (or electrons) per beta particle = $4.5 \times 10^4/35 = 1.3 \times 10^3$

Also, 0.001μCi = 37 dps.

Thus, current flow for 0.001μCi ^{14}C= $\dfrac{1.3 \times 10^3 \times 37}{6.28 \times 10^{18}} = 7.5 \times 10^{-15}$ amp.

and $E = IR = 7.5 \times 10^{-15} \times 10^{12} = 7.5 \times 10^{-3}$ volts
$= 7.5$ millivolts

In practice, the observed potential will be somewhat less than the theoretical value. Efficiency is a function of the construction of the ionization chamber as well as its size. Although each chamber must be calibrated individually, a general idea of anticipated efficiency may be obtained by reference to figure 8.15 which shows the approximate efficiency to be anticipated from the same type of commercial chamber but of different sizes.

The *rate-of-charge method* is more sensitive than the high-resistance-leak method and is applicable to the measurement of very low levels of radioactivity. Resistance R is removed from the circuit, thus the charge which accumulates on capacitance C_v does not leak off. Since an equilibrium potential is not established, the potential continues to increase at a rate proportional to the ionization rate of the gas within the chamber. As full-scale deflection of the meter is attained, the shorting switch S_W is closed. The meter then returns to zero and the cycle is repeated.

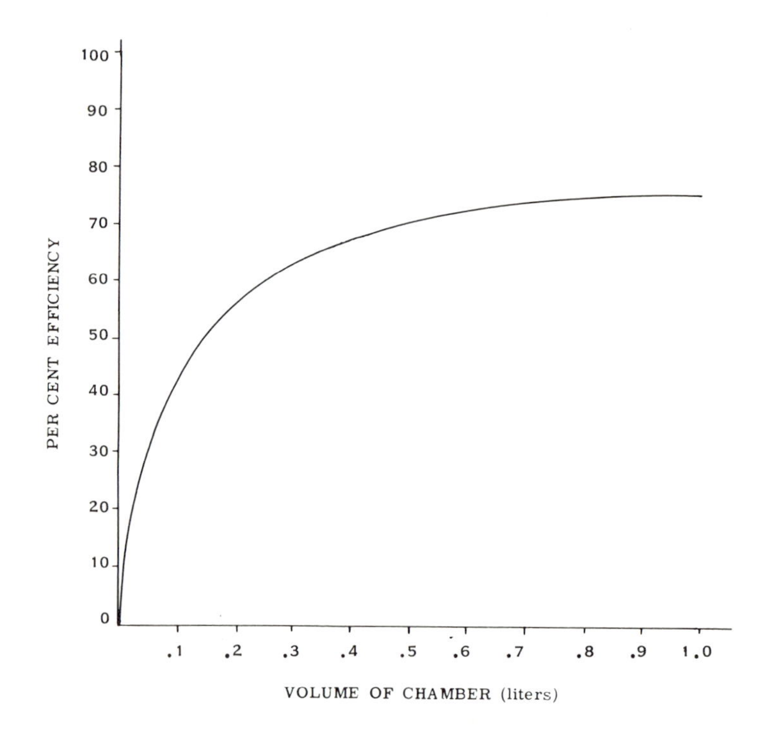

Figure 8.15—APPROXIMATE EFFICIENCY OF IONIZATION CHAMBERS AS A FUNCTION OF VOLUME.

The sensitivity of the rate-of-charge method can be illustrated by use of the same data applied in the calculation for the high-resistance-leak method above, except the 10^{12} ohm resistor is removed. Assume also that the capacitance of C_v and associated stray circuit capacity is 10 picofarads ($10\,\mu\mu f$).
The current flow for $0.001\ \mu Ci\ ^{14}C = 7.5 \times 10^{-15}$ amp. $= 7.5 \times 10^{-15}$ coulombs per second.
The voltage E across capacitance C_v is given by $E = Q/C_v$
Thus, the change in voltage $\Delta E = \Delta Q/C_v$
where ΔQ is the change in charge per second on capacitance C_v.

$$\Delta E = 7.5 \times 10^{-15}/10 \times 10^{-12} = 7.5 \times 10^{-4} \text{ volts per second}$$
$$= 0.75 \text{ mv/sec} = 45 \text{ mv/min}$$

Contamination of the chamber by radon and other alpha emitting nuclides can be bothersome when low level beta activity is measured. A single alpha particle with energy of 5 MeV can produce $5 \times 10^6/35 = 1.4 \times 10^5$ ion pairs, equivalent to a charge of 2.3×10^{-14} coulombs. When this charge is suddenly introduced into a capacitance of 10 picofarads, the observed voltage change is 2.3 millivolts. Such alpha pulses are normally observed at the rate of about one every one or two minutes.

Typical recordings for the high-resistance-leak and the rate-of-charge methods are depicted in figure 8.16.

GENERAL OPERATING INSTRUCTIONS:

1. The shorting switch S_w should be closed and the range switch turned to the highest range.

2. The power switch is now turned on and the instrument is allowed to "warm up" for at least fifteen minutes.

3. The meter is made to read "zero" by adjustment of the "zero" knob. The range switch is then turned to the proper working range and the "zero" is readjusted.

4. The instrument should be checked for drift. If the high-resistance-leak method is used, the switch S_w is opened and the potential allowed to equilibrate. The drift should be less than about 0.01 mv/sec. When the rate-of-charge method is used, replicate observations should show an agreement to within 10^{-16} amperes. If instrument drift is not within these tolerances, a longer warm up period is required.

EQUIPMENT AND REAGENTS:

Vibrating-reed electrometer; resistors (including 10^{12} ohms) for electrometer; vacuum pump; 45 volt to 90 volt battery; batteries providing 1½ volt increments up to about 20 volts; small gamma source; calibrated $Ba^{14}CO_3$ or standard $Na_2{}^{14}CO_3$ solution; gas generator (such as Cary-Tolbert); silicone lubricant; concentrated sulfuric acid; assorted beakers.

PROCEDURE:

PART A—CHARACTERISTIC PLATEAU CURVE

1. To the outside of the ionization chamber, which has been flushed and filled with air, is attached a small radioactive source. A "button" type radium check source is suitable for the purpose. The ionization chamber is then placed in position on the electrometer head.

2. A 10^{12} ohm resistor should be used to provide a high-resistance leak. The electrometer and recorder are also connected to the electrometer head.

3. A variable potential is supplied by means of batteries. A potentiometer should not be used to provide a continuous range of potentials unless capable of the same voltage stability as a battery. Potentials should be measured by means of a sensitive (20,000 ohms/volt) voltmeter.

4. The ionization current is measured at each of about 10 different collection voltages suitably spaced between 1 and 90 volts to provide the necessary data to plot the characteristic plateau curve. Ionization current is calculated from the observed potential and the value of the leak resistor used by means of Ohm's law.

5. A plateau curve is prepared by plotting ionization current as the ordinate and values of collection voltage as abscissa.

6. From the plateau curve, the minimum voltage required to provide a saturation current is ascertained.

PART B—THE HIGH-RESISTANCE-LEAK METHOD

1. Without disturbing the ionization chamber, source or 10^{12} ohm resistor, the variable potential supply is replaced with the regular battery which provides sufficient potential to operate the instrument on the plateau.

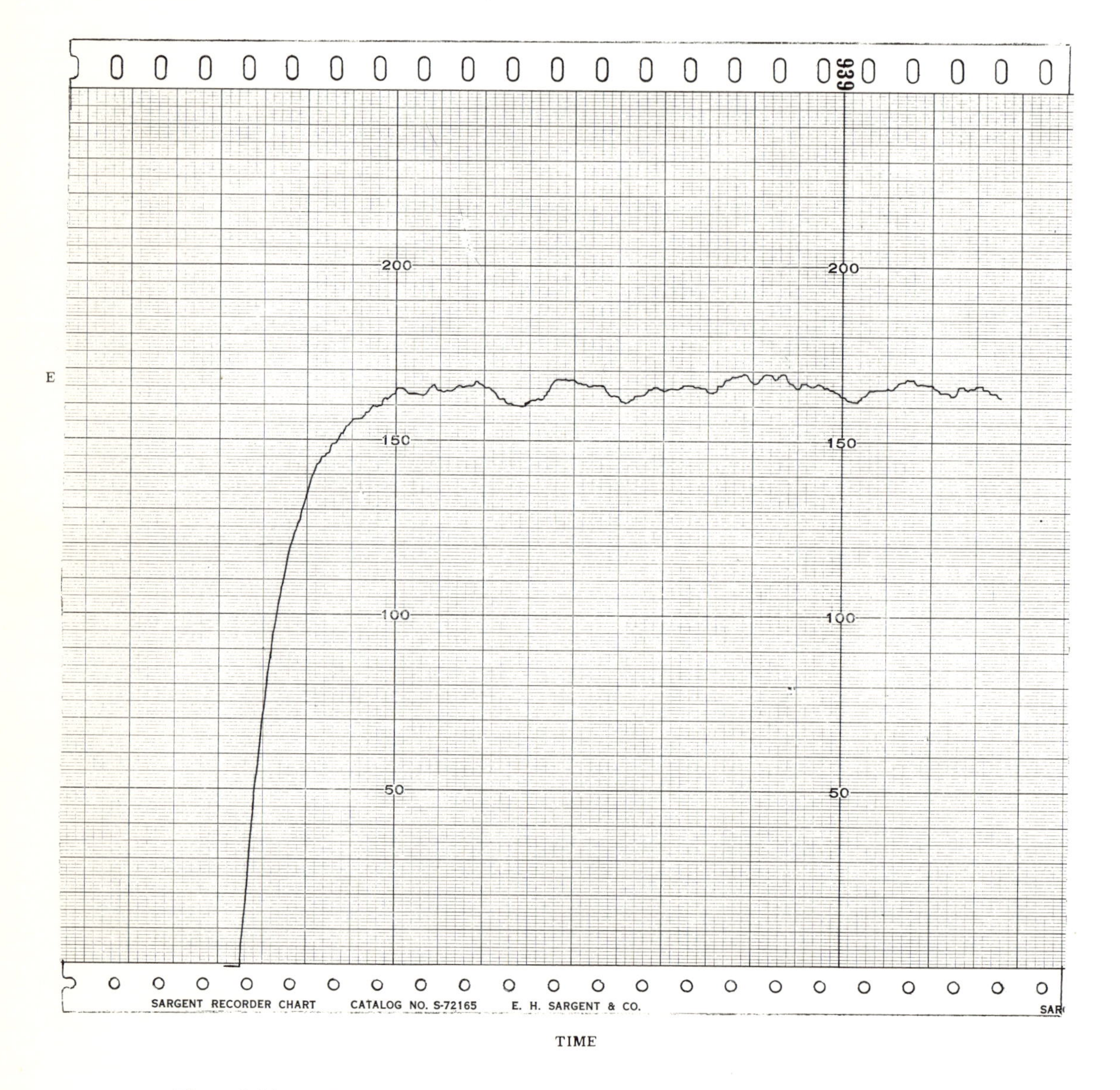

Figure 8.16(a)—TYPICAL RECORDING OF HIGH-RESISTANCE-LEAK METHOD

2. The shorting switch is turned to the open position and the potential, indicated by the recorder, is allowed to come to equilibrium. By means of Ohm's law the ionization current is calculated.

3. The radioactive source should now be removed and the background current measured on the 10 mv range of the instrument. The background current is subtracted from the ionization current calculated in step 2.

4. While measuring the background current on the 10 mv range, alpha pulses will be observed. These occur as very abrupt increases in potential, usually of from 1 to 2 mv. These pulses are easily identified. The chart should be read to eliminate the alpha pulses from the background as well as from readings of sample activities.

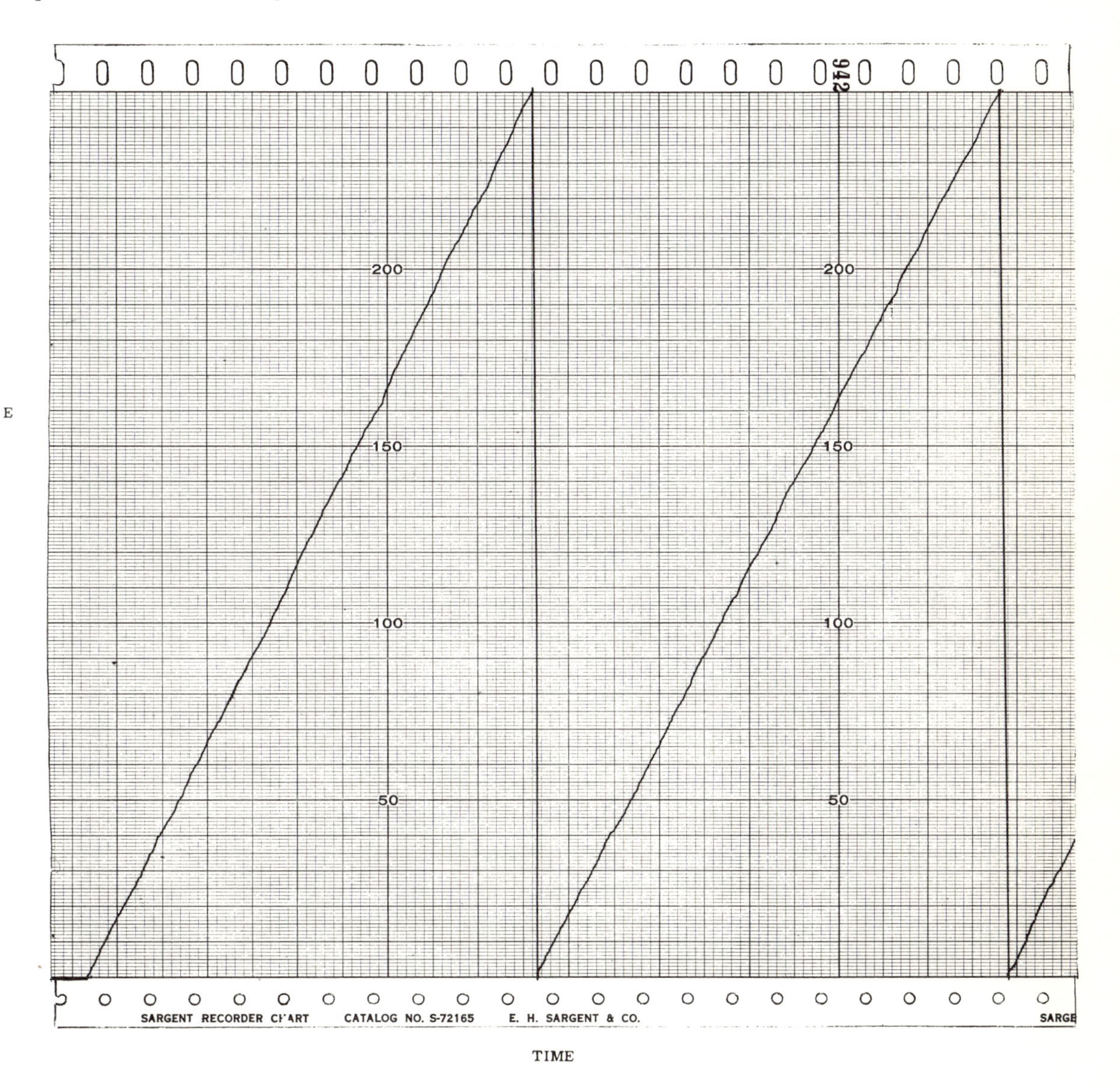

Figure 8.16(b)—TYPICAL RECORDING OF RATE-OF-CHARGE METHOD

PART C—THE RATE-OF-DRIFT METHOD

1. The feedback line is removed from the resistor and the resistor with its holder is removed from the instrument. (This operation does not remove the feedback line to the switch S_w shown in figure 8.12).

2. To measure the instrument drift, the ionization chamber is removed and the dust cap inserted in its place. With the instrument in the 10 mv range and the zero adjusted so the recorder is at about the mid-scale, the shorting switch is opened. Assuming an input capacitance of 10 picofarads, the current (in amperes) is calculated from

$$I = \frac{C\Delta E}{t}$$

where $C = 10^{-11}$ farads and ΔE is the observed potential change (in volts) during time t (in seconds).

3. To measure the background drift rate, the ionization chamber is replaced but without the radioactive source. The meter and recorder are adjusted to zero with the shorting switch closed. With the shorting switch open, the current is again calculated as outlined in step 2. The appropriate range of the instrument should be used.

4. The radioactive source is now replaced exactly as it was in the first two parts of the experiment. The drift rate is again measured and the ionization current calculated. After subtraction of the background, the result should be similar to that obtained using the high-resistance-leak method.

5. Again pulses produced by alpha particles may be observed, especially on the lower ranges. A correction for these pulses should be applied to the calculation.

PART D—ASSAY OF $^{14}CO_2$ (CALIBRATION OF THE IONIZATION CHAMBER)

1. Prior to use, the ionization chamber should be thoroughly flushed with air to remove all traces of the previous radioactive gas sample. This is done by repeated evacuation of the chamber on a vacuum line and refilling with air until a normal background level is observed. Finally, evacuate the chamber.

2. An amount of calibrated barium carbonate-^{14}C, sufficient to provide an activity of 0.01 microcuries, is accurately weighed into the bottom section of a gas generator such as the Cary-Tolbert gas generator shown in figure 8.17. (A standard solution of sodium carbonate-^{14}C may be used if 0.01 microcurie is contained in no more than about 100 microliters of solution.) The stopcock should be closed and the glass joint lubricated with silicone grease.

3. A wad of glass wool is inserted into the upper section of the generator. The lower portion of the glass wool is moistened with concentrated sulfuric acid to assure that air entering the chamber is absolutely dry and to remove any solid barium carbonate which might be carried mechanically by the flow of gas into the ionization chamber. Assemble the generator and connect to the evacuated ionization chamber as illustrated in figure 8.17.

4. The valve on the ionization chamber is now opened and the tip of the generator is immersed in concentrated sulfuric acid. Slowly the stopcock of the generator is opened to allow entrance of a small amount of acid into the generator. The amount of acid

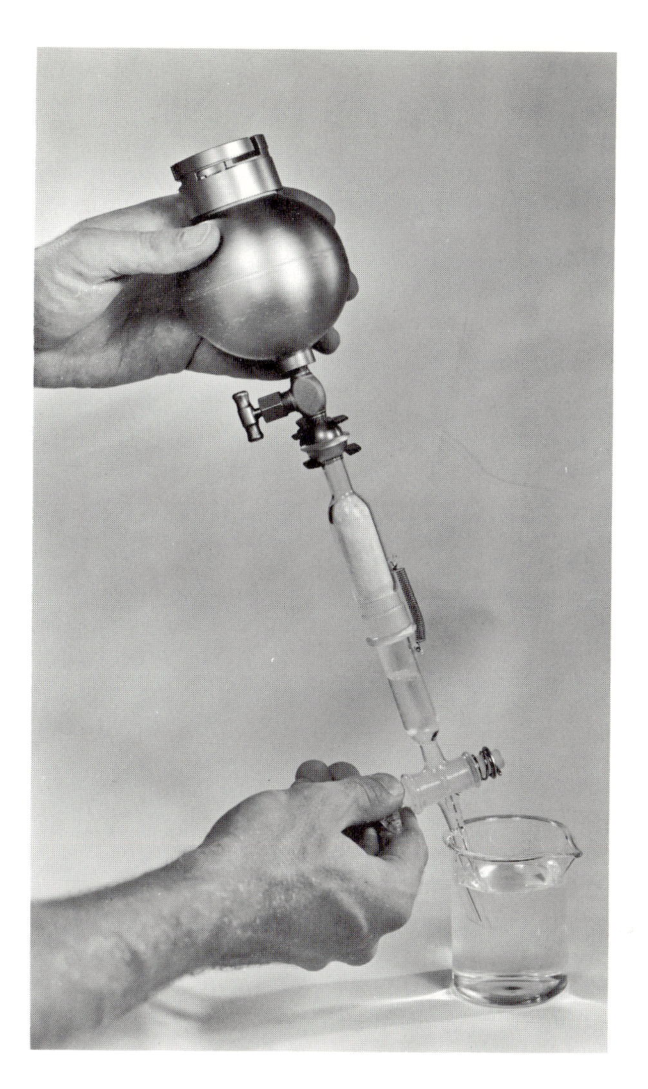

Figure 8.17—CARY-TOLBERT GAS GENERATOR FOR FILLING IONIZATION CHAMBER. (*Courtesy Applied Physics Corp.*)

admitted must be controlled carefully to regulate the rate of release of carbon dioxide. Finally, about two inches of acid are admitted to the generator and warmed by immersion of the generator in a beaker of hot water or by carefully warming it with a small flame.

5. Air is now admitted through the stopcock of the generator to sweep the carbon dioxide out of the generator into the ionization chamber. *CAUTION!* The stopcock must be opened very slowly or sulfuric acid will be carried up into the chamber by the surge of air. (If this should happen, dismantle the chamber immediately, wash with water and dry.) Air is allowed to enter the chamber until the pressure in the chamber is equal to that of the atmosphere. The stopcock on the ionization chamber is immediately closed to prevent diffusion of the radioactive gases out of the chamber.

6. The chamber is now rotated to mix the gases, then placed in position on the electrometer head. It is necessary to wait ten to fifteen minutes for equilibration before a constant reading will be obtained. The current is measured by the high-resistance-leak method.

7. From the data, the calibration constant for the instrument can be calculated in amperes per curie or in terms of a more convenient working constant such as millivolts per millicurie (for a given value of resistance). In addition, the efficiency of the instrument can be calculated from the activity of the sample and the observed current flow. This value is compared to those shown on figure 8.15.

EXPERIMENT 8.3 PROPORTIONAL COUNTERS

OBJECTIVES:

To observe the characteristics of a proportional counter.
To plot the characteristic plateau curve for a proportional counter.

THEORY:

Proportional counters, like Geiger counters, are made in a variety of types and forms (see experiments 3.2 and 8.4). End-window proportional counters and windowless-flow proportional counters are the types used most extensively. These counters are similar to Geiger counters, but they differ in certain constructional details so that operation in the proportional region is made possible.

First, the anode or central positively charged wire may be shortened, often being formed into the shape of a small loop. The diameter of the wire is made as small as manufacturing techniques and durability in use will permit. The purpose of these changes is to alter the potential gradient between the anode and the cathode so that a better control of the gas amplification can be obtained through regulation of the potential applied to the counter. The objective is to increase the potential gradient (volts/cm) in the vicinity of the anode and at the same time to decrease the gradient near the cathode. Some idea of how this is achieved can be seen in figure 8.18. If the electrodes consist of parallel plates, a uniform field results. If the anode is a wire mounted concentrically in the cathode (as in most Geiger counters), the field is not uniform, a higher potential gradient existing near the central wire. Reducing the diameter of the wire and forming it into a loop further distorts this potential field, especially in the vicinity of the loop. The formation of the avalanche is thus controlled because the electron does not attain sufficient energy to produce new ion pairs until it approaches very close to the anode loop. Even though in this region the electron attains considerable energy, now passing through a relatively large potential drop, the number of atoms with which it can collide is restricted and hence the avalanche is controlled.

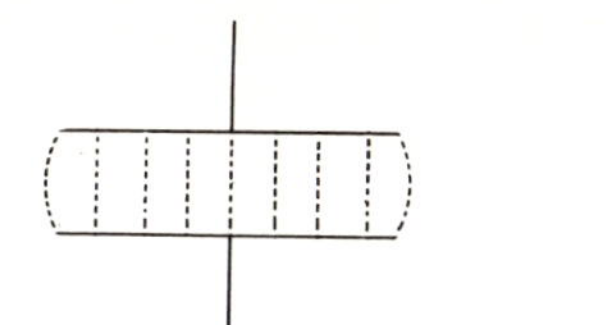

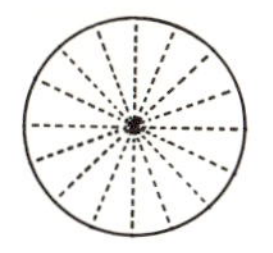

Figure 8.18—EFFECT OF ELECTRODE SHAPE ON POTENTIAL GRADIENT.

Further control of the avalanche is achieved through regulation of the composition of the gas. Proportional gas produces a greater quenching action than does Geiger gas. A typical composition (available from Matheson Co., Inc.) is 90% argon and 10% methane, the methane providing the necessary quenching action.

Figures 8.19 and 8.20 illustrate a typical characteristic curve for a proportional counter. Unlike the Geiger characteristic curve, which has but a single plateau, the proportional curve shows two plateaus—the *alpha plateau* and the *beta plateau.* Alpha particles, having a specific ionization of approximately 1000 times that observed for beta particles, would be expected to produce a pulse proportionately larger than the beta pulse. It is a relatively simple matter to construct an electronic device (a *discriminator*) which will permit large pulses to pass on through, while at the same time blocking the passage of small pulses.

As the voltage on the proportional counter is increased, gas amplification of the tube or counter is likewise increased. At the *alpha threshold*, the alpha pulses, being much larger than the beta pulses, pass through the discriminator or gate and are counted by the scaler. The small beta pulses are blocked. Continued increases in voltage increase the size of both the alpha and beta pulses until finally the beta pulses are also large enough to pass through the discriminator. This point is called the *beta threshold* and both alpha and beta pulses are recorded on the scaler.

The alpha threshold is rather well defined, as compared to the beta threshold. The principal reason for this lies in the energy ranges of the particles. Alpha particles are monoenergetic. Accordingly, they would be expected to produce pulses of nearly equal

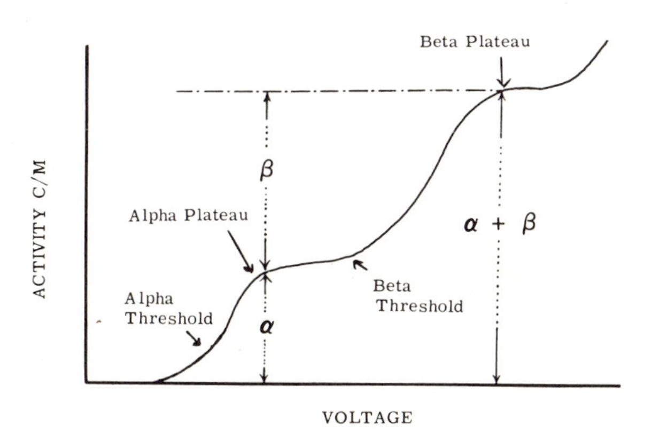

Figure 8.19—TYPICAL PROPORTIONAL CHARACTERISTIC CURVE.

size. Beta particles, on the other hand, have energies ranging from zero to E_m. The corresponding pulse heights range from essentially zero volts to some maximum value equivalent to a beta energy of E_m. With increasing potentials applied to the proportional detector, the largest of the beta pulses would be registered first, the smallest last.

If it is desired to count alpha particles only, the potential is adjusted to the center of the alpha plateau (or that point at which the slope is minimum). On the other hand, it is not possible to count the betas alone. When the potential is adjusted to the operating point of the beta plateau (point of minimum slope), the observed count is the total of the alphas and betas. To obtain the beta count alone it is merely necessary to subtract the alpha count from the count obtained on the beta plateau.

The *resolving time* of a proportional counter is small. At low values of gas amplification, resolving time is less than a microsecond. As the gas amplification increases (at higher applied potentials to the counter), the coincidence loss increases also, having a value of about 10 microseconds in the upper range of the proportional region. It will be recalled that resolving time increases to several hundred microseconds in the Geiger region. Consequently, whereas counting rates in the Geiger region are limited to about 20,000 counts per minute, counting rates of 100,000 to 200,000 counts per minute in the proportional range are quite feasible.

Proportional counters are quite inefficient, as a rule, for counting gamma radiation and are normally not used for this purpose.

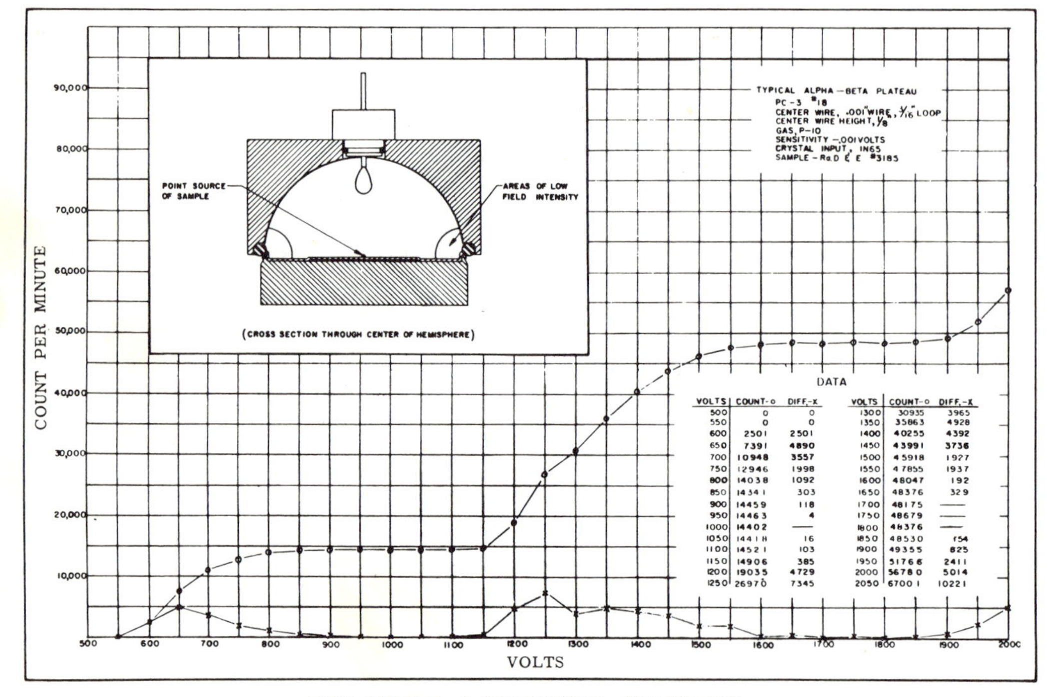

VOLTS	COUNT-o	DIFF.-X	VOLTS	COUNT-o	DIFF.-X
500	0	0	1300	30935	3965
550	0	0	1350	35863	4928
600	2501	2501	1400	40255	4392
650	7391	4890	1450	43991	3736
700	10948	3557	1500	45918	1927
750	12946	1998	1550	47855	1937
800	14038	1092	1600	48047	192
850	14341	303	1650	48376	329
900	14459	118	1700	48175	—
950	14463	4	1750	48679	—
1000	14402	—	1800	48376	—
1050	14418	16	1850	48530	154
1100	14521	103	1900	49355	825
1150	14906	385	1950	51766	2411
1200	19035	4729	2000	56780	5014
1250	26970	7345	2050	67001	10221

FOR NMC 2 pi COUNTING CHAMBERS

Figure 8.20—TYPICAL ALPHA-BETA PLATEAU
(*Courtesy Nuclear Measurements Corp.*)

PROCEDURE:

1. Two characteristic curves will be plotted with the use of the proportional counter, the first for a pure beta emitter and the second for an isotope or mixture of nuclides emitting both alpha and beta particles. For the first curve, use carbon-14, phosphorus-32, or other source of pure betas. For the mixed source of alphas and betas, the use of radium-DEF, a uranium salt, or a thorium salt is suggested.

2. Measure the efficiency of the proportional counter, using the procedure outlined in experiment 8.4.

3. Determine the resolving time by the method of paired sources. The resolving time should be in the order of a few microseconds. A fast scaler must therefore be used, otherwise the measured value of resolving time will be that of the scaler rather than that of the detector. The activity of each half-source should be about 1 million cpm or greater.

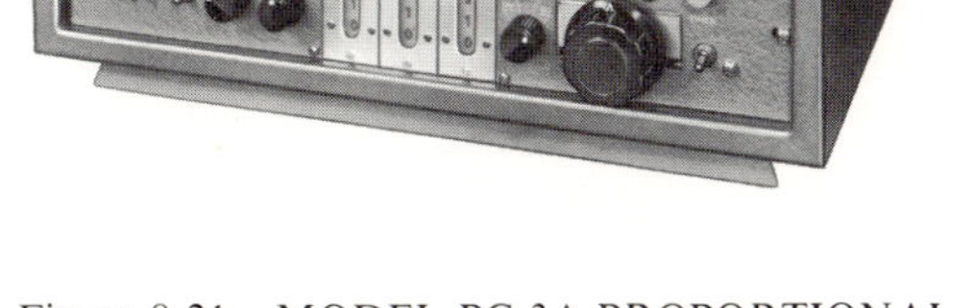

Figure 8.21—MODEL PC-3A PROPORTIONAL COUNTER (*Courtesy Nuclear Measurements Corp.*)

Figure 8.22—MODEL PCC-10 PROPORTIONAL COUNTER CONVERTER (*Nuclear Measurements Corp.*)

OPERATION OF NUCLEAR MEASUREMENTS CORPORATION PROPORTIONAL COUNTER.

1. *Operating the Chamber Slide*—The chamber slide is controlled by the knob in the lower center of the panel. Rotating this knob counterclockwise and then pulling it forward draws the slide from the chamber. A clockwise motion will then life the piston to accept a sample. By rotating the knob counterclockwise again, the piston can be dropped below the level of the slide (which can then be pushed into place). The piston is finally raised into the chamber by another clockwise turn. The knob positions are indexed so that there can be no doubt about the piston

sliding into place. Also, one can easily tell by the feel of the control knob when it is in the correct position.

2. *Introducing the Sample*—Always be sure that the piston is in the down position (the knob turned to the extreme counterclockwise) while the slide is being operated. WARNING: Do not permit the sample to overhang the edge of the piston as it may catch under the "O" ring seal. This precaution is particularly important in the use of large samples.

3. *Purging the Chamber*—After the instrument has been turned on for five minutes, it is ready for operation. The sample is placed on the piston and sealed under the chamber. The "gas-purge" switch is then held down for two minutes, thereby purging the entire gas system of contaminating gases. The Model PCC-12A, because of its larger size, will require about three minutes of purging for optimum results. Normally, when the samples are being changed, the only air which is introduced is that which is carried in with the piston as the sample is put into the chamber.

4. *Checking the Effectiveness of Purging*—If at any time there is doubt concerning the purity of the gas in the system, one should set the high voltage control so that he is counting on the steeply rising characteristic of the counting curve, say 800 volts for alpha counting or 1400 volts for beta counting. Then, purge the chamber normally, let it operate for two minutes at this point and watch for a change in counting rate. If the counting rate varies according to the rate of flow of gas through the chamber, or if counting rate decreases after a short period of continuous flow, either the chamber has not been effectively purged or else a leak has occurred.

5. *Checking Performance*—The performance of the instrument is checked with a sample suitable for the proposed type of counting. If a general purpose instrument (suited for alpha and beta counting) is used, it is best to check it with an alpha-beta emitting sample, such as the National bureau of Standards radium-D and E sample. If the counter is to be used for alphas only, the radium-D and E sample is again quite useful, or a pure alpha source, e.g., radium-F (polonium-210), may be used. At no time should one use radium-226 (common radium) in an alpha counting device unless this is unavoidable as an essential part of a sampling method. Even then the sample must be used carefully and the chamber must be frequently decontaminated in order to prevent a very serious buildup of background activity. For counting betas only, it is best to use a long half-life beta emitter as the standard; a carbon-14 compound is best. However, the Bureau of Standards radium-D and E standard shielded with about 10 mg/cm^2 of metal foil is quite satisfactory.

DATA:

Counter specifications____________________

Scaler specifications____________________

Pure beta source used____________________

Mixed alpha-beta source used____________________

Applied Voltage	Activity of beta source cpm	Activity of mixed source cpm

CALCULATIONS:

1. Plot a graph of the data as illustrated in figure 7.19. Plot the counting rate as the ordinate and the voltage applied to the counter as the abscissa. Plot data for both sources on the same graph.

2. Indicate the best operating voltage for counting on the alpha plateau (________ volts) and for the counting on the beta plateau (________ volts).

3. Calculate the efficiency of the proportional counter.

Activity of standard____________dpm on________________(date).

Activity standard (corrected for decay)____________dpm.

Observed activity— α ____________cpm β ____________cpm

Counter efficiency α ____________% β ____________%

Accepted value of α efficiency—about 51%.

Discuss the effects of backscattering on the α and β efficiencies.

Suggestions for further work:

Using the same technique, plot the curves for other radioactive sources. Examine a gamma emitter in the proportional counter. Is it possible to obtain a gamma plateau?

EXPERIMENT 8.4 WINDOWLESS FLOW-COUNTERS AND G-M COUNTER EFFICIENCY

OBJECTIVES:

To demonstrate the construction, operation, characteristics and use of the windowless flow-counter.

To illustrate a method for measuring the overall efficiency of a G-M counter system and to compare the efficiencies of the windowless flow-counter and end-window G-M tubes.

THEORY:

Windowless flow-counters are often made as 2π-counters, the expression 2π referring to the nature of the counter geometry. They are used for the purpose of increasing counting efficiency, especially for those radioisotopes which emit beta particles of very low energy. This purpose is accomplished by placing the sample inside the tube, thereby improving the physical geometry and at the same time eliminating window absorption.

The construction of a typical windowless flow-counter is depicted in figure 8.23. The counter consists of a stationary member A and a rotating member B which can be turned on a pivot C. The sample is introduced into the depression D and the rotating plate is then turned until the sample is in position beneath the counting chamber. The chamber is then "purged" by a flow of Geiger gas through the inlet tube E. A visible bubble chamber aids in regulating the flow rate of the Geiger gas. An air-tight seal must be maintained between the stationary and rotating members of the counter.

Although the counter will be operated as a Geiger counter in this experiment, windowless flow-counters are available which can be converted quite easily from Geiger operation to proportional operation. To change over to proportional operation, it is necessary to replace the central wire with one shaped into a loop. Also, the gas must be replaced with one having more quenching action and an amplifier must be used between the counter and the scaler.

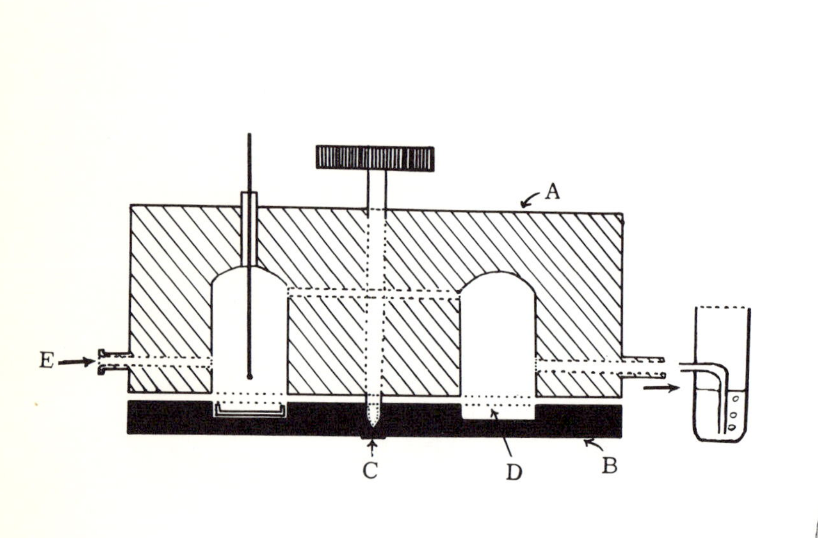

Figure 8.23—WINDOWLESS FLOW-COUNTER (Diagramatic Cross Section)

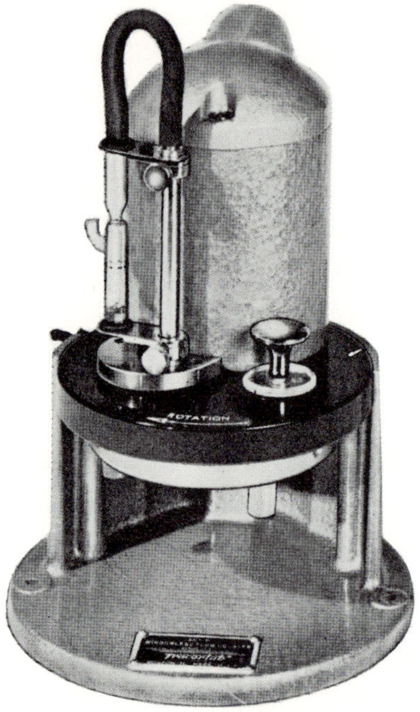

Figure 8.24—WINDOWLESS FLOW-COUNTER (*Tracerlab, Inc.*)

PROCEDURES:

*Operating Instructions for Tracer-lab SC-*16 *Windowless Flow-Counter*—Remove the dust cap and, using the sample tongs, place the sample (mounted in a planchet) in the loading position. Increase the gas flow rate to approximately six bubbles per second. Rotate the lower plate of the counter to the left until the next sample well is in the loading position. Allow the sample to remain in the "flush" position for about 15 seconds before introducing it into the counting position. (While the first sample is being flushed, either a second sample or a background blank may be placed in the loading position.) Again, rotate the lower plate as before to bring the first sample into the count position. Decrease the gas flow to approximately 1 bubble per sec (proper flush rate for operation) during the counting procedure. A third sample may now be placed in the loading position.

COUNTER EFFICIENCY—By use of a calibrated source, the overall counting efficiency of a counter can be calculated. In this experiment, the efficiency for the detection of phosphorus-32 will be measured.

For a calibrated ^{32}P source, one can use the calibrated solutions supplied by Abbott Laboratories, E. R. Squibb and Sons, Mallinckrodt Nuclear and other suppliers.

Prepare a dilution of the calibrated solution so that the activity contained in 25 to 50 lambdas is about 10,000 disintegrations per minute. Place an aliquot of the dilution (containing an activity of about 10,000 dpm) on a planchet (steel, copper or silver) and dry. The planchet used should be "infinitely thick" to the beta particles of ^{32}P.

Observe the counting rate produced by this sample with both the windowless flow-counter and with a standard end-window Geiger tube assembly. Measure the background count of both instruments. Refer to experiment 3.5 for the backscattering factor f_B.

DATA:

Background

Windowless flow-counter__________cpm

End-window Geiger tube__________cpm

Activity of the standard ^{32}P solution

__________dpm/ml at________o'clock on____________(zero date)

__________dpm/____ at________o'clock on____________(assay date)

Calculated activity of standard sample on planchet__________dpm. (A)

	Gross cpm	Net cpm	
Activity of sample in windowless flow-counter	________	________	(B)
Activity of sample with end-window tube	________	________	(C)

Backscattering factor f (for ^{32}P) using________(Fe, Cu or Ag) planchet________

CALCULATION:

1. Correct the sample activities for backscattering. That is, calculate the activities which would be anticipated with zero backscattering.

Windowless flow-counter (B) corrected to "0" backscattering____________(D)

End-window tube (C) corrected to "0" backscattering____________(E)

2. Calculate the efficiencies of the counters.

Efficiency of windowless flow-counter:

Using__________(Fe, Cu or Ag) planchet (100 B/A)____________%

Corrected to "0" backscattering (100 D/A)____________%

Efficiency of end-window counter:

Using__________(Fe, Cu or Ag) planchet (100 C/A)____________%

Corrected to "0" backscattering (100 E/A)____________%

EXPERIMENT 8.5 NEUTRON COUNTERS

OBJECTIVES:

To investigate the methods used to detect neutrons and to observe the characteristics of a neutron detector.

To measure the relative efficiency of a neutron detector for fast and slow neutrons.

THEORY:

The operation of radiation detectors, including neutron detectors, depends upon the production of ions by the incident radiation, either directly or indirectly. Thus, neutrons must undergo a reaction resulting in the formation of ions. This may be accomplished by means of an (n, α) capture reaction with boron-10, by means of neutron fission of ^{235}U or by means of recoil proton production in hydrogenous materials.

Proportional counter for thermal neutrons—Boron, and especially the isotope ^{10}B, presents a large capture cross-section to neutrons. The capture reaction

$$^{10}_{5}B + ^{1}_{0}n \longrightarrow ^{7}_{3}Li + ^{4}_{2}He \tag{8.6}$$

results in the formation of an alpha particle for each neutron captured. It will be recalled from chapter 7 that alpha particles are powerful ionizers. The 2.5 MeV alpha particle released in this reaction produces about 80,000 ion pairs, sufficient to produce a very substantial pulse.

If the neutron flux density is not too great, individual pulses can be detected by means of proportional counting. One type of slow neutron detector (see figure 8.25) consists of a long hollow metallic cylinder filled with boron trifluoride (BF_3), a gas, which has been enriched in the isotope ^{10}B. The cylinder acts as the cathode while a wire located coaxially serves as the anode. A neutron passing through the BF_3 gas is captured by the (n, α) reaction of ^{10}B and an alpha particle is released.

Operation of the detector in the proportional region provides good discrimination against radiation of types other than alpha. Since alpha particles from an external source are incapable of penetrating the metal cylinder, this type of detector is highly specific for neutrons.

By reference to figure 7.48, it is further seen that the cross-section of ^{10}B to fast neutrons is considerably less than the cross-section for slow neutrons. The result is a low detection efficiency for fast neutrons or a high specificity for slow neutrons.

A second reaction suitable for neutron detection is the fission reaction of ^{235}U. The fission fragments, released by ^{235}U when bombarded by neutrons, produce pulses of sufficient height that no problem is encountered in discriminating against pulses from other sources. In these detectors a coating of ^{235}U metal is deposited on the surface of a metal liner within the chamber and the chamber is filled with a gas such as argon. Operation is in the proportional region.

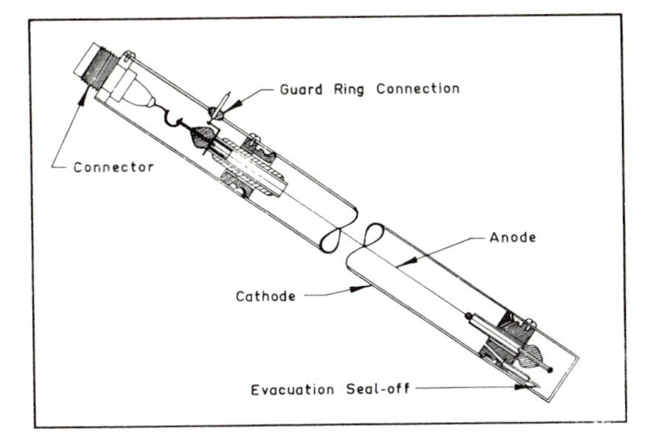

Figure 8.25—THERMAL NEUTRON DETECTOR (*Courtesy Radiation Counter Laboratories, Inc.*)

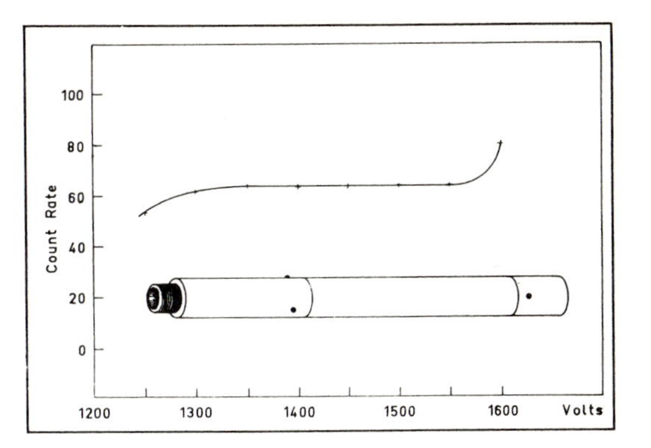

Figure 8.26—CHARACTERISTIC CURVE OF A THERMAL NEUTRON DETECTOR FILLED TO A PRESSURE OF 12 CM OF BF_3 GAS. (*Courtesy Radiation Counter Laboratories, Inc.*)

Proportional detection of fast neutrons—In the design of a fast neutron detector, interactions of neutrons with matter, as described in chapter 7, must once more be considered. It will be recalled that neutrons passing through hydrogenous substances, such as water, paraffin and plastic, undergo collisions with hydrogen nuclei (protons). These protons recoil upon collision with sufficient energy to ionize a gas and, hence, a neutron can again be detected by means of an indirect method.

The fast neutron detector illustrated in figure 8.27 consists of a series of metallic chambers, each lined with 1/16 inch of plastic to provide the recoil protons. The chambers are filled with methane and operated in the proportional region.

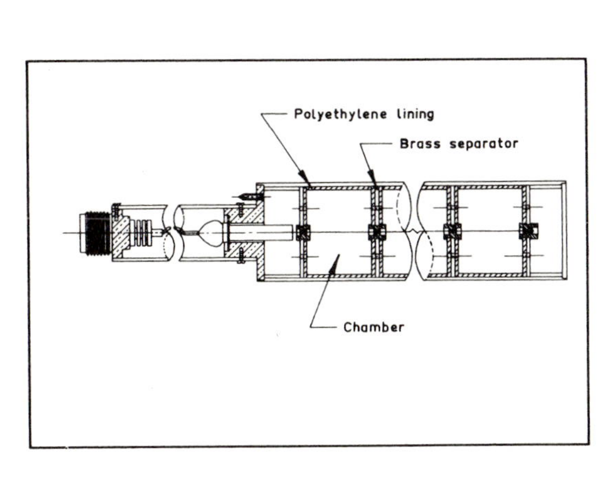

Figure 8.27—CROSS-SECTION OF A FAST NEUTRON DETECTOR. (*Courtesy Radiation Counter Laboratories, Inc.*)

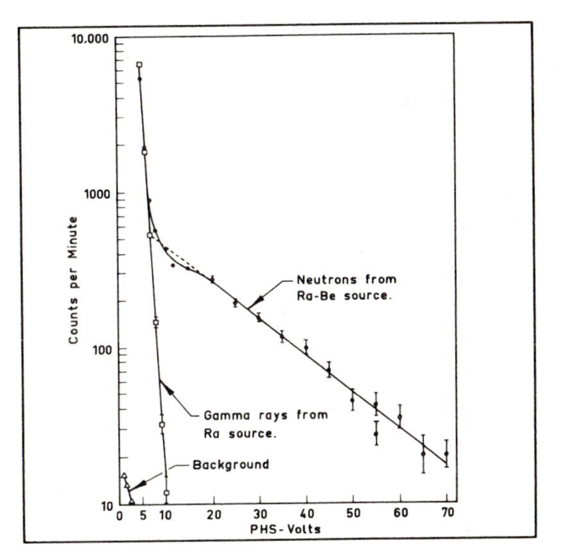

Figure 8.28—PULSE-HEIGHT DISTRIBUTION FOR A FAST NEUTRON DETECTOR. (*Courtesy Radiation Counter Laboratories, Inc.*)

Note that boron is not involved in the construction of the detector. If it were, the instrument would not be specific for fast neutrons. In the detector described, recoil protons resulting in the production of pulses sufficiently large to be accepted by the discriminator are predominately produced by fast neutrons. Alpha and beta radiation will not penetrate the detector and gamma rays produce pulses of insufficient amplitude to be accepted by the discriminator.

Ionization-chamber detectors—Integrating ionization chambers are often used when the flux density is too great for resolution of individual pulses. These chambers often use the (n,α) reaction of boron for neutron detection. Since these chambers are designed primarily for very high neutron flux densities, they are less used in the small laboratory than the proportional counters mentioned previously.

EQUIPMENT:

Neutron source (e.g., neutron howitzer); fast or slow neutron detector; alpha, beta and gamma sources; cadmium foil; boric acid or borax; aluminum or cardboard sheets.

PROCEDURE:

PART A–CHARACTERISTIC CURVE

Both fast- and slow-neutron detectors operating as proportional counters will show a characteristic curve similar to that shown in figure 8.26. As with Geiger and proportional counters, a knowledge of the operating voltage is essential and is best determined from the characteristic curve.

The detector is mounted in front of a port of a neutron howitzer. Data necessary to plot a characteristic curve are obtained by following a procedure similar to that used in experiments 3.2 and 8.2. Only one plateau (the alpha plateau) should be observed. Do not allow the count rate to increase more than 10% above the average plateau level.

The normal operating voltage is selected as the lowest voltage which allows operation in the flat region of the plateau.

PART B—SENSITIVITY TO RADIATION OTHER THAN NEUTRONS

If data resulting from a measurement are to be interpreted and applied accurately, it is essential to know what quantity is being measured. Specifically, one should determine if the neutron counter he is using is sensitive to radiation other than neutrons, and if so, the type of radiation and the detection efficiency should be determined.

The operating voltage as determined in part 1 of the experiment, or as given by the manufacturer, is applied to the neutron detector. The sensitivity of the counter to alpha (or mixed alpha-beta), beta and gamma radiation is measured by positioning appropriate sources close to the tube. Note especially the sensitivity, if any, to a relatively active gamma source. Estimate the efficiency of the neutron counter to alpha, beta and gamma radiation.

PART C—SENSITIVITY TO FAST AND SLOW NEUTRONS

It is difficult to obtain a pure source of either fast or slow neutrons with which to demonstrate the relative sensitivity of the detector. The following procedure will, at best, provide only a general idea of the relative sensitivity of the detector to them.

The detector is mounted in front of the port of a neutron howitzer. As a sheet of cadmium is placed between the source and the detector, the change in activity is observed. A marked change indicates a high sensitivity to slow neutrons. If the decrease in activity is slight, the detector is sensitive to fast neutrons. Refer to figure 6.46 and explain your observations in terms of the cadmium cross-sections.

Repeat the above step using a thin sandwich of boric acid or borax between sheets of aluminum or stiff cardboard. Explain the results using boric acid as compared with those obtained with cadmium.

REFERENCES:

1. Price, W. J., "Nuclear Radiation Detection", McGraw-Hill (1958).
2. Jordan, W. H., "Detection of Nuclear Particles", Ann. Rev. Nuclear Sci., 1, 209-226 (1952).
3. Curtiss, L. F., "Measurements of Radioactivity", Natl. Bur. Stand. Circ. 476, U.S. Dept. Comm. (Oct. 1949).
4. Elmore, W. C., H. Kallman and C. E. Mandeville, "Practical Aspects of Radioactivity Instruments", Nucleonics 8, S1-S32 (June 1951).
5. Buckstein, E., "Basic Electronic Counting", Radio and TV News, p. 122 (March 1958).
6. Glaser, D. A., "The Bubble Chamber", Sci. Amer., p. 46 (Feb. 1955).
7. Borkowski, C. J., "Instruments for Measuring Radioactivity", Anal. Chem., 21, 348-352 (1949).
8. Kohman, T. P., "Measurement Techniques of Applied Radiochemistry", Anal. Chem. 21, 352-364 (1949).
9. Goldsmith, H. H., "Bibliography on Radiation Detection", Nucleonics 4, 142-150 (May 1949).

Ionization Chambers—(Electroscopes and Electrometers)

10. Wilkinson, D. H., "Ionization Chambers and Counters", Cambridge (1950).
11. Rossi, B. B., and H. H. Staub, "Ionization Chambers and Counters", McGraw-Hill (1949).
12. Pinajian, J. J., and J. Christian, "The Assay of Sodium Radio-iodide (I_{131}) Used for Medicinal Purposes", J. A. Ph. A., Sci. Ed., 44, 631-636 (Oct. 1955).
13. Lauritsen, C. C., and T. Lauritsen, "A Simple Quartz Fiber Electrometer", Rev. Sci. Inst., 8, 438-439 (1937).
14. Tolbert, B. M., "Ionization Chamber Assay of Radioactive Gases", University Calif. Rad. Lab. 3499, O.T.S., U.S. Dept. Comm. (March 1956).
15. Tolbert, B. M., "Experiments Using an Ionization Chamber and Vibrating Reed Electrometer for Radioactivity Measurements", Applied Physics Corp. (Feb. 1961).
16. Wyckoff, H. O., and F. H. Attix, "Design of Free-Air Ionization Chambers", Natl. Bur. Stand., Handbook 64, U.S. Dept. Comm. (Dec. 1957).
17. Guinn, V. P., and C. D. Wagner, "A Comparison of Ionization Chamber and Liquid Scintillation Methods for Measurement of Beta Emitters", Atomlight, No. 12, New England Nuclear Corp. (April 1960).
18. Janney, C. D., and B. J. Moyer, "Routine Use of Ionization Chamber Method for C^{14} Assay", Rev. Sci. Instr., 19, 667-674 (Oct. 1948).
19. Reese, H. Jr., "Design of a Vibrating Capacitor Electrometer", Nucleonics 6, 40-45 (March 1950).
20. Glass, F. M., "A Simple Low-Drift Electrometer", Nucleonics 10, No. 2, 36-39 (1952).
21. Facchini, U., and A. Malvicini, "A-N_2 Fillings Make Ion Chambers Insensitive to O_2 Contamination", Nucleonics 13, No. 4, 36-37 (April 1955).
22. Duffy, P. A., "High-Pressure Al-Walled Sensitive Ionization Chamber", Nucleonics 14, No. 9, 132-133 (Sept. 1956).
23. Bylander, E. G., "Ion-Chamber Response in High-Level Radiation Field", Nucleonics 18, No. 5, 102-104 (May 1960).
24. Roberts, D. L., "A-C Ionization Chambers Are Simple and Reliable", Nucleonics 19, No. 53-57 (Feb. 1961).

Proportional Counters

25. Korff, S. A., "Proportional Counters—I", Nucleonics 6, 5 (June 1950); II—Nucleonics 7, 46-52 (Nov. 1950); III—Nucleonics 8, 38-43 (Jan. 1951).
26. Bernstein, W., H. G. Brewer, Jr. and W. Rubinson, "A Proportional Counter X-Ray Spectrometer", Nucleonics 6, 39-45 (Feb. 1950).
27. Sheckler, A. C., "Design Characteristics of Air Proportional Counters", Nucleonics 8, 44-49 (Jan. 1951).
28. Robinson, C. V., "Improved Methane Proportional Counting Method for Tritium Assay", Nucleonics 13, No. 11, 90-91 (Nov. 1955).
29. Miller, D. G., "Proportional Gamma Spectrometer", Nucleonics 13, No. 2, 58-60 (Feb. 1955).
30. Nilsson, G., and G. Aniansson, "Proportional Flow Counter with High-Humidity Gas", Nucleonics 13, No. 2, 38-39 (Feb. 1955).
31. Wolfgang, R., and C. F. Mackay, "New Proportional Counters for Gases and Vapors", Nucleonics 16, No. 10, 69-73 (Oct. 1958).
32. Merritt, W. F., "System for Counting Tritium as Water Vapor", Anal. Chem. 30, 1745 (1958).

Flow Counters

33. Graf, W. L., C. L. Comar and J. B. Whitney, "Relative Sensitivities of Windowless and End-Window Counters", Nucleonics 9, No. 4, 22-27 (1951).
34. Nader, J. S., G. R. Hagee and L. R. Setter, "Evaluating the Performance of the Internal Counter", Nucleonics 12, No. 4, 29-31 (June 1954).
35. Merritt, W. F., and R. C. Hawkings, "The Absolute Assay of Sulfur-35 by Internal Gas Counting", Anal. Chem. 32, 308 (1960).
36. Kelsey, F. E., "An Internal Geiger Counter for the Assay of Low Specific Activity Samples of Carbon 14 and Other Weak Beta Emitters in Biological Samples", Science 109, 566 (1949).
37. Nye, W. N., and J. D. Teresi, "Automatic Sample Changer for Windowless Gas Flow Counters", Anal. Chem. 23, 643 (1951).
38. Karnovsky, M. L., et al, "Correction Factors for Comparing Activities of Different Carbon-14 Labeled Compounds Assayed in Flow Proportional Counter", Anal. Chem. 27, 852 (1955).
39. Perkins, H. J., and M. D. MacDonald, "Gas-Flow Counting of Carbon-14 Compounds: An Improved Technique", Science 138, 1259 (1962).

Solid State Detectors

40. Hofstadter, R., "Crystal Counters—I", Nucleonics 4, 2-27 (April 1949). —II, Nucleonics 4, 29-43 (May 1949).
41. Moos, W. S., and F. Spongberg, "CdS—Crystal Probes Are Convenient for Dosimetry in Body Cavities", Nucleonics 13, No. 6, 88 (June 1955).
42. Hollander, L. E., Jr., "Special CdS Cells Have High X- and Gamma-Ray Sensitivity", Nucleonics 14, No. 10, 68-71 (Oct. 1956).
43. Friedland, S. S., J. W. Mayer and J. S. Wiggins, "Tiny Semiconductor is Fast, Linear Detector", Nucleonics 18, No. 2, 54 (Feb. 1960).
44. Anon., "Putting Semiconductor Detectors to Work", Nucleonics 18, No. 2, 57 (Feb. 1960).
45. Steinberg, R., "Semiconductor Fission Probe", Nucleonics 18, No. 2, 85 (Feb. 1960).
46. Anon., "Semiconductor Detectors", Nucleonics 18, No. 5, 98-100 (May 1960).
47. Jones, A. R., "Uses of Semiconductor Detectors in Health-Physics Monitoring", Nucleonics, 18, No. 10, 86-91 (Oct. 1960).
48. Friedland, S. S., H. S. Katzenstein and M. R. Zatrick, "Semiconductor Detectors for Nuclear Medicine and Biology", Nucleonics 23, No. 2, 57 (1965).
49. Shirley, D. A., "Applications of Germanium Gamma-Ray Detectors", Nucleonics 23, No. 3, 62-66 (1965).
50. Weiss, W. L., and E. M. Whatley, "Surface-Barrier Detectors Make Isotopic Age Measurements", Nucleonics 21, No. 9, 66 (1963).
51. Bilaniuk, O., "Semiconductor Particle-Conductors", Sci. Amer., p. 78 (Oct. 1962).
52. Glos, M. B., "Semiconductors, Scintillators and Data Analysis", Nucleonics 22, No. 5, 50 (May 1964).

Neutron Counters

53. Draper, J. E., "Evaluation of Neutron Counter Efficiency", Nucleonics 6, 32-40 (March 1950).
54. Cohen, B. L., "High-Energy Neutron Threshold Detectors", Nucleonics 8, 29-33 (Feb. 1951).
55. Moyer, B. J., "Survey Methods for Slow Neutrons", Nucleonics 10, No. 4, 14-17 (1952).
56. Thompson, B. W., "Portable Survey Meter for Fast and Slow Neutrons", Nucleonics 13, No. 3, 44-46 (March 1955).
57. Rossi, H. H., G. S. Hurst, W. A. Mills and H. E. Hungerford, Jr., "Inter-comparison of Fast-Neutron Dosimeters", Nucleonics 13, No. 4, 46-47 (April 1955).
58. Nobles, R. G., and A. B. Smith, "Fission Chamber Measures Neutron Distribution Quickly, Accurately", Nucleonics 14, No. 1, 60-62 (Jan. 1956).
59. Kaufmann, S. G., and L. E. Pahis, "Neutron Detectors for Operation at 400° C.", Nucleonics 16, No. 3, 90-93 (March 1958).
60. McKenzie, J. M., "Making Fission Counters for Neutron Monitoring" Nucleonics 17, No. 1, 60-65 (Jan. 1959).

CHAPTER 9
Scintillation Techniques and Nuclear Emulsions

LECTURE OUTLINE AND STUDY GUIDE

I. SCINTILLATION TECHNIQUES
  A. Basic Scintillation Counters (*Experiment* 9.1)
    1. Mode of detection
    2. Phosphor types
    3. Photomultiplier tubes
    4. Characteristics of basic scintillation system
  B. Scintillation Spectrometry (*Experiment* 9.2)
    1. Pulse-height analysis
    2. Integral and differential spectra
    3. Calibration methods
    4. Applications
  C. Liquid Scintillation Counters (*Experiment* 9.3)
    1. Problems of low-energy detection
      a. Selection of fluor
      b. Operation at low temperature
      c. Coincidence techniques
    2. Sample preparation
      a. Simple solution
      b. Common solvent
      c. Compound formation
      d. Suspension
    3. Quenching
      a. Thermal or chemical quenching
      b. Color quenching
    4. Counting techniques
    5. Interpretation of data

II. NUCLEAR EMULSIONS AND AUTO-RADIOGRAPHY (*Experiment* 9.4)
  A. Principles of Technique
  B. Artifacts
  C. Emulsion Types
  D. Exposure and Development Techniques

INTRODUCTION:

In chapter 8 we discussed radiation detection methods which require the formation of ions by the incident radiation and the subsequent collection and measurement of these ions. There are other ways in which radiation can be detected and measured. Some of these are:

1. Formation of photons by the incident radiation and the subsequent collection and measurement of the photons (*Scintillation counters*)

2. Production of chemical or Physico-chemical reactions in systems. The extent of the reaction is determined by:

   a. chemical development of a nuclear emulsion (*Autoradiography, radiography, film-badge dosimetry*)
   b. measurement of fluoresence produced upon exposure of the irradiated system to ultraviolet light (*Glass-rod dosimetry*)

c. measurement of change in optical density, optical rotation or other property of a solution upon irradiation (*Chemical dosimetry*)

3. Vapor condensation by the incident radiation (*Cloud chambers*)
4. Vapor formation by the incident radiation (*Bubble chambers*)

In this chapter several of the radiation detection methods outlined above will be discussed.

EXPERIMENT 9.1 SCINTILLATION COUNTERS

OBJECTIVES:

To observe the mode of operation of a scintillation counter and its operating characteristics.

To plot the integral spectra of several radioisotopes and to determine the effect of the gamma energy on the shape of the curves.

To determine how to calculate the best operating voltage.

THEORY:

When radiation interacts with certain substances called *fluors* (sometimes referred to as *phosphors*), a small flash of visible light (a *scintillation*) is produced. This process constitutes the basis for the operation of all scintillation detectors.

Fluorescent substances—The fluor may consist of any one of a number of suitable chemical substances which have been found to be especially suited to the detection of specific types of radiation.

1. *Alpha fluors*—Zinc sulfide is one of the first phosphors ever used. In the early days the scintillations produced by alpha particles and protons impinging on a screen of zinc sulfide were tediously counted visually with the aid of a low power microscope. Zinc sulfide is still used for the detection of alpha particles, being spread thinly (about 10 mg/cm^2) on a transparent base, but the human eye has been replaced by the photomultiplier tube.

2. *Beta fluors*—Large crystals of anthracene or naphthalene containing a small amount of anthracene have been used for the scintillation detection of beta particles. Recently, plastic fluors have found much use, being much easier to fabricate into large or irregular shapes.

3. *Gamma fluors*—For gamma scintillation, single, large crystals of sodium iodide containing a trace of thallium iodide as an activator are almost exclusively used. This is referred to as a NaI(Tl) crystal. A particular point in favor of the use of sodium iodide over anthracene for gamma ray detection is the much greater density of the sodium iodide. For high detection efficiency the incident radiation must be highly absorbed by the fluor. High absorber densities result in a high rate of absorption.

4. *Neutron fluors*—Lithium iodide crystals doped with europium LiI(Eu) are used for the detection of neutrons. The lithium is enriched in the ^{6}Li isotope in order to promote the reaction ^{6}Li(n,α) ^{3}H. The ionizing radiation produced by this reaction then produces the scintillation.

The *photocell* used to detect these small scintillations must be extremely sensitive and must be connected intimately with the crystal so the light is efficiently transmitted to the *photo-sensitive cathode.* The system used is illustrated in figure 9.1.

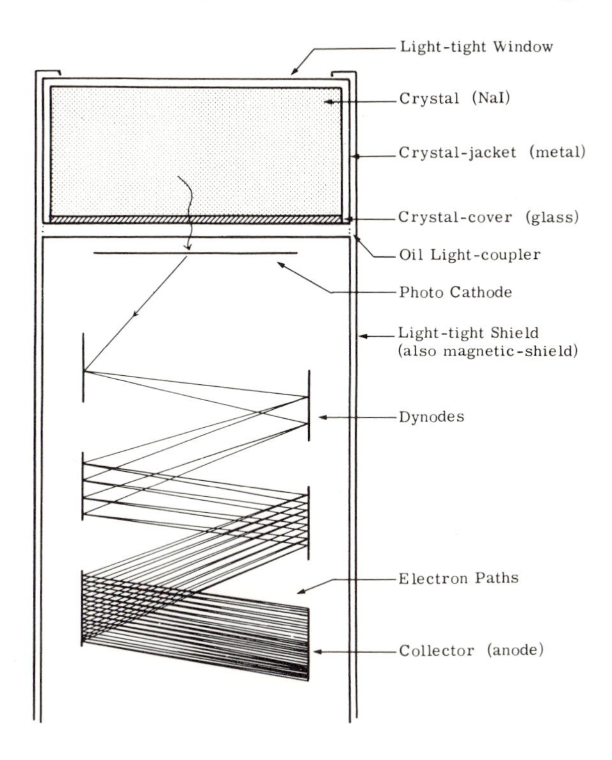

Figure 9.1—GAMMA SCINTILLATION DETECTOR.

A non-aqueous oil or grease having the proper optical properties is used to couple the crystal to the phototube. The index of refraction of the oil must be very nearly equal both to that of the crystal and to the glass of the tube to permit efficient light transmission to the photosensitive cathode.

The photocell used is of the type known as a *photomultiplier*. The photomultiplier is so-called because, for each electron dislodged from the cathode by a photon of light, nearly a million electrons reach the plate of the tube. This process of electron multiplication is made possible by an ingeniously designed system of electrodes (called *dynodes*) within the photomultiplier tube. The result is a gain or amplification of a million within the tube itself.

A *basic scintillation system* consists of a scintillation detector (crystal and photomultiplier), a high voltage supply, a preamplifier and a scaler connected as illustrated in figure 9.2. The inclusion of a linear amplifier and a discriminator (apart from the discriminator incorporated in most scaler inputs), will often be useful.

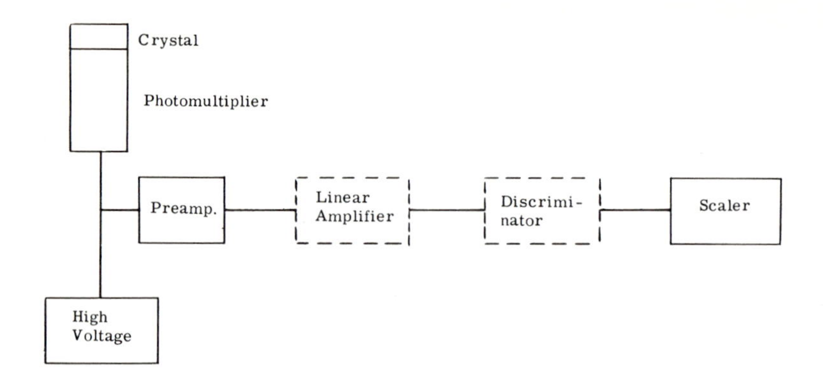

Figure 9.2—BLOCK DIAGRAM OF A BASIC SCINTILLATION COUNTING SYSTEM.

A *high-voltage supply* capable of providing up to about 1200 volts is necessary for the operation of the photomultiplier. Although most photomultipliers are operated at between 900 volts and 1000 volts, the high-voltage supply should be adjustable over the range from 500 volts to 1200 volts. The amplification of the photomultiplier is extremely dependent on the high voltage. For this reason, the high-voltage supply should also be very stable.

The *preamplifier* should be located as close to the photomultiplier as possible. Its function is to convert the weak (high impedance) electrical pulses, produced by the photomultiplier, into strong low impedance pulses capable of being conducted through a cable to the amplifier or scaler.

For the detection of gamma radiation, a scintillation detector of the type described above offers several advantages over Geiger-Müller, proportional and ionization counters.

1. *Sensitivity*—The high density of a NaI(Tl) crystal as compared to a gas provides an enormous increase in sensitivity.

2. *Resolving time*—The resolving time is limited by the electronics of the system rather than by the detector and is of the order of a few microseconds or less as compared to 100 to 500 microseconds for a G-M tube.

3. *Proportional values*—The amplitude of the output pulse is proportional to the energy dissipated by the radiation in the crystal. This advantage is also shared by gas proportional counters and ionization chambers but not by Geiger counters.

When a scintillation detector is used, the activity of a sample as indicated by the observed count on the scaler depends on a number of variables including the following:

1. Number of scintillations occurring in the phosphor.
   *A. Source intensity (including background).
   B. Geometry (including size of phosphor, bremsstrahlung, etc.)
2. Distribution of intensities of scintillations.
   *A. Source type.
   B. Crystal or phosphor size and efficiency.

3. Phototube and optical system.
    A. Efficiency of light transmission to cathode of phototube.
    *B. Amplification by phototube.
        (1) Type of tube—Efficiency of light conversion to electrical energy at cathode.)
        *(2) Potential applied—Controls amplification of electrical energy within phototube.)
4. Associated electrical circuits.
    A. Preamplifier between phototube and scaler.
    B. Discriminator setting on scaler input.

Fortunately, for a particular detector and scaler combination, all of these variables can be kept constant and under normal conditions are held constant, with the exception of those marked with an asterisk. Source intensity and type of radiation will vary with the sample being used. (Source intensity is, of course, the actual quantity we wish to measure). The applied potential and hence the amplification of the phototube is set as accurately as possible with the aid of a meter. A correction is later applied for this and certain other circuit variations by the use of a standard source, such as was used in experiment 5.3. Thus, a relative comparison of the strength of two sources of the same type can be made with considerable accuracy.

If, on the other hand, we wish to measure the true activity, or to compare the disintegration rates of dissimilar samples (say ^{60}Co and ^{131}I), or to compare the activities of two samples of the same type but under conditions where it is difficult to reproduce geometry - e.g., measurements in living animals, iodine uptake studies, etc.—we are then confronted with new problems less easily resolved. Some of these problems may be appreciated and understood more clearly if a "plateau curve" is plotted.

The term "plateau curve" is normally not applied to the scintillation characteristic curve even though the procedure is similar to that used in plotting the plateau curve of a Geiger-Müller counter. The curve produced represents an *integral spectrum* and is so named.

The size of the electrical pulse produced by each gamma ray entering the NaI (Tl) crystal is proportional to the energy dissipated by the gamma ray in the crystal. Consequently, as the voltage is increased, amplification of all the pulses is proportionately increased, and the largest pulses, representing the most energetic gamma rays, are detected first. As the voltage is further increased, amplification is further increased with the result that smaller pulses, representing less energetic gamma rays are detected and counted also. This process continues with increasing voltage until all pulses are detected and recorded including the relatively small noise pulses produced by random electrons in the photomultiplier tube. The result is a curve characterized by a series of steps (figure 9.3), each step representing a particular photopeak produced either by the primary radiation, scattered radiation or noise. Each gamma emitting isotope produces a characteristic curve which can be used for its identification.

Well counters - The purpose of a well counter (figure 9.4) is to improve the counting geometry in order to obtain a higher counting efficiency. This is accomplished by drilling a hole into the center of a crystal to create a well for the sample. The radioactive source is, therefore, almost completely surrounded by the radiation-sensitive crystal.

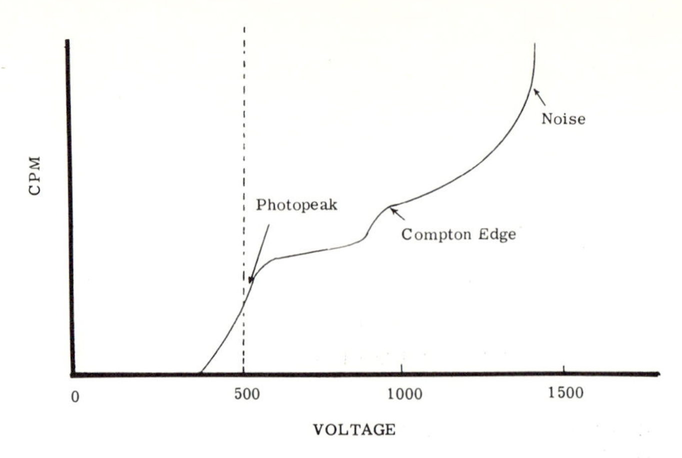

Figure 9.3—INTEGRAL SPECTRUM.

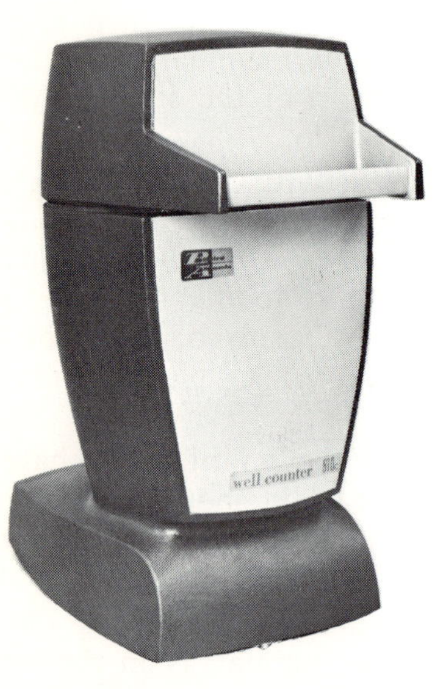

Figure 9.4—WELL COUNTER. (*Courtesy Baird-Atomic, Inc.*)

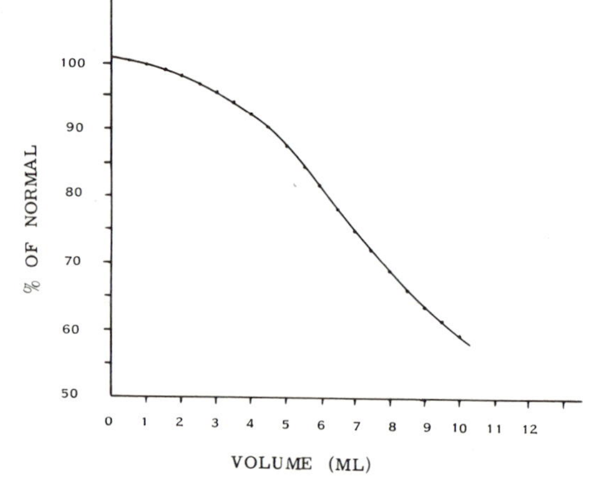

Figure 9.5—CALIBRATION CURVE FOR SILVER-111 IN A WELL COUNTER. Normal volume = 1.0 ml.

When using such counters, the need to use constant sample geometry, or to normalize activity to that for an arbitrary standard geometry, is sometimes overlooked. A typical calibration curve for a well counter is illustrated in figure 9.5. This curve was prepared for liquid samples in standard, plastic test tubes. Activities were measured for 0.5 ml increments of volume while maintaining a constant total activity. Dilution was accomplished by adding non-radioactive solvent to an original 0.5 ml aliquot of the radioactive solution. A volume of 1.0 ml was arbitrarily selected as the normal.

When the volume of the sample is not equal to the standard volume, the measured activity must be normalized.

MATERIALS:

Part A — Scintillation detector and scaler; assorted gamma sources (e.g., ^{60}Co, ^{131}I, ^{226}Ra, ^{137}Cs).

Part B — Well-type scintillation detector and scaler; solution of a gamma emitting nuclide (about 0.01 μCi/ml.

PROCEDURE:

PART A — MEASUREMENT OF SCINTILLATION PLATEAU (INTEGRAL SPECTRUM)

For a particular series of measurements to be used for plotting an integral spectrum, all factors will be held constant except for the potential applied to the phototube. This procedure will be repeated for several different sources thus providing a family of spectra for comparison and analysis.

1. To measure an intergral spectrum, set the high voltage control to its lowest (counter-clockwise) position and insert a suitable source of cobalt-60 into the sample holder. With the counting switch "on," raise the voltage slowly until the counter begins to operate. Measure the counting rate at various voltages, increasing the voltage in increments of 50 volts until the noise pulses are detected. This voltage will be indicated by a sharp increase in the counting rate.

2. Repeat step 1 using other gamma sources (e.g., ^{131}I, ^{226}Ra, ^{137}Cs).

3. Repeat step 1 using no source. This will provide a background curve which must be subtracted from the other curves determined. In making the background correction, be sure to subtract the background from the uncorrected count at similar voltage settings.

CALCULATIONS:

1. To compare the shapes of the curves, corrected for background, they must be normalized to correct for differences in source intensities. This is done by selecting an arbitrary value of voltage near the midpoint of the plateau as a norm.

Determine the factors by which the activity of each sample (measured at normal voltage) must be multiplied in order to equal 1000 cpm. Multiply the activities for a given sample at all voltages by the factor determined for that sample and plot the corrected activities on linear graph paper.

If the geometry has been constant, any differences existing among these curves will be due to the type of the source.

2. Differentiation of the integral spectrum yields a spectrum of peaks. These peaks represent photopeaks and backscatter peaks and occur at points of maximum slope (the inflection points) of the integral spectrum.

From values determined from the literature, determine the relationship between the applied voltage and the gamma ray energy in MeV. Very often, backscattered peaks correspond to the Compton photons after being scattered through angles of either 90° or 180°. Do these values correspond to any peaks of the spectrum? See experiment 7.4 for directions for calculating the energy of the Compton photons.

3. Determine the best operating voltage for the measurement of activity for each source used by use of the "figure of merit." (See equation 5.2) For each source plot a graph of R_s^2/R_b as a function of voltage. The best operating voltage is determined by the maximum of the curve.

PART B—NORMALIZATION OF A WELL COUNTER

1. Iodine-131, gold-198, cobalt-60 or other isotope emitting gamma radiation may be used. The amount used should be adjusted to provide a convenient activity (1,000 to 15,000 cpm) in a volume of 0.5 ml. (About 0.01 μCi/ml will supply this activity.)

2. Into a test tube, designed for use with the well counter, is placed 0.5 ml of the radioactive solution and the activity is measured.

3. Water or carrier solution is added in increments of 0.5 ml. After each addition, the content of the tube is mixed well and the activity is measured. (Note that dilution is made with *non*-radioactive solution so the activity remains constant.)

4. A convenient volume is selected as the normal or standard volume and to it is assigned a relative activity of 100%.

5. The activities observed for other volumes are expressed as a percentage of the normal activity. (For volumes less than the standard volume the relative activity will be greater than 100%.

6. A curve, similar to that in figure 9.5, is plotted from the data.

EXPERIMENT 9.2 SCINTILLATION SPECTROMETRY

OBJECTIVES:

To demonstrate the principles of pulse-height analysis and scintillation spectrometry.

To illustrate applications of scintillation spectrometry for the measurement of radiation energy and for the identification of nuclides.

THEORY:

A scintillation spectrometer is an instrument for the measurement of energy distribution of radiation. It is of great utility in the radioisotope laboratory and finds widespread application in the fields of radiation characterization and nuclear identification and measurement.

The measurement of radiation energy is accomplished electronically by determining the intensity of each scintillation produced by radiation as it impinges on a phosphor. A block diagram of a scintillation system is shown in figure 9.2. The basic mode of detection, the types of fluors used, and the role of the photomultiplier tube, the high voltage supply and the preamplifier were discussed in experiment 9.1. The function of these components in a scintillation spectrometer are basically the same.

In scintillation spectrometry there is one important aspect which must be considered if accurate, quantitative measurements of radiation energy are to be made; namely, the intensity of the scintillation produced by the incident radiation is proportional to the energy *absorbed* by the fluor. If the energy of the incident radiation is totally absorbed, or not absorbed at all, no problem is created. If the energy of the incident radiation is only partially absorbed by the fluor, a part being lost through scattering, a distortion of the facts will result.

Beta fluors— Beta-ray spectrometers which utilize magnetic fields are superior to scintillation spectrometers but these instruments are quite expensive. Beta scintillation detectors do not have the accuracy and resolution available with the magnetic type. This is due largely to backscattering of beta particles from the surface of the fluor. When backscattering occurs, only a part of the beta energy is dissipated in the fluor so the distribution of scintillation intensities is not proportional to the true distribution of beta energies. Beta fluors for scintillation detection should have a low Z to minimize backscattering. It is for this reason that anthracene and other fluorescent carbon compounds are preferred to sodium iodide. In addition, backscattering can be minimized by a careful consideration of the geometry.

Gamma fluors—The energy of electromagnetic radiation is transferred to a fluor—usually a sodium iodide crystal—almost exclusively by two processes, the photoelectric effect and the Compton effect. Pair production becomes predominant at high photon energies. (See experiment 7.4.)

In a photoeolectric interaction, the entire energy of the photon is transferred to an electron—a photoelectron. An electron is a highly ionizing particle and has a path length of less than a millimeter in the crystal. The intensity of the scintillation produced along the path of this photoelectron is proportional to the incident photon energy.

A Compton electron acquires only a part of the incident photon energy, the balance often being lost from the crystal as a scattered photon. See figure 9.6 (a). Sometimes this scattered photon may undergo a collision itself before it emerges from the crystal. If this second collision involves the photoelectric effect (figure 9.6 (b)), then again the entire energy of the incident photon is dissipated in the crystal, and the proportionality of total scintillation intensity to incident photon energy is preserved.

If the incident photon energy is in excess of 1.02 MeV, pair production may also occur. This process is always accompanied by 0.51 MeV annihilation radiation. Thus, a spectral peak will occur not only at an energy $E\gamma$ corresponding to that of the incident photon but also at $E\gamma - 0.51$ MeV and at $E\gamma - 1.02$ MeV due to the loss of either one or both of the annihilation rays.

Two things can be done to assure maximum absorption of photon energy in a gamma fluor. First, the fluor should be as dense as possible. (See experiment 7.5 on properties of gamma rays.) That is why a crystal of sodium iodide is used in preference to anthracene. Secondly, the crystal should be as large as possible. Small crystals are satisfactory for low energy photons but they will produce distorted spectra of high energy photons. Large crystals are always more satisfactory and give much better resolution of photon energies. Their disadvantage is expense.

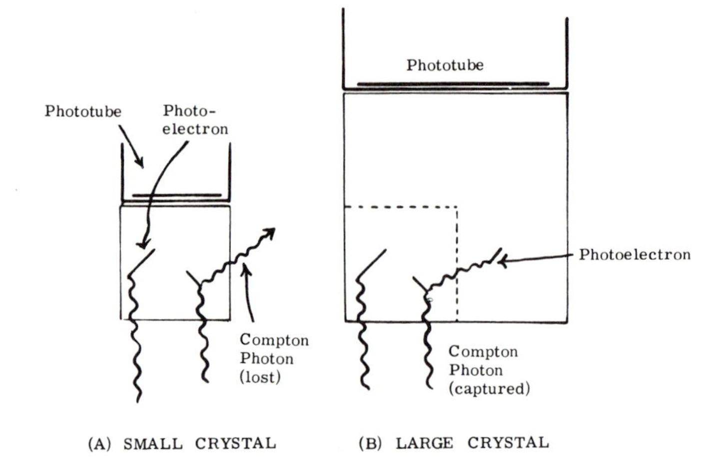

Figure 9.6—ABSORPTION OF GAMMA ENERGY BY A SCINTILLATION CRYSTAL. The larger the crystal the greater the probability of retaining the entire energy of the incident gamma ray within the crystal.

Pulse-Height Analysis—By means of a photomultiplier tube, an electrical pulse is produced by each scintillation in the crystal. If it is assumed that the efficiency of light collection by the photocathode is independent of the location of the pulse in the crystal, and that an electron is emitted from the photocathode for each photon striking it, then the pulse-height is proportional to the scintillation intensity. This is not entirely true, however, and accounts, in part, for the dispersion of spectral peaks.

A preamplifier transmits these small pulses to an amplifier where they are linearly amplified to produce output pulses ranging from 0-100 volts (0-10 volts in transistorized equipment).

As its name implies, the pulse-height analyzer analyzes or sorts pulses according to their height (amplitude or voltage). Since pulse-height is proportional to the incident radiation energy, the pulse-height analyzer is, in fact, analyzing the radiation energy. The *base-line setting* E of the analyzer establishes the lower discriminator level. Pulses with an amplitude less than E are rejected and do not appear at the analyzer output. The *window* ΔE establishes the range of pulse amplitudes greater than E which will be passed by the analyzer and which will consequently appear at the analyzer output. Pulses with an amplitude greater than $E + \Delta E$ are also rejected by the analyzer. Thus, of all the pulses appearing at the analyzer input, only those with an amplitude between E and $E + \Delta E$ pass through to the scaler or rate meter.

Differential Spectra—If the base line (lower level) E is adjusted to 0.0 volt and the window ΔE to 1.0 volt (i.e., 1% of the range of 100 volts), the activity recorded

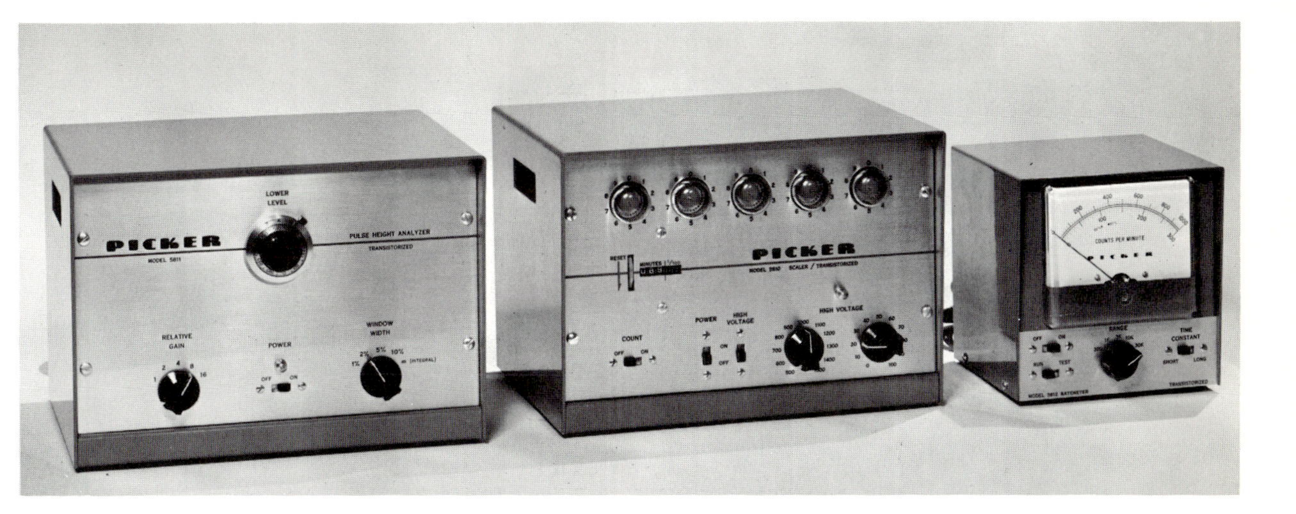

Figure 9.7—PULSE-HEIGHT ANALYZER. An amplifier is built into the student training pulse-height analyzer illustrated with its associated scaler and rate meter. The scintillation detector is not shown. (*Courtesy Picker X-ray Corp.*)

by the scaler will be equal to the number of pulses with amplitudes between 0 and 1 volt. If E is now adjusted to 1 volt, ΔE remaining the same, the activity recorded will be equal to the number of pulses with an amplitude between 1 and 2 volts. This process is repeated, E being increased by increments of 1 volt, until a series of 100 measurements have been made. These activities, plotted as a function of E, will yield a *differential spectrum*.

In the instance cited, 100 observations were recorded, each representing the activity of 1 volt increments. These increments are referred to as *channels* and here 100 channels have been measured. The instrument described, however, is a *single channel analyzer* because the measurements must be made a single channel at a time. *Multichannel analyzers* are also available with up to 400 channels or more. Since multichannel analyzers allow pulses in all channels to be recorded simultaneously, the time requirement is reduced proportionately. If the spectra of very short-lived isotopes are to be measured, a multichannel analyzer is essential.

Analysis of Gamma Spectra— Figure 9.8 depicts the theoretical spectrum for cobalt-60. The spectrum obtained in practice is illustrated in figure 9.9. The obvious differences between these spectra are caused by instrumental and geometrical limitations.

That the 1.17 and 1.33 MeV gamma rays emitted by cobalt-60 are monoenergetic is well established. They should produce the line spectrum of figure 9.8. Such a spectrum would be obtained only if:

1. none of the energy of an incident photon were lost from the crystal through scattering. (If loss were to occur through the use of a crystal of finite dimensions, the theoretical spread of Compton photon energies is given by the area enclosed by the dotted line.)

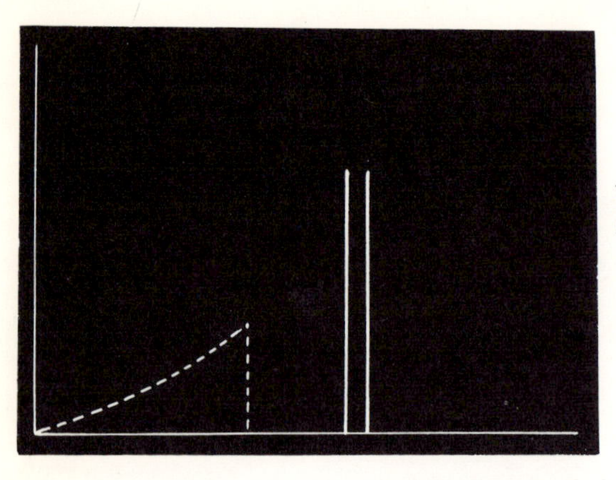

Figure 9.8—THEORETICAL SPECTRUM FOR COBALT-60. The area within the dotted line represents theoretical Compton scattering with finite crystal.

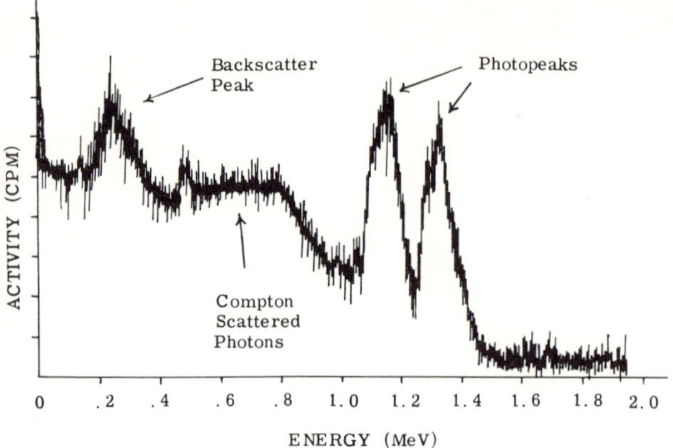

Figure 9.9—PRACTICAL SPECTRUM FOR COBALT-60. Prepared by use of a single channel, recording spectrometer.

2. the scintillation intensities produced in the crystal were strictly proportional to the incident energy dissipated in the fluor,

3. the number of photons reaching the photocathode were strictly proportional to the number of photons produced in a scintillation,

4. the number of electrons emitted from the photocathode were proportional to the number of photons striking it, and

5. perfect proportionality were obtained throughout the electronic section of the pulse-height analyzer.

The first problem of a practical instrument concerns *resolution*. Resolution (See figure 9.10) is calculated as the ratio of the width of a photopeak (measured at a point equal to one-half the amplitude of the peak) to its energy.

$$\% \text{ Resolution} = 100\ \Delta E/E \tag{9.1}$$

Thus, if a photopeak with a maximum at 80 volts had a width of 8 volts at the mid-point of its amplitude, the resolution of the system would be 100 (8/80) =10%. The dispersion of activities about the maximum of a peak is approximately Gaussian.

A second problem concerns Compton scattering, the effects of which were extensively considered in experiment 7.4. Compton scattered radiation accounts for most of the activity observed at energies below the photopeak.

The *Compton edge* (75) corresponds to the maximum energy E_{ce} which can be imparted to an electron in a scintillation crystal through Compton scattering. This occurs when the Compton photon is emitted at an angle of 180° to the incident ray. Thus

$$\frac{0.511}{E'} - \frac{0.511}{E_\gamma} = 1 - \cos\theta = 2 \tag{9.2}$$

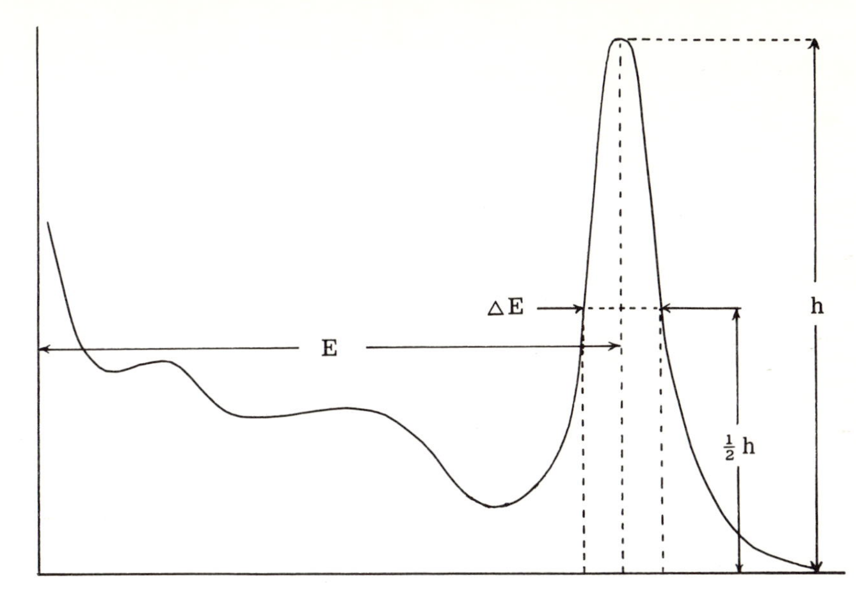

Figure 9.10—PHOTOPEAK RESOLUTION. Resolution is usually expressed as % Resolution = 100 Δ E/E.

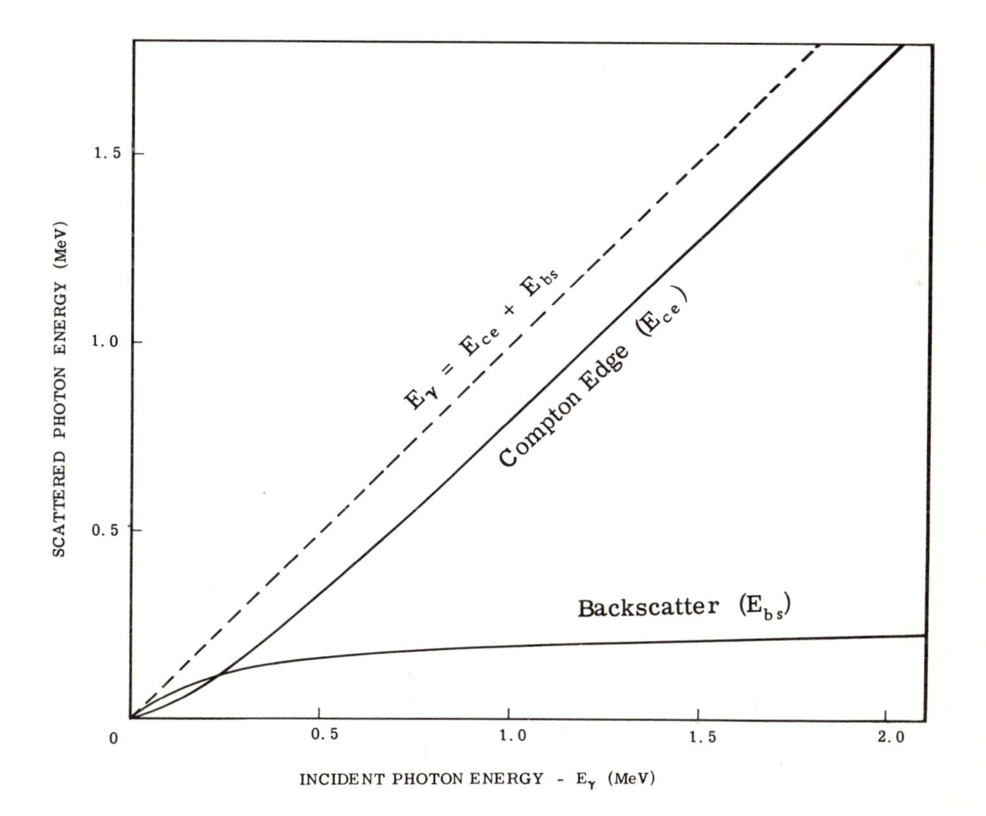

Figure 9.11—COMPTON EDGE AND BACKSCATTER ENERGIES.

Since the Compton photon and the Compton Electron share the incident gamma energy,

$$E_{ce} = E_{\gamma} - E' \tag{9.3}$$

Simulateous solution of equations 9.2 and 9.3 for E_{ce} yields

$$E_{ce} = \frac{E_{\gamma}^{2}}{E_{\gamma} + 0.255} \tag{9.4}$$

A *backscatter peak* will often appear in a spectrum. It is caused by radiation which passes completely through the crystal without interaction and which is then scattered through 180° by the shielding or by parts of the detecting equipment, back into the crystal where it is detected. The theoretical location (energy) of the backscatter peak can be calculated by means of equation 7.34. The energy of the backscatter peak can also be calculated from

$$E_{bs} = E_{\gamma} - E_{ce} \tag{9.5}$$

Base Line Calibration The base line setting of a pulse-height analyzer is frequently indicated in volts. For a given high voltage and amplifier gain adjustment, base line voltage is proportional to the energy of the incident radiation. They may, therefore, be related by a proportionality constant. It is often convenient, however, to adjust the amplifier gain or photomultiplier high voltage so the base line reading and the energy (in MeV) are numerically related by a multiple of 10 to 100. The spectrometer is then calibrated for direct reading.

Calibration is performed by the use of a standard nuclide such as cesium-137 which has a well defined gamma energy of 0.663 MeV. If ΔE is equal to 1 volt or less, a value less than the resolution of the spectrometer, an approximate method for calibration, introducing an error of only 1% or less, is satisfactory for most applications.

The course and fine gain controls are adjusted to about the midpoint of the range, the window is set to infinity and the base line to 66.3 volts to correspond to the 0.663 MeV photopeak of ^{137}Cs. The high voltage is then increased stepwise until counts are just observed on the scaler or ratemeter. Now the window is closed to the desired value and a fine adjustment of amplifier gain or high voltage is made to obtain maximum count rate. The spectrometer is now calibrated so the range of 0-100 volts corresponds to 0-1 MeV. The gain, high voltage and window controls should not be disturbed. The analysis of spectra is now made with the base-line control.

If it is desired to change the range of the spectrometer, reducing the gain by a factor of 2, with the coarse gain control, will double the range to 0-2 MeV. If the amplifier gain is increased by a factor of 2, the range will be halved to 0-0.5 MeV.

It will be observed that in the calibration above, the 0.663 MeV cesium peak was centered in a window from 66.3 to 67.3 volts (Assuming $\Delta E = 1$ volt). For greater accuracy the center of the photopeak should be at 66.3 volts with the window extending from 65.8 to 66.8 volts. That is, the base line should be set at 65.8 volts rather than at 66.3 volts. This consideration is important if wide windows are used.

MATERIALS:

Gamma-ray spectrometer; ^{137}Cs and other gamma sources.

PROCEDURE:

1. Calibrate the spectrometer against ^{137}Cs or other standard sources.
2. Plot a differential spectrum for a gamma emitting isotope.
3. Calculate the gamma energies of the nuclide and compare with the values given in the literature.
4. From the photopeak energies, calculate the energy of anticipated backscatter peaks and determine if peaks appear in the spectrum at these energies.

EXPERIMENT 9.3 LIQUID SCINTILLATION COUNTERS

OBJECTIVES:

To demonstrate the principles of liquid scintillation counting.

To measure carbon-14 and tritium (isotopes which emit weak beta particles) by the use of a liquid scintillation counter.

To study the distribution of pulse-heights produced by weak beta emitters.

To observe the cause and effects of quenching, determine balance point operation and determine the conditions under which the least statistical uncertainty is introduced by background.

THEORY:

Scintillation counters which employ a crystal scintillator such as thallium activated sodium iodide for the detection of radiation have a very limited usefulness for the detection of weak beta particles. Crystal scintillators (see experiment 9.1) are used for the detection of penetrating radiation (gamma rays) and are inefficient for the detection of radiation of low penetrating power (weak beta particles). This is because the radiation of low penetrating power is, absorbed either by self absorption or by the surroundings before it has an opportunity to interact with the scintillator. It is advantageous to incorporate the radioactive sample directly in the scintillator, thereby increasing the efficiency of detection. Theoretically "4π-geometry" is hereby obtained.

In 1950, Kallman (117, 118) and Reynolds (122) independently announced the usefulness of solutions of fluorescent substances in aromatic solvents as radiation detectors. The usefulness of these solutions for the detection of low energy beta parti-

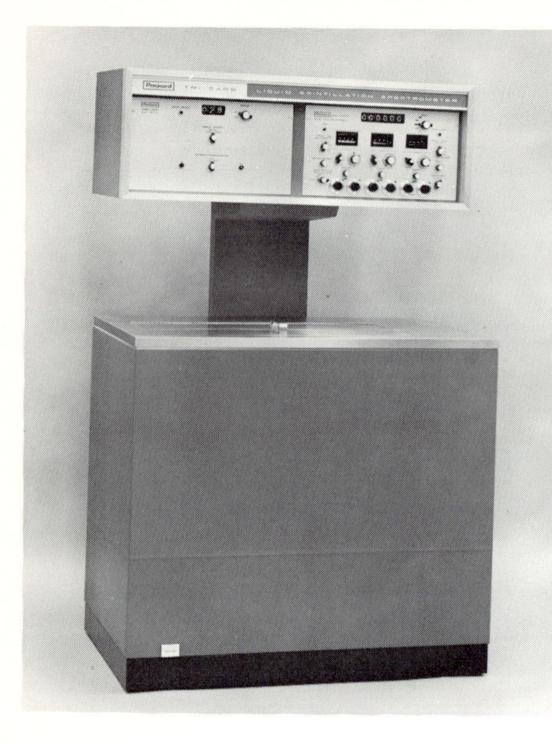

Figure 9.12—PACKARD TRI-CARB LIQUID SCINTILLATION SPECTROMETER MODEL 574. (*Courtesy Packard Instrument Co.*)

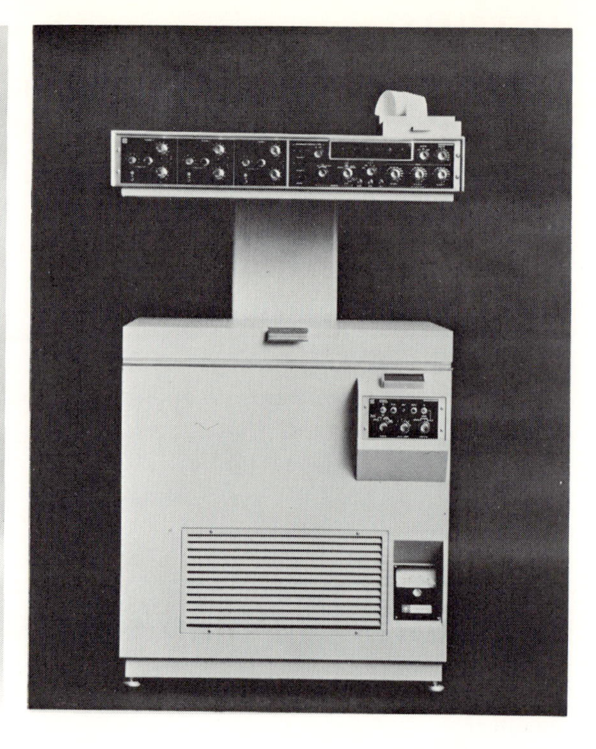

Figure 9.13—NUCLEAR-CHICAGO MODEL 6860 LIQUID SCINTILLATION COUNTING SYSTEM. (*Courtesy Nuclear-Chicago Co.*)

cles is utilized to full advantage in the liquid scintillation counter. In liquid scintillation counting, a solution of a fluorescent substance dissolved in toluene or other suitable solvent replaces the crystal scintillator. This solution is called a *liquid scintillator*. The radioactive substance is dissolved directly in the liquid scintillator or, if the sample is insoluble in the solvent used, it may sometimes be suspended as a fine dispersion. Beta particles, emitted as the radioactive atoms decay, interact with the liquid scintillator to produce very small flashes of light, too faint to be detected by the naked eye. A photomultiplier tube is used to detect the flashes of light in a manner similar to that employed with crystal scintillators. (See experiment 9.1). However, because the flashes of light are so faint, even for detection by a sensitive photomultiplier tube, certain refinements of the method are required over that used for gamma ray measurements.

Three fundamental problems are encountered. These are: (1) The energies of the beta particles are much lower than the usual range of gamma ray energies; (2) Liquid scintillators are less efficient than crystal scintillators; and (3) Photomultiplier tubes produce thermal noise which interferes with the detection of low energy radiation (see experiment 9.1).

Nothing can be done about the first of these problems. The beta energies cannot be changed. In an effort to overcome the second problem, a very large number of fluorescent substances and solvent systems have been tested to find a sensitive liquid

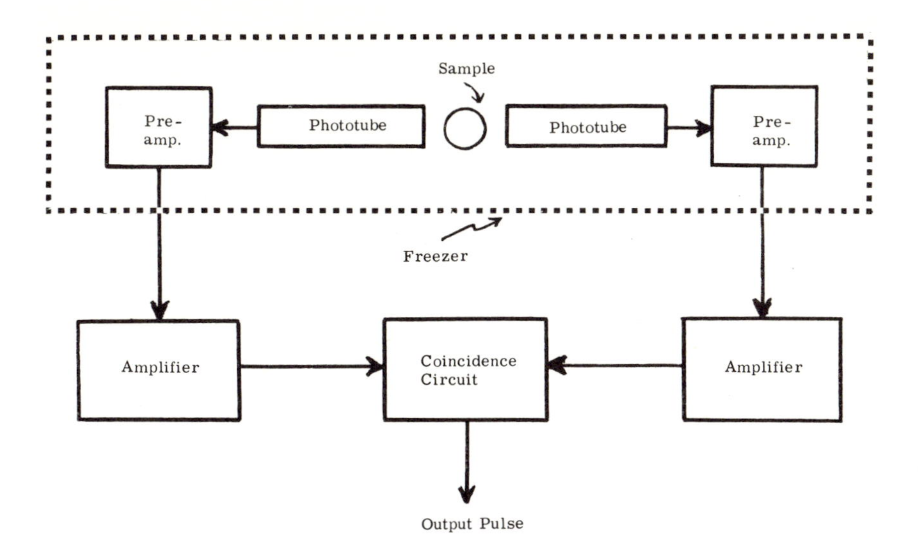

Figure 9.14—SIMPLIFIED COINCIDENCE CIRCUIT.

scintillator. Although none has been found which is as sensitive as a solid crystal for the detection of radiation, several very useful liquid scintillation systems are available.

The third problem, thermal noise from the photomultiplier tubes, can be decreased by reducing the operating temperature of the tubes. To accomplish this a freezer is often included as a standard part of the equipment. The photomultipliers, and thus the sample and the preamplifiers are housed in the freezer. The lower the operating temperature of the phototubes, the less the thermal noise. However, at too low a temperature, the liquid scintillator freezes. As a compromise, a temperature of about $-8°$ C seems to be optimum for many scintillation systems.

A second method for handling the problem of thermal noise (which does not reduce the thermal noise itself) is to use two photomultiplier tubes in conjunction with a coincidence circuit. Each flash of light produced in the liquid scintillator is detected simultaneously by both phototubes. The pulses produced by each tube are amplified and fed into a coincidence circuit. A coincidence circuit rejects any pulse which does not arrive simultaneously with a pulse from the other photomultiplier. Noise pulses are random. Thus, the number of noise pulses from one photomultiplier arriving at the coincidence circuit simultaneously with one from the other photomultiplier is very small.

Some instruments are equipped with newly designed 13 stage photomultiplier tubes, these do not require pre-amplifiers. The pulses derived from both photomultipliers are summed before pulse amplification. Such an electronic arrangement permits a better separation of ^{3}H from ^{14}C, as well as a truer pulse height distribution. Some instruments contain rate-meters which permit easier selection of appropriate instrument settings. Most instruments currently made are equipped with two or more channels and are quite reliable. A block diagram showing the use of a pulse summation circuit is given in figure 9.15.

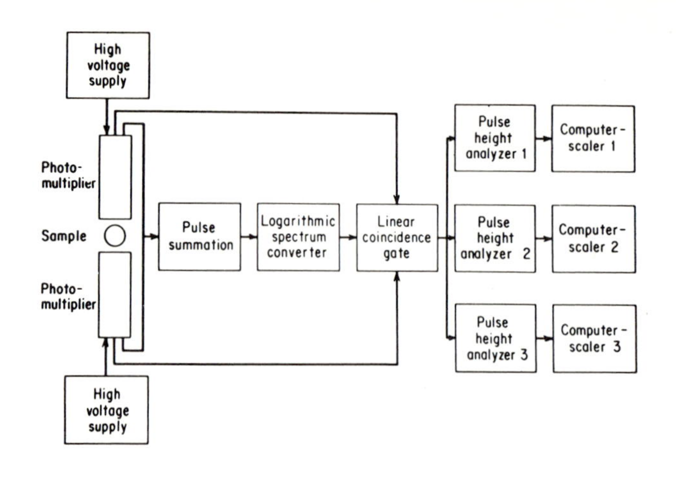

(a) *Ansitron, Ans, Inc.*

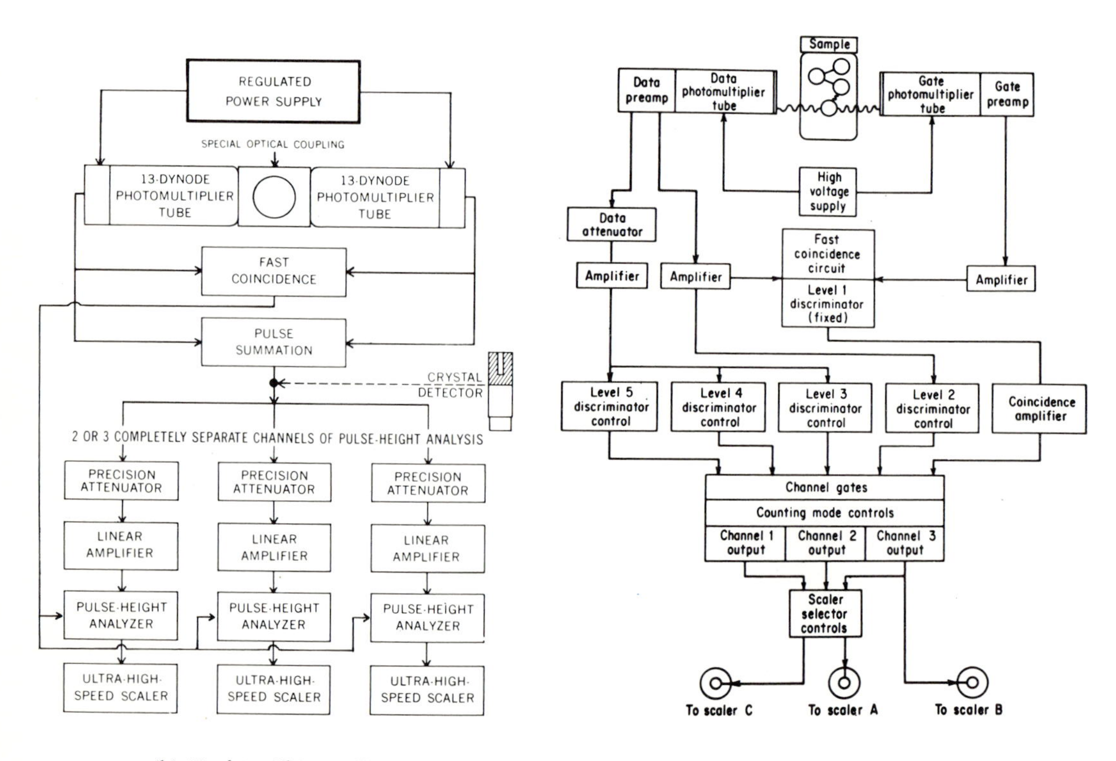

(b) *Nuclear-Chicago Corp.*

(c) *Packard Instrument Co., Inc.*

Figure 9.15—LIQUID SCINTILLATION COUNTING SYSTEMS.
Block diagrams of several commercial instruments.

Other factors, too, must be controlled if a liquid scintillation counter is to operate properly and efficiently. For instance, the efficiency with which the light is transferred from the liquid scintillator to the sensitive cathode of the photomultiplier is important. A high degree of efficiency can be achieved through the careful design of the sample container to prevent the loss of light.

The vial or other container used for the liquid scintillator and sample should be made of a special glass having a low potassium content. Natural potassium contains ^{40}K which is radioactive. Unless the containers are made of low potassium glass, the background count may be excessively high. Sample vials made of materials other than glass are now available; quartz, vicor and polyethylene are used. After exposure of a vial with scintillation solution to strong sunlight or fluorescent light, a high background may be observed. This phosphorescence usually decays in a few hours. Maintaining the sample in the dark or in subdued light to allow decay of phosphorescence to occur is called dark adaption of the sample.

Quenching is a process which results in a decrease in the intensity of the flashes of light produced by the beta particles in the liquid scintillator. *Thermal* or *chemical quenching* is said to occur if the solvent absorbs beta energy. In this case the energy of the beta particles is dissipated as heat rather than as light. *Color quenching* is said to occur if the solvent absorbs the light emitted by the liquid scintillator. Either process causes a decrease in the amount of light received at the photomultiplier.

Sample Preparation. The solvents in general use today are aromatic hydrocarbons or ethers. Solvents of these types are used because they produce the most efficient scintillator solutions. Thus in the preparation of samples for counting there is little flexibility in the choice of solvent or in the composition of the scintillator solution.

The sample itself may be incorporated in the scintillator solution in one of the following ways:

1. Simple solution—Lipids, steroids, hydrocarbons and many other organic compounds are readily soluble in the scintillator solution.

2. Use of common solvent—Small quantities of aqueous systems may be incorporated into toluene and other scintillator solvents by the use of ethanol, anisole, cellosolve, 1,4-dioxane and certain other liquids which act in the role of a common solvent. When used, the common solvent is usually added directly to the stock solution of scintillator.

3. Compound formation—Occasionally when less complicated procedures are not satisfactory, ^{14}C-compounds are oxidized to $^{14}CO_2$ which is then absorbed in scintillator solution containing Hyamine®10-X‡ (p-(diisobutyl-cresoxyethoxyethyl)-dimethylbenzyl-ammonium chloride) a strong base. The resulting hyamine-carbonate is soluble. Ethanolamine has also been used to dissolve carbon dioxide in toluene solution. Dibutyl phosphate is used to complex metallic ions.

4. Suspensions—While it is generally more desirable, from the standpoint of greater efficiency and better reproducibility to dissolve the sample, when suitable solvents systems are not available the finely ground or precipitated sample may be suspended in the scintillator solution. To maintain the sample in suspension the liquid scintillator is converted to a thixotropic gel by addition of Thixin®*, Cab-O-Sil®**, or

‡Hyamine® is available from Rohm & Haas, Inc., Phila., Pa.
*Thixin® is available from Baker Castor Oil Co., Bayonne, N. J.
**Cab-O-Sil® is available from Cabot Corp., Boston, Mass.

other suitable preparation. When given a vigorous shake, the gel becomes liquid but quickly converts to a gel again upon standing. Suspension counting is generally used only with ^{14}C and isotopes producing beta particles of greater energy. With ^{3}H, *self absorption* (not quenching) may be such a serious problem that the counting efficiency is a function of particle size.

One *scintillator system* which has been used extensively is:

PPO (2,5-diphenyloxazole)	3-4 g
POPOP [1, 4-bis-2-(5-phenyloxazolyl)-benzene]	100 mg
Toluene, to make	1000 ml

A *primary fluor* (the PPO) converts the beta particle energy to light energy. A *secondary fluor* (the POPOP) is used to shift the wave length of the light emitted by the primary fluor to a region where the photomultipliers are more sensitive. Minute quantities of secondary fluor may produce very impressive increases in counting efficiency. Best results are attained where strict attention is paid to purity of solvents and scintillators.

Fluors which are sulfur analogues have recently become available. One is 2,5-bis-2-(5-t-butylbenzoxazolyl)-thiophene, usually called BBOT. It is used in a concentration of about 4 g/l in toluene and does not require a secondary fluor.

Nuclides Assayed by Liquid Scintillation

Table 9.1 lists the nuclides conveniently counted by liquid scintillation techniques. In addition, ^{55}Fe, ^{134}Cs, ^{137}Cs, ^{233}U, ^{235}U and ^{239}Pu have also been counted with good results.

Table 9.1—ISOTOPES COMMONLY COUNTED WITH A LIQUID SCINTILLATION SPECTROMETER

Isotope	Half Life	Maximum Beta Energy, MeV
^{3}H	12.3 y	0.018
^{14}C	5770 y	0.15
^{35}S	87.1 d	0.17
^{45}Ca	165 d	0.25
^{65}Zn	245 d	0.33
^{59}Fe	45 d	0.46, 0.27
^{22}Na	2.6 y	β^{+} - 0.54
^{131}I	8.05 d	0.61, 0.25
^{36}Cl	3×10^{5} y	0.71
^{40}K	1.3×10^{9} y	1.32
^{24}Na	15.0 h	1.39
^{32}P	14.3 d	1.71

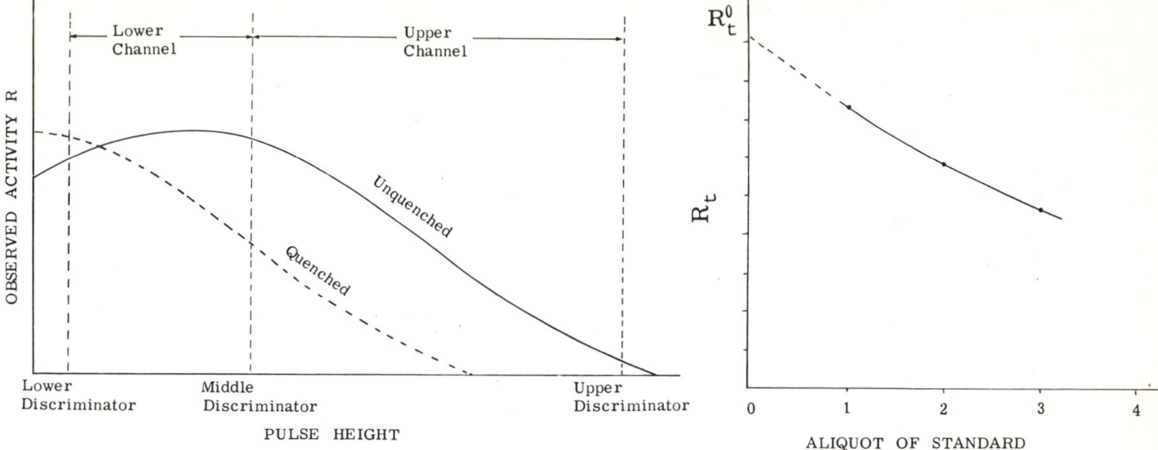

Figure 9.16—BETA SPECTRUM. The effect of quenching on the pulse height distribution is illustrated.

Figure 9.17—QUENCHING CORRECTION. The quenching produced by standard substance is corrected by use of the extrapolated value $R_t°$.

QUENCHING CORRECTION:

Quenching occurs in all liquid scintillator-sample solutions and therefore the extent of quenching must always be determined.

One of the simplest methods for determining the extent of quenching is by use of an *internal standard.* The standard may be any calibrated solution of the nuclide being assayed. After measurement of the unknown sample activity R_u, an aliquot of standard solution is added to the sample and the activity is measured once again. This will be the total activity $R_t = R_u + R_s$. Since the absolute decay rate A_s of the standard is known, the absolute sample activity A_u is given by;

$$A_u = A_s \frac{R_u}{R_s} = \frac{A_s R_u}{R_t - R_u}$$

For improved accuracy, quenching caused by the added reference standard must also be taken into account. After measuring R_u and R_t as described above, an amount of non-radioactive standard, chemically equivalent to the radioactive standard, is now added and the decrease in count rate R_t is noted. The process may be repeated if necessary to provide ample data to extrapolate to $R°_t$, the theoretical activity of the sample plus a massless standard.

To eliminate the time-consuming measurements required when internal standards are used the *external gamma reference* method was developed. In this method a sealed gamma source is moved by a special mechanism from a shielded storage well to a position adjacent to the sample. While the external source provides only a rel- tive correction or normalization of the data rather than a value of absolute d rate as with the internal standard, it is fast and precise. When absolute decay r the sample are required the external source can be calibrated against a standard for the particular scintillation system used.

When the activity in the sample is high and the quenching is not too severe, the *channel ratios method* can be used to correct for counting efficiency. In this method the radioactivity of the sample itself is used to measure the degree of quenching. A change in the spectrum is caused by quenching. This change in spectral shape produces a change in ratio of the net counts in two channels of the pulse-height analyzer adjusted to span certain portions of the spectrum. The counting efficiency, determined by means of an internal standard, is plotted as a function of the channels ratio.

The *counting performance* of liquid scintillation instruments is frequently evaluated by use of a statistical relationship sometimes called the *figure of* merit (see equation 5.2). It is the ratio of the square of the sample activity to the background activity, R_s^2/R_b. The greater the value of the ratio the better the counting conditions.

Balance-Point operation—As the high voltage of a liquid scintillation system is progressively increased, the gain of the photomultiplier tubes is increased thereby producing an overall increase in pulse height. The net effect is somewhat opposite to that produced by quenching (see figure 9.16) since the beta spectra become richer in the higher energies. Over a given range of voltage the activity observed in the upper channel will show an increase while that in the lower channel shows a relatively small change. At one particular voltage the activity observed in the two channels will be approximately equal and the total of the two activities will have reached approximately a maximum value. This voltage roughly corresponds to the condition known as *balance-point operation.* In balance-point operation the decrease in pulse height caused by quenching causes a minimum change in observed activity in each of the channels because the loss of activity from a channel caused by pulses becoming too small to be accepted into a channel will be offset by the addition of pulses to the same channel which were previously too large and fell into the channel above. Balance-point operation is therefore desirable because it corresponds approximately to the condition of highest counting efficiency while at the same time showing a minimum sensitivity to quenching.

Heterogeneous Systems — Homogeneous counting systems in which the sample is mixed directly with the liquid fluor render the sample difficult to recover for further use. If a solution of the sample is poured into a vial containing beads of plastic scintillator which are insoluble in the solution, the radioactive solution can be separated by filtration, after counting, thereby being made available for other purposes. Heterogeneous systems are utilized in *continous flow detectors* used for monitoring the effluent from gas chromatographs. In these instruments anthracene crystals as well as other solid scintillators have been used. When organic solvents are used for the radioactive material, crystal fluors must be covered with a silicone protective coat so they will not dissolve.

...g—After separation of materials by paper-strip chromatog-
...se fractions may be accomplished without the need of in-
...er-strip segment) extractions. Segments of paper-strips (1″
...ced directly into liquid scintillation vials and after the addi-
... assayed for radioactivity. The counting efficiency obtained
... function of the density of the spot. For quantitative evalu-
... standards should be used.

...ation Detectors—To measure radioactivity in humans or
... determination of the ^{40}K content of the body or the body
...gested radioactivity, whole body counters are used. These

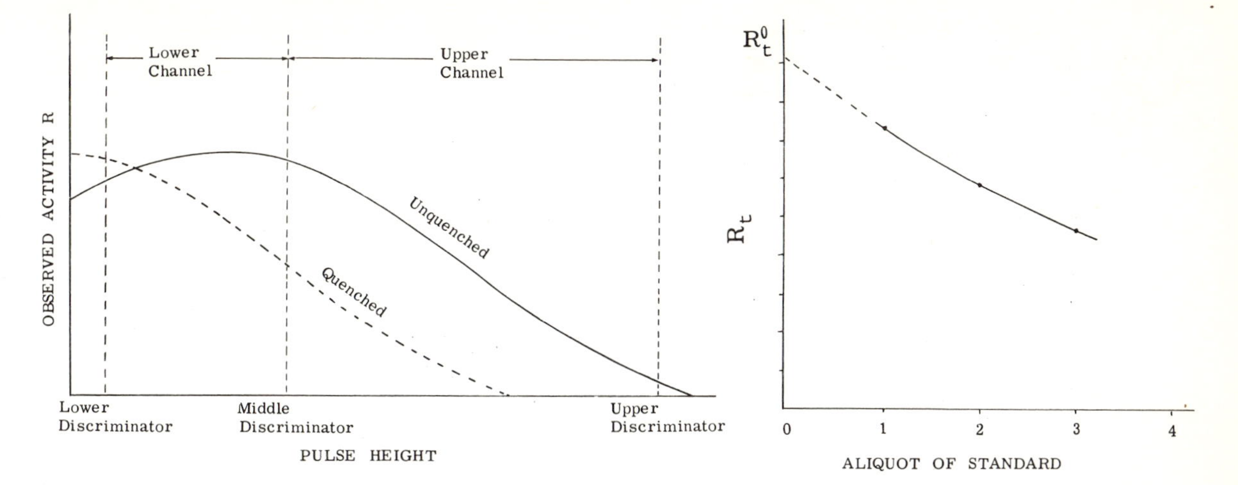

Figure 9.16—BETA SPECTRUM. The effect of quenching on the pulse height distribution is illustrated.

Figure 9.17—QUENCHING CORRECTION. The quenching produced by standard substance is corrected by use of the extrapolated value $R_t°$.

QUENCHING CORRECTION:

Quenching occurs in all liquid scintillator-sample solutions and therefore the extent of quenching must always be determined.

One of the simplest methods for determining the extent of quenching is by use of an *internal standard.* The standard may be any calibrated solution of the nuclide being assayed. After measurement of the unknown sample activity R_u, an aliquot of standard solution is added to the sample and the activity is measured once again. This will be the total activity $R_t = R_u + R_s$. Since the absolute decay rate A_s of the standard is known, the absolute sample activity A_u is given by;

$$A_u = A_s \frac{R_u}{R_s} = \frac{A_s R_u}{R_t - R_u}$$

For improved accuracy, quenching caused by the added reference standard must also be taken into account. After measuring R_u and R_t as described above, an amount of non-radioactive standard, chemically equivalent to the radioactive standard, is now added and the decrease in count rate R_t is noted. The process may be repeated if necessary to provide ample data to extrapolate to $R°_t$, the theoretical activity of the sample plus a massless standard.

To eliminate the time-consuming measurements required when internal standards are used the *external gamma reference* method was developed. In this method a sealed gamma source is moved by a special mechanism from a shielded storage well to a position adjacent to the sample. While the external source provides only a relative correction or normalization of the data rather than a value of absolute decay rate as with the internal standard, it is fast and precise. When absolute decay rates of the sample are required the external source can be calibrated against an internal standard for the particular scintillation system used.

When the activity in the sample is high and the quenching is not too severe, the *channel ratios method* can be used to correct for counting efficiency. In this method the radioactivity of the sample itself is used to measure the degree of quenching. A change in the spectrum is caused by quenching. This change in spectral shape produces a change in ratio of the net counts in two channels of the pulse-height analyzer adjusted to span certain portions of the spectrum. The counting efficiency, determined by means of an internal standard, is plotted as a function of the channels ratio.

The *counting performance* of liquid scintillation instruments is frequently evaluated by use of a statistical relationship sometimes called the *figure of* merit (see equation 5.2). It is the ratio of the square of the sample activity to the background activity, R_s^2/R_b. The greater the value of the ratio the better the counting conditions.

Balance-Point operation—As the high voltage of a liquid scintillation system is progressively increased, the gain of the photomultiplier tubes is increased thereby producing an overall increase in pulse height. The net effect is somewhat opposite to that produced by quenching (see figure 9.16) since the beta spectra become richer in the higher energies. Over a given range of voltage the activity observed in the upper channel will show an increase while that in the lower channel shows a relatively small change. At one particular voltage the activity observed in the two channels will be approximately equal and the total of the two activities will have reached approximately a maximum value. This voltage roughly corresponds to the condition known as *balance-point operation*. In balance-point operation the decrease in pulse height caused by quenching causes a minimum change in observed activity in each of the channels because the loss of activity from a channel caused by pulses becoming too small to be accepted into a channel will be offset by the addition of pulses to the same channel which were previously too large and fell into the channel above. Balance-point operation is therefore desirable because it corresponds approximately to the condition of highest counting efficiency while at the same time showing a minimum sensitivity to quenching.

Heterogeneous Systems—Homogeneous counting systems in which the sample is mixed directly with the liquid fluor render the sample difficult to recover for further use. If a solution of the sample is poured into a vial containing beads of plastic scintillator which are insoluble in the solution, the radioactive solution can be separated by filtration, after counting, thereby being made available for other purposes. Heterogeneous systems are utilized in *continous flow detectors* used for monitoring the effluent from gas chromatographs. In these instruments anthracene crystals as well as other solid scintillators have been used. When organic solvents are used for the radioactive material, crystal fluors must be covered with a silicone protective coat so they will not dissolve.

Paper-Strip Counting—After separation of materials by paper-strip chromatography, radioassay of these fractions may be accomplished without the need of individual "spot" (or paper-strip segment) extractions. Segments of paper-strips (1″ or 1½″ long) may be placed directly into liquid scintillation vials and after the addition of liquid scintillator, assayed for radioactivity. The counting efficiency obtained from the paper-strips is a function of the density of the spot. For quantitative evaluations, similarly prepared standards should be used.

Whole Body Scintillation Detectors—To measure radioactivity in humans or large animals, e.g., for the determination of the ^{40}K content of the body or the body burden of accidentally ingested radioactivity, whole body counters are used. These instruments are usually cylindrical and hold from 40 to 200 gallons of scintillation solution; a toluene solution containing PPO-POPOP is usually used. This solution is monitored by 40 to 100 photomultipliers and is usually surrounded with heavy shielding. For neutron detection, boron salts are added to the solution.

MATERIALS REQUIRED:

Liquid scintillation counter; sample vials; toluene; xylene; PPO; POPOP; benzoic acid-^{14}C; tritiated toluene; 1% eosine solution in ethanol and 1% methylene-blue solution in ethanol; Thixin® (ricinoleic acid); $Ba^{14}CO_3$; Hyamine® hydroxide solution.

PROCEDURE:

In following the general procedural directions outlined below, the operational directions supplied by the manufacturer for each instrument must be carefully followed. The directions, as given below, are generally for a two-channel instrument.

PART A—SELECTION OF COUNTING PARAMETERS

Most liquid scintillation counters are designed to provide considerable flexibility in the selection of counting parameters. Initially some of them must be selected arbitrarily. It is usually desirable, for example, to adjust the amplifier gain to a maximum (or attenuation to a minimum) since maximum electronic gain usually results in a maximum signal-to-noise ratio. For many applications it will also be found useful to set the discriminators for two-channel instruments to the following settings:

0 volts
} Pulses of these amplitudes are rejected
"Lower" discriminator—10 volts
} "Lower" channel
"Middle" discriminator—50 volts
} "Upper" channel
"Upper" discriminator—100 volts
} Pulses of these amplitudes are rejected except when the upper discriminator is switched out of the circuit
∞ volts

Pulse height is normally controlled by changing the gain of the photomultiplier tubes. This is accomplished by changing the high voltage setting.

The *standard scintillator solution* used in the following experiment should contain 3 grams of PPO and 100 milligrams of POPOP per liter of solution in C.P. toluene.

1. Prepare 50 ml of standard sample solution by dissolving a suitable quantity of a toluene soluble ^{14}C-labeled compound (benzoic acid-^{14}C) in standard scintillator solution to make 50 ml, 5 ml of which will produce about 5,000 disintegrations per minute.

2. To each of 5 sample vials add 5 ml of the solution prepared in step 1.

3. Prepare samples for counting by adding to each of the vials from Step 2 the following reagents:

Vial 1—2 ml of standard scintillator solution
Vial 2—2 ml of C.P. Toluene
Vial 3—2 ml of dioxane
Vial 4—2 ml of absolute alcohol
Vial 5—2 ml of 95% alcohol

4. Measure the activity recorded in both the "lower" and "upper" channels as a function of voltage. This should be done for each of the samples prepared in step 3.

5. Measure the background activity of each channel using a vial containing 7 ml of standard scintillator solution.

6. Correct all observed activities for background and compute the channel ratio R_{lower}/R_{upper}. From these values indicate which setting of the high voltage most closely approximates the conditions of balance-point operation.

PART B—EFFICIENCY VARIATION DUE TO PHOSPHOR AND SOLVENT

1. Prepare two liquid scintillator solutions as follows:

A		
PPO		3g
POPOP		100 mg
Toluene		1000 ml

B		
PPO		3g
POPOP		100 mg
Xylene		1000 ml

2. Prepare a solution of benzoic acid-^{14}C in toluene so that each 100 microliters of the solution has an activity of 1000 to 5000 dpm. Transfer 100 microliters of this solution to each of two vials, one containing 15 ml of liquid scintillator A and the other containing 15 ml of liquid scintillator B.

3. Prepare a solution of tritiated toluene in toluene so that each 100 microliters of the solution has an activity of 1000 to 5000 dpm. Transfer 100 microliters of this solution to each of two vials, one containing 15 ml of liquid scintillator A and the other containing 15 ml of liquid scintillator B.

4. To each of two other vials add 15 ml of liquid scintillator A and B respectively. These are the blanks.

5. Measure the activity of each of the six samples in the liquid scintillation counter. Determine the setting of high voltage which provides the optimum combination of balance-point operation and high counting efficiency. Calculate the counting efficiency for each of the samples above.

6. Add 0.1 ml of 1% methylene blue in ethanolic solution to one of the vials containing ^{14}C standard. Shake well and assay for radioactivity.

7. Add 0.1 ml of 1% eosin in ethanolic solution to another of the vials containing ^{14}C standard. Shake well and assay for radioactivity.

8. Repeat steps 6 and 7 for the vials containing the ^{3}H standard.

9. Explain the changes in observed activity in terms of changes in pulse-height distribution caused by quenching. How is counting efficiency influenced by quenching?

PART C—SUSPENSION COUNTING

1. Prepare a standard scintillator solution as described in Part A above.

2. Add sufficient Thixin® (ricinoleic acid) to 50 ml of standard scintillator solution to make a 5% solution.

3. Finely powder 100 mg of $Ba^{14}CO_3$ in a small agate mortar. The activity of this $Ba^{14}CO_3$ should be about 5×10^4 dpm and should be known accurately.

4. Add 10-mg samples of the finely powdered $Ba^{14}CO_3$, accurately weighed, to 15-ml aliquots of solution from step 2 and shake vigorously.

5. Assay for radioactivity. Calculate the counting efficiency and compare with the values obtained above for solution counting.

PART D — COUNTING $^{14}CO_2$ GAS

In chemical and biological research it is frequently necessary to convert carbon-containing samples to CO_2 gas for assay. The apparatus illustrated in figure 12.12 can be used for the conversion of $Ba^{14}CO_3$ to $^{14}CO_2$ or a CO_2 generator can be prepared as follows:

1. Fit a 500-ml Erlenmeyer flask with a 3-hole rubber stopper. A thistle tube is inserted into one of the holes so the stem reaches to the bottom of the flask. A tank of N_2 gas is connected to the flask through the second hole of the stopper. A short piece of glass tubing leads from the third hole to a vial containing 15 ml of liquid scintillator solution to which is added 5 ml of 1% Hyamine hydroxide solution. (Ethanolamine can be used instead of Hyamine hydroxide.)

2. Nitrogen gas is allowed to flow gently through the system. To the Erlenmeyer flask is added 100 mg of $Ba^{14}CO_3$ possessing an activity of 5×10^4 dpm. This is followed by 10 ml of concentrated H_3PO_4, added slowly through the thistle tube. As the $^{14}CO_2$ produced in the reaction is flushed from the generator and through the liquid scintillator solution by the nitrogen, it will be trapped as it reacts to form the carbonate of Hyamine.

3. The activity of the $^{14}CO_2$ trapped in the Hyamine-liquid scintillator solution is compared with that of other methods.

PART E — DOUBLE-LABEL COUNTING

Counting samples which are labeled with two isotopes illustrates a further usefulness of pulse-height analysis. Where the ratio of maximum energies of the two isotopes is four or five to one or greater, the two isotopes can be counted in one operation by adjusting the high voltage so that one isotope is counted efficiently in one channel and inefficiently in the other while the second isotope is counted efficiently and inefficiently in reverse channels. Knowing the efficiency for each isotope in each channel allows their assay through the simultaneous solution of two algebraic equations.

1. Prepare 50 ml of standard benzoic acid-^{14}C in standard scintillator solution (see part A, step 1 above). This is solution 1.

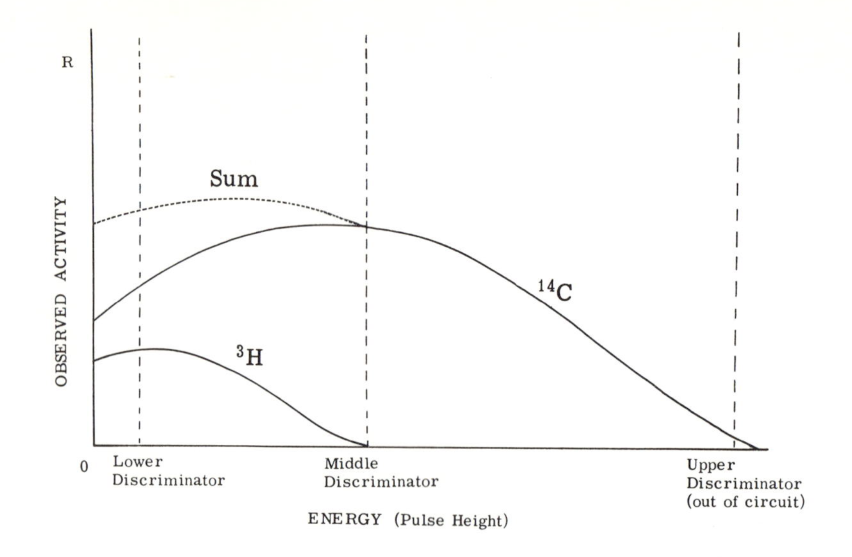

Figure 9.18—DOUBLE-LABEL COUNTING. The isotope emitting the weaker beta particles is measured in the lower channel; the isotope emitting the stronger beta particles is measured in the upper channel.

2. Prepare 50 ml of standard toluene-^{3}H solution, 5 ml of which will provide about 5000 dpm. This is solution 2.

3. Prepare samples for counting as follows:
   Vial A—5 ml of solution 1.
   Vial B—5 ml of solution 2.
   Vial C—5 ml of solution 1 and 5 ml of solution 2.
   Vial D—No radioactive solution is used. (Blank)

To each vial is added enough standard scintillator solution to make a total volume of 15 ml.

4. Place the vial of sample B (tritium) in the liquid scintillation counter.

5. With the lower discriminator set at 10 volts, the middle discriminator set at about 60 volts and the upper discriminator set at infinity, determine the high voltage setting which will provide a maximum activity in the lower channel while at the same time permitting adjustment of the middle discriminator to a value between 50 and 60 at which tritium pulses just fail to appear in the upper channel.

6. Measure the activities of vials A, B and D. Calculate the efficiencies for both the carbon and tritium samples in both the lower and upper channels.

	Typical values are:			Observed values are:		
	Background cpm	Efficiency		Background cpm	Efficiency	
		^{14}C	^{3}H		^{14}C	^{3}H
Lower channel	24.4	ϵ_1 0.254	ϵ_3 0.088		ϵ_1	ϵ_3
Upper channel	11.2	ϵ_2 0.363	ϵ_4 0.000		ϵ_2	ϵ_4

7. Measure the activity of vial C (mixed sample) in both channels. Subtract background and compute the activity of carbon and tritium in the sample. The calculations are made as follows:

Using typical values:

$$A_C = \frac{R_U}{0.363}$$

$$A_T = \frac{R_L - (0.254\, A_C)}{0.088}$$

Using actual values:

$$A_C = \frac{R_U}{\epsilon_2}$$

$$A_T = \frac{R_L - (\epsilon_1 A_C)}{\epsilon_3}$$

In these calculations R_L is the corrected activity recorded in the lower channel and R_U is the corrected activity recorded in the upper channel while A_C and A_T are the calculated activities (dpm) of carbon-14 and tritium respectively. These computed values of activity should be compared with the theoretical values.

EXPERIMENT 9.4 AUTORADIOGRAPHY

OBJECTIVES:

To prepare the fundamentals of autoradiography.
To prepare an autoradiogram of a radioactive specimen.

THEORY:

Becquerel accidently discovered radioactivity in 1896 by a technique similar to autoradiography (actually his was a radiographic technique). In 1911 Reinganum, by microscopic examination, found that alpha particles produced developed silver grains in photographic emulsions placed along their paths. It has since been shown that other ionizing particles also produce developed silver grains along their paths through a photographic emulsion and that the number of developed silver grains per unit of pathlength, and also the shape and size of the path itself, provide much information concerning the nature of the radiation as well as the location of the radioactive atoms emitting it. Autoradiography is concerned principally with the location of the radioactive atoms.

Basically, the autoradiographic technique (sometimes called radioautography) consists of placing a radioactive specimen in contact with a photographic emulsion. The radiation, passing through the film, sensitizes the silver grains which, upon development, will show the location of the radioactive atoms in the specimen. This technique is of importance in biology, mineralogy and other scientific fields.

Technique—If a faithful image of the activity is to be obtained, not only must intimate contact between the specimen and the photographic emulsion be maintained, but other geometric factors as well (illustration in figure 9.19) must be regulated. These factors are:

1. Emulsion thickness.
2. Specimen thickness.
3. Width of gap between specimen and emulsion.
4. Range of the radiation.

In general, the resolution or fidelity of reproduction will be maximum if the above factors are reduced to the minimum practical values.

A moderately thin specimen, such as the leaf of a plant, is autoradiographed by placing a thin layer of plastic film between the emulsion and the specimen. The "sandwich" of emulsion, plastic film and specimen is then wrapped with black paper and either clamped together or weighted down to maintain contact during the exposure period. The plastic film retards the transfer of fluids or vapors to the emulsion which may result in the production of *artifacts*.

Artifacts are images produced by chemical reactions between juices or vapors emitted by the specimen and the emulsion, rather than by ionizing particles emitted from radioactive atoms. The production of artifacts is retarded by the use of a plastic film, as in the case mentioned above. Very often freezing the specimen is helpful, in which case the specimen and emulsion, wrapped in black paper, are placed in a deep freeze during the exposure period.

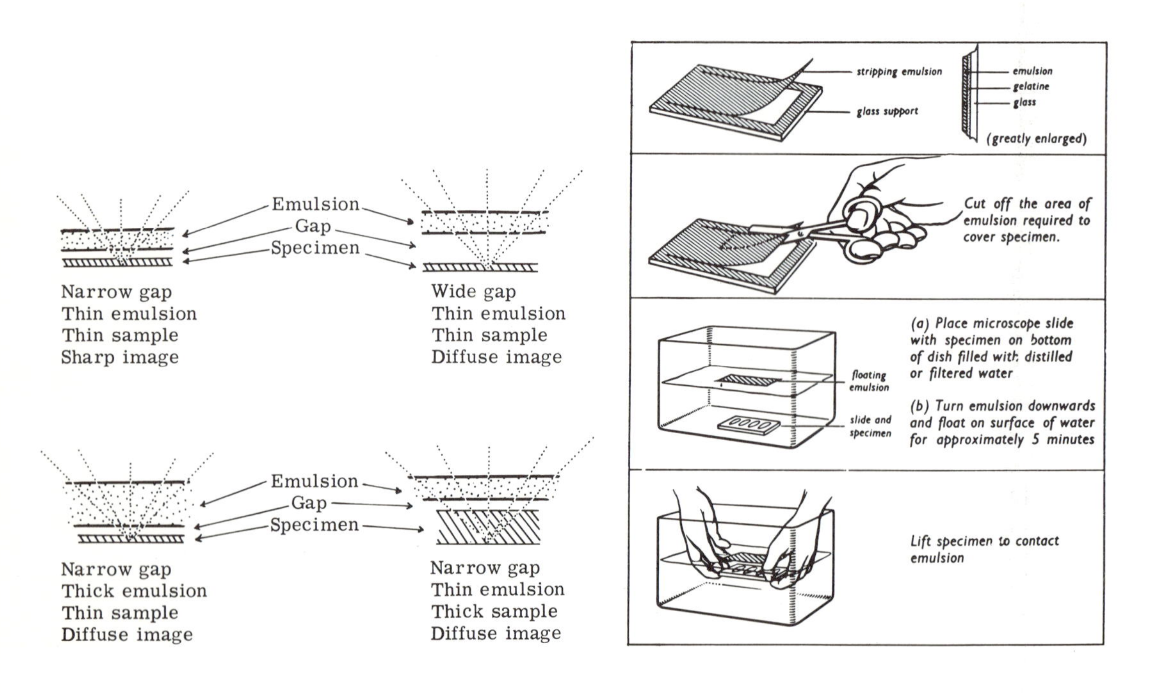

Figure 9.19—GEOMETRY FACTORS IN AUTORADIOGRAPHY.

Figure 9.20—PROCEDURE INVOLVED USING "KODAK" AUTORADIOGRAPHIC STRIPPING FILM. (*Courtesy Kodak Limited, London.*)

A large specimen, such as a rock, mineral or piece of metal, must be smoothed and polished on one surface if it is not so already, so the gap between the emulsion and specimen can be small. Normally a layer of plastic film is not necessary in this case, unless the specimen is moist or is likely to produce vapors.

Very thin sections of tissues can also be examined by this technique. They are first dried (by freeze drying) and imbedded in paraffin, or simply frozen and cut on a microtome by the use of standard histological techniques. The sections are floated on water and scooped up on a photographic plate (available in a 1″ × 3″ size). After exposure the plate is developed. If desired, the specimen may then be stained. This results in the stained specimen being superimposed on the developed image of the radioactivity in the emulsion. The exact location of the radioactivity within the tissue can thus be determined.

Emulsions—There are several categories of emulsions of particular interest for use in autoradiographic techniques.

1. Optical—used for ordinary photography. Can be used for simple demonstrations of autoradiography with large specimens.
2. X-ray—sensitized to ionizing particles.
3. Nuclear—to record tracks of α- and β- particles, protons, etc.
4. Autoradiographic—supplied in form of plates, stripping film and liquid gel.

Optical emulsions—Films used for ordinary photography are generally the least expensive and the most readily available of all types. Roll film, such as the Verichrome type, has been used to demonstrate general autoradiographic techniques with leaves and small plants in simple uptake studies.

Most *x-ray emulsions* are double coated; that is, they have a layer of emulsion on both sides of the supporting film. When used for autoradiography with carbon, sulfur or calcium, the beta particles emitted by these isotopes are so weak they will not penetrate the supporting film, so only one side is exposed. X-ray emulsions contain about 15% of silver halide. The average grain size is about 2 to 8 microns. If a greater sensitivity is desired than is available from optical emulsions, "no-screen" x-ray film may be used to demonstrate autoradiographic techniques with large specimens.

Nuclear emulsions are intended for use by physicists to record tracks of individual particles. Nuclear emulsions contain about 40-45% silver as the halide and the grain size is about 0.2 micron. The small grain size provides a high resolution. Increasing the content of silver halide compensates for the decreased sensitivity associated with small grain size.

The following Kodak nuclear track emulsions are listed in the order of increasing sensitivity:

NTA—for strongly ionizing particles such as α-particles. It is less sensitive to light than type NTB. Available thicknesses—10, 25 and 100μ.

NTB—Background due to cosmic radiation accumulates slower than with types NTB2 or NTB3. Used for very high velocity α-particles, protons, etc.

NTB2—Lower inherent background than NTB3 and gives better differentiation among particle species. Available thicknesses—10, 25 and 100μ.

NTB3—Responds to all charged particles. Will pick up considerable background from cosmic radiation and should be used within three weeks of shipment. This type is an experimental product. Available thicknesses—25 and 100μ.

Autoradiographic emulsions—Sensitized materials used for autoradiography include nuclear track plates, permeable base stripping film, impermeable base stripping film, plates and liquid emulsions.

1. *Nuclear track plates*—Types NTA, NTB, NTB2 and NTB3 listed above are used for autoradiographic techniques, especially when it is desirable to delineate tracks of individual α- or β-particles.

2. *Permeable base stripping film*—With stripping films, the emulsion is attached loosely to a supporting base from which it is removed or stripped prior to use. The technique for using stripping film is illustrated in figure 9.20. A small section of the emulsion, cut to slightly larger dimensions than the specimen, is floated on water, emulsion side down. After it swells and becomes flat, 2 or three minutes more are allowed for complete swelling before it is lifted from the water by raising the slide containing the specimen beneath it. It is dried in a stream of cold air and exposed in a light-tight box for several days to several weeks depending upon the activity of the specimen. It is then developed to bring out the exposed areas.

To prevent *artifacts* a layer of inert material is placed between the specimen and the emulsion. One of the best materials for this purpose is a thin layer of nylon film. Nylon chips are dissolved in hot isobutyl alcohol, care being taken to prevent the introduction of air bubbles into the solution. A drop of this solution is allowed to fall onto the surface of distilled water, carefully adjusted to pH7 with tri-sodium phosphate. After 5 to 10 minutes, the film can be picked up from the surface of the water with a loop of wire and allowed to dry. This film is interposed between the emulsion and the specimen.

Types of stripping film available from Eastman Kodak Company, Rochester, are NTA, NTB, NTB2 and NTB3. In addition, Kodak Limited London supplies:

Type AR.10—A fine-grain autoradiographic stripping plate.
Type AR.50—Fast autoradiographic stripping plate.

3. *Impermeable base stripping film*—Available in type NTB, it is a sensitive autoradiographic film, used largely with β-emitters. To prevent artifacts, it is supported on a film of cellulose ester which serves as an inert film between the emulsion and the specimen.

4. *Autoradiographic plates*—Type A has a fine-grain emulsion for high contrast with β- or γ-emitters, while type "no-screen" has a higher sensitivity at the sacrifice of grain fineness. Both are supplied on $1'' \times 3''$ slides.

5. *Liquid emulsions*—Used for special applications where thick emulsions are desired. It is applied directly to the specimen and dried. Liquid emulsion is not stable for more than a month.

Development—Alkaline developers tend to enlarge the silver grains and to loosen them. This is undesirable if the film is to be examined microscopically. An acid short-stop developer produces distortion by shifting the grains as much as a few microns. Special developers, such as Kodak D-19 or Amidol, are used to develop the emulsions. They are fixed in a 30% sodium thiosulfate solution until clear without the use of an acid short-stop. In special cased, where the specimen (e.g., blood cells) would be destroyed during development by the hypertonicity of the 30% sodium thiosulfate solution, a 10% solution can be used for a longer time.

MATERIALS REQUIRED:

X-ray film, developer, fixer, black paper, film holders, $Na_2H^{32}PO_4$, small plant, 0.01M Na_2HPO_4 carrier, counter with special shield, calibrated ^{32}P solution.

PROCEDURE:

Obtain a small plant of a variety having many small thin leaves, such as a tomato plant. Carefully wash the dirt from the roots without breaking or bruising the rootlets.

Insert the roots of the plant into a 6 inch test tube containing 10 ml of a solution prepared as follows:

Phosphorus-32	10 microcuries
0.01 *M* Na_2HPO_4	10 ml

Allow the plant to stand in the solution for 2 or 3 hours. Survey the foliage for activity.

Carefully remove several leaves from the plant and prepare them for making an autoradiograph. Lay the leaves out flat on a piece of plastic film. Cover with another piece of plastic film and seal the edges with cellulose tape. The leaves should now be sandwiched between plastic film and should lay perfectly flat.

For ample exposure, the film should be in contact with the specimen long enough to accumulate approximately 1×10^7 beta particles per square centimeter. To calculate the proper exposure time, place a lead shield with a 1 cm square opening over the section and determine the activity. The Geiger tube and counter assembly is calibrated by substituting a piece of filter paper for the leaf. An accurate measured quantity of a standard ^{32}P solution, previously spotted on the filter paper, serves as the standard. From the known disintegration rate of the ^{32}P standard, and from the observed counting rates of the standard and leaf, the time required to accumulate 10^7 beta particles per square centimeter of leaf can be calculated.

In the dark room, place the leaf (sandwiched between plastic film) in contact with x-ray film and wrap in black paper. Place on a flat surface, cover with a piece of glass or other flat object, and add sufficient weight on top of the glass to hold the sections flat.

Allow to stand for the calculated exposure time and develop the film.

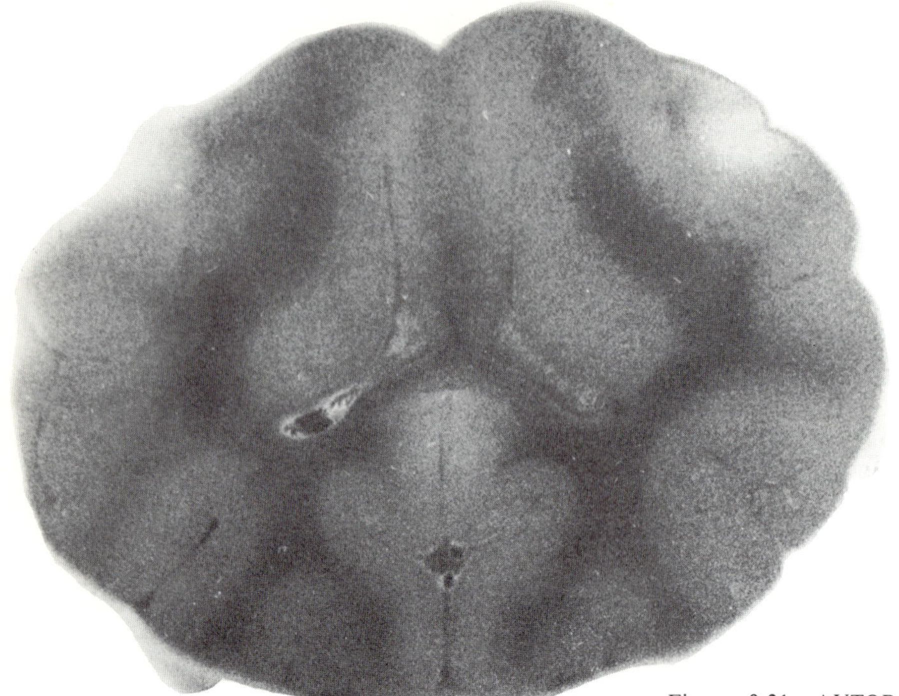

Figure 9.21—AUTORADIOGRAM OF CAT BRAIN. The animal received an injection of ^{14}C-mescaline. During an exposure of one month, the thin section of brain tissue was separated from no-screen x-ray film by 0.25 mil mylar film. In this positive print, light areas indicate the presence of radioactive mescaline. (*Prepared by Norton Neff, Philadelphia College of Pharmacy and Science.*)

Figure 9.22—AUTORADIOGRAM OF JAW OF A DOG. The animal received an injection of 10 μCi of $^{45}CaCl_2$ 48 hours prior to sacrifice. This section, showing a tooth in the jaw, was 100 microns thick. An exposure of three days was used with Kodak Limited type AR. 50 film. Light areas represent regions of increased metabolic rate.

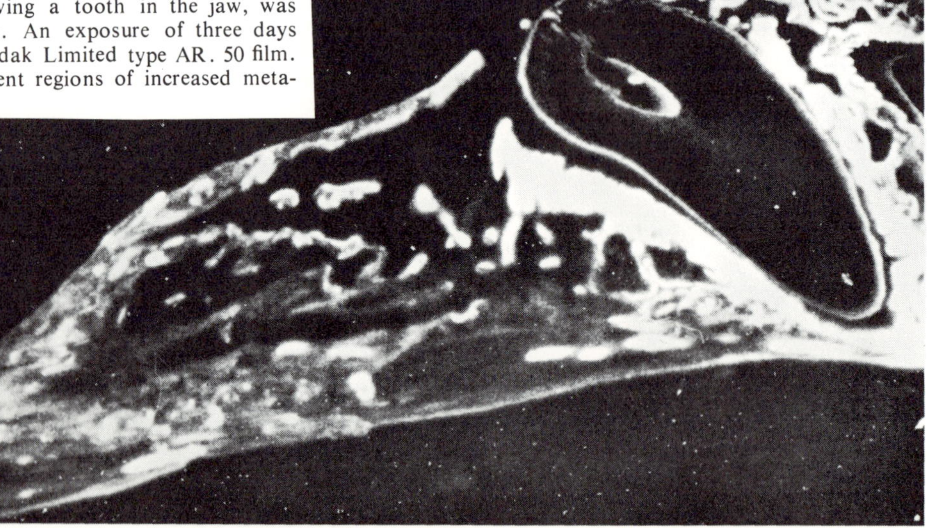

REFERENCES:

Fluors or Scintillators

1. Anon., "New Scintillators", Nucleonics 18, No. 5, 88-89 (May 1960).
2. Anon., "Phosphors and Photomultipliers", Nucleonics 12, No. 3, 26-28 (March 1954).
3. Anon., "Scintillation Response in CsI(Tl)", Nucleonics 21, No. 5, 72 (May 1963).
4. Baskin, R., H. L. Demorest, and S. Sandhaus, "Gamma Counting Efficiency of Two Well-Type NaI Crystals", Nucleonics 12, No. 8, 46-48 (Aug. 1954).
5. Bowen, E. J., "The Luminescence of Organic Substances", Nucleonics 10, No. 7, 14-18 (1952).
6. Brownell, G. L., W. H. Ellett, and H. D. LeVine, "Measuring γ-Ray Spectra with Large Plastic Scintillators", Nucleonics 16, No. 12, 68-70 (Dec. 1958).
7. Brucker, G. J., "Energy Dependence of Scintillating Crystals", Nucleonics 10, No. 11, 72-74 (1952).
8. Buck, W. L., and R. K. Swank, "Preparation and Performance of Efficient Plastic Scintillators", Nucleonics 11, No. 11, 48-52 (1953).
9. Deutsch, M., "Naphthalene Counters for Beta and Gamma Rays", Nucleonics 2, 58 (March 1948).
10. Downs, J. W., and F. L. Smith, "Organic-Glass Scintillators", Nucleonics 16, No. 3, 94-96 (March 1958).
11. Farmer, E. C., and I. A. Berstein, "Molded Multi-Crystalline Stilbene for Scintillation Counting", Nucleonics 10, No. 2, 54-56 (1952).
12. Fischer, J., "Preparing Large-Area Plastic Scintillators", Nucleonics 13, No. 5, 52-54 (May 1955).
13. Frazer, J. F., "An Alpha Scintillation Counter for Solid Samples", Nucleonics 14, No. 6, 88 (June 1956).
14. Friedman, H., and C. P. Glover, "Radiosensitivity of Alkali-Halide Crystals", Nucleonics 10, No. 6, 24-29 (1952).
15. Goodwin, P. N., "Scintillation Counter for Beta-Ray Dosimetry", Nucleonics 14, No. 9, 120-123 (Sept. 1956).
16. Harrison, F. B., "Slow Component in Decay of Fluors", Nucleonics 12, No. 3, 24-25 (March 1954).
17. Hine, G. J., and J. A. Cardarelli, "Conical Plastic Scintillators Show Total Gamma Absorption", Nucleonics 18, No. 9, 92-100 (Sept. 1960).
18. Hine, G. J., and A. Miller, "Large Plastic Well Makes Efficient Gamma Counter", Nucleonics 14, No. 10, 78 (Oct. 1956).
19. Ittner, W. B., III, and M. Ter-Pogossian, "Scintillation Probe for Determining Relative Beta-Ray Intensities", Nucleonics 12, No. 5, 56-57 (May 1954).
20. Kelley, G. G., P. R. Bell, R. C. Davis, and N. H. Lazar, "Intrinsic Scintillator Resolution", Nucleonics 14, No. 4, 53 (April 1956).
21. Lagiss, C., "Thin Plastic Scintillators", Nucleonics 14, No. 3, 66 (March 1956).
22. Lazar, N. H., R. C. Davis, and P. R. Bell, "Peak Efficiency of NaI", Nucleonics 14, No. 4, 52 (April 1956).
23. Leverenz, H. W., "Luminescent Solids (Phosphors)", Science 109, 183-196 (Feb. 25, 1949).
24. Liebson, S. H., "Temperature Effects in Organic Fluors", Nucleonics 10, No. 7, 41-45 (1952).
25. Michel, W. S., G. L. Brownell, and J. Mealey, Jr., "Designing Sensitive Plastic-Well Counters for Beta Rays", Nucleonics 14, No. 11, 96-100 (Nov. 1956).
26. Mitchell, M. L., and L. A. Sarkes, "Detection of S^{35} and Ca^{45} with a Plastic Scintillator", Nucleonics 14, No. 9, 124 (Sept. 1956).
27. Muehlhause, C. O., "Neutron Scintillation Counters", Nucleonics 14, No. 4, 38-39 (April 1956).
28. Sandler, S. R., S. Loshack, E. Broderick, and P. Bernstein, "Dimethyl Styrene Yields More Efficient Scintillators", Nucleonics 18, No. 9, 102-103 (Sept. 1960).
29. Schenck, J., "Neutron-Detecting Phosphors", Nucleonics 10, No. 8, 54-56 (1952).
30. Sun, K. H., P. R. Malmberg, and F. A. Pecjak, "High-Efficiency Slow-Neutron Scintillation Counters", Nucleonics 14, No. 7, 47-49 (July 1956).
31. Swank, R. K., "Recent Advances in Theory of Scintillation Phosphors", Nucleonics 12, No. 3, 14-19 (March 1954).
32. Van Sciver, W., "Spectrum and Decay of NaI", Nucleonics 14, No. 4, 50-51 (April 1956).
33. Williams, F. E., "The Luminescence of Inorganic Crystals", J. of Chem. Educ. 38, No. 5, 242-250 (May 1961).

Photomultipliers

34. Anon., "Photomultiplier Developments", Nucleonics 18, No. 5, 90-91 (May 1960).
35. Caldwell, R. L., and S. E. Turner, "Gain Variation of Photomultiplier Tubes", Nucleonics 12, No. 12, 47-48 (Dec. 1954).
36. Gordon, B. E., and T. S. Hodgson, "Dark Adaptation Reduces Photomultiplier Thermal Noise", Nucleonics 14, No. 10, 64-65 (Oct. 1956).
37. Kinard, F. E., "Temperature Dependence of Photomultiplier Gain", Nucleonics 15, No. 4, 92-97 (April 1957).

38. Linden, B. R., "New Photomultipliers and Operating Data", Nucleonics 12, No. 3, 20-23 (March 1954).
39. Mueller, D. W., G. Beat, J. Jackson, and J. Singletary, "Afterpulsing in Photomultipliers", Nucleonics 10, No. 6, 53-55 (1952).
40. Sharpe, J., "An International Survey of Photomultipliers for Scintillation Counting", Nucleonics 17, No. 6, 82-85 (June 1959).

Scintillation Counters

41. Anon., "Latest Developments in Scintillation Counting", Nucleonics 10, No. 3, 32-41 (1952).
42. Anon., "Scintillation Counting—1956", Nucleonics 14, 36-64 (April 1956).
43. Anon., "Scintillation Counting—1958", Nucleonics 16, No. 6, 54-62 (June 1958).
44. Anon., "Scintillation Crystals", Isomet Corp., Palisades Park, N. J. (1961).
45. Anon., "Scintillation Phosphors", Harshaw Chemical Co. (1958).
46. Anon., "Scintillation Theory", Nucleonics 18, No. 5, 86-87 (May 1960).
47. Anon., "Towards Better Scintillation Counting", Nucleonics 14, No. 4, 46-47 (April 1956).
48. Birks, J. B., "Scintillation Counters", McGraw-Hill (1953).
49. Blau, M., and B. Dreyfus, "Multiplier Photo-Tube in Radioactive Measurements", Rev. Sci. Instr. 16, 245-248 (1945).
50. Borkowski, C. J., "Instruments for Measuring Radioactivity", Anal. Chem. 21, 348-352 (1949).
51. Buckstein, E., "Basic Electronic Counting", Radio and TV News, p. 122 (March 1958).
52. Collins, G. B., "Scintillation Counters", Sci. Amer., p. 36 (Nov. 1953).
53. Curtiss, L. F., "Measurements of Radioactivity", Natl. Bur. Stand. Circ. 476, U.S. Dept. Comm. (Oct. 1949).
54. Elmore, W. C., H. Kallman, and C. E. Mandeville, "Practical Aspects of Radioactivity Instruments", Nucleonics 8, S1-S32 (June 1951).
55. Engstrom, R. W., and J. L. Weaver, "Are Plateaus Significant in Scintillation Counting?", Nucleonics 17, No. 2, 70-74 (Feb. 1959).
56. Glaser, D. A., "The Bubble Chamber", Sci. Amer., p. 46 (Feb. 1955).
57. Goldsmith, G. J., and R. M. Steffen, "Scintillation Counter Symposium", Science 115, 491-492 (May 2, 1952).
58. Herbert, R. J. T., "Discrimination Against Noise in Scintillation Counters", Nucleonics 10, No. 8, 37-39 (1952).
59. Jordan, W. H., "Detection of Nuclear Particles", Ann. Rev. Nuclear Sci. 1, 209-226 (1952).
60. Jordan, W. H., and P. R. Bell, "Scintillation Counters", Nucleonics 5, 30-41 (Oct. 1949).
61. Miller, C. E., L. D. Marinelli, R. E. Rowland, and J. E. Rose, "Reduction of NaI Background" Nucleonics 14, No. 4, 40-43 (April 1956).
62. Post, R. F., "Resolving Time of Scintillation Counters", Nucleonics 10, No. 6, 56-58 (1952).
63. Prener, J. S., and D. B. Sullenger, "Phosphors", Sci. Amer., p. 62 (Oct. 1954).
64. Price, W. J., "Nuclear Radiation Detection", McGraw-Hill (1958).
65. Rawson, E. G., and D. V. Cormack, "A Matrix to Correct for Scintillator Escape Effects", Nucleonics 16, No. 10, 92-97 (Oct. 1958).
66. Reynolds, G. T., "Solid and Liquid Scintillation Counters", Nucleonics 10, No. 6, 46-53 (1952).
67. Solon, L. R., J. E. McLaughlin, and H. Blatz, "Interpreting Background Data from Scintillation Detectors", Nucleonics 12, No. 5, 50-53 (May 1954).
68. Swank, R. K., "Nuclear Particle Detection (Characteristics of Scintillators)", Ann. Rev. Nuclear Sci. 4, 11-140 (1954).
69. Van Dilla, M. A., "Large-Crystal Counting", Nucleonics 17, No. 11, 150-155 (Nov. 1959).
70. Watters, H. J., "Multiple Sodium Iodide Crystal Arrays for Total Body Counting", U.S. Armed Forces Med. Journal 11, No. 4, 381-390 (April 1960).
71. Wonters, L. F., "High-Energy Particles and the Scintillation Counter", Nucleonics 10, No. 8, 48-53 (1952).

Scintillation Spectrometry

72. Anon., "Handbook for Scintillation Spectrometry", Baird-Atomic, Inc. (1958).
73. Beaufait, L. J., Jr., E. E. Anderson and J. P. Peterson, "Development and Preparation of a Set of Gamma Spectrometer Standards", Anal. Chem. 30, 1762 (1958).
74. Brown, B., and E. B. Hooper, Jr., "Plastic Phosphor Matrix for Fast-Neutron Detection", Nucleonics 16, No. 4, 96-103 (April 1958).
75. Crasemann, B., and H. Easterday, "Locating Compton Edges and Backscatter Peaks in Scintillation Spectra", Nucleonics 14, No. 6, 63 (June 1956).
76. De Waard, H., "Stabilizing Scintillation Spectrometers with Counting-Rate-Difference Feedback", Nucleonics 13, No. 7, 36-41 (July 1955).
77. Francis, J. E., P. R. Bell, and C. C. Harris, "Medical Scintillation Spectrometry", Nucleonics 13, No. 11, 82-88 (Nov. 1955).
78. Glenn, W. E., "A Pulse-Height Distribution Analyzer", Nucleonics 9, No. 6, 24-248 (1951).
79. Greenblatt, M. H., M. W. Green, P. W. Davison, and G. A. Morton, "Two New Photomultipliers for Scintillation Counting", Nucleonics 10, No. 8, 44-47 (1952).
80. Gunnink, R., and A. W. Stoner, "Photopeak Counting Efficiencies for 3 × 3 Inch Solid and Well-Type NaI Scintillation Crystals", Anal. Chem. 33, 1311 (1961).

81. Higinbotham, W. A., "Today's Pulse-Height Analyzers", Nucleonics 14, No. 4, 61-64 (April 1956).
82. Hine, G. J., "Beta- and Gamma-Ray Spectroscopy", I—Nucleonics 3, 32-42 (Dec. 1948); II—Nucleonics 4, 56-66 (Feb. 1949).
83. Hine, G. J., "Fast and Simple γ-Ray Scintillation Spectrometry", Nucleonics 11, No. 10, 68-69 (1953).
84. Hine, G. J., B. A. Burrows, L. Apt, M. Pollycone, J. F. Ross, and L. A. Sarkes, "Scintillation Counting for Multiple-Tracer Studies", Nucleonics 13, No. 2, 23-25 (Feb. 1955).
85. Hofstadter, R., and J. A. McIntyre, "Gamma-Ray Spectroscopy with Crystals of NaI(Tl)", Nucleonics 7, 32-37 (Sept. 1950).
86. Hutchinson, G. W., "Applications of Kicksorters", Nucleonics 11, No. 2, 24 (1953).
87. Johnstone, C. W., "A New Pulse-Analyzer Design", Nucleonics 11, No. 1, 36 (1953).
88. Kahn, B., and W. S. Lyon, "Use of a Scintillation Spectrometer in Radiochemical Analysis", Nucleonics 11, No. 11, 61-63 (1953).
89. Kallman, H., M. Furst, and M. Sidran, "Scintillation Counting Techniques", Nucleonics 10, No. 9, 15-17 (1952).
90. Koch, H. W., and R. S. Foote, "Total-Absorption X-Ray Spectrometry (Application to Betatron Experiments)", Nucleonics 12, No. 3, 51-53 (March 1954).
91. Kohl, J., R. E. Nather, and V. P. Guinn, "Simple Instrumentation Determines Several Simultaneous Radioactivities", Nucleonics 14, No. 10, 50-53 (Oct. 1956).
92. Lamonds, H. A., "Fast Differential Analyzer", Nucleonics 14, No. 8, 86-90 (Aug. 1956).
93. Lee, W., "Direct Estimation of Gamma-Ray Abundances in Radionuclide Mixtures. Complement Subtraction Method", Anal. Chem. 31, 800 (1959).
94. Lewis, J. G., J. V. Nehemias, D. E. Harmer, and J. J. Martin, "Analysis of Radiation Fields of Two Gamma-Radiation Sources", Nucleonics 12, No. 1, 40-44 (Jan. 1954).
95. McCollom, K. A., "Discriminators", Nucleonics 17, No. 6, 72-77 (June 1959).
96. Marion, J. B., "γ-Ray Calibration Energies", Nucleonics 18, No. 11, 184 (Nov. 1960).
97. Matlack, G. W., J. W. T. Meadows, and G. B. Nelson, "Low-Energy Standards for Gamma-Ray Spectrometry", Anal. Chem. 30, 1753 (1958).
98. Miller, W. F., and W. J. Snow, "NaI and CsI Efficiencies and Photofractions for Gamma-Ray Detection", Nucleonics 19, No. 11, 174-178 (Nov. 1961).
99. Morrison, G. H., and J. F. Cosgrove, "Determination of Uranium-235 by Gamma Scintillation Spectrometry", Anal. Chem. 29, 1770 (1957).
100. Nicholls, J., "Alpha Scintillation Monitor for Hands and Clothing", Nucleonics 15, No. 3, 80-84 (March 1957).
101. Okada, M., "42 Gamma Spectra of Short-Lived Nuclides", Nucleonics 19, No. 9, 79-81 (Sept. 1961).
102. Owen, R. B., "Pulse-Shape Discrimination Identifies Particle Types", Nucleonics 17, No. 9, 92-95 (Sept. 1959).
103. Reiffel, L., C. A. Stone, and A. R. Brauner, "Fatigue Effects in Scintillation Counters", Nucleonics 9, No. 6, 13-15 (1951).
104. Robinson, K. W., "Pulse-Height Resolution", Nucleonics 10, 34 (March 1952).
105. Roulston, K. I., "A Simple Differential Pulse Height Analyzer", Nucleonics 7, 27-29 (Oct. 1950).
106. Russell, J. T., and H. W. Lefevre, "An F-M Multichannel Pulse-Height Analyzer", Nucleonics 15, No. 2, 76-79 (Feb. 1957).
107. Salutsky, M. L., M. L. Curtis, K. Shaver, A. Elmlinger, and R. A. Miller, "Determination of Protactinium by Gamma Spectrometry", Anal. Chem. 29, 373 (1957).
108. Siegbahm, K. (ed.), "Beta- and Gamma-Ray Spectroscopy", Interscience (1955).
109. Snyder, B. J., and G. L. Gyorey, "Calculating Gamma Efficiencies in Scintillation Detectors", Nucleonics 23, No. 2, 80-86 (Feb. 1965).
110. Spear, W. G., "Scintillation Dose-Rate Meter is Reliable, Easy to Maintain", Nucleonics 15, No. 5, 96-101 (May 1957).
111. Swank, R. K., "Recent Advances in Theory of Scintillation Phosphors", Nucleonics 12, 14-19 (March 1954).
112. Swank, R. K., and W. L. Buck, "Observations on Pulse-Height Resolution and Photosensitivity", Nucleonics 10, No. 5, 51-53 (1952).
113. Thompson, B. W., "Fast-Neutron Scintillation Survey Meter", Nucleonics 12, No. 5, 43-45 (May 1954).
114. Van Rennes, A. B., "Pulse Amplitude Analysis in Nuclear Research", I—Nucleonics 10, 20-27 (July 1952); II—Nucleonics 10, 22-28 (Aug. 1952); III—Nucleonics 10, 32-38 (Sept. 1952); IV—Nucleonics 10, 50-56 (Oct. 1952).
115. Venable, W. H., Jr., "Simple Recording Gamma-Ray Spectrometer", Nucleonics 15, No. 7, 84-86 (July 1957).
116. Wagner, C. D., and V. P. Guinn, "For Low Specific Activity: Use Scintillation Counting", Nucleonics 13, No. 10, 56-59 (Oct. 1955).

Liquid Scintillation
117. Kallman, H., and M. Furst, "Fluorescence of Solutions Bombarded with High Energy Radiation. I and II", Phys. Rev. 79, 853064 (1950).
118. Kallman, H., "Scintillation Counting with Solutions", Phys. Rev. 78, 621-622 (1950).
119. Kallman, H., and M. Furst, "Fluorescence of Liquids Under Gamma Bombardment", Nucleonics 7 (1), 69-71 (1950).

120. Kallman, H., and M. Furst, "Fluorescence of Solutions Bombarded with High Energy Radiation (Energy Transport in Liquids)", Phys. Rev. 79, 864-870 (1950).
121. Reynolds, G. T., "Liquid Scintillation Counters", Nucleonics 6, 68-69 (May 1950).
122. Reynolds, G. T., F. B. Harrison, and G. Salvini, "Liquid Scintillation Counters", Phys. Rev. 78, 488 (1950).
123. Goddu, R. F., and L. B. Rogers, "Counting of Radioactivity in Liquid Samples, Science 114, 99 (1951).
124. Kallman, H., and M. Furst, "Fluorescence of Solutions Bombarded with High Energy Radiation (Energy Transport in Liquids) Part II", Phys. Rev. 81, 853-864 (1951).
125. Raben, M. S., and N. Bloembergen, "Determination of Radioactivity by Solution in a Liquid Scintillator", Science 114, 363-364 (1951).
126. Audric, B. N., and J. V. P. Long, "Measurement of Low-Energy β-emitters by Liquid Scintillation Counting", Research 5, 46-47 (1952).
127. Bahner, C. T., D. B. Zilversmit, and E. McDonald, "The Preparation of Wet Ashed Tissues for Liquid Counting", Science 115, 597-598 (1952).
128. Bowen, E. J., "The Luminescence of Organic Substances", Nucleonics 10, No. 7, 14 (1952).
129. Farmer, E. C., and I. A. Berstein, "Determination of Specific Activities of ^{14}C-labeled Organic Compounds with a Water-Soluble Liquid Scintillator", Science 115, 460 (1952).
130. Furst, M., and H. Kallman, "High Energy Induced Fluorescence in Organic Liquid Solutions (Energy Transport in Liquids) III", Phys. Rev. 85, 816-825 (1952).
131. Harrison, F. B., "Large-Area Liquid Scintillation Counters", Nucleonics 10, No. 6, 40-45 (1952).
132. Hayes, F. N., R. D. Hiebert and R. L. Schuch, "Low Energy Counting with a New Liquid Scintillation Solute", Science 116, 140 (1952).
133. Reynolds, G. T., "Solid and Liquid Scintillation Counters", Nucleonics 10 (7), 46-53 (1952).
134. Blüh, O., and F. Terentiuk, "Liquid Scintillation Beta Counter for Radioactive Solids", Nucleonics 10, No. 9, 48-51 (1952).
135. Belcher, E. H., "Scintillation Counters Using Liquid Luminescent Media for Absolute Standardization and Radioactive Assay", J. Sci. Instr. 30, 286-289 (1953).
136. Birks, J. B., "Scintillation Counters", London, Pergamon Press (1953).
137. Curran, S. C., "Luminescence and the Scintillation Counter", Butterworth (1953).
138. Hayes, F. N., and R. G. Gould, "Liquid Scintillation Counting of Tritium-Labeled Water and Organic Compounds", Science 117, 480 (1953).
139. Hiebert, R. D., and R. J. Watts, "Fast Coincidence Circuit for ^{3}H and ^{14}C Measurements", Nucleonics 11, No. 12, 38 (1953).
140. Jackson, J. A., and F. B. Harrison, "A Slow Component in the Decay of the Scintillation Phosphors", Phys. Rev. 89, 322 (1953).
141. Muehlhause, C. O., and G. E. Thomas, Jr., "Two Liquid Scintillation Neutron Detectors", Nucleonics 11, No. 1, 44-48 (1953).
142. Pringle, R. W., et al., "A New Quenching Effect in Liquid Scintillators", Phys. Rev. 92, 1582-1583 (1953).
143. Reines, R., et al., "Determination of Total Body Radioactivity Using Scintillation Detectors", Nature 172, 521-523 (1953).
144. Schneider, W. A., J. Hershkowitz, and H. Kallman, "Properties of Long Columns of Scintillating Liquids", Nucleonics 11, No. 12, 46-48 (1953).
145. Arnold, J. R., "Scintillation Counting of Natural Radiocarbon: 1. The Counting Method", Science 119, No. 3083, 155-157 (1954).
146. Furst, M., and H. Kallman, "Energy Transfer by Means of Collision in Liquid Organic Solutions Under High Energy and Ultraviolet Excitations", Phys. Rev. 94, 503-507 (1954).
147. Harrison, F. B., "Slow Component in Decay of Fluors", Nucleonics 12(3), 24-25 (1954).
148. Harrison, F. B., C. L. Cowan, Jr., and F. Reines, "Large-Volume Liquid Scintillators: Their Applications", Nucleonics 12, No. 3, 44-47 (March 1954).
149. Rosenthal, D. J., and H. O. Anger, "Liquid Scintillation Counting of Tritium and C^{14}-labeled Compounds", Rev. Sci. Instr. 25, 670-674 (1954).
150. Van Dilla, M. A., R. L. Schuch, and E. C. Anderson, "K-9: A Large 4π Gamma-ray Detector", Nucleonics 12(9), 22-27 (1954).
151. Terentiuk, F., "Solid-Sample Beta Determinations with a Liquid Scintillation Counter", Nucleonics 12, No. 1, 61-62 (Jan. 1954).
152. Anderson, E. C., "A Whole Body Gamma Counter for Human Subjects", U.S. Atomic Energy Commission, LA-1717, 43 p. (1955).
153. Arnold, J. R., "New Liquid Scintillation Phosphors", Science 122, 1139-1140 (1955).
154. Beli, P. R., and R. C. Davis, "Pulse-Height Variation in Scintillation Counters", Rev. Sci. Instr. 26, 726 (1955).
155. Furst, M., and H. Kallman, "Fluorescent Behavior of Solutions Containing More than One Solvent", J. Chem. Phys. 23, 607-612 (1955).
156. Furst, M., and H. Kallman, "High Energy Induced Fluorescence in Organic Liquid Solutions, III", Phys. Rev. 97, 583 (1955).
157. Furst, M., H. Kallman, and F. H. Brown, "Increasing Fluorescence Efficiency of Liquid-Scintillation Solutions", Nucleonics 13, No. 4, 58 (1955).
158. Hayes, F. N., B. S. Rogers, and P. C. Sanders, "Importance of Solvent in Liquid Scintillators", Nucleonics 13, No. 1, 46 (1955).

159. Hayes, F. N., "Liquid Scintillators. 1. Pulse Height Comparison of Primary Solutes", Nucleonics 13(12), 38-41 (1955).
160. Hayes, F. N., D. G. Ott, V. N. Kerr, and B. S. Rogers, "Pulse Height Comparison of Primary Solutes", Nucleonics 13, No. 12, 38 (1955).
161. Ott, D. G., "Argon Treatment of Liquid Scintillators to Eliminate Oxygen Quenching", Nucleonics 13(5), 62 (1955).
162. Pringle, R. W., W. Turchinetz, and B. L. Funt, "Liquid Scintillation Techniques for Radiocarbon Dating", Rev. Sci. Instr. 26, 859 (1955).
163. Swank, R. K., and W. L. Buck, "Decay Times for Some Organic Scintillators", Rev. Sci. Instr. 26, 15 (1955).
164. Ziegler, C., H. H. Seliger and I. Jaffe, "Oxygen Quenching and Wave Length Shifters in Liquid Scintillators", Phys. Rev. 99, 663 (1955).
165. Anderson, E. C., "The Los Alamos Human Counter", Nucleonics 14(1), 26-29 (1956).
166. Basson, J. K., "Absolute Alpha Counting of Astatine-211", Anal. Chem. 28, No. 9, 1472 (1956).
167. Funt, B. L., "Scintillating Gels", Nucleonics 14, No. 8, 83 (1956).
168. Hayes, F. N., "Liquid Scintillators: Attributes and Applications", Intern. J. Appl. Radiation Isotopes 1, 46-56 (1956).
169. Hayes, F. N., D. G. Ott and V. N. Kerr, "Liquid Scintillators. II. Relative Pulse Height Comparisons of Secondary Solutes", Nucleonics 14(1), 42-45 (1956).
170. Hayes, F. N., B. S. Rogers and W. H. Langham, "Counting Suspensions in Liquid Scintillators", Nucleonics 14, No. 3, 48 (1956).
171. Kallman, H. P., M. Furst and F. H. Brown, "Liquid Scintillators with Heavy Elements", Nucleonics 14, No. 4, 48 (April 1956).
172. Lamb, W. R., "Diode Phototube Monitor Uses Liquid Scintillator", Nucleonics 14, No. 8, 84-85 (Aug. 1956).
173. Langham, W. H., W. J. Eversole, F. N. Hayes and T. T. Trujillo, "Assay of Tritium Activity in Body Fluids with Use of Liquid Scintillation System", J. of Lab. and Clin. Medicine 47, No. 5, 819 (1956).
174. Mann, W. B., and H. H. Seliger "Efficiency of 4π-crystal-scintillation Counting. II. Dead-time and Coincidence Corrections", J. of Research of the National Bureau of Standards 57, No. 5, 257-264 (1956).
175. Okita, G. T., J. Spratt and G. V. LeRoy, "Liquid Scintillation Counting for Assay of Tritium in Urine",
176. Seliger, H. H., and C. A. Ziegler, "Liquid-Scintillator Temperature Effects", Nucleonics 14(4), 49 (1956).
177. Seliger, H. H., C. A. Ziegler and I. Jaffe, "Role of Oxygen in the Quenching of Liquid Scintillators", Phys. Rev. 101, 998-999 (1956).
178. Smith, C. C., H. H. Seliger and J. Steyn, "Efficiency of 4π-crystal-scintillation Counting. I. Experimental Technique and Results", J. of Research of the National Bureau of Standards, 57, No. 5, 251-255 (1956).
179. Weinberger, A. J., J. B. Davidson and G. A. Ropp, "Liquid Scintillation Counter for Carbon-14 Employing Automatic Sample Alternation", Anal. Chem. 28, No. 1, 110 (1956).
180. White, C. G., and S. Helf, "Suspension Counting in Scintillating Gels", Nucleonics 14, No. 10, 46 (1956).
181. Williams, D. L., "Preparation of C^{14} Standard for Liquid Scintillation Counter", Nucleonics 14(1), 62-64 (1956).
182. Ziegler, C. A., H. H. Seliger and I. Jaffe, "Three Ways to Increase Efficiency of Liquid Scintillators", Nucleonics 14(5), 84-86 (1956).
183. Agranoff, B. W., "Silica Vials Improve Low-level Counting", Nucleonics 15(10), 106 (1957).
184. Bollinger, L. M., and G. E. Thomas, "Boron-loaded Liquid Scintillation Neutron Detectors", Rev. Sci. Instr. 28, 489-496 (1957).
185. Chleck, D. J., and C. A. Ziegler, "Ultrasonic Degassing of Liquid Scintillators", Rev. Sci. Instr. 28, 466-467 (1957).
186. Davidson, J. D., and P. Feigelson, "Practical Aspects of Internal-Sample Liquid-Scintillation Counting", Intern. J. Appl. Radiation Isotopes 2, 1-18 (1957).
187. Helf, S., and C. White, "Liquid Scintillation Counting of Carbon-14-Labeled Organic Nitro-compounds", Anal. Chem. 29, No. 1, 13 (1957).
188. Kerr, V. N., F. N. Hayes and D. G. Ott, "Liquid Scintillators. III. The Quenching of Liquid Scintillator Solutions by Organic Compounds", Intern. J. Appl. Radiation Isotopes 1, 284-288 (1957).
189. Okita, G. T., et al., "Assaying Compounds Containing H^3 and C^{14}", Nucleonics 15(6), 111-114 (1957).
190. Roucayrol, J., E. Oberhauser and R. Schussler, "Liquid Scintillators in Filter Paper—A New Detector", Nucleonics 15, No. 11, 104-108 (Nov. 1957).
191. Roulston, K. I., and S. I. H. Naqvi, "Gamma Detection Efficiencies of Organic Phosphors", Nucleonics 15, No. 10, 86 (Oct. 1957).
192. Schram, E., and R. Lombaert, "Continuous Determination of ^{14}C and ^{35}S in Aqueous Solutions by a Plastic Scintillator", Biochemical J. 66, 20P (1957).
193. Seaman, W., "Simple Liquid Scintillation Counter for Chemical Analysis with Radioactive Tracers", Anal. Chem. 29, No. 11, 1570 (1957).
194. Vaughan, M., D. Steinberg and J. Logan, "Liquid Scintillation Counting of C^{14} and H^3 Labeled Amino Acids and Proteins", Science 126, No. 3271, 446-7 (1957).
195. Ziegler, C. A., D. J. Chleck and J. Brinkerhoff, "Radioassay of Low Specific Activity Tritiated Water by Improved Liquid Scintillation Techniques", Anal. Chem. 29, No. 12, 1774 (1957).
196. Chin, P. S., Jr., "Liquid Scintillation Counting of C^{14} and H^3 in Plasma and Serum", Proceed. of the Soc. for Exp. Bio. and Med. 98, 546-547 (March 1958).

197. Eisenberg, F., Jr., "Round-Table on Preparation of the Alkaline Absorbent for Radioactive CO_2 in Liquid Scintillation Counting", "Liquid Scintillation Counting", Pergamon Press (1958).
198. Fredrickson, D. S., and K. Ono, "An Improved Technique for Assay of $^{14}CO_2$ in Expired Air Using the Liquid Scintillation Counter", J. Lab. Clin. Med. 51, No. 1, 147 (1958).
199. Greenfield, S., "A Flow-through Cell for Use with Scintillation Counters", Analyst 83, No. 983, 114-116 (1958).
200. Hayes, F. N., "Application of Liquid Scintillation Counters", IRE Transactions on Nuclear Science NS-5, No. 3, 166-170 (1958).
201. Hayes, F. N., et al., "Survey of Organic Compounds as Primary Scintillation Solutes", U.S. Atomic Energy Commission, LA-2176, 24p. (1958).
202. Herberg, R. J., "Phosphorescence in Liquid Scintillation Counting of Proteins", Science 128, 199 (1958).
203. Horrocks, D. L., and M. H. Studier, "Low Level Plutonium-241 Analysis by Liquid Scintillation Techniques", Ana. Chem. 30, 1747 (1958).
204. Lowe, A. E., and D. Moore, "Scintillation Counter for Measuring Radioactivity of Vapours", Nature 182, No. 4628, 133-134 (1958).
205. Radin, N. S., and R. Fried, "Liquid Scintillation Counting of Radioactive Sulfuric Acid and Other Substances", Anal. Chem. 30, 1926 (1958).
206. Ronzio, A. R., C. L. Cowan, Jr. and F. Reines, "Liquid Scintillators for Free Neutrino Detection", Rev. Sci. Instr. 29, 146-147 (1958).
207. Steinberg, D., "Radioassay of Carbon-14 in Aqueous Solutions Using a Liquid Scintillation Spectrometer", Nature 182, 740-741 (1958).
208. Swank, R. K., and W. L. Buck, "Spectral Effects in the Comparison of Scintillators and Photomultipliers", Rev. Sci. Instr. 29, 279-284 (1958).
209. Wilson, A. T., "Tritium and Paper Chromatography", Nature 182, No. 4634, 524 (1958).
210. Funt, B. L., and A. Hetherington, "Spiral Capillary Plastic Scintillation Flow Counter for Beta Assay", Science 129, 1429-1430 (1959).
211. Lutwak, L., "Estimation of Radioactive Calcium-45 by Liquid Scintillation Counting", Anal. Chem. 31, 340 (1959).
212. Miranda, H. A., Jr., and H. Schimmel, "New Liquid Scintillant", Rev. Sci. Instr. 30, 1128-1129 (1959).
213. Ott, D. G., C. R. Richmond, T. T. Trujillo and H. Foreman, "Cab-o-Sil Suspensions for Liquid Scintillation Counting", Nucleonics 17, No. 9, 106-108 (Sept. 1959).
214. Popjak, G., et al., "Scintillation Counter for the Measurement of Radioactivity of Vapors in Conjunction with Gas-Liquid Chromatography", J. Lipd Res. 1, 29-39 (1959).
215. Steinberg, D., "A New Approach to Radioassay of Aqueous Solutions in the Liquid Scintillation Spectrometer", Anal. Biochem. 1, 23-39 (1959).
216. Boyce, I. S., J. F. Cameron and K. J. Taylor, "A Simple Plastic Scintillation Counter for Tritiated Hydrogen", Intern. J. Appl. Radiation Isotopes 9, 122-123 (1960).
217. Domer, F. R., and F. N. Hayes, "Background vs. Efficiency in Liquid Scintillators", Nucleonics 18(1), 100 (1960).
218. Funt, B. L., and A. Hetherington, "Scintillation Counting of Beta Activity on Filter Paper", Science 131, No. 3413, 1608-1609 (1960).
219. Gjone, E., H. G. Vance, and D. A. Turner, "Direct Liquid Scintillation Counting of Plasma and Tissues", International J. of Applied Radiation and Isotopes 8, No. 2/3, 95-97 (1960).
220. Gordon, C. F., and A. L. Wolfe, "Liquid Scintillation Counting of Aqueous Samples", Anal. Chem. 32, No. 4, 574 (1960).
221. Hayes, F. N., E. Hansbury, and V. N. Kerr, "Contemporary Carbon-14. The p-Cymene Method". Anal. Chem. 32, No. 6, 617 (1960).
222. Helf, S., C. C. White, and R. N. Shelley, "Radioassay of Finely Divided Solids by Suspension in a Gel Scintillator", Anal. Chem. 32, 238 (1960).
223. Hendler, R. W., n.d., "Dual Isotope Counting by Channels Ratio Method", Nuclear Chicago Preliminary Tech. Bull., Des Plaines, Ill. (1960).
224. Herberg, R. J., "Backgrounds for Liquid Scintillation Counting of Colored Solutions", Anal. Chem. 32, 1468-1471 (1960).
225. Herberg, R. J., "Determination of Carbon-14 and Tritium in Blood and Other Whole Tissues", Anal. Chem. 32, No. 1, 42 (1960).
226. Jeffay, H., F. O. Olubajo, and W. R. Jewell, "Determination of Radioactive Sulfur in Biological Materials", Anal. Chem. 32, 306 (1960).
227. Karmen, A., and H. R. Tritch, "Radioassay by Gas Chromatography of Compounds Labelled with Carbon-14", Nature 186, 150-151 (1960).
228. Ludwick, J. D., "Liquid Scintillation Spectrometry for the Analysis of Zirconium-95-Niobium-95 Mixtures and Coincidence Standardization of These Isotopes", Anal. Chem. 32, No. 6, 607 (1960).
229. Marlow, W. F., and R. W. Medlock, "A Carbon-14 Beta-ray Standard, Benzoic Acid-7-C^{14} in Toluene, for Liquid Scintillation Counters", J. Res. Nat. Bur. Stand. 64A, 143-146 (1960).
230. Owen, R. B., "Scintillation Counters. A Survey", Nuclear Power 5, 82-86 (1960).
231. Rapkin, E., "The Determination of Radioactivity in Aqueous Solutions", Packard Technical Bulletin (April 1960).

232. Rapkin, E., "Liquid Scintillation Measurement of Radioactivity in Heterogeneous Systems", Packard Technical Bulletin (July 1960).
233. Toporek, M., "Liquid Scintillation Counting of C^{14} Plasma Proteins Using a Standard Quenching Curve", International Journal of Applied Radiation and Isotopes 8, No. 4, 229-230 (1960).
234. Van Dilla, M. A., and E. C. Anderson, "Personnel Monitoring with Large Liquid Scintillation Counters", U. S. Armed Forces Med. Journal 11, No. 5, 526 (1960).
235. Whisman, M. L., B. H. Eccleston, and F. E. Armstrong, "Liquid Scintillation Counting of Tritiated Organic Compounds", Anal. Chem. 32, 484-486 (1960).
236. Williams, D. L., and F. N. Hayes, "Liquid Scintillator Radiation Rate Meters for the Measurement of Gamma and Fast Neutron Rates in Mixed Radiation Fields", U. S. Atomic Energy Commission LA-2375, 59 p. (1960).
237. Badman, H. G., and W. O. Brown, "The Determination of ^{14}C and $_{32}P$ in Animal Tissue and Blood Fractions by the Liquid-Scintillation Method", Analyst 86, No. 5, 342-347 (1961).
238. Berlman, I. B., "Luminescence in a Scintillation Solution Excited by α and β Particles and Related Studies in Quenching", J. Chem. Phys. 34, 598-603 (1961).
239. Brown, W. O., and H. G. Badman, "Liquid-Scintillation Counting of ^{14}C-labelled Animal Tissues at High Efficiency", Biochemical J. 78, 571-578 (1961).
240. Bruno, G. A., and J. E. Christian, "Corrections for Quenching Associated with Liquid Scintillation Counting", Anal. Chem. 33, 650 (1961).
241. Bruno, G. A., and J. E. Christian, "Determination of Carbon-14 in Aqueous Bicarbonate Solutions by Liquid Scintillation TCounting Techniques", Anal. Chem. 33, No. 9, 1216 (1961).
242. Butler, F. E., "Determination of Tritium in Water and Urine. Liquid Scintillation Counting and Rate-of-Drift Determination", Anal. Chem. 33, 409 (1961).
243. Davidson, E. A., "Techniques for Paper Strip Counting in a Scintillation Spectrometer", Packard Tech. Bulletin (June 1961).
244. Daub, G. H., F. N. Hayes, and E. Sullivan, "Proc. Univ. New Mexico Conf. Organic Scintillation Detectors, Albuquerque, 1960", TID-7612. U. S. Government Printing Office, Washington, D. C. 417 p (1961).
245. Hejwowski, J., and A. Szymanski, "Lithium Loaded Liquid Scintillator", Rev. Sci. Instruments 32, 1057-1058 (1961).
246. Hood, D. W., A. F. Isbell, J. E. Noakes, and J. J. Stipp, "Benzene Synthesis Aids C^{14} Dating", Chem. Eng. News (Oct. 1961).
247. Jeffay, H., and J. Alvarez, "Liquid Scintillation Counting of Carbon-14. Use of Ethanolamine-Ethylene Glycol Monomethyl Ether-Toluene", Anal. Chem. 33, 612 (1961).
248. Kepple, R. R., "Sodium Iodide and Sodium Iodide Crystals: Their Use in Scintillation Counting and Spectrometry", A Bibliography, Argonne National Laboratory Report ANL-6446 (1961).
249. Kobayashi, Yutaka, "Liquid Scintillation Counting and Some Practical Considerations", Tracerlab, Waltham, Mass. 20 p. (1961).
250. Lohmann, W., and W. H. Perkins, "Stabilization of the Counting Rate by Irradiation of the Liquid Scintillation Counting Solutions with UV-light", Nuclear Instruments Methods 12, 329-334 (1961).
251. Lukens, H. R. Jr., "The Relationship Between Fluorescence Intensity and Counting Efficiency with Liquid Scintillators", Intern. J. Appl. Radiation Isotopes 12, 134-140 (1961).
252. Ludwick, J. D., and R. W. Perkins, "Liquid Scintillation Techniques Applied to Counting Phosphorescence Emission", Anal. Chem. 33, No. 9, 1230 (1961).
253. Rapkin, E., "Hydroxide of Hyamine 10-X", Packard Tech. Bulletin, Revised June 1960.
254. Tamers, M. A., J. J. Stipp and J. Collier, "High Sensitivity Detection of Naturally Occurring Radiocarbon. I. Chemistry of the Counting Sample", Geochimica et Cosmochimica Acta 24, 266-276 (1961).
255. Tracerlab, Inc., "Quenching in Liquid Scintillation Counting", Waltham, Mass. 7 p. (1961).
256. Whisman, M. L., B. H. Eccleston, and F. E. Armstrong, "Reproducibility of Tritium Analysis of Organic Compounds Using a Liquid Scintillation Spectrometer", U. S. Department of the Interior, Bureau of Mines Report BM-RI 5801 (1961).
257. Bush, E. T., "How to Determine Efficiency Automatically in Liquid Scintillation Counting", Nuclear-Chicago Tech. Bull. No. 13, Des Plains, Ill. 4 p. (1962).
258. Christian, J. E., W. V. Kessler, and P. L. Ziemer, "A 2π Liquid Scintillation Counter for Determining the Radioactivity of Large Samples Including Man and Animals", Intern. J. Appl. Radiation Isotopes 13, 557-564 (1962).
259. Davidson, E. A., "Techniques for Paper Strip Counting in a Scintillation Spectrometer", Packard Tech. Bulletin No. 4, Revised July 1962.
260. Fleishman, D. G., and V. V. Glazunov, "An External Standard as a Means of Determining the Efficience and Background of a Liquid Scintillator", Instruments Exp. Tech. (a translation) No. 3, 472-474 (1962).
261. Funt, B. L., and A. Hetherington, "The Kinetics of Quenching in Liquid Scintillators", Intern. J. Appl. Radiation Isotopes 13, 215-221 (1962).
262. Godfrey, P., and F. Snyder, "A Procedure for *in vivo* $C^{14}O_2$ Collection and Subsequent Scintillation Counting", Analytical Biochemistry 4, No. 4, 310-315 (1962).
263. Higashimura, T., et al., "External Standard Method for the Determination of the Efficiency in Liquid Scintillation Counting", Intern. J. Appl. Radiation Isotopes 13, 308-309 (1962).

264. Jeffay, H., "Oxidation Techniques for Preparation of Liquid Scintillation Samples", Packard Tech. Bulletin No. 10, October (1962).
265. Karmen, A., I. McCaffrey, and R. L. Bowman, "A Flow-through Method for Scintillation Counting of Carbon-14 and Tritium in Gas-liquid Chromatographic Effluents", J. Lipid Res. 3, 372-377 (1962).
266. Levin, L., "Liquid Scintillation Method for Measuring Low Level Radioactivity of Aqueous Solutions. Determination of Enriched Uranium in Urine", Anal. Chem. 34, 1402 (1961[).
267. Mead, E. R. C., and R. A. Stiglitz, "Improved Solvent Systems for Liquid Scintillation Counting of Body Fluids and Tissues", International J. of Appl. Radiation and Isotopes 13, No. 1, 11-14 (1962).
268. Myers, L. S., Jr., and A. H. Brush, "Counting of Alpha- and Beta-Radiation in Aqueous Solutions by the Detergent-anthracene Scintillation Method", Anal. Chem. 34, 342-345 (1962).
269. Piez, K. A., "Continuous Scintillation Counting of Carbon-14 and Tritium in Effluent of the Automatic Amino Acid Analyzer", Anal. Biochem. 4, 444-458 (1962).
270. Plaxco, J. M. Jr., "Cab-O-Sil as a Suspending Agent", American J. Pharm. p. 82, March (1962).
271. Schram, E., and R. Lambaert, "Determination of Tritium and Carbon-14 in Aqueous Solution with Anthracene Powder", Anal. Biochem. 3, 68-74 (1962).
272. Bloom, B., "Use of Internal Scintillation Standards in Heterogeneous Counting Systems", Anal. Biochem. 6, 359-361 (1963).
273. Bonelli, E. J., and H. Hartmann, "Liquid Scintillation Counting of Biological Compounds in Aqueous Solution", Anal. Chem. 35, No. 12, 1982 (1963).
274. Bush, E. T., "General Applicability of the Channels Ratio Method of Measuring Liquid Scintillation Counting Efficiencies", Anal. Chem. 35, 1024-1029 (1963).
275. Cuppy, D., and L. Crevasse, "An Assembly for $C^{14}O_2$ Collection in Metabolic Studies for Liquid Scintillation Counting", Anal. Biochem. 5, No. 5, 462-463 (1963).
276. Johnson, D. R., and J. W. Smith, "Glass Filter Paper Suspension of Precipitates for Liquid Scintillation Counting", Anal. Chem. 35, No. 12, 1991 (1963).
277. Karmen, A., I. McCaffrey, and B. Kliman, "Derivative Ratio Analysis: A New Method for Measurement of Steroids and Other Compounds with Specific Functional Groups Using Radioassay by Gas-liquid Chromatography", Anal. Biochem. 6, 31-38 (1963).
278. Popjak, G., A. E. Lowe, and D. Moore, "Scintillation Counter for Simultaneous Assay of H^3 and C^{14} in Gas-Liquid Chromatographic Vapors", J. of Lipid Res. 3, No. 3, 364-371 (1963).
279. Rapkin, E., "Liquid Scintillation Counting with Suspended Scintillators", Packard Tech. Bulletin No. 11, LaGrange, Ill. 10 p. (1963).
280. Rapkin, E., "Liquid Scintillation Counting: A Review", Intl. J. Appl. Radiation Isotopes 15, 69-87 (1963).
281. Rapkin, E., and J. A. Gibbs, "Polyethylene Containers for Liquid Scintillation Spectrometry", Intl. J. Appl. Radiation Isotopes 14, 71-74 (1963).
282. Rivlin, R. S., and J. Wilson, "A Simple Method for Separating Polar Steroids from the Liquid Scintillation Phosphor", Anal. Biochem. 5, 267-269 (1963).
283. Scales, B., "Liquid Scintillation Counting: The Determination of Background Counts of Samples Containing Quenching Substances", Anal. Biochem. 5, No. 6, 489-496 (1963).
284. Schram, E., "Organic Scintillation Detectors", Elsevier, Amsterdam, 212 p. (1963).
285. Sprokel, G. J., "A Liquid Scintillation Counter Using Anticoincidence Shielding", IBM J. of Res. and Devel. 7, No. 2, 135-145 (1963).
286. Tamers, M. A., and R. Bioron, "Benzene Method Measures Tritium in Rain Without Isotope Enrichment", Nucleonics 21, No. 6, 90 (1963).
287. Willenbrink, J., "On the Quantitative Assay of Radiochromatograms by Liquid Scintillation Counting", Intl. J. of Appl. Radiation and Isotopes 14, No. 4, 237-238 (1963).
288. Zutshi, P. K., "Low-Level Beta Counting and Absorption Measurement with Liquid Scintillators", Nucleonics 21(9), 50-53 (1963).
289. Anonymous, "Calcium Absorption in Man: Based on Large Volume Liquid Scintillation Counter Studies", Science 144, 1155 (1964).
290. Baxter, C. F., and I. Senoner, "Liquid Scintillation Counting of C^{14}-Labelled Amino Acids on Paper, Using Trinitrobenzene-Sulfonic Acid and a Modified Combustion Apparatus", Anal. Biochem. 7, No. 1, 55-61 (1964).
291. Hempel, K., "Uber Die Gleichzeitige Messung von Tritium and ^{14}C in Biologischem Material mit dem Flüssigkeitsszintillationszähler", Atompraxis 10, No. 3, 148-152 (1964).
292. Kornblatt, J. A., Bernath, and J. Katz, "The Determination of Specific Activity of $BaC_{14}O_3$ by Liquid Scintillatiin Assay", Intl. J. of Appl. Radiation and Isotopes 15, No. 4, 191-194 (1964).
293. Rapkin, E., "Liquid Scintillation Counting 1957-1963: A Review", Intl. J. of Appl. Radiation and Isotopes 15, No. 2, 69-87 (1964).
294. Ross, H. H., "Liquid Scintillation Counting of C^{14} Using a Balanced Quenching Technique", Intl. J. of Appl. Radiation and Isotopes 15, 273-277 (1964).
295. Wu, Ray, "Simultaneous Studies of Phosphate Transport and Glycolysis by a Simple Liquid Scintillation Counting Procedure with P^{32}, C^{14}, and H^3 Compounds", Anal. Biochem. 7, 207-214 (1964).
296. Goldstein, G., "Absolute Liquid-Scintillation Counting of Beta Emitters", Nucleonics 23, No. 3, 67 (1965).
297. Hall, T. C., and E. C. Cocking, "High-Efficiency Liquid-Scintillation Counting of ^{14}C-Labelled Material in Aqueous Solution and Determination of Specific Activity of Labelled Proteins", Biochem. J. 96, 626 (1965).
298. Peng, C. T., "Quenching Correction in Liquid Scintillation Counting", Atomlight 44, (March 1965).

Autoradiography

299. Abrahams, A. P., "Autoradiographic Determination of Radioactivity in Rocks", Nucleonics 15, No. 3, 85-86 (Mar. 1957).
300. Bhatragor, A. S., and P. C. Ghosh, "Using Autoradiography for Quantitative Study of U in Ore", Nucleonics 12, No. 4, 58-59 (Apr. 1954).
301. Evans, T. C., "Selection of Radioautography Techniques for Problems in Biology", Nucleonics 2, 52 (Mar. 1948).
302. Faulkenberry, B. H., R. H. Johnson, and C. E. Cole, "Radioautographs Show Quality of Panel Brazing", Nucleonics 19, No. 4, 126-130 (Apr. 1961).
303. Gomberg, H. J., "A New High-Resolution System of Autoradiography", Nucleonics 9, No. 4, 28-43 (1951).
304. Gorbman, A., "Radioautography in Biological Research", Nucleonics 2, 30 (June 1948).
305. Guilbert, J. M., and J. A. S. Adams, "Alpha-Particle Autoradiography with Liquid Emulsion", Nucleonics 13, No. 7, 43 (July 1955).
306. Herz, R. H., "Photographic Fundamentals of Autoradiography", Nucleonics 9, No. 3, 24-39 (1951).
307. Hoecker, F. E., P. N. Wilkinson, and J. E. Kellison, "A Versatile Method of Microautoradiography", Nucleonics 11, No. 12, 60-64 (1953).
308. LaRiviere, P. D., and S. K. Ichiki, "Autoradiographic Method for Identifying Beta-Active Particles in a Heterogeneous Mixture", Nucleonic 10, No. 9, 22-25 (1952).
309. Lotz, W. E., J. C. Gallimore, and G. A. Boyd, "How to Get Good Gross Autoradiographs of Large Undecalcified Bones", Nucleonics 10, No. 3, 28-31 (1952).
310. Lotz, W. E., and P. M. Johnston, "Preparation of Microautoradiographs with the Use of Stripping Film", Nucleonics 11, No. 3, 54 (1953).
311. Preuss, L. E., "Autoradiographic Studies of Surface Detail with Chromium-51", Nucleonics 12, No. 8, 30-32 (Aug. 1954).
312. Towe, G. C., H. J. Gomberg, and J. W. Freeman, "Wet-Process Autoradiography Modified for Studying Metals", Nucleonics 13, No. 1, 54-58 (Jan. 1955).
313. Wainwright, W. W., E. C. Anderson, P. C. Hammer, and C. A. Lehman, "Simplified Autoradiography Exposure Calculatiin", Nucleonics 12, No. 1, 19-21 (Jan. 1954).
314. Williams, A. I., "Method for Prevention of Leaching and Fogging in Autoradiographs", Nucleonics 8, No. 6, 10-14 (1951).

CHAPTER 10
Radiochemical Separation Techniques

LECTURE OUTLINE AND STUDY GUIDE

I. PRECIPITATION METHODS
- A. Standard Analytical Procedures (*Experiment* 10.1)
  1. Semi-micro precipitation techniques
  2. Chemistry of iodide and phosphate separation
  3. Use of a "carrier"
- B. Coprecipitation (*Experiment* 10.2)
  1. Theory of coprecipitation
  2. a. Fajans' precipitation rule
     b. Hahn's precipitation rules
        (1) Isomorphous replacement
        (2) Surface adsorption
        (3) Anomalous isomorphic replacement
        (4) Internal adsorption
  2. Applications of coprecipitation
     a. Separation procedures
     b. Preparation of "carrier-free" isotopes

II. SOLVENT EXTRACTION—Distribution Ratios (*Experiment* 10.3)
- A. Henry's Law
- B. The Partition Law
- C. Counter-current Distribution

III. CHROMATOGRAPHY (*Experiment* 10.4)
- A. Partition Chromatography
- B. Adsorption Chromatography
- C. R_f Value
- D. Column Chromatography
- E. Paper Chromatography
  1. Descending
  2. Ascending

IV. ION EXCHANGE (*Experiment* 10.5)
- A. Mode of Operation
  1. Cation exchangers
  2. Anion exchangers
- B. Types of Exchangers
  1. Natural
  2. Synthetic
- C. Theory of Operation
  1. Mass action law
  2. Order of affinity of ions for resin
  3. Ion exchange chromatography
  4. Effects of pore size and cross linkage
  5. Weak and strong acid resins
  6. Weak and strong basic resins

V. GAS CHROMATOGRAPHY (*Experiment* 10.6)
- A. Definitions
  1. Gas-liquid chromatography
  2. Gas-solid chromatography
  3. Retention time
  4. Peak width
- B. Evaluation of Method
  1. Column efficiencies
  2. Calculation of number of theoretical plates
- C. Applications

VI. DISTILLATION (*Experiment* 10.7)
- A. Chemistry of Iodine
- B. Application of a Carrier
- C. Evaluation of Distillation Procedures

VII. ELECTROCHEMICAL SEPARATIONS
- A. Electrodeposition (*Experiment* 10.8)
  1. Conductors
  2. Electrodes
  3. Electrolytic cells
     a. Cell reactions
     b. Cell potentials
  4. Electroseparation
  5. Electrolytic displacement in electrochemical series
- B. Electrophoresis (*Experiment* 10.9)
  1. Principles
  2. Electrophoresis of proteins
     a. Zwitterions
     b. Isoelectric point
  3. Paper electrophoresis
     a. Tagging proteins
     b. Buffer systems
     c. Location of activity

INTRODUCTION:

By and large, the apparatus found in the average chemical laboratory will also be found in the radiochemical laboratory. Most of the standard chemical techniques and procedures which have been developed over the years have also been applied to radiochemistry. Such commonplace chemical procedures as evaporation, distillation, precipitation, ion exchange, chromatography, electrodeposition, etc., are quite useful in the radiochemical laboratory. In fact, radioisotope techniques have broadened the utility of these basic procedures. For example, the accuracy of quantitative separations of submicrogram quantities has been improved markedly by the addition of radioactive isotopes which permit a calculation of the percentage of recovery.

Prior to the assay of a radioisotope, several preparatory procedures are usually necessary. These include:

1. separation of the nuclide from other radioactive or undesirable substances
2. concentration of the activity to a small volume and
3. preparation of the sample in a form suitable for counting.

The procedures described in this chapter may accomplish one, two or three of these objectives. The electrodeposition of cobalt, for example, serves all three purposes since, in the process of being plated onto a metallic planchet, it is separated from other substances in the solution, it is concentrated as a thin film on the metal disc and, in this form, it is suitable for radioassay. In other instances additional steps of preparation may be required.

In the preparation of a sample for radioassay, radiochemical purity is more important than chemical purity. The problem of attaining radiochemical purity is especially acute when working with carrier-free radioisotopes.

Micro and semi-micro techniques have gained more and more importance in chemistry, especially in the radiochemical laboratory. When one is working with small quantities of materials, these techniques become necessary. They are also desirable because they confine radioactivity to small areas and reduce the volume of radioactive waste. Some of these basic laboratory techniques on a semi-micro scale will be reviewed in the following experiments.

The purpose of this chapter is to present a broad, general coverage of certain useful basic techniques. Emphasis has been placed on fundamental principles and concepts.

EXPERIMENT 10.1 PRECIPITATION

OBJECTIVES:

To conduct selected chemical procedures illustrating semi-micro techniques.

To separate a mixture of radio-iodide and radio-phosphate by precipitation, demonstrating the separation by filtration and by centrifugation.

THEORY:

A radioactive atom possesses the same chemical characteristics as a stable atom of the same element; that is, an element of the same atomic number but different in mass. Although this statement must be modified somewhat in view of the *isotope effect*, particularly in the case of the lighter elements, the general correctness of the statement constitutes the justification for all experimental work involving tracer techniques. Based on this premise, the assumption may be made that radioactive atoms will undergo chemical reactions similar to those of their stable counterparts. This is illustrated by the separation of radio-iodide and radio-phosphate.

Iodide can be separated from phosphate by precipitation as the silver salt:

$$Ag^+ + I^- \longrightarrow \underline{AgI}$$

The precipitation will be conducted in the presence of excess nitric acid which prevents the precipitation of silver phosphate. Separation of the silver iodide precipitate from the supernatant solution containing the phosphate will be accomplished by two methods: (1) by filtration in the first part of the procedure; and (2) by centrifugation in the second part. In each case, the radioactivity of the separated iodide and phosphate will be measured individually and compared against standards as a test of the efficiency of the method and technique used.

Filtration—Several commercial assemblies are available for conducting filtrations of relatively small quantities of material. Those illustrated in figure 10.1 are designed to accommodate a 1″ circle of filter paper. Each consists of two sections. The filter paper, placed on the flat member of the lower section, is held in position when the upper and lower sections are assembled. The precipitate is collected in the funnel by suction by a technique similar to that used with Buchner funnels. After the precipitate is washed, it is dried in the funnel with small quantities of alcohol, acetone or ether to remove the moisture, and then air dried to remove the organic solvent. It

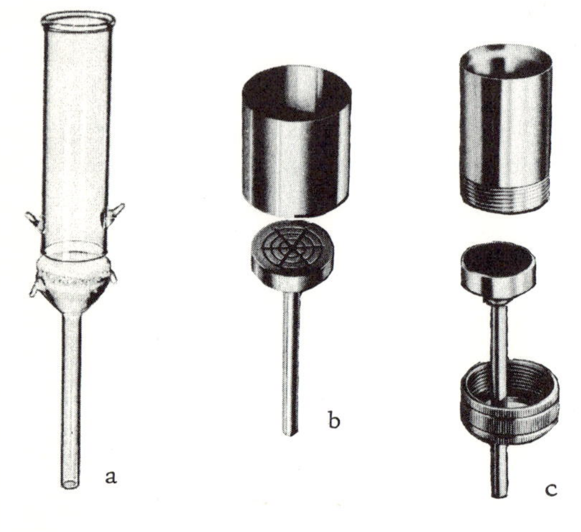

Figure 10.1—FILTER FUNNELS FOR COLLECTION OF PRECIPITATES. (a) *Atomic Accessories, Inc.* (b) and (c) *Tracerlab, Inc.*

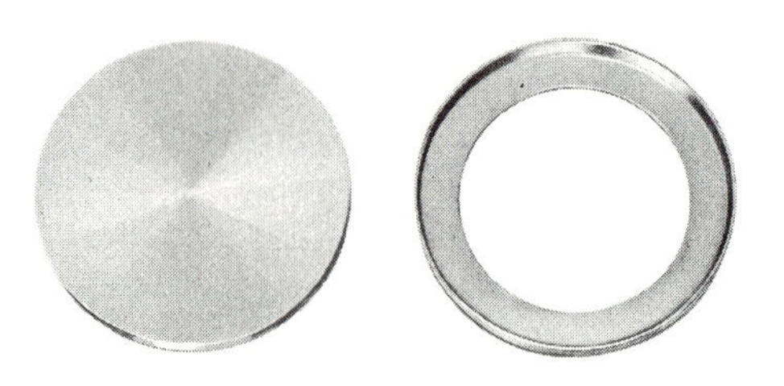

Figure 10.2—DISCS AND RINGS. Discs and rings are used to mount radioactive precipitates deposited on filter paper. After the filter paper has been placed on top of a disc, a ring fits snugly over both paper and disc, holding the paper securely in place. (*Atomic Accessories, Inc.*)

will often be found advantageous to glue the precipitate cake together by passing a small amount of diluted collodion through the precipitate with the aid of suction after it has been dried with solvents.

The precipitate is then transferred, intact, on the filter paper, to a mounting disc (figure 10.2). A small retaining ring holds the filter paper in place.

Trouble is frequently experienced from the adherence of precipitate to the walls of the funnel. If quantitative results are desired, all of the precipitate must be dislodged from the walls and washed down into the filter cake. If the walls of the funnel are first scrupulously cleaned and dried, spraying with silicone compounds will often be found effective because they prevent wetting of the walls. This procedure usually has to be repeated if the precipitate is dried by the use of organic solvents.

Centrifugation—The mixture to be centrifuged, contained in either a test tube or a tapered (centrifuge) tube, is placed in one cup of the centrifuge and another tube containing a similar amount of liquid is placed in the opposite cup. The centrifuge is allowed to run for a sufficient length of time (about 30 seconds), turned off, and stopped smoothly by light pressure of both hands on opposite sides of the revolving head. The operation is completed by very careful removal of the clear liquid with a long-tip dropper and careful transfer to a suitable container. In order not to disturb the residue, be sure to squeeze the dropper-bulb before placing it in the liquid. Be sure to lower the tip of the dropper almost to the level of the precipitate. During this operation, hold the centrifuge tube at approximately a 45 degree angle.

Washing precipitates—Add the proper amount of water or wash liquid to the centrifuged precipitate and, with the aid of a micro-stirring rod, agitate the precipitate until it goes into suspension. Then, centrifuge the mixture and remove the wash liquid with a dropper. Often this operation is carried out more than once on the same precipitate for greater washing efficiency.

Transfer of liquids—When small quantities of liquids are to be transferred from one container to another, the operation is performed most conveniently with a dropper, the nozzle of which has been drawn out to a capillary. If the transfer must be accomplished quantitatively, wash the original container and the dropper with a small quantity of water or wash liquid and use the dropper to transfer the wash liquid to the new container.

Transfer of precipitates—When precipitates are to be transferred from one container to another, add a few drops of water or other suitable reagent to the precipitate. Agitate the precipitate with a capillary dropper until it is finely suspended, then draw the suspension up into the dropper and transfer it to the new container. Use a few additional drops of reagent to wash the original tube and dropper, where quantitative transfers are required.

Water baths—A-100 ml beaker will serve as a water bath for 3-ml and 5-ml test tubes or centrifuge tubes. Heat on a hot plate or over a very low flame.

MATERIALS REQUIRED:

Filtration assembly; 1-inch filter paper circles, semi-micro centrifuge; 3-ml test tubes; centrifuge tubes; droppers; microstirring rods; test-tube rack; 1% silver nitrate solution; 1 molar nitric acid; 500-lambda micropipettes; 1 molar hydrochloric acid; 5% magnesium chloride solution; concentrated ammonium hydroxide; alcohol; ether; diluted collodion; radioactive stock solutions prepared as follows:

Iodine Stock Solution

^{131}I as iodide—about 0.1 microcurie/ml
Sodium iodide carrier—20 mg/ml

Phosphate Stock Solution

^{32}P as phosphate—about 0.1 microcurie/ml
Sodium phosphate (dibasic) carrier—20 mg/ml

PROCEDURE:

PART A—SEPARATION BY FILTRATION

Transfer 500 lambdas of iodine stock solution and 500 lambdas of phosphate stock solution to a 10-ml beaker, rinsing the pipettes with 1 molar nitric acid. Add an additional 5 ml of 1 molar nitric acid and precipitate the iodide by adding 1% silver nitrate dropwise until precipitation is complete. Filter through the filtration apparatus (figure 9.3) collecting the filtrate (which contains the phosphate) in a test tube placed inside the side-arm flask as shown. This eliminates the extra step of transferring the liquid from the flask to the tube and helps to keep radioactive contamination of equipment to a minimum. Rinse the 10-ml beaker with two 1-ml portions of 1 molar nitric acid, using this wash acid to rinse the funnel and precipitate as well. Now remove the test tube containing the phosphate solution and replace the funnel on the side-arm flask. Wash the precipitate with a few ml of water and then dry it by washing first with alcohol and then with ether. Finally, add 1 ml of diluted collodion and draw air through the precipitate until it is dry.

Carefully, without disturbing the precipitate, remove the top piece of the funnel and transfer the filter paper containing the precipitate to a mounting disc. Slip a retaining ring over the paper. It is now ready for measurement of the radioactivity. (This is sample FIM).

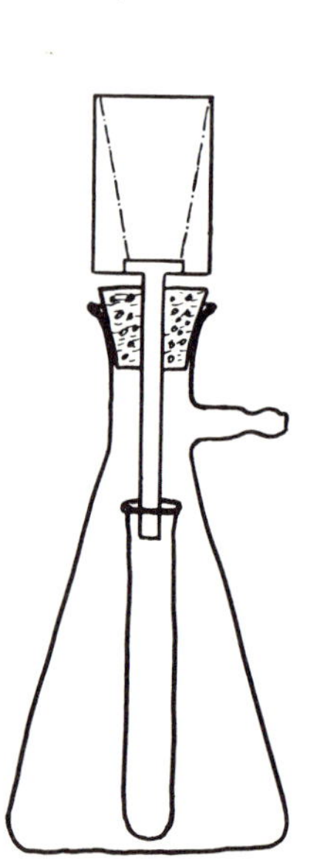

Figure 10.3—ASSEMBLED FILTRATION APPARATUS. A test tube used as shown facilitates the quantitative collection of small volumes of filtrate.

To the filtrate containing the phosphate, add an excess of concentrated hydrochloric acid, drop by drop, to precipitate the excess of silver ion. Filter and wash the precipitate with a few ml of dilute hydrochloric acid. Discard the precipitate. To the filtrate add 1 ml of magnesium chloride solution and add concentrated ammonium hydroxide dropwise until the solution has an alkaline reaction and a distinct odor of ammonia. Mix well and allow to stand for one hour. Filter through the filtration assembly. Use dilute ammonium hydroxide as a wash liquid, dry with alcohol and ether and mount for counting. (This is sample FPM).

Prepare standards for comparison. The iodine standard is prepared by precipitating the iodide in 500 lambdas of iodine stock solution with silver nitrate. The precipitate is collected in the filtering apparatus, washed, dried and counted as previously described. (This is sample FIS).

Prepare the phosphate standard by precipitating the phosphate directly in 500 lambdas of phosphate stock solution with magnesium chloride and ammonium hydroxide, and mounting as above. (This is sample FPS).

PART B—SEPARATION BY CENTRIFUGATION

Transfer 500 lambdas of each stock solution to a 3-ml test tube, rinsing the pipettes with dilute nitric acid. If the solution is not acidic, make it so by the addition of concentrated nitric acid, dropwise. Then, add 1% silver nitrate until precipitation is complete. Mix well and centrifuge. Transfer the supernatant liquid to a clean 3 ml centrifuge tube. This is the phosphate filtrate. Wash the precipitate once with 1 ml of 1 molar nitric acid, centrifuge and add the wash liquid to the centrifuge tube containing the phosphate filtrate.

Transfer the silver iodide precipitate to a plastic planchet (or other suitable planchet) with a few drops of water and dry under the infrared lamp. Measure the activity. (This is sample CIM).

Precipitate the excess silver in the phosphate filtrate with excess hydrochloric acid. Centrifuge and remove the filtrate to a clean 5-ml centrifuge tube. Wash the precipitate once with 1 ml of $1M$ hydrochloric acid and add the wash liquid to the filtrate.

Precipitate the phosphate in the filtrate as $MgNH_4PO_4$ by the addition of 0.5 ml of 5% magnesium chloride solution and an excess of concentrated ammonium hydroxide. Stir vigorously for several minutes and centrifuge. Remove the filtrate, test it for radioactivity and discard it if it is inactive. Suspend the precipitate in a few drops of dilute ammonium hydroxide and transfer it to a planchet with a dropper. Dry under the infrared lamp and measure the activity. (This is sample CPM).

Check your procedure by preparing two standards as follows: Into a 3-ml test tube, place 500 lambdas of iodine stock solution. Add a drop of nitric acid and an excess of silver nitrate. Mix, centrifuge, remove the filtrate and transfer the precipitate to a planchet. Measure the activity. (This is sample CIS).

Into another 3-ml centrifuge tube, place 500 lambdas of phosphate stock solution. Add 0.5 ml of magnesium chloride solution and an excess of concentrated ammonium hydroxide. Mix, centrifuge, remove the filtrate and transfer the precipitate to a planchet as before. Measure the activity. (This is sample CPS).

DATA:

	Mixture	Standards
Filtration		
Iodine	(FIM)__________cpm	(FIS)__________cpm
Phosphorus	(FPM)__________cpm	(FPS)__________cpm
Centrifugation		
Iodine	(CIM)__________cpm	(CIS)__________cpm
Phosphorus	(CPM)__________cpm	(CPS)__________cpm

EXPERIMENT 10.2 COPRECIPITATION

OBJECTIVES:

To demonstrate the use of coprecipitation in the separation of radioisotopes.

To illustrate the use of a "carrier" in radiochemical reactions and to observe the behavior of "carrier-free" solutions.

To become acquainted with the chemistry of a few elements constituting a part of a naturally occurring radioactive series.

To illustrate a growth decay curve.

THEORY:

In experiment 10.1 the separation of radioactive atoms by precipitation was demonstrated. This separation was conducted in the presence of an excess of nonradioactive or stable atoms of the same element in the same chemical form as the radioactive atoms. The stable elements, added in excess, are called carriers. Their purpose will be further illustrated by the separation of the radium-DEF series. See experiment 5.3 for the decay scheme.

Radium-D (lead-210) can be separated from radium-E (bismuth-210) by precipitation as the sulfate.

$$^{210}Pb^{+2} + SO_4^{-2} \longrightarrow \underline{^{210}Pb\,SO_4}$$

On the other hand, radium-E (bismuth-210) can be precipitated as the basic bismuth salt by the addition of an excess of sodium hydroxide to the nitric acid solution of radium-DEF, leaving radium-D in solution as the plumbite.

$$^{210}Bi(NO_3)_3 + 2\,NaOH \longrightarrow \underline{^{210}BiONO_3} + 2\,NaNO_3 + H_2O$$

$$^{210}Pb(NO_3)_2 + 4\,NaOH \longrightarrow Na_2{}^{210}PbO_2 + 2\,NaNO_3 + 2\,H_2O$$

If the solution is carrier-free, no precipitate will form upon addition of the reagents. This is because the ion product is less than the solubility constant. For example, the ion product (I. P.) for lead sulfate is given by the expression,

$$\text{I. P.} = [Pb^{+2}]\ [SO_4^{-2}]$$

where the brackets represent the molar concentrations of the lead and sulfate ions, respectively. In order to produce a precipitate, the product of these concentrations of ions must exceed 1.1×10^{-8}, the value of the solubility product constant for lead sulfate.

If the solution contains sufficient radium-D (lead) to give 10,000 disintegrations per minute per ml, the total amount of lead present, calculated from its 22 year half-life, is only 3×10^{-13} gram atoms per ml, or 3×10^{-10} gram atoms per liter. From this concentration and the solubility product constant, the minimum sulfate ion concentration required to produce precipitation of lead sulfate can be calculated.

$$K_{sp} = [Pb^{+2}]\ [SO_4^{-2}]$$

$$1.1 \times 10^{-8} = 3 \times 10^{-10}\ [SO_4^{-2}]$$

$$[SO_4^{-2}] = 40 \text{ molar}$$

Obviously, it is impossible to obtain such a high concentration of sulfate ion. Even if all the lead could be precipitated as lead sulfate, the quantity which would be obtained from a solution of this concentration would amount to only about 0.1 micrograms per liter.

Similarly, the concentration of 14 day phosphorus providing 10,000 disintegrations per minute per ml is only 5×10^{-13} molar, approximately 1,000 times more dilute than the 22 year lead. Such minute amounts are called *tracer concentrations*.

In the separation of lead and bismuth, the addition of carrier lead and carrier bismuth resolves the problem. The addition of carrier increases the ionic product to a value in excess of the solubility product constant.

If carrier is added only for the substance being precipitated and not for the ions forming soluble compounds, the latter may be *coprecipitated* with the former. Consequently, if separation is to be effected, carrier usually must be added for all radioactive species of ions in the solution. Thus, to effect a clean separation of radium-D and radium-E both carrier lead and carrier bismuth should be added. A carrier added to prevent the coprecipitation of a soluble compound is called a *holdback carrier*.

Radium-F (polonium-210) will be coprecipitated in this experiment because no polonium carrier is available. Of the approximately twenty isotopes of polonium, none are stable. The chemical properties of polonium strongly resemble those of bismuth and tellurium.

The behavior of trace concentrations was illustrated in experiment 5.7, where some of the problems encountered in handling such small quantities were demonstrated. The aforementioned phenomenon of *coprecipitation* of tracer quantities is another consideration of importance in radiochemistry.

Although coprecipitation, at times, creates problems of procedure, coprecipitation phenomena can also be used to advantage, especially in the separation of carrier-free isotopes. A procedure for such a separation is illustrated in part B of this experiment. Rules regarding coprecipitation phenomena were formulated by Fajans in 1913 and extended by Hahn in 1926.

Fajans and Beer (Fajans, "Radioactivity"—Methuen, 1922) established the following rule: "A radio-element is precipitated from exceedingly dilute solutions with the most varied precipitates if the latter are precipitated under conditions where the radio-element in question, if present in weighable quantities, forms a slightly soluble salt." For example, Fajans and Richter found a distinct parallel between the precipitation of radioactive lead with various precipitates and the solubility of the corresponding lead salt, thorium-B (^{212}Pb) being almost completely precipitated with bismuth *sulfide*, manganese *carbonate* or barium *sulfate* and only partially precipitated with silver *iodide* and silver *chloride*. Thus, the less the solubility of the salt of the corresponding trace element, the greater will be the extent of coprecipitation.

Hahn concluded that several processes, rather than a single one, were responsible for the observed precipitation phenomena. He catalogued these phenomena as isomorphous replacement, surface adsorption, anomalous mixed crystal formation and internal adsorption.

Isomorphous replacement—If two salts crystallize in the same crystal system and in the same form, they are isomorphous and one salt may be incorporated into the crystal lattice of the other to form an isomorphous mixed crystal. This will occur, particularly, if macro quantities of tracer elements of long half-life are present.

If the distribution of the tracer element throughout the crystal is *homogeneous* and there is established an equilibrium between the ions in solution and the ions in the interior of the crystal, then the *Nernst* distribution law applies.

$$K = \frac{[\text{ tracer in crystal }]}{[\text{ tracer in solution }]}$$

If, on the other hand, equilibrium is established between the solution and the crystal surface only, then as the crystals grow, a series of different equilibria will be established as the solution is slowly depleted of its ions. As a result, the distribution of the tracer throughout the crystals will assume a logarithmic pattern rather than a homogeneous one. *Doerner* and *Hoskins* have developed the mathematics for such a logarithmic distribution.

Surface adsorption of a tracer will occur if a precipitate possesses a surface charge opposite to the charge of the tracer ion to be adsorbed. The presence of other ions of charge similar to that of the tracer will cause interference. The efficiency of adsorption is also dependent upon the surface area of the precipitate and upon the solubility of any tracer compounds which can be formed from the ions present.

Anomalous mixed crystals are apparently formed at times, even though the salts of the tracer and carrier are not isomorphous. This process may be similar to *internal adsorption* wherein the tracer is adsorbed on the surface of a growing crystal so that as the crystal grows the adsorbed layer becomes engulfed. Repetition of this process produces a crystal with the tracer element distributed throughout. Through studies of the Nernst and Doerner-Hoskins distribution coefficients, Hahn attempts to distinguish between the process of anomalous mixed crystal formation and that of internal adsorption.

MATERIALS REQUIRED:

Microcentrifuge; centrifuge tubes; radium-DEF solution (about 1μCi per ml); lead nitrate solution; bismuth nitrate solution; 10% sulfuric acid; sodium hydroxide solution;

10% HCl; carrier-free phosphorus-32 (about 0.1μCi/ml) and iodine-131 (about 0.1μCi/ml); ferric chloride solution; ammonium hydroxide solution; isopropyl ether; well counter; end-window G-M counter; scintillation detector.

PROCEDURE:

PART A—COPRECIPITATION PHENOMENA

CAUTION: BE PARTICULARLY CAREFUL NOT TO CONTAMINATE EQUIPMENT, DESKS, OR YOURSELF WITH Ra-D, HALF-LIFE 22 YEARS!

Method 1. *Precipitation of Radium-D as Sulfate*

To each of three test tubes (3-ml size), each containing 1½ ml of 10% nitric acid, add 100 lambdas of Ra-DEF solution. Treat each tube, respectively, as follows:

Tube A-1. Add nothing—"carrier free"
Tube A-2. Add 3 drops of lead-carrier solution
Tube A-3. Add 3 drops of lead-carrier solution and 3 drops of bismuth-carrier solution

To each tube add 10 drops of 10% sulfuric acid. Mix the contents of each tube and allow to digest for 1 hour in a hot water bath. Tubes A-2 and A-3 should contain a precipitate. Cool the tubes and centrifuge. Transfer 1 ml of supernatant from each tube to a planchet and measure the activity of each solution with an end-window G-M tube.

Tube A-1. ________________ cpm for 1 ml of supernatant
Tube A-2. ________________ cpm for 1 ml of supernatant
Tube A-3. ________________ cpm for 1 ml of supernatant

Carefully remove the balance of the supernatant solution from each tube and wash the precipitates once with 1 ml of water. Transfer the precipitates quantitatively to planchets and dry them. Prevent loss of the dry precipitate by applying a drop of diluted collodion. Finally, evaporate all solvents thoroughly under the infrared lamp.

(Quantitative transfer of small volume precipitates can be made as water suspensions with the help of a dropper or pipette and a rubber bulb. Add several drops of water to the precipitate. Mix well to suspend the precipitate, using the tip of the long slender dropper. Using the same dropper, draw up the liquid and transfer it to the planchet. Repeat at least once, using several drops of fresh water. If necessary, repeat the process several times to effect quantitative transfer. The liquid is then evaporated.)

Elapsed time (hours)	Tube (A-1, A-2, or A-3)	Activity Alpha Plateau	Activity Beta Plateau	Alpha Activity	Beta Activity

The activity of each precipitate is measured in the proportional counter on both the alpha and beta plateaus. Repeat the measurement of activity in 1 day, 2 days, 4 days, and 8 days.

Method 2. *Precipitation of Radium-E as a Basic Salt*

To each of three test tubes (3-ml size), each containing 1½ ml of water, add 100 lambdas of Ra-DEF solution. Treat each as follows:

Tube B-1. Add nothing—"carrier-free"
Tube B-2. Add 3 drops of bismuth nitrate solution
Tube B-3. Add 3 drops of bismuth nitrate solution and 3 drops of lead nitrate or lead acetate solution

To each tube add NaOH solution dropwise until no further precipitate forms, adding exactly the same amount to each tube. Mix and centrifuge. Transfer 1 ml of supernatant liquid from each tube to a planchet and measure the activity of each solution with an end-window G-M tube.

Tube B-1. ____________ cpm for 1 ml of supernatant
Tube B-2. ____________ cpm for 1 ml of supernatant
Tube B-3. ____________ cpm for ml of supernatant (SAVE)

To the planchet containing the supernatant solution from tube B-3 (from part B of the experiment), add very carefully an excess of conc. HCl to neutralize the NaOH. Then, dry the solution under the infrared lamp until no excess HCl remains. Measure the activity of this sample each day for several days in the same manner used to calculate the half-life.

Elapsed Time	Background cpm	Ra-DEF Standard cpm	Unknown cpm	Unknown Corrected cpm

CALCULATIONS AND REPORT:

1. Plot the activities of the sample on linear graph paper.
2. Identify the nuclide or nuclides responsible for each of the observed activity curves plotted in step 1. Refer to the decay scheme in experiment 5.3.
3. For each precipitation conducted above, determine the fate of each component (Ra-D, Ra-E and Ra-F), indicating the extent to which it was precipitated or left in solution.
4. Explain the behavior of each component on the basis of the precipitation rules.

Tube	Ra-D		Ra-E		Ra-F	
	Ppt.	Filt.	Ppt.	Filt.	Ppt.	Filt.
A-1						
A-2						
A-3						
B-1						
B-2						
B-3						

Key: 0 = none; x = trace; xx = moderate; xxx = most or all

PART B—PREPARATION OF A CARRIER-FREE ISOTOPE BY CO-PRECIPITATION

The separation of radiophosphorus and radioiodine was accomplished in experiment 10.1 by simple precipitation. The separation was efficient, but the radioactive substances were no longer carrier-free. In this experiment separation will be effected by coprecipitation without resort to a carrier.

Place 500 lambdas each of carrier-free ^{32}P solution (about 0.1μCi/ml) and carrier-free ^{131}I solution (about 0.1μCi/ml) in a 5-ml centrifuge tube. Rinse the micropipettes with dilute HNO_3 and add these rinsings to the centrifuge tube. Measure the activity in a well-type scintillation counter. This is solution (A).

Prepare two standards by placing 500 lambdas of each carrier-free solution into separate centrifuge tubes. Measure the activity of each in the well counter.

Prepare two additional standards by placing 500 lambdas of each carrier-free solution into separate planchets. Add a drop of iodine carrier to the ^{131}I standard to fix the activity and dry both standards under the infrared lamp. Measure the activity of each. Use a scintillation detector for measuring the ^{131}I activity and an end-window G-M tube with a 200 mg/cm^2 absorber between the source and the tube for measuring the ^{32}P activity.

To the mixed sample, solution (A), in the centrifuge tube, add one drop of ferric chloride solution and sufficient ammonium hydroxide to precipitate the ferric hydroxide. Stir the sample. The phosphate tracer will coprecipitate with the $Fe(OH)_3$, and the iodine, though initially oxidized to free iodine by the ferric ion, will remain in solution. Centrifuge the sample and transfer the supernatant liquid to another 5-ml centrifuge tube. Measure the activities of the precipitate (B) and the supernatant (C) in the well counter. Where is the ^{131}I? (In ____________________).

To the supernatant (C) add one drop of ferric chloride solution and sufficient ammonium hydroxide to precipitate the ferric hydroxide. Mix well, centrifuge and transfer the supernatant to a clean 5-ml centrifuge tube. Measure the activities of the precipitate (D) and the supernatant (E).

Transfer supernatant (E) to a second plastic planchet and evaporate to dryness. Measure the activity of this residue (1) with the scintillation detector and also, through a 200 mg/cm^2, with an end-window tube for phosphorus.

Combine the ferric hydroxide precipitates (B) and (D). Dissolve the precipitates in a minimum amount of 5% hydrochloric acid and reprecipitate the $Fe(OH)_3$ by the addition of ammonium hydroxide. Centrifuge and remove the supernatant. Measure the activities of the precipitate (F) and the supernatant (G) in the well counter. Transfer the supernatant (G) to a planchet, evaporate and measure the activities of the residue (2) with the scintillation detector and, through a 200-mg/cm^2 absorber, with an end-window tube for phosphorus.

Dissolve the precipitate (F) in 500 lambdas of 10*M* hydrochloric acid and add 500 lambdas of isopropyl ether which has been saturated with 10*M* hydrochloric acid. Shake well; centrifuge if necessary to separate the layers. Transfer the isopropyl ether layer to a second plastic planchet. Repeat the extraction three times. (Iron should be extracted by the isopropyl ether, leaving the carrier-free ^{32}P activity in the aqueous phase).

Combine the three extracts and evaporate to dryness in the plastic planchet. Measure the activities of this residue (3) for iodine and phosphorus.

Measure the activity of the aqueous phase (H) in the well counter. Then, transfer quantitatively to a plastic planchet. Dry and measure the activities of this residue (4) for iodine and phosphorus.

DATA:

Measurements with well-type scintillation counter:

	Activity cpm	Estimated % iodine
Standards — Phosphorus		xxxx
Iodine		100
Original solution (A)		
Precipitate (B)		
Supernatant (C)		
Precipitate (D)		
Supernatant (E)		
Precipitate (F)		
Supernatant (G)		
Aqueous phase (H)		

Measurements on Planchets:

Sample	Phosphorus-32*		Iodine-131**	
	cpm	%	cpm	%
(1) Evaporated supernatant (E)				
(2) Evaporated supernatant (G)				
(3) Isopropyl ether extracts				
(4) Evaporated aqueous phase				
Standards		100		100

*Phosphorus-32 activities are measured with a 200 mg/cm² absorber between sample and end-window G-M tube.

**Iodine-131 activities are measured with a scintillation detector.

EXPERIMENT 10.3 SOLVENT EXTRACTION

OBJECTIVES:

To demonstrate the separation of a substance from a mixture for the purpose of purification or concentration.

To illustrate the use of a radioisotope for the determination of a physical constant—the partition coefficient.

To determine the partition coefficient of iodine for the system chloroform-water.

THEORY:

As early as 1872, M. Berthelot noted that if to a system consisting of two immiscible solvents is added a third component (a solute) which is soluble in both, then the solute will distribute itself between the two solvents in a definite manner. The law was clearly formulated by Nernst in 1891. If C_1 and C_2 represent the concentrations of the solute, respectively, in the two solvents, then

$$C_1/C_2 = K \text{ (a constant)} \qquad (10.1)$$

This is known as the *partition law* or *distribution law* and the constant K is the *partition coefficient* or the *distribution coefficient*.

Justification for equation 10.1 may be demonstrated by reference to Henry's law, a form of the partition law. Henry's law refers to the solubility of a gas or volatile substance in a solvent and relates the concentration C of the gas in the solvent to the vapor pressure of the gas

$$C = k\,p \tag{10.2}$$

If one considers a closed system (figure 10.5) consisting of two chambers containing two different solvents, and connected in such a way that the vapor or gas is common to both, then two equilibria can be proposed,

$$C_1 = k_1 p \tag{10.3}$$

and

$$C_2 = k_2 p \tag{10.4}$$

where C_1 and C_2 are the concentrations of the vapor in the two solvents, p is the pressure of the vapor and is the same for both systems, and k_1 and k_2 are constants. Dividing equation (10.3) by equation (10.4) gives the relationship:

$$C_1/C_2 = k_1/k_2 = K \tag{10.5}$$

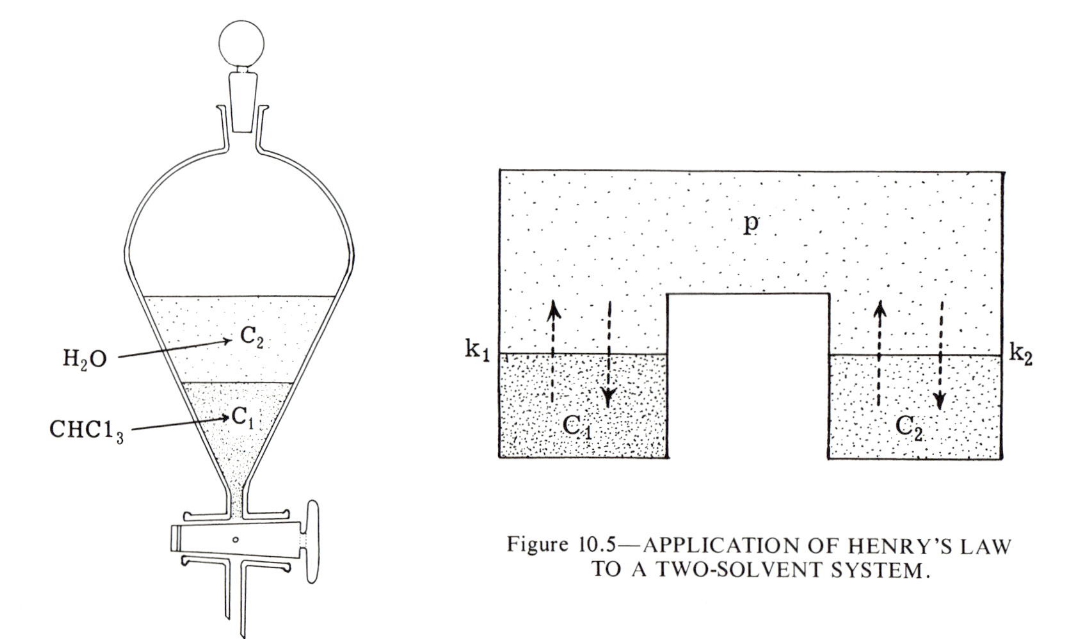

Figure 10.4—SOLVENT EXTRACTION. The ratio C_1/C_2 is constant for a given solute and solvent system.

Figure 10.5—APPLICATION OF HENRY'S LAW TO A TWO-SOLVENT SYSTEM.

This relationship (10.5) is identical to (10.1).

If, now, the vapor phase is removed from between the two liquid phases so that they come into contact, the same relationship is applicable and, in fact, becomes even broader in scope, for now the solute need not be volatile.

If the solute is *associated* or *dissociated* in either or both of the liquid phases, the relationship (10.1) or (10.5) must be modified. If α_1 and α_2 represent the fraction of the total solute associated or dissociated in each of the solvents, then

$$\frac{C_1(1-\alpha_1)}{C_2(1-\alpha_2)} = K \qquad (10.6)$$

Example: 10 ml of an immiscible solvent (solvent 1) are equilibrated with 50 ml of a solution (solvent 2) containing 0.5 g of solute. The partition coefficient is 20, the solute being more soluble in solvent 1 than in solvent 2. Calculate the weight of solute in each solvent at equilibrium. Assume that no association or dissociation occurs.

$$C_1/C_2 = K = 20$$

Let X = weight of solute remaining in solvent 2.

$0.5 - X$ = weight of solute extracted by solvent 1.

Then $C_1 = (0.5 - X)/10$ g/ml *

$C_2 = X/50$ g/ml *

$$\frac{(0.5 - X)/10}{X/50} = 20$$

$$X = 0.1\ \text{g}$$

$$0.5 - X = 0.4\ \text{g}$$

Thus, $C_1 = 0.4/10 = 0.04$ g/ml *

$C_2 = 0.1/50 = 0.002$ g/ml *

and $C_1/C_2 = 0.04/0.002 = 20 = K$

PROCEDURE:

Stock Iodine Solution—Dilute an aliquot of radioactive iodide solution with potassium iodide solution (10 mg KI per ml) so that the resulting solution contains an activity of about 10 microcuries per ml. Add an equal volume of chloroform to this solution and oxidize the iodide to free iodine by adding 5% ferric chloride solution, dropwise. Shake well to extract the iodine into the chloroform layer. Discard the aqueous phase into the radioactive waste container.

Determination of Partition Coefficient—Transfer 1 ml of the stock iodine solution in chloroform to a 3-ml glass stoppered test tube, or to a 2-ml volumetric flask. The latter is available in the shape of a test tube and is convenient to use. Add 1 ml of distilled water, stopper and shake thoroughly. Allow the layers to separate, centrifuging if necessary, remove and discard the aqueous phase into the radioactive waste container. Wash again two or three times to remove traces of unreduced potassium iodide or unreacted ferric chloride which might otherwise cause erroneous results.

* Although molarity is normally used to express concentration, other concentration units may be used if the constant K is dimensionless; that is, if concentration units cancel out.

Now, add 1 ml of water, equilibrate the chloroform-iodine-131 solution by shaking thoroughly and allow the layers to separate. It is imperative that both phases be clear and free of any indication of emulsification. Centrifuge if necessary to clear the layers.

Measurement of Radioactivity—

1. *By means of a G-M counter*—Transfer 0.5 ml of the aqueous phase to a planchet, add 0.5 ml of thiosulfate solution to convert the iodine to iodide and dry under the infrared lamp.

Carefully remove the balance of the aqueous phase from the test tube and discard it into the waste receptacle indicated. Transfer 0.5 ml of the chloroform layer to another test tube and add 0.5 ml of $Na_2S_2O_3$ solution. Shake thoroughly so the free iodine in the chloroform layer will be extracted completely as iodide by the aqueous phase. Centrifuge, if necessary, to separate the layers and measure an aliquot of the aqueous layer into a planchet containing a few drops of water to distribute the sample evenly over the bottom. Dry under the infrared lamp.

Determine the activity of each sample with a G-M counter and calculate the partition coefficient. Correct for resolving time if necessary.

2. *By means of a well scintillation counter*—Transfer 0.5 ml of the aqueous phase to a test tube designed for the well counter and measure the activity.

Transfer 100 μl of the chloroform layer to a 5-ml volumetric flask and make up to volume with a solution containing 1% KI and 1% Na_2SO_3. Shake until the chloroform is dissolved and the solution is colorless. Transfer 0.5 ml of this solution to a test tube and measure the activity in a well counter.

DATA AND CALCULATIONS:

Background	______	cpm	
Activity of aliquot of $CHCl_3$ phase (observed)	______	cpm	
(corrected)	______	cpm	
Activity of 1 ml of $CHCl_3$ phase (calculated)	______	cpm	(A)
Activity of aliquot of aqueous phase (observed)	______	cpm	
(corrected)	______	cpm	
Activity of 1 ml of aqueous phase (calculated)	______	cpm	(B)

Partition coefficient = (A)/(B) ______

EXPERIMENT 10.4 CHROMATOGRAPHY

OBJECTIVES:

To present the basic principles of chromatography and to illustrate the varieties of chromatographic procedures in current use, with particular emphasis on paper chromatography.

To demonstrate the significance of the term "Radio-chemical purity."

To conduct a chromatographic analysis for measuring the radio-chemical purity of ^{131}I by the U. S. P. method.

THEORY:

Chromatography is primarily a technique for the separation of molecular mixtures of essentially similar compounds which takes advantage of differences in their distribution (or partition) coefficients. The various chromatographic processes are named according to the nature of the phases.

<table>
<tr><th colspan="2" rowspan="2"></th><th colspan="2">Mobile Phase</th><th rowspan="2">Laws which are obeyed</th></tr>
<tr><th>Gas</th><th>Liquid</th></tr>
<tr><td rowspan="3">Stationary Phase</td><td>Liquid (Partition)</td><td>Gas-liquid Chromatography (GLC)</td><td>Paper Chromatography
----
Column-partition Chromatography</td><td>Henry & Nernst</td></tr>
<tr><td>Solid (Adsorption)</td><td>Gas-solid Chromatography (GSC)</td><td>Column-adsorption Chromatography
----
Thin-layer Chromatography (TLC)</td><td>Freundlich & Langmuir</td></tr>
<tr><td>Solid (Exchange)</td><td></td><td>Ion Exchange Chromatography</td><td>Donnan & Mass Action</td></tr>
</table>

The term *adsorption chromatography* implies that one of the phases is essentially a surface phase and competes for the "solute" by a process of *adsorption* rather than dissolution, while the second phase is a liquid in which the "solute" exhibits a solubility within a specific range. The term partition chromatography implies that both phases are liquids or that one is a gas and the other is a liquid.

In all chromatographic procedures, one phase moves relative to the second phase, the second usually being designated as the *stationary phase* and the first as the *moving phase.*

In *column adsorption chromatography,* the stationary solid phase (starch, talc, alumina, etc.) is packed into a glass column (see figure 10.6) while the liquid, moving phase, in which are dissolved the substances to be separated, is allowed to flow slowly downward through the column. Those substances having a greater affinity for the solid phase move more slowly, those having a greater affinity for the liquid phase move more rapidly and appear in the first fractions of eluate to be collected in the receiving vessel. In *column partition chromatography,* the solid which fills the column is first wetted with a suitable

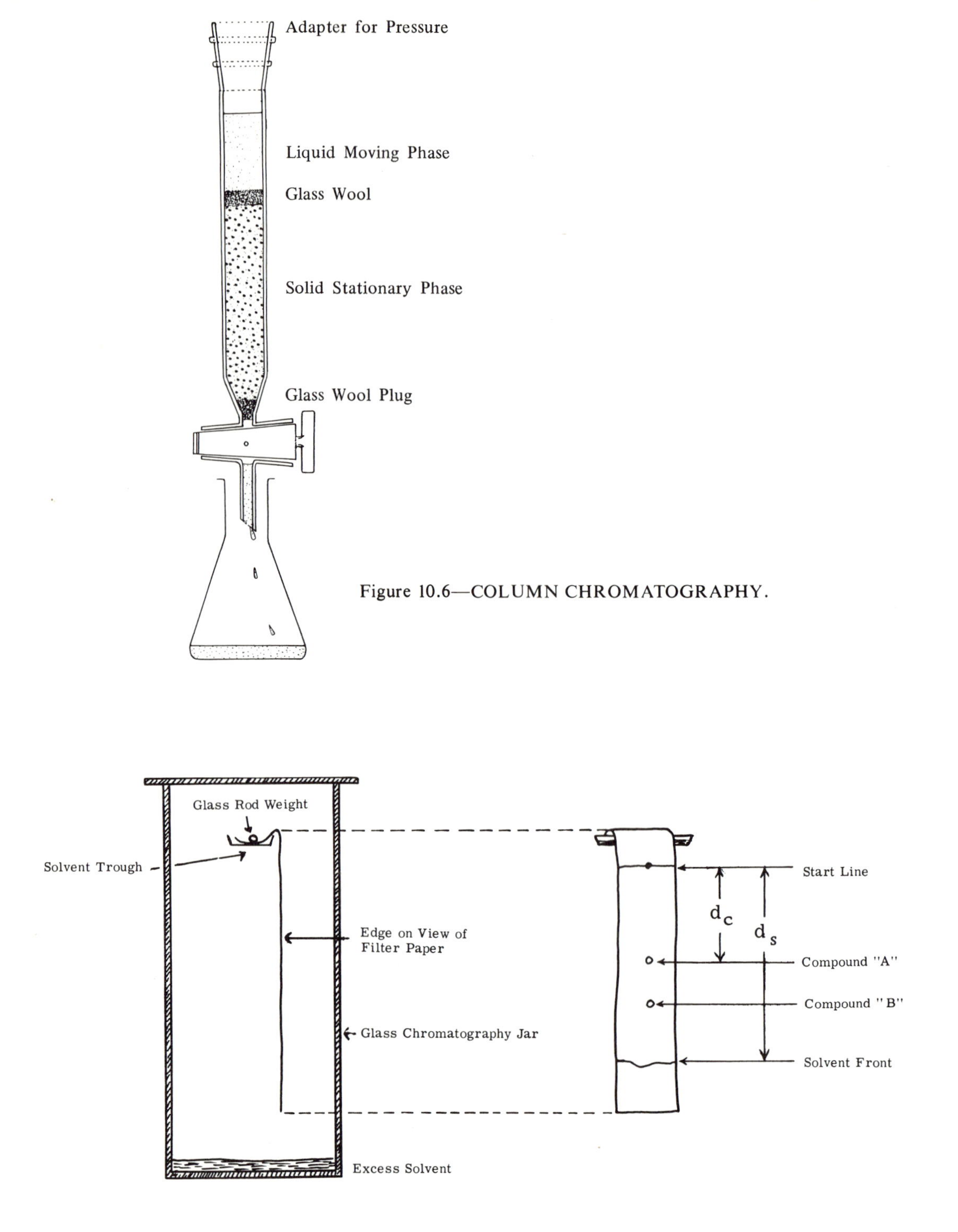

Figure 10.6—COLUMN CHROMATOGRAPHY.

Figure 10.7—DESCENDING PAPER CHROMATOGRAPHY.

liquid. This liquid becomes the stationary liquid phase. It must be essentially insoluble in the moving liquid phase and both liquids must be "equilibrated" with each other before use by previous intimate contact. Separation of the solutes present in this case is a function of their relative solubilities in the two liquids.

In some ways *paper chromatography* is similar to column chromatography. A piece of filter paper replaces the packed column. *Descending paper chromatography* is illustrated in figure 10.7. In this procedure a light pencil line (the *starting line*) is made near one end of a strip of filter paper. A small quantity of a solution of the compounds to be separated is "spotted" on the start line and allowed to evaporate. The paper strip is then hung in a tall glass chromatography jar where it is supported at the upper end in a solvent trough. A thick glass rod is used as a weight to prevent the paper from slipping out of the trough. Solvent is poured into the trough and a small amount of solvent is poured into the jar. The jar is sealed to prevent solvent evaporation. Attention is paid to the progress of the *solvent front* as it passes down the filter paper strip, the process usually being interrupted before the solvent reaches the end of the paper. It is necessary to remove the paper strip and immediately mark the position of the solvent front before the solvent evaporates. The paper is then hung up to dry.

If the proper choice of solvents is made, the compounds will occupy different positions along the strip. If the compounds are colored they will be readily visible. If they are colorless, their positions can usually be determined by examination of the strip under an ultra violet lamp or by spraying the strip with reagents which will react with the compounds to produce colored products.

The R_f value is useful for the identification of a compound. It is defined as the ratio of the distance d_c traveled by the compound to the distance d_s traveled by the solvent front. That is:

$$R_f = d_c/d_s \tag{10.7}$$

In *ascending paper chromatography,* the solvent is placed in the bottom of the jar and the paper strip is supported so that its lower edge dips into the solvent. In this case, the starting line is situated near the bottom, rather than at the top of the strip. Simple apparatus suitable for ascending paper chromatographic procedures is illustrated in figure 10.9.

The principle involved in chromatography is readily understood by analogy with a liquid-liquid extraction process. (See experiment 10.3). Let the aqueous phase (figure 10.4) represent the stationary phase, which was used to saturate the column packing or the filter paper, and let the chloroform be the moving phase. Let it further be supposed that a mixture of substances A and B is to be separated and that the respective partition coefficients K_A and K_B are equal to 1 and 4, respectively. Then,

$$K_A = \frac{C_s}{C_m} = 1 \quad \text{and} \quad K_B = \frac{C_s}{C_m} = 4$$

By countercurrent extraction, B moves less rapidly than A and gradually becomes separated from A. (See figure 10.8). This multiple extraction process, if carried out 50, 100 or 200 times, constitutes the countercurrent extraction procedure which has been extensively studied by Craig and Craig. In chromatography, the number of extractions accomplished is theoretically in the thousands.

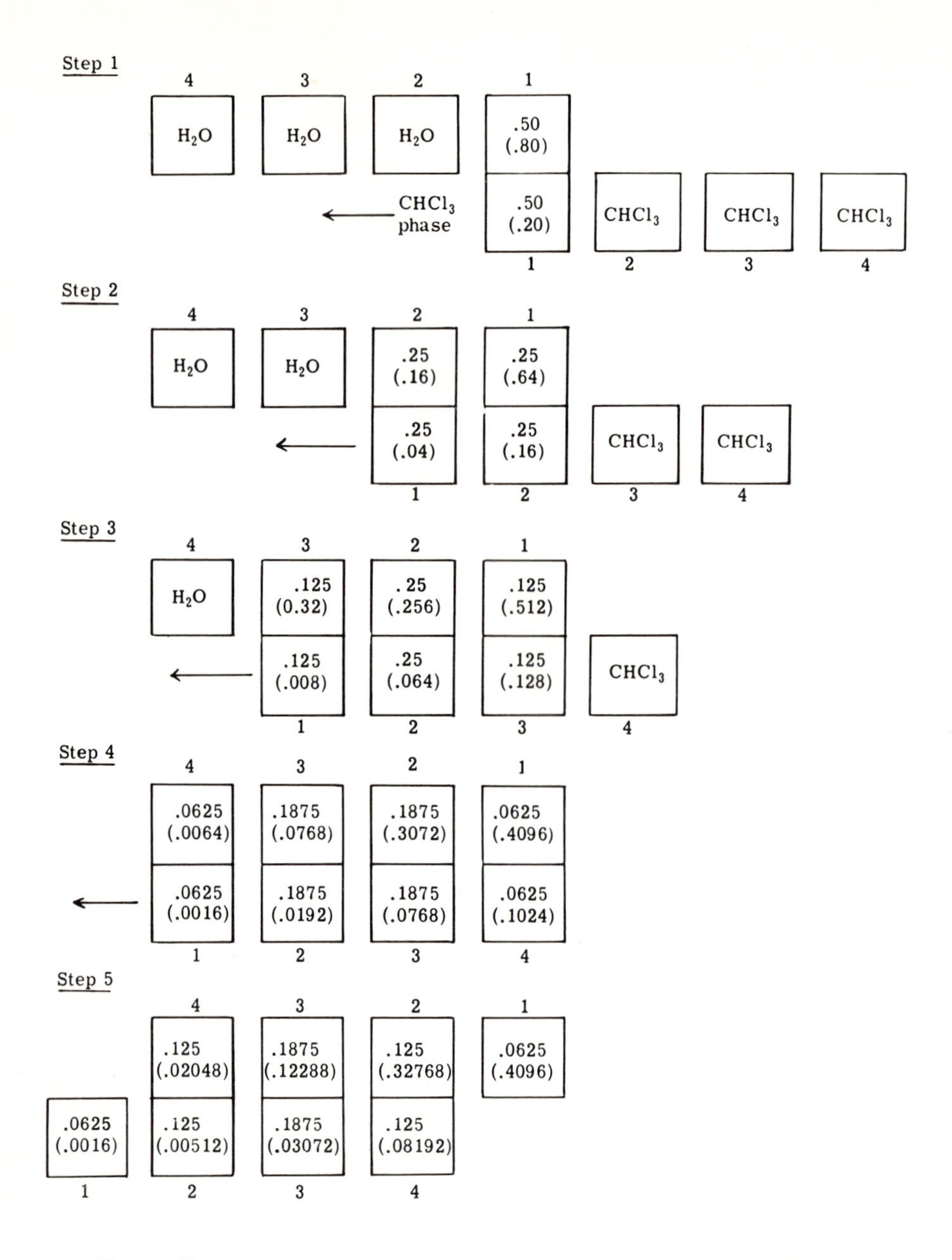

Figure 10.8—COUNTERCURRENT EXTRACTION OF TWO COMPOUNDS. The partition coefficient for compound A = 1 and for compound B = 4. The fraction of A in each phase is given in plain numbers, that of B is in parentheses. Complete equilibration is assumed for each extraction.

The location and relative amounts of radioactivity along the length of a chromatogram may be determined by one of several methods:

1. the dry strip is placed in contact with a photographic film to produce an autoradiogram of the activity,

2. the strip is cut into 1-cm pieces and the activity of each piece is measured by placing it beneath a G-M counter, a scintillation counter or in a liquid scintillation counter,
3. the intact strip is moved slowly and at constant speed past a detector, the output of which passes through a ratemeter to a recorder where a permanent record of the activity is made.

Equipment for the latter method is illustrated in figures 10.10 and 10.11. Choice of the detector will depend upon the type of activity on the strip.

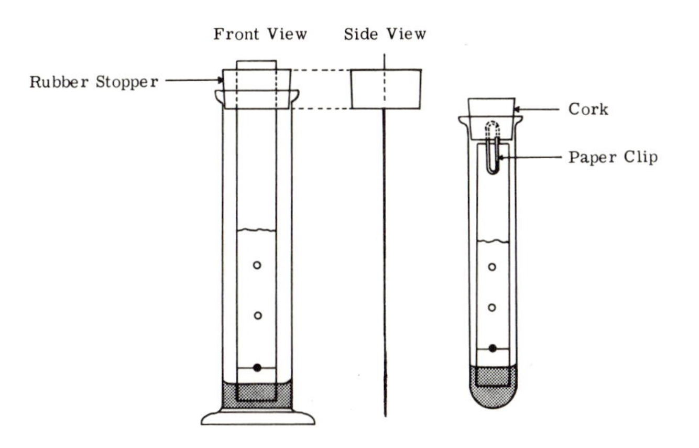

Figure 10.9—APPARATUS FOR ASCENDING PAPER CHROMATOGRAPHY.

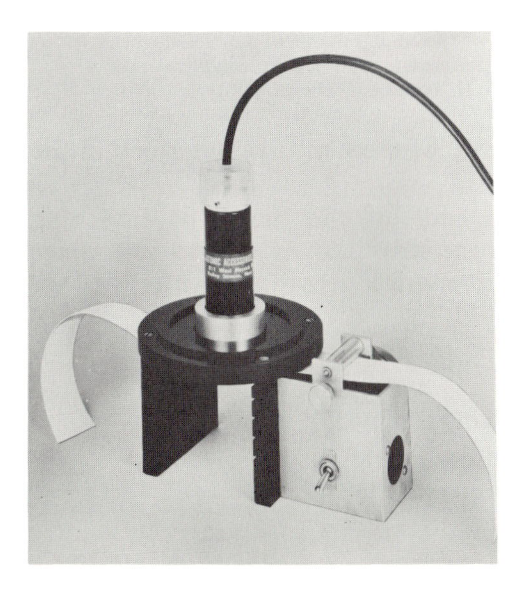

Figure 10.10—EDUCATIONAL RADIOCHROMATOGRAM SCANNER. (*Atomic Accessories, Inc.*)

Figure 10.11—WINDOWLESS-FLOW, 4-π RADIOCHROMATOGRAM SCANNER. (*Atomic Accessories, Inc.*)

PROCEDURE:

PART A—TEST FOR RADIOCHEMICAL PURITY OF SODIUM IODIDE (^{131}I) SOLUTION

In the monograph for Sodium Radio-iodide (^{131}I) Solution in the U. S. P., it is stated, "Iodine-131 activity as iodate must not exceed 5 per cent of the iodide activity. Other forms of radioactivity are absent."

As originally produced, either as a uranium fission product or by neutron bombardment of tellurium, the ^{131}I may exist in any of its oxidation states and may be mixed with any number of other elements and compounds. It must, therefore, undergo a purification process before it is suitably pure for use. However, even though the iodine may be removed successfully from other contaminants and be reduced essentially to iodide, it may still contain a substantial quantity of ^{131}I in the form of iodate. Needless to say, if iodate is used indiscriminately in place of iodide, consistent results of certain experiments and procedures will not be obtained. For this reason, the U. S. P. has placed the 5 per cent limit on iodate content.

The test is carried out by the ascending chromatographic technique with 75 per cent methanol as the developing liquid. The R_f of iodide is about 0.76, of iodate about 0.45.

The official procedure as described in the U. S. P. XVI is as follows:

"Place one drop of a solution containing 0.1 per cent of potassium iodide, 0.2 per cent of potassium iodate, and 1 per cent of sodium bicarbonate 25 mm from one end of a 25 × 300 mm strip of chromatographic paper and allow the paper to dry. To the same area add approximately 0.01 ml of a solution of Sodium Radio-iodide Solution, representing an activity of about 20,000 counts per minute per 0.01 ml and allow to dry. Develop the chromatogram over a period of about 4 hours by ascending chromatography, using 75 per cent methanol. Dry the chromatogram in air and determine the radioactivity of succeeding 1-cm lengths, using a shielded counter (at least 1.6 mm of brass or the equivalent) with an opening in the shield having a width of 10 mm and a length of 25 mm. The total activity of the iodate band does not exceed 5 per cent of the total activity of the iodide band. No other radioactive bands are present. The R_f values for the iodate and iodide bands fall within plus or minus 5 per cent of the values found for iodine-131 sample of known purity when both are determined under parallel conditions."

Prepare a chromatogram and measure the activity of the bands as described in the official procedure.

Using a similar chromatogram, extend both ends of the paper by attaching additional strips of paper with cellulose tape. Feed the strip into an automatic scanner of the type illustrated in figure 10.10 or 10.11 and prepare a record of the activity on the chart.

Calculate the area under the peaks on the chart for both iodide and iodate. Calculate the percentage of iodate.

The spots can be located and identified by spraying to develop a color.

a. For iodide, spray with H_2O_2 and follow with starch solution.
b. For iodate, spray with ascorbic acid and starch solutions.

PART B—SEPARATION OF RADIUM-DEF

A mixture of radium-D (^{210}Pb), radium-E (^{210}Bi) and radium-F (^{210}Po) is readily separated by ascending chromatography. The apparatus may consist simply of a hydrometer jar or a test tube as depicted in figure 10.9.

Whatman #1 filter paper has been used successfully. The Ra-DEF solution is applied to the start line in 10 μl quantities, allowing each to dry before applying the next, until about 0.05 μCi of activity has been applied.

The chromatogram is developed by use of butanol saturated with 1*N* HCl. This developing solution is prepared by shaking butanol and 1*N* HCl together in a separatory funnel for about 5 minutes. The pases are allowed to separate and the aqueous phase is discarded.

Development of the chromatogram will take several hours. The anticipated R_f values are approximately 0.1 for ^{210}Pb, 0.6 for ^{210}Bi and 0.8 for ^{210}Po.

EXPERIMENT 10.5 ION EXCHANGE

OBJECTIVES:

To demonstrate the application of ion exchange methods to the separation and purification of elements in small amounts.

THEORY:*

Ion exchange has attained considerable importance as a tool in both fundamental and industrial chemistry. The method has been of extreme importance in the separation of radioelements and especially so in the separation of the rare earths and the actinides. *Ion exchange* is the reversible interchange of ions between a liquid phase (solution) and a solid material which does not involve a substantial change in the structure of the solid. By proper manipulation of the exchange process, separation of the desired ions can be effected. Once the ions are held on the column of the exchanger they can be removed by passage of a specially prepared solution, called the *eluant,* through the column.

The solid material is a polymerized, high molecular weight, insoluble electrolyte called the *"ion exchange resin."* This electrolyte usually has one very large, heavy ion and an oppositely charged, small, simple ion which can be "exchanged" for ions in the liquid in contact with the resin. Both natural and synthetic materials are available as ion exchange resins. Ion exchange resins are also classified as either cation exchange resins or as anion exchange resins, depending upon their mode of action.

Cation exchange resins consist of small, simple cations which can be exchanged and large, high molecular weight anions. Resins of this type include sulfonic acid derivatives, carboxylic acid derivatives and silicates. A typical cation exchange reaction is

$$\text{Resin-H} + \text{NaCl} \longrightarrow \text{Resin-Na} + \text{HCl} \qquad (10.8)$$

* Based on lecture notes of Dr. Louis A. Reber, Professor of Chemistry, Philadelphia College of Pharmacy and Science.

Anion exchange resins consist of small, simple anions which can be exchanged and large, high molecular weight cations. Those of the quaternary ammonium compounds are strong bases; those of the polyamine type are weak bases. A typical anion exchange reaction is

$$\text{Resin-OH} + \text{NaCl} \longrightarrow \text{Resin-Cl} + \text{NaOH} \quad (10.9)$$

Ion exchange equilibria should obey the mass action law. Thus, for the exchange reaction

$$RX + C^+ = RC + X^+ \quad (10.10)$$

where R represents the resin and X^+ and C^+ the exchangeable ions. The equilibrium expression, assuming all activity coefficients to equal unity, is

$$K = \frac{[RC]\ [X^+]}{[RX]\ [C^+]} \quad (10.11)$$

and the relative concentrations of the two ions on the exchanger is expressed by

$$\frac{[RC]}{[RX]} = K\,\frac{[C^+]}{[X^+]} \quad (10.12)$$

From this, it is seen that the relative quantities of the two ions associated with the resin depend both on the equilibrium constant K and on the concentrations of the ions in solution.

Selected applications of ion exchange include:

1. Water softening

$$2\ \text{Resin-Na} + CaCl_2 \longrightarrow (\text{Resin})_2\text{-Ca} + 2\ \text{NaCl}$$

2. De-ionization (de-mineralization). The solution is first passed through a cation exchanger

$$\text{Resin-H} + \text{NaCl} \longrightarrow \text{Resin-Na} + \text{HCl}$$

The effluent from the cation exchanger is then passed through an anion exchange column

$$\text{Resin-OH} + \text{HCl} \longrightarrow \text{Resin-Cl} + H_2O$$

In mixed bed de-ionizers, a cationic resin (Resin-H) and anionic resin (Resin-OH) are mixed together in a single column. De-ionization is thus accomplished in a single pass of the solution.

3. Isolation and concentration of electrolytes. If ions are in dilute solution, the solution may first be shaken with a small amount of resin to concentrate it and then eluted with a suitable solvent.

4. Chromatography by ion exchange.

Ion Exchange Chromatography—A solution containing the ions to be separated is passed through a column containing a suitable ion-exchange resin. In the usual case, all the ions of similar charge (e.g. cations) enter into ionic exchange and are held back on the column. The next step is to bring about a separation of these ions. This may be accomplished by the use of a suitable wash liquid (eluant) which will permit a repeated exchange of the ions between the resin and the flowing liquid. In this way, the effect of small differences in the affinity of the resin for different ions is multiplied until the ions become localized at different levels on the resin column, or are washed out of the column in sequence, completely separated from each other.

The *eluant* may be a buffer solution whose pH is made to change constantly by the addition of acid or alkali, as used successfully in amino acid separations; or the eluant may be a complexing agent, such as the citrate or glycolate buffers used in the separation of the rare earths and other fission products, and in the isolation of the transuranium elements.

The affinity between the resin and different ions shows certain regularities which permit prediction of ion behavior up to a point. For example,

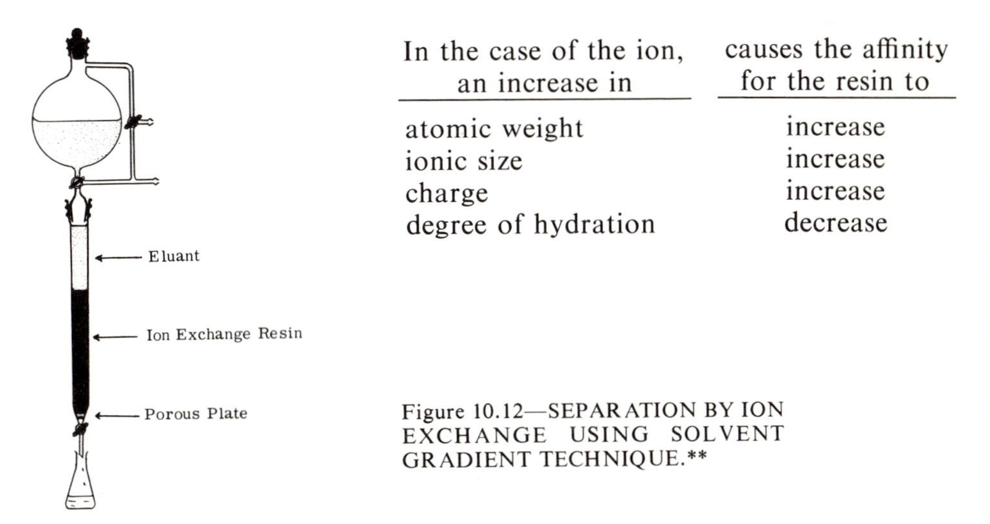

In the case of the ion, an increase in	causes the affinity for the resin to
atomic weight	increase
ionic size	increase
charge	increase
degree of hydration	decrease

Figure 10.12—SEPARATION BY ION EXCHANGE USING SOLVENT GRADIENT TECHNIQUE.**

Effects of pore size and cross-linkage—Resins may be considered as having a porous or sponge-like structure which ions must penetrate in order to be exchanged. Pore size is controlled by the degree of cross-linkage in essentially linear polymers. For example, in resins (Dowex) which are co-polymers of styrene and divinylbenzene, the greater the percentage of divinylbenzene, the greater the amount of cross-linkage, the less permeable the resin and the more difficult is the exchange of large ions. Thus, a certain degree of selectivity and separation is possible on the basis of ion-size.

Weak (-COOH) vs. strong (-SO_3H) acid resins—The hydrogen ions of a resin acid must not be held too firmly by the acidic group if they are to be exchanged for other ions. In other words, the equilibrium

$$\text{R-COOH} \rightleftharpoons \text{RCOO}^- + \text{H}^+$$

**A column similar to that illustrated is available from California Laboratory Equipment Co., 98 Rincon Road, Berkeley 7, Calif.

must be displaced to the right in order for ionic exchange to take place. If RCOOH is a weak acid, this will occur only in solutions of relatively high pH, and the resin will be ineffective in acid media. Such a resin will not enter into exchange with the cation of a weak base and may be useful in separating a weak base from a strong base.

On the other hand, sulfonic acid resins are of the strong acid type and usually function over a wide pH range, entering into exchange with the cations of both weak and strong bases. Therefore, they are useful for the removal of all cations from a solution rather than for the selective removal (or separation) of such ions.

Weak (polyamines) vs. strong (quaternary) basic resins—By similar reasoning, it can be shown that an anion exchange resin must not attract hydroxyl ions too strongly if these ions are to be exchanged for other anions. This means the equilibrium

$$\text{R-OH} \rightleftharpoons \text{R}^+ + \text{OH}^-$$

must be displaced to the right in order for ionic exchange to take place. If the resin is a weak base, it will require relatively low pH (acid) solutions to function. Such a resin will not enter into exchange with the anion of a weak acid and may be useful in separating a weak acid from a strong acid.

Anion exchange resins of the quaternary ammonium type are strong bases and usually function over a wide pH range, entering into exchange with anions of both weak and strong acids. Therefore, they are useful for the removal of all anions from a solution rather than for the selective removal (or separation) of such ions.

*Separation of Cobalt and Nickel by Ion Exchange**

In the laboratory, a separation of cobalt and nickel is effected by ion exchange. This separation is based on the following considerations. In the presence of high concentrations of chloride ion, the following equilibrium is set up and displaced far to the right:

$$\text{Co}^{++} + 4\,\text{Cl}^- \rightleftharpoons [\text{CoCl}_4]^{-2}$$

Nickel does not form a similar complex anion. Therefore, when a mixture of cobalt ions and nickel ions is applied to an anion exchange resin, the nickel passes through, while the cobalt (as $[CoCl_4]^{-2}$) enters into ion exchange and is held back as a colored ring on the column. After all the nickel has been eluted with 8*N* HCl, the column is then eluted with water. The addition of water decreases the chloride ion concentration and displaces the above equilibrium toward the formation of simple Co^{++} ions in the water phase. This procedure therefore elutes the cobalt.

MATERIALS REQUIRED:

Micro ion-exchange column (see figure 10.13) filled with Dowex 1-X8 resin (an anion exchange resin) in the chloride form, medium porosity, 100-200 mesh; dropper bottle of 1% dimethylglyoxime in alcohol; dropper bottle of concentrated NH_4OH;

*This procedure is based on a similar one in reference 100.

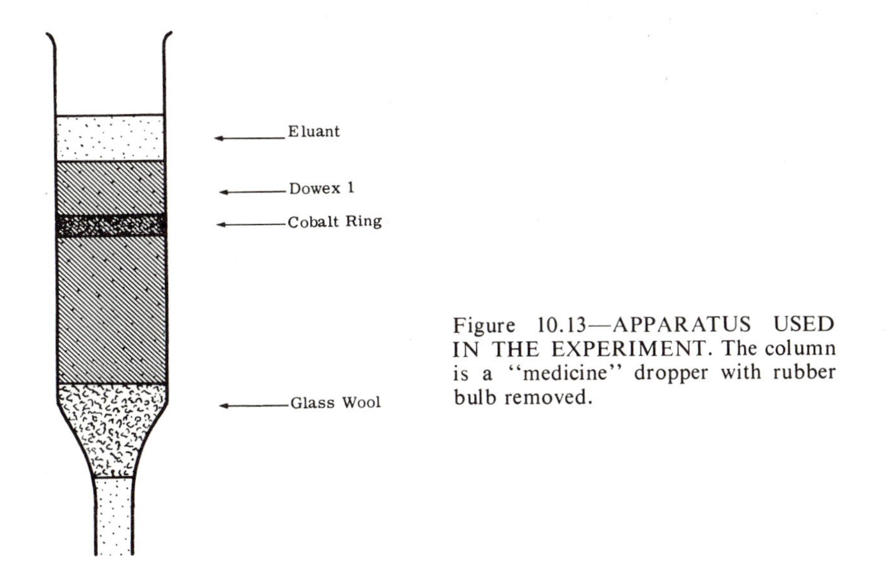

Figure 10.13—APPARATUS USED IN THE EXPERIMENT. The column is a "medicine" dropper with rubber bulb removed.

Ni-^{60}Co solution containing approximately one mg of each element per ml in 8*N* HCl, with approximately 10^6 dpm of ^{60}Co activity per ml; watch crystals—1 inch in diameter; planchets.

PROCEDURE:

The micro ion-exchange columns should be prepared as follows: Cover the Dowex-1 resin with water and allow it to stand until it is fully expanded. This avoids breaking of the column, which may occur if the dry resin is used. Pour the suspension of the resin into the column to form a bed about 3 cm high. Then, treat this column with 8*N* HCl.

Remove any solution remaining above the resin.

Add 20 lambdas of a solution containing a mixture of ^{60}Co and Ni directly above the resin. Let this run through and elute with 8*N* HCl.

Test the effluent for Ni by the following method: Collect each group of 1 or 2 consecutive drops in a separate watch glass (do *not* use planchets) and add to each watch glass, 2 drops of dimethylglyoxime and NH_4OH to excess. Number the watch glasses and observe which watch glass shows the deepest red color. Continue until a treated sample is obtained which is colorless.

When the effluent no longer gives an indication of Ni, pipette off the liquid above the resin. Elute the Co with water. Count the drops coming from the column. When the Co band is approximately at the middle of the column, start collecting each group of 1 or 2 drops in a separate planchet. Number the planchets in sequence. Dry each planchet and make a count with an end-window tube of the G-M type.

Plot an elution curve of the activity of each planchet versus its ordinal number.

EXPERIMENT 10.6 GAS CHROMATOGRAPHY

OBJECTIVES:

To demonstrate the application of gas chromatography to the separation and purification of small amounts of isotopically-labeled substances.

THEORY:

Gas chromatography, abbreviated GC, includes all chromatographic procedures in which the moving phase is a gas or a vapor. The separation of components is achieved by selective partitioning between the gaseous phase and a stationary phase. *Gas-liquid chromatography,* abbreviated GLC, utilizes a liquid as the stationary phase. This liquid, supported on a finely divided, inert solid, acts as a solvent for the components of the gaseous phase. On the other hand, *gas-solid chromatography,* abbreviated GSC, utilizes an active solid as the stationary phase which selectively adsorbes components of the gaseous phase.

The stationary phase is packed into a *column.* A typical column consists of a ¼-inch diameter copper or aluminum tube from one to 10 or more feet in length. The column need not be straight or vertical but may be coiled to fit into a heated compartment.

Samples are introduced at the head of the column. This may be done by injecting the sample from a syringe, the needle of which has been inserted into the column through a septum. If the sample is a liquid the section of the column at the point of injection is often heated so the sample will be flash vaporized. The temperature at which the entire column is held will also be determined by the nature of the sample.

Carrier gas constitutes the major constituent of the moving phase. Carrier gas is propelled through the column by pressure and is used to sweep (or elute) the sample components through (or from) the column. The column effluent is gaseous, even for liquid samples.

In gas chromatography a detector and a recorder are integral parts of the system. Many types of detector are available to sense composition changes in a gaseous stream. One commonly used is the *thermal-conductivity detector* which senses differences in the conduction of heat through pure carrier gas and through the mixture of carrier gas and eluted components. The recorder plots these differences as a function of time. The result is a *chromatogram* or elution curve similar to that illustrated in figure 10.14.

The area under the elution curve of a particular component is approximately proportional to the weight per cent of that component. The *retention time* is characteristic of the nature of the component and depends on its volatility, solubility, polarity and other properties. As the retention time increases there is a general increase in peak width. This is caused by diffusion of the vapor molecules. Both retention time and peak width are related to the efficiency of a column. The efficiency is proportional to the number of theoretical plates N where,

$$N = 16 \frac{\text{Retention time}}{\text{Peak width}}$$

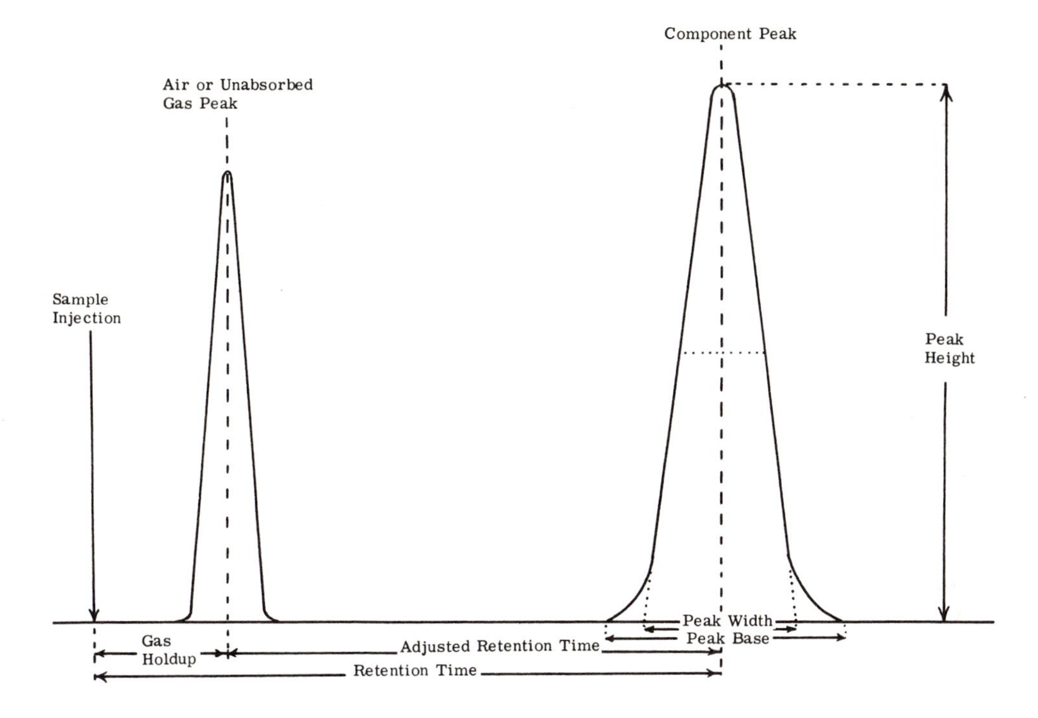

Figure 10.14—TYPICAL GAS CHROMATOGRAM.

This expression is useful for appraising the ability of a column to resolve or separate components of a mixture.

PROCEDURE:

In the following procedure ^{14}C-tagged sterols are prepared biosynthetically in a mouse, extracted from the tissues of the mouse and separated and purified by means of gas chromatography.

One ml of a solution containing 100 μCi of ^{14}C-2-sodium acetate is injected intraperitoneally into a 20-g mouse. After 24 hours the animal is sacrificed and the liver and gall bladder are removed. These tissues are finely minced by means of small scissors and placed into an Erlenmeyer flask containing 50 ml of 40% NaOH solution and 50 ml of methanol. The mixture is digested overnight at room temperature or for two hours with gentle heating to disintegrate the tissues and release the lipids. The sterols are extracted from this mixture with 5 successive 50-ml portions of ether. The ether extracts are combined and dried with 50 g of anhydrous $MgSO_4$. After shaking well to remove the water, the $MgSO_4$ is filtered off. The dry ether is evaporated at reduced pressure until the volume has been reduced to less than 10 ml. The solution of sterols is then transferred to a pre-weighed 10-ml volumetric flask and the ether is completely evaporated at 25° under a stream of nitrogen. The sterols are then dried to constant weight in a desiccator. Finally they are made up to a volume of 10 ml with toluene and stored at 0° until used.

An aliquot of the toluene stock solution containing between 0.01 and 1.0 mg of sterols is mixed with 1 ml of "Tri-sil"*. After shaking for 30 seconds and waiting for 5 minutes conversion of the sterols to the corresponding tri-methylsilyl ethers is complete. The boiling point of the silyl ethers is much lower than that of the original sterols. Separation at lower column temperatures is therefore possible with less danger of thermal decomposition of the components.

Any conventional type of GC equipment may be employed. While various columns may be used for the separation, a ¼-inch coiled stainless steel column, 8 feet long and packed with 6.6 g of succinate polyester of ethylene glycol on 20 g of 60 mesh Chromosorb is suitable. The column is operated at 220° C. Helium is used as the carrier gas at a pressure of about 30 psi and at a flow rate of about 75 ml per minute.

Samples are introduced by means of a micro-syringe into a heated injection chamber so that flash vaporization of the sample occurs. Provision should also be made to heat the exit tube from the detector to avoid condensation of the components before they arrive at a condensing trap where they are collected.

The percentage composition can be estimated by calculating the ratio of the area under each peak to the sum of the areas under all of the component peaks.

Each peak is identified by repeating the above procedure with known samples. Identification is made through a comparison of the retention times. Knowing the retention times, each component can be collected in a separate tube as it leaves the gas chromatograph. The percentage composition can then be checked by adding liquid scintillator to each collection tube and measuring the activity in a liquid scintillation counter.

EXPERIMENT 10.7 DISTILLATION

OBJECTIVES:

To demonstrate the use of distillation as a procedure for the separation and purification of a radioisotope.

To illustrate the effect of a carrier on the efficiency of such distillation.

To demonstrate the application of a radioactive standard for measuring the quantitative yield of a distillation of microgram quantities of radioisotopes.

THEORY:

If the vapor pressures of two substances are considerably different, it is frequently possible to separate and purify one or the other from a mixture of the two by simple distillation. This technique has found widespread use in the standard chemical laboratory,

* Hexamethyl-di-silazane and trichloro-methyl-silane in pyridine solution, product of Pierce Chemical Co., Post Office Box 117, Rockford, Illinois.

as have variations of the simple procedure, including fractional distillation, steam distillation, and distillation under reduced pressures.

Volatilization techniques are also applicable to the separation of radioactive substances, although limitations of the method are often encountered with "carrier-free" distillations. Typical separations by volatilization include the removal of ruthenium from fission products following its oxidation to the volatile ruthenium tetroxide.

Carbon-14 can frequently be removed from contaminating substances by oxidation to $^{14}CO_2$, in which form it is flushed over into a collecting vessel containing NaOH or $Ba(OH)_2$.

Iodine-131 is one of the most widely used isotopes. In the free state it can readily be volatilized from numerous contaminating substances. The usual procedure in iodine distillation is to oxidize the iodine to iodate. This procedure will liberate "bound" iodine and destroy organic matter. The iodate is then reduced to free iodine and the iodine distilled.

Oxidation to iodate can be accomplished by the use of the following reagents, singly or in mixtures: Dichromate, perchlorate, permanganate, bismuthate, hydrogen peroxide and ozone. For the oxidation of iodine in mixtures containing organic compounds, concentrated sulfuric acid and sometimes mixtures of sulfuric and phosphoric acids are used to promote the reaction.

One of the best known mixtures for oxidizing carbon is the Van Slyke-Folch mixture. This consists of 25 grams of chromium trioxide, 5 grams of potassium iodate, 167 ml of phosphoric acid (density 1.7) and 333 ml of fuming sulfuric acid (20% free SO_3). This mixture cannot be used for iodine determinations because of the presence of iodate, but a similar formula with iodate omitted is quite useful.

Another useful oxidizing solution for the determination of iodine is made by adding 20 grams of potassium dichromate to 100 ml of 50% sulfuric acid. Still another consists of a mixture of potassium chlorate and nitric acid.

Reduction of iodate is accomplished by the addition of oxalic acid or other suitable reducing agent to the iodate. A ferrous salt is also useful for reduction, being itself oxidized to a ferric salt which prevents the reduction of iodine to iodide. This precaution is not necessary, however, in the presence of an excess of concentrated sulfuric acid, which will also oxidize iodide to iodine.

The quantity of iodine to be distilled is an important consideration. With milligram quantities, no difficulty is normally experienced. When distilling microgram quantities, one must take considerable precaution to prevent loss if quantitative results are desired. The difficulties which arise in the distillation procedure for protein bound iodine in blood, wherein the quantity of iodine involved is approximately 0.1 microgram, are typical. Distillation of carrier-free iodine involves submicrogram quantities.

The apparatus used is likewise important. For distillations of micro-quantities, it is advisable to maintain the exposed surface to a minimum. A very useful distilling arrangement is shown in figure 10.15. Distillation is conducted in a test tube equipped with a standard tapered joint and a delivery tube. The delivery tube should extend to the bottom of a small receiving tube immersed in ice water.

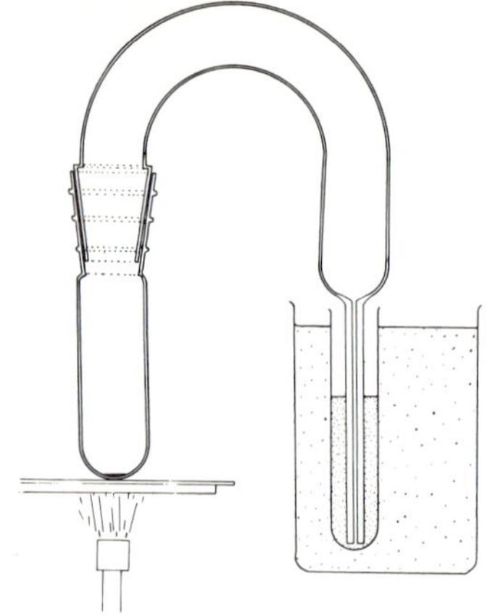

Figure 10.15—SEMIMICRO DISTILLATION APPARATUS.

PROCEDURE:

PART A—DISTILLATION WITH CARRIER

Use the apparatus shown in figure 10.15, or a similar distilling apparatus. Place 1 ml of water in the tube and add 50 lambdas of carrier-free iodide solution (having an activity of about 50,000 counts per minute per 50 lambdas). Add approximately 1 mg of KI carrier (1 drop of a solution containing 20 mg per ml). Add 0.2 gram of potassium dichromate and 1 ml of concentrated sulfuric acid. Allow the mixture to digest for 15 minutes over a low flame. Allow to cool, then chill in ice water. Add 5 ml of water and 0.2 gram of oxalic acid and immediately attach the delivery tube. The end of the delivery tube should be immersed in 1 ml of 0.1 N sodium hydroxide solution containing about 5 mg of sodium sulfite. Distill the mixture until the total volume in the receiving tube equals at least 2 ml. Dilute the distillate to exactly 5 ml. This is *solution A*.

PART B—DISTILLATION OF MICROGRAM QUANTITIES

Rinse the apparatus with water. Then rinse with small portions of 1% KI solution and finally rinse with distilled water. Check the apparatus for activity.

Repeat the procedure as outlined in *part A,* but use only 1 microgram of KI carrier instead of 1 mg. The distillate is again diluted to exactly 5 ml. This is *solution B.*

PART C—CARRIER-FREE DISTILLATION

Again wash the apparatus carefully, first with water, then with 1% KI solution, and finally with distilled water. Check for activity.

Repeat the procedure as outlined in *part A,* but omit the carrier altogether. Dilute the distillate to exactly 5 ml. This yields *solution C.*

Standard Solution—To about 4 ml of water containing a small amount of KI carrier, add 50 lambdas of the same radioiodide solution used above. Dilute this to exactly 5 ml without distillation. This is the *standard solution* or radioiodide.

DATA AND CALCULATIONS:

Measure the relative activities of aliquots of each of the solutions above. Compare their activities with that of the standard solution and calculate the per cent recovery in each part.

Solution	Activity cpm	% Recovery
Standard		x x x x
A		
B		
C		

EXPERIMENT 10.8 ELECTRODEPOSITION

OBJECTIVES:

To illustrate the separation of a metal from contaminating substances by electrolytic deposition.

To demonstrate a method for concentrating a radioactive metal.

To illustrate the electrolytic preparation of a radioactive metal in a form suitable for the measurement of its radioactivity.

THEORY:

Electrodeposition is the process of separation and deposition (plating out) of an element or compound at an electrode by passage of a direct electric current between electrodes immersed in an electrolyte. Electrochemical methods are found quite useful under the proper circumstances as a means for (1) separating an element from a number of contaminating substances; (2) concentrating a radioactive element; and (3) preparing a radioactive element in a convenient form for measuring its radioactivity.

Conductors—Two types of conductors are found in electrolytic cell systems—(1) *metallic,* in which the current is conducted by electrons, and (2) *electrolytic,* in which the current is conducted by ions. Metallic conductors are common. A piece of wire, (copper, iron, etc.) for example, is a metallic conductor. Metallic conductors have a negative coefficient of conductivity. That is, their resistance increases with increasing temperature. An electrolytic conductor consists of a solution of a suitable solute (an electrolyte) in a solvent (commonly water) in which ions are produced. Electrolytic conductors have positive coefficients of conductivity. Metals, in general, are better conductors than electrolytes.

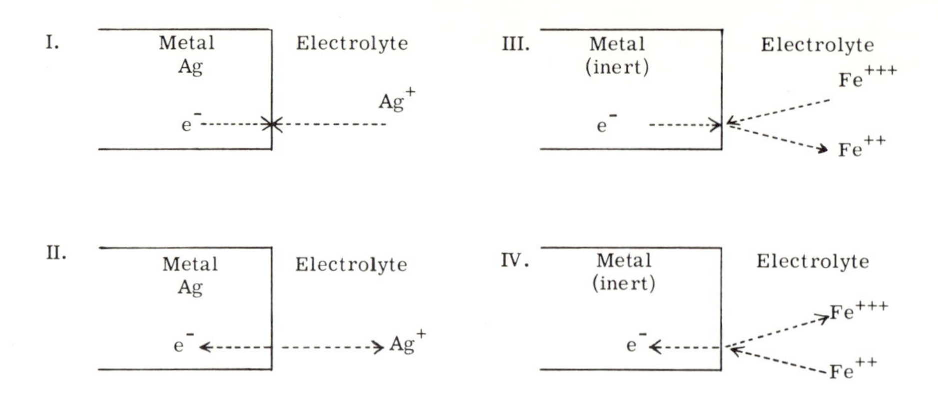

Figure 10.16—CONDUCTION AT AN INTERFACE.

Conduction at an interface—If, in order to complete the circuit, the current must flow through both metallic and electrolytic conductors, the junction where the metallic and electrolytic conductors come into contact is called an *interface.* At the interface the mode by which the current is conducted must change—i.e., from electrons to ions, or from ions to electrons. This change in the mode of current conduction necessitates a chemical reaction to bring about the required change from ion flow to electron flow, or vice versa. All reactions at an interface between different types of conductors (electrode reactions) involve a change of valence.

Figure 10.16 depicts four different examples of current flow at an interface. In examples I and III the electron flow takes place from the metal to the electrolyte; in examples II and IV electron flow is reversed. The following half-reactions occur at the electrodes:

I. $e^- + Ag^+ \longrightarrow Ag$
II. $Ag \longrightarrow e^- + Ag^+$
III. $Fe^{+3} + e^- \longrightarrow Fe^{+2}$
IV. $Fe^{+2} \longrightarrow Fe^{+3} + e^-$

Electrodes—Electrodes are the means by which the current enters and leaves an electrolytic conductor. The *cathode* is the electrode by which electrons enter a cell (κατὰ = down; ὁδὸς = road; hence cathode = "road down"). *Reduction* occurs at the cathode. The *anode* is the electrode by which electrons leave a cell (ἀνὰ = up; hence anode = "road up"). *Oxidation* occurs at the anode.

There are three types of electrodes:

1. Inert—those which do not enter into reaction—e.g., platinum—(See examples III and IV above).

2. Simple metal electrodes—electrodes composed of metals which participate in reactions—(See examples I and II above).

3. Composite electrodes:

a. metal and gas—e.g., ((Pt)) H_2
b. metal and solid—e.g., Ag: <u>AgCl</u>

Electrolytic cells—An electrolytic cell consists of two electrodes immersed in an electrolyte. In a typical cell, consisting, for example, of a homogeneous electrolyte and two electrodes of dissimilar metals, a potential difference will be produced between the electrodes. Figure 10.17 shows two such cells connected together externally. The cell producing the greater potential becomes the *driving cell* and determines the direction of current flow. The "weaker" cell—the one having the lower potential—becomes the *driven cell.* Attention is directed to the following facts and conventions of definition:

1. Electrons leave the negative pole (anode) of the generating cell and enter at the negative pole (cathode) of the driven cell.

2. The negative pole is that electrode which is negative with respect to the other, because of an excess of electrons. "Cathode" and "negative pole" are, therefore, *not* synonymous.

3. Within the cells, the anions (negative ions) move in the same continuous direction as do the electrons in the external circuit, while the cations (positive ions) migrate in the opposite direction. Cations always migrate toward the cathode and anions toward the anode.

4. The electron current flows in the direction opposite to that of the "current" as spoken of in reference to D.C. electrical equipment and apparatus.

5. By convention, the sign of the potential of a cell is defined as (a) positive if the anode is drawn on the left and (b) negative if the cathode is on the left.

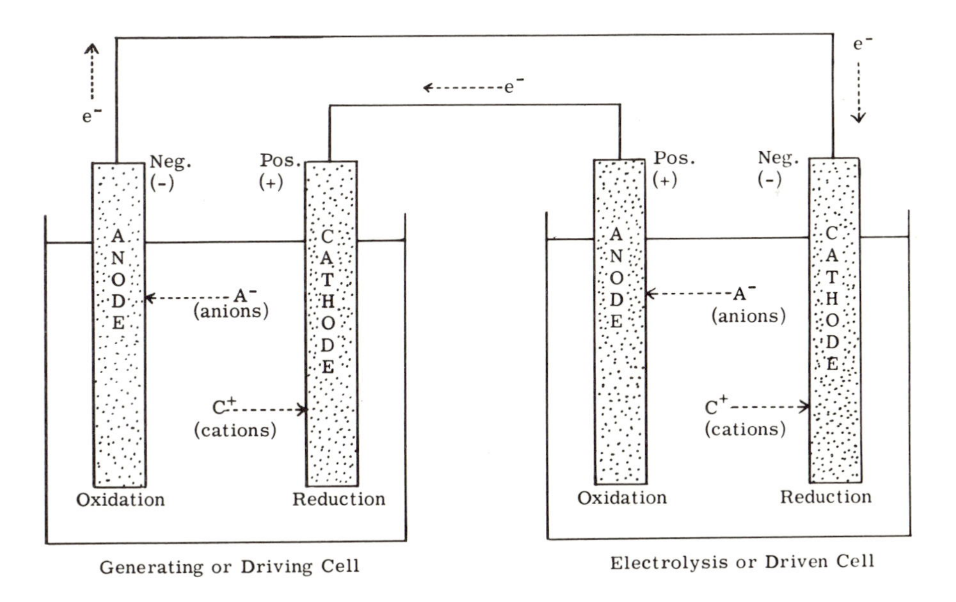

Figure 10.17—ELECTROLYTIC CELLS.

Cell reactions—The nature of a cell reaction is determined by (1) the solvent, (2) the solute and (3) the electrodes. Generalized cell reactions for driven cells made up of inert electrodes and aqueous solutions of electrolytes are given in table 10.1.

Table 10.1

Electrolyte	Cathode Reaction	Anode Reaction
Dilute acids (e.g., HCl)	$e^- + H^+ \longrightarrow \frac{1}{2}H_2$	$\frac{1}{2}H_2O \longrightarrow \frac{1}{4}O_2 + H^+ + e^-$
Conc. halogen acid (HCl)	$e^- + H^+ \longrightarrow \frac{1}{2}H_2$	$Cl^- \longrightarrow \frac{1}{2}Cl_2 + e^-$
Strong bases (e.g., NaOH)	$e^- + H_2O \longrightarrow \frac{1}{2}H_2 + OH^-$	$OH^- \longrightarrow \frac{1}{4}O_2 + \frac{1}{2}H_2O + e^-$
Dilute salts (e.g., NaCl)	$e^- + H_2O \longrightarrow \frac{1}{2}H_2 + OH^-$	$\frac{1}{2}H_2O \longrightarrow \frac{1}{4}O_2 + H^+ + e^-$
Dilute salts of weak bases		
(e.g., $AgNO_3$)	$e^- + Ag^+ \longrightarrow Ag$	$\frac{1}{2}H_2O \longrightarrow \frac{1}{4}O_2 + H^+ + e^-$
(e.g., $CuSO_4$)	$e^- + \frac{1}{2}Cu^{++} \longrightarrow \frac{1}{2}Cu$	$\frac{1}{2}H_2O \longrightarrow \frac{1}{4}O_2 + H^+ + e^-$
(e.g., $CoCl_2$)	$e^- + \frac{1}{2}Co^{++} \longrightarrow \frac{1}{2}Co$	$\frac{1}{2}H_2O \longrightarrow \frac{1}{4}O_2 + H^+ + e^-$

The *complete reaction* for a cell is obtained by addition of the two electrode reactions. For a dilute cobalt chloride solution, the complete reaction (for the exchange of a single electron) would be

$$\tfrac{1}{2}Co^{++} + \tfrac{1}{2}H_2O \longrightarrow \tfrac{1}{2}Co + \tfrac{1}{4}O_2 + H^+$$

Cell potential—The total potential of a cell E_c is equal to the sum of the single electrode potentials.

$$E_c = E_{anode} + E_{cathode} \tag{10.13}$$

The single electrode potential E is related to the ease of the oxidation or reduction processes occurring at the metal-electrolyte interface. If, for simplicity, molar concentrations are used instead of activities, it is expressed by the Nernst equation

$$E = E^0 + \frac{RT}{n\mathfrak{F}} \ln \frac{C''}{C'} \tag{10.14}$$

where E^0 is the standard electrode potential (the potential the cell will have when all ion concentrations are equal to unity), T is the absolute temperature, n is the number of Faradays required to accomplish the reaction (the hydrogen equivalent or the number of electrons transferred as indicated by the equation for the cell reaction), and $\mathfrak{F}$ is the Faraday (96,496 coulombs* per equivalent), ln is the symbol for the natural logarithm, and C' and C'' are the molar concentrations of the reactants and products, respectively. For a temperature of 25° C, the Nernst equation can be written in common logarithms, as follows:

$$E = E^0 + \frac{0.0591}{n} \log \frac{C''}{C'} \tag{10.15}$$

Values of E^0, the standard electrode potential, are given in table 10.2. These standard electrode potentials are referred to hydrogen, which is taken as zero.

* 96,496 coulombs = 6.023×10^{23} electrons (see Avogadro number).

Table 10.2

Electrode Reaction*	Potential (E^0)
$K = K^+ + e^-$	+2.92
$Na = Na^+ + e^-$	2.71
$\frac{1}{2}Mg = \frac{1}{2}Mg^{+2} + e^-$	2.40
$\frac{1}{3}Al = \frac{1}{3}Al^{+3} + e^-$	1.70
$\frac{1}{4}U = \frac{1}{4}U^{+4} + e^-$	1.40
$\frac{1}{2}H_2 + OH^- = H_2O + e^-$	0.83
$\frac{1}{2}Zn = \frac{1}{2}Zn^{++} + e^-$	0.76
$Pu^{+++} = Pu^{++++} + e^-$	0.72
$\frac{1}{2}Fe = \frac{1}{2}Fe^{++} + e^-$	0.44
$Tl = Tl^+ + e^-$	0.34
$\frac{1}{2}Co = \frac{1}{2}Co^{++} + e^-$	0.28
$\frac{1}{2}Ni = \frac{1}{2}Ni^{++} + e^-$	0.23
$\frac{1}{2}Sn = \frac{1}{2}Sn^{++} + e^-$	0.14
$\frac{1}{2}Pb = \frac{1}{2}Pb^{++} + e^-$	0.12
$\frac{1}{2}H_2 = H^+ + e^-$	0.0000
$\frac{1}{3}Bi = \frac{1}{3}Bi^{+3} + e^-$	−0.23
Saturated calomel	−0.2415
Calomel (1.0 *N* KCl)	−0.2800
Calomel (0.1 *N* KCl)	−0.3338
$\frac{1}{2}Cu = \frac{1}{2}Cu^{++} + e^-$	−0.34
$OH^- = \frac{1}{4}O_2 + \frac{1}{2}H_2O + e^-$	−0.40
$\frac{1}{4}Po = \frac{1}{4}Po^{+4} + e^-$	−0.40
$Fe^{++} = Fe^{+++} + e^-$	−0.77
$Ag = Ag^+ + e^-$	−0.80
$\frac{1}{2}Pd = \frac{1}{2}Pd^{++} + e^-$	−0.90
$\frac{1}{2}H_2O = \frac{1}{4}O_2 + H^+ + e^-$	−1.23
$Cl^- = \frac{1}{2}Cl_2 + e^-$	−1.36

Electrodeposition—It has been shown that cathodic deposition of a metal is accomplished in driven cells containing as an electrolyte a dilute solution of a salt of a weak base. (See table 10.1). This is the basis for most electrodeposition or electroplating procedures. Any convenient source of direct current producing ample potential can be used as the driving cell, such as a battery, a D.C. generator or a rectifier. Figure 10.18 illustrates the use of a battery B as a driving cell. The current flow is regulated by the resistance *R* and is measured by the ammeter A. The voltage applied to the cell is indicated on the voltmeter V. For certain measurements, it will be found convenient to use a rectifier power supply as a variable potential source because different voltages are required for the deposition of different metals.

The applied potential is bucking the potential generated by the cell. If the applied potential is just equal to the cell potential, no current will flow. Thus, the net potential responsible for current flow is the difference, $E_{applied} - E_c$, and the current I flowing in the cell is given by

*Reactions are written as they would occur at the anode. At the cathode, the reactions are reversed as well as the sign of the potential, e.g., at the cathode, $Ag^+ + e^- = Ag$ and the potential is +0.80 volts.

$$I = \frac{E_{\text{applied}} - E_c}{R_i} \tag{10.16}$$

where R_i is the internal resistance of the cell. This current causes chemical reactions, and under suitable conditions, the deposition of the desired metal will occur.

In the cell shown in figure 10.18, the cathode reaction is

$$\tfrac{1}{2}Co^{++} + e^- \longrightarrow \tfrac{1}{2}Co \quad (-0.28 \text{ volts})$$

and the anode reaction is

$$\tfrac{1}{2}H_2O \longrightarrow \tfrac{1}{4}O_2 + H^+ + e^- \quad (-1.23 \text{ volts})$$

For a one molar solution of $CoCl_2$, assuming unit activities, the reversible emf of the cell is

$$\begin{aligned} E &= E_{Co^{++}} + E_{O_2} \\ &= -0.28 + (-1.23) = -1.5 \text{ volts} \end{aligned}$$

In practice, however, the reaction is not reversible and the applied potential must be in excess of the reversible cell potential by an amount ω, the *overpotential* or *overvoltage.* Thus,

$$I = \frac{E_{\text{applied}} - (E_c + \omega)}{R_i} \tag{10.17}$$

The exact value of ω depends upon temperature, ionic concentration, and the applied potential, but it is approximately equal to 0.2 volts or less.

Equation 10.17 is expressed graphically in figure 10.19. It is seen that the slope of the curve is equal to $1/R_i$, where R_i is the cell resistance.

Electroseparation of metals is possible by (1) controlled voltage, (2) by anodic deposition, and (3) by complexing. If the potentials of the metals (table 10.2) differ by 0.2 volts or more they can ordinarily be separated by *controlled voltage.* In figure 10.20, cation 1 will be reduced at a potential of E_1, and will be plated alone up to a potential of E_2, when cation 2 will likewise become reduced. By this means, copper (−0.34 V) can easily be separated from nickel (0.23 V). On the other hand, it would be difficult to separate cobalt (0.28 V) from nickel by this process.

Anodic deposition is frequently used for preparing deposits of lead as PbO_2. The electrolysis is conducted in the presence of nitrate ion which prevents the cathodic deposition of lead. The anode reaction is given by

$$\tfrac{1}{2}Pb^{++} + H_2O = \tfrac{1}{2}PbO_2 + 2H^+ + e^- \quad (-1.46 \text{ volts})$$

Under suitable conditions, oxides of manganese, molybdenum and uranium can also be deposited at the anode.

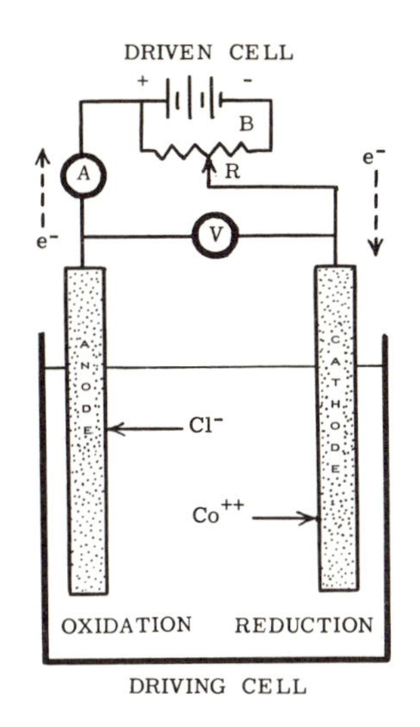

Figure 10.18—ELECTRODEPOSITION OF COBALT. The electrodes used are inert.

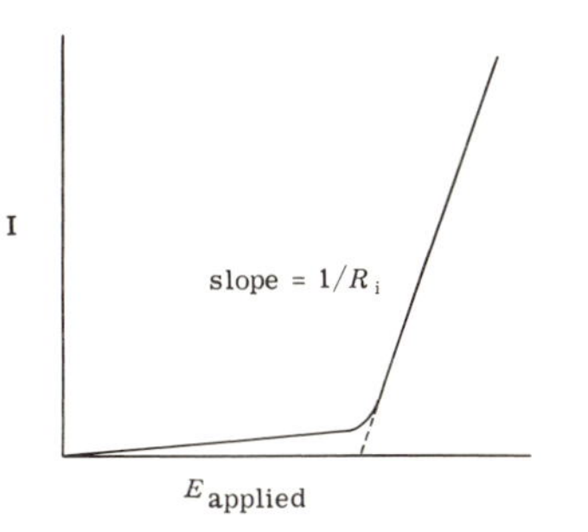

Figure 10.19—CURRENT-VOLTAGE RELATIONSHIP IN A DRIVEN CELL.

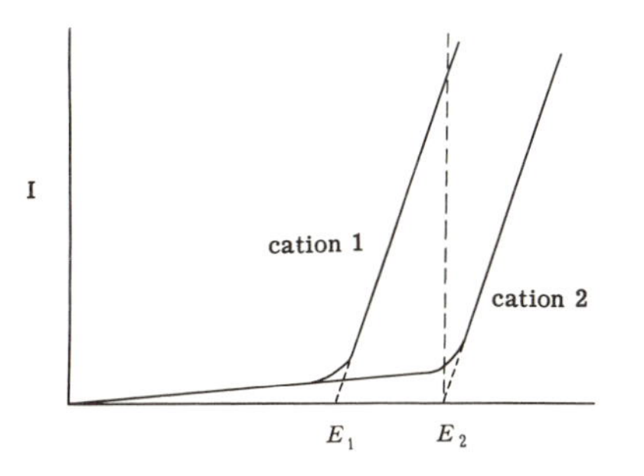

Figure 10.20—CURRENT-VOLTAGE RELATIONSHIP FOR ELECTROSEPARATION.

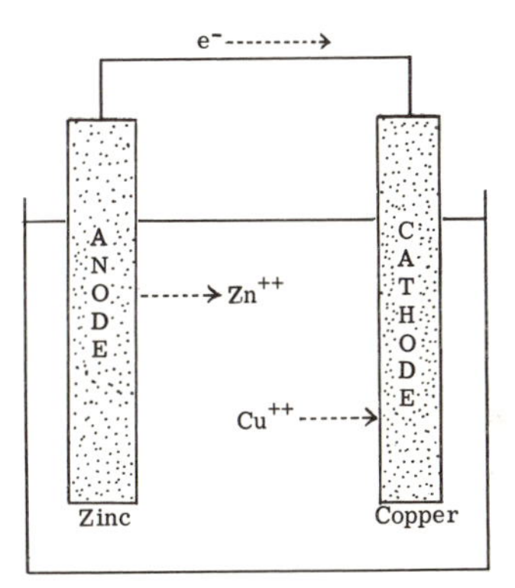

Figure 10.21—ELECTROLYTIC DISPLACEMENT.

Complexing is often a possible means for separating two metals where potentials differ by less than 0.2 volts. If one of the metals can be converted to a complex ion (cyanide or ammonia complex) a greater voltage will be required for its deposition, and thus separation may be possible.

Electrolytic displacement is simply the plating out of a more noble element on a less noble one without the external application of a current. The process is analogous to the case of a driving cell with its electrodes connected directly. (See figure 10.21). If copper and zinc electrodes are used, copper will deposit on the cathode, while the anode of zinc metal will dissolve, forming zinc ions. Any metal in table 10.2 will deposit in this manner on an electrode prepared from any free metal above it in the table.

PROCEDURE:

PART A—ELECTRODEPOSITION OF COBALT

The following excerpt is from an article by C. L. Comar, et al., "Cobalt Metabolism Studies: Radioactive Cobalt Procedures with Rats and Cattle," Arch. Biochem., *9,* 149-158 (1946).

"The body of the electrolytic cell is a piece of Pyrex glass tubing with the ends ground square, 25 mm outside diameter, 1.5 mm wall thickness, and about 125 mm in length. The cathode, which forms the bottom of the cell, is a copper disk, 25.4 mm in diameter. The disks are washed in petroleum ether, dipped in a solution of sulfuric and nitric acids, rinsed thoroughly and dried, and kept protected from dust until used. The disk is fastened to the glass tube with Pyseal sealing wax which makes a leakproof joint. The cell is mounted in a brass holder.

"The plating solution contains 100 g ammonium sulfate, 180 ml concentrated ammonium hydroxide, and 5 g ammonium hypophosphite per liter. About 30 ml of this plating solution containing 10 mg cobalt and the labeled cobalt to be plated out are used in the electrolytic cell. When the whole sample is used, the total cobalt content is essentially 10 mg, since that amount of carrier has been previously added. When aliquots are used, however, a proportional amount of inert cobalt is added to bring the cobalt content to this value.

"A Braun Electrolytic Outfit, which accommodates six cells, was found convenient for the plating operation although individual motor stirrers can be used. Platinum anodes serve as stirrers and the current is supplied by storage batteries and regulated by rheostats. Experiment has shown that practically complete recovery is attained by plating at a current density of 27 milliamperes per sq. cm. for 5½ to 6 hours."

Set up the electrolysis cell and prepare two identical samples, taking precautions to obtain equal amounts of activity in each sample. Use about 0.01μCi of the ^{60}Co solution and 30 ml of the electrolytic fluid. Dry the samples and count the activity of each, using the end-window G-M counter.

Compare the two samples as an indication of the precision of this method of sample preparation. Calculate the odds that the difference in counts observed on the two samples is due to statistics rather than to experimental error. (See chapter 4).

An exploded and cut-away view of the cell is shown in figure 10.22. When you assemble, be sure to place the rubber gasket *between* the glass cell and the planchet to prevent leakage. Be sure, also, to use a planchet of the proper size. The metal planchets used in chapter 11 (Simulated Reference Source) are a different size than those intended for use with the electrolysis cell and do not give a tight fit with the gasket.

It will also be convenient to use an electronic battery eliminator (Heath Company, Benton Harbor, Michigan) in place of storage batteries and an external milliameter to provide greater sensitivity of current measurement.

PART B—ANODIC DEPOSITION OF LEAD

Place 30 ml of 0.1*M* nitric acid in the electrolytic cell (figure 10.22) and add 2 or 3 drops of lead carrier solution representing about 5 milligrams of lead. Now introduce about 0.1 μCi of radium-DEF solution. Connect the electrolytic cell so the planchet is the anode (to positive pole of battery) and the stirrer is the cathode. Allow current to flow through the cell for about 1 hour. When electrolysis is complete, remove the planchet (containing a brown deposit of PbO_2) and wash it in distilled water.

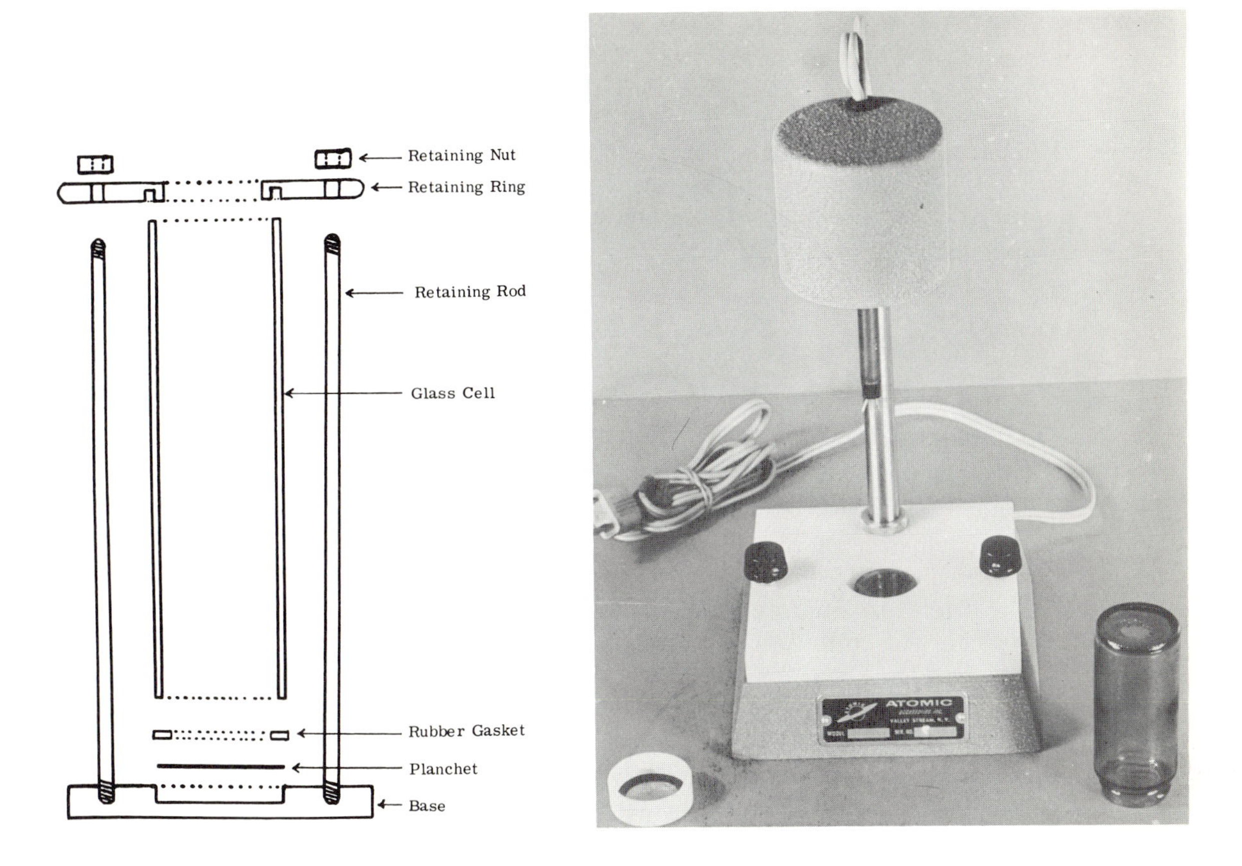

Figure 10.22—EXPLODED VIEW OF ELECTROLYTIC CELL.

Figure 10.23—EDUCATIONAL ELECTRODEPOSITION UNIT. (*Courtesy Atomic Accessories, Inc.*)

Measure the activity with an end-window G-M counter or, if available, a proportional counter on both the alpha and beta plateaus. Measure the alpha and beta activities of the sample periodically over a 5-day period. Plot the alpha and beta activities as a function of time.

PART C—ELECTROLYTIC DISPLACEMENT OF POLONIUM

To each of three 10-ml beakers, add 5 ml of water and about 0.01μCi of radium-DEF solution. Place small sheets of silver (a dime), copper and tin (or nickel, zinc or iron) in each of the three beakers, respectively. Allow the metals to contact the solutions for about two hours, then wash and dry them.

The radium-DEF solution contains lead (Ra-D), bismuth (Ra-E), and polonium (Ra-F). Refer to the activity series in table 10.2. Predict which elements will plate out and which ones will not for each of the metals used.

Prove your predictions by making suitable measurements of the activities deposited on each metal. Explain your results.

EXPERIMENT 10.9 ELECTROPHORESIS

OBJECTIVES:

To demonstrate the basic principles of electrophoresis.

To apply electrophoresis, in particular paper electrophoresis, as a means for separating blood proteins.

THEORY:

Electrophoresis may be defined as the migration of ions or colloidal particles in an electric field. In general, the ions or particles undergoing migration are charged electrically, either positively or negatively.

A simplified version of the process is depicted in figure 10.24. If a positively charged colloid is placed in the U-tube, under the influence of the potential gradient, the particles will tend to migrate toward the negative electrode. On the other hand, if a negatively charged colloid is used, the particles will be found to migrate toward the positive electrode.

In the biological and medical fields, this process has been used to separate mixtures of proteins. If blood serum is used, the blood proteins can be separated into the albumin, alpha globulin, beta globulin and gamma globulin fractions.

For the sake of simplicity, let us consider the amino acids from which proteins are made. The amino acids are amphoteric; that is, they can act either as acids or as bases according to the pH of the environment by virtue of the carboxyl and amino

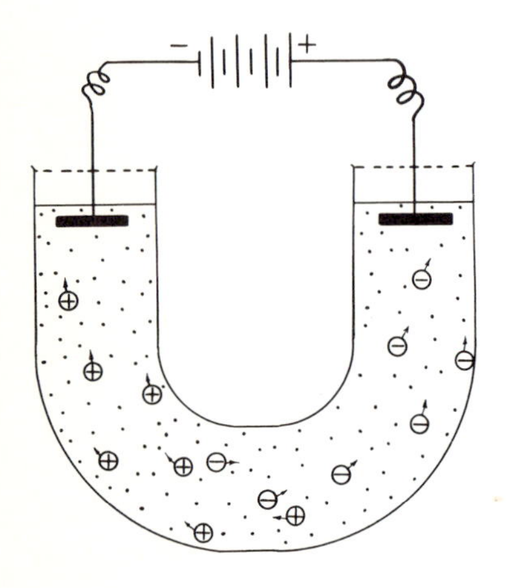

Figure 10.24—APPARATUS FOR ELECTROPHORESIS.

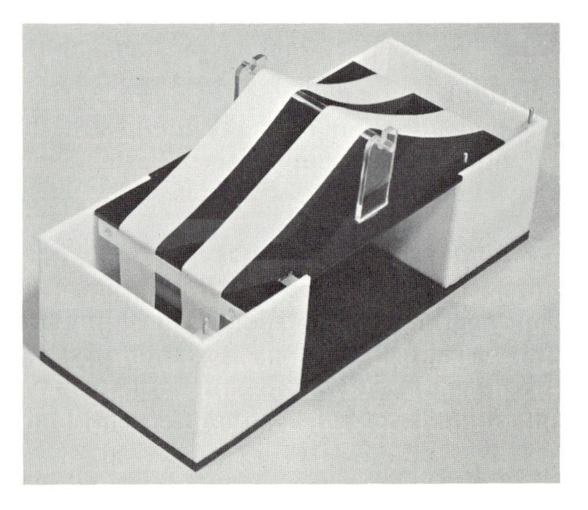

Figure 10.25—APPARATUS FOR ELECTROPHORESIS. Shown with cover removed. (*Courtesy Atomic Accessories, Inc.*)

groups which they contain. In solution, these molecules, called "zwitterions," have both positive and negative charges. The net charge on the molecule can be made more positive or more negative by a change in the pH of the solution. The charges on the "zwitterion" of the amino acid glycine and the changes in these charges with the addition of hydrogen or hydroxyl ions are illustrated below.

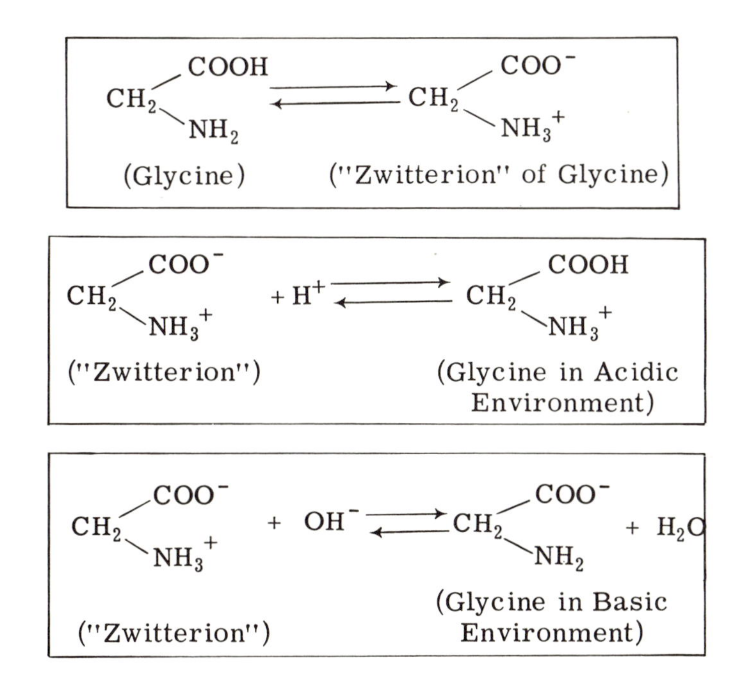

At a particular median value of pH, the acidic and basic ionizations of the molecule will be equal in extent, and the molecule will be electrically neutral. This pH is known as the *"isoelectric point."* Amino acids, amphoteric electrolytes in general and the dispersed phase of colloidal systems will fail to migrate in an electric field if the pH has been adjusted to correspond to the isoelectric point. If the pH is decreased below the isoelectric point, the molecules will become positively charged and migrate toward the negative electrode. If the pH is increased above the isoelectric point, the molecules will become negatively charged and migrate toward the positive electrode.

The pH at which the isoelectric point occurs depends upon the acidic and basic ionization constants of the amphoteric compounds. The values of these constants will differ from one compound to another. It therefore becomes possible to find a pH at which one species of molecule will be positively charged and another negatively charged. If electrophoresis is conducted at this particular pH, one compound will migrate toward the positive electrode and the other toward the negative electrode. Thus, a separation is effected.

Proteins show the same behavior as the amino acids. At the isoelectric point there is no drift towards an electrode. In more acidic solutions, proteins become positively charged and in more alkaline solutions they become negatively charged. Each protein is characterized by a specific value for the isoelectric point. Consequently, mixtures of proteins can be fractionated by electrophoresis in a solution buffered at the proper pH.

Paper electrophoresis is a variation of the classical method in which the solution is supported by a strip of filter paper. When an electrical potential is applied across the paper, charged ions and molecules are caused to migrate along the paper strips.

The apparatus used for paper electrophoresis is illustrated in figure 10.25. Strips of filter paper, approximately 1 inch wide, are moistened with a buffer solution. They are then suspended between two troughs, each containing the same buffer solution used to moisten the paper strips. About 5 or 10 microliters of the protein solution are placed on the strips at the "start line," previously drawn on the paper with a pencil. The entire assembly is enclosed in a box to retard evaporation, and an electrical potential is applied to electrodes immersed in the troughs of buffer solution. Electrophoresis is allowed to proceed until separation of the fractions occurs.

MATERIALS:

Electrophoresis apparatus, filter paper strips, blood serum or heparinized plasma, iodine-131 (iodide), paper chromatogram scanner, 0.2% ninhydrin solution in ethanol, pH meter, sodium barbital, sodium acetate, acetic acid, 3% hydrogen peroxide, potassium iodide, potassium iodate, 1*N* hydrochloric acid.

PROCEDURE:

Tagging the protein—From time to time some difficulty may be experienced in tagging the protein with radioactive iodine. This may be caused by the presence of sulfite in the ^{131}I solution sometimes added by the supplier to prevent oxidation of iodide. Increasing the amount of oxidant in the following procedures will usually remedy the problem. Two methods, modifications of those given by McFarlane (151), are presented.

Method 1—To 0.5 ml of serum or plasma add 100 μl of buffer solution (pH 8.5) and 25 to 100 μl of radioactive iodide, carrier-free, representing an activity of about 10 μCi of ^{131}I. (The same volume of a more concentrated ^{131}I solution will provide a higher and more satisfactory activity in the labeled protein.) This is followed by 0.05 ml of 3% hydrogen peroxide to liberate the iodine. Mix the serum well and allow it to incubate at 37° C. for 15 to 30 minutes. During this time the iodine becomes fixed to the blood proteins.

Method 2—To 25 to 100 μl of radioactive iodide, carrier-free, representing an activity of about 10 μCi (or more) add 0.1 ml of a solution containing 0.5 mg of KI and 0.1 ml of a solution containing 0.5 mg of KIO_3, followed by one drop of 1*N* HCl. Approximately 0.5 mg of iodine is liberated. This is added to 0.5 ml of serum or plasma. Mix the serum well and incubate at 37° C. for 15 to 30 minutes. During this time the iodine becomes fixed to the blood proteins.

Assembling the apparatus—Clean the apparatus and place the chambers in position in the plastic box. The cotton wicks which are placed through the holes in the wall of the electrode chambers serve as electrical connections but prevent electrode products from contaminating the buffer. Place about 100 ml of the buffer in each of the large compartments and fill the electrode compartments until the platinum electrodes are covered.

Preparation of the buffer

Sodium barbital	4.4 g
Sodium acetate	4.7 g
Acetic acid, 0.1*N*	30.0 ml
Water, q. s. ad	1000.0 ml

This buffer should have a pH of 8.5. Measure the pH and adjust it, if necessary, by adding 10% NaOH or 36% acetic acid, as required. Only a few drops of either reagent should be needed.

Preparation of Paper Strips—Cut 1 inch wide paper strips into lengths sufficiently long to reach from one trough to the other, allowing about 1 inch at each end to be submerged below the buffer solution in the troughs. Mark a light pencil line on each strip as the starting point, as well as other identifying marks. The polarity (+ or −) should also be indicated at one end.

Immerse the paper strips in a beaker filled with the buffer. Then remove the strips from the solution and lay them flat on a paper towel to remove the excess buffer. Place the moist strips in the apparatus with the ends immersed in the buffer and with their centers supported by the glass rod. Make sure the strips are taut and that there is no sag in the middle.

Apply 5 lambdas of the tagged blood serum or plasma (at the start line) to the strips in the cabinet. Replace the cover on the cabinet and plug in the leads from the power supply, observing the polarity marked on the strips. Turn on the power supply and allow current to flow until adequate separation has occurred. Normally the following voltages and times are adequate:

16 hours at 50 volts, or
8 hours at 100 volts, or
4 hours at 200 volts

Better results are obtained by the use of lower voltages because less power is dissipated in the paper strip. Thus, less heat is produced and evaporation is decreased. If time permits, apply 50 volts for 16 hours.

At the termination of electrophoresis, remove the strips. Suspend them vertically or lay them on a towel to dry. A *moderately* increased temperature may be used to shorten the time required for drying. An infrared lamp at least 1 foot from the strips may be used for this purpose.

When the strips are dry, determine the location of the activity by the use of a chromatogram scanner similar to that used in experiment 10.4.

To locate the protein bands visually, spray the strips with 0.2% ninhydrin in ethyl alcohol. Full color development is attained at room temperature in about 24 hours. The color can be developed in as little as one-half hour at elevated temperatures, but caution must be used since overheating will cause darkening of the entire strip. (If this happens, respraying with ninhydrin and redrying at a lower temperature will frequently bleach out the undesired color, but the results are of a poorer quality).

Identify the albumin, alpha globulin, beta globulin and gamma globulin fractions from the color produced by the ninhydrin spray. Compare the recorder chart with the developed strip. Which fractions were tagged with iodine?

REFERENCES

Separation Techniques

1. Choppin, G. R., "Experimental Nuclear Chemistry", Prentice Hall (1961).
2. Cohen, K., "Fundamentals of Isotope Separation", Nucleonics 2, 3 (June 1948).
3. Cunningham, B. B., "Microchemical Methods Used in Nuclear Research", Nucleonics 5, 62-85 (Nov. 1949).
4. DeVoe, J. R., and W. W. Meinke, "Radiochemical Separations of Cadmium", Anal. Chem. 31, 1428 (1959).
5. Dunn, R. W., "Methods of Producing Radioiron", Nucleonics 10, No. 7, 8-13 (1952).
6. Dunn, R. W., "Techniques Related to Production of Fe^{59}", Nucleonics 10, No. 8, 40-43 (1952).
7. Faires, R. A., and B. H. Parks, "Radioisotope Laboratory Techniques", Pitman, London (1958).
8. Finston, H. L., and J. Miskel, "Radiochemical Separation Techniques", Ann. Rev. Nuclear Sci. 5, 269-296 (1955).
9. Goldin, A. S., R. J. Velten, and G. W. Frishkorn, "Determination of Radioactive Strontium", Anal. Chem. 31, 1490 (1959).
10. Kirk, P. L., "Quantitative Ultramicroanalysis", Wiley (1950).
11. Overman, R. T., and H. M. Clark, "Radioisotope Technique", McGraw-Hill (1960).
12. Seaman, W., and G. L. Roberts, "Radiotracer Method for Determination of Adsorption of Surfactants on Copper Phthalocyanine", Anal. Chem. 33, 414 (1961).
13. Sunderman, D. N., and W. W. Meinke, "Evaluation of Radiochemical Separation Procedures", Anal. Chem. 29, 1578 (1957).
14. Wilkinson, G., and W. E. Grummitt, "Chemical Separation of Fission Products", Nucleonics 9, No. 3, 52-62 (1951).
15. Williams, R. R., "Principles of Nuclear Chemistry", Van Nostrand (1950).

Precipitation

16. Abers, E. L., "An Inexpensive Filter-Cup for Collecting and Counting Active Precipitates", Nucleonics 3, 43 (Oct. 1948).
17. Broadhead, K. G., and H. H. Heady, "Radiochemical Precipitation Studies of Rare-Earth Oxalates", Anal. Chem 32, 1603 (1960).
18. Handley, T. H., and C. L. Burros, "Determination of Radioactive Cesium", Anal. Chem. 31, 332 (1959).
19. Heyn, A. H. A., and H. L. Finston, "Separation of Magnesium from Sodium and Potassium, A Tracer Study", Anal. Chem. 32, 328 (1960).
20. Meadows, J. W. T., and G. M. Matlack, "Determination of Radioactive Zirconium in Fissioned Plutonium", Anal. Chem. 32, 1607 (1960).
21. Migicovsky, B. B., and W. A. Evans, "Filtration and Mounting Device for Ca^{45}", Nucleonics 9, No. 3, 77-78 (1951).
22. Pruitt, M. E., R. R. Rickard, and E. I. Wyatt, "Radiochemical Determination of Yttrium and Promethium. A Precipitation Technique", Anal. Chem. 34, 283 (1962).
23. Sacks, J., "All-Glass Filtration Apparatus for Radioactive Tracer Experiments", Anal. Chem. 21, 876-877 (1949).
24. Sodd, V. J., A. S. Goldin, and R. J. Velten, "Determination of Radioactivity of Saline Waters", Anal. Chem. 32, 25 (1960).
25. Stang, L. G., Jr., W. D. Tucker, H. O. Banks, Jr., R. F. Doering, and T. H. Mills, "Production of Iodine-132", Nucleonics 12, No. 8, 22-24 (Aug. 1954).
26. Sugihara, T. T., H. I. James, and E. J. Troianello, "Radiochemical Separation of Fission Products from Large Volumes of Sea Water. Strontium, Cesium, Cerium, and Promethium", Anal. Chem. 31, 44 (1959).

Coprecipitation

27. Goldin, A. S., "Determination of Dissolved Radium", Anal. Chem. 33, 406 (1961).
28. Gordon, L., and K. Rowley, "Coprecipitation of Radium with Barium Sulfate", Anal. Chem. 29, 34 (1957).
29. Lima, F. W., "Calculation of Amount of Tracer Carried with Precipitates of its Radioactive Parent", Anal. Chem. 25, 1924-1925 (1953).
30. Myers, R. J., D. E. Metzler, and E. H. Swift, "The Distribution of Ferric Iron Between Hydrochloric Acid and Isopropyl Ether Solutions", J. Am. Chem. Soc. 72, 3767-3771 (1950).
31. Rickard, R. R., and E. I. Wyatt, "Radiochemical Determination of Fission Ruthenium in Aqueous Solutions. A Nondistillation Technique", Anal. Chem. 31, 50 (1959).

Solvent Extraction

32. Bagget, B., and L. L. Engel, "The Use of Countercurrent Distribution for the Study of Radiochemical Purity", J. Biol. Chem. 229, 443-450 (1957).
33. Beard, H. C., and L. A. Lyerly, "Separation of Arsenic from Antimony and Bismuth by Solvent Extraction", Anal. Chem. 33, 1781 (1961).

34. Eberle, A. R., and M. W. Lerner, "Separation of Uranium from Thorium, Bismuth and Ores with Tributyl Phosphate. Spectrophotometric Determination with 8-Quinolinol", Anal. Chem. 29, 1134 (1957).
35. Grahame, D. C., and G. T. Seaborg, "The Distribution of Minute Amounts of Material Between Liquid Phases", J. Am. Chem. Soc. 60, 2524-2528 (1938).
36. Kirby, H. W., "Preparation of Radiochemically Pure Cerium by Solvent Extraction", Anal. Chem. 29, 1599 (1957).
37. Kirk, P. L., and M. Danielson, "A Liquid-Liquid Microextractor for Solvents Lighter than Water", Anal. Chem. 20, 1122-1123 (1948).
38. Lowe, R. W., S. H. Prestwood, R. R. Rickard, and E. I. Wyatt, "Determination of Radioantimony by Extraction into Diisobutylcarbinol", Anal. Chem. 33, 874 (1961).
39. Maeck, W. J., G. L. Booman, M. C. Elliott, and J. E. Rein, "Determination of Neptunium in Uranium-Fission Product Mixtures. Initial Extraction with Methyl Isobutyl Ketone", Anal. Chem. 32, 605 (1960).
40. Maeck, W. J., M. E. Kussy, and J. E. Rein, "Radiochemical Determination of Molybdenum by Solvent Extraction", Anal. Chem. 33, 237 (1961).
41. Maeck, W. J., M. E. Kussy, and J. E. Rein, "Solvent Extration Method for the Radiochemical Determination of Chromium", Anal. Chem. 34, 1602 (1962).
42. Maeck, W. J., M. E. Kussy, and J. E. Rein, "Solvent Extraction Method for the Radiochemical Determination of Zinc", Anal. Chem. 33, 235 (1961).
43. Marsh, S. F., W. J. Maeck, G. L. Booman, and J. E. Rein, "Solvent Extraction Method for Radiocerium", Anal. Chem. 34, 1406 (1962).
44. Matuszek, J. M., Jr., and T. T. Sugihara, "Low-Level Radiochemical Separation of Manganese", Anal. Chem. 33, 35 (1961).
45. Meadows, J. W. T., and G. M. Matlack, "Radiochemical Determination of Ruthenium by Solvent Extraction and Preparation of Carrier-Free Ruthenium Activity", Anal. Chem. 34, 89 (1962).
46. Moore, F. L., "Separation and Determination of Neptunium by Liquid-Liquid Extraction", Anal. Chem. 29, 941 (1957).
47. Moore, F. L., and S. A. Reynolds, "Determination of Protactinium-233", Anal. Chem. 29, 1596 (1957).
48. Moore, F. L., and S. A. Reynolds, "Radiochemical Determination of Uranium-237", Anal. Chem 31, 1080 (1959).
49. Morrison, G. H., and H. Freiser, "Solvent Extraction in Analytical Chemistry", Wiley (1957).
50. McCown, J. J., and R. P. Larsen, "Radiochemical Determination of Cerium by Liquid-Liquid Extraction", Anal. Chem. 32, 597 (1960).
51. McCown, J. J., and R. P. Larsen, "Radiochemical Determination of Total Rare Earths by Liquid-Liquid Extraction", Anal. Chem. 33, 1003 (1961).
52. Smith, G. W., and F. L. Moore, "Separation and Determination of Radiocerium by Liquid-Liquid Extraction", Anal. Chem. 29, 448 (1957).
53. Sunderman, D. N., I. B. Ackermann, and W. W. Meinke, "Radiochemical Separations of Indium", Anal. Chem. 31, 40 (1959).
54. Velten, R. J., and A. S. Goldin, "Simplified Determination of Strontium-90. Preferential Extraction of Yttrium with Tributyl Phosphate", Anal. Chem. 33, 128 (1961).

Paper Chromatography

55. Aronoff, S., "A Two-Dimensional Scanner for Radio-Chromatograms", Nucleonics 14, No. 6, 92-94 (June 1956).
56. Berliner, D. L., O. V. Dominguez, and G. Westenskow, "Determination of Carbon-14 Steroids on Paper Chromatograms", Anal. Chem. 29, 1797 (1957).
57. Cassidy, H. G., "Fundamentals of Chromatography", Interscience (1957).
58. Clegg, D. L., "Paper Chromatography", Anal. Chem. 22, 48-59 (1950).
59. Cohn, D. V., G. W. Buckaloo, and W. E. Carter, "Automatic Paper-Strip Scanner for Detecting Radioactivity", Nucleonics 13, No. 8, 48 (Aug. 1955).
60. Frierson, W. J., and J. W. Jones, "Radioactive Tracers in Paper Partition Chromatography of Inorganic Ions", Anal. Chem. 23, 1447-52 (1951).
61. Harrison, A., and F. P. W. Winteringham, "4π Beta Counter for Scanning Paper Chromatograms", Nucleonics 13, No. 3, 64-68 (Mar. 1955).
62. Jones, A. R., "Instrumental Detection of Radioactive Material on Paper Chromatograms", Anal. Chem. 24, 1055 (1952).
63. Lederer, E., and M. Lederer, "Chromatography", 2nd Ed., Van Nostrand (1957).
64. Lowenstein, J. M., and P. P. Cohen, "A Windowless Flow Counter for Paper Strips", Nucleonics 14, No. 5, 98-100 (May 1956).
65. Muller, R. H., and E. N. Wise, "Use of Beta-Ray Densitometry in Paper Chromatography", Anal. Chem. 23, 207-208 (1951).
66. Lewin, S. Z., "Chemical Instrumentation — 14. Chromatographic Equipment", J. Chem. Educ. 38, A713-A739 (1961).
67. Pinajian, J. J., and J. E. Christian, "The Determination of Iodide-Iodate Activity in Sodium Radio-Iodide (I^{131}) by Automatic Scanning of Paper Chromatograms", J. Amer. Pharm. Assoc., Sci. Ed., 44, 107-109 (1955).
68. Price, T. D., and P. B. Hudson, "Fluoroscope and Geiger Counter for Measuring Ultraviolet Absorption of Chromatograms", Anal. Chem. 26, 1127-32 (1954).

69. Rockland, L. B., J. Lieberman, and M. S. Dunn, "Automatic Determination of Radioactivity on Filter Paper Chromatograms", Anal. Chem. 24, 778-782 (1952).
70. Smith, (ed.), "Chromatographic Techniques", Interscience (1958).
71. Steenberg, K., and A. A. Benson, "A Scintillation Counter for Soft-β Paper Chromatograms", Nucleonics 14, No. 12, 40-43 (Dec. 1956).
72. Soloway, S., F. J. Rennie, and De W. Stetten, Jr., "An Automatic Scanner for Paper Radiochromatograms", Nucleonics 10, No. 4, 52-53 (1952).
73. Stein, W. H., and S. Moore, "Chromatography", Sci. Amer. p. 35-41 (March 1951).
74. Strain, H. H., "Chromatography", Anal. Chem. - Annual Reviews 32, 3R (April 1960).
75. Winteringham, F. P. W., A. Harrison, and R. G. Bridges, "Radioactive Tracer Techniques in Paper Chromatography", Nucleonics 10, No. 3, 52-57 (1952).

Ion Exchange Chromatography

76. Banerjee, G., and A. H. A. Heyn, "Separation of Uranium from Bismuth by Anion Exchange Resins", Anal. Chem. 30, 1795 (1958).
77. Bryant, E. A., J. E. Sattizahn, and B. Warren, "Strontium-90 by an Ion Exchange Method", Anal. Chem. 31, 334 (1959).
78. Bunney, L. R., N. E. Ballou, J. Pascual, and S. Foti, "Quantitative Radiochemical Analysis by Ion Exchange. Anion Exchange Behavior of Several Metal Ions in Hydrochloric, Nitric, and Sulfuric Acid Solutions", Anal. Chem. 31, 324 (1959).
79. Cohn, W. E., G. W. Parker, and E. R. Tompkins, "Ion-Exchangers to Separate, Concentrate and Purify Small Amounts of Ions", Nucleonics 3, 22 (Nov. 1948).
80. Doering, R. F., W. D. Tucker, and L. G. Stang, Jr., "A Simple Device for Milking High-Purity Y^{90} from Sr^{90}", Brookhaven Nat. Lab. Report, BNL 5454 (1960).
81. Evans, H. B., C. A. A. Bloomquist, and J. P. Hughes, "Anion Exchange Separation and Spectrophotometric Determination of Microgram Quantities of Rhodium in Plutonium-Uranium-Fissium Alloys", Anal. Chem. 34, 1692 (1962).
82. Faris, J. P., and J. W. Warton, "Anion Exchange Resin Separation of the Rare Earths, Yttrium, and Scandium in Nitric Acid-Methanol Mixtures", Anal. Chem. 34, 1077 (1962).
83. Freiling, E. C., J. Pascual, and A. A. Delucchi, "Quantitative Radiochemical Analysis by Ion Exchange. Anion Exchange Equilibrations in Phosphoric Acid Solutions", Anal. Chem. 31, 330 (1959).
84. Fritz, J. S., and B. B. Garralda, "Anion Exchange Separation of Thorium Using Nitric Acid", Anal. Chem. 34, 1387 (1962).
85. Fritz, J. S., and J. E. Abbink, "Cation Exchange Separation of Vanadium from Metal Ions", Anal. Chem. 34, 1080 (1962).
86. Kitchener, J. A., "Ion Exchange Resins", Methuen; Wiley (1957).
87. Korkisch, J., and F. Tera, "Separation of Thorium by Anion Exchange", Anal. Chem. 33, 1264 (1961).
88. Krau, K. A., and F. Nelson, "Radiochemical Separations by Ion Exchange", Ann. Rev. Nuclear Sci. 7, 31-46 (1957).
89. Kressin, I. K., and G. R. Waterbury, "The Quantitative Separation of Plutonium from Various Ions by Exchange", Anal. Chem. 34, 1598 (1962).
90. Kunin, R., and A. F. Preuss, "Ion Exchange in the Atomic Energy Program", Ind. Eng. Chem. 48, 30A-35A (Aug. 1956).
91. Kunin, R., "Ion Exchange Resins", 2nd Ed., Wiley (1958).
92. Lederer, E., and M. Lederer, "Chromatography", Elsevier (1957).
93. MacNevin, W. M., and E. S. McKay, "Separation of Rhodium from Platinum, Palladium, and Iridium by Ion Exchange", Anal. Chem. 29, 1220 (1957).
94. Maeck, W. J., and J. E. Rein, "Determination of Fission Product Iodine. Cation Exchange Purification and Heterogeneous Isotopic Exchange", Anal. Chem. 32, 1079 (1960).
95. Nachod, F. C., (ed.), "Ion Exchange—Theory and Application", Academic Press (1949).
96. Nachod, F. C., and J. Schubert (eds.), "Ion Exchange Technology", Academic Press (1956).
97. Nelson, F., and K. A. Kraus, "Anion Exchange Studies. XI. Lead (II) and Bismuth (III) in Chloride and Nitrate Solutions", J. Am. Chem. Soc. 76, 5916 (1954).
98. Newacheck, R. L., L. J. Beaufait, Jr., and E. E. Anderson, "Isotope Milker Supplies Ba^{137} from Parent Cs^{137}", Nucleonics 15, No. 5, 122.
99. Osborn, G. H., "Synthetic Ion Exchangers", Chapman and Hall (1955).
100. Overman, Coffey and Muse, "Separation of Co and Ni by Ion Exchange", J. Chem. Ed. 35, No. 6, 296-298 (1958).
101. Pascual, J., and E. C. Freiling, "Rare-Earth Solutions for 4π Counting", Nucleonics 15, No. 5, 94 (May 1957).
102. Perkins, R. W., "Filtration-Precipitation Separation of Barium-140 from Lanthanum-140", Anal. Chem. 29, 152 (1957).
103. Porter, C., D. Cahill, R. Schneider, P. Robbins, W. Perry, and B. Kahn, "Determination of Strontium-90 in Milk by an Ion Exchange Method", Anal. Chem. 33, 1306 (1961).
104. Power, W. H., H. W. Kirby, W. C. McCluggage, G. D. Nelson, and J. H. Payne, Jr., "Separation of Radium and Barium by Ion Exchange Elution", Anal. Chem. 31, 1077 (1959).
105. Samuelson, O., "Ion Exchangers in Analytical Chemistry", Wiley (1953).

106. Smit, J. Van R., W. Robb, and J. J. Jacobs, "AMP—Effective Ion Exchanger for Treating Fission Waste", Nucleonics 17, No. 9, 116-123 (Sept. 1959).
107. Stanley, C. W., and P. Kruger, "Determination of Sr^{90} Activity in Water by Ion-Exchange Concentration", Nucleonics 14, No. 11, 114-118 (Nov. 1956).
108. Tompkins, E. R., "Laboratory Application of Ion-Exchange Techniques", J. Chem. Educ. 26, 32-38, 92-100 (1949).
109. Tucker, W. D., "Radioisotopic Cows", Brookhaven Nat. Lab. Report, BNL 4908 (1960).
110. Walton, H. F., "Ion Exchange", Sci. Amer. p. 48-51 (Nov. 1950).
111. Wade, M. A., and H. J. Seim, "Ion Exchange Separation of Calcium and Strontium. Application to Determination of Total Strontium in Bone", Anal. Chem. 33, 793 (1961).
112. Winsche, W. E., L. G. Stang, Jr., and W. D. Tucker, "Production of Iodine-132", Nucleonics 8, No. 3, 14-18, 94 (1951).
113. Wish, L., "Quantitative Radiochemical Analysis by Ion Exchange. Calcium, Strontium, and Barium", Anal. Chem. 33, 53 (1961).
114. Wish, L., "Quantitative Radiochemical Analysis by Ion Exchange. Sodium and Cesium", Anal. Chem. 33, 1002 (1961).
115. Wish, L., "Quantitative Radiochemical Analysis by Ion Exchange. Anion Exchange Behavior in Mixed Acid Solutions and Development of a Sequential Separation Scheme", Anal. Chem. 31, 326 (1959).

Gas Chromatography

116. Cacace, F., "Labeled Organics in Gas Chromatography", Nucleonics 19, No. 5, 45-50 (May 1961).
117. Gudzinowicz, B. J., and W. R. Smith, "New Radioactive Gas Chromatographic Detector for Identification of Strong Oxidants", Anal. Chem. 35, 465 (1963).
118. James, A. T., and E. A. Piper, "A Compact Radiochemical Gas Chromatograph", Anal. Chem. 35, 515 (1963).
119. Johnson, H. W., and F. H. Stross, "Terms and Units in Gas Chromatography", Anal. Chem. 30, 1586 (1958).
120. Jones, W. L., and R. Kieselbach, "Units of Measurement in Gas Chromatography", Anal. Chem. 30, 1590 (1958).
121. Karmen, A., I. McCaffrey, J. W. Winkelman, and R. L. Bowman, "Measurement of Tritium in the Effluent of a Gas Chromatography Column", Anal. Chem. 35, 536 (1963).
122. Koch, R. C., and G. L. Grandy, "Xenon—Krypton Separation by Gas Chromatography", Nucleonics 18, No. 7, 76-80 (July 1960).
123. Lee, J. K., E. K. C. Lee, B. Musgrave, Yi-Noo Tang, H. W. Root, and F. S. Rowland, "Proportional Counter Assay of Tritium in Gas Chromatographic Streams", Anal. Chem. 34, 741 (1962).
124. Winkelman, J., and A. Karmen, "Use of an Ionization Chamber for Measuring Radioactivity in Gas Chromatography Effluents", Anal. Chem. 34, 1067 (1962).
125. Wolfgang, R., and F. S. Rowland, "Radioassay by Gas Chromatography of Tritium- and Carbon-14-Labeled Compounds", Anal. Chem. 30, 903 (1958).

Distillation

126. Kahn, M., A. J. Freedman, and C. G. Shultz, "Distillation of 'Carrier-Free' Iodine-131 Activity", Nucleonics 12, No. 7, 72-75 (July 1954).

Electrodeposition

127. Beacom, S. E., and B. J. Riley, "Tracer Follows Leveler in Electroplating Bath", Nucleonics 18, No. 5, 82-84 (May 1960).
128. Caldwell, P. A., and J. D. Graves, "Secondary-Standard Co^{60} Sources Prepared by Electrodeposition", Nucleonics 13, No. 12, 49-52 (Dec. 1955).
129. DeFord, D. D., "Electroanalysis", Anal. Chem. 26, 135-140 (1954).
130. Huff, J. B., "Electrodecontamination of Metals", Nucleonics 14, No. 6, 70-77 (June 1956).
131. Ko, R., "Americium Electrodeposition", Nucleonics 14, No. 7, 74 (July 1956).
132. Ko, R., "Electrodeposition of the Actinide Elements", Nucleonics 15, No. 1, 72-77 (Jan. 1957).
133. Kolthoff, I. M., J. Jordan, and A. Heyndricks, "Voltammetric Determination of Lead as Lead Dioxide at the Rotated Platinum Wire Electrode", Anal. Chem. 25, 884-887 (1953).
134. Lingane, J. J., "96493 Coulombs", Anal. Chem. 30, 1716 (1958).
135. Maletskos, C. J., and J. W. Irvine, Jr., "Quantitative Electrodeposition of Radiocobalt, Zinc and Iron", Nucleonics 14, No. 4, 84-93 (Apr. 1956).
136. Miller, H. W., and R. J. Brouns, "Quantitative Electrodeposition of Plutonium", Anal. Chem. 24, 536-538 (1952).
137. Mitchell, R. F., "Electrodeposition of Actinide Elements at Tracer Concentrations", Anal. Chem. 32, 326 (1960).
138. Moore, F. L., and G. W. Smith, "Electrodeposition of Plutonium", Nucleonics 13, No. 4, 66-69 (Apr. 1955).
139. Peterson, "Separation of Radioactive Iron from Biological Materials", Anal. Chem. 24, No. 11, 1850-1852 (1952).
140. Van Cleve, A., and F. D. McDonough, "Electroplating Technique for Tl^{204}", Nucleonics 12, No. 12, 53 (Dec. 1954).
141. Wilson and Langer, "Electrodeposition of Uranium Oxide on Aluminum", Nucleonics 11, No. 8, 48 (1953).

Amalgam Exchange

142. DeVoe, J. R., H. W. Nass, and W. W. Meinke, "Radiochemical Separation of Cadmium by Amalgam Exchange", Anal. Chem. 33, 1713 (1961).
143. Silker, W. B., "Separation of Radioactive Zinc from Reactor Cooling Water by an Isotope Exchange Method", Anal. Chem. 33, 233 (1961).

Electrophoresis

144. Aronsson, T., and A. Grönwall, "Improved Separation of Serum Proteins in Paper Electrophoresis—A New Electrophoresis Buffer", Science Tools, 5, No. 2, (Aug. 1958).
145. Formusa, K. M., R. R. Benerito, W. S. Singleton, and J. L. White, "Quantitative Determination of Serum Proteins by Paper Electrophoresis. Rapid Dyeing Method", Anal. Chem. 29, 1816 (1957).
146. Garvin, I. E., "Student Experiment with Filter Paper Electrophoresis", J. Chem. Educ. 38, 36-37 (1961).
147. Gilmore, R. C., Jr., M. C. Robbins, and A. F. Reid, "Labeling Bovine and Human Albumin with I^{131}", Nucleonics 12, No. 2, 65-68 (Feb. 1954).
148. Gray, G. W., "Electrophoresis", Sci. Amer. p. 45-53 (Dec. 1951).
149. Grunbaum, B. W., and P. L. Kirk, "Design and Use of a Refined Microelectrophoresis Unit", Anal. Chem. 32, 564 (1960).
150. Lederer, M., "Introduction to Paper Electrophoresis and Related Methods", Elsevier (1955).
151. McFarlane, A. S., "Labelling of Plasma Proteins with Radioactive Iodine", Biochem. J. 62, 135-143 (1956).
152. Strain, H. H., and J. C. Sullivan, "Analysis by Electromigration Plus Chromatography", Anal. Chem. 23, 816-823 (1951).
153. Yoon, C. H., "Electrophoretic Analysis of the Serum Proteins of Neurological Mutations in Mice", Science 134, 1009-1010 (1961).

CHAPTER 11
Nuclear Analysis (Identification and Calibration of Nuclides)

LECTURE OUTLINE AND STUDY GUIDE

I. QUALITATIVE IDENTIFICATION OF NUCLIDES (*Experiment* 11.1)
- A. Criteria for Identification
  1. Half-life
  2. Type of radiation (alpha, beta, gamma, etc.)
  3. Energy of radiation
  4. Chemical properties
- B. Determination of Half-Life (Chapter 6)
- C. Identification of Type of Radiation
  1. Interaction with magnetic and electrostatic fields
  2. Characteristics of path produced in cloud chamber or in photographic emulsion
  3. Absorption measurements
  4. Scattering characteristics
  5. Use of detectors with selective sensitivity (e.g., proportional and scintillation counters)
- D. Determination of Energy of Radiation
  1. From range and absorption measurements
  2. By means of spectrometry

II QUANTITATIVE CALIBRATION OF RADIOACTIVE SOURCES
- A. Objectives
- B. Definitions
  1. Types of standards
  2. Units of radioactivity
  3. Units of dose
  4. Units of dose rate
  5. Units of radiation intensity
  6. Units of energy absorption
- C. Standards
  1. Primary
  2. Secondary
- D. Primary Standardization Methods
  1. Coincidence method
  2. Calorimetry
  3. 4 counting
  4. Absolute alpha counting (*Experiment* 11.2)
  5. Absolute beta counting by defined geometry
  6. Methods utilizing parent-daughter equilibria (*Experiment* 11.6)
- E. Calibration by Comparison with a Standard
  1. Standard prepared from same isotope as unknown (*Experiment* 11.3)
     - a. Precautions required
       - (1) Same position of standard and sample
       - (2) Same area of standard and sample
       - (3) Same backing material for standard and sample
     - b. Available standards
       - (1) Long-lived standards
       - (2) Short-lived standards
  2. Standard (Beta) prepared from different isotope (*Experiment* 11.4)
     - a. Criteria
       - (1) A beta ray or positron must be emitted for every disintegration
       - (2) Beta-ray spectrum of standard must not be too different from that of sample
       - (3) Absorption correction must be applied
     - b. Types of standards available
  3. Simulated standards (*Experiment* 11.5)
     - a. Phosphorus-32
     - b. Iodine-131
  4. Gamma ray standards and methods
     - a. Using the electroscope (*Experiment* 8.1)
     - b. Using the scintillation detector

INTRODUCTION:

The roles of qualitative and quantitative analysis in the field of chemistry are well defined. It may be convenient to consider a parallel disciplinary subdivision in the field of nucleonics. Thus, nuclear analysis may be either qualitative or quantitative, the former dealing with the identification of nuclides and the latter with the measurement of the quantity of a particular nuclide present in a sample.

QUALITATIVE IDENTIFICATION OF NUCLIDES:

In qualitative analysis the identification of an ion or chemical substance is based upon the observed chemical and physical properties. The identification of a nuclide is also based upon its observed properties. These properties may include the half-life, the type of radiation emitted and the energy of the emitted radiation. In addition, it is sometimes possible to determine the chemical identity of an unknown nuclide by means of a chemical analysis. A chemical analysis could prove to be difficult, however, unless the scope of possibilities has been narrowed to somewhat fewer than the more than 100 chemical elements known. A more complete discussion of the qualitative identification of nuclides will be found in experiment 11.1.

QUANTITATIVE CALIBRATION OF NUCLIDES:

Quantitative measurements of radioactive sources may be performed for any of several basic purposes:

1. To measure the ratio of activities of two or more sources of the same nuclide. (i.e., to measure R_1/R_2).
2. To measure the true decay rate of a source. (i.e., to measure A).
3. To measure the dose rate or radiation intensity at a given place.

When only the ratio of activities need be known the procedure of measurement can usually be simplified. It is only necessary to maintain a constant value of counting efficiency since the true decay rate A is related to R, the observed activity corrected for background, by the relation $R = A\epsilon$. If the efficiency ϵ is constant, then

$$\epsilon = \frac{R_1}{A_1} = \frac{R_2}{A_2} = \frac{R_3}{A_3} = \ldots\ldots$$

and

$$R_1{:}R_2{:}R_3{:}\ldots\ldots\ldots = A_1{:}A_2{:}A_3{:}\ldots\ldots\ldots$$

Such measurements are made, for example, in certain uptake and distribution studies wherein the results are reported in terms of a percentage of the total activity. The total activity should always be known approximately for the sake of safety but need not be known with great accuracy. The majority of radioactive measurements in chemistry, biology and medicine are of this type.

To measure the true decay rate of a source requires additional information. The decay rate is related to the decay constant λ and the number of radioactive atoms N present in the sample.

$$A = -dN/dt = \lambda N$$

If any two quantities are known the third can be calculated. Normally the value of the decay constant is readily available from the half-life. Thus in practice it is only necessary to evaluate either the decay rate $A = -dN/dt$ or to measure the number of atoms N present in the sample. The determination of one of these quantities constitutes the objective of a radioactive standardization. The accuracy of a standardization depends upon the accuracy with which these quantities can be measured. From the previous paragraph it can be seen that one approach would be to evaluate the efficiency ϵ from which the true decay rate A could be calculated from the observed value R.

When the interest lies not so much in knowing the disintegration rate of a sample as in knowing what effect the radiation will have on the surroundings, a measure of the dose rate delivered to a given place in space or the dose absorbed by a given tissue is of greater importance.

DEFINITIONS:

An *absolute* or *primary standard* is one which has been calibrated in terms of the accepted standards of mass, length and time. The calibration process is known as an *absolute standardization.*

A *relative standard* (a *derived* or *secondary standard*) is one which has been calibrated against an absolute or primary standard.

A *reference source* or *performance standard* is usually a long-lived source used to check the constancy (or changes in efficiency) of radiation detection equipment. The decay rate of such a source need not be known.

RADIOLOGICAL UNITS

UNITS OF RADIOACTIVITY (Quantity of material, activity or decay rate)

Curie—The quantity of any radioactive nuclide in which the number of disintegrating atoms per second is 3.7×10^{10}.

*Rutherford**—The quantity of any radioisotope which disintegrates at the rate of 10^6 atoms per second.

UNITS OF DOSE

Roentgen (r)—*Exposure dose of X- or gamma radiation* at a certain place is a measure of the radiation that is based upon its ability to produce ionization.

The unit of exposure dose of X- or gamma radiation is the *roentgen.* One roentgen is an exposure dose of X- or gamma radiation such that the associated corpuscular emission per 0.001293 g of air produces, in air, ions carrying 1 electrostatic unit of quantity of electricity of either sign. (0.001293 g of air occupies a volume of 1 cc at STP.)

$$\begin{aligned} 1\ r &= 1.61 \times 10^{12} \text{ ion pairs per gram of air (1 g of air = 711.5 cc)} \\ &= 83.8 \text{ ergs per gram of air} \\ &= 93.1 \text{ ergs per gram of water} \end{aligned}$$

*These units are obsolete or are used only rarely.

Roentgen-equivalent-physical (rep)* — The roentgen-equivalent-physical applies to any type of radiation. (Note — The roentgen applies only to X- or gamma radiation) The dose is one rep if the energy lost by ionization in the tissues is the same as the energy loss for one r of gamma radiation absorbed in air. That is:

1 rep = 83 ergs per gram of tissue

*Gram-roentgen** — The gram-roentgen is the energy dissipated in one gram of air by one roentgen of gamma radiation. Thus:

1 gram-roentgen = 83.8 ergs in an unspecified amount of tissue

or 1 rep = 1 gram-roentgen per gram of tissue

*Energy-unit** — In considering the energy dissipated in tissue by fast neutrons, Gray and his associates used an "energy-unit" defined as the same energy in one gram of tissue as is dissipated by gamma rays in one gram of water, about 93 ergs. Thus:

1 energy-unit = 93 ergs per gram of tissue

Roentgen-equivalent-man (rem)* — One rem is the estimated amount of energy absorbed in tissue which is biologically equivalent in man to one roentgen of X-or gamma radiation. By definition:

1 rem = 83/RBE ergs per gram of tissue

The relative biological effectiveness (RBE) is an estimate of the relative effectiveness of different types of radiation on tissue as compared to the same dose of X- or gamma radiation.

Rad — Absorbed dose of any ionizing radiation is the energy imparted to matter by ionizing particles per unit mass of irradiated material at the place of interest.

The unit of absorbed dose is the rad. One rad is 100 ergs/g.

Gram rad — Integral absorbed dose in a certain region is the energy imparted to matter by ionizing particles in that region.

The unit of integral absorbed dose is the gram rad. One gram rad is 100 ergs.

RBE dose — RBE dose is equal numerically to the product of the dose in rads and an agreed conventional value of the RBE with respect to a particular form of radiation effect. The standard of comparison is X- or gamma radiation having a LET in water of 3 keV/μ delivered at a rate of about 10 rad/min.

The unit of RBE dose is the *rem*. It has the same inherent looseness as the RBE.

UNITS OF DOSE RATE

Exposure dose rate is the exposure per unit of time. The unit of exposure dose rate is the *roentgen per unit of time.* e.g., mr/hr

Absorbed dose rate is the absorbed dose rate per unit of time. The unit of absorbed dose rate is the *rad per unit of time.*

Specific gamma-ray emission (specific gamma-ray output) Γ of a radioactive nuclide is the exposure dose rate produced by the unfiltered gamma-rays from a point source of a defined quantity of nuclide at a definite distance. The unit of specific gamma-ray emission is the *roentgen per millicurie hour* (r/mCih) *at 1 cm.*

UNITS OF RADIATION INTENSITY

Intensity of radiation (radiation energy flux density) at a given place is the energy per unit time entering a small sphere of unit cross-sectional area centered at that place. The unit of intensity of radiation may be *ergs per square centimeter second,* or *watts per square centimeter.*

*These units are obsolete or are used only rarely.

UNITS OF ENERGY ABSORPTION

Linear Energy Transfer (LET) is the linear-rate of energy loss (Locally absorbed) by an ionizing particle traversing a material medium. Linear energy transfer may be expressed in kilo electron volts per micron (keV/μ)

STANDARDS:

A *primary standard* is one which has been calibrated in absolute measure in terms of the accepted standards of mass, length and time. A good primary standard should have a relatively long half-life to eliminate problems arising from changes in activity during the standardization procedure and to eliminate the need for carrying out repeatedly the necessarily tedious process of standardization at frequent intervals. The decay scheme for the nuclide chosen for the preparation of a primary standard must also be known accurately.

International primary standards of radium are based on mass. These international standards have been intercompared and are maintained in Washington, Paris and other cities throughout the world in order to provide a means for the comparison of results from laboratories in one country with results from those in another. The first standard, consisting of 21.99 mg of pure anhydrous radium chloride, was prepared in 1911 by Mme. Curie in Paris. In Vienna, Professor Otto Hönigschmid also prepared several radium standards which were compared with that of Mme. Curie. For the next twenty years the Curie radium standard in Paris and one of the Hönigschmid standards in Vienna served as the international standards. Then the International Radium Standards Committee, being concerned about the deterioration of these standards, invited Hönigschmid to prepare a set of new ones. In 1934, Hönigschmid prepared twenty new primary standards from highly purified radium chloride. One of these, containing 22.23 mg of radium chloride was selected as the new international primary standard while the others were designated as international secondary standards. These are the radium standards currently used.

When artificially prepared radioactive substances become available in large quantities it was considered necessary to have standards for nuclides other than radium-226. In 1951, Canada, the United Kingdom and the United States exchanged samples of carbon-14, sodium-24, phosphorus-32, cobalt-60, bromine-82, strontium-90-yttrium-90, iodine-131, gold-198 and thallium-204 for the purpose of intercomparison. Responsibility for international standards of radioactivity has now been assumed by The International Commission on Radiological Units and Measurements (ICRU). Recommendations of the ICRU have been published in National Bureau of Standards Handbooks 47, 62, 78, 84, 85, 86, 87, 88 and 89. A complete list of these handbooks appears in appendix J.

Secondary standards (*derived* or *relative* standards) are available from the National Bureau of Standards and from many commercial suppliers of radioisotopes. They are often classified as alpha, beta or gamma standards, several activities and physical forms usually being available for each of the most commonly used nuclides. Those available from the National Bureau of Standards at the time of publication of Circular 594 are listed in Table 11.1. A very complete list of currently available standards will be found in "The Isotope Index"*.

*Published by Scientific Equipment Corp., 23 N. Hawthorne Lane, Indianapolis 19, Indiana.

PRIMARY STANDARDIZATION METHODS:

The standardization of a radioactive source requires the evaluation of at least one of the following three quantities:

1. Decay rate. ($A = -dN/dt$)
2. Number (or mass) of radioactive atoms in the sample. ($N = N_0 m$)
3. Dose rate or radiation intensity at a given place—(r/h for gamma or x-rays)

Absolute methods in use by the National Bureau of Standards are listed in table 11.2. Some of these methods will be discussed in the following paragraphs.

COINCIDENCE COUNTING:

Coincidence counting is a unique and ingenious approach to calibration. The emission of a particle and a photon or of two photons during radioactive-decay is required. Possible combinations are α-γ, β-γ, γ-γ and x-γ coincidences. The technique is illustrated here by a β-γ coincidence.

The sample is placed between two counters, one of which is sensitive to beta particles and the other to gamma rays. If A is the true activity of the sample, the quantity we wish to determine, then the total activity R_β registered on the beta counter will be

$$R_\beta = A\,\epsilon_\beta \tag{11.1}$$

where ϵ_β is the efficiency of the beta detector system, including the instrument efficiency and the geometry factor. Similarly, the recorded activity of the gamma scaler R_γ will be

$$R_\gamma = A\,\epsilon_\gamma \tag{11.2}$$

Pulses from both the beta detector and the gamma detector are also fed into a coincidence circuit, the output from which is recorded on the coincidence scaler. There will be an output to the coincidence scaler only if two pulses from the beta and gamma detectors, respectively, arrive at the coincidence circuit during the coincidence time. The activity R_c thus recorded is given by the expression

$$R_c = A\,\epsilon_\beta \epsilon_\gamma \tag{11.3}$$

If both equations 11.1 and 11.2 are solved for the efficiency and substituted into equation (11.3), one obtains the basic uncorrected expression

$$A = \frac{R_\beta R_\gamma}{R_c} \tag{11.4}$$

Table 11.1—*Primary alpha, beta, and gamma radioactivity standards maintained by the National Bureau of Standards*

Sample No.	Radiation	Nuclide	Nominal activity [a]	Volume	Chemical form of standard	Method of primary standardization	Method of secondary standardization
4900	α	Polonium-210[b]	200 dps	(e)	Polonium plated from polonium-chloride solution.	$2\pi\alpha$-prop. counting	$2\pi\alpha$-prop. counting.
4901	α	Polonium-210[b]	500 dps	(e)	do	do	Do.
4902	α	Polonium-210[b]	1,000 dps	(e)	do	do	Do.
4903	α	U_3O_8[d]	15 dps	(e)		do	Do.
4910	$\beta(\alpha)$	RaD+E[f]	200 dps	(c)	Lead peroxide	Quantitative extraction from pitchblende.	Defined-geometry G.-M. counting.
4911	$\beta(\alpha)$	RaD+E[f]	500 dps	(c)	do	do	Do.
4912	$\beta(\alpha)$	RaD+E[f]	1,000 dps	(c)	do	do	Do.
4913	$\beta(\gamma)$	Cobalt-60	10^4 dps/ml	(g)	Chloride in HCl solution	$4\pi\beta$-prop. counting; coinc. counting.	$4\pi\gamma$-ion. chamber; formamide counting.
4914	$\gamma(\beta)$	Cobalt-60	10^5 dps	5.0 ml	do	do	Do.
4915	$\gamma(\beta)$	Cobalt-60	10^6 dps	5.0 ml	do	do	Do.
4916	β	Phosphorus-32	10^5 dps/ml	(g)	Phosphoric acid solution	$4\pi\beta$-prop. counting	$2\pi\beta$-ion. chamber; formamide counting.
4917	$\beta(\gamma)$	Iodine-131	10^5 dps/ml	(g)	Sodium iodide solution	$4\pi\beta$-prop. counting; coinc. counting.	$2\pi\beta$-$4\pi\gamma$-ion. chambers; formamide counting.
4918	$\beta(\gamma)$	Gold-198	10^5 dps/ml	(g)	Auric cyanide solution	do	Do.
4919	β	Strontium-90 / Yttrium-90	10^4 dps/ml	(g)	Chloride in HCl solution	$4\pi\beta$-prop. counting	$2\pi\beta$-ion. chamber; formamide counting.
4920	β	Thallium-204	10^4 dps/ml	(g)	Thallic nitrate in HNO_3 solution.	do	Do.
4921	$\beta(\gamma)$	Sodium-22	10^4 dps/ml	(g)	Chloride in HCl solution	$4\pi\beta$-prop. counting; coinc. counting; Co^{60} comparison.	$4\pi\gamma$-ion. chamber.
4922	$\gamma(\beta)$	Sodium-22	10^6 dps	5.0 ml	do	do	Do.
4923	$\beta(\gamma)$	Sodium-24	10^5 dps/ml	(g)	do	$4\pi\beta$-prop. counting; coinc. counting.	Do.
4924	β	Carbon-14	10^3 dps/ml	25.0 ml	Sodium carbonate solution	Compensated gas counting	Formamide counting; gas ion. chamber; liq. scint. counting.
4925	β	Carbon-14[h]	10^4 dps/g	(g)	Benzoic acid in toluene	do	Do.
4926	β	Hydrogen-3	10^4 dps/ml	25.0 ml	Tritiated water solution	Calorimeter; compensated gas counting.	Liq. scint. counting; gas ion. chamber.
4927	β	Hydrogen-3	10^5 dps/ml	(g)	do	do	Do.
4928	β	Sulfur-35	10^4 dps/ml	(g)	Sodium sulfate solution	Calorimeter; $4\pi\beta$-prop. counting.	Liq. scint. counting; formamide counting.
4929	K	Iron-55	10^5 dps/ml	(g,i)	Ferric chloride in HCl solution.	Liq. scint. counting	Liq. scint. counting; X-ray counting.
4930	$K(\gamma)$	Zinc-65	10^5 dps/ml	(g)	Chloride in HCl solution	Liq. scint. counting; coinc. counting.	Do.
4931	$\gamma(\beta)$	Cesium-137 / Barium-137	10^4 dps	5.0 ml	do	$4\pi\beta$-prop. counting; coinc. counting.	$4\pi\gamma$-ion. chamber.
4932	$\gamma(\beta)$	Mercury-203	10^6 dps	5.0 ml[i]	Mercurous nitrate solution	do	Do.
4933	$\gamma(\beta)$	Potassium-42	10^5 dps/ml	(g)	Chloride in HCl solution	do	Do.
4934	$\gamma(\beta)$	Tantalum-182	10^5 dps/ml	(g)	Fluoride in HCl solution	$4\pi\beta$-prop. counting	Do.
4935	β	Krypton-85	10^4 dps	(j)	Gaseous krypton	Compensated gas counting	Gas ion. chamber; gas counting.

[a] The disintegration rate as of the reference data is given on a certificate accompanying the standard.
[b] Samples consist of a practically weightless deposit of polinium-210 on a silver disk 1 in. in diameter, 1/16 in. thick, and faced with 0.002 in. of palladium.
[c] Deposited source.
[d] Samples consist of U_3O_8 deposited on a 0.1-mm-thick platinum foil and mounted on an aluminum disk, 1¼ in. in diameter and 1/32 in. thick. The alpha-ray disintegration rate as of the date of calibration and in the forward hemisphere is indicated on the certificate accompanying the standard.
[e] Evaporated source.
[f] Standards consist of Pb-210 and Bi-210 in equilibrium, deposited on a silver disk 1 in. in diameter, 1/16 in. thick, and faced with 0.002 in. of palladium.
[g] Approximately 3 ml of low-solids carrier solution containing the active nuclide in a flame-sealed ampoule.
[h] Benzoic acid in toluene for use in liquid scintillation counters.
[i] In preparation.
[j] Approximately 10 ml of krypton at STP in a break-seal ampoule.

Radium standards (for radon analysis)

Sample No.	Radium content (in grams)	Volume (in ml)
4950	10^{-9}	100
4951	10^{-11}	100
4952	Blank solution	100

Samples are sealed in glass containers. They were prepared by determining the radium content of a purified sample of radium chloride by direct comparison with the U. S. Primary Radium Standards by means of the gold-leaf electroscope and radiation balance (Mann, Stockmann, Youden, Schwebel, Mullen, and Garfinkel, 1958). This radium chloride was then transferred quantitatively to a carrier solution consisting of 0.2 percent by weight $BaCl_2 \cdot 2H_2O$ in a 5 percent by weight solution of HCl.

Radium gamma-ray standards

Sample No.	Radium content (in grams)	Volume (in ml)
4955	0.1×10^{-6}	5
4956	0.2	5
4957	0.5	5
4958	1.0	5
4959	2.0	5
4960	5.0	5
4961	10	5
4962	20	5
4963	50	5
4964	100	5

Samples are contained in flame-sealed glass ampoules. They were prepared by determining the radium content of a purified sample of radium bromide by direct comparison with the U. S. Primary Radium Standards by means of the gold-leaf electroscope. The radium bromide was then dissolved quantitatively in a 5 percent by weight solution of HNO_3.

Reproduced from National Bureau of Standards Circular 594, "Preparation, Maintenance, and Application of Standards of Radioactivity", June 11, 1958.

Table 11.2—*Absolute methods for the preparation of standards of radioactive nuclides*

Method	Nuclides to which method is applicable	Activity of measured source	Special corrections, limitations, etc.	Accuracy under most favorable conditions*
1. *4π Counting*				
(a) Geiger-Müller counter	α emitters, β emitters, with or without γ rays.	0.01μc or less	Accuracy usually limited by uncertainty in corrections for loss of β particles in source material and source mount, particularly for β particle energies <0.5 Mev.	±2%
(b) Proportional counter	α emitters, β emitters, with or without γ rays.	0.1μc or less	Accuracy usually limited by uncertainty in corrections for loss of β particles in source material and source mount, particularly for β particle energies <0.5 Mev.	±2%
(c) High pressure proportional counter.	Electron-capturing nuclides	0.1μc or less	Method assumes knowledge of K-fluorescence yield and K/L capture ratio.	±5%
(d) Scintillation counting; solid phosphor.	β emitters, with or without γ rays.	0.1μc or less	Suitable for medium and high energy β emitters; low energy cut-off limits accuracy for low energy β emitters.	±2%
(e) Scintillation counting; liquid phosphor.	β emitters, with or without γ rays, α emitters.	0.1μc or less	Suitable for medium and high energy β emitters; low energy cut-off limits accuracy for low energy β emitters.	±2%
2. *Defined Solid Angle*				
(a) End window Geiger-Müller counter.	β emitters	0.1μc or more	Complex corrections for absorption and scattering effects	±5%
(b) Proportional counter	α emitters	0.1μc or more		±0.5%
	β emitters	0.1μc or more		±5%
(c) Zinc sulphide screen	α emitters	0.1μc or more		±0.5%
3. *Coincidence Counting*				
(a) βγ	β+γ emitters having simple decay schemes.	1μc or less	Uncertainty in the correction for the gamma sensitivity of the β detector usually limits accuracy.	±2%
(b) 4πβ-γ	Many β+γ emitters	0.1μc or less	Corrections, in general, very small and hence method is more accurate than βγ method.	±1%
(c) γγ	So far chiefly used for Co^{60}		Requires careful adjustment of efficiencies of counters, and allowance for angular correlation of γ rays.	±1%
(d) Xγ and 4πX-γ	Electron-capturing nuclides	0.5μc or more		±5%
4. *Internal gas counting*	Low-energy β emitters. Electron-capturing nuclides.	0.01μc or less	Limited to radionuclide samples having suitable gaseous form.	±2%
5. *Measurement of loss of charge*	β emitters	0.1mc	Application limited by experimental difficulties	±5%
6. *Calorimetry*	α emitters	1mc or more		±0.5%
	β emitters	1mc or more	Involves a knowledge of the mean energy of the β particles.	±2%
	β+γ emitters	1mc or more	Involves a knowledge of the mean energy of the β particles, the quantum energies of the γ rays, the number of γ quanta of each energy per disintegration and the fraction of the total γ radiation absorbed in the calorimeter.	±5%
7. *Ionization measurement of dose rate.*	γ emitters having relatively simple γ-ray spectra.	1mc or more	Accuracy of disintegration rate determination limited by uncertainty in *W* (energy per ion pair) for air and in some cases by incomplete knowledge of γ-ray spectra.	±4%
	β emitters		Accuracy limited by uncertainty in *W* and in average energy of β particles.	±5%
8. *Weighing*	Radium and long-lived radionuclides.	10 mg	High chemical and radio-chemical purity of radionuclide or parent essential.	±0.2%†
9. *Mass spectrometer*	Has been used for H^3			±1% (routinely)

*The figures given in this column are estimates of the extreme limits of total error in the disintegration-rate value, and the errors which contribute to this total depend on various factors which are not the same for each nuclide; the limits given indicate the accuracy believed to be obtainable under the most favorable conditions presently available.
†Radium standards.

Reproduced from "Radioactivity", National Bureau of Standards Handbook 86, November 29, 1963.

From the three scaler readings one is able to determine the absolute activity of the sample. The unique thing about this method is that the efficiencies of the counting systems do not enter into the final calculation.

In the derivation of equation 11.4, counting corrections were neglected to avoid confusing the basic concept. In practice, however, corrections must be applied. These are of two types: (1) corrections for background, resolving time and other factors

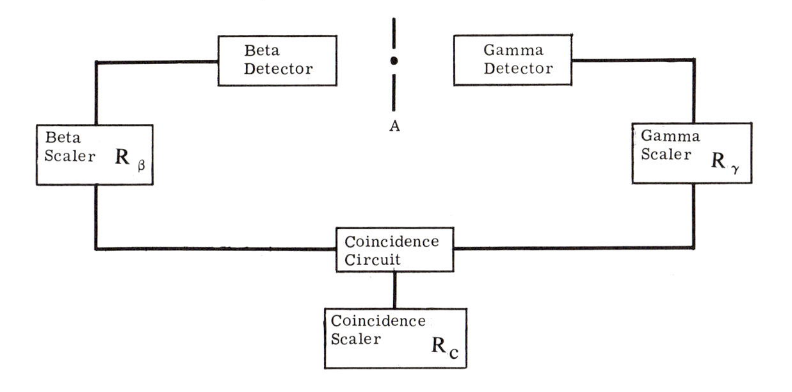

Figure 11.1—BLOCK DIAGRAM OF APPARATUS FOR β-γ COINCIDENCE COUNTING.

concerning the electronic circuitry and (2) corrections concerning the decay scheme. For background the following relationships must be considered:

$$r_\beta = R_\beta + r_{b\beta} = A\epsilon_\beta + r_{b\beta} \quad (11.5)$$

$$r_\gamma = R_\gamma + r_{b\gamma} = A\epsilon_\gamma + r_{b\gamma} \quad (11.6)$$

$$r_c = R_c + r_{bc} = A\epsilon_c + r_{bc} \quad (11.7)$$

where r_β, r_γ and r_c are the observed beta, gamma and coincidence activities, respectively and $r_{b\beta}$, $r_{b\gamma}$ and r_{bc} are the corresponding background activities.

The potential complexities of this type of measurement, as fewer assumptions and approximations are made, is illustrated by the following relationship*:

$$A = \frac{R_\beta R_\gamma}{R_c}\left[1 + R_c T_D + \frac{2\, r_\beta r_\gamma - r_c(r_\beta + r_\gamma)}{r_c} T\right] \quad (11.8)$$

In this equation T is the resolving time of the coincidence circuit and T_D is the dead time of each circuit.

An empirical expression**, similar to equation 11.8 but of lesser accuracy, is given by

$$A = \frac{R_\beta R_\gamma}{R_c}\left(1 + T'\frac{R_\beta R_\gamma}{R_c}\right) \quad (11.9)$$

Here the parameter T′ must be evaluated for each nuclide and for each particular apparatus.

*NBS Handbook 80
**NBS Handbook 86

CALORIMETRY:

Calorimetric methods are used to measure the heat Q evolved by the nuclear reaction. The basic principles involved are similar to those of ordinary calorimetry, but the problems encountered are somewhat different. Because of the smallness of the weight of the sample, and the correspondingly small amount of heat to be measured, one must use a liquid nitrogen calorimeter or other extremely sensitive type, such as the Peltier-effect microcalorimeter. On the other hand, the amount of the sample required is in the order of one curie, which poses a considerable problem from the standpoint of handling the sample. Nevertheless, calorimetric methods have proved their usefulness in nuclear studies.

4-PI COUNTER (PROPORTIONAL OR GEIGER):

The four-pi counter was designed to eliminate backscattering errors by producing 100% geometry. These counters consist essentially of two 2-pi counters mounted face to face, with the sample mounted at the geometrical center between them on an extremely thin film. In this way, all particles emitted from the sample, regardless of the direction of emission, will be counted equally. Several commercial models of the 4-pi counter are now available.

ABSOLUTE BETA COUNTING BY DEFINED GEOMETRY:

This method, which was discussed briefly in chapter 3, consists essentially of measuring accurately each term of the equation relating disintegrations per second, A, to counts per second, R.

$$A = \frac{R}{G\, f_A f_B f_H f_S f_W} \tag{11.10}$$

where G = the physical geometry factor (see experiment 5.4)
f_A = factor for forescattering in air
f_B = backscattering factor (See experiment 5.5)
f_H = factor for scattering by walls and housing
f_S = self absorption and self scattering factor (See experiment 5.6)
f_W = factor for absorption due to air, window and sample cover

Following the evaluation of each of these factors according to the methods outlined in chapter 5, the calculation of the disintegration rate from the observed activity resolves itself to the simple calculation of equation 11.10.

EXPERIMENT 11.1 QUALITATIVE IDENTIFICATION OF NUCLIDES

OBJECTIVES:

To demonstrate a procedure for the identification of nuclides.

To identify an unknown nuclide.

THEORY:

The systematic identification of a nuclide consists of (1) measurement of the half-life, (2) identification of the type or types of radiation emitted, (3) measurement of the energy of the radiation, (4) separation and identification by chemical means, if appropriate, and finally (5) comparison of the observed data with that in tables and other literature. The first three of these steps have been discussed, at least in part, in previous experiments.

Unless some knowledge of the sample activity is available, suitable protective measure should be taken. A cutie-pie type ionization chamber is especially well suited to determine the radiation level from the source. If necessary, the source should be handled by means of tongs and other remote handling equipment behind a lead shield sufficiently thick to reduce radiation to a safe level. If the source is very active, it must be subdivided to provide a source of the proper activity for counting.

Half-life—Certain techniques used for the determination of half-life were discussed in chapter 6. For the most part these same techniques are applicable in this experiment as well.

Upon receipt of the radioactive sample, half-life measurements should be started as soon as possible in order that very short-lived nuclides might be detected and their half-lives measured. This is particularly true of samples which have been recently activated by subjection to neutron bombardment. Quite often, recently activated samples contain a mixture of nuclides, in which case a combined decay curve of the type indicated in experiment 6.2 will be obtained.

When the half-life is very short (less than 10 minutes), its measurement by the method used in experiment 6.1 is difficult. Frequent observations of activity must be made. A ratemeter and recorder are usually much more useful than a scaler in such cases since a continuous record of activity is obtained. Also the results are more meaningful statistically because the maximum number of disintegrations has been measured. Still better, for short lived isotopes, is the technique of recording the output of the detector on a tape recorder. The information is then permanently recorded exactly as it occurred and can be analyzed at the leisure of the individual.

On the other hand, when the half-life is relatively long (more than two weeks) the time required for an accurate measurement is extended. If this is the case, the determination of types and energy of the radiation becomes more important, especially if immediate results are required.

Type of radiation—The following simple screening procedure for the identification of the type of radiation may be found useful:

Table 11.3—BETA PARTICLE ENERGIES AND HALF-LIVES

β-Energy \ Half-life	< 1 hour	1-6 hours	6-24 hours	1-5 days	5-30 days	30 days-1 year	> 1 year
0-0.3 MeV				Au^{199}	Os^{191}	*S^{35}, *Ca^{45} (2)Fe^{59}, Nb^{95} Ru^{103}, Hg^{203}	*H^{3}, *C^{14} *Ni^{63}, *Tc^{99} (2)Cs^{134}, *Pm^{147} Eu^{155}
				Br^{82}, *Sn^{121} (1)Yb^{175}	Xe^{133}, *Er^{169} Lu^{177}	Sc^{46}, (1)Fe^{59} Zr^{95}, Ce^{141} Ce^{144}, Hf^{181} *W^{185}	Co^{60}
0.5-0.7 MeV		(3)Mn^{56}, (2)Ni^{65}	(1)Cu^{64}, Te^{127} (2)Gd^{159}, (1)W^{187} Pt^{197}	As^{77}, (1)Rh^{105}	I^{131}	(1)Sb^{124}, Ir^{192}	Na^{22}, Kr^{85}, *Sr^{90} (1)Cs^{134}, Cs^{137}
0.7-1.0 MeV	Zn^{69}	Kr^{85m}	(1)Ga^{72}, (1)Gd^{159}	Au^{198}	*Pr^{143}, (1)Nd^{147}	Tm^{170}	*Cl^{36}, *Tl^{204}
1.0-1.5 MeV	(2)Cl^{38}	*Si^{31}, A^{41} (2)Mn^{56}, (3)Ni^{65} Nb^{97}, Ru^{105} Nd^{149}	Na^{24}, Pd^{109} Er^{171}	Mo^{99}, (1)Sb^{122} Ce^{143}, Pm^{149} (1)Re^{186}, Os^{193} *Bi^{210}	Ag^{111}	*Sr^{89}	
1.5-3.0 MeV	Al^{28}, Ga^{70} (2)Rb^{88}, Mg^{27}	(1)Mn^{56}, (1)Ni^{65} Br^{80}, Ba^{139}	(2)K^{42}, (2)Ga^{72} Zr^{97}, Pr^{142} Re^{188}, Ir^{194}	As^{76}, *Y^{90} (2)Sb^{122}, Ho^{166}	*P^{32}, (1)Rb^{86}	Y^{91}, In^{114} Cd^{115m}, (2)Sb^{124}	
>3.0 MeV	(1)Cl^{38}, (1)Rb^{88}		(1)K^{42}, (3)Ga^{72}			Ce^{144}/Pr^{144}	Ru^{106}/Rh^{106}

Numbers in parentheses indicate that more than one β-group exists. The most abundant group is numbered (1), the second most abundant (2) and so on.
*Pure β-emitter.

(From R. A. Allen, R. J. Millet and D. B. Smith, "Radioisotope data" reproduced with permission of United Kingdom Atomic Energy Authority.)

Table 11.4—GAMMA ENERGIES AND HALF-LIVES

γ-Energy \ Half-life	<1 hour	1-6 hours	6-24 hours	1-5 days	5-30 days	30 days-1 year	>1 year
0-0.3 MeV		(1)Ba^{139}	Er^{171}, Re^{188}	Xe^{133m} Sm^{153}, Au^{199}	Xe^{133}, (2)Nd^{147}	Ce^{141}, Tm^{170} (2)Hf^{181}, Hg^{203}	Eu^{155}
0.3-0.5 MeV	(2)In^{116m}	Sr^{87m}	Zn^{69m}, Gd^{159}	Rh^{105}, (2)La^{140} Au^{198}	Cr^{51}, I^{131}	Be^{7}, Sn^{113} Hf^{175}, (1)Hf^{181}	Ba^{133}
0.5-0.7 MeV		Br^{80}, Nb^{97}	$Cu^{64}(\beta^{+})$	(1)As^{76}, (1)Br^{82} Sb^{122}	$Mn^{52}(\beta^{+})$ (1)Nd^{147}	$Co^{58}(\beta^{+})$, Ru^{103} (1)Sb^{124}	$Na^{22}(\beta^{+})$ (1)Cs^{134}, Cs^{137}
0.7-1.0 MeV	(2)Rb^{88}, Mg^{27}	(1)Mn^{56}, Ru^{105} (1)I^{132}	(1)Ga^{72}, Zr^{97}	(2)Br^{82}	(1)Mn^{52}	(1)Sc^{46}, Mn^{54} Co^{58}, Zr^{95}, Nb^{95}	(2)Cs^{134}
1.0-1.5 MeV	(1)In^{116m}	A^{41}, Ni^{65} (2)Ba^{139}	(1)Na^{24}	(2)As^{76}, (3)Br^{82}	(2)Mn^{52}, Rb^{86}	(2)Sc^{46}, Fe^{59} Zn^{65}	Na^{22}, Co^{60}
1.5-3.0 MeV	Al^{28}, Cl^{38} (1)Rb^{88} (3)In^{116m}	(2)Mn^{56}	(2)Na^{24}, K^{42} (2)Ga^{72}, Pr^{142}	(1)La^{140}		(2)Sb^{124}	

Only isotopes having relatively simple γ-decay schemes are listed.
Numbers in parentheses indicate that more than one γ-group exists. The most abundant group is numbered (1), the second most abundant (2) and so on.
(From R. A. Allen, R. J. Millet and D. B. Smith, "Radioisotope data", reproduced with permission of United Kingdom Atomic Energy Authority.)

1. *Gamma rays*—A one-quarter inch sheet of plastic is placed between the source and a sodium iodide crystal scintillation detector. Only gamma radiation (including possibly a small amount of bremsstrahlung) will pass through the plastic. Alpha and beta radiation will be absorbed.

2. *Beta particles*—With the source held at a distance of about two inches or more from the window of a G-M tube, the activity with and without a one-quarter inch sheet of plastic interposed between the source and the tube is observed. The G-M tube has an efficiency of only 1-2% for gamma radiation and at two inches or more, alpha radiation is absorbed by the air and window and is not detected. A pronounced increase in activity upon removal of the plastic absorber is indicative of beta radiation. It should be noted, however, that soft beta radiation such as that emitted by ^{14}C and ^{3}H may not be detected by this method.

3. *Alpha particles*—Alpha radiation is most conveniently identified by means of a scintillation probe equipped with a zinc sulfide screen, or by means of a proportional counter biased to discriminate against beta and gamma radiation.

The methods outlined above are presented only as a guide. Many other procedures will be found useful also, the method used often depending upon the equipment which may be available. In addition, the procedures outlined below for the measurement of the radiation energy will further serve to confirm the type of radiation.

Energy of radiation—After the type of radiation has been identified, the energy is measured. The following are suggested methods by which this may be done:

1. *Alpha energy*—The range of alpha particles can be measured by means of the method outlined in experiment 7.1. The energy is then obtained by reference to figure 7.3.

For greater accuracy, a scintillation counter (e.g., liquid scintillation counter) can be calibrated by the use of known alpha sources. Standard pulse height analysis is then used to determine the energy of the unknown alpha particle. Other more sophisticated methods are also available.

2. *Beta energy*—Unless a beta-ray spectrometer is available, the Feather analysis (experiment 7.3) provides the most direct method for the determination of beta energies with reasonable accuracy.

3. *Gamma energy*—From the absorption coefficient (experiment 7.5), one can estimate the gamma energy. If accurate results are to be obtained by this method, the geometry must be carefully designed to provide "narrow-beam" data.

A better method for measuring gamma energies is scintillation spectrometry (experiment 9.2). Spectrometry is not only faster but usually provides more accurate information than the absorption method.

Chemical separation—If the source appears to contain a mixture of nuclides, and if a tentative identification has been made, a chemical separation may be helpful. Carriers for the suspected nuclides are added prior to their separation, the choice of separation method depending upon the chemical nature of the nuclides (see chapter 10). If the separation is successful—the nuclides having exhibited the expected chemical behavior—additional proof of identity has been provided. Furthermore, the measurements of radiation energies will no longer be complicated by the presence of alien nuclides.

Identification—As an aid to the identification of the nuclides, beta particle energies and half-lives, and gamma energies and half-lives have been listed in tables 11.3

and 11.4 respectively. By means of these or other tables (1-4), a tentative identification can be made or, at least, the search may be narrowed to but a few possibilities. Punched cards for the identification of radioisotopes (5, 6) are also very useful. Such a set of cards, called the "Isotope Datadex" is shown in figure 11.2. Finally half-lives and other nuclear data can be checked on the "Chart of the Nuclides" and as one of the final steps in the identification of a nuclide emitting gamma radiation, the spectra can be compared directly with standard spectra (7, 8) prepared using known isotopes.

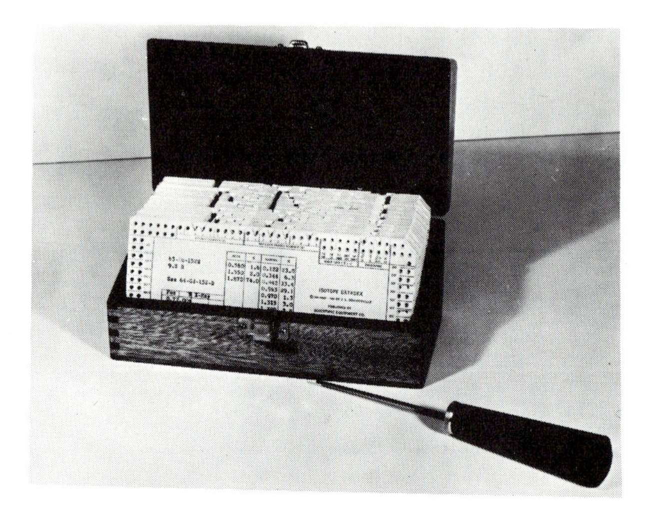

Figure 11.2—PUNCHED CARDS FOR THE IDENTIFICATION OF RADIOISOTOPES. The "Isotope Datadex" illustrated lists approximately 265 isotopes. (*Courtesy Scientific Equipment Co.*)

EQUIPMENT AND REAGENTS:

Required: Geiger and scintillation detectors; aluminum and lead absorbers; ¼″ sheet of plastic; unknown source (any available sources may be used; a neutron howitzer provides the means for several short-lived samples).

Useful: Proportional counter; gamma-ray spectrometer; liquid scintillation counter.

PROCEDURE:

In this experiment an unknown radioactive source provided by the instructor is identified. The general procedure to be followed is that which has been outlined in the section on "Theory." The specific procedure to be followed will depend upon the nature of the unknown and must be left to the good judgment of the student.

EXPERIMENT 11.2 ABSOLUTE ALPHA COUNTING

OBJECTIVES:

To illustrate a method for the absolute calibration of an alpha source using 2π proportional counting.

To check the calibration of a standard alpha source.

THEORY:

To calibrate a source by defined geometry it is necessary to evaluate all terms of equation 5.1. For an infinitely thin alpha source, deposited on a smooth surface, the geometry factor G is 0.5—this corresponds to 2π geometry—and the self absorption factor f_S is unity. By measuring the activity in a windowless flow-counter the scattering and absorption factors f_A, f_H and f_W are also reduced to unity. Thus the only term remaining to be evaluated is the backscatter factor f_B.

Rutherford scattering is the principal process by which alpha particles are backscattered. It will be recalled from chapter 1 (see equation 1.10) that the observed scattering of alpha particles through an angle of 90° or more by a thin gold foil 0.00004 cm thick was about one alpha particle in 20,000. While this represents but a very small fraction of the incident particles, the increase in observed activity caused by backscattering can be up to about 2% above that anticipated without backscattering.

One approach to the solution of this backscatter problem was suggested by Curtiss, Heyd, Olt and Eichelberger (23). Using very thin aluminum absorbers, absorption curves are plotted. The absorbers are placed directly on the alpha source which should be securely deposited or plated on a smooth metal disc. With this particular geometry, the observed activity is a *linear* function of the absorber thickness. That it is not a logarithmic function as experienced in the cases of beta- and gamma-ray absorption can be explained by reference to figure 11.3.

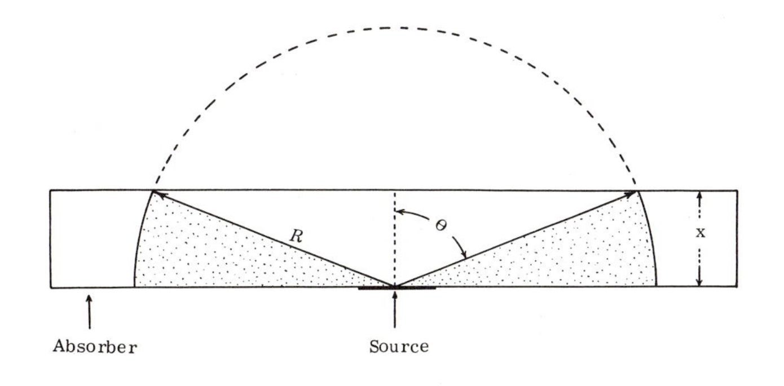

Figure 11.3—ABSORPTION OF ALPHA PARTICLES. Alpha particles within the shaded region are absorbed. (After Reference 23.)

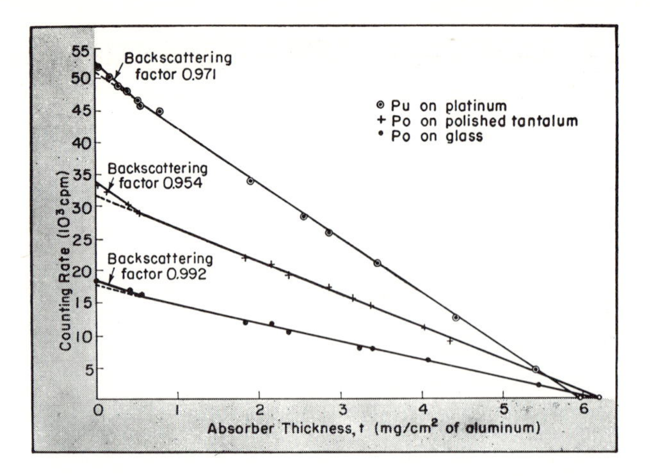

Figure 11.4—ABSORPTION CURVES FOR ALPHA PARTICLES. As zero absorber thickness is approached the lines bend upward since backscattered alpha particles are able to penetrate the thin absorber. (*Reproduced with permission from Nucleonics, May* 1955.)

Only those particles emitted within the angle θ to the perpendicular will be counted and a count rate R (corrected for background) will be observed for an absorber thickness x. At zero absorber thickness all alpha particles are counted and the activity is R_0. The fraction of particles counted will be equal to the ratio of the solid angle ω to 2π, or

$$\frac{R}{R_0} = \frac{\omega}{2\pi} \tag{11.11}$$

By reference to equation 5.15 it is seen that

$$\omega = 2\pi(1 - \cos\theta) = 2\pi(1 - \frac{x}{R}) \tag{11.12}$$

Therefore

$$\frac{R}{R_0} = 1 - \frac{x}{R} \tag{11.13}$$

or

$$R = R_0 - R_0 \frac{x}{R} \tag{11.14}$$

This is the equation for a straight line, the intercepts of which are the true count rate R_0 and the range R. The slope of the curve is R_0/R.

EQUIPMENT AND REAGENTS:

2π Gas-flow proportional counter; thin, calibrated aluminum foils or uncalibrated foils and micrometer; ring and disc mounts; standard alpha source, e.g., $^{210}PbO_2$ deposited on a silver disc.

PROCEDURE:

In this experiment the calibration supplied with a standard source is verified by means of alpha counting. In Ra-DEF sources more than a year old the ^{210}Pb, ^{210}Bi and ^{210}Po can be assumed to have attained a state of secular equilibrium. Therefore the decay rate for each of these three nuclides should be equal.

The alpha activity of the source is determined by the absorption method described in the theory section above, the extrapolated value at zero absorber thickness R_0 being taken as half of the activity. That is, $A = 2R_0$ since 2π geometry has been used. If suitable calibrated absorbers are not available, thin sheets of aluminum (or mica or other uniform material) can be calibrated by means of the micrometer caliper. It is especially important that the foils remain snug against the source during the measurements.

After making decay corrections, the measured value of activity A is compared with that given by the supplier of the standard source.

The backscatter factor f_B should be calculated from the absorption curve.

EXPERIMENT 11.3 BETA CALIBRATION WITH STANDARD OF SAME ISOTOPE

OBJECTIVES:

To illustrate the calibration of a radioactive source against a standard of the same isotope.

To calibrate a solution of an isotope of known identity but of unknown strength against a standard of the same isotope.

DISCUSSION:

When a radioactive standard is prepared from the same isotope as that to be measured, only three simple precautions are required to secure reliable results. These are: (1) readings must be made with the standard in the same position as that at which readings are made on the sample; (2) the sample must be uniformly distributed over approximately the same geometrical area as the standard; and (3) the sample must be supported on a layer of material identical with that supporting the standard, or at least one producing the same backscattering effect.

Standards of short-lived isotopes (^{24}Na, ^{32}P, ^{42}K, ^{131}I and ^{198}Au) are distributed by the Nuclear-Chicago Corporation on a semi-annual schedule. They are supplied as solutions sealed in ampules and have an activity of approximately 10^5 dps at time of shipment.

Standards of long-lived isotopes are available from a number of manufacturers. The problems of preparation and distribution contingent on a short half-life are, of course, not true of these long-lived standards which may be kept on hand in the laboratory and require only a simple calculation for the determination of their current activity.

Solution standards—To calibrate a solution of an isotope when the standard is supplied in the form of a solution, aliquots of both standard and unknown solution are accurately measured into identical planchets. If there is a possibility of loss of the radioactive solution through volatilization, a small quantity of a fixing or carrier solution is added to aliquots in the planchets before drying. The activities of both standard and unknown are then measured under identical conditions of geometry.

If either the standard or unknown solution is too concentrated, an accurate dilution must be made. An aliquot of the diluted solution is then measured into the planchet. Dilutions should be made with carrier solution, the carrier usually being an approximately 0.001*M* solution of a salt containing the stable isotope in the same chemical form as the radioactive isotope. In addition, it is also helpful to add a drop of wetting agent to each planchet prior to drying the samples to assist in providing an even distribution of activity over a defined area.

"Sodium Radio-iodide (^{131}I) Solution" has become official in the United States Pharmacopeia because of its widespread medical applications. The following is an excerpt from the U.S.P. XVI monograph describing the official assay procedure. It is a typical example of the use of a solution standard of the same isotope.

ASSAY FOR RADIOACTIVITY:

Using a beta-sensitive Geiger-Müller assembly, determine the activity of a calibrated iodine-131 standard. Dilute an accurately measured aliquot of Sodium Radio-iodide (^{131}I Solution with 0.001*M* sodium iodide so that each 0.01 ml has an activity of 1000 to 20,000 counts per minute. Add one drop of dioctyl sodium sulfosuccinate solution (1 in 100) or other surface active agent to a sample mount identical to that used for the standard. Spread the detergent solution uniformly over an area approximately equivalent to that of the standard, dry, and add from a micropipet exactly 0.01 ml of the diluted Sodium Radio-iodide (^{131}I) Solution. Rinse the pipet twice with water, adding the rinse water to the sample. Add one drop of silver nitrate solution (1 in 1000), and dry under an infrared lamp at a temperature not above 50°. Determine the activity, with the Geiger-Müller assembly, of the sample and of the standard at approximately the same time, and under identical geometrical conditions. Correct all activity determinations for coincidence and background. Calculate the activity A_0 of the sample, in millicuries per ml, by the formula,

$$A_0 = A/B \times S \times (D/0.01) \tag{11.15}$$

in which A and B are the activities expressed in counts per minute of the sample and the standard, respectively, S is the millicurie strength of the standard, and D is the dilution factor.

Mounted standards—Many long-lived standards are supplied already mounted on planchets with the radioactive deposit covered by a thin aluminum foil for protection. A kit of these sources is illustrated in figure 11.5. A single source is shown in cross-section in figure 11.6. When this type of standard is used, the aliquot of the unknown solution must be placed on an identical planchet and spread out over approximately the same area. Planchets, mounting blocks and retaining rings for this purpose are supplied with the standards or may be purchased separately. Standard and unknown are then compared under identical conditions of geometry. If the standard is covered with an aluminum foil the unknown must be similarly covered. The activity of the unknown is then calculated.

$$\frac{A_s}{A_u} = \frac{R_s}{R_u} \tag{11.16}$$

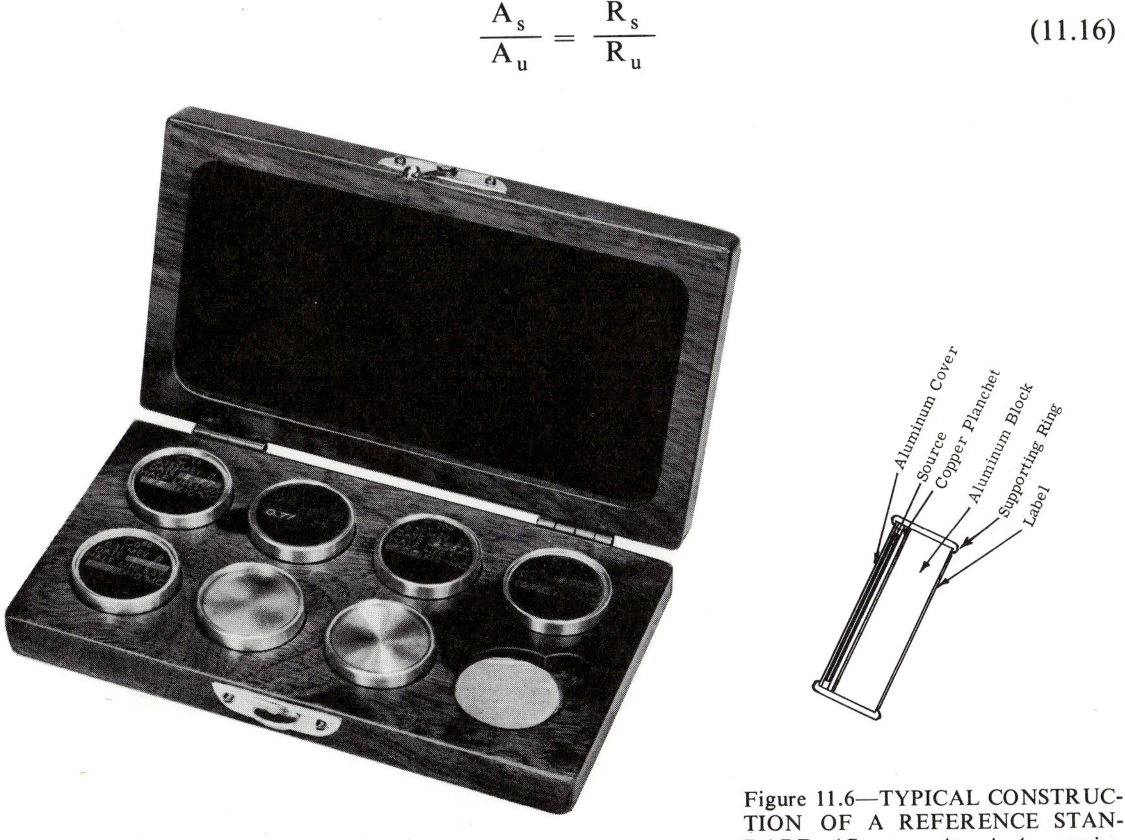

Figure 11.5—CALIBRATED BETA REFERENCE SOURCES. (*Courtesy Atomic Accessories, Inc.*)

Figure 11.6—TYPICAL CONSTRUCTION OF A REFERENCE STANDARD. (*Courtesy Atomic Accessories, Inc.*)

A_s and A_u are the activities (in dps or millicuries) of the standard and unknown sources respectively, while R_s and R_u are the observed decay rates (in cps or cpm) of the standard and sample respectively.

When standard and unknown are isotopically identical, no correction for air and window absorption is required. This is not true if they are different nuclides.

EQUIPMENT AND REAGENTS:

End-window G-M tube and sample support; scaler; standardized solution of same isotope to be calibrated*; radioactive solution to be calibrated**.

PROCEDURE:

PART A—SOLUTION STANDARDS

In this experiment a solution of either phosphorus-32, iodine-131 or gold-198 will be assayed for activity, the procedure for each being similar except for the diluting fluid. For phosphorus-32 solutions, 0.001*M* phosphoric acid is used for dilution, and for gold-198 use 0.0001*M* gold chloride.

Previous to usage, a planchet is scrupulously cleaned and then rinsed in distilled water. The type of planchet used is unimportant so long as the same type is used for all solutions. Place one drop of a dilute detergent solution (Duponal and other wetting agents are also satisfactory) on the planchet. Spread uniformly. A known aqueous aliquot of the unknown solution, or a suitable dilution is now placed on the planchet with a micropipette. This aliquot is then evaporated under an infrared lamp.

Prepare similar mounts for the standard and the unknown sample, using, if possible, the same pipette for measuring both solutions. A radioassay is then made under identical geometrical conditions.

DATA:

Activity of standard

________ dps/ml at ________ o'clock on ________ (zero date)

________ dps/_____ at ________ o'clock on ________ (assay date)

millicuries in aliquot ________ (S)

	Gross Activity (cpm)	Net Activity*** (cpm)
Sample		(A)
Standard		(B)

Dilution factor of sample ________________ (D)

*A commercial standard may be used but an accurate dilution (by the instructor) of the sample to be calibrated will serve to instruct the student in the use of this type of standard. The prepared standard should have an activity of 10^4 to 10^5 dps/ml.

**An exercise in which the student checks the manufacturers calibration of a shipment of an isotope (a standard having been prepared from this same solution by the instructor) has proven to be both instructive and interesting to the student.

***Gross activity corrected for background and resolving time yields net activity.

CALCULATIONS:

1. Calculate the strength of the sample (mCi/ml) by the use of equation 11.15 ________mCi/ml
2. Check the results with the accepted value for the "unknown" sample.

________mCi/ml at ________o'clock on ________(date calibrated)

________mCi/ml at ________o'clock on ________(assay date)

PART B—MOUNTED STANDARDS

A dilution of the unknown solution is prepared so that 25 μl will provide an activity similar to that of the standard. Before placing the aliquot on the planchet supplied with the standard, this activity can be estimated by measuring the activity of 25 μl in a steel or nickel planchet, aluminum foil or other disposable mount. If necessary the dilution is adjusted to provide an activity within 50% of that of the standard. The final preparation is then made on a planchet supplied with the standard. This planchet should be thoroughly cleaned as described in part A of the experiment. Dry the 25-μl aliquot of solution under an infrared lamp, cover the foil and mount on a block with a retaining ring. Measure accurately the activity of the standard and unknown.

DATA:

Activity of standard

____________mCi on ____________(Date indicated on label of standard)

____________mCi on ____________(Assay date) This is A_s.

	Gross Activity (cpm)	Net Activity (cpm)
Standard		(R_s)
Sample		(R_u)

Dilution factor for unknown solution ____________

CALCULATIONS:

By the use of equation 11.16 calculate the millicurie activity A_u of the unknown in the planchet. This is the activity in 25 μl of diluted unknown solution. Calculate the activity per ml of the original unknown solution.

Activity calculated ______________ mCi/ml

Activity reported by manufacturer ______________ mCi/ml

Figure 11.5 illustrates a set of standards of the type depicted in the sketch of figure 11.6. In these kits, beta standards are prepared from isotopes selected to provide a range of different energy maxima. Standards included in one commercially available set are listed in the following table.

Table 11.5—TYPICAL SET OF BETA STANDARDS

Isotope	Half-Life (years)	Maximum Beta Energy (MeV)	Thickness of Protective Foil (mg/cm^2)
^{14}C	5700	0.155	1.5
^{60}Co	5.2	0.310	11.5
^{204}Tl	4.0	0.78	11.5
^{210}Bi (Ra-E)	25	1.17	11.5
^{234}Pa (Uranium)	5×10^9	2.32	150.0

When using these standards for a direct calibration by the method described above, it is necessary to use additional absorber for the sample measurement so the total absorber thickness for both the standard and the unknown are equal. When thin aluminum foils are used to cover the unknown also, their thickness can be measured with a micrometer.

EXPERIMENT 11.4 BETA CALIBRATION WITH STANDARD OF DIFFERENT ISOTOPE

OBJECTIVES:

To illustrate the calibration of a radioactive source against a standard of a different nuclide.

DISCUSSION:

Ra-E beta standards, previously distributed by the National Bureau of Standards, are now available from the Nuclear-Chicago Corporation, 223 West Erie Street, Chicago 10, Illinois. These consist of electrolytic deposits of lead containing Ra-D in the form of PbO_2 on the palladium surface of a palladium-clad silver disc 1/16 of an inch thick. This provides saturation backscattering of the beta particles so that beta sources compared with these standards should have similar mountings.* The deposit of PbO_2 has an area of 1.13 cm^2. All silver discs, those containing the standard deposit of PbO_2 as well as the blank discs used for mounting the unknown sample, are exactly one inch in diameter.

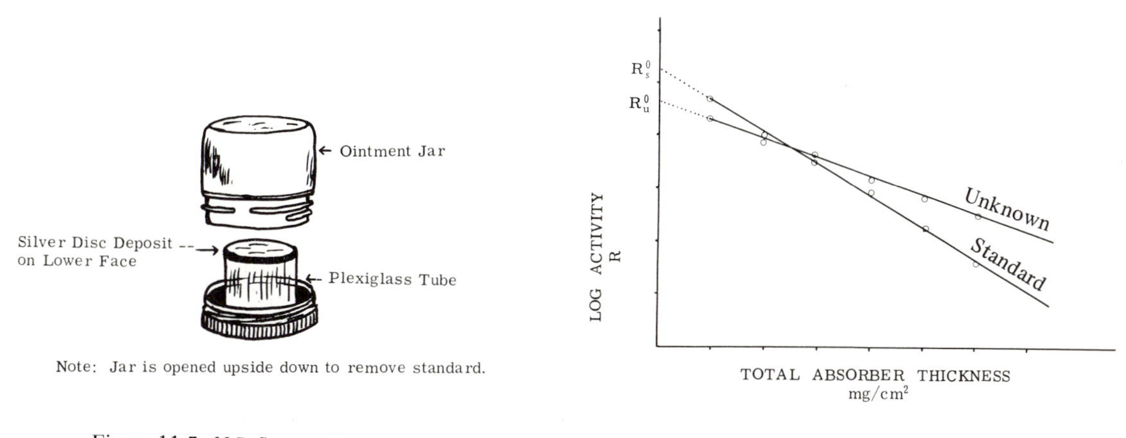

Figure 11.7–N.B.S. RADIUM-DEF STANDARD.

Figure 11.8—TYPICAL ABSORPTION CURVES. Slope of an absorption curve is determined by the maximum beta energy. In general, the standard selected should have a beta energy approximating that of the unknown.

The standards available have disintegration rates of approximately 200, 500 and 1000 disintegrations per second. The exact disintegration rate is given on a certificate which accompanies each standard. The radioactive decay processes taking place in the standard are shown in figure 5.6 (experiment 5.3), beginning with the decay of Ra-D (^{210}Pb). Secular equilibrium is established between the Ra-D and Ra-E (^{210}Bi) in about 2 months, while about 3 years are required to establish secular equilibrium with Ra-F (Po^{210}).

Because these standards emit weak beta particles from Ra-D as well as polonium alpha particles, additional absorbers to make a total of at least 6 mg/cm^2 must be placed between the counter and the source to absorb all radiation other than the Ra-E (^{210}Bi) beta particles which have a maximum energy of 1.17 MeV.

It should be noted that these standards are, strictly speaking, only standards for ^{210}Bi, but they supply satisfactory calibrations involving beta-emitting isotopes which have a beta spectrum similar to that of Ra-E, namely, isotopes which emit beta particles with a maximum energy in the range from 0.8 MeV to 2.0 MeV.

*Blank discs are available from Engelhard Industries, Inc., Baker Platinum Division, 113 Astor Street, Newark 2, N. J.

Radium-E standards, as well as many others suitable for calibration of unknown beta sources, are included among the reference standards of the type shown in figure 11.5. The reference source selected for use should be the one with a maximum beta energy approximating that of the unknown.

For calibration, the unknown source is mounted on a planchet, composed of the same material as that used for mounting the standard, to eliminate the need to correct for backscattering. Activities over a range of absorber thicknesses are measured for both standard and unknown and extrapolated to zero total absorber. Extrapolated activities, at zero total absorber, are used in the calculation of the unknown activity. All absorbing substances, including the tube window, air and foils covering either the standard or unknown, must be considered when extrapolating.

PROCEDURE:

1. *Selection of the Standard*—When available, the standard should be one prepared from the same nuclide as that comprising the source being calibrated. When such a standard is not available the next best choice is one having a similar beta energy distribution. The choice can be made simply by reference to table 11.5.

2. *Preparation of Dilution and Test Deposit*—Prepare an accurate dilution of the unknown sample so that 25 microliters of the diluted solution will have an activity within 50% of the disintegration rate of the standard. Pipette 25 microliters of the diluted sample onto the center of a plastic planchet, microscope cover glass or other expendable support. Do not use a silver (or copper) disc. Evaporate to dryness under the infrared lamp and compare the counting rate of the sample with that of the standard using the same geometry. The counting rates of the standard and the sample should be between 2500 and 4000 counts per minute with no added absorber. If necessary, prepare another dilution of the isotope so that the counting rate will be within 50% of that for the standard to eliminate resolving time corrections. Use this dilution for preparing the final mount. Use 0.001 M H_3PO_4 for making the dilutions of ^{32}P solutions. Use an appropriate carrier solution for other types of activity.

3. *Preparation of Final Deposit for Counting*—The deposit of the sample must be on a support of the same composition as that used to mount the standard. This may be a disc of silver, copper or other metal. Check the activity of the blank disc. If residues from previous usage give a count more than about 10% above background, the disc should be cleaned.

To remove any deposit remaining from a previous test, clean the disc as follows: Rinse with ethyl ether; scrub with dilute HCl, using a swab; rinse with water; scrub with dilute NH_4OH, using a swab; polish with a mild abrasive, such as diatomaceous earth, optical polishing powder or equivalent; rinse; immerse in boiling, distilled water; remove and allow to dry. Recheck the activity of the disc. If less than 10% above background, pipette 25 microliters of the proper dilution, as described in step 1, onto the clean disc and dry under the infrared lamp.

4. *Determination of Absorption Curves*—Obtain data sufficient to plot the absorption curves for both standard and unknown (as in experiment 7.3). Start with a total absorber thickness of 6 or 7 mg/cm^2 (including window and air). The density of air is 1.2 mg/cm^3.

Separate background counts should be made for the standard and the unknown, especially if the unknown is a gamma emitter. Background should be measured with the sample in position and with an absorber placed between the source and the Geiger tube. The absorber should be sufficiently thick to absorb all beta particles (about 1000 mg/cm^2).

Backgrounds

Standard ______________________ cpm
Unknown ______________________ cpm

Standard

Total Absorber mg/cm^2	Observed Activity r (cpm)	Corrected Activity R (cpm)

Unknown

Total Absorber mg/cm^2	Observed Activity r (cpm)	Corrected Activity R (cpm)

5. *Extrapolation to Zero Total Absorber*—Plot the above data on semi-logarithmic paper. Extend the plots for the standard and for the unknown to zero total absorber. Read the intercepts which correspond to net counts per second for the standard at zero total absorber and net counts per second for the sample at zero total absorber. These values are used for calculating the activity of the sample.

Extrapolated activities at zero total absorber:

Standard R_s^0 ______________________ cpm
Sample R_u^0 ______________________ cpm

Activity of standard from certificate or label:

A_s ______________________ cpm

6. *Calculation of Activity of Sample*—The activity of the sample A is calculated in terms of the number of millicuries per milliliter at reference time. Compare this value with that given by the supplier.

Activity of sample (Calculated) ______________________ mCi/ml
Activity of sample (manufacturer's calibration) ______________________ mCi/ml

EXPERIMENT 11.5 BETA CALIBRATION WITH SIMULATED REFERENCE SOURCES

OBJECTIVES:

To illustrate the calibration of a radioactive source against a simulated radioactive reference used as a standard.

To calibrate a solution of ^{32}P or ^{131}I.

DISCUSSION:

Simulated standards are available for phosphorus-32 and iodine-131. Only the simulated source for phosphorus-32 is described here.

Phosphorus-32 is a pure beta emitter with a continuous beta spectrum terminating at a maximum energy level of 1.71 MeV. It is apparent that what is required as a simulated source is a long lived isotope, or a combination of long lived isotopes, such that the spectral distribution of the respective beta emissions will be essentially identical to that of ^{32}P over a significant range of absorber thicknesses. Uranium-238 in secular equilibrium with thorium-234 (UXI) and protoactinium-234 (UXII) is such a combination and is used as the standard source in this experiment. This is due to the fact that UXI emits beta particles whose maximum energy is 0.2 MeV, while UXII emits beta particles whose maximum energies are 2.32 (95%) MeV and 1.52 (5%) MeV. This mixture also emits 5% gamma rays of energy 0.8 MeV. However, the efficiency of the beta Geiger tubes as gamma detectors is less than 2%, so that incremental contribution of the gamma ray to the counting rate can be neglected.

In the absorber range of 120 mg/cm² to 200 mg/cm² of aluminum, the slopes of the UXII standard curve and an unknown ^{32}P curve are identical, and hence, within this region the spectral distribution of the respective beta particles must be essentially equivalent. Therefore, if a "standard" 130-140 mg/cm² aluminum absorber is placed over each radioactive sample the ratio of the ^{32}P counting rate to the UXII counting rate will render an accurate calculation of the former, provided the latter is known.

To facilitate a standardization of this type, a number of manufacturers supply "Simulated Radioactive Reference Source" kits complete with a calibrated source, standard absorbers and standard mounts for the unknown sample. (See figure 11.9). When using one of these reference source kits, it is important to mount the sample on the planchets provided with the kit in the prescribed manner in order to reproduce the geometry of the standard.

PROCEDURE:

The ^{32}P reference source kit must be used for the calibration of ^{32}P and the ^{131}I reference kit must be used for the calibration of ^{131}I. They are not interchangeable.

Figure 11.9—SIMULATED RADIOACTIVE-REFERENCE SOURCE KIT. (*Tracerlab, Inc.*)

Previous to usage, one of the copper planchets supplied with the reference source kit is scrupulously cleaned, then washed with hot benzene, warm 2*N* HCl, and finally with distilled water. This assures the removal of grease and any surface contamination which may have been present.

The planchet is removed from the distilled water and one drop of a dilute detergent solution (Duponal or other suitable wetting agent) is placed on the surface. This is spread uniformly over the surface by means of a rubber policeman. A known aqueous aliquot of the unknown solution is now placed on the planchet with a pipette. This aliquot is then evaporated under an infrared lamp. When evaporation is complete, the planchet is placed in a sample mount and a "standard" dural foil absorber is placed over it and locked into position by means of the absorber ring. The "standard" absorber is of the same weight (± 1 mg) as the dural foil which protects the simulated standard. It will now be noted that the unknown and the simulated source are identical in every geometric respect. A radioassay is then made under identical geometric conditions by measurement of the counting rates of the standard and of the sample. In each case the "standard" absorber is interposed between the sample being counted and the G-M tube.

To be certain that the curve for the standard and the unknown are parallel (i.e., of equal slope) at the thickness of the "standard" absorber, obtain data for other absorber thicknesses and plot their respective curves in this region.

Background ______________________ cpm

Simulated Source

Total Absorber mg/cm²	Observed Activity r (cpm)	Corrected Activity R (cpm)

Unknown

Total Absorber mg/cm²	Observed Activity r (cpm)	Corrected Activity R (cpm)

CALCULATIONS:

1. Plot the above data on semilogarithmic paper. Draw a tangent to each of the curves at the "standard" absorber thickness. If everything has been done properly these two tangent lines will be parallel to each other. This serves both as a check on your work and, also, on the equipment.

2. From the ordinate, read the activity of the simulated and unknown sources at the standard absorber thickness.

Activity of simulated source ____________ cpm (A)

Activity of unknown source ____________ cpm (B)

3. Record the activity represented by the simulated source. This will be found inscribed on the back of the mount. ____________________ μCi (C)

4. Calculate the activity on the unknown mount from the relationship

$C \times B/A =$ ________________________ μCi

5. From the dilutions used, calculate the activity of the orginal solution.

Activity __________ μCi/ml at __________ (time)

__________ (date)

6. By use of the half-life, calculate the concentration of the original solution at "zero" time.

Activity __________ μCi/ml at __________ (time)

__________ (date)

7. If the "unknown" has been calibrated by the distributer or other individual, calculate the activity to the same "zero" time.

Activity __________ μCi/ml at __________ (time)

(at "zero" time) __________ (date)

8. Compare the results obtained in step 6 with those in step 7. Calculate the per cent error. ________________ %.

EXPERIMENT 11.6 BETA CALIBRATION BY MEANS OF A PARENT-DAUGHTER EQUILIBRIUM

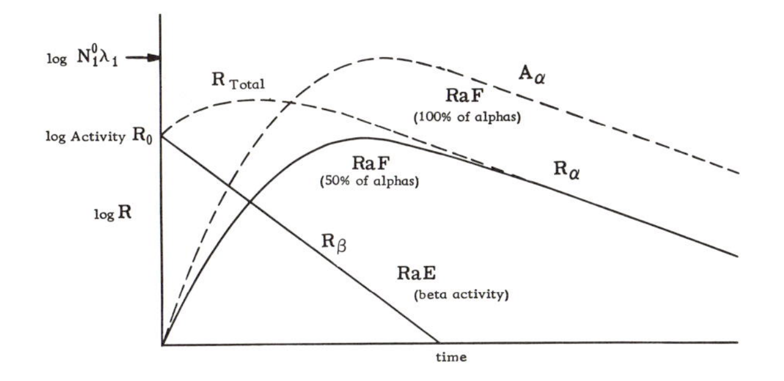

Figure 11.10—GROWTH-DECAY CALIBRATION METHOD.

OBJECTIVES:

To illustrate the use of a growth-decay method for the beta calibration of a proportional counter.

To measure the beta efficiency of a proportional counter using a Ra-DEF source in secular equilibrium.

THEORY:

Growth decay provides a means for the beta calibration of a proportional counter. Starting with a Ra-DEF equilibrium mixture, pure radium-E (bismuth-210) is separated and immediately placed in a proportional counter wherein the beta activity of the sample, due to the decay of Ra-E to Ra-F, is measured over a period of several days. At the same time the alpha activity of the Ra-F is measured. These activities the beta and the alpha, respectively are plotted versus time on semi-log paper and a graph similar to that shown in figure 11.10 obtained. Theoretically, in a 2-pi counter, 50% of the alpha particles should be detected. Actually, backscattering of the alpha particles increases the efficiency of the counter to about 52%. By use of this value for the alpha efficiency, the number of atoms of Ra-F present at any time t can be calculated. Use of the Bateman solution then permits the calculation of the number of atoms of Ra-E present at zero time.

$$\underset{\text{(Ra-D)}}{^{210}\text{Pb}} \xrightarrow[\beta^-]{} \underset{\substack{\text{(Ra-E)} \\ N_\beta}}{^{210}\text{Bi}} \xrightarrow[\beta^-,\ 1.17\ \text{MeV}]{\lambda_\beta} \underset{\substack{\text{(Ra-F)} \\ N_\alpha}}{^{210}\text{Po}} \xrightarrow[\alpha,\ 5.3\ \text{MeV}]{\lambda_\alpha} \underset{\substack{\text{(Ra-G)} \\ \text{(stable)}}}{^{206}\text{Pb}}$$

The growth-decay method, following the chemical separation of bismuth-210, can be outlined as follows:

1. Measure the alpha activity r_α at time t and correct for background and resolving time to obtain R_α.

2. Determine the alpha efficiency of the proportional counter by means of the procedure described in experiment 11.2 or assume $\epsilon_\alpha = 0.52$.

3. Calculate the alpha decay rate A_α at time t: $A_\alpha = R_\alpha/\epsilon_\alpha$.

4. Since the alpha activity A_α is related to the number or atoms N_α of ^{210}Po present in the source at time t by $A_\alpha = -dN_\alpha/dt = N_\alpha\lambda_\alpha$, N_α can be evaluated.

5. By means of the Bateman equation the number of atoms N^0_β of ^{210}Bi present at zero time is calculated.

$$N_\alpha = N^0_\beta \frac{\lambda_\beta}{\lambda_\alpha - \lambda_\beta}\left(e^{-\lambda_\beta t} - e^{-\lambda_\alpha t}\right)$$

6. The absolute beta decay rate A^0_β can now be calculated from the relation $A^0_\beta = N^0_\beta\lambda_\beta$.

7. From a plot of R_β, the corrected values of the beta activity, one can determine R^0_β, the corrected activity at zero time.

8. The beta efficiency ϵ_β can now be calculated.

$$\epsilon_\beta = \frac{R^0_\beta,}{N^0_\beta \lambda_\beta} = \frac{R^0_\beta}{A^0_\beta}$$

Secular equilibrium provides another approach to the beta calibration of a proportional counter. If the half-life of the parent nuclide is very long compared to that of the daughter(s), then the decay rates can be considered equal.

In the case of Ra-DEF the ^{210}Po decays be alpha emission. The decay rate of the polonium is determined by use of the procedure outlined in experiment 11.2 or, if desired, by assuming an alpha efficiency of 0.52. Since both ^{210}Pb and ^{210}Bi decay by beta emission, the beta activity should be twice the alpha activity. That is, $A_\beta = 2A_\alpha$. The beta efficiency is then calculated from A_β and the observed beta activity R_β.

PROCEDURE:

The beta efficiency of a proportional counter is determined by one of the above methods. When time is a factor it is suggested that the method utilizing secular equilbrium be used.

The Plateau characteristics of the counter must be determined in order that the high voltage can be adjusted to the alpha or beta plateau as required.

It will be observed that while $A_\beta = 2A_\alpha$ it does not follow that $R_\beta = 2R_\alpha$. This is, of course, due to a difference between ϵ_α and ϵ_β. Explain the value obtained for ϵ_β (or the relationship between A_β and R_β) in terms of G, f_A, f_B, f_H, f_S and f_W.

REFERENCES

Identification of Nuclides

1. Hallden, N. A., "Beta Emitters by Energy and Half-Life", Nucleonics 13, No. 6, 78-79 (June 1955).
2. Anonymous, "Beta Emitters by Half-life and Energy", Nucleonics 19, No. 2, 70 (February 1961).
3. Smith, G. W., and D. R. Farmelo, "Radionuclides Arranged by Gamma-Ray Energy", Nucleonics 16, No. 2, 80-81 (February 1958).
4. Anonymous, "Gamma Emitters by Half-Life and Energy", Nucleonics 18, No. 11, 196-197 (Nov. 1960).
5. Neissel, J. P., "Keysort Cards for Nuclear Data", Nucleonics 16, No. 11, 142 (November 1958).
6. Lukens, H. R. Jr., E. E. Anderson and L. J. Beaufait, Jr., "Punched Card System for Radioisotopes", Anal. Chem. 26, 651-652 (1954).
7. Heath, R. L., "Scintillation Spectrometry Gamma-Ray Spectrum Catalogue", AEC Research and Development Report IDO-16408, Office of Technical Service, U. S. Dept. of Commerce (1957).
8. Heath, R. L., "Scintillation Spectrometry Gamma-Ray Spectrum Catalogue", (IDO-16880) Office of Technical Services, U. S. Department of Commerce, Washington, D. C. (1964).

Calibration

9. Anonymous, "Measurement of Radioactive Isotopes", Nucleonics 3, 55 (Sept. 1948).
10. Axton, E. J., "Neutron-Source Calibrations: A Review", Nucleonics 19, No. 3, 90-94 (March 1961).
11. Curtiss, L. R., "Measurement of Radioactive Isotopes", Chem. Eng. News 26, 3560-62 (Nov. 29, 1948).
12. Evans, R. D. "Radioactivity Units and Standards", Nucleonics 1, 32-43 (October 1947).
13. Kinsley, M. T., J. B. Cumming, and H. L. Finston, "Application of Conversion X-Ray Spectra to Isotopic Analysis. Resolution of Cesium-134/Cesium-137 Mixtures", Anal. Chem. 32, 1081 (1960).
14. Lee, D. A., "Microdetermination of Deuterium by Effusiometry", Anal. Chem. 30, 1296 (1958).
15. Lyon, W. S. "Determination of Neptunium-239 Counting Efficiency", Anal. Chem. 29, 1048 (1957).
16. Mandeville, C. E., and M. V. Scherb, "Nuclear Disintegration Schemes and the Coincidence Method", Nucleonics 3, 2 (October 1948).
17. Mann, W. B. and H. H. Seliger, "Preparation, Maintenance, and Application of Standards of Radioactivity", Nat. Bur. Stand. Circ. 594, Supt. Documents (June 1958).
18. Myers, O. E., "Calorimetric Radioactivity Measurements", Nucleonics 5, 37-47 (November 1949).
19. Paneth, F. A., "Radioactive Standards and Units", Nucleonics 8, No. 5, 38-41 (1951).
20. Rosholt, J. N., Jr., and J. R. Dooley, Jr., "Automatic Measurements and Computations for Radiochemical Analyses", Anal. Chem. 32, 1093 (1960).
21. Rudstam, S. G., "Preparing Samples for Absolute Counting", Nucleonics 13, No. 2, 64-66 (Feb. 1955).
22. Tolmie, R. W., "Dual Detectors Key to Radioisotope Identifier", Nucleonics 18, No. 10, 92 (Oct. 1960).

Alpha Calibration

23. Curtis, M. L., J. W. Heyd, R. G. Ott, and J. F. Eichelberger, "Absolute Alpha Counting", Nucleonics 13, No. 5, 38-41 (May 1955).
24. Kirby, H. W., "Radium Determination by Alpha Counting", Anal. Chem. 25, 1238-1241 (1953).
25. Miller, B. L., and F. E. Hoecker, "Quantitation of Alpha Emitters in Bone", Nucleonics 8, No. 5, 44-52 (1951).
26. Miller, D. G., and M. B. Leboeuf, "Effect of High Beta Backgrounds on Precision Alpha Counting", Nucleonics 11, No. 4, 28-31 (1953).
27. Miller, D. G., and M. B. Leboeuf, "Pulse-Distribution Evaluation of Alpha-Counting Instruments", Nucleonics 11, No. 11, 56-60 (1953).
28. Pietri, C. E., "Plutonium Sulfate Tetrahydrate, a Proposed Primary Analytical Standard for Plutonium", Anal. Chem. 34, 1604 (1962).

Beta Calibration

29. Baker, R. G., and L. Katz, "Absolute Beta Counting of Thick Planar Samples", Nucleonics 11, No. 2, 14 (1953).
30. Bayhurst, B. P., and R. J. Prestwood, "A Method for Estimating Beta-Counting Efficiencies", Nucleonics 17, No. 3, 82-85 (March 1959).
31. Beischer, D. E., "Absolute Beta Measurement of Tritium Monolayers", Nucleonics 11, No. 12, 24-27 (1953).
32. Bizzell, O. M., W. T. Burnett, Jr., P. C. Tompkins, and L. Wish, "Properties of Phosphorus-Bakelite Beta-Ray Sources", Nucleonics 8, No. 4, 17-27 (1951).
33. Blau, M. and J. E. Smith, "Beta-Ray Measurements and Units", Nucleonics 2, 67 (June 1948).
34. Burtt, B. P., "Absolute Beta Counting", Nucleonics 5, 28-43 (Aug. 1949).
35. Gleason, G. I., J. D. Taylor, and D. L. Tabern, "Absolute Beta Counting at Defined Geometries",
36. Gunnink, R., L. J. Colby, Jr., and J. W. Cobble, "Absolute Beta Standardization Using 4 Pi Beta-Gamma Coincidence Techniques", Anal. Chem. 31, 796 (1959).

37. Jeffay, H., F. O. Olubajo, and W. R. Jewell, "Determination of Radioactive Sulfur in Biological Materials", Anal. Chem. 32, 306 (1960).
38. Kalmon, B., "Experimental Method for Determination of Counting Geometry", Nucleonics 11, No. 7, 56-59 (1953).
39. Libby, W. F. "Simple Absolute Measurement Technique for Beta Radioactivity. Application to Naturally Radioactive Rubidium", Anal. Chem. 29, 1566 (1957).
40. Loeoinger, R., and S. Feitelberg "Simplified Beta Counting, Using Bremsstrahlung Detection by a Scintillator", Nucleonics 13, No. 4, 42-45 (April 1955).
41. McIsaac, L. D., and E. C. Freiling, "Counting Efficiency Determinations of Pa^{233}", Nucleonics 14, No. 10, 65-67 (October 1956).
42. Merritt, J. S., J. G. V. Taylor, W. F. Merritt, and P. J. Campion, "The Absolute Counting of Sulfur-35", Anal. Chem. 32, 310 (1960).
43. Reynolds, S. A., and W. A. Brooksbank, Jr., "Thallium-204 as a Standard for Radioassays", Nucleonics 11, No. 11, 46-47 (1953).
44. Seliger, H. H., and A. Schwebel, "Standardization of Beta-Emitting Nuclides", Nucleonics 12, No. 7, 54-63 (July 1954).
45. Tompkins P. C., L. Wish, and W. T. Burnett, Jr., "Estimation of Beta-Activities from Bremsstrahlung Measurements in an Ionization Chamber", Anal. Chem. 22, 672-676 (1950).
46. Wish, L., "Counting Efficiency for Np^{239} Beta Rays", Nucleonics 14, No. 5, 102-106 (May 1956).
47. Zumwalt, L. R., C. V. Cannon, G. H. Jenks, W. C. Peacock, and L. M. Gunning, "Comparison of the Determination of the Disintegration Rate of Radio-phosphorus by Absolute Beta Counting and Calorimetric Measurement", Science, p. 47 (January 9, 1948).

Gamma Calibration

48. Anonymous, "Mock Iodine for I^{131} Calibrations", Nucleonics 16, No. 8, 134 (August 1958).
49. Axtmann, R. C., and J. S. Stutheit, 'Scintillation Counting of Natural Uranium Foils", Nucleonics 12, No. 7, 52-53 (July 1954).
50. Colby, L. J., Jr., and J. W. Cobble, "Precision Calibration of Large 4 Pi Scintillation Crystals for Routine Gamma Counting", Anal. Chem. 31, 798 (1959).
51. Covell, D. F., "Determination of Gamma-Ray Abundance Directly from the Total Absorption Peak", Anal. Chem. 31, 1785 (1959).
52. Heath, R. L. "Scintillation Spectrometry Gamma-Ray Spectrum Catalogue", AEC R & D Report IDO-16408, Office of Tech. Serv., U. S. Dept. of Commerce (1957).
53. Pinajian, J. J., and J. E. Christian, "The Assay of Sodium Radio-iodide (I^{131}) Used for Medical Purposes", J. A. Ph.A., Science, Ed. XLIV, 631-636 (1955).
54. Upson, U. L., R. E. Connally, and M. B. Leboeuf, "Analyzing for Low-Energy Gamma Emitters in a Radionuclide Mixture", Nucleonics 13, No. 4, 38-41 (April 1955).

PART III

APPLICATIONS OF RADIOISOTOPES

CHAPTER 12
Radioisotopes in Chemistry

LECTURE OUTLINE AND STUDY GUIDE

I. PHYSICAL CHEMISTRY
- A. Solubility Measurements (*Experiment* 12.1)
  - 1. Law of mass action
  - 2. Law of chemical equilibrium
  - 3. Solubility product constant
  - 4. Complex ion formation
- B. Exchange Kinetics (*Experiment* 12.2)
  - 1. Types of reactions
    - a. Direct
      - (1) Ionic bond reactions
      - (2) Covalent bond reactions
    - b. Indirect with intermediate complex formation
    - c. Electron transfer reactions—Redox reactions
  - 2. Isotope effects
  - 3. Kinetic mechanisms
    - a. Derivation of the rate equations
    - b. Reaction half-time
    - c. Activation energy

II. ANALYTICAL CHEMISTRY
- A. Isotope Dilution Analysis (*Experiment* 12.3)
  - 1. General principles
  - 2. Direct isotope dilution
  - 3. Inverse isotope dilution
  - 4. Modified inverse isotope dilution
  - 5. Applications of isotope dilution
- B. Radiometric Analysis (*Experiment* 12.4)
  - 1. General principles
  - 2. "Direct titration" method
  - 3. "Indirect titration" method
  - 4. Other methods
- C. Activation Analysis (*Experiment* 12.5)
  - 1. Activation of the sample
  - 2. Activity calculations
  - 3. Identification and/or isolation of the nuclide
  - 4. Quantitative assay for the nuclide

III. ORGANIC CHEMISTRY
- A. Synthesis of Labeled Compounds (*Experiment* 12.6)
  - 1. Selection of a method for synthesis
    - a. Yield of the method
    - b. Radiation hazards
  - 2. Choice of glassware
  - 3. Need for preliminary "dry run"
  - 4. Typical techniques
- B. Degradation Methods (*Experiment* 12.7)
  - 1. Purpose of degradation
  - 2. Schmidt reaction
  - 3. Van Slyke-Folch wet oxidation

IV. NUCLEAR CHEMISTRY
- A. Recoil Reactions (Szilard-Chalmers Process) (*Experiment* 12.8)
  - 1. Hot atom chemistry
  - 2. Mechanics of the process
  - 3. Applications
    - a. Preparation of carrier-free isotopes
    - b. Calibration of weak neutron sources
    - c. Preparation of labeled compounds
- B. Radiation Chemistry (*Experiment* 12.9)
  - 1. Chemistry of reactions induced by radiation
    - a. Excitation and ionization
    - b. Free-radical formation
    - c. Formation of final compounds
      - (1) By alpha radiation in pure water
      - (2) By beta and gamma radiation
      - (3) In aerated water
      - (4) In solutions of inorganic ions
  - 2. Measurement of radiation effect
    - a. G value
    - b. Calculation of dose

EXPERIMENT 12.1 SOLUBILITY MEASUREMENTS

OBJECTIVES:

To demonstrate the use of a radioactive tracer for measuring the solubility of a difficultly soluble compound and for determining the degree of formation of a soluble complex ion.

To measure the solubility product constant for silver iodide and the dissociation constant for the silver diammino ion.

THEORY:

The *law of mass action* is a statement of the rate at which a chemical reaction takes place, expressed in terms of the concentrations of the reacting substances. Thus, if substances A and B react together to form C and D, the chemical equation is

$$aA + bB \rightleftharpoons cC + dD \tag{12.1}$$

where a, b, c and d are the coefficients in the balanced chemical equation. Then the rate R_1 at which A and B react to form C and D is given by the expression

$$R_1 = k_1 [A]^a [B]^b \tag{12.2}$$

Note: Brackets represent molar concentrations.

Similarly, in the reverse reaction, the rate R_2 at which C and D react to form A and B is

$$R_2 = k_2 [C]^c [D]^d \tag{12.3}$$

When sufficient time has elapsed for the system to establish equilibrium, $R_1 = R_2$ and thus,

$$\frac{[C]^c [D]^d}{[A]^a [B]^b} = \frac{k_1}{k_2} = K_e \tag{12.4}$$

where K_e is the equilibrium constant. Equation 10.4 is a statement of the *law of chemical equilibrium.*

The law of chemical equilibrium can be applied to the quantitative measurement of solubility for the so-called "insoluble compounds," more accurately referred to as difficultly soluble compounds. For example, if silver iodide and water are shaken together, a very small amount of it will dissolve. In time, an equilibrium will be established between the solid solute (i.e., solid AgI) in contact with the solvent and the ions in solution. The chemical reaction is

$$AgI_{(solid)} \rightleftharpoons Ag^+ + I^- \tag{12.5}$$

and the equilibrium expression is

$$K_e = \frac{[Ag^+]\ [I^-]}{[AgI_{solid}]} \tag{12.6}$$

Since the concentration of AgI in solid silver iodide is a constant, the expression can be simplified to

$$K_{sp} = [Ag^+]\ [I^-] \tag{12.7}$$

where K_{sp} is the solubility product constant which for silver iodide has a value of 1.5×10^{-16}. From this the solubility of AgI in pure water can readily be calculated. From equation (12.5) it is seen that the concentrations of silver ion and iodide ion will be equal and the concentration of one or the other will be equal to the concentration of silver iodide in solution. Thus, if X = molar concentration of silver iodide in solution, then

$$X = \sqrt{K_{sp}} \tag{12.8}$$

Therefore, the molar concentration of silver iodide is numerically equal to

$$X = (1.5 \times 10^{-16})^{1/2} = 1.2 \times 10^{-8} \text{ molar}$$

Complex ions are produced if silver iodide is treated with ammonia.

$$AgI + 2\,NH_3 = Ag(NH_3)_2^+ + I^- \tag{12.9}$$

The ammonia complex, $Ag(NH_3)_2^+$, called the silver diammino ion, dissociates to some extent into its constituents

$$Ag(NH_3)_2^+ = Ag^+ + 2\,NH_3 \tag{12.10}$$

This reaction, too, obeys the law of chemical equilibrium,

$$\frac{[Ag^+]\ [NH_3]^2}{[Ag(NH_3)_2^+]} = K_i = 6.8 \times 10^{-8} \tag{12.11}$$

The constant K_i is called the *instability constant*. From this information it is possible to calculate how much silver iodide will dissolve in a solution of ammonia (ammonium hydroxide) of known concentration. Let X = concentration of silver iodide dissolved. Thus,

$$X = [I^-] = [Ag(NH_3)_2^+] \tag{12.12}$$

If we substitute X into equations (12.7) and (12.11), we obtain

$$[Ag^+]\,(X) = K_{sp} \tag{12.13}$$

$$\frac{[Ag^+]\ [NH_3]^2}{X} = K_i \tag{12.14}$$

The equilibria represented by equations (10.13) and (10.14) occur in the same solution. Therefore, the silver ion concentrations represented in both equations are equal.

$$[Ag^+] = \frac{K_{sp}}{X} = \frac{K_i(X)}{[NH_3]^2} \tag{12.15}$$

Hence,

$$X^2 = \frac{K_{sp}\,[NH_3]^2}{K_i} \tag{12.16}$$

Replacing K_{sp} and K_i with their numerical equivalents gives

$$X = 4.7 \times 10^{-5}\,[NH_3] \tag{12.17}$$

Therefore, the concentration of silver iodide in solution will be proportional to the concentration of NH_3.

In this experiment, a precipitate of radioactive silver iodide ($Ag^{131}I$) will be prepared from radioactive sodium iodide. The solubility of the radioactive silver iodide in water and in known concentrations of ammonium hydroxide will be determined by measurements of the radioactivity of the solution in contact with the precipitate. From this data, the solubility product constant of AgI and the dissociation constant of $Ag(NH_3)_2^+$ will be calculated.

MATERIALS REQUIRED:

$^{131}I^-$ solution*; sodium iodide solution; 1% silver nitrate solution; ammonium hydroxide; 10% nitric acid; centrifuge; centrifuge tubes; scintillation well counter or other efficient beta or gamma detector.

PROCEDURE:

To exactly 1 ml of 0.05*M* potassium iodide in a 3-ml centrifuge tube, add an accurately measured quantity of carrier-free* radioiodide containing 200 to 300 microcuries of activity. Then add 1 drop of 10% nitric acid followed by $AgNO_3$ solution dropwise until there is a slight excess of silver ion present. Mix and centrifuge. Check a small amount of the supernatant liquid for activity. It should be essentially zero.

*Iodotope Solution (Squibb) contains 0.2% sodium thiosulfate with 0.9% benzyl alcohol as a preservative and sodium chloride for isotonicity. The presence of AgCl which will precipitate with the AgI will produce serious alterations in the equilibria. AgCl can be removed from the AgI by washing the precipitate with 5% NH_4OH solution.

The potassium iodide solution and the radioiodide solution must be measured accurately. These volumes will be used to calculate the specific activity S of the silver iodide. (S = cpm/mg of AgI).

Remove the supernatant liquid from the AgI precipitate and wash the AgI three times with CO_2-free distilled water to remove all excess silver ions. Equilibrate the AgI with 2 ml of distilled water by mixing or shaking for 15 minutes.* Centrifuge the mixture and transfer exactly 1 ml to another test tube. Measure the activity in a well counter. Repeat this procedure until agreement of activity measured on successive 1 ml portions of saturated supernatant are obtained. Record these activities.**

Remove the excess water from the AgI precipitate and wash it with 0.1M NH_3 (NH_4OH). Now, add 2 ml of 0.1M NH_3 and mix or shake as above until the solution is saturated. Then centrifuge the mixture and measure the activity of exactly 1 ml of supernatant liquid. Repeat, using a fresh quantity of 0.1M NH_3 to make certain that the mixture was not contaminated and that saturation was attained.

Repeat the equilibration and measurement of activity, using increasing strengths of ammonia. The use of 0.5M, 1.0M and 5.0M solutions is suggested.

As the strength of the ammonia is increased, the measured activity may increase excessively. It may be found advisable to make a 1:10 or 1:100 dilution. To make a 1:10 dilution, transfer exactly 1 ml of supernatant to a 10 ml volumetric flask. Dilute to the mark using ammonia of strength similar to that used in the extraction and containing a drop or two of iodide carrier. Mix and transfer exactly 1 ml of this dilution to a tube for the measurement of activity.

If the exact strength of the 0.05M potassium iodide solution is not known, it will be necessary to determine its strength by standard gravimetric or volumetric techniques.

The exact quantity of ^{131}I incorporated in the AgI must be known but because the activity is so great it cannot be measured directly. Prepare a dilution of the stock radioiodide solution such that the activity of exactly 1 ml of the diluted solution can be measured in the well counter. From these data the activity in the 200 lambdas of stock solution used to prepare the AgI can be calculated.

Carry out the following calculations:

DATA AND CALCULATIONS:

1. *Specific activity of AgI*

		Item
Exact molarity of 0.05M KI	________ M	(A)
Volume of 0.05M KI used	________ ml	(B)
Molecular weight of AgI	________	(C)
Weight AgI (milligrams) = (A) (B) (C)	________ mg	(D)
Volume of radioiodide stock solution used	________ ml	(E)
Dilution factor (Total volume/E)	________	(F)
Activity of 1 ml of diluted ^{131}I solution	________ cpm	(G)
Activity of volume (E) of stock ^{131}I = (E)(F)(G)	________ cpm	(H)
Specific activity of AgI = H/D	________ cpm/mg	(I)

*Much longer periods for equilibration may be required where precise results are desired.

**The formation of a colloidal dispersion of AgI will produce high results. Heating, addition of electrolyte and other techniques often used to break colloids cannot be used here. If allowed to stand too long decomposition of the AgI will also result in error.

2. *Solubility of AgI in water*

Activity in 1 ml of water	________ cpm	(J)
Wt AgI per ml = J/I	________ mg	(K)
Molarity of AgI = K/C	________ *M*	(L)

3. *Solubility of AgI in 0.1M* NH_3

Activity in 1 ml of 0.1*M* NH_3	________ cpm	(M)
Wt AgI per ml = M/I	________ mg	(N)
Molarity of AgI = N/C	________ *M*	(O)

4. *Solubility of AgI in 0.5M NH_3*

Activity of 1 ml of 0.5*M* NH_3	________ cpm	(P)
Wt AgI per ml = P/I	________ mg	(Q)
Molarity of AgI = Q/C	________ *M*	(R)

5. *Solubility of AgI in 1.0M NH_3*

Activity in 1 ml of 1.0*M* NH_3	________ cpm	(S)
Wt AgI per ml = S/I	________ mg	(T)
Molarity of AgI = T/C	________ *M*	(U)

6. *Solubility of AgI in 5.0M NH_3*

Activity in 1 ml of 5.0*M* NH_3	________ cpm	(V)
Wt AgI per ml = V/I	________ mg	(W)
Molarity of AgI = W/C	________ *M*	(Y)

7. *Calculation of solubility product constant K_{sp}*

Use equation (12.8). The molar concentration of AgI in solution, represented by "X" in equation (12.8), is item (L) recorded in data above.

K_{sp} (calculated) ________________

K_{sp} (literature) ________________

8. *Calculation of instability constant for $Ag(NH_3)_2^+$*

Solve equation (10.16) for K_i.

$$K_i = K_{sp}\left(\frac{[NH_3]}{X}\right)^2$$

From the data in steps 3 to 6, inclusive, plot $[NH_3]$ vs X. Draw the best straight line through the points and from the slope determine the ratio $[NH_3]/X$. Using the value of K_{sp} measured in step 7, calculate the value of K_i.

K_i (calculated) ________________

K_i (literature) ________________

EXPERIMENT 12.2 EXCHANGE KINETICS

OBJECTIVES:

To demonstrate the existence of exchange reactions.

To illustrate the use of tracers for the determination of the mechanism of an exchange reaction and for the measurement of the rate constant for such reaction. In particular, to measure the rate of exchange of iodide ^{131}I with the iodine in an alkyl iodide.

To measure the activation energy of this reaction.

THEORY:

Since 1920, when Hevesy first demonstrated the exchange of lead ions between $Pb(NO_3)_2$ and $PbCl_2$, numerous investigations of exchange reactions have been made with radioisotopes as atomic tracers. It has been shown, for example, that a very rapid exchange of atoms occurs between Cl^- and Cl_2, Br^- and Br_2, I^- and I_2, MnO_4^{-1} and MnO_4^{-2}, and between ions of soluble salts of the nature investigated by Hevesy.

It is interesting that little or no exchange has been observed for manganese between Mn^{+2} and MnO_4^{-1}, or for sulfur between S^{-2} and SO_4^{-2} or SO_3^{-2} and SO_4^{-2}, as well as for many other examples mentioned in the literature.

It will be noted that exchange reactions involving different types of bonds have been investigated. Where *ionic bonds* exist between the element in question and the rest of the compound, exchange is extremely rapid. Where *covalent bonds* exist, rates of exchange are slow or exchange has not been observed.

The use of tracer elements is of particular importance in the study of reaction mechanisms. This is especially true in organic chemistry, where two or more possible mechanisms have often been proposed to explain a particular reaction.

Electron transfer reaction studies are now possible with tagged atoms. The reaction

$$^*Fe^{++} + Fe^{+++} = Fe^{++} + ^*Fe^{+++}$$

can be investigated only by the use of radioactive iron. Investigations of this sort have greatly increased our knowledge of chemical reactions.

Rate equations—Consider the exchange of iodine between inorganic iodide and an alkyl iodide, in particular, the reaction between n-butyl iodide and sodium radioiodide.

$$CH_3(CH_2)_3I + NaI^* = CH_3(CH_2)_3I^* + NaI$$

This type of reaction can be expressed by a generalized equation.

$$AX + BX^* = AX^* + BX$$

where X* represents a radioactive atom of X.

1. *Reaction Rate*—R. B. Duffield and M. Calvin [J. A. C. S., *68*, 557 (1946)] have found the reaction to be first order with respect to the activity and have developed the rate equation for this type of reaction.

Let

$$[AX] + [AX^*] = a$$
$$[BX] + [BX^*] = b$$
$$[AX^*] = x \text{ and } x = x_\infty \text{ at } t_\infty$$
$$[BX^*] = y \text{ and } y = y_\infty \text{ at } t_\infty$$

Brackets are used to indicate molar concentrations, and R′ is the reaction rate constant. The rate of appearance of x is some function of a and b, namely:

$$dx/dt = R'(y/b - x/a) = (R'/ab)(ay - bx) \quad (12.18)$$

Since

$$x + y = x_\infty + y_\infty$$
$$y = x_\infty + y_\infty - x$$

also

$$ay_\infty = bx_\infty \text{ and thus, } y_\infty = x_\infty (b/a)$$

therefore,

$$y = x_\infty + x_\infty (b/a) - x \quad (12.19)$$

Substitution in the rate equation gives

$$dx/dt = R'/ab\,[a(x_\infty + x_\infty (b/a) - x) - bx]$$

or

$$dx/dt = (R'(a + b)/ab)(x_\infty - x)$$

Hence,

$$dx/(x_\infty - x) = [R'(a + b)/ab]dt \quad (12.20)$$

Integration gives

$$-\ln (x_\infty - x) = [R't(a + b)/ab] + C$$

Since at t_0, $x = 0$, then the integration constant $C = -\ln x_\infty$

and

$$-\ln (x_\infty - x) = R't(a + b)/ab - \ln x_\infty$$

or

$$-\ln (1 - x/x_\infty) = R't(a + b)/ab \quad (12.21)$$

For ease in calculation and in plotting the data, the terms of equation (12.21) are rearranged and an arbitrary constant k is defined by equation (12.22).

$$k = \frac{R'(a + b)}{2.303\,(ab)} = -\left[\log\left(1 - \frac{x}{x_\infty}\right)\right]\left(\frac{1}{t}\right) \quad (12.22)$$

Thus, if $-\log (1 - x/x_\infty)$ is plotted against time, the slope of the curve is equal to the constant k which is related to R′ by equation (12.22).

2. *Half-Time of the Reaction*—The half-time of the reaction is the time required for the butyl iodide to receive one-half of the activity it will ultimately receive in infinite time, that is, when $x/x_{\infty} = ½$. Thus, from equation (12.18)

$$- \ln (1 - ½) = R't_{½} (a + b)/ab$$
$$\ln 2 = R't_{½} (a + b)/ab$$
$$t_{½} = 0.693\ ab/(a + b)R'$$

3. *Activation Energy*—In reactions involving two or more molecules, it is assumed that a collision between the molecules must occur before a reaction takes place. On the other hand, it has been calculated that the number of collisions occurring between molecules exceeds the number of molecules undergoing reaction by many powers of ten. The explanation for this apparent discrepancy is that the molecules will not react unless they collide with an energy in excess of a certain critical value ($E_{activation}$ in figure 12.1). The energies of the molecules have a Maxwellian distribution, as

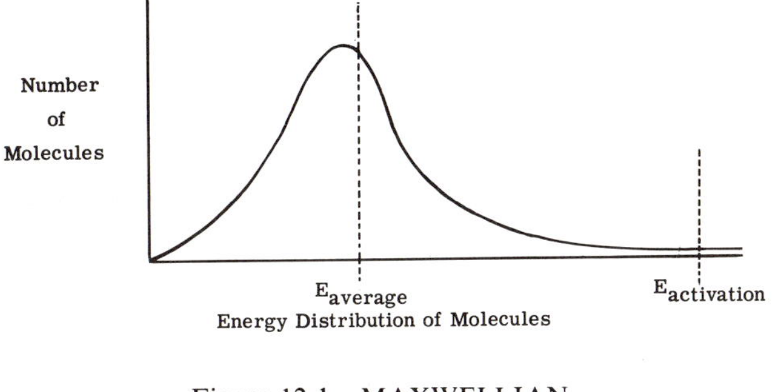

Figure 12.1—MAXWELLIAN
ENERGY DISTRIBUTION.

illustrated, with only a small fraction possessing sufficient energy to react. Calculation of the activation energy E, through the use of reaction rate constants, is made possible by the van't Hoff equation

$$\frac{d \ln k}{dT} = \frac{E}{RT^2}$$

which, when integrated and solved for E, becomes the useful working equation

$$E = 2.303\ R \frac{T_2 T_1}{T_2 - T_1} \log \frac{k_2}{k_1}$$

where k_1 is the rate constant at temperature T_1 and k_2 is the rate constant at temperature T_2, absolute temperature being used. R is the gas constant.

PROCEDURE:

Group A carries out the experiment at the lower temperature.________°C.
Group B carries out the experiment at the higher temperature.________°C.

1. Place five 2-ml volumetric flasks in a beaker filled with ice and add the following reagent to each flask:

1 ml 0.04*M* NaI aqueous solution
1 ml benzene

Reserve these flasks (chilled) for use in step 3.

2. Place one 5-ml volumetric flask in a thermostatically controlled bath of the desired temperature and add, in the order listed:

1 ml 0.04*M* NaI (in 90% acetone)
100 lambda ^{131}I solution
1.1 ml 0.04*M* n-butyl iodide (in 90% acetone)

THIS IS THE STARTING TIME OF THE EXPERIMENT. NOTE TIME t_0.

3. Shake the 5-ml flask gently for about one minute. At appropriate time intervals (about 15 minutes for experiments at higher temperatures and 20 - 25 minutes for experiments at lower temperatures) take out 100 lambdas of sample from this flask and add it to the benzene-NaI mixture in one of the 2-ml volumetric flasks prepared in step 1.

4. Shake each of the 2-ml flasks one to two minutes and proceed as follows: Take out *carefully* all the benzene solution with a pipette and deposit it in a plastic cup. (Some of the aqueous solution may be drawn up into the pipette to insure getting all the benzene and the aqueous layer returned to the flask.) Place the cup on a counting card. *Cover immediately* and count the benzene fraction. This is the activity of x.

5. Transfer the aqueous solution to a plastic cup and rinse the flask with a few drops of water, using a pipette for transfer and washing. Add the rinsings to the cup. Cover and count. This is the activity of y.

DATA:

Background________________cpm

GROUP A

Sample	Time min	Activity of "x" cpm	Activity of "y" cpm	Corr. Activity "x"	Corr. Activity "y"	x_∞
1	______	______	______	______	______	______
2	______	______	______	______	______	______
3	______	______	______	______	______	______
4	______	______	______	______	______	______
5	______	______	______	______	______	______

T_1 ______ °A $x_\infty = (x + y)/2$ when $a = b$

x_∞ ______ (average)

R' ______

k_1 ______

$t_{1/2}$ ______

GROUP B

Sample	Time min.	Activity of "x" cpm	Activity of "y" cpm	Corr. Activity "x"	Corr. Activity "y"	x_∞
1	______	______	______	______	______	______
2	______	______	______	______	______	______
3	______	______	______	______	______	______
4	______	______	______	______	______	______
5	______	______	______	______	______	______

T_2 ______ °A

x_∞ ______ (average)

R' ______ Activation Energy:

k_2 ______ E ______ cal/mole

$t_{1/2}$ ______

CALCULATIONS:

1. Plot $[-\log(1 - x/x_\infty)]$ vs time on linear graph paper. The slope of the curve is equal to the velocity constant k.

$$k = R'(a + b)/(2.303\ ab)$$

Calculate k at each temperature.

2. Calculate the value of R' at each temperature.
3. Calculate the half-time $t_{1/2}$ of the reaction at each temperature.
4. Using the data from both groups, calculate the activation energy E for the reaction.

EXPERIMENT 12.3 ISOTOPE DILUTION TECHNIQUES

OBJECTIVES:

To illustrate the principles of isotope dilution methods and their application to analytical procedures.

To analyze an "unknown" solution by means of an isotope dilution technique.

THEORY:

Isotope dilution analysis was introduced by Hevesy and Hofer (55) in 1934, but it was not until 1940 that the technique was revived by Rittenberg and Foster (62). Subsequent to 1940 the usefulness of isotope dilution methods has been reported frequently and is now accepted as an indispensable technique for performing various analyses which would otherwise be extremely tedious or impossible. Volumetric analysis depends upon the use of a titrant which is specific for the substance sought, and gravimetric analysis depends upon the ability to separate the substance sought in pure form for subsequent weighing. Often it is impossible to accomplish either of these aims, and one must resort to isotope dilution methods for analysis.

There are three general types of isotope dilution methods. These are (a) direct, (b) inverse, and (c) double isotope dilution. These methods are based on the same fundamental principles but they differ both in technique and procedure and are applied under different circumstances.

Direct Isotope Dilution—(Determination of an Inactive Compound by Dilution with an Active Compound). The following technique is used to determine the quantity of a nonradioactive or untagged constituent in a mixture of closely related compounds which are difficult to separate quantitatively by conventional methods. It is applicable to both inorganic and organic mixtures.

The technique consists of:

Step 1. Addition of a "spike" consisting of a known amount of isotopically labeled compound with a known specific activity S' to the unknown mixture containing the same compound made up of stable isotopes. The two are thoroughly mixed to obtain a uniform distribution.

Step 2. Suitable treatment of the mixture to isolate the same compound in pure form. It is essential that the isolated compound be pure, but it is not necessary that all of the compound be recovered from the mixture. Thus, it is possible to avoid tedious processes required for quantitative separations and frequently to carry out analyses otherwise impractical.

Step 3. Determination of the isotope content of the isolated portion by measurement of its specific activity S. The ratio of active and inactive molecules depends on the relative masses of (active) substance added m' and (inactive) substance originally present m.

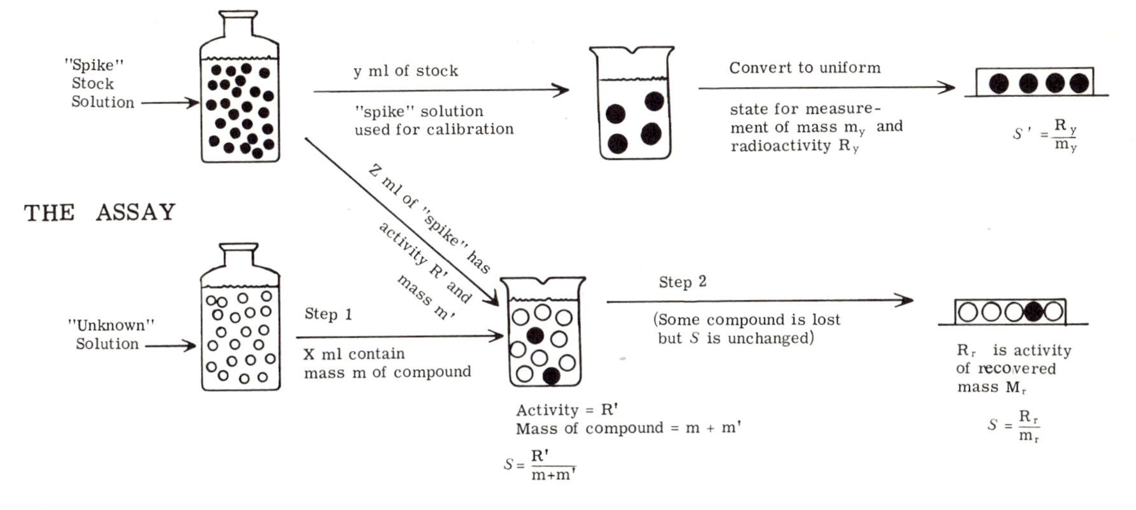

If

R′ = activity (cps) in spike used for assay
m′ = mass of spike used for assay
m = mass of inactive substance in unknown used in assay then

the specific activity of the spike S' is given by

$$S' = R'/m' \tag{12.23}$$

and the specific activity of the substance isolated in pure form from the mixture is

$$S = R'/(m' + m) = R_r/m_r \tag{12.24}$$

It is desired to know the mass m of substance in the unknown in terms of easily measurable quantities. This may be accomplished by dividing equation 12.23 by equation 12.24 and solving for m. Thus,

$$\frac{S'}{S} = \frac{R'/m'}{R'/(m' + m)} = \frac{(m' + m)}{m'} = 1 + \frac{m}{m'} \tag{12.25}$$

$$\frac{m}{m'} = \frac{S'}{S} - 1 \tag{12.26}$$

$$m = m' [(S'/S) - 1] \tag{12.27}$$

Example: A sample of mass m was spiked with 0.05 grams of sodium phosphate having an activity of 1950 cps.

$$S' = R'/m' = 1950/0.05 = 39{,}000 \text{ cps/g}$$

0.1 grams (i.e., m_r) of pure sodium phosphate was recovered having an activity R_r of 60 cps.

$$S = 60/0.1 = 600 \text{ cps/g}$$
$$m = m' [(S'/S) - 1]$$
$$m = 0.05 [(39{,}000/600) - 1] = 3.20 \text{ grams}$$

Inverse Isotope Dilution—(Determination of Radioactive Compound by Dilution with Inactive Compound). As the name implies, inverse isotope dilution is essentially the reverse of direct isotope dilution. It is applicable when a system contains an unknown amount of an isotopically labeled substance of known specific activity S'. To determine the amount m′ of the labeled substance present, a measured quantity m of unlabeled substance is added and allowed to equilibrate. A small quantity of the substance is then removed from the system, purified and measured for its specific activity S. Then, from equation (12.26)

$$\frac{m}{m'} = \frac{S' - S}{S} \tag{12.28}$$

thus

$$m' = m\left(\frac{S}{S' - S}\right) \tag{12.29}$$

The purity of the isolated compound is of utmost importance, particularly if more than one isotopically labeled compound is present. In general, the greater the amount of unlabeled substance added, the more accurate will be the results so long as the activity of the sample can still be measured with accuracy.

Double Isotope Dilution Analysis—Block and Anker (48) describe an isotope dilution procedure which does not require a knowledge of the specific activity S' of the unknown substance. In this case two aliquots of the sample are removed and different amounts of carrier m and m″ added to each. A small quantity of the substance to be determined is extracted from each sample and purified. The specific activities S and S'' respectively, are measured. For the first sample, the amount of unknown substance m′ as given by equation (10.29) is

$$m' = \frac{m\,S}{S' - S} \quad \text{or} \quad S' = \frac{mS}{m'} + S \tag{12.30}$$

and for the second sample

$$m' = \frac{m''\,S''}{S' - S''} \quad \text{or} \quad S' = \frac{m''\,S''}{m'} + S'' \tag{12.31}$$

For this case, where the value of S'' cannot be measured, we equate 12.30 and 12.31 and solve for m′

$$\frac{m\,S}{m'} + S = \frac{m''\,S''}{m'} + S'' \tag{12.32}$$

$$m\,S + m'\,S = m''\,S'' + m'\,S'' \tag{12.33}$$

$$m'\,S'' - m'\,S = m\,S - m''\,S'' \tag{12.34}$$

$$m' = \frac{m\,S - m''\,S''}{S'' - S} \tag{12.35}$$

Similarly, the specific activity can be calculated by the simultaneous solution of equations 12.30 and 12.31 for S′,

$$\frac{\text{m}\,S}{S' - S} = \frac{\text{m}''\,S''}{S' - S''} \tag{12.36}$$

$$S' = \frac{S\,S''\,(\text{m} - \text{m}'')}{\text{m}\,S - \text{m}''\,S''} \tag{12.37}$$

The nature of the double isotope dilution method does not allow the degree of precision attainable with the direct and inverse isotope dilution methods. Normally, however, it does permit more than a rough estimation of quantities not measurable by other means.

Modified Inverse Isotope Dilution—This procedure is the same as the inverse isotope dilution method, except that a radioactive substance is determined by a second radioactive substance. This method is used when a stable isotope is unavailable or its use is undesirable. Procedures involving polonium and palladium are typical examples.

MATERIALS REQUIRED:

Ferric chloride solution containing about 15 mg Fe per ml and a small amount of ^{59}Fe; ammonium hydroxide; "unknown" iron solution.

PROCEDURE:*

To calibrate the tracer (spike) solution, place into one 5-ml centrifuge tube (Tube A):

1.0 ml of radioactive iron solution
3.0 ml of distilled water

To assay the unknown solution, place into a second 5-ml centrifuge tube (Tube B):

1.0 ml of "unknown" iron solution
0.20 ml of radioactive iron solution
3.0 ml of distilled water

Precipitate ferric hydroxide in each tube by adding, slowly with stirring, 6*N* NH_4OH. Centrifuge. Decant the supernatant liquids and redisperse each precipitate in about 3 ml of water plus 1 or 2 drops of 6*N* NH_4OH. Centrifuge and decant the supernatant liquids. Transfer most of each precipitate to a tared planchet using acetone. Completely dry the ferric oxide residues cautiously under the heat lamp.

Determine the mass (nearest tenth mg) of the ferric oxide in each case.

Count the samples using an aluminum absorber (approximately 200 mg/cm^2) between the sample and the counter. The absorber permits only the gammas to be counted and minimizes errors caused by self absorption of the betas.

*Adapted from Overman, et al., J. Chem. Ed., *35*, 296 (1958).

DATA:

	(m) Mass Fe_2O_3 mg	(R) Activity cps	Specific Activity $S = R/m$
Tube A (known)	________ (m_y)	________ (R_y)	________ (S')
Tube B (unknown)	________ (m_r)	________ (R_r)	________ (S)
Mass of spike (i.e., mass of iron in 0.200 ml)			________ (m')

CALCULATIONS:

Assuming the radioactive iron solution to contain 15 mg Fe per ml, calculate the mg of Fe per ml in the "unknown" solution.

The mass of iron (m) in 1 ml of unknown solution is given by the formula

$$m = [(S'/S) - 1]\, m'$$

EXPERIMENT 12.4 RADIOMETRIC ANALYSIS

OBJECTIVES:

To investigate the principles of radiometric analysis.

To analyze an unknown by means of a radiometric analysis.

THEORY:

Radiometric analysis is a quantitative analytical procedure involving a reaction between an unknown and a radioactive reagent (or a radioactive unknown and a stable reagent) resulting in the stoichiometric production of a radioactive compound easily separated from the original radioactive substance by filtration, volatilization or other procedure.

The essence of the radiometric method of quantitative chemical analysis is the use of a reagent whose radioactivity has been standardized in terms of chemical equivalence. The number of equivalents of the reagent consumed in a given reaction is determined by measuring the radioactivity of the reaction product or the radioactivity of the unconsumed reagent when the reaction is completed. In a purely radiometric analysis, the activity measurement is the only measurement necessary. In some cases, measurement of the radioactivity is analogous to weighing the reaction product in a gravimetric analysis but is usually simpler and less time consuming.

It is not anticipated, however, that radiometric analysis will replace the more standard analytical techniques. The value of radiometric methods lies in the high degree of sensitivity often attainable and in the time saved in certain techniques where drying to constant weight can be eliminated.

Direct titration—If a substance produces a radioactive precipitate upon titration with a radioactive titrant, the end-point of the reaction will be indicated by (1) the inflection point of the activity curve for the precipitate (figure 12.2), the activity of which will plateau at the end-point, and (2) the inflection point of the activity curve for the supernatant (figure 12.3), which will show an abrupt increase at the end-point. In either case, separation of the precipitate is necessary. This requirement often imposes a cumbersome restriction on the method.

In the direct-titration method, measurements of activity are made after each addition of titrant, and the end-point is determined through the construction of a curve as illustrated in figure 12.2 or 12.3. If sufficient sample is available, problems associated with separation of the precipitate can be minimized by measuring equal aliquots of unknown solution into each of several centrifuge tubes. To each tube, in sequence, is added a measured amount of titrant, the volumes of titrant being selected to give a series of values on either side of the equivalence-point. The content of each tube is mixed and centrifuged, and an aliquot of supernatant is measured for its radioactivity.

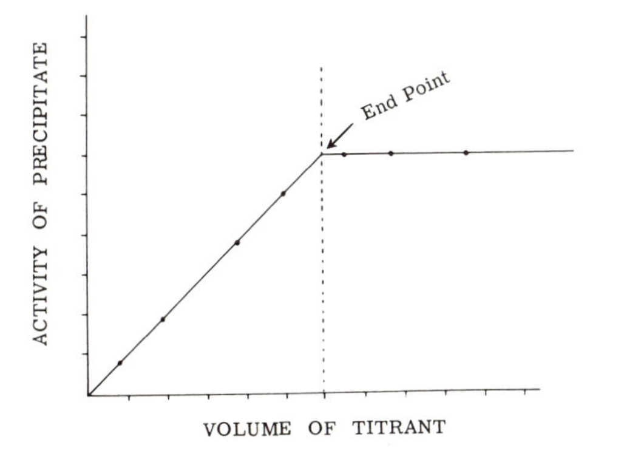

Figure 12.2—END-POINT DETERMINED FROM ACTIVITY OF PRECIPITATE.

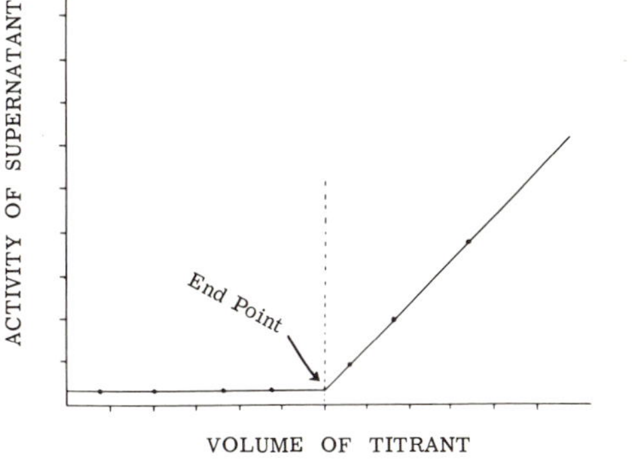

Figure 12.3—END-POINT DETERMINED FROM ACTIVITY OF SUPERNATANT.

Indirect titration—A method comparable to the standard indirect-titration or back-titration technique has been used. A measured excess of titrant is added to the sample. After mixing, the solution is centrifuged and the activity of an aliquot of the supernatant is determined. A comparison of the activity of the supernatant to that of the original titrant measures the amount of unknown present. To illustrate this method, let

R_s = Activity of the standard solution, cpm
V_c = Volume of standard solution counted, ml
N_y = Normality of the supernatant

N_s = Normality of the standard solution
V_x = Volume of unknown solution used for analysis, ml
V_s = Volume of standard solution added to unknown, ml
V_y = Volume of supernatant counted, ml
R_y = Activity of supernatant counted, cpm
N_x = Normality of unknown solution
V_r = Total volume of all solutions ($V_s + V_x +$ necessary reagents)

When an excess of standard solution is added to a solution of the unknown, the milliequivalents of standard consumed will be equal to the milliequivalents of unknown $(\text{meq})_x$ present in the aliquot V_x, and this will be the difference between the amount added $(\text{meq})_s$ and that remaining at the completion of the reaction. Thus,

$$(\text{meq})_x = (\text{meq})_s - (\text{meq})_y \left[\frac{\text{(total volume of all solutions)}}{\text{(volume of supernatant counted)}} \right] \tag{12.38}$$

$$(\text{meq})_x = N_s V_s - (N_y V_y)(V_r/V_y) \tag{12.39}$$

The normality of the supernatant N_y is measured by comparing its radioactivity R_y to that of the standard solution R_s. The term (V_c/V_y) corrects for any difference in volumes of these solutions used for measuring radioactivity. Thus,

$$(\text{meq})_x = N_s V_s - N_s \left[\frac{R_y V_c}{R_s V_y} \quad V_y (V_r/V_y) \right] \tag{12.40}$$

or

$$(\text{meq})_x = N_s \left[V_s - \frac{R_y V_c V_r}{R_s V_y} \right] \tag{12.41}$$

Because the activities of the standard solution R_s and the aliquot of supernatant R_y appear as the ratio (R_y/R_s) in equation 12.41, correction for radioactive decay is not necessary if both are measured at about the same time.

If equal volumes of standard and supernatant are counted to maintain constant geometry, V_c and V_y are then equal, and the ratio V_c/V_y is unity. Thus,

$$(\text{meq})_x = N_s \left[V_s - \frac{R_y V_r}{R_s} \right] \tag{12.42}$$

If desired, the normality of the unknown solution can also be calculated.

$$N_x = (\text{meq})_x / V_x \tag{12.43}$$

Example: Chloride was determined by precipitation as AgCl by the addition of $0.001N$ $^{110}AgNO_3$. The activity of 0.2 ml of $0.001N$ $^{110}AgNO_3$ was 10,000 cpm. To 2.5 ml of the chloride solution was added 0.5 ml of dilute HNO_3 and 3.0 ml of the standard $^{110}AgNO_3$ solution. The solutions were mixed well, heated for a few minutes to coagulate the AgCl, cooled well, heated for a few minutes to coagulate the AgCl, cooled and centrifuged. When 1.0 ml of the supernatant was counted, the activity found was 4,000 cpm. Thus,

$$(meq)_{Cl^-} = N_s\left[V_s - \frac{R_y V_c V_r}{R_s V_y}\right] = 0.001\left[3.0 - \frac{(4{,}000)(0.2)(2.5 + 0.5 + 3.0)}{(10{,}000)(1.0)}\right]$$

$$(meq)_{Cl^-} = 2.5 \times 10^{-3} \text{ (in 2.5 ml of chloride solution).}$$

$$N_{Cl^-} = (2.5 \times 10^{-3})/2.5 = 0.001N$$

APPARATUS AND REAGENTS:

End-window G-M tube, sample holder and scaler; calibrated $0.001N$ Na_2HPO_4 solution containing about 0.1 μCi ^{32}P per ml; 10% NH_4OH; "unknown" solution of approximately $0.001N$ $MgCl_2$.

PROCEDURE:

In this experiment, magnesium is determined by precipitation as magnesium ammonium phosphate with ^{32}P-tagged dibasic sodium phosphate.

1. To exactly 2.0 ml of "unknown" magnesium solution in a 10-ml centrifuge tube, add 3.0 ml of standard, radioactive $0.001N$ Na_2HPO_4 solution and 1.0 ml of 10% NH_4OH. Mix well and allow to stand fifteen minutes or longer, scratching the walls of the tube vigorously to promote the formation of the precipitate.

2. Centrifuge. Transfer 1.0 ml of the supernatant to a planchet and, without drying, measure the activity of the aliquot.

3. Repeat steps 1 and 2 until constant activity of the supernatant is achieved.

4. Using the same geometry, measure the activity of 1.0 ml of the standard phosphate solution. If this activity is too high, coincidence losses may be excessive. In this case, use a smaller aliquot of standard phosphate solution diluted to 1.0 ml with carrier solution or water.

Note: The quantity of precipitate obtained in this procedure is extremely small. It is therefore important that attention be paid to good technique if satisfactory results are to be obtained.

EXPERIMENT 12.5 ACTIVATION ANALYSIS

OBJECTIVES:

To demonstrate the basic principles of neutron activation analysis.

To measure the neutron flux produced by a neutron source.

To analyze a sample for its content of a specific nuclide.

THEORY:

The principle of neutron activation analysis is explained by Guinn[101] by analogy with optical emission spectroscopy. In the optical spectroscope, the sample is caused to emit light by excitation with an electric arc. This light is separated into a spectrum which is recorded on a photographic plate. Each element produces a characteristic spectrum by which it can be identified.

In activation analysis, neutrons replace the electric arc. Nuclei which have been activated (excited) by neutron capture (see experiment 7.6) emit characteristic gamma rays by which they can be identified.

In optical emission spectroscopy, activation and spectrometry occur simultaneously. In neutron activation analysis, they are separate steps. A complete analysis by neutron activation may include the following steps:

1. Activation of the sample (see experiment 7.6).

2. Identification and/or isolation of the nuclide (see experiment 11.1). Identification very often requires a chemical separation.

3. Quantitative assay for the nuclide (see chapter 11).

Activation of the sample—The activity A produced in a specific nuclear species in an irradiation time t is given by

$$A = N\phi\sigma(1 - e^{-0.693t/t_{1/2}}) \qquad (12.44)$$

where N = the number of atoms of the nuclide in the sample, capable of forming the radioisotope in question

ϕ = the neutron flux (n cm^{-2} sec^{-1})

$t_{1/2}$ = the half-life of the nuclear species produced, in the same units as t

σ = the isotopic thermal neutron capture cross section (cm^2)

In most cases, activation of the sample occurs by an (n,γ) reaction. Thus, the atomic number, and hence the chemical identity, of the element does not change, but the mass number increases by one unit.

If the nuclide produced by activation were stable, the amount present in the sample would increase during irradiation as shown by the dotted, diagonal line of figure 10.4, but because it is radioactive, it accumulates at a decreasing rate, as shown by the solid, curved line, until an equilibratory concentration is attained. When this occurs, saturation is said to have been achieved, and the nuclide is now decaying as fast as it is being produced. It should be noted that this curve is exactly the reverse of the decay curve for the nuclide produced. After an irradiation time of $t = t_{1/2}$, the activity of the sample is 50% of the saturation value. There is no need, therefore, to irradiate for more than several half-lives since little additional sample activity will be acquired.

Values for $t_{1/2}$ and σ will be found in the chart of the nuclides. Values of particular interest where sources of low neutron flux are used are given in table 12.1. It should be noted that σ used in equation 12.44 is the isotopic cross section. Thus, N in this equation is equal to the number of atoms in the sample capable of forming the radioisotope in question. N is equal to the total number of atoms of a particular element in the sample only when the isotopic abundance is 100%.

Table 12.1—PARTIAL LIST OF MORE READILY ACTIVATED NUCLIDES

Target Nuclides	% Natural Abundance	Isotope Produced	Half-Life	Isotopic Cross Section (barns)	For $\phi = 10^5$ Approximate Sensitivity (mg nuclide)
^{45}Sc	100	^{46}Sc	84d	13	2.0
^{51}V	99.76	^{52}V	3.77m	4.5	
^{55}Mn	100	^{56}Mn	2.58h	13.3	0.6
^{59}Co	100	^{60m}Co	10.5m	18	
^{63}Cu	69.09	^{64}Cu	12.9h	4.5	7.0
^{65}Cu	30.09	^{66}Cu	5.1m	2.3	
^{75}As	100	^{76}As	26.5h	4.3	2.0
^{79}Br	50.54	^{80}Br	18m	8.5	
		^{80m}Br	4.5h	2.9	3.0
^{81}Br	49.46	^{82}Br	35.7h	3	
^{108}Pd	26.71	^{109}Pd	13.6h	12	5.0
^{107}Ag	51.82	^{108}Ag	2.4m	40	
^{109}Ag	48.18	^{110}Ag	249d	2.8	
		^{110}Ag	24s	82	
^{115}In	95.72	^{116m}In	54m	150	0.1
		^{116}In	14s	50	
^{121}Sb	57.25	^{122}Sb	2.8d	6	4.0
^{123}Sb	42.75	$^{124m_2}Sb$	21m	30	
		$^{124m_1}Sb$	1.3m	30	
^{127}I	100	^{128}I	25m	6.4	2.0
^{139}La	99.911	^{140}La	40.2h	8.9	2.0
^{141}Pr	100	^{142}Pr	19.2h	11	2.0
^{152}Sm	26.72	^{153}Sm	46.7h	220	0.6
^{151}Eu	47.82	^{152}Eu	9.3h	1700	0.03
^{159}Tb	100	^{160}Tb	73d	46	4.0
^{164}Dy	28.18	^{165}Dy	2.3h	800	0.03
		^{165m}Dy	75s	2000	
^{165}Ho	100	^{166}Ho	27.2h	64	0.4
^{169}Tm	100	^{170}Tm	127d	125	2.0
^{174}Yb	31.84	^{175}Yb	4.2d	60	2.0
^{175}Lu	97.41	^{176m}Lu	3.7h	18	1.0
^{186}W	28.41	^{187}W	24.0h	40	3.0
^{185}Re	37.07	^{186}Re	90h	110	0.7
^{187}Re	62.93	^{188}Re	17h	70	0.6
^{191}Ir	37.3	^{192}Ir	74d	750	0.4
		^{192m}Ir	1.4m	250	
^{193}Ir	62.7	^{194}Ir	19h	130	0.3
^{197}Au	100	^{198}Au	64.8h	96	3.0

Example: Calculate the number of microcuries of ^{56}Mn produced by the irradiation of 2.0 grams of MnO_2 for 5.2 hours in a neutron flux of 10^5 n cm^{-2} sec^{-1} by the reaction

$$^{55}Mn(n, \gamma)^{56}Mn$$

^{55}Mn is 100% abundant, σ = 13.3 barns, $t_{1/2}$ for ^{56}Mn = 2.6 hours, the target (2.0 g of MnO_2) consists of 1.265 g ^{55}Mn and the number of atoms of ^{55}Mn in the target is

$$N_{^{55}Mn} = (6.025 \times 10^{23})(1.265/55 = 1.385 \times 10^{22})$$

Thus, the activity, calculated by means of equation 10.44, is

$$A = (1.385 \times 10^{22})(10^5)(13.3 \times 10^{-24})(1 - e^{-0.693(5.2)/2.6})$$

$$A = 1.39 \times 10^3 \text{ dps}$$

or, since $\mu Ci = dps/(3.7 \times 10^4)$

$$\frac{1.39 \times 10^3}{3.7 \times 10^4} = 0.376 \ \mu Ci \text{ of } ^{56}Mn$$

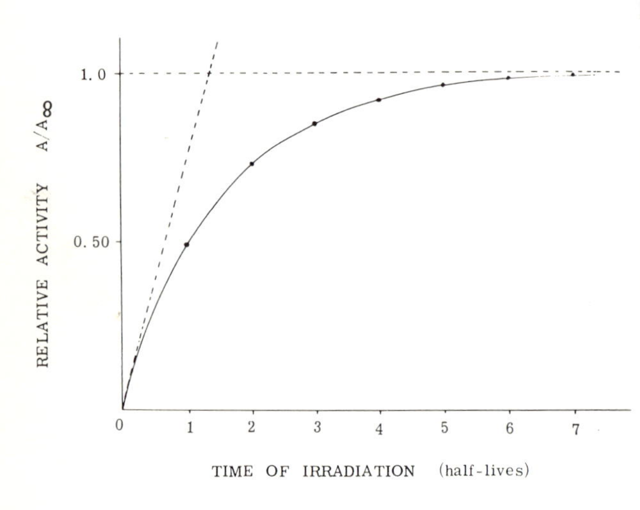

Figure 12.4—ACTIVATION CURVE FOR IRRADIATION OF A SAMPLE BY NEUTRONS.

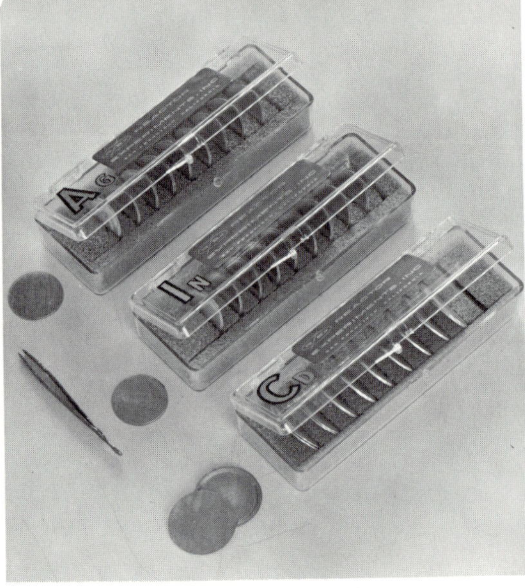

Figure 12.5—FOILS FOR NEUTRON ACTIVATION MEASUREMENTS. (*Courtesy Reactor Experiments, Inc.*)

Identification of the nuclide—Where a quantitative analysis for a specific nuclide is desired, it is common procedure to separate, by chemical means, the nuclide sought from other interfering nuclides. If the element sought is present in trace quantities only, a carrier must be added prior to chemical separation. On the other hand, if it is not known which nuclide may be present, it is necessary to make a complete identification as outlined in experiment 11.1.

Quantitative assay for the nuclide—If the absolute decay rate A of the nuclide is measured, and extrapolated back to the time when irradiation was discontinued, the quantity of target nuclide can be estimated by use of equation 12.44. In this case N is the unknown. This method is only good as an estimate, however, since it is limited by the accuracy with which ϕ and σ are known.

When a more exact analysis is desired, it is necessary to irradiate a known quantity of the element sought simultaneously with the unknown. Any variation in either ϕ or σ will then be the same for both. A simple comparison of the relative activities of the unknown and the known is all that is required for the calculation.

APPARATUS AND REAGENTS:

Neutron source (neutron howitzer, sub-critical assembly or small accelerator); indium foil for calibration; "unknown" sample (e.g., gold foil, compound of manganese or other nuclide selected from table 12.1*); end-window G-M tube and counter assembly; plastic test tubes for irradiating samples.

PROCEDURE:

In this experiment the principles of activation analysis are demonstrated by use of a neutron howitzer or other small neutron source even though these sources, limited by a relatively low neutron flux, may not all find wide-spread use for activation analysis.

In the method given, several assumptions and approximations are made which will influence the accuracy of the results. This sacrifice is made rather than risk losing sight of the basic principle by complicating the procedure.

PART A.—CALCULATION OF NEUTRON FLUX

1. Place an indium foil in a holder and irradiate the foil for at least two half-lives if a neutron howitzer is used, or for a correspondingly shorter time if a more intense neutron source is available. The exact time of irradiation should be noted.

2. Note the time of removal of the foil from the source and measure its activity. (13 minute ^{116}In will decay to a negligible amount in a few minutes.) The activity is measured by comparison with a beta reference source as outlined in experiment 11.4. A Ra-DEF source is suggested. While measuring the activities as a function of absorber thickness, the activity of the ^{116m}In will decrease owing to its 54 minute half-life. It is important, therefore, to note the exact time of each measurement so a decay correction can be applied. The time used for decay correction should be calculated from the mean between the beginning and end of a count. Finally, correct the absolute decay rate to the time of removal of the foil from the source.

3. From the mass of the foil, calculate the number of atoms N of ^{115}In irradiated.

4. From the values of A and N as calculated above, and from the value of σ obtained from table 12.1, calculate the flux ϕ.

*Available from Fairmount Chemical Co., Inc., 136 Liberty St., New York 6, N. Y.

PART B.—ACTIVATION ANALYSIS OF A NUCLIDE

1. Two grams of sample are measured into a plastic test tube and irradiated for at least two half-lives, care being taken to position the sample in the same location occupied by the indium during its irradiation.

2. The time is noted when irradiation is terminated. The sample is spread out in a planchet to duplicate the geometry of the indium foil and the standard source as accurately as possible. An absorption curve is prepared, and the activity at zero total absorber is corrected for decay from the time of termination or irradiation.

3. Using the value of A measured in step 2, the value of ϕ measured in part A of the experiment, and the value of σ from table 12.1, calculate N for the sample.

4. From a knowledge of the chemical identity of the sample and its mass, calculate N and compare to the value obtained in step 3.

EXPERIMENT 12.6 SYNTHESIS OF LABELED COMPOUNDS

OBJECTIVES:

To illustrate a few general techniques for the manipulation of radioactive compounds in the process of organic synthesis.

To prepare several selected labeled compounds requiring the use of typical techniques.

DISCUSSION:

When radioisotopes are prepared in a nuclear reactor, they are normally recovered, either from fission products or from irradiated substances, as free elements or in the form of simple inorganic compounds. When isotopically labeled compounds of a more complex nature are desired, they must be synthesized from the simple compounds or elements obtained from irradiation. Carbon-14, for example, is initially produced as the carbonate. Tritium is readily available as the oxide (tritiated water) or as the free element.

Standard methods of synthesis may be used, but the method used should be selected with discretion. Very often the most commonly used or most direct method is not suited to the preparation of labeled compounds because of low yield. Thus, a method satisfactory for the production of an unlabeled compound may be found uneconomical when using relatively expensive radioactive materials. In order to obtain satisfactory yields, more sophisticated methods are often necessary.

A second important feature to be considered in the design of labeling methods is contamination, especially through the loss of volatile radioactive products or by-products. The use of closed systems employing cold traps to condense and retain volatile materials is a common practice. The apparatus illustrated in figure 12.6 is such a closed system. Starting material is placed at A in the tube on the left. A reagent which must be added during the course of the reaction is placed in the side tube at B from which it can be transferred to A when desired merely by twisting the standard-taper joint. By warming tube A and chilling tube C, the product can be distilled from A to C where it is collected. The entire operation can be conducted *in vacuo* by evacuation through the stopcock while tube A is chilled to prevent loss of starting material. With a sealed system such as this, hazards to personnel from radioactive vapors are minimized.

A good rule, especially when synthesizing radioactive compounds, is to design the apparatus to fit the method rather than to force a method to fit existing equipment. Improved yields of expensive radioactive products through the use of proper equipment will more than offset the expense of special glassware. Hazards to personnel are also reduced when the proper equipment is used.

"Dry runs" should always be made before using a radioactive reactant. A dry run is necessary to check both the equipment and the operator. It is suggested that three steps be taken in the approach to a new synthesis.

1. A dry run using non-radioactive compounds on a scale about 10 times that contemplated for the radioactive synthesis. Radioactive syntheses are often run on such a small scale that products, on occasion, are unaccountably lost. A dry run on a 10 fold scale will help to predict such an occurrence and details can be worked out to eliminate these problems.
2. Dry run using actual quantities of materials to be used in the final synthesis.
3. Synthesis using radioactive materials.

Before proceeding to the next, each step should be repeated until the operator gains proficiency. The importance of the procedure from the standpoint of safety cannot be overemphasized.

Before a container of a liquid radioactive starting material is opened, it should be chilled with ice, acetone-dry ice or liquid nitrogen, depending upon its vapor pressure. If the vapor pressure of the radioactive liquid is relatively low, it can be transferred to the reaction vessel by means of a syringe (see figure 12.7), the needle of the syringe being inserted through a rubber stopper of the type used to seal multiple-dose vials. Reagents and solvents required for the reaction are placed in the reaction vessel prior to addition of the radioactive compound. The apparatus in figure 12.7 is shown with a condenser, for reactions requiring reflux. Should there be a possibility of solids condensing and plugging the condenser tube, a deflegmator may be used to break up the solid and return it to the reaction vessel.

Volatile radioactive liquids are often received in break-seal tubes (figure 12.8). These are opened by breaking the fragile tip of the tube. Before opening, the tube is carefully placed within a larger tube and chilled in an acetone-dry ice bath, or in liquid nitrogen, as required. With the break-seal assembly connected to the reaction vessel (figure 12.9), the magnetic breaker (a small magnetic-stirrer magnet) is pulled out of the side arm by means of another magnet held outside of the assembly, and is allowed to fall on the glass tip. As the break-seal tube is allowed to warm slowly, the vapors pass over into the reaction mixture in the three-neck flask through a bubbler tube. Last traces of the product can be removed from the break-seal assembly by gentle warming.

Reagents are added to the 3-neck flask through a separatory funnel equipped with a pressure equalizer tube. When necessary to conduct the reaction in a vacuum, the

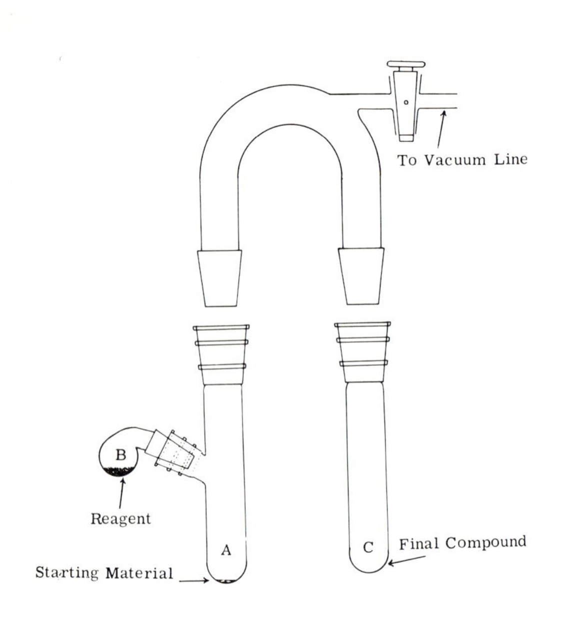

Figure 12.6—REACTION VESSEL AND DISTILLING APPARATUS. For a volume of about 1 ml of starting material.

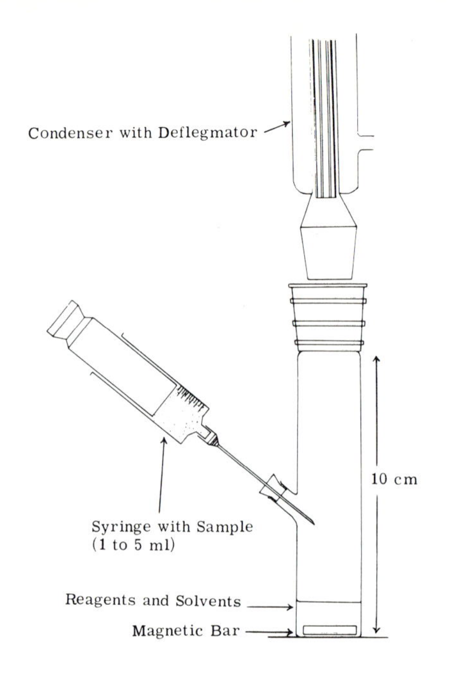

Figure 12.7—REACTION VESSEL WITH CONDENSER FOR REFLUX. Radioactive reactant is injected by means of a syringe.

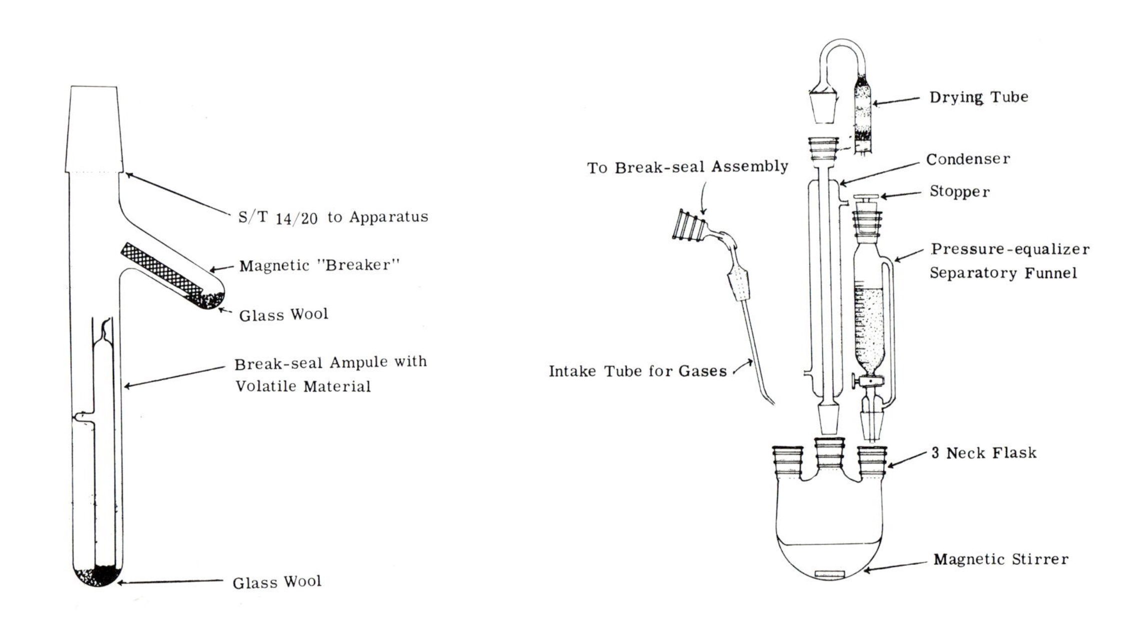

Figure 12.8—BREAK-SEAL ASSEMBLY.

Figure 12.9—REACTION ASSEMBLY.

system can be evacuated through the drying tube at the top of the condenser.

A convenient beaker for use in syntheses involving the formation of a precipitate is shown in figure 12.10. The beaker itself serves as the reaction vessel. A precipitate produced by the reaction is separated from the solution by filtration simply by inverting the beaker to bring the funnel into the proper position. The filtrate is collected in a second beaker of the same type, the male joint of the funnel fitting into the female joint of the second beaker. Either pressure or vacuum can be applied to accelerate filtration.

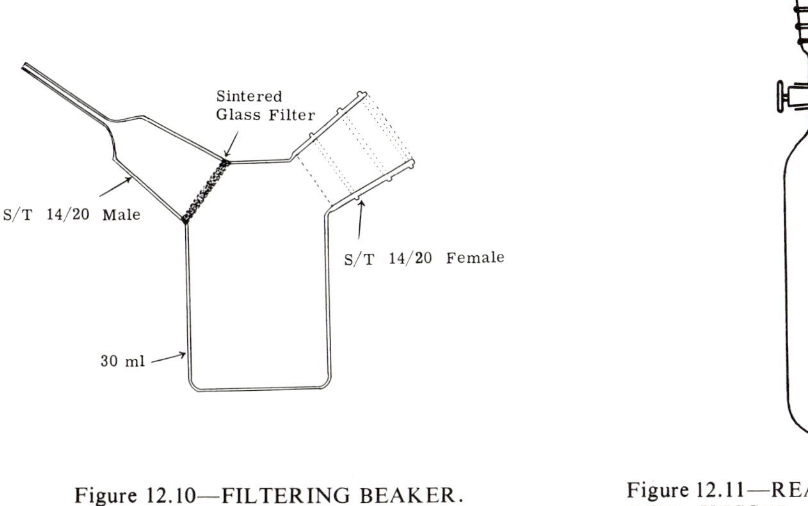

Figure 12.10—FILTERING BEAKER.

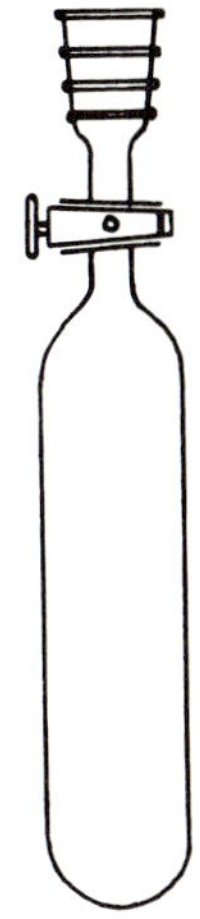

Figure 12.11—REACTION OR STORAGE VESSEL FOR VOLATILE COMPOUNDS.

MATERIALS:

PART A—Adam's catalyst (PtO_2) reduced with tritium; tritiated water; lauric acid; KOH; dry ice; 100% phosphoric acid (H_3PO_4); ether; acetone.

PART B-1—$Ba^{14}CO_3$; Ottawa sand; liquid nitrogen; 85% phosphoric acid (H_3PO_4); lithium aluminum hydride ($LiAlH_4$); tetrahydrofurfuryloxytetrahydropyran; dry ice; compressed nitrogen gas; tetrahydrofurfuryl alcohol.

PART B-2—Hydroiodic acid (density 1.7); liquid nitrogen; $^{14}CH_3OH$; soda lime; P_2O_5.

PART B-3—KCN; $^{14}CH_3I$; dry ice; acetone.

PART B-4—Acetonitrile ($^{14}CH_3CN$); NaOH; H_2SO_4; Ag_2SO_4.

PROCEDURE:

The syntheses given as examples were selected because (1) they represent initial syntheses which must be conducted with pile-produced isotopes prior to the preparation of more complex labeled material, and (2) the techniques required present a greater challenge and have more to offer in the way of training than do the less complex precipitation reactions.

CAUTION: Dry runs should always be made before use of labeled starting compounds.

PART A—PREPARATION OF ^{3}H-LAURIC ACID

75 mg of Adam's catalyst (PtO_2), previously reduced with $^3H_2$, is suspended in 0.75 ml of tritium enriched water contained in a long-neck flask. To this suspension is added 0.75 g of lauric acid and 10 mg of KOH. The flask is cooled with dry ice, evacuated and sealed. It is then agitated gently for six days at 130° C, after which the flask is again chilled and opened. The water is removed by vacuum distillation and the residue is acidified by addition of a few drops of 100% phosphoric acid. The ^{3}H-lauric acid is then extracted from the residue with ether. It is then extracted from the ether with dilute alkali and re-extracted with ether after acidification. Crystals of lauric acid are obtained upon evaporation of the ether. On recrystallization from acetone, 0.7 g of randomly labeled ^{3}H-lauric acid, m. p. 45°, are obtained.

This is a general method for other fatty acids as well. Since the hydroxl hydrogen is not labeled, the tritium introduced by this method is not removed by dilute acid or alkali even at 85°. Thus, this compound is suitable for most biological applications.

PART B-1—PREPARATION OF ^{14}C-METHYL ALCOHOL

1 millimole (197 mg) of finely powdered $Ba^{14}CO_3$, mixed with fine Ottawa sand, is placed in a small reaction vessel (figure 12.6) and 1 ml of concentrated H_2SO_4 is placed in the tilting side arm. To ensure purity of the $^{14}CO_2$ produced, the apparatus is evacuated. The receiving vessel C, which should be provided with a stopcock just below the joint (figure 12.11), is chilled in liquid nitrogen. When the side arm is tilted, the acid will liberate $^{14}CO_2$ from the $Ba^{14}CO_2$. Solid $^{14}CO_2$ will condense in vessel C. Tube A is finally dipped into warm water to ensure completion of the reaction. The stopcock on tube C is now closed to retain the $^{14}CO_2$. Chilling is no longer required. (1 millimole of CO_2 will occupy about 25 ml at 1 atmosphere.) In a reaction vessel (figure 12.9), equipped with a magnetic stirrer, is placed 250 ml of 0.8 molar $LiAlH_4$ in tetrahydrofurfuryloxytetrahydropyran (prepared by refluxing 7.6 g of $LiAlH_4$ in the solvent for 6-8 hours). The apparatus is evacuated and the $^{14}CO_2$ (1-15 millimoles) is admitted at the intake tube.

$$4\,CO_2 + 3\,LiAlH_4 \longrightarrow LiAl(OCH_3)_4 + 2\,LiAlO_2 \qquad (12.45)$$

After an hour, absorption is complete and nitrogen is admitted through the vacuum stopcock line. The addition-line is removed and a receiving tube is substituted. Under nitrogen atmosphere, 2 moles of tetrahydrofurfuryl alcohol are added dropwise to the stirred mixture which is then heated for 2-4 hours at 110° under reflux.

$$LiAl(OCH_3)_4 + 4\,ROH \longrightarrow LiAl(OR)_4 + 4\,CH_3OH \qquad (12.46)$$

The methanol is then distilled from the mixture, using a gentle stream of nitrogen, and collected in a dry ice trap. This is a mixture of $^{14}CH_3OH$ and solvents. It is then vacuum distilled in the apparatus of figure 12.6, at room temperature into a receiver cooled with liquid nitrogen. Yield—65-85%.

PART B-2—PREPARATION OF METHYL IODIDE-^{14}C

$$^{14}CH_3OH + HI \longrightarrow {}^{14}CH_3I + H_2O \tag{12.47}$$

In a glass reaction vessel (figure 12.6), is placed 7 ml of HI (density 1.7), and the tube is frozen in liquid nitrogen. By distillation, 2 to 5 millimoles of methanol-^{14}C are placed in the opposite tube, which contains a glass enclosed magnetic stirrer, and is chilled with liquid nitrogen. The vessels are connected by means of the U-tube and the apparatus is evacuated. The temperature is allowed to increase to room temperature and the HI is added to the methanol by tilting the tube. The mixture is then stirred, finally being heated to 80-85° for 2 hours. After cooling to room temperature, an absorption tube containing soda lime and P_2O_5 on glass wool is inserted between the reaction vessel and a receiving vessel. Methyl iodide is vacuum distilled, the receiving vessel (figure 10.11) being chilled with dry ice while the reaction vessel is warmed in hot water. Yield—90-99%.

PART B-3—PREPARATION OF ACETONITRILE-2-^{14}C

$$^{14}CH_3I + KCN \longrightarrow {}^{14}CH_3CN + KI \tag{12.48}$$

Into one vessel (figure 12.11) is placed 1.7 g of KCN and 3 ml of water; another tube contains 2.6 g of $^{14}CH_3I$. Under reduced pressure, the $^{14}CH_3I$ is distilled into the KCN by warming one tube and chilling the other. The stopcock of the receiving tube is closed and the mixture is held overnight at room temperature with occasional shaking. The mixture of acetonitrile-2-^{14}C and unreacted $^{14}CH_3OH$ is fractionally distilled in a vacuum.

(By a similar process other long-chain fatty acid nitriles can be prepared.)

PART B-4—PREPARATION OF SODIUM ACETATE-2-^{14}C

The acetonitrile-2-^{14}C from the previous reaction is vacuum distilled into a 60-ml flask containing 2 ml of carbonate-free 20 *N* NaOH. The flask is sealed and heated for 4 hours at 80°. After chilling, the flask is opened and the solution mixed with 100 ml of water. Steam is then flushed through the hydrolysate mixture to remove traces of impurity. The alkaline solution is then acidified with concentrated H_2SO_4 and 0.4 g of Ag_2SO_4 is added. The acetic acid-2-^{14}C is separated by steam distillation and then titrated with 1*N* NaOH solution to pH 8.8 by the use of a glass electrode pH meter. The solution is then evaporated to dryness and the resulting granular powder dried under vacuum at 250°. Yield—1.5g, 90-95%.

EXPERIMENT 12.7 DEGRADATION METHODS

OBJECTIVES:

To demonstrate the Schmidt reaction and the wet-oxidation of a sample.
To degrade acetic acid "carbon by carbon."

THEORY:

In the determination of metabolic pathways and in other studies, it is often necessary to establish the position of a radioactive atom within a molecule. To do this the compound must be degraded. That is, some of its atoms must be separated from the rest and radioassayed. Sometimes a "carbon by carbon" degradation is required.

The degradation method employed depends on the chemical nature of the compound. A methyl group next to a carbonyl, for example, can be separated easily from the remaining carbon skeleton by use of haloform reaction.

$$CH_3COR \xrightarrow[NaOH]{NaOI} CHI_3 + NaOCOR \qquad (12.49)$$

If it is desired to determine the radioactivity of the methyl carbon only in 2-butanone ($^{14}CH_3COCH_2\text{-}CH_3$), the compound is treated with NaOI. The resulting $^{14}CHI_3$ (iodoform) is separated by distillation from the solution and assayed for its radioactive content. The remaining carbon chain will be in the form of CH_3CH_2COONa (propionate).

One of the most useful assays is the determination of the radioactivity of a carboxyl carbon. Separation of the carboxyl group may be accomplished by means of the Schmidt reaction.

$$R\text{-}^{14}COOH \xrightarrow[H_2SO_4]{NaN_3} RNH_2 + {}^{14}CO_2 \qquad (12.50)$$

The CO_2 that is formed is easily collected and assayed directly or as $BaCO_3$ or Na_2CO_3. A useful apparatus for this assay is depicted in figure 12.12. This apparatus may also be used for the total wet-oxidation of the remainder of the sample.

EQUIPMENT AND REAGENTS:

All glass apparatus as shown in figure 12.12; sodium azide (NaN_3); H_2SO_4; CH_3COONa (either 1 or 2-^{14}C); saturated $Ba(OH)_2$ solution; toluene; N_2 gas; dry ice; water bath; thermometer; drying tubes loaded with $CaCl_2$ and KOH; Van Slyke-Folch oxidizing mixture—[25 g CrO_3, 5 g KIO_3; 167 ml H_3PO_4 (density 1.7, made by boiling 85% acid) and 333 ml fuming (20% SO_3) H_2SO_4. Heat to 150° and cool.]

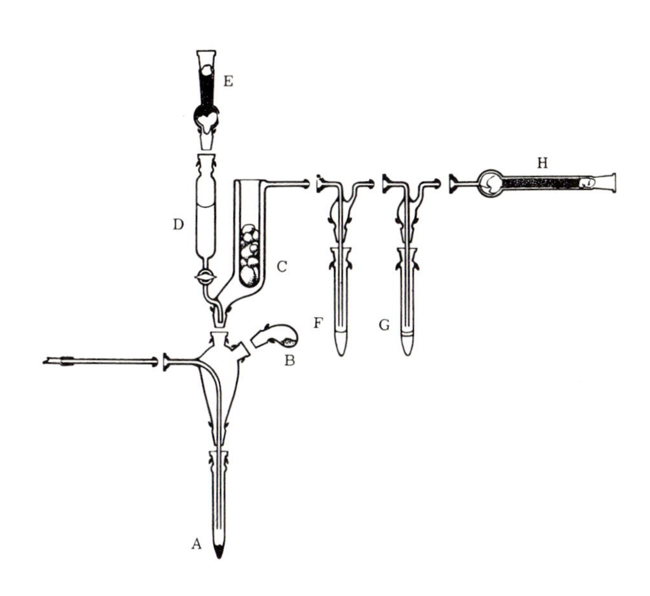

Figure 12.12—APPARATUS FOR CONVERTING ORGANIC MATERIALS TO BARIUM CARBONATE FOR CARBON-14 COUNTING. (*Anal. Chem.*, 29, 982, 1957, *J. L. Rabinowitz.*)

PROCEDURE:

Place the sample (1-5 mg) of ^{14}C-acetate and 10-15 mg of dry NaN_3 into test tube A and pass a gentle stream of nitrogen through the system. Place 10-15 ml of 97% H_2SO_4 in separatory funnel D. Load dry ice into finger trap C, and place CO_2-free $Ba(OH)_2$ saturated solution, layered with toluene, in bubble-tubes F and G. Place the calcium chloride drying tubes E and H in position. Place beaker with ice around test tube A and chill for a few minutes. Let the H_2SO_4 fall dropwise into tube A until 10-12 ml have been added. Remove the beaker of ice from around A and replace it with a water bath at 95-100°. The reaction is completed in 1-2 hours. The CO_2 formed is flushed from the reaction by the N_2 stream and reacts with the $Ba(OH)_2$ to precipitate as $BaCO_3$. Tubes F and G are removed at this time. The precipitates, while in the test tubes, are washed free of adhering $Ba(OH)_2$ with two 5 ml portions of water. This is accomplished by centrifugation and removal of the supernatant with an eye dropper. The washed $BaCO_3$ is dried overnight at 120°, and is assayed for its radioactive content. This carbon is the carboxyl carbon of the acetate sample.

Test tube A now contains $CH_3NH_2 \cdot \frac{1}{2}H_2SO_4$. This compound may be oxidized to CO_2 in the same apparatus. For this purpose Van Slyke-Folch oxidizing mixture is added through the separatory funnel into test tube A or by turning addition tube B. Test tube A is inserted in an oil bath at 150° for one to two hours. The amine carbon is thus quantitatively converted into CO_2. This CO_2 represents the carbon of the methyl group of the acetate sample. The recovery of CO_2 and assay are carried out as before.

EXPERIMENT 12.8 RECOIL REACTIONS (Szilard-Chalmers Process)

OBJECTIVES:

To demonstrate the Szilard-Chalmers reaction.

To investigate the results of recoil reactions in several chemical systems.

THEORY:

Investigations of unusual chemical reactions, occurring as a result of neutron bombardment, were conducted by L. Szilard and T. A. Chalmers (165) as early as 1934. These reactions are known as recoil reactions. The process by which they occur is known as the Szilard-Chalmers process and the term "hot atom chemistry" has been applied to this area of chemical research.

Neutron reactions of the (n,γ) type occur much more frequently than other types of capture reactions. When an (n,γ) reaction occurs, the new nucleus produced exists in a very excited state. In somewhat less than a microsecond, this excess energy of the excited nucleus is emitted as a photon. The nucleus remaining is still radioactive and will decay at a rate regulated by its half-life. A typical reaction of particular interest in hot atom chemistry is the activation of iodine.

$$^{127}\mathrm{I}\,(\mathrm{n},\gamma)^{128}\mathrm{I} \tag{12.51}$$

Let a sample of ethyl iodide be exposed to a source of neutrons. As the neutron velocity decreases, through elastic collisions with hydrogen and other nuclei, the probability of capture by iodine increases. Upon capture, the excited iodine nucleus emits a gamma ray with an energy of about 8 MeV. The recoil, analogous to that experienced when firing a gun, imparts kinetic energy to the iodine atom equivalent to a temperature of about a million degrees. Hence, the term "hot atom chemistry" is applied. Since the chemical bond energy holding the iodine atom in the compound is only about 5 eV, this bond is easily ruptured and the ^{128}I atom speeds off through the ethyl iodide, colliding with other atoms and rupturing many other bonds in the surrounding molecules.

When irradiated ethyl iodide is extracted with aqueous sodium sulfite or sodium thiosulfate, about 60% of the radioactive atoms are found in the aqueous layer. This means that about 40% have recombined with organic fragments. Most of this organic radioactive iodine is found as ethyl iodide and a small part is found in the form of diiodoethane.

In other hot atom reactions, manganese is ejected from $NaMnO_4$ in aqueous solution and is precipitated as MnO_2. Iron is stripped from potassium ferrocyanide from which solutions it can be precipitated as the hydroxide. It should be remembered, however, that the quantities produced are extremely small from a chemical point of view. Nevertheless, when suitable extraction procedures are applied, the Szilard-Chalmers reaction becomes a source of carrier-free isotopes.

Table 12.2—SZILARD-CHALMERS PROCESS

Element	References	Element	References
Antimony	168, 169	Iodine	192-196
Arsenic	170, 171	Iridium	197-199
Bromine	172-183	Iron	200-202
Calcium	184	Maganese	203-206
Cobalt	185-190	Osium	207, 208
Holmium	191	Rhenium	209

APPARATUS AND REAGENTS:

Neutron source; ethyl iodide; 0.1 *N* $Na_2S_2O_3$ or Na_2SO_3; $Ca(MnO_4)_2$ or $NaMnO_4$; well counter or end-window G-M tube with associated scaler; separatory funnel; filter funnel.

PROCEDURE:

PART A—SZILARD-CHALMERS PROCESS WITH ETHYL IODIDE

1. About 50 ml of fresh ethyl iodide in a plastic bottle is irradiated for 30 minutes to 1 hour in a neutron source. If necessary, extract with a dilute solution of sodium sulfite or sodium thiosulfate prior to irradiation, especially if the sample is old, to remove any decomposition products such as HI or free iodine.

2. Transfer the irradiated ethyl iodide to a separatory funnel and extract with three 5 ml portions of 0.1 *N* sodium thiosulfate.

3. Combine the aqueous extracts and transfer a 10-ml aliquot to a plastic test tube and count in a well counter, or transfer the aliquot to a small petri dish and count with the end-window G-M counter.

4. Add a drop of tincture of iodine to the ethyl iodide remaining in the separatory funnel as a carrier and repeat step 3.

5. Transfer a similar aliquot of the ethyl iodide to a counting tube or dish and measure its activity.

6. From the volume of each phase, the volume of each aliquot counted and the observed activities, calculate the percentage of radioactive iodine in each phase. What was the effect of adding carrier solution?

Note: Because the half-life of ^{128}I is so short, it is necessary to record the time of each measurement and to correct for decay.

PART B—SZILARD-CHALMERS REACTION WITH PERMANGANATE

The use of a permanganate solution is especially interesting since this reaction serves as a basis for the calibration of neutron sources. The source is placed at the center of a large bath of $KMnO_4$, $NaMnO_4$ or $Ca(MnO_4)_2$. The MnO_2 produced by the Szilard-Chalmers reaction is filtered off after a measured exposure time. By applying a decay correction, the number of neutrons emitted by the source can be calculated from the activity of the MnO_2. For a small sample, the sodium and calcium salts are used in preference to $KMnO_4$ because of their greater solubility. [e.g., the solubility of $Ca(MnO_4)_2$ is 331 g/100 g of water.]

1. A concentrated solution (ca. 50%) of calcium or sodium permanganate is prepared and filtered through a glass filter into a clean plastic container.

2. After irradiation over night in the neutron source, 10 ml of the solution is transferred to a plastic tube and its activity is measured in a well counter. (The solution can also be placed in a small petri dish and counted with an end-window counter.)

3. The balance of the solution is filtered through a fine filter to remove MnO_2, and 10 ml of the filtrate is counted by the same method used in step 2.

4. Compare the activities of the two solutions and explain the results.

EXPERIMENT 12.9 RADIATION CHEMISTRY

OBJECTIVES:

To illustrate a few of the basic principles of radiation chemistry.

To measure the radiation-induced oxidation of ferrous ion.

THEORY:

In chapter 7 it was seen that the principal interactions of charged particles with matter are coulombic and result in the excitation and ionization of atoms. Each individual collision of a particle along its path involves an average energy loss of about 30 to 35 eV. On the other hand, electromagnetic radiation may pass through a great amount of matter without interaction or loss of any energy but its ultimate collision is a violent one. In a photoelectric interaction or in pair production all of the photon energy is transferred to a charged particle or pair of particles while in a Compton interaction a large part of the photon energy is transferred to a charged particle. The energetic particles produced by these collisions then, in turn, produce excited and ionized atoms through coulombic interactions.

When the excited or ionized atom is a part of a molecule, the molecule is raised to an excited state, is ionized or even split into fragments such as free radicals. By means of a series of reactions, thermal and chemical equilibria are once again restored. This over-all process is called *radiolysis*.

STAGES IN RADIOLYSIS—It is convenient to divide the overall effect of radiation into three stages (254):

1. *Physical stage*—A period of about 10^{-13} seconds or less (231) during which the energy of the incident radiation is dissipated in the system. A variety of excited molecular species is produced in the process. The spacial distribution of those species is inhomogeneous since the ionizing particle produces "spurs" or clusters of excited ions along its path.

2. *Physicochemical stage*—A period of about 10^{-11} seconds during which thermal equilibrium is established in the system. In the process stable molecules may be found and free radicals are produced.

3. *Chemical stage*—A period of about 10^{-8} seconds or longer during which a variety of chemical reactions occur resulting in the establishment of chemical equilibrium.

RADIOLYSIS OF WATER—Because of the unique role of water in chemical and biological systems the irradiation of pure water is of special interest. Early experiments demonstrated hydrogen, oxygen and hydrogen peroxide as products. These reactions have been explained in terms of free-radical reactions.

1. *Physical stage*—In this first step (239) ionization of the water molecules takes place. If the incident radiation is electromagnetic the photoelectric effect, Compton effect or pair production is involved.

$$H_2O \rightsquigarrow H_2O^+ + e_S^- \qquad (12.52)$$

The secondary radiation (i.e., the electron e_S^-) may be a photoelectron or a Compton electron. It is a high energy electron and travels up to about a millimeter through water producing additional ionized water molecules.

2. *Physico—Chemical stage*—The secondary radiation e_s^- produces ionized water molecules by means of the reaction:

$$H_2O + e_S^- \longrightarrow H_2O^+ + e^- + e_S^- \qquad (12.53)$$

The interactions in this reaction are coulombic and although the electron e_S^- emerges from the collision with less energy (by about 30 eV) than it possessed prior to the collision it is still a high-energy electron. On the other hand, the average energy of the excited electron e^- is comparatively low.

A. H. Samuel and J. L. Magee (259) estimate that a 10 eV electron would travel only about 20 Angtroms before becoming thermal. At this distance it would still be within the sphere of influence of the positive ion and would be attracted back to it forming an excited water molecule H_2O^* which dissociates into a hydrogen and a hydroxy l radical.

$$H_2O^+ + e^- \longrightarrow H_2O^* \longrightarrow H\bullet + OH\bullet \qquad (12.54)$$

It is also possible for the H_2O^+ ion to react with a water molecule before it is neutralized by the excited electron.

$$H_2O^+ + H_2O \longrightarrow H_3O^+ + \bullet OH \qquad (12.55)$$

Neutralization then occurs

$$H_3O^+ + e^- \longrightarrow H\bullet + H_2O \qquad (12.56)$$

and its overall reaction is the same as equation 12.54.

Alternatively (246), a more energetic excited electron might escape greater distance from the positive ion reacting with a molecule of water

$$e^- + H_2O \longrightarrow H_2O^- \longrightarrow H\bullet + OH^- \quad (12.57)$$

The intermediate product H_2O^- is a hydrated electron and may be written e^-_{aq}.

E. J. Hart (239) has shown that the hydrated electron can be distinguished from the hydrogen radical only by measurement of rate constants, spectrophotometric methods, etc., but not on the basis of reaction products. They differ only by a proton

$$e^-_{aq} + H^+ \longrightarrow H\bullet \quad (12.58)$$

or

$$e^-_{aq} + H_3O^+ \longrightarrow H\bullet + H_2O \quad (12.59)$$

Their equivalence is illustrated by the reactions:

$$Cu^{++} + H\bullet \longrightarrow Cu^+ + H^+ \quad (12.60)$$

or

$$Cu^{++} + e^-_{aq} \longrightarrow Cu^+ \quad (12.61)$$

in which copper (II) is reduced to copper (I) and by the reactions of the hydroperoxy radical

$$H\bullet + O_2 \longrightarrow \bullet HO_2 \quad (12.62)$$

$$e^-_{aq} + O_2 \longrightarrow O_2^- \quad (12.63)$$

where $\bullet HO_2$ is equivalent to O_2^- since

$$O_2^- + H^+ \longrightarrow \bullet HO_2 \quad (12.64)$$

3. *Chemical stage*—In the third step the free radicals H• and OH• react to form final compounds. The sequence of these reactions will depend upon the nature of the radiation and the nature of the solution.

Alpha radiation and pure water—The specific ionization in the initial step is very great if the ionization is produced by alpha particles. This results in the free radicals being densely populated in a small volume of liquid along the path of the alpha particle. Because of this high concentration, a high percentage of interactions among the •H and •OH radicals in all possible combinations will occur.

$$\bullet H + \bullet H \longrightarrow H_2 \quad (12.65)$$

$$\bullet H + \bullet OH \longrightarrow H_2O \quad (12.66)$$

$$\bullet OH + \bullet OH \longrightarrow H_2O_2 \quad (12.67)$$

Beta (and gamma) radiation in pure water—Since the specific ionization for beta particles is much less than for alpha particles, reactions 12.65, 12.66 and 12.67 take place to a lesser extent. Thus, more free radicals remain to react with the hydrogen and hydrogen peroxide which have diffused throughout the solution.

$$H_2 + \bullet OH \longrightarrow \bullet H + H_2O \quad (12.68)$$

$$H_2O_2 + \bullet H \longrightarrow \bullet OH + H_2O \quad (12.69)$$

and

$$\bullet H + \bullet OH \longrightarrow H_2O$$

Thus, for beta radiation, the buildup of hydrogen peroxide is not so great as it was in the case of alpha radiation.

Aerated water—If dissolved oxygen is present in the water, the hydroperoxy radical will be produced.

$$\bullet H + O_2 \longrightarrow \bullet HO_2 \quad (12.70)$$

This radical is a strong oxidizing agent and will initiate many diverse chemical reactions.

Reactions of inorganic ions—When other ions are present, as in the ferrous ammonium sulfate solution used in this experiment, free radical reactions with these ions are possible also.

$$Fe^{++} + H^+ + \bullet HO_2 \longrightarrow Fe^{+++} + H_2O_2 \quad (12.71)$$

To avoid the need for a three-body collision, equation (12.71) can be written in two steps:

$$Fe^{++} + \bullet HO_2 \longrightarrow Fe^{+++} + \bullet HO_2^-$$

$$\bullet HO_2^- + H^+ \longrightarrow H_2O_2 \quad (12.73)$$

Additional ferrous ions are then oxidized:

$$Fe^{++} + H_2O_2 \longrightarrow Fe^{+++} + OH^- + \bullet OH$$

$$Fe^{++} + \bullet OH \longrightarrow Fe^{+++} + OH^-$$

MEASUREMENT OF RADIATION EFFECT—The tendency of a compound or ion to undergo a radiation-induced reaction is measured by its G value. G is defined as the number of molecules (or ions) produced or destroyed for each 100 eV absorbed. The value of G for the oxidation of ferrous ion is about 15.5 which means that it requires approximately 100/15.5 = 6.5 eV per ion oxidized.

The energy required to cause the reaction of one mole of ions or molecules is

$$(100/G)(6.02 \times 10^{23}) = 6.02 \times 10^{25}/G \text{ electron volts}$$

and the energy required for one micromole is

$$6.02 \times 10^{19}/G \text{ electron volts} = 6.02 \times 10^{13}/G \text{ MeV}$$

Since 1 MeV = 1.60×10^{-6} ergs, this energy is equivalent to $9.6 \times 10^7/G$ ergs. Thus,

$$1 \text{ micromole/g} = 9.6 \times 10^7/G \text{ erg/g}$$

and, since 1 rad = 100 erg/g,

$$1 \text{ micromole/g} = 9.6 \times 10^5/G \text{ rad}$$

For the oxidation of ferrous ion, where G = 15.5,

$$1 \text{ micromole } Fe^{+++}/g = 9.6 \times 10^5/155 = 6.2 \times 10^4 \text{ rads}$$

The dose may be delivered by an external source of radiation or by an internal source such as ^{32}P which has been mixed directly with the solution containing ferrous ion. In the latter case, the dose is calculated as follows:

$$1 \text{ rad} = 100 \text{ ergs per g of solution} = 6.25 \times 10^7 \text{ MeV/g}$$

The average energy of ^{32}P beta particles is 0.7 MeV. Therefore, the number of beta particles N_β required to deliver one rad is

$$N_\beta = 6.25 \times 10^7/0.7 = 8.9 \times 10^7 \text{ beta particles/g}$$

If the concentration of ^{32}P is one millicurie per gram, then

$$\begin{aligned} 1 \text{ mc } {}^{32}P/g &= \frac{3.7 \times 10^7}{8.9 \times 10^7} = 0.42 \text{ rad/sec} \\ &= 25 \text{ rad/min} = 1500 \text{ rad/hour} \end{aligned}$$

In these calculations it is assumed that the decrease in activity of the ^{32}P due to decay is negligible. From these figures it follows that

$$1 \text{ mc } {}^{32}P/g = \frac{1.5 \times 10^3}{6.2 \times 10^4} = 0.0242 \text{ micromoles } Fe^{++}/\text{hour/g of solution}$$

APPARATUS AND REAGENTS:

Beckman Spectrophotometer Model DU, hydrogen lamp and 1 cm quartz cells; ferrous ammonium sulfate; sodium chloride; sulfuric acid; carrier-free $Na_2H^{32}PO_4$ solution.

PROCEDURE:

1. Prepare a solution containing 2×10^{-3} *M* ferrous ammonium sulfate, 2×10^{-3} *M* sodium chloride and 1.6*N* sulfuric acid. Just before use this solution is saturated with oxygen by bubbling air through it.

2. Equal volumes of aerated ferrous ammonium sulfate solution (from step 1) and radioactive sodium phosphate are mixed and placed in a spectrophotometer cell. The total volume (about 3 ml) should be sufficient to fill the cell to the proper level. The concentration of the activity in the final solution should be about 1 mCi/ml.
CAUTION: The quantity of radioactivity used in this experiment requires the use of shielding. The area should be carefully monitored and warning signs posted where necessary.

3. Using a slit width of 0.5 mm, the absorbance (optical density) at a wavelength of 305 millimicrons is measured at intervals of 15 to 30 minutes. Maximum absorption of ferric ion occurs at 305 mμ while the absorbance of ferrous ion at this wavelength is negligible.

4. Calibrate the spectrophotometer with dilutions of ferric sulfate from 0.0001*M* to 0.001*M* in 0.5*M* sulfuric acid.

5. Construct a calibration curve and by use of the curve, calculate the equivalents of ferrous ion oxidized as a function of time from the data of step 3.

6. Calculate the dose rate in rads/hour and the value of G for the oxidation of ferrous ion.

REFERENCES

1. Abraham, B. M., H. E. Flotow and R. D. Carlson, "Particle Size Determination by Radioactivation", Anal. Chem. 29, 1058-1060 (1957).
2. Alder, M. G., and H. Eyring, "Kinetics of Radiation-Induced Decomposition of Certain Organic Molecules in Solution", Nucleonics 10, No. 4, 54-57 (1952).
3. Anonymous, "The Stability of Labelled Organic Compounds", RCC Review 3, The Radiochemical Centre, Amersham, Bucks, England (April 1965).
4. Arnstein, H. R. V., and R. Bentley, "Isotopes in the Study of Chemical Reactions", Nucleonics 6, 11-27 (June 1950).
5. Bayly, R. J., and H. Weigel, "Self-Decomposition of Compounds Labelled with Radioactive Isotopes", Nature 188, 384-87 (October 29, 1960).
6. Cahn, L. and W. J. Cadman, "New Microtechniques", Anal. Chem. 30, 1580 (1958).
7. Claycomb, C. K., Hutchens, T. T., and Van Bruggen, J. T., "Techniques in the Use of C^{14} as a Tracer. I. Apparatus and Technique for Wet Combustion of Non-Volatile Samples. II. Preparation of $BaCO_3$ Plates by Centrifugation. III. Synthesis Labeled Acetic Acid", Nucleonics 7, 38-48 (September 1950).
8. Conway, J. G., and M. F. Moore, "Spectrographic Analysis of Radioactive Materials", Anal. Chem. 24, 463, 464 (1952).
9. Corless, J. T., "Determination of Calcium in Sea Water—Analytical Experiment Using the Radionuclide Ca^{45}", J. Chem. Ed. 42, 421-23 (1965).
10. Cunningham, B. B., "Ultramicrochemistry", Sci. Amer., p. 76-82 (February 1954).
11. Dunn, H. W., "X-Ray Absorption Edge Analysis", Anal. Chem. 34, 116 (1962).
12. Eastwood, T. A., "Recent Advances in Nuclear and Radio-Chemistry", Nucleonics 19, No. 4, 132-137 (April 1961).
13. Feldman, C., M. B. Hawkins, M. Murray and D. R. Ward, "Spectrochemical Analysis of Radioactive Solutions", Anal. Chem. 22, 1400-1403 (1950).
14. Finston, H. L., and E. Yellin, "Determination of Carbon-Hydrogen Ratios by Neutron Scattering", Anal. Chem. 35, 336 (1963).
15. Gray, P. R., D. H. Clarey, and W. H. Beamer, "Interaction of Beta Particles with Matter. Analysis of Hydrocarbons by Beta-Ray Backscattering", Anal. Chem. 31, 1065 (1959).
16. Hahn, O., "Applied Radiochemistry", Cornell University Press (1936).
17. Heemstra, R. J., J. W. Watkins, and F. E. Armstrong, "Laboratory Evaluation of Nine Water Tracers", Nucleonics 19, No. 1, 92-96 (January 1961).
18. Helmholz, H. R., and R. A. Schneider, "Analysis of Trace Activities of Cobalt-60 in Hanford Treated Waste Solutions", Anal. Chem. 31, 1151 (1959).
19. Hudgens, J. E., Jr., "Analytical Applications of Radiochemical Techniques", Anal. Chem. 24, 1704-1708 (1952).
20. Kelly, M. T., "The Analytical Chemist in Nuclear Technology", Anal. Chem. 29, 21A-27A (1957).
21. Leboeng, M. B., D. G. Miller, and R. E. Connally, "Gamma Absorptrometry: A New Tool for Analytical Chemistry", Nucleonics 12, No. 8, 18-21 (August 1954).
22. Lemmon, R. M., "Radiation Decomposition of Carbon-14-Labeled Compounds", Nucleonics 11, No. 10, 44-45 (1953).
23. Meinke, W. W., W. H. Beamer and D. C. Stewart, "Nuclear Technique Training for Analysts Lagging", Anal. Chem. 29, 19A-33A (1957).
24. Newman, David S., "Radio Tracer Studies of Metal-Metal Ion Exchange", J. Chem Ed. 42, 424-426 (1965).
25. Peck, G. E., J. E. Christian and G. S. Banker, "Determination of Moisture Content of Lactose by Neutron Thermalization", J. Pharm. Sci. 53, 632-635 (1964).
26. Raben, M. S., "Microdetermination of Iodine Employing Radioactive Iodine", Anal. Chem. 22, 480-482 (1950).
27. Reynolds, S. A., and G. W. Leddicotte, "Radioactive Tracers in Analytical Chemistry", Nucleonics 21, No. 8, 128-142 (August 1963).
28. Rodden, C. J., "Recent Developments and Current Problems in Inorganic Analytical Chemistry—Nuclear Materials", Anal. Chem. 31, 1940-1945 (1959).

29. Ropp, G. A., "Effect of Isotope Substitution on Organic Reaction Rates", Nucleonics 10, No. 10, 22-27 (1952).
30. Ropp, G. A., and O. K. Neville, "A Review of the Uses of Isotopic Carbon in Organic Chemical Research", Nucleonics 9, No. 2, 22-37 (1951).
31. Rosenblum, C., "The Chemistry and Application of Tritium Labeling", Nucleonics 17, No. 12, 80-83 (December 1959).
32. Ross, C. P., "Particle Size Analysis by Gamma-Ray Absorption", Anal. Chem. 31, 337 (1959).
33. Schulman, J., and M. Falkenheim, "Review of Conventions in Radiotracer Studies", Nucleonics 3, 13 (October 1948).
34. Tolbert, B. M., "Radiation Self-Decomposition of Labeled Compounds", Atomlight, New England Nuclear Corp. (February 1960).
35. Tolbert, B. M., N. Garden, and P. T. Adams, "Special Equipment for C^{14} Work", Nucleonics 11, No. 3, 56-58 (1953).
36. Wahl, A. C., and N. A. Bonner (eds.), "Radioactivity Applied to Chemistry", Wiley (1951).
37. Zeigler, C. A., L. L. Bird, and D. J. Chleck, "X-Ray Rayleigh Scattering Method for Analysis of Heavy Atoms in Low Z Media", Anal. Chem. 31, 1794 (1959).

Solubility Measurements

38. Banks, J. E., "The Equilibria of Complex Formation", J. Chem. Ed. 38, 391-393 (1961).
39. Butler, J. N., "An Approach to Complex Equilibrium Problems", J. Chem. Ed. 38, 141-143 (1961).
40. Butler, J. N., "Calculating Molar Solubilities from Equilibrium Constants", J. Chem. Ed. 38, 460-463 (1961).
41. Corsaro, G., "Ion Strength, Ion Association and Solubility", J. Chem. Ed. 39, 622-626 (1962).
42. Firsching, F. H., "Selective Precipitation of Silver Halides from Homogeneous Solution", Anal. Chem. 32, 1876-1878 (1960).
43. Lochmüller, C., and M. Cefola, "Solubility in Mixed Solvents—A radiochemistry experiment", J. Chem. Ed. 41, 604 (1964).
44. Ramette, R., "Solubility and Equilibria of Silver Chloride", J. Chem. Ed. 37, 348-354 (1960).
45. Sheppart, J. C., and R. C. Jensen, "The Solubility of Thallium (I) Bromide", J. Chem. Ed. 40, 34 (1963).

Exchange Kinetics

46. Duffield, R. B., and M. Calvin, "The Stability of Chelate Compounds. III. Exchange Reactions of Copper Chelate Compounds", J.A.C.S., 68, 557 (1946).
47. Frost, A. A., and R. G. Pearson, "Kinetics and Mechanisms", New York, John Wiley & Sons, Inc., (1961), p. 192.

Isotope Dilution

48. Bloch, K., and H. S. Anker, "An Extension of the Isotope Dilution Method", Science 107, 228 (1948).
49. Breckinridge, C. E. Jr., and J. E. Christian, "Study of Inverse Isotope Dilution Analyses of Chlortetracycline", J. Pharm. Sci. 50, 777-779 (1961).
50. Christian, J. E., and J. J. Pinajian, "The Isotope Dilution Procedure of Analysis. II. Procedure", J. Amer. Pharm. Assoc., Sci. Ed., 42, 304-307 (1953).
51. Clingman, W. H., Jr., and H. H. Hammen, "Determination of Hydrocarbon Oxidation Products—Reverse Isotope Dilution Analysis", Anal. Chem. 32, 323-325 (1960).
52. Craig, J. T., and P. F. Tryon, "Determination of the Gamma Isomer Content of Benzene Hexachloride (By Chlorine-36 Isotope Dilution Method)", Anal. Chem. 25, 1661-1663 (1953).
53. Gordon, M., A. J. Virgona, and P. Numerof, "An Isotope Dilution Assay for Total Penicillins", Anal. Chem. 26, 1208-1210 (1954).
54. Goris, P., W. E. Duffy, and F. H. Tingey, "Uranium Determination by the Isotope Dilution Technique", Anal. Chem. 29, 1590 (1957).
55. Hevesy, G. von, and E. Hofer, "Elimination of Water from Human Body", Nature 134, 879 (Dec. 1934).
56. Moyer, J. D., and H. S. Isbell, "Structural Analysis of Clinical Dextrans by Periodate Oxidation and Isotope Dilution Techniques", Anal. Chem. 29, 1862 (1957).
57. Overman, R. T., D. L. Coffey, and L. A. Muse, "Radioisotope Experiments in the ORINS Summer Institute Programs", J. Chem. Ed. 35, 296 (1958).
58. Pinajian, J. J., J. E. Christian, and W. E. Wright, "The Isotope Dilution Procedure of Analysis. I. Historical and Literature Survey", J. Amer. Pharm. Assoc., Sci. Ed., 42, 301-304 (1953).
59. Quimby, O. T., A. J. Mabis, and H. W. Lampe, "Determination of Triphosphate and Pyrophosphate by Isotope Dilution", Anal. Chem 26, 661-667 (1954).
60. Raben, M. S. "Microdetermination of Iodine Employing Radioactive Iodine", Anal. Chem. 22, 480-482 (1950).
61. Ralph, W. D., Jr., T. R. Sweet, and I. Mencis, "Determination of Small Amounts of Cobalt by Isotope Dilution Analysis", Anal. Chem. 34, 92 (1962).
62. Rittenberg, D., and G. L. Foster, "New Procedure for Quantitative Analysis by Isotope Dilution, with Application to Determination of n-amino Acids and Fatty Acids", J. Biol. Chem. 133, 737 (1940).
63. Rosenblum, C., "Improving Quantitative Analysis", Nucleonics 14, No. 5, 58-59 (May 1956).
64. Rosenblum, C., "Principles of Isotope Dilution Assays", Anal. Chem. 29, 1740 (1957).
65. Swartz, H. A., and J. E. Christian, "An Inverse Isotope Dilution Analysis of Salicylic Acid", J. Amer. Pharm. Assoc., Sci. Ed. 47, 701-702 (1958).

66. Trenner, N. R., R. W. Walker, B. Arison, and C. Trumbauer, "Stable Isotope Dilution Method for Nicotinic Acid Determination", Anal. Chem. 23, 487-490 (1951).

Radiometric Analysis

67. De Lange, P. W., "Radiometric Analysis of Leached Uranium-Thorium Ore Samples with the Beta-Gamma-Gamma Method", Anal. Chem. 31, 812 (1959).
68. Hommel, C. O., F. J. Brousaides, and R. L. Bersin, "Determination of Gaseous Fluorine Using Radioactive Clathrates", Anal. Chem. 34, 1608 (1962).
69. Langer, A., "Radiometric Titration with Radioactive Silver as End-Point Indicator", Anal. Chem. 22, 1288-1290 (1950).
70. Regier, R. B., "Radiometric Determination of Krypton-85", Anal. Chem. 31, 54 (1959).
71. Richter, H. G., and A. S. Gillespie, Jr., "Thallium-240 Radiometric Determination of Dissolved Oxygen in Water", Anal. Chem. 34, 1116 (1962).
72. Scott, B. F., and W. Driscoll, "Radiometric Chemistry for Automatic Process Control", Nucleonics 19, 49 (1960).

Activation Analysis

73. Abraham, B. M., H. E. Flotow, and R. D. Carlson, "Particle Size Determination by Radioactivation", Anal. Chem. 29, 1058 (1957).
74. Amiel, S., "Analytical Applications of Delayed Neutron Emission in Fissionable Elements", Anal. Chem. 34, 1683 (1962).
75. Anders, O. U., "Identification of a Previously Unassigned 5-Second Bromine Activity and Its Use in Neutron Activation Analysis", Anal. Chem. 34, 1678 (1962).
76. Anders, O. U.,"Neutron-Activation Sensitivaties", Nucleonics 18, No. 11, 178-179 (November 1960).
77. Anders, O. U., and W. H. Beamer, "Resolution of Time-Dependent Gamma Spectra with a Digital Computer and Its Use in Activation Analysis", Anal. Chem. 33, 226 (1961).
78. Anonymous, "Analysis, Tracing and Radiation Processing", Nucleonics 17, No. 11, 182-199 (Nov. 1959).
79. Beard, D. B., R. G. Johnson, and W. G. Bradshaw, "Photon Activation Measures Oxygen, Carbon in Beryllium", Nucleonics 17, No. 7, 90-96 (July 1959).
80. Benson, P. A., W. D. Holland, and R. H. Smith, "Determination of Iron and Uranium in High Purity Lead Foil by Neutron Activation Analysis", Anal. Chem. 34, 1113 (1962).
81. Bogart, D., "Estimation of Neutron Energy for First Resonance from Thermal-Neutron Absorption Cross Sections", Nucleonics 10, No. 10, 35-39 (1952).
82. Bopp, C. D., and O. Sisman, "How to Calculate Gamma Radiation Induced in Reactor Materials", Nucleonics 14, No. 1, 46-50 (January 1956).
83. Boyd, G. E., "Method of Activation Analysis", Anal. Chem. 21, 335-347 (1949).
84. Brooksbank, W. A., Jr., Leddicotte, G. W., and J. A. Dean, "Neutron Activation Analysis of Aluminum-Base Alloys", Anal. Chem. 30, 1785 (1958).
85. Corth, R., "Determination of Tantalum in Tungsten by Activation Analysis", Anal. Chem. 34, 1607 (1962).
86. Cosgrove, J. F., R. P. Bastian, and G. H. Morrison, "Determination of Traces of Mixed Halides by Activation Analysis", Anal. Chem. 30, 1872 (1958).
87. Cosgrove, J. F., and G. H. Morrison, "Activation Analysis of Trace Impurities in Tungsten Using Scintillation Spectrometry", Anal. Chem. 29, 1017 (1957).
88. Davis, M. V., and D. T. Hauser, "Thermal-Neutron Data for the Elements", Nucleonics 16, No. 3, 87-89 (March 1958).
89. De, A. K., and W. W. Meinke, "Activation Analysis with an Antimony-Beryllium Neutron Source", Anal. Chem. 30, 1474 (1958).
90. Delbecq, C. J., L. E. Glendenin, and P. H. Yuster, "Determination of Thallium by Radioactivation", Anal. Chem. 25, 350-351 (1953).
91. Eichholz, G. G., "Activation Assaying for Tantalum Ores", Nucleonics 10, No. 12, 58-61 (1952).
92. Freiling, E. C., "Nomogram for Radioactivity Induced in Irradiation", Nucleonics 14, No. 8, 65 (August 1956).
93. Geiger, R. C., and R. C. Plumb, "Slow-Neutron Capture Radioisotopes Arranged by Half-Life", Nucleonics 14, No. 2, 30-31 (February 1956).
94. Gillespie, A. S., Jr., and W. W. Hill, "Sensitivities for Activation Analysis with 14-MeV Neutrons", Nucleonics 19, No. 11, 170-173 (November 1961).
95. Gilmore, J. T., and D. E. Hull, "Nitrogen-13 in Hydrocarbons Irradiated with Fast Neutrons", Anal. Chem. 34, 187 (1962).
96. Girardi, T., and R. Pietra, "Neutron Activation Analysis of Aluminum. Determination of Gamma-Emitting Impurities with Long Half Lives", Anal. Chem. 35, 173 (1963).
97. Goldberg, M. D., "Neutron Cross Sections in the MeV Region", Nucleonics 11, No. 5, 42-45 (1953).
98. Greenfield, M. A., R. L. Koontz, A. A. Jarrett, and J. K. Taylor, "Measuring Flux Absolutely with Indium Foils", Nucleonics 15, No. 3, 57-61 (March 1957).
99. Gruverman, I. J., and W. A. Henninger, "Neutron Activation Analysis of Alloy Steel and Electro-Etch Residues for Sixteen Elements", Anal. Chem. 34, 1680 (1962).
100. Guinn, V. P., "Instrumental Neutron Activation for Rapid, Economical Analysis", Nucleonics 19, No. 8, 81-84 (August 1961).

101. Guinn, V. P., "Neutron Activation Analysis", Internat. Sci. and Tech., Prototype Issue, 74-76, 80, 82, 84 (August 1961).
102. Guinn, V. P., and C. D. Wagner, "Instrumental Neutron Activation Analysis", Anal. Chem. 32, 317 (1960).
103. Harbottle, G., "Activation of Nuclear Isomers by Gamma Rays", Nucleonics 12, No. 4, 64-67 (April 1954).
104. Haskin, L. A., H. W. Fearing, and F. S. Rowland, "Neutron Activation Analysis for U^{235}, Especially in Limestones, by Measurement of Xe^{133}", Anal. Chem. 33, 1298 (1961).
105. Hoste, J., F. Bouten, and F. Adams, "Minor-Constituent Analysis with Neutron Activation", Nucleonics 19, No. 3, 118-123 (March 1961).
106. Hudgens and Cali, "Determination of Antimony by Radioactivation", Anal. Chem. 24, 171 (1952).
107. Hudgens, J. E., Jr., and H. J. Dabagian, "Radioactivation Determination of Zirconium in Zr-Hf Mixtures", Nucleonics 10, No. 5, 25-27 (1952).
108. Hudgens, J. E., and L. C. Nelson, "Determination of Small Concentrations of Indium by Radioactivation", Anal. Chem. 24, 1472-1475 (1952).
109. Hume, D. H., "Radiochemical Activation Analysis", Anal. Chem. 21, 322-326 (1949).
110. Kaplan, L., and K. E. Wilzbach, "Lithium Isotope Determination by Neutron Activation", Anal. Chem. 26, 1797-1798 (1954).
111. Koontz, R. L., and A. A. Jarrett, "Estimating Radioactivity of Irradiated Uranium", Nucleonics 12, No. 6, 26-28 (June 1954).
112. Kroeger, H. R., "Thermal Neutron Cross Sections and Related Data", Nucleonics 5, 51-54 (October 1949).
113. Laing, K. M., R. E. Jones, D. E. Emhiser, J. V. Fitzgerald, and G. S. Bachman, "Neutron Activation and Autoradiography as Coupled Techniques in Tracer Experiments", Nucleonics 9, No. 4, 44-46 (1951).
114. Leddicotte, G. W., and S. A. Reynolds, "Activation Analysis with the Oak Ridge Reactor", Nucleonics 8, 62-65, 78 (March 1951).
115. Leddicotte, G. W., "Activation Analysis—What Can It Do for You?", Nucleonics 14, No. 5, 46-47 (May 1956).
116. Lightowlers, E. C., "Determination of Submicrogram Quantities of Aluminum in Natural Diamonds by Neutron Activation Analysis", Anal. Chem. 34, 1398 (1962).
117. Lowe, L. F., H. D. Thompson, and J. P. Cali, "Neutron Activation Analysis of Silicon Carbide", Anal. Chem. 31, 1951 (1959).
118. Lukens, H. R., Jr., "The Activation of Antimony in Adhesive Products", Nucleonics 17, No. 1, 83 (January 1959).
119. Meinke, W. W., "Pneumatic Tubes Speed Activation Analysis", Nucleonics 17, No. 9, 86-89 (September 1959).
120. Meinke, W. W., "Sensitivity Charts for Neutron Activation Analysis", Anal. Chem. 31, 792 (1959).
121. Meinke, W. W., and R. E. Anderson, "Activation Analysis Using Low Level Neutron Sources", Anal. Chem. 25, 778-783 (1953).
122. Meinke, W. W., and R. E. Anderson, "Activation Analysis of Several Rare Earth Elements—A Comparison with Spectrophotometric Procedures", Anal. Chem. 26, 907-909 (1954).
123. Meinke, W. W., and R. A. Maddock, "Neutron Activation Cross-Section Graphs", Anal. Chem. 29, 1171 (1957).
124. Mesler, R. B., "Rapid Assessment of Neutron Activation", Nucleonics 18, No. 1, 73-75 (January 1960).
125. Moljk, A., R. W. P. Drever, and S. C. Curran, "Neutron Activation Applied to Potassium-Mineral Dating", Nucleonics 13, No. 2, 44-46 (February 1955).
126. Muehlhause, C. O., and G. E. Thomas, "Use of the Pile for Chemical Analysis", Nucleonics 7, 9-17, 59 (July 1950).
127. Peterson, D. C., S. H. Webster, and E. J. Liljegren, "Activities Induced in Some Common Foods by Thermal-Neutron Exposure", Nucleonics 10, No. 1, 33-36 (1952).
128. Plumb, R. C., "Measuring Trace Elements by Activation Analysis", Nucleonics 14, No. 5, 48-49 (May 1956).
129. Plumb, R. C., and J. E. Lewis, "How to Minimize Errors in Neutron Activation Analysis", Nucleonics 13, No. 8, 42-46 (August 1955).
130. Ricci, E., and W. D. Mackintosh, "Neutron Activation Method for the Determination of Traces of Cadmium in Aluminum", Anal. Chem. 33, 230 (1961).
131. Schroeder, G. L., and J. W. Winchester, "Determination of Sodium in Silicate Minerals and Rocks by Neutron Activation Analysis", Anal. Chem. 34, 96 (1962).
132. Senftle, F. E., and W. Z. Leavitt, "Activities Produced by Thermal Neutrons", Nucleonics 6, 54-63 (May 1950).
133. Smales, A. A., and B. D. Pate, "Determination of Submicrogram Quantities of Arsenic by Radioactivation", Anal. Chem. 24, 717-721 (1952).
134. Smith, H., "Estimation of Manganese in Biological Material by Neutron Activation Analysis", Anal. Chem. 34, 190 (1962).
135. Steele, E. L., and W. W. Meinke, "Determination of Oxygen by Activation Analysis with Fast Neutrons Using a Low-Cost Portable Neutron Generator", Anal. Chem. 34, 185 (1962).
136. Szekely, G. "Determination of Traces of Copper in Germanium by Activation Analysis", Anal. Chem. 26, 1500-1502 (1954).

137. Thompson, B. A., "Analysis of Thin Metal Films by Neutron Activation", Anal. Chem. 31, 1492 (1959).
138. Veal, D. J., and C. F. Cook, "A Rapid Method for the Direct Determination of Elemental Oxygen by Activation with Fast Neutrons", Anal. Chem 34, 178 (1962).
139. Westcott, C. H., "Effective Cross Sections for Thermal Spectra", Nucleonics 16, No. 10, 108-111 (October 1958).
140. Winchester, J. W., "Determination of Potassium in Silicate Minerals and Rocks by Neutron Activation Analysis", Anal. Chem. 33, 1007 (1961).

Synthesis Labeled Compounds

141. Catch, J. R., "Preparation and Analysis of Tracer Compounds", Anal. Chem 29, 1726 (1957).
142. Cox, J. D., H. S. Turner, and H. J. Warne, "Synthesis with Isotopic Tracer Element. Part I. The Preparation of Methanol and Sodium Acetate Labeled with Carbon Isotopes", J. Chem Soc. 3167 (1950).
143. Cox, J. D., and R. J. Warne, "Preparation of Methanol Labeled with Isotopic Carbon", Nature 165, 563 (1950).
144. Crompton, C. E., and N. H. Woofrugg, "Chemical Synthesis of Radioisotope-Labeled Compounds", Part I—Nucleonics 7, 49 (September 1950); Part II—Nucleonics 7, 44 (October 1950).
145. Melville, D. B., J. R. Rachele, and E. B. Keller, "A Synthesis of Methionine Containing Radiocarbon in the Methyl Group", J. Biol. Chem. 169, 419 (1947).
146. Moyer, J. D., and H. S. Isbell, "Preparation and Analysis of Carbon-14-Labeled Cyanide", Anal. Chem. 29, 393 (1957).
147. Murray, A., and A. R. Ronzio, "Micro-synthesis with Tracer Elements. VII. The Synthesis of 2-methyl-C^{14}-1, 4-naphthoquinone (Pro-vitamin K)", J. Am. Chem. Soc. 74, 2408 (1952).
148. Murray, A., and D. L. Williams, "Organic Syntheses with Isotopes", Interscience 1, 861 (1958).
149. Murray, A., and D. L. Williams, "Organic Syntheses with Isotopes", Interscience 1, 40 (1958).
150. Murray, A., and D. L. Williams, "Organic Syntheses with Isotopes", Interscience 2, 1269 (1958).
151. Van Heyningen, W. E., et al., "The Preparation of Fatty Acids Containing Deuterium", J. Biol. Chem. 125, 495 (1938).
152. Woodruff, N. H., and E. E. Fowler, "Biological Synthesis of Radioisotope-Labeled Compounds", Nucleonics 7, 26-41 (August 1950).

Degradation Methods

153. Calvin, M., C. Heidelberger, J. C. Reid, B. M. Tolbert, and P. E. Yankwich, "Isotopic Carbon", Wiley (1949).
154. Gold, L., "Distribution of Radioactivity in Labeled Polymers", Nucleonics 11, No. 7, 48-50 (1953).
155. Phares, E. R., and M. V. Long, "The Complete Degradation of C^{14}-labeled Succinic Acid and Succinic Anhydride by the Schmidt Reaction", J. Am. Chem. Soc. 77, 2556-2557 (1955).
156. Pregl, F., and J. Grant, "Quantitative Organic Microanalysis", 4th ed., Blakiston (1946).
157. Rabinowitz, J. L., "Apparatus for Wet-Oxidation of Organic Samples and Carbon Dioxide Trapping for Subsequent Radioactive Assay", Anal. Chem. 29, 982-984 (1957).

Recoil Reactions

158. Dodson, R. W., M. Goldblatt, and J. H. Sullivan, "Measurement of Weak Neutron Intensities by Szilard-Chalmers Reaction in Calcium Permanganate Solution", U. S. Atomic Energy Commission Document MDDC 344 (1956).
159. Edge, R. D., "Neutron Experiments with a Sensitive Szilar-Chalmers Detector", Australian J. Phys. 9, 429 (1956).
160. Levey, G., and J. E. Willard, "The Influence of Structure, Phase and Added Iodine on the Organic Yields of the I^{127} (n,γ) I^{128} Reaction in Alkyl Iodide", J. Am. Chem. Soc. 74, 6161-6167 (1952).
161. Libby, W. F., "Hot Atom Chemistry", Sci. Amer., p. 44, (March 1950).
162. Libby, W. F., "Reactions of High Energy Atoms Produced by Slow Neutron Capture", J. Am. Chem. Soc. 62, 1930-1942 (1940).
163. Schuler, R. H., R. R. Williams, and W. H. Hamell, "Laboratory Exercises in Nuclear Chemistry—III. Preparation and Properties of Several Halogen Activities", J. Chem. Educ. 26, 667-670 (1949).
164. Sun, K. H., and F. A. Pecjak, "Recoil Separation of Isotopes", Nucleonics 14, 122-126 (November 1956).
165. Szilard, L., and T. A. Chalmers, "Chemical Separation of the Radioactive Element from its Bombarded Isotope in the Fermi Effect", Nature 134, 462 (1934).
166. Willard, J. E., "Chemical Effects of Nuclear Transformations", Annual Rev. Nuclear Sci. 3, 193-220 (1953).
167. Willard, J. E., "Hot-Atom Reactions", Nucleonics 19, No. 10, 61-64 (1961).
168. Melander, L., "Szilard-Chalmers Reaction on Antimony", Acta Chem. Scand. 2, 290-291 (1948). (In English.)
169. Kahn, M., "Enrichment of Antimony Activity Through the Szilard-Chalmers Separation", JACS 73, 479-480 (1951).
170. Laurent, H. and P. Simonnin, "Preparation of $Arsenic^{76}$ by the Szilard-Chalmers Process", J. Phys. 14, 294-298 (1953).
171. Miller, H. and E. Broda, "The Szilard-Chalmers Effect on the Oxygen Acids of Arsenic", Monal. 82, 48-52 (1951).
172. Maddock, A. G. and H. Muller, "Chemical Effects of Radiative Thermal Neutron Capture. VII Calcium Bromate", Trans. Faraday Soc. 56, 509-518 (1960).

173. Harbottle, G., "Hot-Atom Chemistry of Br Atoms in Crystalline $KBrO_3$" J. Am. Chem. Soc., 82, 805-809 (1960).
174. Nesmeyanov, A. N. and A. C. Borisov, "Chemical Action of Radioactive Bromine Atoms, obtained in the reaction of Bromine with Neutrons in Chlorobromomethane, Dichlorobromomethane, and Chlorodibromomethane", Radiokhemeya, No. 1, 86-90 (1959).
175. Campbell, I. G., "Some Properties of the Unstable Intermediate Produced by Isomeric Transition or (n) Reaction in Bromates", J. Inorg. and Nuclear Chem. 15, 46-49 (1960).
176. Milman, M. and P. F. D. Shaw, "Szilard-Chalmers Reaction in Ethyl Bromide", J. Chem. Soc., 1303-1310 (1957).
177. Milman, M., "Szilard-Chalmers Effect in Solid Ethyl Bromide", J. Am. Chem. Soc. 80, 5592-5595 (1958).
178. Milman, M., "Scavenger Effect in Solid Ethyl Bromide", J. Am. Chem. Soc., 79, 5581-5582 (1957).
179. Milman, M., P. F. D. Shaw, and I. B. Simpson, "Szilard-Chalmers Reaction in Ethylene Dibromide", J. Chem. Soc., 1310-1317 (1957). (England.)
180. Herr, W., G. Stocklin, and F. Schmidt, "Effect of Amine Addition on the Amount and Composition of Br^{82} Retention in Neutron Irradiated n- and Isopropyl Bromide", Z. Naturforsch 14b, 693-699 (1959).
181. Berne, E., "The Szilard-Chalmers Reaction with Potassium Bromate", Acta. Chem. Scand. 6, 1106-1115 (1952). (In English.)
182. Cupion, P. C. and E. Crevicoeur, "Difference in the Retention on Nuclear Isomers of $Bromine^{80}$ in the Radioactive Capture of Thermal and Resonance Neutrons", J. Chem. Phys. 49, 29-45 (1952).
183. Levey, G. and J. Willard, "Some Reactions of Elemental Bromine, Hydrogen Bromide and Szilard-Chalmers Bromine with Ethylene Bromide", J. Am. Chem. Soc. 73, 1866-1867 (1951).
184. Bruno, M. and U. Belluco, "The Possibility of Using Complexes with Ethylenediamine-Tetracetic Acid in Szilard-Chalmers Reaction", Recerca Sci. 26, 2085-2089 (1956).
185. Kayas, G. and P. Sne, "The Behavior of Certain Cobalt Complexes in Szilard and Exchange Effects", J. Chem. Phys. 45, 188-190 (1948).
186. Nath, A., and J. Shankar, "Possibility of Preparation of High Specific Activity Cobalt-60 Source for Radio-Therapeutic Use", Current Sci. (India) 24, 267 (1955).
187. Beiser, W., and U. Imobirsteg, "Exchange Reaction and Szilard-Chalmers Effect on Cobalt and Hexacyano-Cobaltate", Experientia 9, 288-289 (1953).
188. Nath, A., and J. Shankar, "Szilard-Chalmers Reaction with Cobalt Phthalocyanine", Current Sci. (India) 22, 372-373 (1953).
189. Ikeda, N., K. Yoshihora, and N. Mishio, "Hot-Atom Chemistry of Ammine Complex Salts. I. Preparation of Co^{60m} and Co^{60} of High Specific Activity", Radioisotopes (Tokyo) 8, 242-245 (1959).
190. Rauscher, H. and G. Harbottle, "Szilard-Chalmers Reaction in Potassium Cobalticyanide", J. Inorg. and Nuclear Chem. 4, 155-170 (1957).
191. Vobecky, M., "Szilard-Chalmers Effect on Holmium Oxide. Preliminary Communication", Collection Czechoslav Chem. Communs 25, 1506 (1960).
192. Brestad, T. and J. Baarli, "Szilard-Chalmers Reaction on CH_3I", J. Chem. Phys. 22, 1311-1313 (1954).
193. Alimarin, I. P. and K. F. Svoboda, "Some Peculiarities in the Yields of the Szilard-Chalmers Process in Alkyl Iodides", Soviet J. At. Energy 5, No. 1, 73-75 (1958).
194. Zlotowski, I., A. Halpern and A. Polaczek, "Radioactive Isotopes of Iodine and Bromine Enriched by the Szilard-Chalmers Method", Zeszyty Nauk. Univ. Jgiellonskiego, No. 3, Ser. Mat.—Przyr. No. 1, 65-81 (1955).
195. Cleavy, R., W. Hamill, and R. Williams, Jr., "Szilard-Chalmers Reaction with Inorganic Compounds of Iodine", J. Am. Cham. Soc. 74, 4675-9 (1952).
196. Campbell, I. G., "Effects of Amine Scavengers on Produces of Neutron Irradiation of Organic Halides", J. Inorg.& Nuclear Chem. 15, 37-45 (1960).
197. Herr, W. and K. Heine, "Nuclear Recoil in Crystalline Hexachloro Complexes of Quadravalent Iridium", Z. Naturforsch. 15a, 323-325 (1960).
198. Croatto, U., G. Giacomello, and A. G. Maddock, "New Methods for the Study of the Szilard-Chalmers Processes as Applied to Sodium Chloroiridate", Ricerca Sci. 21, 1788-1790 (1951).
199. Croatto, U., G. Giacomello, and A. G. Maddock, "Difference in Yield of Iridium 192 and Iridium 194 in the Szilard-Chalmers Separation by Neutron Bombardment of Sodium Chloroiridate", Ricerca Sci. 22, 265-269 (1952).
200. Jack, J. and N. Stuin, "The Szilard-Chalmers Reaction in Ferricinium Picrate", J. Inorg. and Nuclear Chem. 7, 5-7 (1958).
201. Sutin, M. and R. W. Dodson, "The Szilard-Chalmers Reaction in Ferrocene", J. Inorg. and Nuclear Chem. 6, 91-98 (1958).
202. Henry, R., C. Ambertin, C., and J. Valade, "Szilard-Chalmers Effect on Potassium Ferrocyanide", Comm. Energie at. (France), Rappt. No. 1360, 1-8 (1959).
203. Sano, H. T., "Retention of Mn^{56} in Neutron-Irradiated $KMnO_4$," Bull. Chem. Soc. Japan 33, 1738-1739 (1960).
204. Rieder, W., "Szilard-Chalmers Effect With Slow and Fast Neutrons", Acta. Phys. Austriaca 4, 290-303 (1950).
205. Jordan, P., "Exchange Reactions Between Inorganic Compounds of Different Valence", Helv. Chem. Acta. 34, 699-714 (1951) Switzerland.
206. Dodson, R. W., M. Goldblatt and J. H. Sullivan, "Measurement of Weak Neutron Intensities by Szilard-Chalmers Reaction in Calcium Permanganate Solution", U.S. Atomic Energy Commission Document MDDC 344, (1956).

207. Mitchell, R. F. and D. S. Martin, Jr., "Szilard-Chalmers Process for Osmium from Hexachloroosmate (IV) Targets", J. Inorg. Nuclear Chem. 2, 286-289 (1956).
208. Herr, W., and R. Dreyer, "Production of Osmium 191 and 193 Preparations of Very High Specific Activity", Inorg. M. Allgem. Chem. 293, 1-4 (1957).
209. Schweitzer, G. K., and D. L. Wilhelm, "Szilard-Chalmers Reactions in Some Rhenium Compounds", J. Inorg. Nuclear Chem.

Radiation Chemistry

210. Allen, A. O., "The Radiation Chemistry of Water and Aqueous Solutions", Van Nostrand (1961).
211. American Society for Testing Materials, "Tentative Method of Test for Absorbed Gamma Radiation Dose by Chemical Dosimetry", ASTM Method D 1671-59T. ASTM Bull. No. 239, 30 and 52, (July 1959).
212. Anonymous, "Advances in Radiation Effects", Nucleonics 18, No. 9, 64-86 (September 1960).
213. Anonymous, "Hot-Atom Chemists Study Nuclear-Reaction Effects", Nucleonics 19, No. 1, 90-91 (January 1961).
214. Anonymous, "Radiation Chemistry", (A special report), Nucleonics 19, No. 10, 37-68 (October 1961).
215. Anonymous, "Radiation Processing of Chemicals", Chem. & Eng. News, 80-91 (April 22, 1963).
216. Anonymous, "Exposing Polymers for Radiation-Effects Testing", Nucleonics 21, 86 (1963).
217. Anonymous, "Commercial Radiation Synthesis of Ethyl Bromide", Nucleonics 21, 74 (1963).
218. Baxendale, J. H., E. M. Fielden and J. P. Keene, "Pulse Radiolysis of Dioxane Solutions", Science 148, 637 (1965).
219. Beckmann, R., "Radiation Effects in Quartz—A Bibliography", Nucleonics 16, No. 3, 122-138 (March 1958).
220. Bernstein and Katz, "Fluorine 18: Preparation, Properties, Uses", Nucleonics 11, No. 10, 46 (1953).
221. Billington, D. S., "How Radiation Affects Materials—Basic Mechanisms", Nucleonics 14, No. 9, 54-57 (September 1956).
222. Box, H. C., and H. G. Freund, "Paramagnetic Resonance Shows Radiation Effects", Nucleonics 17, No. 1, 66-76 (January 1959).
223. Burr, J. G., "Radiation Chemistry—Organics", Nucleonics 19, No. 10, 49 (October 1961).
224. Burton, M., "Development of Current Concepts of Elementary Processes in Radiation Chemistry", J. Chem. Ed. 36, 273-278 (1959).
225. Cercek, B., M. Ebert, J. P. Keene and A. J. Swallow, "Pulse Radiolysis of Potassium Bromide Solutions", Science 145, 919-920 (1964).
226. Cermak, V., and Z. Herman, "Molecular Dissociation in Charge-Transfer Reactions", Nucleonics 19, No. 9, 106-114 (September 1961).
227. Chapiro, A., "Radiation Chemistry—Polymerization", Nucleonics 19, No. 10, 65 (October 1961).
228. Charlesby, A., "Beneficial Effects of Radiation on Polymers", Nucleonics 14, No. 9, 82-85 (September 1956).
229. Charlesby, A., "How Radiation Affects Long-Chain Polymers", Nucleonics 12, No. 6, 18-25 (June 1954).
230. Choppin, G. R., "Experimental Nuclear Chemistry", Chapter 12, Prentice-Hall (1961).
231. Dewhurst, H. A., A. H. Samuel and J. L. Magee, "A Theoretical Survey of the Radiation Chemistry of Water and Aqueous Solutions", Radiation Research 1, 73 (1954).
232. Dorfman, L. M., "Radiation Chemistry—Intermediates in Liquids", Nucleonics 19, No. 10, 54 (October 1961).
233. Edwards, R. R., and T. H. Davies, "Chemical Effects of Nuclear Transformations", Nucleonics 2, 40 (June 1948).
234. Garrison, W. M., M. E. Jayko and W. Bennett-Corniea, "Radiation-Chemical Oxidation of Peptides in the Solid State", Science 146, 250-252 (1964).
235. Glockler, G., "Early Contributions of S. C. Lind to the Radiation Chemistry of Gases", J. Chem. Ed. 36, 262-266 (1959).
236. Hamill, W. H., "Ion-Molecule Reactions", J. Chem. Ed. 36, 346-349 (1959).
237. Hart, E. J., "Development of the Radiation Chemistry of Aqueous Solutions", J. Chem. Ed. 36, 266-272 (1959).
238. Hart, E. J., "Radiation Chemistry—Water", Nucleonics 19, No. 10, 45 (October 1961).
239. Hart, E. J., "The Hydrated Electron", Science 146, 19-25 (1964).
240. Harteck, P., and S. Dondes, "Producing Chemicals with Reactor Radiations", Nucleonics 14, No. 7, 22-25 (July 1956).
241. Harteck, P., and S. Dondes, "Radiation Chemistry of the Fixation of Nitrogen", Science 146, 30-35 (1964).
242. Karpov, V. L., et al., "Radiation Makes Better Wood and Co-polymers", Nucleonics 18, No. 3, 88-90 (March 1960).
243. Kuppermann, A., "Theoretical Foundations of Radiation Chemistry", J. Chem. Ed. 36, 279-285 (1959).
244. Kuppermann, A., "Radiation Chemistry—Diffusion Kinetics", Nucleonics 19, No. 10, 38-42 (October 1961).
245. Lampe, F. W., "Irradiation Reactions in Hydrocarbon Gases", Nucleonics 18, No. 4, 60-65 (April 1960).
246. Lea, D. E., "Action of Radiation on Living Cells", Cambridge Univ. Press (1946).
247. Levy, B., L. M. Epstein, and G. Handler, "Hydrazine from Irradiation of Solid and Liquid Ammonia", Nucleonics 18, No. 9, 128-130 (September 1960).

248. Lind, S. C., "Radiation Chemistry—Gases", Nucleonics 19, No. 10, 43-44 (October 1961).
249. Matheson, M. S., "Radiation Chemistry—Intermediates in Solids", Nucleonics 19, No. 10, 57 (October 1961).
250. Murphy, C. B., and J. A. Hill, "Detection of Irradiation Effects by Differential Thermal Analysis", Nucleonics 18, No. 2, 78-80 (February 1960).
251. Myers, L. S., et al., "Radiolysis of Thymine in Aqueous Solution: Change in Site of Attack with pH", Science 148, 1234 (1965).
252. Odian, G., "Organic-Irradiation Yields", Nucleonics 18, No. 11, 185 (November 1960).
253. Platzman, R. L., "What is Ionizing Radiation", Sci. Amer., p. 74-83 (September 1959).
254. Platzman, R. L., "The Physical and Chemical Basis of Mechanisms in Radiation Biology", in "Radiation Biology and Medicine", pp. 15-72, W. D. Clause, ed. Addisson-Wesley, N.Y. (1956).
255. Rice, W. L. R., "Effects of Gamma Radiation on Organic Fluids and Lubricants", Nucleonics 16, No. 10, 112-113 (October 1958).
256. Rowland, F. S. and R. Wolfgang, "Tritium Recoil Labeling of Organic Compounds", Nucleonics 14, No. 8, 58-61 (July 1956).
257. Ryan, J. W., "Effect of Gamma Radiation on Certain Rubbers and Plastics", Nucleonics 11, No. 8, 13 (1953).
258. Saller, H. A., "Beneficial Effects of Radiation on Metals", Nucleonics 14, No. 9, 86-88 (September 1956).
259. Samuel, A. H., and J. L. Magee, "Theory of Radiation Chemistry. II Track Effects in Radiolysis of Water", J. Phys. Chem. 21, 1080 (1953).
260. Schulke, A. A., "Radiation Effects in Diamonds", Nucleonics 21, No. 2, 68-70 (February 1963).
261. Ward, R. A., "Irradiated Polyethylene Comes of Age", Nucleonics 19, No. 8, 54-57 (August 1961).
262. Weiss, J., "Chemical Dosimetry Using Ferrous and Ceric Sulfates", Nucleonics 10, No. 7, 28 (1952).
263. Weiss, J., A. O. Allen, and H. A. Schwartz, "Use of Fricke Ferrous Sulfate Dosimeter for Gamma Ray Doses in the Range 4 to 40 kr", Proceedings of the International Conference on the Peaceful Uses of Atomc Energy 14, 179 United Nations (1956).
264. Whetcher, S. L., M. Rotheram and N. Todd, "Radiation Chemistry of Cysteine Solutions", Nucleonics 11, No. 8, 30-33 (1953).
265. Wiebe, A. K., W. P. Conner, and G. W. Kinzer, "Fission Fragments Initiate Ethylene Glycol Synthesis", Nucleonics 19, No. 2, 50-52 (Feb. 1961).
266. Willard, J. E., "Radiation Chemistry—Hot Atom Chemistry", Nucleonics 19, No. 10, 61 (October 1961).
267. Wippler, C., "Irradiation Mechanism in Polyvinyl Chloride", Nucleonics 18, No. 8, 68-72 (August 1960).

CHAPTER 13
Isotopic Tracers in Biology

LECTURE OUTLINE AND STUDY GUIDE

I. GENERAL CONSIDERATIONS
- A. The Tracer Concept
  - 1. Isotopic versus non-isotopic tags
  - 2. Nature of the isotopic label
  - 3. Information supplied by tracers
- B. Sensitivity of Tracer Methods
- C. Biological and Effective Half-lives (*Experiment* 13.6)
- D. Dosage
  - 1. Activity required
  - 2. Total amount of element or compound required
- E. Health Physics Measures
  - 1. Safety of laboratory personnel
  - 2. Safety of community
    - a. Waste disposal
    - b. Air-borne contamination

II. LOCATION AND ASSAY OF BIOLOGICAL TRACERS
- A. Animal Sacrificed
  - 1. *In-vitro* measurements (*Experiment* 13.1)
    - a. Sample preparation
      - (1) Wet ashing
      - (2) Dry ashing
      - (3) Direct assay
    - b. Standardization and calibration curves
  - 2. Autoradiography
- B. Animal Not Sacrificed
  - 1. Use of metabolic cages for measurements on excretion products (*Experiment* 13.2)
  - 2. *In-situ* measurements (*Experiment* 13.3)
  - 3. Scanning techniques (*Experiment* 13.4)
  - 4. Total body counters (*Experiment* 13.5)
- C. Intermediary metabolism (*Experiment* 13.6)

III. KINETIC STUDIES
- A. Reaction Rate Determination (*Experiment* 13.7)
- B. Metabolic Equilibria

IV. BIOSYNTHESIS OF LABELED COMPOUNDS
- A. In a Tissue Homogenate (*Experiment* 13.8)
- B. In a Living Plant (*Experiment* 13.9)

V. SOURCES OF ERROR IN TRACER STUDIES
- A. Chemical Effects, Due to
  - 1. Extraneous elements
  - 2. Principal activity
- B. Radiochemical Purity
  - 1. Extraneous elements
  - 2. Chemical state
  - 3. Distribution in other species of molecules
  - 4. Radiocolloids
  - 5. Parent-daughter activities
- C. Radiation Effects
  - 1. Physiological effects
  - 2. Chemical effects
- D. Isotope Effects
  - 1. Cause and effects
  - 2. Examples
- E. Exchange Reactions
  - 1. Loss of label by exchange
  - 2. Catalyst and enzyme studies

INTRODUCTION:

It would be extremely difficult to calculate the value of the radioisotope as a tool in biological research. Certainly much of the progress achieved in this field in recent years would have been impossible prior to the development of the atomic pile. Two basic facts account for the utility of these radioactive elements. We can tell *where they are (location),* and *how much is present (quantity),* even after they have become intermixed with large numbers of stable atoms of the same element.

THE TRACER CONCEPT:

"Tracers" or "tags" of sorts have been used for many years. The tagging of birds and fish by affixing a small metal band to the animal has been used as a means for studying their migratory habits. In this way one is able to determine not only where they go, but also how many migrate to a particular location. To study the migratory habits of flies and mosquitoes by this technique would pose a problem with respect to tagging the animals. Radioisotopes have solved this problem. Even bacteria are not too small to accept a radioactive tag. Indeed, even a single molecule can be identified in this manner.

To tag a molecule an isotope of one of the component atoms of the molecule must be used. That is, a stable atom in the molecule is replaced by a radioactive one of the same element. Such a tag is called *isotopic.* To tag objects of complex chemical composition it is often not necessary to use an isotopic tag. Paper money, boxes, mosquitoes and many other things can be tagged simply by touching with a drop of any radioactive solution. Such a tag is said to be *non-isotopic.*

If the *location* and *quantity* of glucose is to be determined in a biological system, the carbon-14-labeled compound may be used. The labeled glucose, being chemically indistinguishable from native glucose, will mix completely with the available glucose pools in the body of the organism studied. Both the *quantity* and *location* of this sugar present in tissues can then be determined by radioactive assay of biopsies.

For proper tagging, the physical and chemical nature of the compound must be known. The tag must be in such a position that it will render maximum information without being needlessly lost. Glycine and other amino acids may be synthesized with ^{14}C, ^{18}O, ^{3}H and ^{15}N, but in order to classify an amino acid as "glucogenic" (glycogen formers) or "ketogenic" (lipid formers), the fate of the carbon skeleton must be known. Therefore only ^{14}C-labeled amino acids may be used for this purpose. On the other hand, the contribution of an amino acid to the "amine pool" can only be measured with ^{15}N-tagged material, while amino acids labeled with ^{3}H or ^{18}O must be used for the evaluation of kinetic rates of peptide bond formation. Thus, for each type of study a different labeled material may be required.

A molecule may contain more than one atom of a particular element. If all atoms of this element are tagged, the compound is *totally labeled* for that element. If the specific activity of each radioactive atom is equal, the compound is also *uniformly labeled.* Biological systems, when not in "substrate equilibrium," may give totally labeled compounds that are not uniformly labeled. Nevertheless these compounds may render much information about the metabolic "pathways" of the substrate.

In many studies it is advantageous to label only a single atom of the carbon chain (e. g. acetate may be ^{14}C-tagged in either of its two carbons.) If acetic acid were prepared by the following method

$$CH_3MgI + {}^{14}CO_2 \xrightarrow[\text{b. } [H^+]\text{, hydrolysis}]{\text{a. ether, } I_2} CH_3{}^{14}COOH$$

the acetic acid formed would be ^{14}C-labeled in the carboxy position (1-^{14}C). If the label were desired in the methyl position, ^{14}C-labeled methyl Grignard reagent would first be prepared and subsequently treated with stable CO_2. Totally labeled ^{14}C-acetic acid would result if both the methyl Grignard reagent and the CO_2 were labeled.

If two different species of radioactive atoms are employed, *doubly labeled* compounds can be prepared. Thus, the hydrolysis of $K^{14}CN$ by tritiated water produces doubly labeled formic acid:

$$K^{14}CN + 2\,{}^{3}H_2O + 2\ HCl \xrightarrow[\text{heat}]{\text{Catalist}} {}^{3}H^{14}COOH + KCl + NH_4Cl$$

In this reaction some of the tritium label would also appear in the ammonium chloride.

In the discussion which follows on "Sensitivity of Tracer Methods" it is pointed out that for an activity of 10 millicuries per millimole of ^{14}C activity, only 1 carbon atom in 38.4 carbon atoms is radioactive. Thus, the probability of both tritium and ^{14}C appearing in the same molecule of doubly labeled formic acid is very slight except at unusually high specific activities. At low specific activities, the "double" label is actually a mixture of the singly labeled molecules $^{3}HCOOH$ and $H^{14}COOH$. When any iodine atom in iodoform [CHI_3] is labeled, all iodine atoms are effectively labeled since they are chemically equivalent. This is also true of the chlorine atoms in trichloroacetic acid [Cl_3CCOOH]. Similarly, when a methyl group of acetone [$(CH_3)_2CO$] is carbon labeled, it is equivalent to having them both tagged. It is not possible to have only 1-^{14}C-acetone; it is actually 1, 3-^{14}C-acetone. The specific activity of the methyl group is halved but since there are two, the total activity remains the same.

When labeled compounds are administered to an organism they become a part of the general metabolic pool and undergo reactions characteristic of the metabolism of that particular organism. The original compound may remain unchanged, or in the course of the metabolism it may undergo changes to become part of a new compound, or it may be degraded to simpler compounds. When such changes occur, the radioactive label becomes a part of the new compound formed. It is often important, therefore, to determine the chemical identity of radioactive compounds isolated in the course of a tracer experiment.

SENSITIVITY OF TRACER METHODS:

The weight of carbon-14 which possesses one millicurie of activity is 0.227 milligrams. This has an activity of 3.7×10^7 disintegrations per second. Liquid scintillation counters can easily measure activities of 3.7 disintegrations per second. Thus, the presence of 10^{-8} milligrams of carbon-14 would present no problem of detection. With care, this lower limit of detection could be extended to 10^{-9} milligrams (10^{-6} micrograms). For shorter lived isotopes this limit of detection is even further extended (e. g., 14 day ^{32}P, less than 10^{-10} micrograms).

Consider the activity of ^{14}C-labeled compounds in which all the carbon is the mass 14 isotope. If the activity of 0.218 milligrams of ^{14}C is 1 millicurie, then the activity of pure ^{14}C is 64 millicuries per millimole. From this it is seen that the activity of a completely labeled compound, expressed in millicuries per millimole, will equal to 64 multiplied by the number of carbon atoms in the compound. Thus for benzene—$^{14}C_6H_6$—

the maximum possible activity is $64 \times 6 = 384$ millicuries per millimole (mM). That is, the activity of benzene would be 384 mCi/mM if every carbon atom in every molecule were ^{14}C. Such high specific activities are not yet possible. One supplier of benzene-^{14}C quotes an activity of 1 mCi/mM. This would mean that of every 384 carbon atoms only 1 is radioactive, or 1 out of 64 benzene molecules contains a radioactive atom. Specific activities of 10 mCi/mM are also available in which the ratio is 1 radioactive atom per 6.4 molecules or 1 radioactive atom per 38.4 carbon atoms. On special order, slightly higher specific activities are available.

BIOLOGICAL AND EFFECTIVE HALF-LIVES:

If loss of radioactivity from an organism is assumed to occur through radioactive decay only, the concentration of activity C remaining in a tissue at time t is given by:

$$C = C^0 e^{-\lambda t} \qquad (13.1)$$

where C^0 is the initial tissue concentration.

In a living organism depletion of radioactive materials occurs not only by radioactive decay but through excretion as well. Excretion usually occurs by way of several routes and by means of many complex mechanisms. However, the rate of excretion of a particular element or substance is found to be roughly proportional to its concentration. Therefore the mathematics describing the depletion of radioactivity through excretion is similar to that for radioactive decay and the rate can be expressed by a biological decay constant λ_b.

Since depletion occurs simultaneously through the parallel pathways of radioactive decay and excretion the mathematics representing this process is analogous to that for branching decay. Thus,

$$\lambda_e = \lambda_r + \lambda_b \qquad (13.2)$$

where λ_e is the effective decay constant, λ_r is the radiological decay constant and λ_b is the biological decay constant. The tissue activities and concentrations are therefore related by:

$$\frac{A}{A^0} = \frac{R}{R^0} = \frac{C}{C^0} = e^{-\lambda_r t} e^{-\lambda_b t} = e^{-(\lambda_r + \lambda_b)t} = e^{-\lambda_e t} \qquad (13.3)$$

where $e^{-\lambda_e t}$ is the fraction of activity remaining at time t.

It is often more convenient to express the depletion of radioactivity in terms of half-life rather than decay constants. By use of the relationship $\lambda = 0.693/t_{1/2}$ equation 13.2 can be written

$$\frac{0.693}{t_e} = \frac{0.693}{t_r} + \frac{0.693}{t_b} \qquad (13.4)$$

Division by 0.693 yields

$$\frac{1}{t_e} = \frac{1}{t_r} + \frac{1}{t_b} \tag{13.5}$$

or

$$t_e = \frac{t_r t_b}{t_r + t_b} \tag{13.6}$$

DOSAGE CALCULATIONS:

In answer to the question—What microcurie strength of dose should be used?—the sensitivity of the measuring instrument for the isotope used should first be considered. Liquid scintillation, gas ionization chambers and Bernstein-Ballentine counters offer the highest degree of efficiency—from 50 to 90%. Thus, if 111 counts per minute is considered an acceptable minimum counting rate, taking into account the statistics and available counting time per sample, then for an instrument of 50% efficiency one needs only 222 decaying atoms per minute (3.7 per second) in the sample under observation. Thus, a sample activity of 10^{-4} microcuries is required for counting.

Secondly, the *dilution factor* must be estimated. Suppose the dose is to be administered to a 500 gram animal and that a 1 ml sample of blood is to be withdrawn for the determination of activity. Some knowledge of the disposition of the dose in the animal tissues is presupposed.

Case I—Uniform distribution of dose throughout the animal.—The dilution factor here is 500. Thus 500×10^{-4} μCi or 5×10^{-2} μCi must be administered to attain the minimum count desired in the sample. Administration of 10 times this dose would, of course, provide 10 times the minimum counting rate.

Case II—Concentration of the dose in a particular tissue or organ.—If the dose is administered intravenously as tagged macromolecules unable to permeate through the blood vessels, the blood volume of the animal and the volume of sample (1 cc of blood) taken will determine the dilution factor. Assuming 10% of the body weight as blood, the dilution factor is 50. Thus 50×10^{-4} μCi must be administered.

These criteria determine the microcurie level of the dose, but there is one additional factor which must always be considered. That is, the specific activity required or the quantity of non-radioactive material which should be added either as a carrier, as a diluent for ease in handling, or as a necessary metabolite. This is the *chemical carrier* factor.

Occasionally it is desirable to administer a carrier-free radioisotope (e.g., thyroid metabolism studies). When this is so, the quantity of the dose with respect to its content of chemical substances is essentially zero and the specific activity is extremely high.

On the other hand, it may be desirable to add a considerable quantity of diluent to the radioactive substance to obtain the desired results. This would result in a relatively low specific activity, even though the total activity may not be affected. (e.g., iodinated oleic acid used for metabolic investigations.)

Example I—A dose of 10 μCi of carrier-free radioiodine (as NaI) is administered to a patient to measure the thyroid uptake. What is the total weight of NaI and the specific activity?

$t_{1/2} = 8.0 \text{ days} = 6.9 \times 10^5 \text{ seconds}$

$A = 10\mu Ci = 3.7 \times 10^5 \text{ dps} = \lambda N = (0.69/t_{1/2})\, N = (0.69/6.9 \times 10^5)\, N$

$N = 3.7 \times 10^{11}$ atoms of I or molecules of NaI

$$\text{Weight of NaI} = 3.7 \times 10^{11} \times \frac{154}{6.02 \times 10^{23}} = 9.4 \times 10^{-11} \text{ grams}$$

$$= \underline{9.4 \times 10^{-5} \text{ micrograms}}$$

$$\text{Specific Activity} = 10\mu\text{Ci}/9.4 \times 10^{-5}\,\mu\text{g} = 1.06 \times 10^{5}\,\mu\text{Ci}/\mu\text{g}$$

$$= \underline{1.06 \times 10^{5} \text{ curies/gram}}$$

Example 2—A dose of 10 μCi of iodine-tagged oleic acid is administered to a patient. The total weight of oleic acid is 100 mg. Calculate the specific activity.
Note: Total activity is the same as in example 1.

$$\text{Specific Activity} = 10\,\mu\text{Ci}/100 \text{ mg}$$

$$= \underline{100 \text{ microcuries/gram}}$$

Dilution with an additional 900 milligrams of oleic acid would result in a specific activity of only 10 microcuries per gram. (Total activity is still the same.)

Note the tremendous difference between the specific activities in examples 1 and 2.

Specific activity may be reported in terms of activity per unit of volume, per unit of weight or per unit of chemical equivalence. For example, if a 10 mg sample of $Ba^{14}CO_3$ assays 100 cpm, the specific activity may be reported as:
(a) 10 cpm/mg $BaCO_3$, (b) 44.7 cpm/mg CO_2, (c) 164.1 cpm/mg C or (d) 1970 cpm/μmole.

HEALTH PHYSICS MEASURES:

General procedures of safety are discussed in chapter 3. A further word of caution is indicated regarding the handling of laboratory animals for the protection of the laboratory personnel and the disposal of waste and animal carcasses with regard to the safety of the community.

Precautions should be taken to prevent contamination of the laboratory by confining the work area and by the generous use of absorbent paper to cover the desks and tables. Animals should be kept in cages especially designed for ease in the removal and disposal of excreta. Adequate ventilation should be maintained, especially if the radioisotope used is volatile or if it forms volatile compounds (^{131}I, ^{14}C, etc.). Normally, animal carcasses should be buried and other waste disposed of in accord with the Atomic Energy Commission Regulations* unless special permission is granted by the A. E. C. to follow an alternate procedure.

In most laboratories engaged in low-level work, protection from direct radiation does not pose so serious a problem as do surface and air borne contamination and proper waste disposal.

SOURCES OF ERROR IN TRACER STUDIES:

In biological experiments involving radioactive tracers there are a number of pitfalls of which the experimenter must be mindful lest a serious error in the design or technique of the experiment lead to incorrect conclusions. These sources of error may by placed generally into the following categories:

1. Chemical effects
2. Radiochemical purity
3. Radiation effects
4. Isotope effects
5. Exchange reactions

* Federal Register, November 17, 1960.

Chemical effects may be caused either by extraneous elements or by the principal activity itself. *Extraneous elements* or trace elements, present as impurities, have caused considerable difficulties where enzyme reactions are involved. Poisoned by these trace elements, the enzymes may cease to function or may function in an abnormal manner.

The quantity of the *principal activity* administered will often alter its biological distribution, e.g., the uptake of iodine by the thyroid gland. The greater the mass of iodine administered, the smaller the percentage of uptake. As the amount of iodine is increased the thyroid becomes saturated with the element and, although the total amount of iodine (both radioactive and stable) continues to rise with increased doses, the percentage decreases.

Radiochemical purity is affected not only by the presence of extraneous elements, but also by the chemical state of the principal activity, its colloidal state, and the production of radioactive daughters. Some methods of radioisotope production result in the formation of complex mixtures of radioactive substances which must undergo many steps of separation and purification. Thus it is not surprising, especially in the early days of pile-produced isotopes, that an occasional contaminant consisting of an *extraneous radioactive element* should be found in a sample of radioisotope. Living organisms exert a concentrating action on many elements and although the contaminants may be present in seemingly insignificant amounts, reports are available which show spurious data resulting from this concentrating action of the organisms on extraneous elements.

Of more importance, because of its greater frequency, is the complication involving the *chemical state* of the radioactive element. A typical example is the presence of iodate in iodide for which the United States Pharmacopeia has established a permissible maximum of 5% (see experiment 10.4). Other examples are the presence of ferric ion in ferrous compounds and of phosphate in phosphites. The chemistry of the carbon

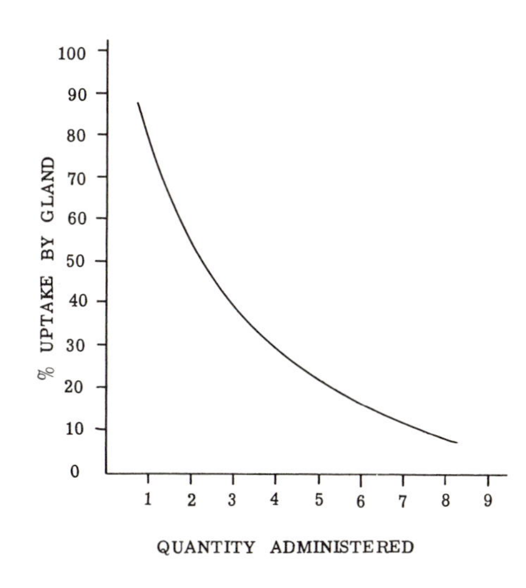

Figure 13.1—EFFECT OF QUANTITY OF DOSE ADMINISTERED ON % UPTAKE OF ^{131}I.

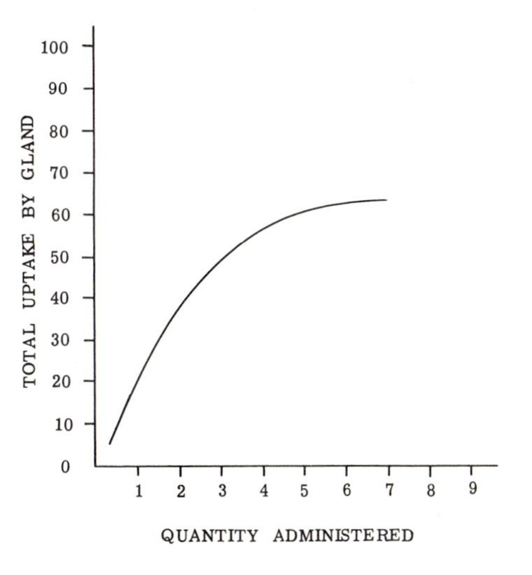

Figure 13.2—EFFECT OF QUANTITY OF DOSE ADMINISTERED ON TOTAL UPTAKE OF ^{131}I.

compounds becomes even more complex and the problem of radiochemical purity with respect to the chemical state of carbon-14 and tritium-labeled compounds is still more acute. Where small quantities only are desired, chromatography offers a useful procedure for purification. Counter-current distribution has also been used. Not only is a starting compound of known purity and identity important, but, as previously pointed out, it is equally important for one to know the identity of labeled compounds isolated from the metabolic pool of organisms in the course of tracer experiments. The label is frequently found on a species of molecule other than the originally labeled substance.

Labeled amino acids, citric acid and acetate may yield totally different compounds, such as labeled glycogen, fatty acids and steroids through metabolic transformations.

Solutions of the radioelements and their compounds may be true solutions, in which the solute is in ionic form or is reduced to units of single molecules. The elements may also exist as *radiocolloids*. Some solutions previously thought to be true solutions have since been shown to be radiocolloids. The action of the reticulo-endothelial system on colloidal preparations of gold, chromium phosphate and zirconyl phosphate depends on the particle size of the colloid. The enzymatic behavior of suspended cholesterol also depends on its particle size.

In tracer experiments employing strontium-90 and a large number of other radioisotopes, the complication presented by the formation of a radioactive daughter can not be avoided.

$$^{90}\text{Sr} \xrightarrow{28\text{ y}} {}^{90}\text{Y} \xrightarrow{64\text{ h}} {}^{90}\text{Zr (stable)}$$

When this problem arises, it is often necessary to wait for the mixture to achieve secular equilibrium before making a measurement of activity. Even though the parent and daughter may have been in a state of equilibrium before administration to the organism, such combinations, easily separated by chromatography, adsorption, etc., are easily separated in animal bodies by similar mechanisms.

Radiation effects are important in some tracer experiments where the total amount of radioactive material required may be of such magnitude that the radiation is detrimental to the organism. The physiological and biochemical status of the organism may be influenced and the observed results invalidated. To determine if the radioactivity itself is influencing a reaction, it is necessary to repeat the same experiment with varying amounts of radioactivity while maintaining the same total chemical concentrations. Lists of maximum permissible radiation levels are available. These levels are continually being downgraded as better biochemical assays become available. Maximum permissible doses are different for each isotope and species of organism. Levels at which biological effects of radiation from the tracer element have been reported are

^{131}I	0.045 μCi/g of body weight in mice
^{226}Ra	0.03 μCi/g of body weight
^{32}P	0.05 μCi/ml of solution on mosquito laryae
^{32}P	0.8 μCi/g of body weight in rats
^{32}P	2.0 μCi/liter of nutritional fluid for plants

These effects are considered to be physiological rather than pathological.

Carbon-14-labeled compounds have shown a greater rate of decomposition on the shelf than the corresponding carbon-12 compounds. The reason for this increased rate of decomposition is associated with the rupture of chemical bonds by the alpha, beta or gamma radiation emitted by the radioactive substance. If the specific activity is very low, radiation effects are negligible. At higher specific activities, this phenomenon becomes appreciable, necessitating frequent repurification of such compounds. Decomposition from radiation can be reduced by dilution of the activity with carrier. Dilution with the non-radioactive form of the substance produces a permanent dilution of the activity. If dilution is accomplished by dissolving the substance in a suitable solvent, then evaporation of the solvent will restore the initial activity. Another ingenious procedure is to add ground glass or very fine glass beads to the substance if it is a liquid, or to concentrated solutions of solids. The glass absorbs much of the radiant energy and the original substance or solution is obtained merely by draining it from the beads; evaporation is unnecessary.

Isotope effects—Radioisotopes are utilized as tracers on the assumption that the radioactive atom behaves in the same manner as its stable counterpart. This may be true or almost true in most cases, but it is not true in certain instances, especially with the lighter elements. This difference in behavior is known as the *isotope effect* and is caused by the difference in the masses of the respective atoms. In some instances, tritium is not even considered the same element as hydrogen because of the great difference. For example, Weinberger and Porter (50) observed the uptake of tritium oxide by algae to be only 45 to 49% of that for hydrogen oxide because of the isotope effect. In large molecules, however, it is legitimate to ignore the mass difference if the tritium is somewhat removed from the reactive group of the molecule.

```
  H T H H              H H H H
  | | | |              | | | |
H-C-C-C-C-OH         H-C-C-C-C-OT
  | | | |              | | | |
  H H H H              H H H H

Negligible effect    Marked effect
```

Ropp (40) has shown that plants can utilize ^{12}C about 17% faster than ^{14}C and Rabinowitz, *et al.* (27) have reported a carbon isotope effect of about 20%. In general, the isotope of greater mass reacts at a slower rate.

In otherwise symmetrical carbon compounds the introduction of a tag may produce asymmetry and thus create optical activity.

```
          H                H³
          |                |
- - - - R-C-ϕ - - - - -  R-C-ϕ - - -  plane of symmetry
          |                |
          H                H

     symmetric        asymmetric
```

A radioisotope may also have a mass which is intermediate between the masses of the natural, stable isotopes with the result that no isotope effect is observed. For example, copper-63 and copper-65 are natural isotopes; copper-64 is radioactive.

In addition to causing kinetic changes, isotopes also produce changes in the infrared spectra of compounds. When crystalline substances are tagged, differences in crystal cell size and cell angles are observed. Differences in chromatographic behavior have also been reported between untagged and tagged amino acids.

Exchange Reactions. Much interest has been centered on exchange reactions suitable for tagging compounds.

The method of Wilzbach (52) is a simple approach to random labeling or organic molecules with tritium. The compound is sealed in an ampoule with tritium gas and allowed to stand for one or two weeks, during which time tritium exchanges with and replaces some of the hydrogen atoms in the compound. Not all hydrogen atoms in the compound undergo this exchange with equal ease, the rate of exchange depending upon the bond energy as well as upon the stereoconfiguration of the molecule. Unfortunately the Wilzbach method has also been reported to cause internal molecular rearrangements. Thus the product obtained requires rigorous purification.

It should be remembered that atoms which become a part of a molecule by exchange can also be lost by exchange. Deuterium and tritium labels attached to oxygen, nitrogen, sulfur and active carbon atoms are lost by exchange with water almost instantly. When attached to inactive carbon atoms, deuterium and tritium labels are relatively permanent. A label should not be used which will be lost too readily and thereby appear in the general metabolic pool of the host.

Occasionally, confusion may arise from the exchange of label between a reactant and a catalyst in a chemical reaction, as illustrated by the Friedel-Crafts reaction

$$C_6H_6 + CH_3COCl^* \xrightarrow{AlCl_3} C_6H_5COCH_3 + HCl^*$$

When chlorine is labeled in the acetyl chloride, the radioactive chlorine atom will also be found in the aluminum chloride even though the $AlCl_3$ was initially unlabeled. When aluminum chloride is labeled, the radioactive chlorine atom will also be found in the acetyl chloride. This exchange makes the determination of the role of $AlCl_3$ in the reaction exceedingly difficult.

In some metabolic processes the tag may be transferred to the enzyme or cofactor, or to an unrelated material present in the reaction. If this transfer does not occur in the presence of an enzyme poison (CN^-, FCH_2COO^-, AsO_4^{-3}) it may be assumed to be the result of an enzymatic process since true exchange is neither affected by protein denaturing agents nor by enzyme poisons.

EXPERIMENT 13.1 *IN-VITRO* MEASUREMENTS (With Sacrifice of Animal)

OBJECTIVES:

To illustrate a typical procedure for the radioassay of biological materials.

To demonstrate wet-ashing procedures, a method for the measurement of activity in liquid preparations, and various ways in which to report results.

In particular, to measure the relative uptake of ^{32}P by various tissues of the rat.

THEORY:

Radioisotopes find widespread use as tracers in biological studies. In metabolism studies, the fate of specific substances frequently cannot be determined without resort to isotopic tagging of the compound. And even with the availability of isotopically tagged compounds there are numerous problems which must be resolved. One of these involves the manner in which a sample of the radioactive substance is assayed.

General Considerations. For determining the distribution of a substance in a biological system, one should carry out the following steps:

1. Administer a dose of known activity.
2. Wait for a specific interval of time to allow the dose to become distributed throughout the animal or biological system.
3. Obtain samples by sacrifice of the animal or by collection of excreta or withdrawal of fluids.
4. Convert all samples to a uniform physical state for counting so as to permit a comparison of samples under identical conditions of counting geometry.
5. Assay the samples.
6. Compute the per cent of original dose in each sample by comparison with a calibration curve prepared from dilutions of an aliquot of the same material used for the dose.

Obtaining the Samples. Very often an animal is sacrificed to permit the removal of specific organs or tissues for radioassay. This procedure has the obvious disadvantage that further observations cannot be made on the same animal. If continuous or periodic observations must be made on the same animal, samples are limited to blood, breath, excreta, biopsy specimens, and other fluids or tissues which can be removed without upsetting the life processes of the organism. Continuous observations of activity can sometimes be made *in situ* by the use of hard beta-or gamma-emitting tracers (see experiment 13.4).

Tissues removed from a sacrificed animal must be reduced to a uniform state for counting. The reasons for preparing uniform samples for the measurement of radioactivity have been discussed in chapter 5. Tissues are converted to a uniform condition by a variety of ashing procedures, following which they can be counted either as liquids or solids or, in some cases, as gases (e. g., $^{14}CO_2$).

Wet-ashing procedures. Conventional wet-ashing procedures employ concentrated acids in combination with oxidizing agents and catalysts to convert the sample to a liquid of uniform composition. Most common of these procedures is the Kjeldahl method, which uses concentrated sulfuric acid with potassium sulfate and phosphoric acid to elevate the digestion temperature, and copper or mercury salts as catalysts.

Simple digestion with concentrated nitric acid will often be found satisfactory, except when the tissue contains large amounts of fat. Comar (9) suggests the use of isoamyl alcohol for fat extraction from the nitric acid digestion mixture. The activities of each phase are then measured and totalled. Others have suggested the use of dioxane or acetone as a common solvent for the aqueous and fatty phases, the final solution containing 60 to 80% of solvent.

Dry-ashing procedures. Planchets of stainless steel and nickel are convenient for ashing small samples, the radioassay being made on the sample in the same container. For larger samples, tared crucibles can be used. In most cases, the sample is placed in a cold muffle furnace. The furnace is heated slowly to prevent spattering, the temperature finally being raised to about 600° C. To avoid spattering, samples may be dried before ignition.

Wet ashing provides several advantages over dry ashing. Most important is the reduced opportunity for loss of sample through volatilization. Also, wet ashing is usually less time consuming than dry ashing.

Some samples may not require ashing prior to radioassay. This depends upon the circumstances. Blood, urine, milk, and certain other liquids may fall into this category. Such samples usually require only simple dilution to a specific volume prior to the measurement of activity. A simple dilution is not sufficient if weak beta emitters are to be measured. These may be assayed by liquid scintillation techniques after decoloration by H_2O_2.

The Calibration Curve. Using an aliquot of the same stock solution or preparation used to provide the aliquot for the "dose," make a series of dilutions covering the normal or anticipated activity range of samples. From these data construct a calibration curve (figure 13.3). The geometry of the standard dilutions must be identical with the geometry of the prepared and diluted samples. The calibration curve then permits a direct conversion of "observed activity" to "per cent of dose," automatically correcting for background and dead time as well as for all other geometry factors. If the calibration curve is prepared just prior to the radioassay of the samples, then changes in sample activity caused by radioactive decay are also taken into account.

Effects of Dose Activity and Animal Weight on Distribution. Discounting deleterious effects of the radiation itself on the animal system which may be incurred through use of abnormally high activities, table 13.1, illustrates the effect of changing the total activity of the dose and/or the weight of the animal used in making a distribution study. It should be noted that:

1. the *specific activity* in the tissues is a function of both the total activity of the dose (compare animals #1 and #2) and the weight of the animal (compare animals #l and #3).

2. use of *% of dose per gram of tissue* avoids the variation caused by differences in administered dose (animals #1 and #2) but not that caused by differences in animal weight (animals #1 and #3).

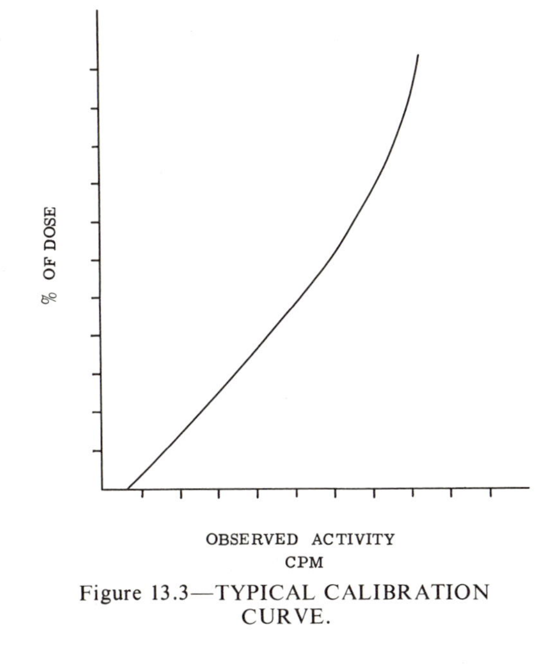

Figure 13.3—TYPICAL CALIBRATION CURVE.

Errors due to variation in animal weight can be reduced to some extent by using only those animals within a certain weight range. A further correction can be made by normalizing the data to an arbitrary animal weight. To do this multiply "% of dose per gram of tissue" by the weight of the animal and divide by the normal weight.

Table 13.1—EFFECTS OF DOSE ACTIVITY AND ANIMAL WEIGHT ON DISTRIBUTION

Animal No.	A Total Activity of Dose	B Weight of Animal	C Specific Activity in Tissues	D % of Dose per Gram of Tissue	×	B Body Weight	=	100%	E Normalized Specific Activity*
#1	100 cpm→	100 g Animal	→ 1 cpm/g	1 %/g	×	100 g	=	100%	0.66 cpm/g
#2	200 cpm→	100 g Animal	→ 2 cpm/g	1 %/g	×	100 g	=	100%	0.66 cpm/g
#3	100 cpm→	200 g Animal	→ 0.5 cpm/g	0.5 %/g	×	200 g	=	100%	0.66 cpm/g

*Normalized to an animal weight of 150 g. (E = C × B/150 g)

MATERIALS REQUIRED:

Rats (or mice or guinea pigs); radioactive sodium phosphate (carrier-free); 1-ml syringes and hypodermic needles; dissecting tools; balance; beakers; concentrated nitric acid; volumetric flasks; 60-mm petri dishes; asbestos gloves.

PROCEDURE:

Inject* a known quantity (60 to 100 μCi) of ^{32}P as phosphate in 0.5 ml of solution into a rat's hind leg. The rat may be anesthetized with ether or other suitable anesthetic for ease in handling. After a specific length of time, sacrifice the rat by over-anesthetizing it with ether.

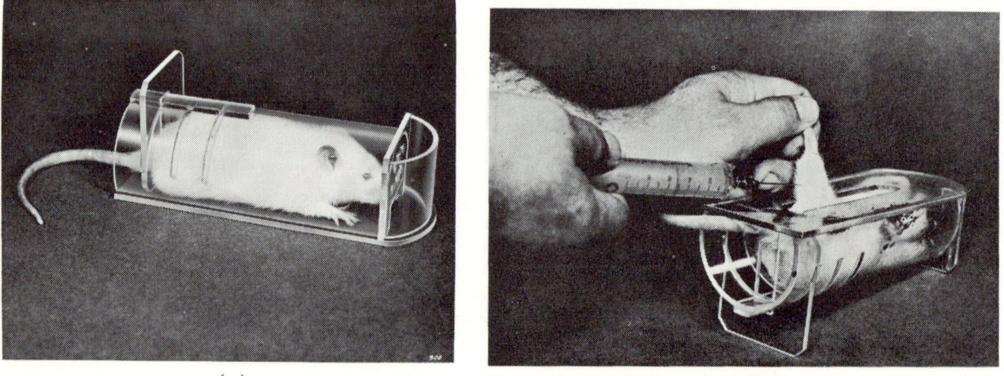

(a) (b)

Figure 13.4—RESTRAINING CAGES FOR RATS AND MICE.
(a) *Courtesy A. H. Thomas Co.*
(b) *Courtesy Econo-Lab Division of Maryland Plastics Co.*

(Place a small piece of absorbent cotton, saturated with ether in the bottom of a small beaker. Hold the rat in one hand, grasping him firmly from the back, thumb and index finger so placed behind his head that he cannot turn. With the free hand place the beaker over his nose).

Remove five or more organs of interest, such as the heart, lungs, spleen, etc. (Do not take any samples from the hind leg, which was the site of the injection).

Weigh each organ to the nearest 10 mg in a tared 50-ml beaker. Record the total weight. Remove a portion of tissue from each organ and record the weight of tissue used. The portion of tissue used should weigh between 0.5 and 1.0 gram.

Add a few ml of concentrated HNO_3 and allow the tissue to soak for about 5 minutes. (Do not use any more acid than necessary). Heat gently and cautiously. Frothing usually occurs in the process of digestion, but frothing can be controlled by flaming the froth directly with the flame of a microburner, or by adding a few drops of silicone anti-foam.

When digestion is complete (frothing ceases and all solids are decomposed), allow to cool. Dilute to 25 ml with water or 0.001*M* H_3PO_4. Place 10-ml aliquots in a small petri dish and count. (DO NOT DRY.)

* It is advisable to wear *asbestos* gloves when handling rats, especially at the time of injection since some rats are prone to bite. Matted asbestos resists penetration by sharp teeth; cloth and leather offer little protection.

Preparation of calibration curve. Dilute 100 μl of the solution injected (solution A) to 100 ml with $0.001M$ H_3PO_4. This is solution B. Prepare dilutions of solution B as follows:

Solution No.	Volume of Solution B (ml)	Volume of $0.001M$ H_3PO_4 added (ml)	Total Volume (ml)	Per Cent of Dose*	
				In 25 ml	In 10 ml
C_1	0	25	25	0	0
C_2	5	20	25	1	0.4
C_3	10	15	25	2	0.8
C_4	15	10	25	3	1.2
C_5	20	5	25	4	1.6
C_6	25	0	25	5	2.0

Measure the activities of 10 ml aliquots of these dilutions in petri dishes. (DO NOT DRY). Prepare a calibration curve for the particular instrument to be used for the radioassays, plotting $\mu Ci/ml$ versus cpm, and a second curve plotting % of dose versus cpm. These curves can then be used for evaluation of the samples.

DATA:

Original Solution (from supplier)

________ mCi/ml at ________ o'clock on ________ (Date standardized)

________ mCi/ml at ________ o'clock on ________ (Date of experiment)

Solution A—(to be injected)

Prepared by diluting ________ ml of original solution to ________ ml with $0.001M$ H_3PO_4. Contains ________ $\mu Ci/ml$ or ________ μCi in the total dose. Inject ________ ml.

Solution B—Prepared by dilution 100 μl of solution A to 100 ml with $0.001M$ H_3PO_4 as directed above.

Solutions C_1 to C_6—Prepared as directed above.

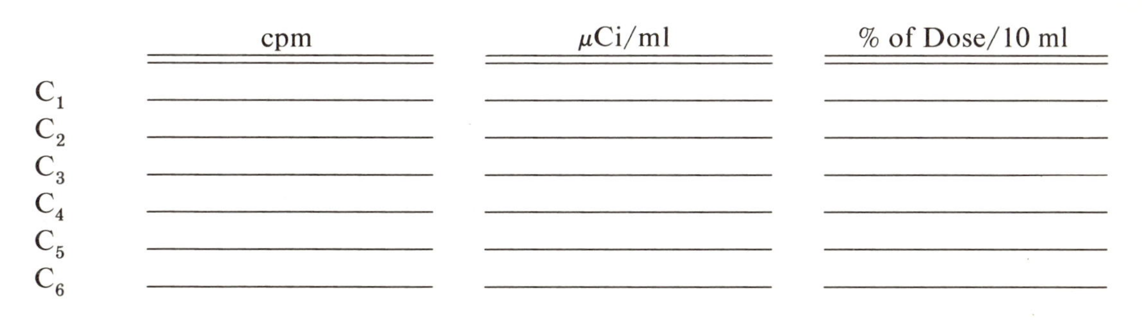

	cpm	$\mu Ci/ml$	% of Dose/10 ml
C_1			
C_2			
C_3			
C_4			
C_5			
C_6			

*Based on dose of 0.5 ml.

	A	B	C	D	E	F	G	H	I	J
Organ	Total Wt.	Wt. Used	cpm per 10 ml	μCi/ml	% of Dose in 10 ml Aliquot	% of Dose in Sample	% of Dose per g of tissue	% of Dose per g of tissue (normalized)	mg P/g of tissue	% of Dose per mg P
Bone									50.5	
Blood									0.5	
Brain									1.3	
Eye										
Heart									2.7	
Intestine									1.0	
Kidney									1.4	
Liver									2.1	
Lung									1.2	
Muscle									2.2	
Spleen									3.8	

Weight of rat ________ grams.

CALCULATIONS:

1. Refer to the calibration curve to convert "cpm/10 ml" of sample aliquot (column C) to "μCi/ml" (column D) and to "% of dose in 10 ml aliquot" (column E).

2. Because the digested sample was diluted to 25 ml and only a 10 ml aliquot of this dilution was used for the assay, then

$$F = \frac{25}{10} E = 2.5 E$$

where F is the % of dose in the sample.

3. To calculate the % of dose per gram of tissue, divide the % of dose in the sample (F) by the weight of sample used (B). Thus $G = F/B$.

4. Correct for the effect of animal weight on the distribution of dose by normalizing to an arbitrary animal weight. The value used should be approximately equal to the average weight of the animals used in the experiment. A weight of 300 g is used here for illustration.

$$H = G \frac{\text{Wt of rat used}}{300}$$

5. The uptake of phosphorus by a tissue should be expected to depend on the amount of phosphorus normally present in a tissue. For example, a tissue which normally contains no, or little, phosphorus would not be expected to take up much radioactive phosphorus. On the other hand, a tissue-like bone should take up a proportionately larger amount if time permits. Column I lists the milligrams of phosphorus per gram of certain tissues. Calculate the % of dose per mg of phosphorus (column J).

$$J = H \div I$$

$$\frac{\%\ \text{Dose}}{\text{mg P}} = \frac{\%\ \text{Dose}}{\text{g tissue}} \div \frac{\text{mg P}}{\text{g tissue}}$$

This is a measure of the rate of exchange of radioactive phosphate for stable phosphate in these tissues.

EXPERIMENT 13.2 INTRODUCTION TO METABOLIC STUDIES

OBJECTIVES:

To demonstrate the use of a metabolic cage.

To measure the percentage of carboxyl carbon of acetate excreted in the breath, urine and stools of an animal.

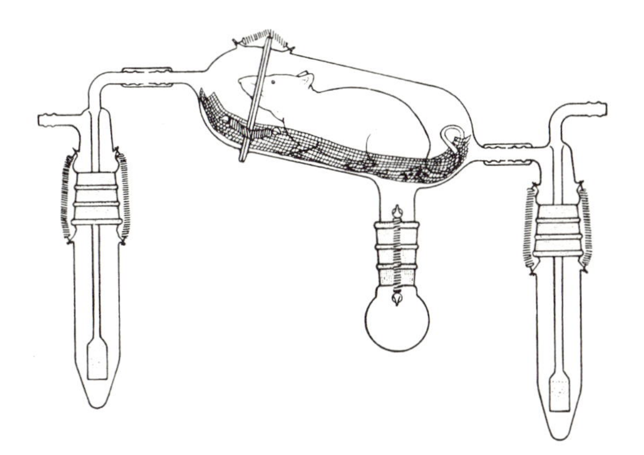

Figure 13.5—WEINHOUSE METABOLIC CAGE (*Courtesy J. Biol. Chem.* 191, 1951.)

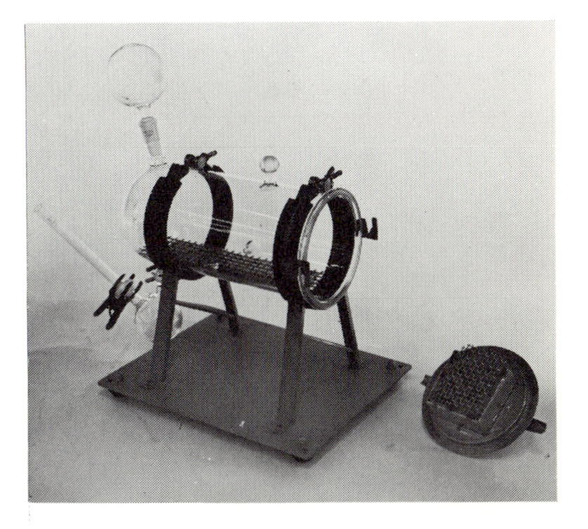

Figure 13.6—CARY METABOLIC CAGE (*Courtesy Applied Physics Corp.*)

THEORY:

Metabolic cages are for the purpose of collecting the urine, feces and breath of an animal while it is metabolizing tagged materials. If these metabolic products are collected at regular intervals, rates of detoxication, metabolic oxidation and other metabolic processes can be determined.

In the Weinhouse cage, air is drawn in through the tube on the right where carbon dioxide is removed by bubbling the air through a solution of sodium hydroxide. This CO_2-free air is drawn through the cage for the animal to breathe. Exhaled carbon dioxide is then absorbed in sodium hydroxide or barium hydroxide solution in the tube on the left. Urine and feces are separated by means of a screen in the bottom of the cage. The Cary cage is similar but also provides water and food facilities for the animal.

In the Roth cage, air is drawn in through an absorption tube, through the cage proper and finally through an absorption tube where exhaled CO_2 is collected. Both urine and feces fall through the floor of the cage. Feces are collected in a flask directly beneath the cage while urine flows down the funnel-shaped walls of the lower section into a channel from which it flows into a small flask containing a small amount of toluene to retard decomposition and evaporation. In this cage the animal can also be supplied with food and water for long term studies.

In the use of either cage it is preferable to draw air through the animal chamber by means of a slight negative pressure on the outlet side. In this way labeled carbon dioxide will not be lost should a small leak develop in the cage.

Most organisms readily convert acetate into acetyl-coenzyme A, a very reactive compound. Thus, labeled acetate which has been injected or fed to an animal quickly becomes activated as acetyl-coenzyme A. In this form it undergoes many reactions and

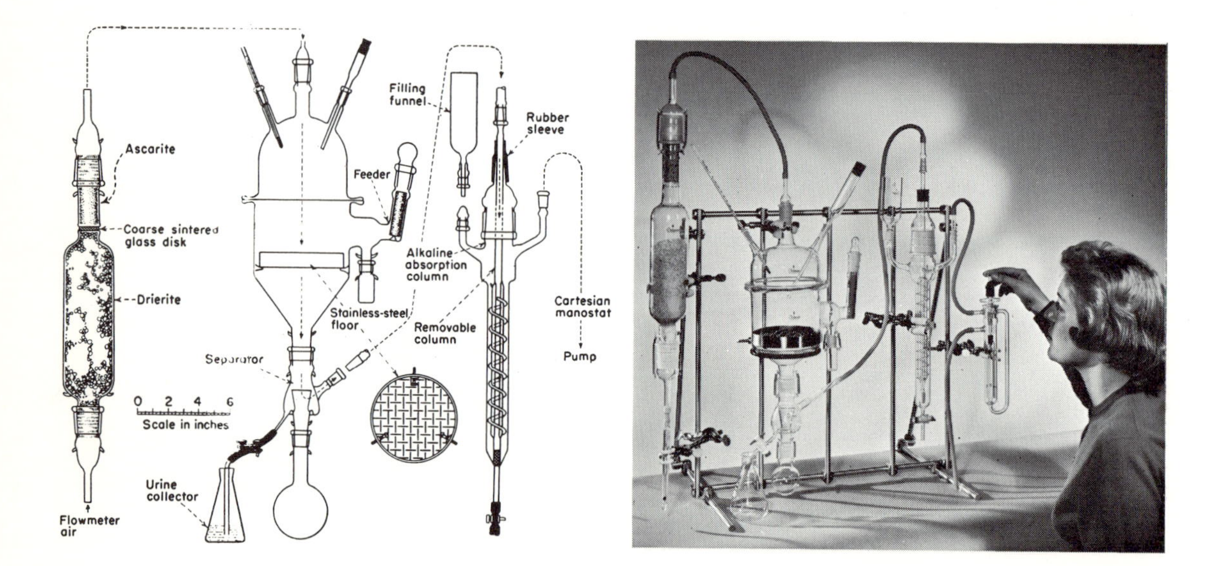

Figure 13.7—ROTH METABOLIC CAGE. (*Courtesy Delmar Scientific Laboratories.*) (See Reference 41.)

mixes with the acetate pool of the animal. Acetate is a known precursor of fatty acids, hydrocarbons and steroids. It may also transfer the ^{14}C tag to other materials such as amino acids and nucleic acids by indirect pathways.

Some of the acetate will be oxidized as indicated by the radioactivity of the CO_2 of breath. The radioactivity found in the first stools and urine samples is a measure of the unused portion of acetate (i. e. that in excess of the threshold). Subsequent samples may contain radioactive materials other than acetate. These materials represent degradation products of compounds synthesized from the acetate.

MATERIAL:

Metabolic cage; mice or rats (depending on size of available cage); rat chow; 20-50 mg of 1-^{14}C-acetate (1 mCi/mM); syringe; saturated NaOH solution (for collecting CO_2); toluene (for preservation of stools and urine); chloroform; #20 hypodermic needles; normal saline solution; vibrating-reed electrometer; liquid scintillation counter or windowless-flow counter; volumetric flasks.

PROCEDURE:

The air pump is turned on and the metabolic cage is inspected for leaks and the proper flow of air. The dry sodium acetate to be used is dissolved in a minimal volume (1-1.5 ml) of normal saline solution. A volume of solution containing 0.5-1.0 mg of ^{14}C-sodium acetate (1 mCi/mM) per gram of animal weight is injected intraperitoneally or subcutaneously and the animal is placed in the cage. Breath, urine and stool samples are collected after 5 min, 1 hr and 24 hrs, and radioassayed. More frequent samples can be taken if desired. In order to calibrate the measuring instrument, an aliquot of the dose is also radioassayed by the same method. Results are reported as percentage of original dose.

EXPERIMENT 13.3 *IN-VIVO* AND *IN-SITU* MEASUREMENTS

OBJECTIVES:

To illustrate selected procedures for conducting *in vivo* and *in situ* measurements of radioactive substances.

To measure simultaneously the relative blood level and the thyroid uptake of $Na^{131}I$ in the rat.

THEORY:

There are several advantages to *in vivo* and *in situ* measurements of radioactivity where such measurements are possible. These are: (1) The organism or animal does not have to be destroyed to obtain a sample for assay; (2) numerous or even continuous measurements of activity can be made on the same animal, and thus errors caused by differences among the animals themselves are eliminated; (3) distribution rates and elimination rates can be measured more accurately; and (4) such measurements are applicable to human beings.

In making *in vivo* and *in situ* measurements there are four major factors which must be considered. These are:

1. The geometry factor G which defines the solid angle subtended by the source and the sensitive volume or window of the counter. Other factors being constant, the inverse square law which applies here for gamma rays will limit the sensitivity of detection at a given depth of location of the source in the tissue.

2. Scattering in the tissue not only decreases the sensitivity of detection but also produces an apparently more diffuse source. Pulse height discrimination is useful for rejecting scattered radiation and tends to "focus" the detector when some estimate of the size and shape of the radioactive area is desired.

3. The nature and energy of the radiation play an important role. It is impractical to measure internal alpha sources with an external detector, but this is not a serious handicap since the only important alpha emitters are those elements with an atomic number greater than 82.

Only those beta emitters producing very energetic beta particles can be measured with an external detector and then only if they are located relatively near the surface of the tissue. ^{32}P has been used in limited cases. For example, the relative blood level in a mouse or rat can be measured with a detector placed above the animal's tail, even though counting efficiency is low.

Gamma emitters are by far the most useful generally because of the very penetrating nature of the gamma ray. Even sources located in the deep tissues can be observed, although scattering of the radiation, especially by Compton scattering, can cause serious problems, as mentioned above, by causing the source to appear diffuse.

4. Radiation from radioactive material in adjacent tissues adds considerably to the background and may even overshadow the radiation being received from the source of interest. Directional counters (figure 13.9), designed with heavy shielding so that radiation is preferentially detected from a single direction, are quite useful for reducing this problem.

MATERIALS REQUIRED:

Scintillation detector, G-M detector; ratemeters; recorders; shields; rat; $Na^{131}I$. (Scalers may be substituted for ratemeters and recorders).

PROCEDURE:

Fasten a large, anesthetized rat to an animal board with the animal on its back. Secure the rat's tail in the "tail shield" for a G-M tube. This is illustrated in figure 13.8. The end-window G-M tube, also in a shield, is placed on top of the "tail shield" which is designed with a circular depression into which the tube shield fits snugly.

The window of the tube is then in close proximity to the tail. Strips of cellophane tape are used, if necessary, to prevent the tail from touching the tube window. The G-M tube is then connected through a ratemeter to a recorder. The relative blood level will be recorded by this system.

Position a scintillation counter above the neck of the animal to record the activity in the thyroid. A simple lead shield placed over the end of the detector will be quite helpful in reducing background. Figure 13.9 illustrates such a shield.

Connect the scintillation counter to a second ratemeter and record on a second recorder, if available. This will permit a simultaneous record of both thyroid level and blood level. Adjust the high voltage to the scintillation detector so the iodine gamma peak is detected but the scattered radiation is rejected.

Fill a syringe with about 5 microcuries of $Na^{131}I$ solution. Use of the activity contained in the syringe as a source will assist in selection of the proper ratemeter ranges. Inject the ^{131}I solution intraperitioneally.

An additional lead shield placed across the animal's body just below the neck region will also reduce the background. The shield, notched as shown in figure 13.10, is held in a vertical position. The animal's body fits into the slot.

If ratemeters and recorders are not available, scalers may be used, one minute counts being made at 10 minute intervals. The activities are then plotted against time to obtain the curves.

All observations should be continued for a minimum of two hours, if time permits. At the end of the period, survey the rest of the rat's body for activity.

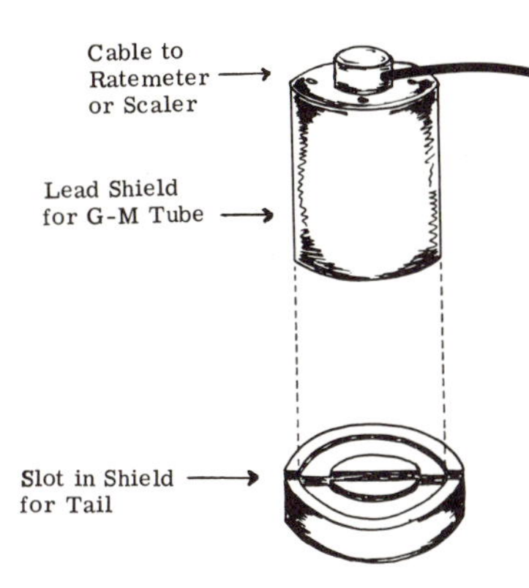

Figure 13.8—SHIELD FOR TAIL OF A RAT OR MOUSE.

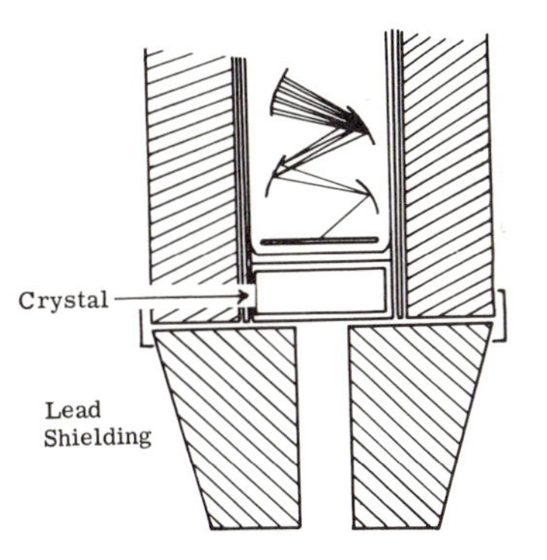

Figure 13.9—SCINTILLATION DETECTOR WITH COLLIMATED LEAD SHIELD.

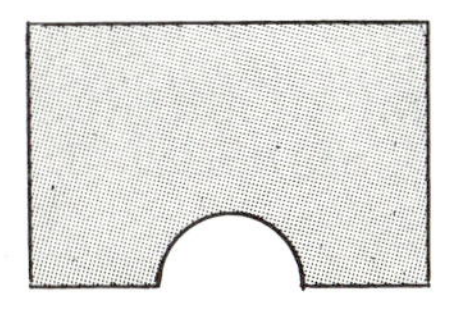

Figure 13.10—LEAD SHIELD USED IN THE EXPERIMENT.

TYPICAL DATA

Reading	Time (min.)	G-M Tube on Tail Observed cpm	G-M Tube on Tail Corrected cpm	Scintillation on Thyroid Observed cpm	Scintillation on Thyroid Corrected cpm
1	00	19	0	200	0
2	10	56	37	2300	2100
3	20	117	98	4300	4100
4	30	205	186	5500	5300
5	40	234	215	6000	5800
6	50	284	265	6150	5950
7	60	289	270	6200	6000
8	70	282	263	6000	5800
9	80	332	313	5900	5700
10	90	339	320	5700	5500
11	100	348	329	5750	5550
12	110	318	299	5800	5600

Rat Expired t = 95 min.
Groin Scan t = 102 min. A = 2,169
Mid-Section t = 104 min. A = 10,875
Lower Chest t = 106 min. A = 732

Measured with G-M Tube

REPORT:

1. From the graph of thyroid activity determine the time required for iodine to first appear in the thyroid.

2. Was iodine concentrated in any other part of the rat's body? Note especially if there was any concentration in the bladder, but do not be misled in your conclusion by activity remaining at the site of injection.

3. What value should be used as the background count for each detector?

4. Calculate the ratio of the uptake *rate* 15 minutes after administration of the ^{131}I to the uptake *rate* one hour after administration.

EXPERIMENT 13.4 SCANNING TECHNIQUES

OBJECTIVES:

To illustrate a method for the localization and mapping of radioactivity in an animal.

To demonstrate a procedure for locating metastatic thyroid tissue.

THEORY:

Metabolism is primarily controlled by the thyroid gland. The thyroid exerts this control through the secretion of thyroxine, an iodine-containing hormone. If an animal is given inorganic ^{131}I, e.g., $Na^{131}I$, iodine pickup by the thyroid can be detected by scanning the neck region as demonstrated in experiment 13.3. Maximum localization of radioactivity in the thyroid should occur within four to six hours.

For this experiment it is desirable to use the same rat used in experiment 13.3 since the approximate time for localization of the ^{131}I has already been determined. The procedure consists of surgical removal of one lobe of the thyroid and its transplantation to the upper part of one of the hind legs. The thyroid tissue will continue to function after the transplant.

Spontaneous relocations of malignant tissue are called metastases. When this occurs with thyroid tissue the metastases can be located by administration of radioactive iodine followed by scanning.

PROCEDURE:

1. Anesthetize the rat. This can be done by means of an intraperitoneal injection of 3 ml of nembutal solution (50 mg/ml). Place the animal on its back.

2. An incision is made in the inner surface of the upper part of one of the hind legs. The excised thyroid tissue will be placed in this incision.

3. An I-shaped incision is carefully made in the neck over the trachea. Muscle tissue will be revealed by this procedure. Cut through the muscle tissue being careful not to cut too deeply. A smooth sheath of muscle will now be visible. The thyroid gland is immediately under this sheath and lies on either side of the trachea. Applying light pressure on the scalpel, make an incision in the sheath. This will expose the trachea. On either side of the white-colored trachea a tiny pinkish mass of tissue will be found. These are the thyroid lobes. Remove one of the lobes and place in the incision previously made in the leg. Suture the incision.

4. Allow the animal a day or two to recover. Infection of the wound is not likely to occur.

5. An intraperitoneal dose of approximately 5 μCi of $Na^{131}I$ is administered. A successful surgical procedure will be indicated by concentration of radio-activity in the leg as well as in the neck. It is advisable to anesthetize the animal before scanning. A probe providing good collimation should be used to discriminate against activity from the bladder.

EXPERIMENT 13.5 WHOLE BODY COUNTERS (Applied to Measurement of Uptake Rate and Biological Half-Life)

OBJECTIVES:

To demonstrate the use of whole-body counting techniques for the measurement of total or average radioactivity in the body of a small animal.

To illustrate the application of whole-body counting to the measurement of biological half-life.

DISCUSSION:

In experiment 13.2 the rate of excretion of radioactive material was determined by use of a metabolic cage. If the amount of radioactive material administered to the animal is known, the amount retained can be determined by difference.

Whole body counters provide a direct means for measuring the amount of radioactive material retained by an animal or a human being. To be efficient a whole-body counter must have a very large sensitive volume at the center of which is placed the radioactive specimen. These counters are limited, however, to the measurement of gamma-emitting nuclides since it is apparent that alpha and beta radiation are readily absorbed by matter and cannot be detected in large masses of intact tissue.

It will be recalled from chapter 9 that solid scintillators such as NaI(Tl) afford the most efficient materials for the detection of gamma radiation. When applied to whole-body counters, however, the size and number of crystals required make crystal whole-body counters expensive. Consequently, plastic and liquid scintillators are very popular for this application.

Figure 13.11—SMALL ANIMAL OR ARM COUNTER. (*Courtesy Metrix, Inc.*)

PROCEDURE:

If small tropical fish such as neon tetras or glow tetras are used as the experimental animal a sodium iodide crystal well counter can be used for the whole-body counter. Between two and five milliliters of water are added to the test tube so the fish will survive the counting procedure without harm.

PART A—EFFECT OF DIMETHYLSULFOXIDE ON ABSORPTION RATE

Dimethylsulfoxide is a membrane activator and affects absorption rates. Its effect on the absorption of diiodofluorescene will be measured.

The fish are divided into two groups. To one group, contained in one liter of water, is added 10 μCi of ^{131}I-diiodofluorescene (or other radioactively-labeled dye). The radioactivity of the fish is measured at intervals to determine the rate at which the compound is taken up and to determine the time required for maximum uptake.

To the second group of fish is added the same amount of labeled diiodofluorescene plus enough dimethylsulfoxide to give a 1% solution. The radioactivity of these fish is also measured at intervals to determine the rate of uptake and the time required for maximum uptake.

The average activity of the fish in each group is plotted as a function of time. Any difference noted between the two groups should be explained.

PART B—DETERMINATION OF BIOLOGICAL HALF-LIFE

The fish from part A of the experiment are placed into clean containers filled with fresh water. The radioactive compound will now be excreted and desorbed by the fish into the water. The water should be changed frequently to prevent the buildup of radioactivity in it.

The radioactivity of the fish is measured at intervals as before. The logarithm of the activity is plotted as a function of time. From the slope of the curve the effective half-life can be determined. From the effective half-life and the radiological half-life the biological half-life is calculated.

Both parts A and B of the experiment can be repeated using other radioactively tagged compounds. If desired, the group receiving dimethylsulfoxide can be eliminated. Small animals such as mice can be used in place of fish. They are restrained during the measurement of radioactivity by placing them in a beaker covered with screening. In this case the beaker would have to be placed on top of the crystal and counting efficiency would be reduced.

EXPERIMENT 13.6 INTERMEDIARY METABOLISM

OBJECTIVES:

To illustrate a general method suitable for the isolation of intermediary compounds in the investigation of metabolic pathways.

To isolate ^{14}C-β-hydroxy-β-methylglutaric acid (HMG), a biologically short-lived intermediary in the synthesis of cholesterol from ^{14}C-acetate.

THEORY:

To establish the metabolic pathway by which simple compounds are converted into complex ones by enzyme systems, it is necessary to identify metabolic intermediaries (e.g. mevalonic, acetoacetic and farnesenic acids) that are usually found in very small amounts in biological systems. The following radioactive tracer method is usually employed for the isolation of intermediaries.

A ^{14}C-labeled precursor is added to tissue slices, an organ perfusion, or a tissue homogenate. The biological system must, of course, be capable of utilizing the precursor. Immediately after its addition, a large amount of a non-labeled chemical is added, a chemical known or postulated to be an intermediary. The biological system will metabolize the radioactive precursor. Therefore all the intermediaries made from this precursor will at some time be present as labeled compounds in the system. The addition of a very large quantity of unlabeled intermediary traps the radioactive substance by dilution. Because the biological system can only convert a small amount of this large pool to the next intermediary, most of the radioactive intermediary will remain unchanged. This large pool of both labeled and unlabeled material can now be isolated and purified. If a constant specific activity is obtained after repurification, the chemical is presumed to be the trapped intermediary or a closely related compound. Many of the intermediaries in the metabolism of food have been determined by means of this technique.

If another carrier is used, repetition of the above procedure will result in the isolation of the corresponding intermediary. Similarly all intermediaries may be identified and thus a metabolic pathway suggested. The field of intermediary metabolism blossomed only after radioactive materials and the means for their assay became available. Metabolic pathways and pool levels (threshold) have contributed to the basic knowledge of the biochemistry and physiology of organisms. This information in turn has been useful for the establishment of clinical, diagnostic, and therapeutic techniques.

MATERIALS:

Rats 150 g (starved for 24 hrs); cages; petri dishes; ice; 25-ml Erlenmeyer flasks; corks; constant temperature bath at 37° C provided with a shaker; oxygen; dissection kit; 0.1*M* potassium phosphate buffer at pH 7.2; potassium chloride; magnesium chloride; 5 mg potassium 1-^{14}C-acetate with a specific activity of 50 μCi/mg; 100 mg β-hydroxy-β-methylglutaric acid (HMG); ethyl ether; petroleum ether; centrifuge tubes; metaphosphoric acid pellets; anhydrous magnesium sulfate; filter paper; funnels; Norite absorbent charcoal; planchets; flow counter or liquid scintillation counter.

PROCEDURE:

1. *Preparation of rat liver slices*—A rat is sacrificed by cervical fracture and the liver is rapidly excised. The liver is placed in a petri dish chilled in a tray of ice. It is cut into thin slices (0.1 to 0.2 mm thick) which are transferred to a tared 25-ml Erlenmeyer flask by means of forceps and the weight of eight to ten slices is determined. This weight should be about 0.5 g.

2. *Incubation*—Eight ml of 0.1*M* phosphate buffer (pH 7), 5 mg of potassium chloride, 2 mg of magnesium chloride, 1 mg of the radioactive substrate (50 μc of 1-^{14}C-acetate) and 10 mg of HMG carrier are added to the liver slices. Oxygen is bubbled through the solution for a few minutes. The flask is then stoppered and the mixture is shaken for 45 minutes in a constant temperature bath at 37° C.

3. *Isolation*—After incubation, 100 to 200 mg of metaphosphoric acid pellets is added to the mixture which is digested for one-half hour on a water bath to effect complete mixing of "cold" and "hot" HMG after the tissues are denatured. The mixture is extracted with 10 successive 25-ml portions of ethyl ether for 4 hours in a continuous ether extracter. The combined ether extracts are dried overnight with 5 to 7 g of anhydrous magnesium sulfate. The magnesium sulfate is filtered off and the ether solution is evaporated to a 10-ml volume and transferred to a small centrifuge tube. It is then taken to dryness and finally placed in a vacuum desiccator over KOH for 24 hours to remove radioactive acetate. The solids in the centrifuge tube are a mixture of carrier HMG and radioactive HMG along with tracer quantities of other intermediaries in the synthesis of cholesterol.

4. *Purification*—To the dry mixture in the centrifuge tube is added 2 to 3 ml of absolute ether and 5 to 10 mg of Norite absorbent charcoal. The tube is heated to 30°; the contents are mixed well and filtered. By judicious addition of petroleum ether to the clear ethyl ether filtrate, fine crystals of HMG will be formed. The solution is chilled and centrifuged and the supernatant is carefully removed. The crystals of HMG remaining in the centrifuge tube are dried in a desiccator, weighed on a planchet and assayed for radioactivity.

DATA AND CALCULATIONS:

Calculate the specific activity of the 1-^{14}C-acetate precursor.

Calculate the specific activity of the ^{14}C-HMG recovered.

Determine the efficiency with which the HMG was produced by liver tissue. Assume it is possible to recover all 10 mg of carrier HMG from the mixture.

EXPERIMENT 13.7 REACTION RATE OF ENZYME-CATALYZED REACTIONS

OBJECTIVES:

To review the nature of enyzme-catalyzed reactions and tracer methods used to investigate them.

To determine the reaction rate constant for the enzymatic decarboxylation of formic acid.

THEORY:

The rate of a chemical reaction is determined by:

1. Concentration (or pressure) as defined by the *law of mass action.*
2. Temperature as described, for example, by the Arrhenius equation:

$$k = Ae^{-E_a/RT} \quad \text{or} \quad \ln k = -\frac{E}{RT} + \ln A \tag{13.7}$$

3. Catalysts, which can be classified according to:

   a. behavior —as positive or negative
   b. composition—as inorganic or organic
   c. phases —as homogeneous or heterogeneous

Enzymes are homogeneous, organic, positive catalysts. Of the more than 150 which have been prepared in crystalline form, all are proteins. Some enzymes require a coenzyme or a metal ion in order to be active.

Enzymes exhibit the properties of proteins. They are therefore especially sensitive to pH and temperature. A maximum activity is usually observed at a particular pH. At both higher and lower values of pH the enzyme activity decreases. This effect is illustrated in figure 13.12.

An increase in temperature results in an increase in reaction rate. If the increase in rate were in accord with the Arrhenius equation (equation 13.7) the plot of ln k versus 1/T would be linear (figure 13.13). However, the increase is limited by the amount of enzyme present and denaturization of the enzyme at temperatures above 65° C will result in a sharp decrease in the reaction rate.

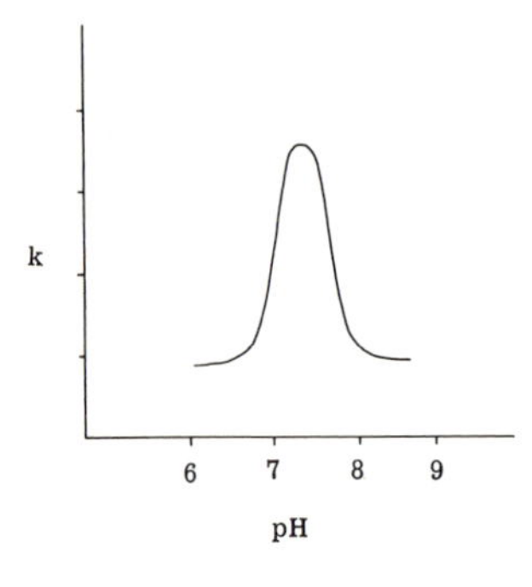

Figure 13.12—EFFECT OF pH ON ENZYME ACTIVITY.

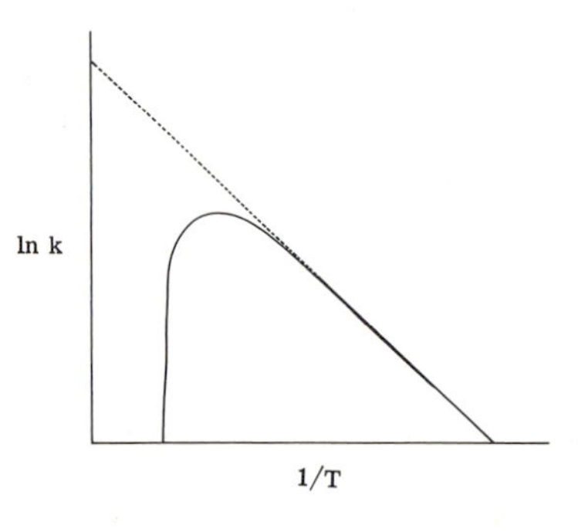

Figure 13.13—EFFECT OF TEMPERATURE ON ENZYME ACTIVITY.

Enzyme activity may also be inhibited by certain compounds produced in the enzyme-catalyzed reaction itself. Because of this problem and others associated with concentration changes, reaction-rate studies are often restricted to about the first 10% of the reaction.

A general enzyme-catalyzed reaction can be written

$$\underset{\text{(Substrate)}}{S} \xrightarrow[\text{(Enzyme)}]{E} \underset{\text{(Products)}}{P} \qquad (13.8)$$

or, if the formation of an intermediate complex X is assumed:

$$S + E \underset{k_2}{\overset{k_1}{\rightleftharpoons}} X \underset{k_4}{\overset{k_3}{\rightleftharpoons}} P + E \qquad (13.9)$$

In addition, a *steady state* of X is generally assumed. Thus,

$$\frac{d(X)}{dt} = 0 = k_1(S)(E) - (k_2 + k_3)(X) + k_4(P)(E) \qquad (13.10)$$

If only the initial phase of the reaction is considered, (P) is very small and the last term can be neglected. Also, let the total concentration of enzyme, both free and in the form of a complex be $(E_0) = (E) + (X)$. Thus,

$$k_1(S)[(E_0) - (X)] - (k_2 + k_3)(X) = 0 \qquad (13.11)$$

Solution of equation 13.11 for the concentration of intermediate complex gives:

$$(X) = \frac{(E_0)}{1 + \dfrac{k_2 + k_3}{k_1(S)}} \qquad (13.12)$$

From equation 13.9 it can be seen that the rate of formation of product is

$$\frac{d(P)}{dt} = k_3(X) - k_4(P)(E) \qquad (13.13)$$

Again since (P) is initially very small the second term can be neglected and substitution of equation 13.12 into equation 13.13 yields

$$\frac{d(P)}{dt} = \frac{k_3(E_0)}{1 + \dfrac{k_2 + k_3}{k_1(S)}} = \frac{k_3(E_0)}{1 + k_m/(S)} \qquad (13.14)$$

In this expression, k_3 is known as the "*turnover number*" and

$$\frac{k_2 + k_3}{k_1} = k_m$$

is called the *Michaelis* constant.

If the rate of product formation d(P)/dt is plotted as a function of substrate concentration the curve in figure 13.14 is obtained. It can be seen that the denominator approaches unity as (S) becomes very large. Thus $k_3(E_0)$ represents the limiting rate for the reaction.

If a linear plot of the data is desired, equation 13.14 can be rearranged to the form:

$$\frac{1}{d(P)/dt} = \frac{1}{k_3(E_0)} + \frac{1}{(S)} \cdot \frac{k_m}{k_3(E_0)} \qquad (13.15)$$

which will be recognized as the equation for a straight line. A plot of this relationship is illustrated in figure 13.15.

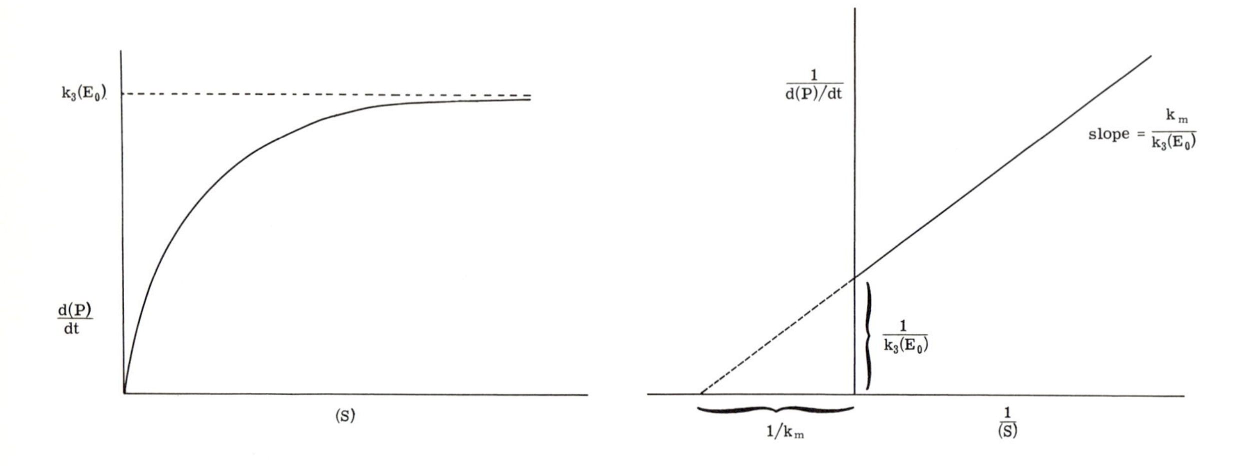

Figure 13.14—INFLUENCE OF SUBSTRATE CONCENTRATION ON RATE OF PRODUCT FORMATION. (See Equation 13.14.)

Figure 13.15—GRAPH ILLUSTRATING STANDARD TREATMENT OF DATA TO OBTAIN A LINEAR PLOT. (See Equation 13.15.)

PROCEDURE:

1. *Preparation of the enzyme*—To prepare a cell-free enzyme extract, 100 grams of split green peas are soaked in water overnight. The water is decanted and the peas are blended with 200 ml of 0.1*M* Na_2HPO_4 in a Waring blender for two minutes. The mixture is allowed to stand at room temperature for two hours and is then strained through cheesecloth to remove coarse material. A light green supernatant containing the enzyme is obtained by centrifuging at 2000 rpm for 30 minutes.

2. *Decarboxylation of formic acid*—Ten milligrams of potassium nitrate and 3 mg of DPN are dissolved in 30 ml of the enzyme solution. This solution is then added to a weighed quantity of $H^{14}COONa$ reaction vessel contained in a thermostated bath at 37° C. The $^{14}CO_2$ released by the decarboxylation of the formate is flushed from the reaction vessel with CO_2-free oxygen. The oxygen is bubbled through the reaction mixture at about 100 ml per minute and the gases are then passed through a manifold to a CO_2 collection tube.

3. *Collection of the carbon dioxide*—The tube which carries the CO_2-oxygen mixture from the reaction vessel should extend to the bottom of a centrifuge tube containing barium hydroxide solution. The top of the solution is layered with toluene to prevent CO_2 from the atmosphere from reacting with the barium hydroxide. Dilution with atmospheric CO_2 would reduce the specific activity.

4. *Preparation of the samples for counting*—The tubes from step 3 are centrifuged and the precipitate of $BaCO_3$ produced in the experiment is washed several times with CO_2-free water (e.g., boiled water) to remove the excess barium hydroxide. The $BaCO_3$ is then transferred to a filter with a small quantity of acetone. By means of suction, the filter cake is given a preliminary drying. The samples are then transferred to an oven and kept overnight at 115° C. They are then weighed and the activity is measured with an end-window or windowless flow counter. Self absorption must be considered as described in experiment 5.6.

Alternatively, the weighted samples of $BaCO_3$ can be transferred to a Cary-Tolbert gas generator and the CO_2 regenerated for counting in a vibrating-reed electrometer (see experiment 8.2). A liquid scintillation method (experiment 9.3) can also be used, if desired.

5. *Treatment of data*—By means of a series of tubes containing barium hydroxide, samples of CO_2 are collected at intervals. None of the CO_2 should be allowed to escape.

The total CO_2 liberated up to a time is plotted as a function of time. From the slope of this curve one can obtain the value of d(P)/dt. If the experiment is repeated using a different concentration of formate, (i.e., a different value of (S)), data required to plot the curves shown in figures 13.14 and 13.15 can be obtained.

The reaction can also be followed by plotting the total activity accumulated up to a given time rather than the total weight of CO_2. The results will be slightly different due to an *isotope effect*.

If the specific activity, normalized for self absorption, is determined for each sample it will be seen that the specific activity changes as the reaction proceeds. This is caused by the isotope effect mentioned above. The lighter $^{12}CO_2$ tends to be liberated at a slightly higher rate than the $^{14}CO_2$.

EXPERIMENT 13.8 BIOSYNTHESIS OF A LABELED COMPOUND IN A TISSUE HOMOGENATE

OBJECTIVES:

To illustrate a method for the production of a labeled compound normally too complex to prepare by chemical synthesis.

To measure the conversion of ^{14}C-acetate to cholesterol.

THEORY:

The production of radioactively tagged compounds by standard techniques or organic synthesis in sometimes impractical, especially in the case of complex molecules. As an alternative, biosynthesis can sometimes by used.

Either animal or plant systems can be used for this purpose. In this experiment an homogenate of an animal tissue is used to prepare tagged cholesterol from tagged acetate. Although the tissues may be considered destroyed, the enzymes are still active. In experiment 13.9 a living plant is used for the biosynthesis of mescaline.

MATERIALS:

Constant-temperature shaker (Warburg of similar apparatus); 3-5 mice or rats; Adenosine monophosphate (AMP); Diphosphopyridine nucleotide (DPN); ^{14}C-sodium acetate (1 mCi/mM); metaphosphoric acid (HPO_3) pellets; KCl; ether; vacuum desiccator; 0.1*M* KH_2PO_4; 0.1*M* K_2HPO_4; 0.6*M* $MgCl_2$; 3*M* nicotinamide; saturated $NaHCO_3$ solution; acetic acid; digitonin; ethanol; acetone; trays with ice shavings; glass (Potter-Elvehjem) homogenizer; small angle centrifuge; Warburg flasks (or 25-ml Erlenmeyer flasks with a 2-hole rubber stopper with glass tubing for inlet and exit line); compressed oxygen; centrifuge tubes; dissecting kit; planchets; eye droppers.

PROCEDURE:

Sacrifice a 15-g mouse by placing the thumb firmly at the base of the skull and pulling up and out on the tail to sever the spinal column. Immediately place it on a tray of ice. Quickly remove the entire liver and place it in chilled buffer solution (30). The buffer is 42 ml—0.1*M* KH_2PO_4; 67 ml—0.1*M* K_2HPO_4; 1 ml—0.6*M* $MgCl_2$; 1 ml—3*M* nicotinamide; and 1 ml—saturated $NaHCO_3$ solution mixed well and kept refrigerated). Maintain a 5:2 volume ratio of buffer to tissue. Homogenize the buffer-tissue mixture in a loose-fitting, glass Potter-Elvehjem homogenizer, being careful to introduce no body heat. Homogenization time should not exceed 20 to 30 seconds. To obtain an homogenate free of debris, centrifuge at ⅓ the maximum speed of a small angle centrifuge for 5 minutes maintaining a temperature of 0-5° C. An homogenate is a cell-free suspension of nuclei, mitochondria and cytoplasmic particles prepared from tissue. (It is important in these first steps to work rapidly, and to maintain a temperature of 0-5° C).

Transfer the liver homogenate to a 50-ml Warburg flask containing 1 mg each of the following; (a) *Cofactors:* AMP and DPN: (b) *Precursor:* ^{14}C-NaOAc (1 mCi/mM); (c) *Carrier:* cholesterol. Incubate this mixture in a constant-temperature shaking bath, with medium agitation, under O_2 for 3 hours at 37° C. At the end of the incubation period, add a few pellets of HPO_3 and 0.5 g KCl to the mixture. Extract the mixture for 24 hours by means of a continuous ether extractor (or extract 5 times with 50-ml portions of ether). The pooled ether solutions are poured into a beaker and evaporated to dryness on a water bath. A few drops of glacial acetic acid are then added to wet the dry residue. The beaker is placed in an evacuated desiccator containing KOH pellets in the base. The KOH will trap unused ^{14}C-acetate. After dryness is attained (24 hours later) add 5 ml of absolute ethanol to the residue in the beaker and warm gently. Filter the extract into a centrifuge tube. Add 0.5 to 1 ml of 1% digitioin in 90% ethanol. Allow the solution to stand overnight at 0° C to effect complete precipitation of cholesterol-digitonide. The tube is then centrifuged at room temperature at medium speed for 45 minutes. Decant the mother liquor. Add 1 ml of ethanol to the precipitate in the tube, shake, centrifuge for 10 minutes and decant the solvent. The precipitate is then washed by the above method with 1 ml of acetone and then with 1 ml of anhydrous ether. The washed and dried precipitate is the radioactive cholesterol in the form of cholesterol-digitonide. It is dissolved in 0.5 ml of methanol and the resulting solution is then placed in a tared planchet by means of a dropper. The solvent is evaporated under a heat lamp, and when dry, the cholesterol-digitonide is weighed and assayed for its radioactivity.

CALCULATIONS:

A = Specific activity of recovered ^{14}C-cholesterol, dpm/mg C
B = Specific activity of ^{14}C-NaOAc precursor, dpm/mg C

$$\frac{A \times 100}{B} = \text{\% efficiency of conversion of acetate into cholesterol}$$

If pure cholesterol is desired, the cholesterol-digitonide is dissolved in 0.2 ml of pyridine. To this solution is added 2 ml of ether. After filtering, the solution is poured into 0.5 ml of water and extracted twice with 2 ml ether. The ether extracts are dried over $CaCl_2$. The $CaCl_2$ is then removed by filtration, and the ether is evaporated. The residue is multilabeled ^{14}C-cholesterol.

EXPERIMENT 13.9 BIOSYNTHESIS OF A COMPOUND IN A LIVING PLANT

OBJECTIVES:

To illustrate a procedure for the biosynthesis of a radioactively-labeled compound in an intact plant.

To prepare mescaline-^{14}C by growing a cactus, peyotl, in an atmosphere containing $^{14}CO_2$.

THEORY:

Peyotl (Lophophora Williamsii) is a cactus indigenous to Mexico and the southwestern areas of the United States. The dried tops of the cactus, often referred to as "mescal buttons" have long been used by natives of these areas to promote trances and hallucinations. Mescaline is the component which produces these results. This same cactus is also available in many botanical gardens and shops. It is easily recognized by its button shape.

PROCEDURE:

A small peyotl cactus, about two or three inches high, is placed inside a glass bell jar. The bell jar must be equipped with some means for releasing $^{14}CO_2$. One arrangement uses a mixing "Y" tube as shown in figure 13.16. One arm of the "Y" tube contains 2 to 3 μCi of $Na_2{}^{14}CO_3$ while the other contains about 2 ml of concentrated H_2SO_4. By rotating the tube the acid is mixed with the labeled carbonate to generate $^{14}CO_2$.

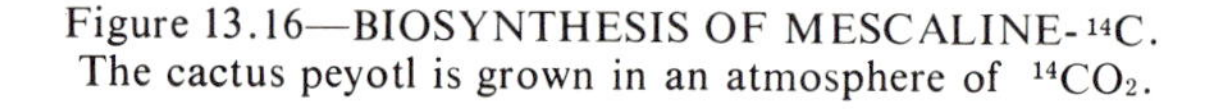

Figure 13.16—BIOSYNTHESIS OF MESCALINE-^{14}C. The cactus peyotl is grown in an atmosphere of $^{14}CO_2$.

The bell jar is kept near a window for about five days while the $^{14}CO_2$ is metabolized by the plant. The bell jar is then transferred to a hood and the plant is removed. The button or the head is carefully cut off from the main body of the cactus. Mescaline will be extracted from the button while the rest of the cactus may be set aside.

The button is carefully minced and transferred to a glass homogenizer. About 10 mg of carrier mescaline hydrochloride is added and the mixture is homogenized in 95% ethanol. The homogenate is then acidified with concentrated hydrochloric acid to pH 2 and allowed to stand for about eight hours. The digest is then centrifuged at high speed and the alcoholic supernatant containing the mescaline is carefully removed. The solid residue is discarded.

To extract the mescaline the alcoholic supernatant is evaporated to a small volume and made basic (pH 10.5) by addition of ammonium hydroxide. The free mescaline base is now extracted with several 15-ml portions of chloroform. Evaporation of the chloroform leaves a residue of the crude alkaloid which is then dissolved in alcohol for chromatography.

The alcoholic solution is spotted on a strip of filter paper (Whatman #1) with a micropipette or capillary tube and dried in a stream of warm air. The paper is then chromatographed by ascending chromatography using one of the following solvent systems:

1. 1-butanol, glacial acetic acid, water (4:1:5)
2. benzene, glacial acetic acid, water (2:1:1)
3. isopropanol, ammonium hydroxide, water (20:1:2)

Twelve or more hours may be required for chromatography. The mescaline is identified by spraying the sheet with 0.1% ninhydrin in isopropanol and warming the paper to 35° C. The strip is also passed through a scanner to locate the radioactivity.

REFERENCES

Tracer Uses

1. Aaronoff, S., "Techniques of Radiobiochemistry", Iowa State College Press (1956).
2. Amber, H., J. A. Watson, and T. B. Grucci, "Procedure for Digestion and Radioassay of Animal Tissues", Nucleonics 12, No. 8, 40-41 (August 1954).
3. Anderson, E. C., R. L. Schuck, J. D. Perrings, and W. H. Langham, "The Los Alamos Human Counter", Nucleonics 14, No. 1, 26-29 (January 1956).
4. Anonymous, "Tritium Tracing—A Rediscovery", Nucleonics 16, No. 3, 62-67 (March 1958).
5. Bogash, M., J. M. Schneeberg, G. Perch, and J. L. Rabinowitz, "The Dissemination of Radioactive Carbon Particles by Transurethral Surgery", Am. J. Urol. 91, 586 (1964).
6. Bruner, H. D., and J. D. Perkinson, Jr., "A Comparison of Iodine-131 Counting Methods", Nucleonics 10, No. 10, 57-61 (1952).
7. Calvin, M., C. Heidelberger, J. C. Reid, B. M. Tolbert, and P. E. Yankwich, "Isotopic Carbon", Wiley (1949).
8. Comar, C. L., "Radioisotopes in Biology and Agriculture", McGraw-Hill (1955).
9. Comar, C. L., "Radioisotopes in Nutritional Trace Element Studies", Nucleonics 3, Part I, p. 32 (September 1948); Part II, p. 30 (October 1948); Part III, p. 34 (November 1948).
10. Edelmann, A., "Radioactivity for Pharmaceutical and Allied Research Laboratories", Academic Press (1960).
11. Gloyna, E. F., and B. B. Ewing, "Uranium Recovery from Saline Solutions by Biological Slimes", Nucleonics 15, No. 1, 78-81 (January 1957).
12. Gorbman, A., "Radioautography in Biological Research", Nucleonics 2, 30 (June 1948).
13. Hansard, S. L. "Radioisotope Procedures with Farm Animals", Part I. Nucleonics 9, No. 1, 13-25 (1951); Part II. Nucleonics 9, No. 2, 38-45 (1951).
14. Hansard, S. L., and C. L. Comar, "Radioisotope Procedures with Laboratory Animals", Nucleonics 11, No. 7, 44 (1953).
15. Hassitt, C. C., W. H. Summerson, and F. Solomon, "A Biosynthesis of C^{14}-Labelled Glycogen", Nucleonics 12, No. 4, 59-60 (April 1954).
16. Hevesy, G., "Radioactive Indicators", Interscience (1948).
17. Jenkins, D. W., and C. C. Hassett, "Radioisotopes in Entomology", Nucleonics 6, 5-14 (March 1950).
18. Juda, A., J. Sklaroff, D. W. Cohen, and J. L. Rabinowitz, "Tooth Movement Studies in Dogs Using Ca-45", I.A.D.R. (Abstract #195) Pittsburgh, Pa. (1963).
19. Kamen, M. D., "Isotopic Tracers in Biology", 3rd Ed., Academic Press, Inc. (1957).
20. Mayr, G., H. D. Bruner, and M. Bruner, "Boron Detection in Tissues Using the (n,α) Reaction", Nucleonics 11, No. 10, 21-25 (1953).
21. Melander, L., "Isotope Effects on Reaction Rates", Ronald Press (1960).
22. Neff, N., G. V. Rossi, G. D. Chase, and J. L. Rabinowitz, "Distribution and Metabolism of Mescaline-C-14 in the Cat Brain", J. Pharmacol. & Therap. 144, 1 (1964).
23. Perkinson, J. D. Jr., and H. D. Bruner, "Preparation of Tissues for Iodine-131 Counting", Nucleonics 10, No. 11, 66-67 (1952).
24. Poddar, R. K., "Quantitative Measurement of S^{35} in Biological Samples", Nucleonics 15, No. 1, 82-83 (January 1957).
25. Preuss, L. E., "Rugged Scintillation-Counter Housing for Biological Application", Nucleonics 11, No. 6, 74 (1953).
26. Rabinowitz, J. L., "The Biosynthesis of Radioactive Beta-hydroxy-isovaleric Acid in Rat Liver", J. Am. Chem. Soc., 77, 1295 (1955).
27. Rabinowitz, J. L., and G. D. Chase, et al., "Studies on Isotope Effects with Carbonic Anhydrase Using C^{14} Sodium Bicarbonate", Atompraxis 6, 432 (1960).
28. Rabinowitz, J. L., and R. M. Dowben, "The Biosynthesis of Radioactive Estradiol, I.—Synthesis by Surviving Tissue Slices and Cell-free Homogenates of Dog Ovary", Biochimica et Biophysica Acta 16, 96 (1955).
29. Rabinowitz, J. L., and S. Gurin, "Biosynthesis of Cholesterol and Beta-hydroxy-beta-methyl-Glutaric Acid by Extracts of Liver", J. Biol. Chem. 208, 307 (1954).
30. Rabinowitz, J. L., and S. Gurin, "The Biosynthesis of Radioactive Cholesterol, beta-methylglutaconic Acid and beta-methylcrotonic Acid", J. Am. Chem. Soc. 76, 5169 (1954).
31. Rabinowitz, J. L., J. S. Lafair, H. D. Strauss, and H. C. Allen, "Carbon Isotope Effects in Enzyme Systems. II. Studies with Formic Acid Dehydrogenase", Biochimica et Biophysica Acata 27, 544 (1958).
32. Rabinowitz, J. L., R. M. Myerson, G. T. Wohl, "The Deposition of C-14 Labeled Cholesterol in the Atheromatous Aorta", Proc. Soc. Expl. Med. 105, 241 (1960).
33. Rabinowitz, J. L., and T. Sall, "Studies on a Glucuronolactone Decarboxylase", Biochimica et Biophysica Acta 23, 1289 (1957).

34. Rabinowitz, J. L., T. Sall, J. Bierly, Jr., and O. Oleksyshyn, "Carbon Isotope Effects in Enzyme Systems. I.—Biochemical Studies with Urease", Arch. Biochem. Biophys. 63, 437 (1956).
35. Rabinowitz, J. L., P. V. Skerrett, and R. W. Riemenschneider, "Fat Composition and Fat Biogenesis in Various Areas of a Single Human Aorta—A Preliminary Report", J. of A.M.A. 179, 153 (January 13, 1962).
36. Rabinowitz, J. L., and H. J. Teas, "Biosynthesis of Rubber from C-14 Acetate by Stem Slices and Latex Ensyme from Hevea", Proc. Nat. Rubber Res. Conf. 777, (1961).
37. Radin, N. S., "Isotope Techniques in Biochemistry-V (Chemical and Biological Methods for Labeling)", Nucleonics, p. 33 (February (1948).
38. Radin, N. S., "Isotope Techniques in Biochemistry-IV (Kinetic Approach to Problems of Intermediary Metabolism)", Nucleonics 2, 50 (January 1948).
39. Reid and Wilson, "Concomitant *in Vivo* Measurement of Regional Erythrocyte and Plasma Concentrations Using I-131 and P-32", Nucleonics 11, No. 5, 64 (1953).
40. Ropp, G. A., "Effect of Isotope Substitution on Organic Reaction Rates", Nucleonics 10, 22-27 (October 1952).
41. Roth, L. J., L. Leifer, J. J. Hogness, and W. H. Langham, "Studies on the Metabolism of Radioactive Nicotinic Acid and Nicotinamide in Mice", J. Biol. Chem. 176, 249-257 (1948).
42. Schulman and Falkenheim, "Review of Conventions in Radiotracer Studies", Nucleonics 3, 13 (October 1948).
43. Schubert, J., and E. E. Conn, "Radio-colloidal Behavior of Some Fission Products", Nucleonics 4, 2 (June 1949).
44. Serlin, D., and Rabinowitz, J. L., "The Distribution of Radioactivity in Dog Tissues Following the Administration of the ThioTEPA-^{32}P as an Adjuvant to Surgery", Atompraxis 11, 122 (1965).
45. Staple, E., and Rabinowitz, J. L., "Formation of Tri-hydroxycoprostanic Acid from Cholesterol in Man", Biochim. et Biophys. Acta. 59, 735 (1962).
46. Stokinger, H. E., "Size of Dose: Its Effect on Distribution in the Body", Nucleonics 11, No. 4, 24 (1953).
47. Tait, I. F., and E. S. Williams, "Assay of Mixed Radioisotopes", Nucleonics 10, No. 12, 47-51 (December 1952).
48. Thompson, R. C., "Biological Applications of Tritium", Nucleonics 12, No. 9, 31-35 (September 1954).
49. Veall, N., and H. Vetter, "Radioisotope Techniques in Clinical Research and Diagnosis", Butterworth (1958).
50. Weinberger, D., and J. W. Porter, "Incorporation of Tritium Oxide into Growing Chlorella Pyrenoidasa Cells", Science 117, 636-638 (1953).
51. Weinhouse, S., and B. Friedmann, "Metabolism of Labeled Carboxylic Acids in the Intact Rat", J. Biol. Chem. 191, 707-717 (1951).
52. Wilzbach, K. E., "Tritium-Labeling by Exposure of Organic Compounds to Tritium Gas", J. Am. Chem. Soc. 79, 1013 (1957).
53. Wolstenholme, G. E. W. (ed.), "Isotopes in Biochemistry", Ciba Foundation Symposium, Blakiston (1952).
54. Wood, H. G., "Symposium on Use of Isotopes in Biology & Medicine", Edited by H. T. Clarke, Univ. of Wisconsin Press (1959).
55. Zilversmit, D. B., and M. L. Shore, "A Hydrodynamic Model of Isotope Distribution in Living Organisms", Nucleonics 10, No. 10, 32-34 (1952).

CHAPTER 14
Radiation Biology

LECTURE OUTLINE AND STUDY GUIDE

I. SOURCES OF RADIATION
   A. Natural
   B. Artificial

II. TYPES OF RADIATION AND RELATIVE EFFECT
   A. External Source Produces a Biological Effect Dependent Upon:
      1. Penetration or range of radiation
      2. Dose delivered
      3. Tissue involved
      4. Area involved
      5. Time for accumulation of dose
   B. Internal Source Produces a Biological Effect Dependent Upon:
      1. Amount assimilated
      2. Type and energy of radiation
      3. Effective half-life
      4. Concentration in specific tissues

III. MOLECULAR MECHANISMS OF RADIATION DAMAGE
   A. Target Theory (Direct Action)
      1. Elementary target theory
      2. Multiple-hit theory
   B. Diffusion Theory (Indirect Action)
   C. Combined or "Fuzzy-Target" Theory
   D. Chemical Reactions of Macromolecules

IV. BIOLOGICAL EFFECTS OF RADIATION
   A. Somatic Effects
      1. Acute vs. chronic dose and relative recovery
      2. Dose levels and lethal doses
      3. Effects on systems
         a. Enzymes (*Experiment* 14.1)
         b. Cells (*Experiment* 14.2)
         c. Tissues (*Experiment* 14.3)
   B. Genetic Effects
      1. Genetic structure—chromosomes, genes, nucleotides, DNA and RNA
      2. Mutant types
         a. Chromosome abberations
         b. Point mutations
      3. Results of mutations
         a. Dependence on rest of message
         b. Dependence on environment
         c. Genetic diseases
      4. Factors limiting mutations
      5. Factors causing or sustaining mutations
         a. Spontaneous mutations
         b. Radiation (*Experiment* 14.4)
         c. Chemical agents—radiomimetics, mutagens

V. ANTIRADIATION DRUGS

SOURCES OF RADIATION:

Man has always lived with radiation and always will—he has no choice. Radioisotopes were present in the earth when it was created and everything which has come from the earth, including our own bodies, is therefore naturally radioactive. Every second over 7,000 of the atoms in the body of an average adult undergo radioactive decay. In the same time about 300 cosmic rays have passed through his body. From drinking water one receives an average of 10^{-9} to 10^{-10} μCi/ml of uranium and/or radium. Milk contains potassium and has an activity of 6×10^{-8} μCi/ml as ^{40}K. This does not mean that one should refrain from drinking milk. The ^{40}K activity has always been and presumably always will be part of normal, wholesome milk. Sources of natural radiation are summarized in table 14.1.

In recent years a number of artificial sources of radiation have also been created. Estimates of doses received from artificial sources are summarized in table 14.2.

Table 14.1
NATURAL RADIATION SOURCES*

Local gamma	45%	
from ground		60 mr/yr
wood house		10 mr/yr
brick house		40 mr/yr
Cosmic rays	30%	50 mr/yr at sea level 560 mr/yr at 20,000 ft.
Body K-40	20%	20 mr/yr (5,000 dps)
Other sources	5%	
Body C-14		(2,200 dps)
Ra-226 in water		6 mr/yr 200 dps-250
Totals	100% ~	140 mr/yr

Table 14.2
ARTIFICIAL RADIATION SOURCES*

Fallout	1-15 mr/yr
(3% of natural radiation)	5 mr/yr
Occupational	
Radiologists	Variable
Atomic Energy workers	Variable
Medical diagnosis	100 mr/yr
Lumbar x-ray	5000 mr/x-ray
Dental x-ray (jaw)	5000 mr/x-ray
(gonads)	5 mr/x-ray
Use of devices	
Luminous-dial watch	30 mr/yr
Shoe-fitting x-ray	

*All figures are estimates and are intended to provide only an idea of the order of magnitude of the dose.

EXPOSURE FROM EXTERNAL SOURCES OF RADIATION:

The biological effect produced upon exposure to an external source of radiation depends upon several factors. These include:

1. The penetration or range of radiation.
2. The total dose received.
3. The tissue involved.
4. The area or volume involved.
5. The energy of the radiation.
6. The time for accumulation of the dose.

Type of Radiation—Alpha particles cannot penetrate more than about 0.1 mm of tissue and would normally not be expected to penetrate even into the germinal layers of skin. Externally, alpha sources do not generally constitute a hazard but it must be remembered that the situation is quite different for alpha sources within the body and that the *potential* hazard posed by the possibility of their ingestion is therefore great.

Beta particles may penetrate up to a few millimeters in tissue depending upon their energy. Since the air range of beta particles can be considerable, the possibility of external exposure from beta particles is likely from an unshielded source. In an estimate of dose from high intensity beta sources, even though shielded, the contribution from bremsstrahlung must be considered.

Gamma and x-rays are, of course, very penetrating and even those which originate from an external source can cause deep-seated radiation damage. The calculated dose rate at one foot from a point-source gamma emitter is given in figure 14.2. It can be seen that the dose rate is a function of source strength, gamma ray energy and distance from the source as given by equation.

Neutrons also penetrate deeply. Tissue damage results from ionization caused by recoil of the nuclei struck by the neutron.

Dose Delivered—The dose delivered to a given volume of tissue is equal to the product of dose rate and time. That is, in terms of exposure dose,

$$\text{Dose}_{(\text{roentgens})} = \text{Dose Rate}_{(\text{r/hr})} \times \text{Time}_{(\text{hr})}$$

The dose rate is influenced by the distance from the source and by shielding. For a point gamma or neutron source the inverse square law applies ($I \sim 1/d^2$). For a source which is spread out $I \sim 1/d$. The composition of a radiation shield is governed by the type of radiation, thus for maximum absorption of radiation the following materials should be used:

for β particles — plastic
for γ rays — lead
for neutrons — hydrogenous material (paraffin, water), boron, cadmium

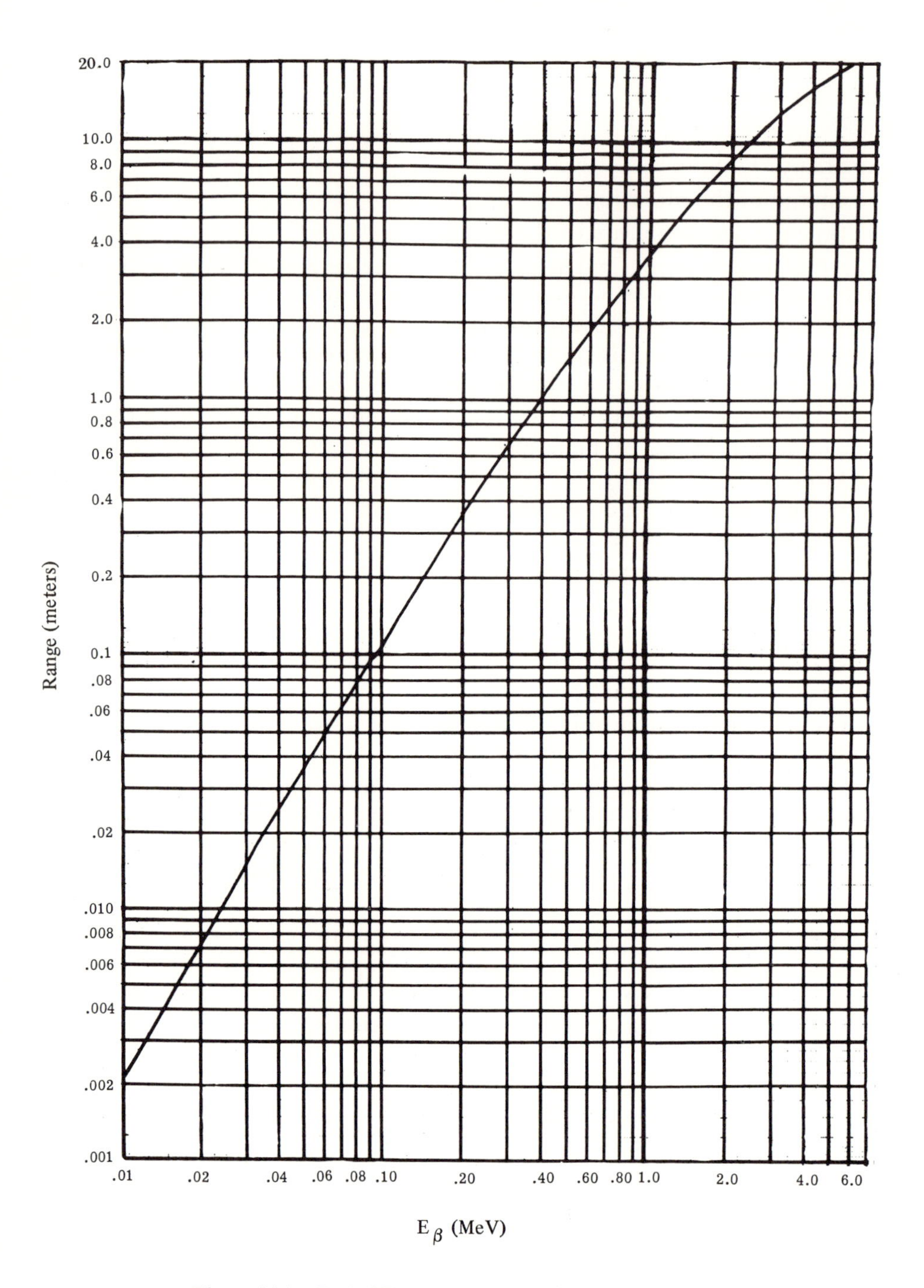

Figure 14.1—RANGE OF BETA PARTICLES IN AIR.

Table 14.3
RANGE OF BETA PARTICLES IN AIR

E_{MeV}	Maximum Range (Meters of Air)
.01	.0022
.02	.0072
.03	.015
.04	.024
.05	.037
.06	.050
.07	.064
.08	.080
.09	.095
.10	.11
.15	.21
.2	.36
.3	.65
.4	1.0
.6	1.8
.8	2.8
1.0	3.7
1.5	6.1
2.0	8.4
3.0	13
4.0	16
5.0	19

Table 14.4
CALCULATED DOSAGE RATE AT ONE FOOT FROM POINT-SOURCE GAMMA EMITTERS

E_{MeV}	r/hr/mCi at 1 foot*
0.1	0.5
0.2	1.1
0.3	1.8
0.5	3.1
0.8	5.0
1.0	6.0
1.5	8.2
2.0	10
2.5	11.6
3	13
4	15

* One photon per disintegration

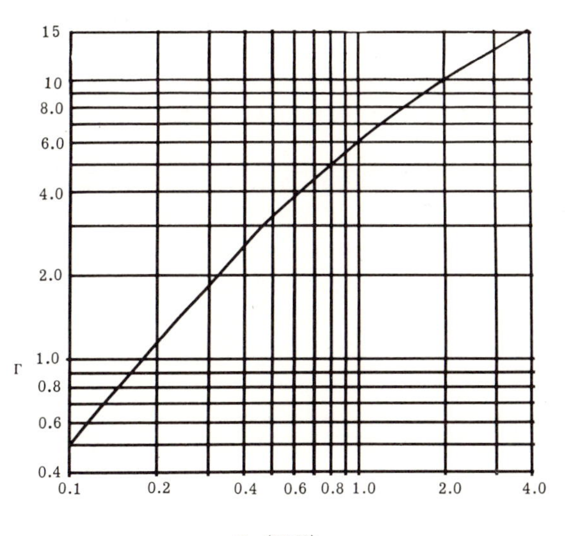

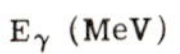

Figure 14.2—SPECIFIC GAMMA RAY OUTPUT. Values are based on emission of one gamma ray per disintegration.

Tissue Involved—Tissues differ in their sensitivity to radiation. Those most sensitive comprise the hematopoetic system including the lymphatic system and bone marrow. Next in order of decreasing sensitivity are the gastro-intestinal system, the kidneys and the central nervous system. The skin, liver and lungs are least sensitive.

Cells in the order of decreasing sensitivity are lymphocytes, bone marrow cells (erythroblasts and megakaryocytes), sperm and egg cells, cells of the lining of the small intestine, sebacious cells, sweat gland, eye, cartilage, liver cells, kidney cells, nerve cells and muscle cells.

Area Involved—It can simply be said that the less the body area exposed to radiation the greater the tolerance to a given dose of radiation. For example, total body exposure to 1000 r would almost certainly result in the death of the individual. A similar dose restricted to a hand or foot would not.

Time Involved—If a given radiation dose is delivered over a long period of time it will produce less permanent somatic damage than if delivered over a short period of time. For example, half of the people exposed to an acute total body dose of 400 r will die; if exposed over a period of 20 years to the same dose it is doubtful if any effect would be observed at all.

EXPOSURE FROM INTERNAL SOURCES OF RADIATION:

Internal radiation sources, ingested by swallowing radioactive material or by breathing radioactive dust or vapors can be very serious. It is difficult to calculate or estimate the amount of radioactive material present and the dose rate and elimination rate cannot be appreciably altered. Furthermore, some elements tend to concentrate in specific organs or tissues and will therefore produce extensive local damage to a specific tissue.

Types of Radiation—If equal millicurie quantities of radioactive materials are compared, nuclides which decay by alpha particle emission cause the greatest amount of internal damage. An average alpha particle might deliver 5 MeV or more of energy within a range of 100 microns or less. Specific ionization is very great along the path of an alpha particle. The result is extensive local damage to a cell with low tissue recovery.

In the case of beta particles, energy absorption is also complete but beta particle energies rarely exceed 2 MeV and are generally less than 1 MeV. For beta particles the average energy $\bar{E}$ is between ⅓ and ½ of E_m.

Per particle or ray, gamma radiation constitutes the least hazard since some rays which originate within the body will escape without interaction with body tissues at all. Only radiation energy expended in a tissue is harmful.

MOLECULAR MECHANISMS OF RADIATION DAMAGE:

To be acceptable a theory must explain observed phenomena. The effects of radiation on living systems have been measured and attempts have been made to correlate the facts. Because biological systems are so varied and complex a single unified theory, which explains all phenomena for all systems in a satisfactory way has not yet been presented. Several theories, each capable of explaining certain experimentally observed facts, have been proposed. These are (1) the target theory (direct hit); (2) the diffusion theory (indirect); (3) the combined or "fuzzy target" theory.

Target theory—When cultures of bacteria are irradiated it has been observed that a plot of the number of surviving organisms as a function of dose is usually logarithmic and is independent of the rate at which the dose is administered. Such a logarithmic relationship can be explained by assuming an "all or none" type process involving a target of volume V. Then the change in the number of survivors dN with increments of dose dD can be given by

$$-dN/dD = kNV \tag{14.1}$$

or

$$N = N_0e^{-kVD} \tag{14.2}$$

Careful selection of the units for V and D could eliminate the need for the proportionality constant k thus leading to a simplification of the expression. The target volume can then be calculated from the slope.

As the organisms become more complex the nature of the survival curve changes, usually to a sigmoid form. A sigmoid curve can be explained by assuming either (1) that a single target must receive more than one hit to be inactivated or (2) that more than one target must be hit to inactivate an organism or process.

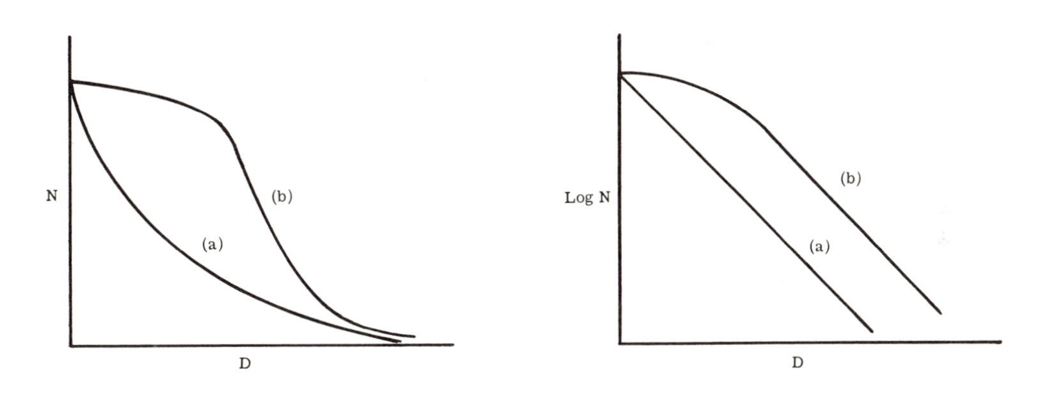

Figure 14.3—SURVIVAL CURVES. (a) simple logarithmic function
(b) Sigmoid curve

The simple target theory predicts that larger doses are required to inactivate smaller targets and vice versa. By assuming a value of 100 electron volts required to produce a primary ionization, Hutchinson (38) arrives at a simple relationship between the dose D_0 in rads necessary to reduce the activity to 37 percent* of the initial value and the molecular weight M of the target.

$$D_0M = 7 \times 10^{11} \tag{14.3}$$

*D_0 is called the inactivation dose and represents the dose which gives kVD = 1 so that $N = N_0e^{-1}$ or $N = 0.37\ N_0$.

This relationship holds for a number of compounds of biological importance when they are irradiated in the dry state.

Diffusion Theory—Living cells are about 70 to 80% water. Thus when a cell is irradiated it is reasonable to assume that most of the radiation will interact with water and that only a very small percentage of the incident radiation will involve a direct hit with a vital center as predicted by the target theory. In the diffusion theory it is assumed that free radicals, peroxides and other chemical substances produced in the water or through subsequent reactions spread through cells and tissues. Reactions for the radiolysis of water and mechanisms in the formation of certain compounds will be found in Experiment 12.9.

The Combined or "Fuzzy-Target" Theory—Although the radiation inactivation of dry molecules of biological importance could be made to fit an equation (see equation 14.3), these same molecules, when dissolved in water required a much smaller dose. The conclusion, of course, is that the diffusion theory plays an important role in the aqueous systems. For enzymes in yeast cells (38) the ratio of dry to wet dose was found to be 2 to 1 for invertase, 20 to 1 for alcohol dehydrogenase and about 100 to 1 for coenzyme A.

Hutchinson (38) has proposed a theory which combines, in effect, the target and diffusion theories. To simplify the picture, spherical molecules of radius r are assumed, each surrounded by a layer of water of thickness d where d represents the average distance radicals diffuse before they are inactivated. Thus, radicals produced within this volume of water contribute to the distruction of the molecule. Calculations show that the average value of d is about 30 angstroms. Hutchinson also points out that this dimension is roughly the thickness of water layer required to account for a cellular water content of 80%. While both the target and diffusion theories are combined it is seen that this combined theory results in a "fuzzy target."

Chemical Reactions in Macromolecules—Reactions of radiation with macromolecules, either directly or indirectly, may result in a variety of reactions. Types of reactions include simple degradation, oxidation, crosslinkage and disulfide interchange.

A *direct reaction* between an incident gamma ray and a macromolecule is illustrated by the following protein reaction.

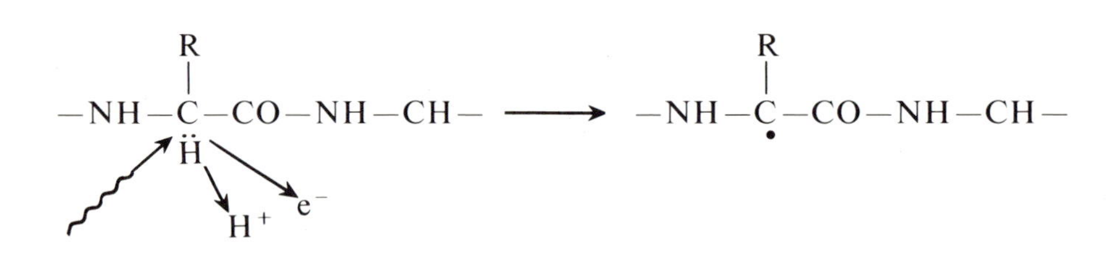

An electron (a photoelectron or a compton electron) is ejected from the molecule. The energy also ejects a proton (H^+) leaving a protein chain in which one carbon atom has an unshared electron. Such a molecule is highly reactive.

An *indirect reaction* may result in the formation of the same product molecule. The agent causing the reaction might be a hydroxyl radical produced through an interaction of radiation with water.

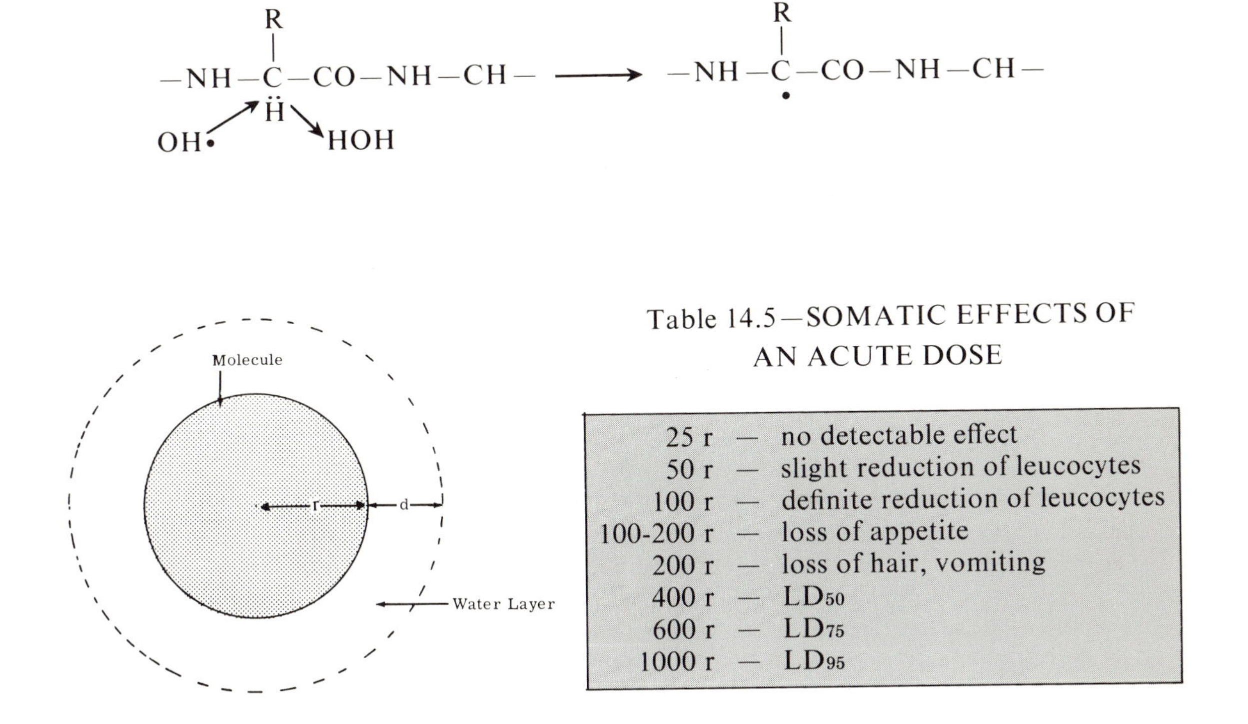

Table 14.5—SOMATIC EFFECTS OF AN ACUTE DOSE

Dose		Effect
25 r	—	no detectable effect
50 r	—	slight reduction of leucocytes
100 r	—	definite reduction of leucocytes
100-200 r	—	loss of appetite
200 r	—	loss of hair, vomiting
400 r	—	LD_{50}
600 r	—	LD_{75}
1000 r	—	LD_{95}

Figure 14.4—COMBINED TARGET AND DIFFUSION THEORIES. For simplicity a spherical molecule of radius r is assumed.

SOMATIC EFFECTS OF RADIATION EXPOSURE:

The importance of the rate of delivery of a dose was pointed out above. If delivered over a sufficiently long time no effect can be measured. This raises the question of a possible "*threshold.*" Is it possible that there is a level of radiation below which no harm is done? This question will probably not be answered until more sensitive methods are devised to detect radiation damage.

The effects noted in a population receiving an acute whole body dose of gamma radiation are listed in table 14.5.

Chronic, low-level doses produce no detectable harm at first but in time may result in tissue necrosis, changes in the pigmentation of the skin, loss of hair and other symptoms. One of the most famous examples of chronic radiation damage concerns the radium watch-dial painters who pointed their paint brushes by twisting the bristles between pursed lips. Many of these individuals required surgery on the jaw and other regions of the face to remove necrotic bones. Some presumably died as a result of the exposure.

GENETIC STRUCTURE:

The key to genetic information is *deoxyribonucleic acid* (DNA). DNA is found in the chromosomes within cell nuclei and in the mitochondria. It is a macromolecule in the form of a double helix and contains a backbone of deoxyriboses linked by diphosphate ester bonds. An organic base is attached to each deoxyribose. These organic bases are adenine, thymine, cytosine and guanine. Millions of these base moieties are required to produce a DNA molecule in which they are arranged in a specific sequence. The particular sequence constitutes the genetic message in which the bases are the characters of a four-letter "alphabet" arranged in a series of three-letter "words." This is called the *triplet code.*

Duplication is the process by which DNA reproduces itself in order to supply the genetic needs of new cells. In the process, parent DNA serves as a template for the production of a new molecule of identical composition and structure. In this way the "instructions" in the form of a DNA "sentence" of three-letter "words" are provided for the new cell so its function will be identical to that of the parent.

Enzymes govern all metabolic activity in the cell. Enzymes are proteins and consist of one or more high molecular weight polypeptide chains. The amino acids in these chains occur in an orderly and specific sequence. When additional enzyme molecules are required for cell function, messenger ribonucleic acid (m-RNA) is the pattern for their production.

Transcription is the process by which m-RNA is produced through the action of polymerase using DNA as the template. The proper sequence of bases in m-RNA is obtained through *base pair matching.* The following transcriptions occur:

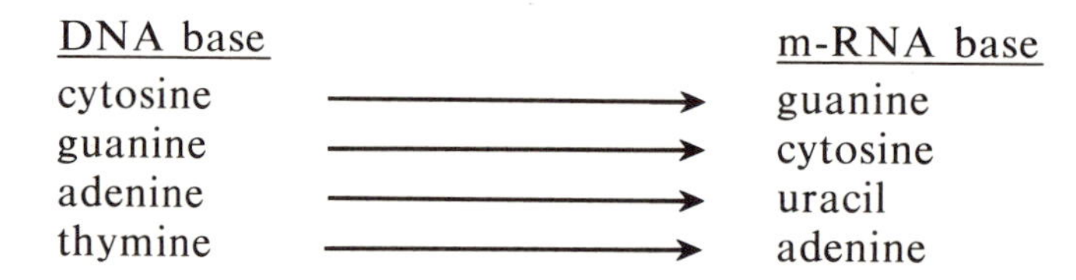

DNA base		m-RNA base
cytosine	→	guanine
guanine	→	cytosine
adenine	→	uracil
thymine	→	adenine

Messenger RNA, produced in the nucleus in the presence of DNA primer, migrates to the surface of a cytoplasmic ribosome where protein synthesis occurs. Protein production requires an interpretation of the triplet code, a process called *translation.* Translation determines the amino acid sequence of the protein molecule.

A *gene* is considered to be the segment of DNA responsible for a complete polypeptide chain. It is estimated that a cell contains about three million genes.

GENETIC MUTATIONS:

Mutations are of two types: (1) chromosome abberations and point mutations. When cells are irradiated, scissions in chromosomes are observed during metaphase. The majority of such breaks undergo spontaneous repair by joining at the broken ends but if the radiation dose is quite large the number of broken chromosomes increases and the probability that fragments of two different chromosomes may combine is also proportionately greater. Such *chromosome abberations* are discernable microscopically. Chromosome abberations occur only infrequently because large doses of radiation are normally required to produce them.

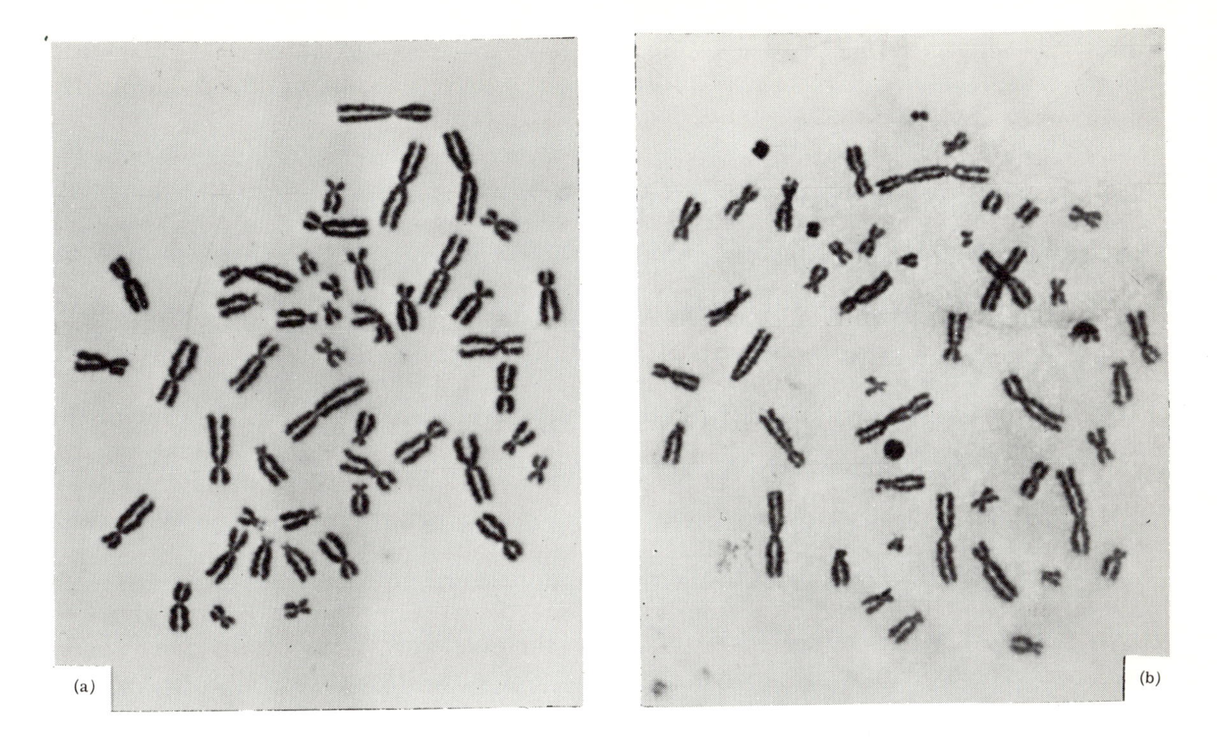

Figure 14.5—GENETIC ABBERATIONS IN HUMAN MITOTIC FIGURES.
(*Courtesy P. Nowell, U. of Penna. Hosp.*)
(a) chromosomes of normal female
(b) chromosomes of patient treated with x-rays

Point Mutations are more common. They involve a change in the genetic code in one of four ways: (1) addition of a nucleotide to the chain, (2) removal of a nucleotide, (3) transposition of the code or (4) replacement of one nucleotide for another. The first two processes result in a change in the entire code to one side of the mutation since in one case a "letter" has been added, and in the other case a "letter" has been removed from the "sentence."

Most mutations produce undesirable changes in the cell or new individual. Examples of genetic diseases are alkaptonuria (black urine), diabetes, mongolism (associated with the presence of an extra chromosome), glaucoma and club foot. One might also include myopia, baldness and brown eyes. Only a very small number of mutants result in the production of a freak.

Some mutations can be beneficial. Individuals (heterozygotes) with the sickle-cell anemia trait are less sensitive to the fulminating type of malaria. In countries where malaria is endemic the life expectancy is thus greater among the population with the sickle-cell anemia trait.

The perpetuation of highly undesirable mutants is limited by *genetic death.* Genetic death simply means the individual possessing a particular mutant does not reproduce due to his premature death or inability to reproduce. Genetic death may be caused by the inability of a defective sperm cell to fertilize an ovum or vice versa.

Mutations are caused by (1) radiation, (2) chemical agents or (3) they may occur spontaneously. It is estimated that only about 10% of mutations are caused by radiation. The molecular mechanisms involved have been discussed earlier. Chemical agents causing mutations are called *mutagens*. Agents producing effects similar to those caused by radiation are termed *radiomimetic*.

ANTIRADIATION DRUGS:

Normally a person would not be forewarned of an impending accident resulting in his receipt of a hazardous dose of radiation. It would therefore be desirable to have an antiradiation drug which is effective even when administered after the radiation exposure. Although some effects of radiation damage are delayed, a review of the mechanism of radiation damage will reveal that the discovery or development of such a drug is not likely.

One of the most rewarding approaches in the development of antiradiation compounds has been the pre-irradiation administration of drugs which are highly reactive with free radicals. Such drugs presumably compete for the free radicals with radiation sensitive compounds in the body tissues. For example, the sulfhydryl group —SH, required by a number of enzymes for their activity, is especially sensitive to oxidation. The oxidation may result from reaction with a hydroxyl radical.

$$-SH + OH\bullet \longrightarrow -S\bullet + HOH$$
$$-S\bullet + -S\bullet \longrightarrow -S-S-$$

A protective action is shown by cysteine and other compounds containing the —SH group. Some compounds containing this same group, however, are not active and so it is not possible to make a general statement concerning chemical structure and antiradiation activity. Many compounds containing the —SH group have been tested. Those showing some degree of protection include cysteine, cysteamine, glutathione and aminoisothiouronium salts. These compounds may be considered as free radical scavengers.

Some degree of radiation protection is also provided by reducing the oxygen tension. The mechanism is not clear but may be the result of (1) decreased formation of the hydroperoxy radical or (2) a general decrease in the metabolic rate. Compounds which lower the oxygen tension and thus afford a degree of radiation protection include tryptamine, p-aminopropiophenone, malononitrile, thiosulfates and dimethylsulphoxide derivatives.

As suggested above, a third approach to the control of radiation damage is through a lowering of the general metabolism of the organism. Compounds producing this effect may be considered antithyroid or hypometabolic drugs and include rutin, thiazolidines, N-2-mercaptoethylpiperazines and aminoalkylthiosulfates.

The radioprotective action of a compound is usually measured in terms of its ability to prevent mortality following whole body x- or gamma irradiation of the animal at a dose which is lethal to 50% of the unprotected animals in 30 days. This dose is called the $LD_{50/30}$.

EXPERIMENT 14.1 EFFECT OF RADIATION ON ENZYMES

OBJECTIVES:

To demonstrate the inactivation of an enzyme by radiation.

THEORY:

Saliva contains a very active enzyme called $\alpha_T$amylase which splits the glycosidic linkages in starch. Because it is found in saliva it is often called salivary amylase. The potency of salivary amylase can be measured by the rate at which a given volume of saliva hydrolyzes a given amount of starch solution. Amylase activity varies with species and individuals.

Irradiation inactivates the enzyme. The extent of radiation inactivation can be determined by measuring the potency of the amylase through its action on starch as mentioned above.

PROCEDURE:

1. Collect about 10 ml of saliva. To stimulate flow it is helpful to chew on a piece of paraffin. Instead of saliva, crystalline alpha amylase* from bacterial sources may be used in an 0.01% solution in an $0.02M$, pH 7 phosphate buffer.**

2. Place one gram of starch in a beaker containing 100 ml of water and boil. Cool and filter the resulting starch solution. Save the supernatent liquid.

3. *First approximation of enzymic activity:* Place 10 ml of the 1% starch solution into a test tube and add 1 ml of saliva. Mix thoroughly and place in a beaker of water kept at 37° (± about 3°). Every 30 seconds, test a drop of the digest with a drop of $0.1N$ iodine solution*** (use a porcelain test plate for these tests). When the blue-black color does not develop, complete hydrolysis of the starch is indicated. The digestion time is recorded.

4. *Standardization of Amylase:* Prepare a dilution of saliva or of the pure α-amylase solution with saline solution such that it requires 3 to 5 minutes digestion time at 37° to obtain a negative iodine test.

5. Put 0.5 ml of the standardized amylase solution into each of four containers and irradiate the samples for various time intervals (10, 20, 30 and 50 min). The radiation source may be ^{60}Co, ^{85}Sr, an x-ray machine or other source of such strength that the solutions receive over 5 r/hr.

* Nutritional Biochemicals Corp., Cleveland 28, Ohio, or Calbiochem, 3625 Medford St., Los Angeles, Calif.

** 3.471 g Na_2HPO_4 and 2.118 g KH_2PO_4 in 2000 ml of water.

***To prepare $0.1N$ iodine solution dissolve 3 g KI in 25 ml H_2O, add 1.27 g I_2 and dilute to 100 ml with water.

The enzyme is very sensitive to all forms of radiation, so that even a strong UV lamp may be used to irradiate the samples. For this purpose 25-ml petri dishes (uncovered) are convenient to hold the samples for irradiation.

6. After irradiation each of the solutions is transferred to a test tube containing 10 ml of the 1% starch solution and determine the time required for the complete disappearance of the starch-iodine color.

7. Plot the time required for the blue color to disappear vs the time interval the enzyme was being irradiated. (If the enzyme is totally inactivated, the blue color will not change).

EXPERIMENT 14.2 EFFECT OF RADIATION ON CELLS

OBJECTIVES:

To observe the effect of radiation on living cells.

To measure the change in blood hematocrit in a rat after treatment with a large dose of a "boneseeker" like ^{32}P.

THEORY:

Some atoms exchange easily with the atoms in bone and are incorporated into the bone structure. Among these are calcium, magnesium and phosphorus. When sodium phosphate is introduced into a vertebrate, the phosphorus is absorbed by the bones. If the phosphate contains ^{32}P, the beta radiation will destroy some of the hematopoietic centers, since blood is manufactured continuously in the bone marrow. As a result of this destruction or attenuation, the ratio of the volume of erythrocytes (red blood cells) to plasma will be affected.

The blood hematocrit is the ratio of the volume of red blood cells per unit volume of circulating blood. Microhematocrit determinations require the use of only a few drops of blood. In the micro-method, the blood is picked up by capillary force into small capillary tubes previously treated with heparin or sodium oxalate to prevent coagulation of the blood. The capillary is then sealed and spun in a small centrifuge to pack the cells rapidly and uniformly.

Polycythemia vera is a disease in which the afflicted individual over-produces erythrocytes. Since the blood volume is approximately constant in an individual, the contribution to this volume by red blood cells becomes very great. Hematocrits of from 70% to 80% have been recorded. If the hematocrit is above 60% the patient is in a dangerous condition for there is a likelihood of capillaries breaking throughout the body, resulting in the formation of clots. The skin of these patients is usually quite red and they suffer associated pain in all joints because the capillaries in the joints are subjected to so much trauma. The treatment prior to the advent of radioactive materials consisted of regular phlebotomies, approximately 500 ml of blood being removed at regular intervals. With the availability of ^{32}P, cells are now destroyed by radiation. A patient may receive from 2 to 5 mc of ^{32}P in the form of Na_2HPO_4, the dose being regulated according to the individual's hematocrit. After treatment, patients usually need no further therapy for one or two years. While ^{32}P can be very useful in polycythemia vera, myelogenous leukemia and other dyscrasias, it can cause severe damage when improperly used.

MATERIALS:

Two to four 250 g rats; large shears; absorbent cotton; large glass funnel; large beaker and cover (or a large desiccator); oxalated or heparinized capillary tubes*, 1 ½″-2″ in length; microhematocrit apparatus* (or small centrifuge, centrifuge cups and millimeter rules and corks); one large metallic spatula; burner; $Na_2H^{32}PO_4$ solution, 1 mCi per animal; No. 20 hypodermic needle; 2-ml syringe; rubber gloves.

PROCEDURE:

1. Place a pad of cotton, soaked with ethyl ether, in the bottom of a large beaker or desiccator and cover tightly for 30 to 60 seconds to create an ether atmosphere.

2. Quickly put a rat into the container and replace the cover until the rat just stops moving. Do not over anesthetize the animal or it will die.

3. Remove the rat from the chamber and, to assure anesthetization throughout the procedure, place an ether-soaked cotton cone (glass funnel with a piece of ether-soaked cotton wedged in the orifice) close to but not quite touching the rat's nose.

4. Working quickly, snip away a 1 cm portion of the rat's tail with large, sharp shears. Half-fill each of six oxalated or heparinized capillary tubes by gingerly touching each tube to the freshly exuded blood.

5. After filling the capillaries, heat a spatula to red heat over a burner and place the hot spatula firmly against the wounded end of the tail to stop the flow of blood (cauterization).

6. Seal the end of each capillary tube—the end away from the blood—by holding it in a small pin-point flame, i.e., micro-burner, alcohol lamp or safety match. This forms a small centrifuge tube. Measure the hematocrit in the micro-hematocrit apparatus following the manufacturer's directions. If this apparatus is not available, an effective measure of the hematocrit can be made in the following way: Fit corks tightly into the tops of six small centrifuge cups—make a small hole in the center of each cork with a small drill or similar instrument—place the capillary tubes in the holes and spin the centrifuge at maximum speed for two minutes. Measure the length of the packed red blood cell column and the length of the total volume of blood with a millimeter rule. Determine the percentage of the volume of packed cells to the total volume for each tube. Average the values obtained for each of the six tubes for each animal. This is the rat's hematocrit.

*Drummond Scientific Co., Philadelphia, Pa.

7. Don rubber gloves to protect the hands from contact with ^{32}P. Inject 1 mCi of the ^{32}P solution intraperitoneally into the rat. Place the rat in a cage, being careful that its position allows for free breathing. (CAUTION: The urine and feces of the animal will be highly radioactive from this point on. The excreta must be carefully collected and left to decay in a thick jar.)

8. Take blood samples and measure the hematocrit at 48-hour intervals for a period of 4 to 8 days. Use the technique outlined above.

9. To dispose of the rats, inject each with 2 ml of a saturated solution of $MgSO_4$ and let the radiation decay while the rats are stored in a formalin solution in a protected jar.

DATA:

Blood hematocrits:		After ^{32}P		
	Before ^{32}P	2 Days	4 Days	6 Days
Capillary #1				
2				
3				
4				
5				
6				
Average				

SUGGESTIONS FOR FURTHER WORK:

The experiment may be repeated with the use of a lower dosage range of ^{32}P. It should be observed that there is a dose range for ^{32}P which causes no perceptible change in the blood hematocrit value. In this manner, maximum permissible doses for each type of experimental animal may be determined.

Conversely, by increasing the ^{32}P dosage, with the use of the dose range resulting in no significant further decrease in hematocrit.

A "clinically" safe dose for an animal can be ascertained for a specific isotope from observations of blood damage and total histopathological changes for each tissue. Data on many animals of a given species would be required for the results to be of significance.

EXPERIMENT 14.3 SOMATIC EFFECTS OF RADIATION

OBJECTIVES:

To demonstrate a somatic effect of radiation.

To determine the changes produced in the tissues of the rat salivary gland by beta radiation.

To demonstrate that calcium concentrates in the salivary gland.

THEORY:

The salivary gland is a radio-sensitive target and its cells are large and well defined. Radiation damage to the salivary gland is therefore readily demonstrable by means of histological techniques. The salivary gland concentrates calcium ion to a greater extent than adjacent tissues. It is therefore possible to deliver a dose of beta radiation to the gland through the administration of $^{45}Ca^{++}$. Morphological changes can be observed in about three days.

MATERIALS:

Six male Wistar rats, each weighing between 150 and 200 g and each in an individual cage; 25 μCi $^{45}CaCl_2$ solution adjusted to pH 7 with 0.1*N* Na_2HPO_4; 25 μCi $Na_2H^{32}PO_4$ solution; planchets; flow counter; large shears; dissecting kit; paraffin imbedding apparatus; histological microtome; stain jars with hematoxylin; eosin; washing solutions; microscope.

PROCEDURE:

1. *Administration of the dose*—Six rats are divided into three groups of two each. Each rat receives an intramuscular injection as follows:

Group I —5 to 10 μCi of $^{45}Ca^{++}$ solution.
Group II —5 to 10 μCi of $H^{32}PO_4^{-2}$ solution.
Group III—(controls)—non-radioactive solution chemically equivalent to that given animals in Group I.

2. *Calibration of dose*—Aliquots of the ^{45}Ca and ^{32}P solutions injected above are assayed to determine the exact dose administered. Alternately, it is usually adequate to calculate the dose from the manufacturer's data, decay of the nuclide and dilutions prepared.

3. *Dissection of tissues*—After 3 days the animals are sacrificed and the salivary glands are excised. The parotids are large sack-like structures on each side of the neck, near the ears. Each whole parotid is carefully freed of adhering tissue and weighed. The weight should be about 500 mg.

4. *Assay of radioactivity*—A thin section of the parotid gland is placed in a tared planchet and weighed. The activity of the tissue is then determined. If measured with a liquid scintillation counter, homogenize the weighed section with 1 ml of ethanol in a glass homogenizer and transfer quantitatively to a vial with 19 ml of liquid scintillator solution.

For comparison, the activity of a weighed sliver of muscle or lung tissue is determined in a similar fashion. The extent of calcium concentration in the salivary glands can then be estimated from these activities.

5. *Histological examination*—The remaining section of each parotid gland is placed in 10% neutral formalin, each in an individual flask. Twenty four hours is allowed for fixation in the formalin. A portion of each section is then selected for paraffin imbedding.

The sections are imbedded in paraffin and sections are sliced on a microtome. To assure obtaining a useful section the imbedded tissues are turned through 90° and a few additional sections are cut and mounted on glass slides. The paraffin is removed from each section and the tissues are stained with hematoxylin and eosin. The sections are now examined microscopically.

DATA AND RESULTS:

Dose injected: $^{45}Ca^{++}$ ________ μCi ^{32}P __________ μCi

Rat	Type of Activity Administered	Weight of Whole Parotid	Weight of Tissue Assayed	Activity of Tissue Assayed	Activity Per g of Tissue (μCi/g)	Weight of Muscle Assayed	Activity of Muscle Assayed	Activity Per g of Muscle (μCi/g)
I a	^{45}Ca							
I b	^{45}Ca							
II a	^{32}P							
II b	^{32}P							
III a	Control							
III b	Control							

Calculate the per cent of dose per gram of tissue and compare the uptake of calcium activity with the phosphorus activity in the parotid glands as compared with muscle (or lung tissue). Do the salivary glands concentrate calcium?

Histological evaluation: Parotid glands of the control rats and those receiving ^{32}P should be normal. Parotid glands of the rats receiving ^{45}Ca will show interstitial fibrosis, inflammation, atrophy, pseudoglandular arrangement and some necrotic cells.

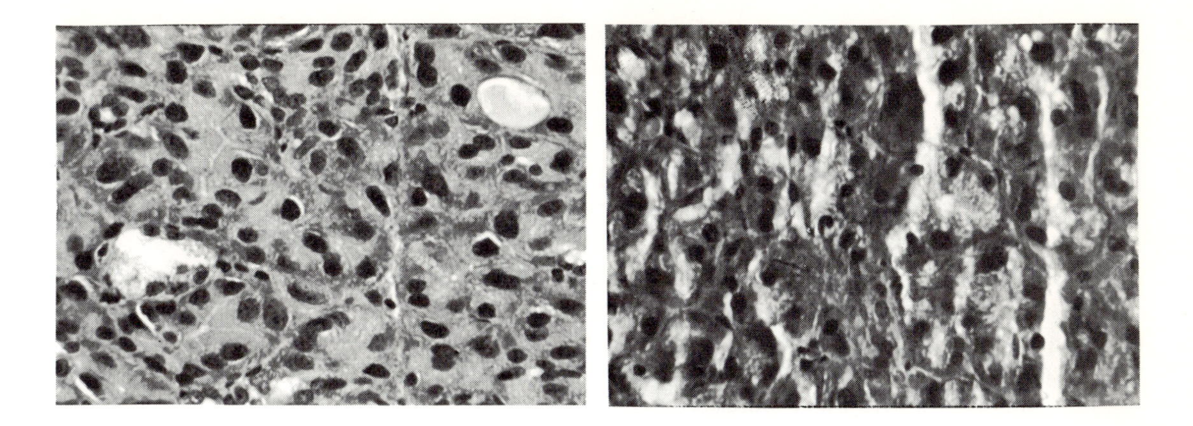

Figure 14.6—RADIATION DAMAGE TO THE SALIVARY GLAND.
(a) control
(b) cells exposed to radiation from ^{45}Ca.
The tissues have been stained with hematoxylin-eosin.

EXPERIMENT 14.4 GENETIC EFFECTS OF RADIATION

OBJECTIVES:

To investigate the effect of radiation on chromosomes.

To illustrate a technique for visualizing chromosomes.

To observe the morphological changes in chromosomes in the cells of *Vicia faba* root tips produced by exposure to radiation.

THEORY:

Rapidly dividing cells show the least resistance to radiation. For this reason cells of the meristematic region of the root of plants are very sensitive to radiation. Intense radiation will produce pyknotic nuclei containing adhering chromosomes. If the tissues are stained with feulgen dye, these defective chromatic forms can be observed. Roots of the flat bean, *Vicia faba,* grow rapidly and are well suited for radiation experiments.

MATERIALS:

Vicia faba beans*; wire screen or sand; pan; 0.1% colchicine solution; fixative solution (95% alcohol-acetic acid, 3:1); 1 *N* HCl; feulgen stain**; glass slides; cover

* Available from Burpee Seed Co., Philadelphia, Pa.
** Feulgen reagent is available as Schiff's reagent (Harleco #2818).

glasses; absorbent paper; dry ice; dissecting needle and scissors; 0.02% fast green dye in 95% ethanol; 95% ethanol; absolute alcohol; diaphane mounting media.

PROCEDURE:*

1. *Stimulation of lateral root growth*—About two dozen beans are softened by soaking in water overnight. The seed coats are removed and they are suspended in water on a wire screen or by placing in sand. The beans are allowed to grow in the dark for 3 or 4 days during which time the main root should grow to a length of about one inch. The tip of the root is cut off to stimulate the growth of lateral roots. In another 3 or 4 days the lateral roots should be sufficiently developed.

2. *Irradiation of the roots*—A dose of 200 to 250 r is needed to cause visible damage to the root cells. This can be delivered by means of an x-ray machine or by exposure to a suitable radioactive source. Some of the beans are not irradiated to serve as the controls. The beans are returned to the pan of water.

3. *Preparation of the roots*—Lateral roots are removed from the control beans and from the irradiated beans after 0, 6, 12, 24 and 48 hours. The roots are immersed in 0.1% colchicine solution for exactly 3 hours before transfer to the alcohol-acetic acid fixative. The roots are now ready for staining.

4. *Staining*—

a. Place root in 1*N* HCl at 60° for 12 minutes.
b. Wash in cold water.
c. Place in Feulgen reagent for 45 minutes at room temperature.
d. Place root tip in a drop of 45% acetic acid on a glass slide. Cut off and save about 2 to 3 mm of the root tip.
e. Place a cover glass on the root tip and tap gently to produce a monocellular layer. Press between layers of absorbent paper to flatten cells.
f. Freeze the slide on a block of dry ice.
g. Remove the cover glass with a needle. If properly frozen it should snap off.
h. Place slide in 0.02% fast green dye in 95% alcohol for 3 minutes.
i. Wash in absolute alcohol.
j. Mount in diaphane media. Diaphane media requires a few days to harden.

5. *Examination of the slide*—Using low power, examine the slide with a microscope. The region of small cuboidal cells with large nuclei is the meristematic region.

When the meristematic cells are located and identified, the magnification is changed to high power.

The number of cells in mitoses should be noted. The number of abnormalities in cells of the controls is compared with those of the irradiated cells.

* Adapted from Casarett and Davis (reference 2).

REFERENCES

1. Bellamy, W. D., and E. J. Lawton, "Problems in Using High-Voltage Electrons for Sterilization", Nucleonics 12, No. 4, 54-57 (April 1954).
2. Casarett, A. P., and T. P. Davis, "Elementary Experiments in Radiation Biology", AEC Research and Development Report UR-627, Off. Tech. Services, Department of Commerce, Washington 25, D.C.
3. Comar, C. L., "Biological Aspects of Nuclear Weapons", Amer. Scientist 50, No. 2, 339-353 (1952).
4. Errera, M., and A. Forssberg, "Mechanisms in Radiobiology", Vol. 1 & 2, Academic Press (1960).
5. Goldfeder, A., "Ionizing Radiation as a Tool in Biological Research", Trans. of N.Y. Acad. Sciences, Sect. Biol., Ser. II, Vol. 20, No. 8, 809 (June 1958).
6. Hahn, P. F., "Therapeutic Use of Artificial Radioisotopes", Wiley (1960).
7. Hassett, C. C., and D. W. Jenkins, "Use of Fission Products for Insect Control", Nucleonics 10, No. 12, 42-46 (1952).
8. Jefferson, M. E., "Irradiated Males Eliminate Screwworm Flies", Nucleonics 18, No. 2, 74-76 (February 1960).
9. Lemmon, R. M., "Radiation Decomposition of Carbon-14 Labeled Compounds", Nucleonics 11, No. 10, 44 (1953).
10. O'Meara, J. P., "Radiation Chemistry and Sterilization of Biological Materials by Ionizing Radiations", Nucleonics 10, No. 2, 19-23 (1952).
11. Pochin, E. E., "Therapeutic Use of Radioisotopes", P. F. Hahn, ed., Wiley (1956).
12. Proctor, B. E., and S. A. Goldblith, "Effect of High-Voltage X-Rays and Cathode Rays on Vitamins (Niacin)", Nucleonics 3, 32 (August 1948).
13. Rubin, B. A., "Radiation Microbiology: Problems and Procedures", Nucleonics 7, 5-20 (September 1950).
14. Sparrow, A. A., and E. Christensen, "Improved Storage Quality of Potato Tubers after Exposure to Co^{60} Gammas", Nucleonics 12, No. 8, 16-17 (July 1954).

Somatic Effects

15. Allen, B., "Anemia in Mice Resulting from Irradiation of the Exteriorized Spleen", Nucleonics 11, No. 3, 44 (1953).
16. Brennan, J. T., P. S. Harris, R. E. Carter, and W. H. Langham, "The Biological Effectiveness of Thermal Neutrons on Mice—II", Nucleonics 12, No. 4, 31-35 (April 1954).
17. Fritz, M., D. W. Cohen, and J. L. Rabinowitz, "Lipogenesis and Cholesterolgenesis in the Wistar Rat Parotid Gland after Uptake of Ca-45", Radiation Res. 22, 543-547 (1964).
18. Grosch, D. W., R. L. Sullivan, and L. E. LaChance, "Biological Response to Mixed Radiations", Nucleonics 15, No. 12, 64-66 (December 1957).
19. Lawrence, C. A., L. E. Brownell, and J. T. Graikoski, "Effect of Cobalt-60 Gamma Radiation on Microorganisms", Nucleonics 11, No. 1, 9 (1953).
20. Osgood, E. F., A. J. Seaman, and H. Tivey, "Comparative Survival Times of X-ray Treated vs. P^{32} Treated Patients with Chronic Leukemia Under the Program of Titrated, Regularly Spaced Total-Body Irradiation", Radiology 64, 373-381 (1955).
21. Rugh, R., "Effect of X-Irradiation on Tissue Hydration in the Mouse", Nucleonics 12, No. 1, 28-33 (January 1954).
22. Stokinger, H. E., "Size of Dose: Its Effect on Distribution in the Body", Nucleonics 11, No. 4, 24-27 (1953).
23. Sullivan R. L., and D. S. Grosch, "The Radiation Tolerance of an Adult Wasp", Nucleonics 11, No. 3, 21 (1953).
24. Winthrop, M. M., "Clinical Hematology", 5th ed., Lea & Febiger, Phila.

Genetic Effects

25. Anonymous, "DNA Regenerates Cells Damaged by X-Rays", Chem. Eng. News, 40 (April 26, 1965).
26. Bacq, A., and P. Alexander, "Fundamentals of Radiobiology", Pergamon Press, New York (1961).
27. Darlington, C. D., and L. F. LaCour, "The Handling of Chromosomes", Allen and Unwin, London (1962).
28. Hollaender, A., ed., "Radiation Biology" Volume 1, Chapters 9 and 10, McGraw-Hill, New York (1954).
29. Kelly, E. M., L. B. Russel and W. L. Russel, "Radiation Dose Rate and Mutation Frequency", Science 128, 1546-1550 (1958).
30. Muller, H. J., "The Prospects of Genetic Change", Amer. Scientist 47, No. 4, 551 (December 1959).
31. Osborne, T. S., "Experiments with Radiation on Seeds", Booklets #1 and #2. USAEC, Div. of Technical Information, P.O. Box 62, Oak Ridge, Tenn.
32. Plough, H. H., "Radiation Tolerances and Genetic Effects", Nucleonics 10, No. 8, 16-20 (1952).
33. Read, J., "Radiation Biology of *Vicia Faba* in Relation to the General Problem", Charles C. Thomas, Springfield, Illinois (1959).
34. Russel, W. L., "First Generation Effect of Radiation", Science 125, 1135 (1957).
35. Spalding, J. F., V. G. Strang and W. L. Stourgeon, "Heritability of Radiation Damage in Mice", Genetics 46, 129-133 (1961).

36. Spalding, J. F., V. G. Strang, and W. L. Stourgeon, "Characteristics of Offspring from Ten Generations of X-Irradiated Mice", Genetics 48, 341-346 (1963).

Mechanism of Radiation Damage

37. Anonymous, "New Theory Explains Radiation Damage", Chem. Eng. News 39, No. 14, 25 (April 3, 1961).
38. Hutchinson, F., "Molecular Basis for Action of Ionizing Radiation", Science 134, 533 (1961).

Radiomimetic Agents

39. Alexander, P., "Radiation-Imitating Chemicals", Scientific Amer., 99 (January 1960).
40. Swinyard. E. A., and S. C. Harvey, "Antineoplastic Drugs", Chapter 73 of "Remington's Pharmaceutical Sciences", 13th edition, E. Martin, ed., Mack Publishing Co., Easton, Pa. (1965).
41. Teas, H. J., H. J. Sax and K. Sax, "Cycasin: Radiomimetic Effect", Science 149, 541 (1965).

Antiradiation Drugs

42. Alexander, P., Z. Bacq, S. F. Cousins, M. Fox, A. Serve, and J. Lajor, "Mode of Action of Some Substances to Protect Against the Lethal Effects of X-rays", Radiation Res. 2, 392 (1955).
43. Anonymous, "Government Seeks Antiradiation Drugs", Chem. Eng. News, 42 (November 23, 1959).
44. Ashwood-Smith, M. J., "The Radioprotective Action of Dimethyl Sulphoxide and Various Other Sulphoxides", Int. J. Rad. Biol 3:1, 41 (1961).
45. Bond, V. P., E. P. Cronkite, and R. A. Conard, "Radiation Illness: Its Pathogenesis and Therapy", Atomic Medicine, 3rd Ed., C. F. Behrens, ed., (1960).
46. Bridges, B. A., "Radiation Protection by Some Sulfhydryl Derivatives of Pyridoxine and a New BAL Preparation", Int. J. Rad., Biol. 3:1, 49 (1961).
47. Cipriani, A. J., "Prevention and Treatment of Radiation Injury", Canad. Serv. M. J. 11:7, 497-501 (1955).
48. Crouch, B. G., and R. R. Overman, "Chemical Protection Against X-Radiation Death in Primates", Science 125, 1092 (1957).
49. Doherty, D. G., and W. T. Burnett, "Protective Effect of S-B, AET, Br. HBr, and Related Compounds Against X-Radiation Death in Mice", Proc. Soc. Exper. Biol. and Med. 89, 312-314 (1955).
50. Doherty, D. G., W. T. Burnett, Jr., and R. Shapiro, "Chemical Protection Against Ionizing Radiation II. Mercaptoethylamines and Related Compounds with Protective Activity", Radiation Research 7, 13-21 (1957).
51. Doherty, D. G., W. T. Burnett, Jr., and R. Shapiro, "Chemical Protection Against Ionizing Radiation III. Mercaptoalkylguanidines and Related Isothiuronium Compounds with Protective Activity", Rad. Res. 7, 22 (1957).
52. Eldjarn, L., and A. Pihl, "On the Mode of Action of X-ray Protective Agents; Fixation *in vivo* of Cysteamine and Protein", J. Biol. Chem. 223, 341 (1956).
53. Fowler, J. M., ed., "Biological Effects of Radiation" and "Protective Treatment", *Fallout; A Study of Superbombs, Strontium* 90, *and Survival* (1960).
54. Foye, W. O., and D. H. Kay, "Antiradiation Compounds III—N-2-Mercaptoethylpiperazines", J. Pharm. Sciences 51, 1098-1101 (1962).
55. Foye, W. O., and R. H. Zaim, "Antiradiation Compounds V—α-Amino Acid Esters of 2-Mercaptoethylamine", J. Pharm. Sciences 53, 906-908 (1964).
56. Foye, W. O., and J. Mickles, "Antiradiation Compounds VI—Metal Chelates and Complexes of 2-Mercaptoethylamine and 2-Mercaptoethylguanidine", J. Pharm. Sciences 53, 1030-1033 (1964).
57. Gray, J. L., J. T. Tew, and H. Jensen, "Protective Effect of Serotonin and of P-Aminopropiophenone Against Lethal Doses of X-Radiation", Proc. Soc. Exp. Biol 80, 604-607 (1952).
58. Haley, T. J., "Radiation Injury and the Present Status of Therapeutics", Indust. M. and S. 22:11, 569-572 (1953).
59. Holmberg, B., and B. Sorbo, "Protective Effect of B-aminoethylthiosulfuric Acid Against Ionizing Radiation", Nature 183, 832 (1959).
60. Jones, M. M., "An Alternating Mechanism for Chemical Protection Against Radiation Damage", Nature 185, 96-97 (1960).
61. Kalkwarf, D. R., "Chemical Protection from Radiation Effects", Nucleonics 18, No. 5, 76-81, 130 (May 1960).
62. Kaluszyner, A., P. Czerniak, and E. D. Bergman, "Thiazolidines and Aminoalkylthiosulfuric Acids as Protecting Agents Against Ionizing Radiation", Rad. Res. 14, 23 (1961).
63. Lindop, P. J., and J. Rotblat, "Protection Against the Acute Effects of Radiation by Hypoxia", Nature 185, 593-594 (1960).
64. Maisin, J. R., and D. G. Doherty, "Chemical Protection of Mammalian Tissue", Federation Proc. 19:2, 564 (1960).
65. Needleman, P., "Treatment and Prophylaxis of Radiation Damage", Am. J. Pharm. 133, 166 (1961).
66. Patt, H. M., "Protective Mechanisms in Ionizing Radiation Injury", Physiol. Rev. 33, 35 (1953).
67. Patt, H. M., "The Modification of Radiation Effects by Chemical Means", Progress in Radiation Therapy, ed. by F. Buschke, (1958).
68. Patt, H. M., "Chemical Approaches to Radiation Protection in Mammals", Federation Proc. 19:2, 549 (1960).
69. Pihl, A., and N. Eldjarn, "Pharmacological Aspects of Ionizing Radiation and of Chemical Protection in Mammals", Pharmacological Review 10:4, 437-474 (1958).

70. Smith, A. D., and D. Lowan, "Radioprotective Action of Methoxamine", Nature 184, 1729 (1959).
71. Smith, W. E., "Drugs and Techniques Used in Radiation Damage", Am. J. Pharm. 128:12, 410-420 (1956).
72. Van Bekkum, D. W., "The Protective Action of Dithiocarbamates Against the Lethal Effects of X-Irradiation in Mice", Acta. Physiol. Pharmacol. Neerl. 4, 508 (1956).

CHAPTER 15
Nuclear Medicine (Clinical and Veterinary)

LECTURE OUTLINE AND STUDY GUIDE

I. DOSAGE FORMS
  A. Official in U.S.P. XVII
  B. Non-official
  C. Applications or Use
    1. As tracers (diagnostic)
    2. As radiation sources
      a. Utilizing effect of tissue on radiation (diagnostic x-ray)
      b. Utilizing effect of radiation on tissue (therapeutic)
  D. Properties of Medical Isotopes

II. THERAPEUTIC APPLICATIONS (Used as Radiation Sources)
  A. External Sources
    1. Teletherapy sources
    2. Surface sources
  B. Internal Sources
    1. Infusion
    2. Interstitial implant
    3. Selectively absorbed or concentrated

III. DIAGNOSTIC APPLICATIONS
  A. Isotope Dilution Techniques
    1. Blood and plasma volume
      a. Radio-iodinated serum albumin method (*Procedure* 15.1)
      b. Sodium radiochromate method (*Procedure* 15.2)
    2. Total body water
    3. Sodium space
    4. Potassium space
    5. Chloride space
  B. Rate of Transfer
    1. Circulation time
    2. Cardiac output
  C. Rate of Disappearance
    1. Test of circulation in tubed pedicles in skin grafts
    2. RBC survival time (*Procedure* 15.3)
    3. Sequestration of blood cells (*Procedure* 15.4)
    4. Test of gastrointestinal bleeding (*Procedure* 15.5)
  D. Concentration and Metabolism of Materials
    1. Thyroid function tests
      a. Thyroid uptake (*Procedure* 15.6)
      b. Urinary excretion (*Procedure* 15.7)
      c. PBI conversion ratio (*Procedure* 15.8)
        (1) Trichloroacetic acid method
        (2) Ion exchange method
      d. Salivary activity
      e. Thyroid clearance
      f. Triiodothyronine *in vitro* uptake methods (*Procedure* 15.9)
        (1) RBC-T_3 method
        (2) Sephadex method
    2. Blood dyscrasias
      a. Schilling test for pernicious anemia (*Procedure* 15.10)
      b. Kinetics of iron metabolism (*Procedure* 15.11)
    3. Fat absorption
    4. Liver function
    5. Renal function (*Procedure* 15.13)
    6. Differential uptake studies
    7. Scanning techniques

IV. ESTABLISHMENT OF A MEDICAL RADIOISOTOPE PROGRAM
  A. Personnel
  B. Program
  C. Facilities and Work Areas
  D. Instrumentation
  E. Licensure and Reports—AEC, state and city
  F. Records
  G. Isotopes—Procurement, storage and disposal

INTRODUCTION:

Radium has the distinction of being the first radioisotope used in medicine, having been employed as early as 1901. This nuclide was the most important medical radioisotope in use up to about 1946, when artificially produced isotopes became available in quantity. Since that date, growth in the medical applications of radioisotopes has been very rapid as their usefulness has become more and more apparent in diagnosis, therapy and medical research and as greater numbers of physicians and other scientific personnel have been trained in their use.

DOSAGE FORMS AND MODES OF USE:

Current medical procedures employ more than a dozen radioisotopes in a wide variety of chemical and physical forms. Table 15.1 lists eleven radioactive preparations of sufficient importance to warrant their inclusion in the Pharmacopeia (XVII) of the United States and one in the National Formulary. It also lists a large number of non-official preparations supplied by pharmaceutical manufacturers.

Radioisotopes are used in medicine in two different ways. They may be used (1) as radiation sources or (2) as radioactive tracers. As *radiation sources* their principal role is in therapy. Here, the choice of the isotope for a given application is governed largely by the properties of the radiation required for treatment; type and energy of the radiation and range in tissues are prime considerations. Except in special cases, the chemical properties or chemical form of a given isotope are relatively unimportant.

As a *radioactive tracer* the chemical identity and form of the nuclide are most important since, with but few exceptions, the tracer must be isotopic with the element being traced or must otherwise be capable of being incorporated as a part of a particular molecule. The nature of the radiation emitted by a tracer radioisotope is important primarily from the standpoint of its ease of detection. Radioactive tracers are used in medicine principally for diagnostic purposes.

Where radioisotopes are used as external sources or as sealed sources implanted in a tissue, the dose is terminated by removal of the source. When they are administered internally as an unsealed source, the dose administered to the patient, either deliberately in therapy or incidentally in diagnosis, cannot be terminated at will by removal of the source. In therapeutic applications the total dose must be calculated from a knowledge of the effective half-life of the isotope, the type and energy of the radiation emitted and the concentration of the isotope in the tissue.

When radioisotopes are used for diagnosis, the radiation dose delivered to the patient is maintained at as low a level as possible. This is accomplished through the judicious choice of isotope for the best combination of minimum half-life, minimum retention in the body and minimum quantity of isotope which will permit its detection and accurate measurement. Accordingly, certain isotopes, such as ^{90}Sr, ^{226}Ra and many others, are never used as unsealed, internal sources or tracers. In order to reduce the radiation dose to the population there is a trend toward the use of shorter lived isotopes, when available, for diagnostic purposes. It is for this reason that ^{57}Co and ^{58}Co are often used in place of ^{60}Co, where possible, in diagnostic procedures.

Table 15.1—RADIOPHARMACEUTICALS

Official in the U. S. Pharmacopeia XVII

Cyanocobalamin Co 57 Capsules (Racobalamin 57, Rubratope-57 Capsules)
Cyanocobalamin Co 57 Solution (Racobalamin 57, Rubratope-57 Oral Solution)
Cyanocobalamin Co 60 Capsules (Racobalamin 60, Rubrotope-60 Capsules)
Cyanocobalamin Co 60 Solution (Racobalamin 60, Rubrotope-60 Oral Solution)
Iodinated I 131 Serum Albumin (Albumotope, RISA, IHSA)
Rose Bengal Sodium I 131 Injection (Robengatope)
Sodium Chromate Cr 51 Injection (Chromitope Sodium, Rachromate)
Sodium Iodide I 131 Capsules (Iodotope Diagnostic, Radiocaps)
Sodium Iodide I 131 Solution (Iodotope Diagnostic, Oriodide, Theriodide)
Sodium Iodohippurate I 131 Injection (Hippuran, Hipputope, Radio-Hippuran)
Sodium Phosphate P 32 Solution (Phosphotope)

Official in the National Formulary XII

Gold Au 198 Solution (Aureotope, Aurcoloid)

Official in the British Pharmacopeia 1963

(Not included in USP XVII nor in NF XII)

Sodium Iodide (^{131}I) Injection
Sodium Phosphate (^{32}P) Injection

Non-official

Chlormerodrin Hg 197 (Neohydrin-197, Radio Chlormerodrin Hg 197)
Chlormerodrin Hg 203 (Neohydrin-203, Radio Chlormerodrin Hg 203)
Chromic Cr 51 Chloride (Chromitope Chloride)
Chromic Phosphate P 32 Suspension, Colloidal Chromic Radiophosphate P 32 (Phosphocol, Chromphosphotope)
Cobalt 60 (Actaloy)
Cobalt Co 57 Chloride (Cobatope 57)
Cobalt Co 58 Chloride (Cobatope 58)
Cobalt Co 60 Chloride (Cobatope 60)
Cyanocobalamin Co 58 (Rubratope 58)
Ferric Fe 59 Chloride
Ferrous Fe 59 Citrate (Ferrutope)
Ferrous Fe 59 Sulfate
Iodinated I 131 Diatrizoate Methylglucamine (Radio-Renografin)
Iodinated I 131 Iodipamide Sodium (Radio-Cholografin)
Iodinated I 131 Iodopyracet Injection (Radio-Diodrast)
Iodinated I 131 Oleic Acid (Raoleic Acid, Oleotope)
Iodinated I 125 Oleic Acid (Oleotope I-125 Capsules, Oleotope I-125 Oral Solution)
Iodinated I 131 Glyceryl Trioleate (Raolein, Trioleotope)
Iodinated I 125 Serum Albumin (RISA-125, Albumotope I-125 Injection)
Iridium Ir 192 Seed Ribbons (Iriditope Ribbons)
Iron Fe 59 Globulin Complex
Polymetaphosphate P 32
Rose Bengal Sodium I 125 Injection (Robengatope I-125 Injection)
Sodium Na 22 Chloride (Natritope Chloride Injection)
Selenomethionine Se 75 (Sethotope)
Sodium Arsenate As 74
Sodium Iodide I 125 (Iodotope I-125 Injection, Iodotope I-125 Oral Solution)
Sodium Iodohippurate I 125 Injection (Hipputope I-125 Injection)
Strontium Sr 85 Nitrate (Strotope)
Triiodothyronine I 131 (T-3, Triomet)
Tritiated Water T_2O (Tritiotope)

THERAPEUTIC APPLICATIONS OF ISOTOPES:

For therapy, isotopes are used as radiation sources, not as tracers. These sources may be used either externally or internally. Their use may be summarized as follows:

External Sources
- Teletherapy sources — ^{60}Co, ^{137}Cs
- Surface sources — ^{90}Sr, ^{32}P

Internal Sources
- Infusion — ^{198}Au
- Interstitial implant — ^{192}Ir
- Selectively absorbed or concentrated — ^{32}P, ^{131}I

The therapeutic area of clinical chemistry is basically justified by the fact that radioactive material, when present in a tissue or organ in sufficient quantity, will produce emanation capable of destroying existing cells and preventing the formation of new tissue. For this reason, isotopic therapy is generally applied only to those diseases in which there exists extensive cellular metabolic malfunction or to those conditions in which an organ or tissue produces physiological harm through overactivity.

Radiotherapy, involving internal sources, is largely confined to treatment with four different radioisotopes:

1. *Radiogold ^{198}Au,* introduced as a colloidal gold suspension into a fluid-containing serious cavity, will initially diffuse rapidly throughout the fluid; it will then localize on the surface of the cavity as large aggregates of precipitate. Used in this way, it has found wide and successful use in the treatment of peritoneal and pleural effusions associated with malignant tumors in those cases in which fluid has accumulated in the abdomen or chest without the presence of large masses or of severe constitutional effects from the tumor. The tumor itself is generally destroyed only superficially, or not at all. A side effect of radiation sickness has been noted occasionally, more frequently in intraperitoneal than in intrapleural administration (1).

^{198}Au has been used experimentally in the treatment of prostate and cervix uterine carcinoma and bladder tumors. The evaluation of results is as yet incomplete (2, 3, 4).

2. *Iridium ^{192}Ir Seed Ribbons,* consisting of ^{192}Ir seeds spaced at intervals along a nylon ribbon, are used for removable interstitial implant therapy of tumors. The procedure is a surgical one which must be conducted in an operating room.

3. *Radiophosphate ^{32}P* may be used in the treatment of polycythemia vera to decrease the rate of formation of the erythrocytes. Since ^{32}P is metabolized in a manner similar to naturally occurring phosphorus, the isotope is readily distributed to all tissues and is concentrated in those tissues where proliferation is most rapid. Thus, cancerous tissues concentrate the greatest amount of the isotope. A large dose of ^{32}P — 3 to 5 millicuries — will concentrate in the bone marrow but will suppress erythrogenesis only partially. In severe cases of polycythemia a phlebotomy is necessary in conjunction with ^{32}P therapy. Colloidal chromic phosphate is used in general abdominal fluid control where the production of fluid is due to carcinomatous conditions.

^{32}P may also be utilized in the treatment of chronic granulocytic and lymphatic leukemia. This treatment, however, cannot achieve a cure, but can serve only to

Table 15.2—PROPERTIES OF ISOTOPES USED FREQUENTLY IN MEDICINE

Isotope	Radiological Half-life	Beta Energies* (MeV)	Gamma Energies (MeV)
^{3}H	12.26 y	0.018(100%)	----
^{14}C	5,770 y	0.156(100%)	----
^{22}Na	2.58 y	β^+ 0.54(89%), EC(11%)	1.28(100%), 0.51**
^{24}Na	15.0 h	1.39(100%)	1.37(100%), 2.75(100%)
^{32}P	14.3 d	1.71(100%)	----
^{35}S	86.7 d	0.168(100%)	----
^{42}K	12.4 h	3.53(82%), 2.01(18%)	1.52(18%)
^{45}Ca	165 d	0.25(100%)	----
^{47}Ca-^{47}Sc	4.5 d	0.66(83%), 1.96(17%)	1.31(16%)
(Daughter ^{47}Sc)	3.4 d	0.44(74%), 0.60(26%)	0.16(74%)
^{51}Cr	27.8 d	EC(100%)	0.32(8%)
^{59}Fe	45 d	0.46(54%), 0.27(46%)	1.10(57%), 1.29(43%), 0.19(2.8%)
^{57}Co	267 d	EC(100%)	0.122(93%), 0.137(7%)
^{58}Co	71 d	β^+ 0.48(14%), EC(86%)	0.81(100%), 0.51**
^{60}Co	5.27 y	0.31(100%)	1.17(100%), 1.33(100%)
^{64}Cu	12.9 h	EC(43%), 0.57(38%) β^+ 0.66(19%)	0.51**
^{65}Zn	245 d	EC(98.5%), β^+ 0.325(1.5%)	1.11(45%), 0.51**
^{74}As	18 d	1.36(17%), 0.69(16%) β^+ 0.90(26%), EC(38%)	0.60(63%), 0.64(16%), 0.51**
^{82}Br	36 h	0.44(100%)	0.78(87%), 0.55(70%), 0.62(43%), 1.04(30%), 0.70(29%), 1.32(28%), 0.83(26%), 1.48(18%)
^{86}Rb	18.7 d	1.77(91%), 0.7(9%)	1.08(9%)
^{90}Sr-^{90}Y	28 y	0.54(100%)	-----
(Daughter ^{90}Y)	64.2 h	2.26(100%)	-----
^{125}I	57.4 d	EC	0.035
^{131}I	8.08 d	0.61(87%), 0.33(9%) 0.25(3%)	0.36(80%), 0.64(9%), 0.28(5%), 0.72(3%)
^{192}Ir	74 d	0.67(48%), 0.53(41%), 0.26(7%), EC(3.5%)	0.296(29%), 0.31(28%), 0.60(11%), 0.61(7%), 0.48(6%), 0.59(6%)
^{198}Au	64.8 h	0.97(99%), 0.28(1%)	0.411(96%), 0.674(1%)
^{203}Hg	47 d	0.21(100%)	0.279(83%), IC(17%)
^{226}Ra	1620 y	α4.78(94.3%) α4.59(5.7%)	0.187(4%)
(^{226}Ra daughters)		Many	Many

* Unless otherwise specified, the energies given are for negatrons.
** Annihilation radiation from positron emission.
Events occuring in less than 1% of the decays have been omitted.

alleviate the symptoms of the disease. When ^{32}P is used in conjunction with local x-ray treatment, some phases of the disease may be controlled in the earlier stages.

Radiophosphorus rarely induces a side effect of radiation sickness, but excessive doses can result in serious effects on the hematopoietic system (5).

4. *Radio-iodine* ^{131}I has several therapeutic applications. In cases of hyperthyroidism, therapeutic doses of ^{131}I will destroy thyroid tissue by means of radiation produced from within the gland. This procedure provides a more desirable mode of therapy than external roentgen-ray treatment since there is less radiation danger to the surrounding tissues. Also, a much greater tissue dose can be administered. Dosage must be regulated carefully in the treatment of hyperthyroidism since too large a dose may induce hypothyroidism (6).

^{131}I is used along the same lines in the management of euthyroid cardiac diseases, including congestive heart failure and angina pectoris. The control of cardiac disease is based on the ability of the isotope to reduce thyroid activity by radiation thyroidectomy, thereby lowering the total metabolic rate of the body and thereby reducing the stress on the heart. Dosage range from 25 to 75 millicuries, given as a single dose or extended over a period of a few weeks. Some cases of thyroid carcinoma with metastasis respond to ^{131}I therapy.

DIAGNOSTIC APPLICATIONS OF ISOTOPES:

Except as a source of radiation for diagnostic x-rays (figure 15.1), isotopes are used diagnostically as radioactive tracers and not as radiation sources. If results are to be meaningful, the tagged substances must be handled by the body in a manner similar to that of the untagged substance.

Radioisotope studies may be divided into four categories. These are: (1) Isotope dilution; (2) Rate of isotope transfer; (3) Rate of isotope disappearance; and (4) Degree of isotope concentration, or metabolic rate.

1. *Isotope Dilution.* The principles of isotope dilution are discussed in experiment 12.3. The clinical application of this technique is illustrated by its use for the measurement of blood volume. The more popular procedure uses radio-iodinated human serum albumin injected intravenously (7, 8). Ten minutes after injection, a time sufficient to allow adequate mixing of the radio-iodine in the intravascular pool, yet not long enough for metabolic activity or seepage into extravascular pools to occur, a blood sample is withdrawn. The blood volume is calculated from the measured radioactivity of the injected sample of serum and the measured radioactivity of the sample of blood withdrawn. Red blood cell volume and plasma volume are related to the blood volume by the peripheral venous hematocrit. RBC volumes can also be determined by the use of cells labeled with ^{51}Cr in the form of sodium chromate.

Radioactive hydrogen ^{3}H in the form of tritiated water can be used to determine total body water (9). Total body potassium, sodium and chloride, usually referred to as "spaces," can be determined by the use of the radioactive isotopes of these elements (10-12). In the case of chloride, ^{52}Br is usually used instead of ^{36}Cl because of the long half-life of the latter (13).

2. *Rate of Isotope Transfer.* In these procedures, a labeled substance is injected into one part of the vascular system and the time required for its arrival at another

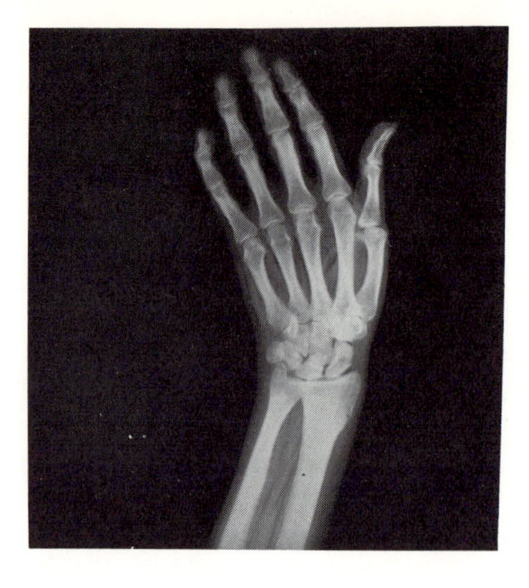

Figure 15.1—USE OF ISOTOPES FOR RADIOGRAPHY. Iodine-125 x-rays were used in producing this radiograph. The use of radioisotopes in place of heavy electronic equipment results in a light-weight portable x-ray unit. (*Courtesy W. G. Meyers, The Ohio State University Health Center and The Ohio State Medical Journal, Vol.* 58, *July* 1962.)

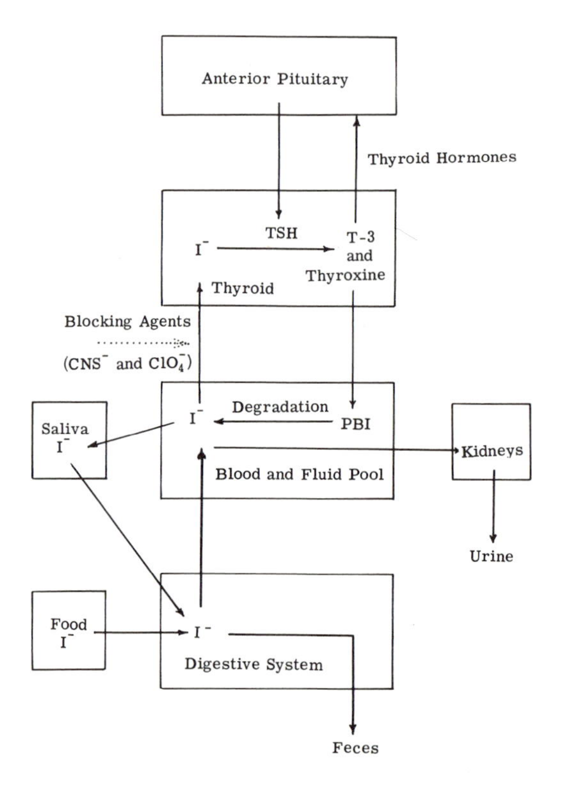

Figure 15.2—SIMPLIFIED SCHEMATIC OF IODINE CYCLE.

part is determined. This technique has been used widely to determine circulation times, especially in the extremities. ^{24}Na is well suited for this purpose, since it has a short half-life, is a normal body constituent, is not selectively absorbed by any tissue and is readily detected (14, 15).

An extension of these methods has been used in the measurement of cardiac output. The passage of a radioisotope through the heart and lungs, following intravenous injection, can be recorded either by assay of serial arterial blood samples or by external counting over the heart. It is important that the radioactive material does not diffuse into the tissues during the studies. Radio-iodinated serum albumin has been found most satisfactory; measurements are possible with as little as 25 microcuries injected intravenously (16.17).

3. *Rate of Isotope Disappearance.* The rate at which an isotope disappears from a tissue into which it has been injected is a measure of the circulation in that tissue. This test has been used successfully to determine the extent of circulation in tubed skin grafts in plastic surgery (18, 19). A small amount of an isotope, e.g., radiosodium chloride, is injected directly into the tissue. The disappearance rate of the isotope is measured by means of a counter placed directly over the site of injection.

The RBC destruction mechanism and RBC half-life are measured by means of a disappearance-rate technique. If erythrocytes are labeled in vitro with ^{51}Cr and then re-injected, the fate of the tagged cells can be followed by assay of serial blood samples taken every two or three days for at least two weeks. Since the labeled cell group contains cells of all ages, only a mean survival time can be determined for these cells. The rate of decrease in circulating ^{51}Cr is approximately exponential and can thus be characterized by a half-clearance-time. The normal RBC half-life is about twenty-six days. This study is a valuable aid in the diagnosis of hemolytic anemias.

4. *Isotope Concentration or Metabolic Rate.* Most of the more familiar radioisotope studies are in this category. The concentration of a particular radioisotope in normal or abnormal tissue or in an organ provides data from which the function of the tissue or the metabolic condition of the organ can be evaluated.

Several studies of thyroid function can be carried out with the aid of radioactive ^{131}I. These studies include: (1) The rate of deposition of iodine in the gland *in vivo;* (2) The total accumulation of iodine in the gland within a specified period of time; and (3) The output of thyroid hormone into which radioactive iodine has been incorporated.

Because the thyroid has such an avidity for iodine, precautions must be taken to prevent exposure to or ingestion of even small amounts of iodine by the patient if valid results are to be obtained. Treatments to be avoided include the external application of iodine, ingestion of iodine-containing medicaments (these may influence thyroid uptake for weeks), the use of x-ray contrast media containing organic iodine compounds (these may produce an effect on the thyroid for months), and myelography and bronchography (these may have a permanent effect). The radio-iodine thyroid uptake may, in addition, be lowered by such substances as thyroid preparations, antithyroid drugs, thiocyanate and perchlorate, corticotropin and corticosteroids, phenylbutazone, sulfonamides and p-aminosalicylic acid, arsenic, lead and mercury. Malabsorption syndromes and renal disease will also lower thyroid uptake.

For *thyroid uptake* the most commonly employed technique involves the oral administration of from 5 to 25 microcuries of $Na^{131}I$, followed by a measure of the radioactivity of the thyroid after the elapse of a given time. The most reliable and widely used test remains the 24-hour uptake, with the 48-hour uptake providing a useful double check. Normal uptake values vary from clinic to clinic, but usually range from 15% to 45%. Some laboratories use a 6-hour uptake.

The *thyroid clearance* test measures the rate of clearance of ^{131}I from the plasma. First, a small amount of the isotope is given intravenously, then the rate of uptake over the thyroid gland is measured for thirty minutes. At the end of this time, the urinary excretion of ^{131}I is measured. The iodine collected by the thyroid per minute is divided by the average plasma concentration of ^{131}I (μCi/ml) during the elapsed time. The result is thyroid clearance in ml/min. The normal clearance is about 25 ml/min. In hyperthyroidism the value may rise to 250 ml/min., while for hypothyroidism the clearance may fall to about 2 ml/min.

The *protein-bound iodine conversion ratio* is a measure of thyroid activity. It is the fraction of inorganic iodide converted to thyroid hormone and bound to the plasma proteins in 24 hours. Sodium radio-iodide is given by mouth and after 24 hours the protein-bound ^{131} iodine ($PB^{131}I$) and the total ^{131}I in the plasma are determined. The ratio of $PB^{131}I$ to total ^{131}I is the conversion ratio. Normal values are usually in the range of 13% to 42% with hypothyroidism being indicated by values below and hyperthyroidism by values above these limits.

Thyroid activity can be measured by two *in vitro* methods in which ^{131}I in the form of labeled triiodothyronine (T-3) is bound to the red cells or to plasma proteins. The first of these is the Hamolsky T-3 RBC uptake (20) which measures the percentage of T-3 absorbed on the surface of the cells. The uptake is increased in hyperthyroidism and decreased in hypothyroidism. Simplified methods have been developed (21). A newly developed Sephadex method (22) determines the capacity of plasma proteins for binding with labeled T-3. A high capacity of plasma proteins indicates hypothyroidism; a low capacity indicates hyperthyroidism. In contrast to *in vivo* studies with ^{131}I, test results with T-3 appear to be but little affected by exogenous iodine or iodides, anxiety, hypertension, congestive heart failure and polycythemia.

One method of isotope concentration is designed to detect pernicious anemia and to differentiate it from other macrocytic anemias by the use of vitamin B_{12} labeled with ^{60}Co, ^{58}Co and ^{57}Co. Pernicious anemia is associated with the absence of the intrinsic factor of Castle. The absence of this factor also prevents the absorption of vitamin B_{12}. Therefore, in pernicious anemia there will always be low excretion of B_{12}in the urine. A urinary excretion in 24 hours of less than 5% of an administered dose of labeled B_{12} is suggestive of pernicious anemia. After the administration of intrinsic factor, the urinary excretion of labeled B_{12} will increase from two-fold to eight-fold in a patient with pernicious anemia. In normal individuals and in patients with steatorrhea or other gastrointestinal disorders, this increase will not occur (23).

Iron-deficiency anemias can be studied by oral administration of ^{59}Fe and subsequent determination of ^{59}Fe excreted in the stools and incorporated in the red cells. ^{59}Fe may also be given intravenously and the plasma clearance rate or RBC uptake then determined. In hypochromic anemias the RBC uptake and plasma clearance are rapid; in aplastic anemias they are slow (24, 25).

^{14}C has been used in many metabolic studies, such as those involving cholesterol and steroids. The use of radiocarbon and other radioisotopes for intermediary metabolism studies is for the most part experimental (26, 27). Limitations on the usefulness of tagging have been mentioned previously (Chapter 13) and should be considered for accurate measurement (28).

Radio-iodinated triolein is used for the measurement of fat digestion and absorption. The thyroid is first blocked with Lugol's solution. The patient is then given a test meal followed by oral administration of the labeled fat. Venous blood samples are drawn at intervals up to 24 or 48 hours to prepare a curve of plasma radioactivity. Levels of 12% are normally obtained in six hours. Fecal radioactivity may also be calculated. The normal individual excretes less than 2% of the administered labeled fat in the 48 hours following injestion (29-33). The rate of gastric emptying and gastrointestinal mobility will influence excretion and absorption patterns. In malabsorption states plasma levels are lower and fecal excretion is increased. Abnormal patterns have been described in individuals with coronary artery disease (34, 35).

Liver function is measured by intravenous injection of Rose Bengal labeled with ^{131}I and external counting over the area of the liver. Some laboratories measure blood disappearance of the ^{131}I-dye rather than direct liver uptake. Normally, radioactivity over the liver reaches a peak 15 to 20 minutes after injection and falls off gradually as excretion takes place. In liver disease there is a decreased uptake of the dye by the liver. In biliary obstruction, the uptake remains normal in the absence of parenchymal involvement, but the excretion rate is diminished (36). Liver function has also been studied by measurement of the rate of excretion of various labeled radioactive materials such as iodipamide (37).

In certain types of hemolytic anemias, red cells disappear rapidly from the blood stream, being trapped and eventually destroyed by the spleen. The extent of RBC uptake by the spleen can be determined by tagging the cells in vitro with ^{51}Cr, reinjecting them intravenously, counting externally over the spleen and liver and calculating the ratio of spleen: liver radioactivity. A high ratio, associated with decreased RBC survival, may indicate the need for a splenectomy (33).

The use of radio-iodinated hippuric acid provides an advantageous method for the evaluation of renal function, since this compound is rapidly and selectively absorbed and excreted by the kidneys. The isotope is given intravenously and excretion from the renal area is measured externally by means of a pair of radiation detectors. The radioactivity of each kidney is recorded graphically to provide a pattern of the renal function.

^{32}P has been of value in the detection and delineation of eye tumors (39, 40). Yet when used in attempts to detect and localize gastrointestinal, gastrourinary, pulmonary and breast tumors, the results have not been too favorable. Experiments have been undertaken recently to localize prostatic metastases and lesions by use of ^{32}P. The isotope is administered orally and the radioactive uptake by the prostate is determined by means of an internal radiation detector (41). Some real palliation is observed in some cases from ^{32}P therapy. In the field of brain tumors, the isotope may be administered preoperatively and at surgery the marginal limit of the tumor may be delineated by the use of detectors in the brain while the skull is open (42).

Scanning techniques—Where standard x-ray techniques fail to provide sufficient information because tissues do not exhibit sufficient contrast, it is often possible to delineate an organ or to localize a tissue by radioisotopic-scanning techniques. To make a scanning technique feasible, a radioisotope must be preferentially absorbed by the tissue.

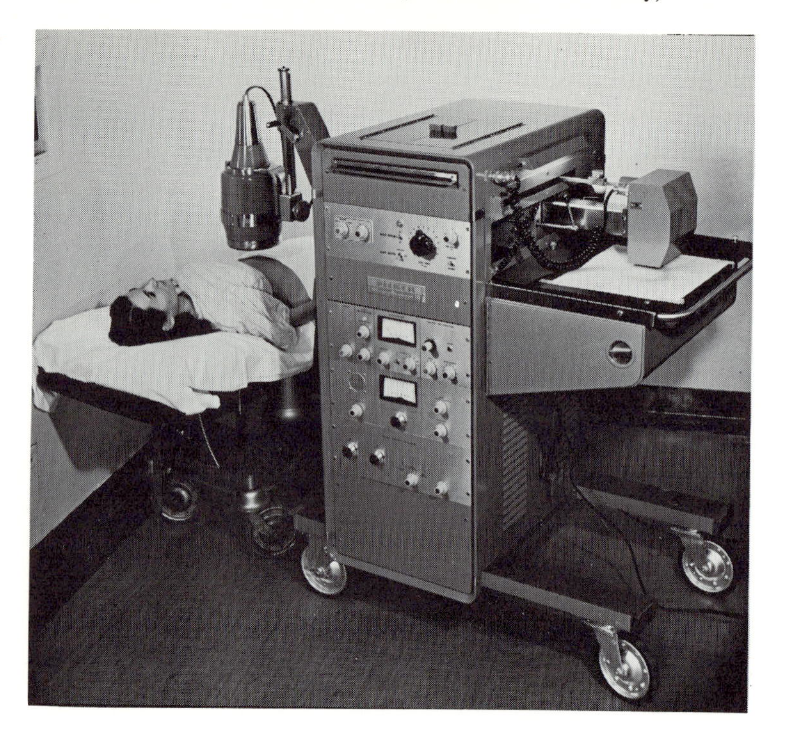

Figure 15.3—SCANNER. (*Courtesy Picker X-ray Corp.*)

A scanner (figure 15.3) consists of a sensitive, collimated detector or probe which is mechanically coupled to a recording device. As the probe moves back and forth over an area of the patient's body, the radioactivity detected by the probe is indicated on the recording by a pattern of dots made by a stylus. A dot scan of the thyroid of a patient who has received radioactive ^{131}I is shown in figure 15.4. In more recent scanners recording is also accomplished by the action of a light beam on sensitized paper, the intensity of the light being controlled by the level of radioactivity.

Successful brain tumor localization has been accomplished recently by use of chlormerodrin ^{203}Hg (neohydrin ^{203}Hg), also used for kidney scanning. For brain tumor detection a dose of meralluride is given first to minimize renal uptake of the labeled chlormerodrin (43). The latest technique for brain tumor localization employs technetium (^{99m}Tc) as the pertechnitate. By use of suspensions of ^{131}I-radioalbumin (macroaggregates), Taplin and associates (44) have successfully visualized the lung and spleen.

ESTABLISHMENT OF A MEDICAL RADIOISOTOPE PROGRAM:

Types of Programs—There are two types of clinical radioisotope programs:

(a) Institutional program sponsored by a medical organization or hospital and carried out under the guidance of a medical isotope committee, and

(b) Private medical practice programs.

Among the latter, a distinction may be made between private practice programs confined primarily to the use of radioactive materials in a private office and private

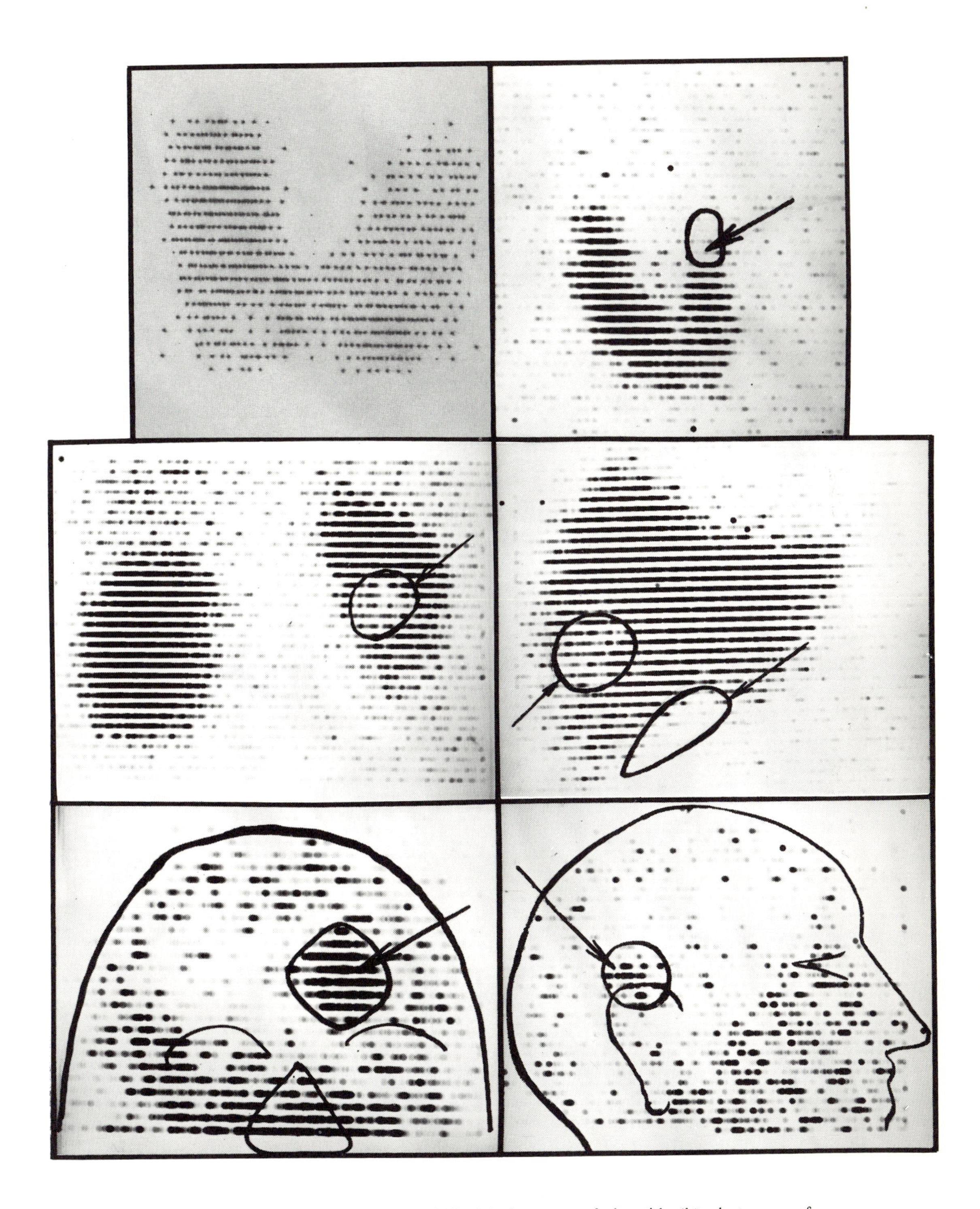

Figure 15.4—TYPICAL SCANS. (a) dot scan of thyroid; (b) photoscan of thyroid; (c) photoscan of kidney; (d) photoscan of liver; (e) photoscan of brain; (f) photoscan of brain. Arrows point to malignancies.

practice programs where the materials are used within a medical facility. For radiological safety patients receiving other than a tracer dose of radioactive material must be hospitalized. In most cases physicians have found it desirable to carry out their treatments within a hospital rather than in their private offices. The responsibility still remains with the individual physician, however, the hospital merely providing the facilities.

A number of publications are available which are of great value to institutions organizing a radioisotopes program (45-48).

Personnel—Radioactive compounds used in humans for therapy and diagnosis are radioactive drugs and should be treated as such. The administration of any drug, including radioactive ones, to humans, must be supervised by a physician specially trained in their use. Thus to organize a medical radioisotope program there must be at least one qualified physician. His training must be ample to satisfy the AEC that he can use the radioactive material properly and safely for the particular procedures proposed. Suggested minimum experience requirements for the physician for specific diagnostic and therapeutic procedures have been published by the AEC (45).

The organization of an isotope committee is highly recommended. Membership on the committee may include physicians, physicists, chemists, pharmacists, business administrators or others as required for the proper functioning of the committee. One member of the isotope committee should be designated as *Health Officer.* His specific responsibility is to see that all radioisotopes are handled with minimum hazard to personnel and that all regulations governing the use of radioisotopes are observed.

Program—One of the first duties of the isotope committee is to outline the radioisotope program for the hospital. Although hospital programs may differ, one from the other, they all have certain common problems. Some of the subjects to be considered by the committee are:

Diagnostic or therapeutic procedures to be performed

Facilities and work areas required (treatment room, laboratory, counting area, etc.)

Instrumentation (selection, cost)

Licensure and reports (AEC, state, city)

Records (Licenses, isotopes ordered and received, disposition of isotopes, patients records, personnel safety records)

Isotopes (procurement, storage, waste disposal)

Facilities and Work Area—An isotope storage area, a laboratory for the manipulation of isotopes and the preparation of prescribed dosage forms, a counting area for calibration of the dose and a treatment room are among the facilities and work areas which should be considered. With the availability of pre-calibrated dosage forms of most radiopharmaceuticals, elaborate facilities are not required for many diagnostic procedures. An existing, standard chemical laboratory will normally provide all necessary facilities for the preparation and handling of small amounts of radioactive drugs.

In some hospitals radiopharmaceuticals are dispensed by the hospital pharmacist (49). Non-radioactive pharmaceuticals, including stable forms of most of the radiopharmaceuticals listed in Table I, have been dispensed by pharmacists for years. Adding the property of radioactivity to these substances modifies them but does not cause them to be less of a drug than they were before. It therefore seems quite logical

Figure 15.5—MEASUREMENT OF A RADIOACTIVE SOLUTION.

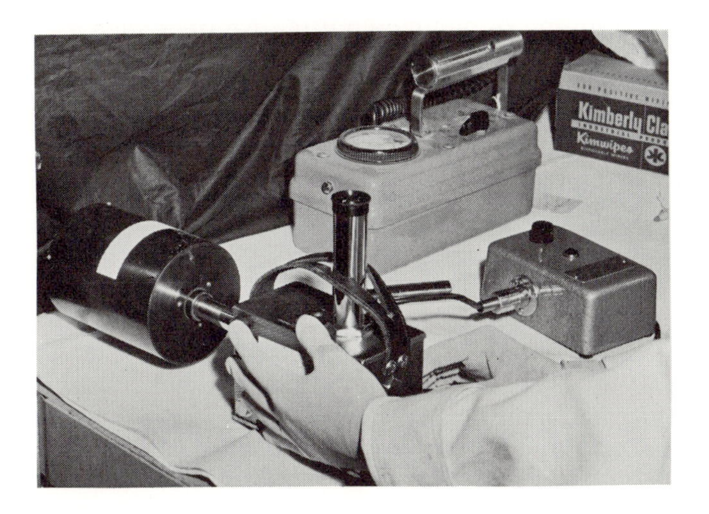

Figure 15.6—THE BRAESTRUP CHAMBER. This radiation detector is an ionization chamber useful for checking the calibration of radioisotope shipments.

that these products be dispensed by the hospital pharmacist where he has taken the initiative to acquire the necessary knowledge and training.

In the selection of the counting area, the location of x-ray equipment and other radiation sources must be considered lest the background in the counting area be erratic or excessively high.

Instrumentation—The majority of clinical studies and analyses can be performed with the aid of but a few different types of radiation detection equipment. Of these the scintillation detector will usually be found most useful.

Because the 24-hour radio-iodide thyroid uptake is one of the routine clinical procedures most commonly employed, the instruments necessary to carry out this study may be considered essential equipment for a clinical radioisotope laboratory. The necessary components of the system are an adjustable gamma scintillation probe attached to a scaler equipped with an elapsed-time clock. With this system the uptake of gamma-emitting isotopes can be measured in any organ of the body. The uptake of ^{131}I by the thyroid and the uptake of ^{51}Cr-labeled red blood cells by the spleen are examples of techniques possible with this basic equipment.

Use of a scintillation counter is greatly enchanced by a pulse-height analyzer. The resulting spectrometer can be "tuned" to a particular isotope. Such a system not only allows the detection of one isotope in the presence of others but also permits the use of a smaller amount of radioactive material in a diagnostic procedure because of the significant reduction in background and the improved statistics resulting therefrom. A further advantage to the use of a spectrometer, again resulting from the possible high selectivity or "tuning" effect, is the resulting discrimination against Compton scattered radiation which could otherwise cause serious errors in the measurements, especially should changes in the counting geometry occur. With a gamma-ray spectrometer, studies utilizing ^{51}Cr, ^{59}Fe, ^{60}Co, ^{131}I and ^{137}Ca, as well as other isotopes, may be performed.

For the *in vitro* counting of gamma-emitting isotopes such as those above, a well-type detector will provide detection efficiencies as high as 60% or more. Accordingly, the addition of a well counter to the basic scintillation system, or better, to a scintillation spectrometer system, will provide the necessary instrumentation for performing numerous other diagnostic procedures. Included are blood volume measurements as well as other isotope dilution techniques, RBC survival time, test of gastrointestinal bleeding, PBI conversion ratio, triiodothyronine *in vitro* uptake studies, Schilling test, fat absorption and others.

Measurements to determine the relative concentration of an isotope throughout an organ can be accomplished by means of scintillation scanning. A scintillation probe moving evenly and uniformly over the radioactive site detects differences in activity concentrated in the organ. The impulses detected by the probe are amplified and made to modulate a small beam of light which is directed onto a sheet of x-ray film. Since the detector and the light are connected mechanically, visible evidence of the pattern of radiation concentration in the area under examination is produced in the form of a photoscan. Dense regions in the scan, produced by greater exposure to the light, are indicative of regions of high activity. During the recording of the film scan, a mechanical dot scan can also be produced by means of a stylus which transmits the pattern of activity detection to a sheet of electrically sensitive paper. Variations in activity may be observed throughout the scanning procedure by means of a ratemeter and an audio signal. A scintillation scanner such as this permits accurate mapping of radioactivity uptake in such organs as the thyroid, brain, liver, kidneys, spleen and cardiac blood pool (50, 51).

Renal function may be evaluated by means of radioactive renograms. A useful instrument for this study is a dual-probe kidney scintillation scanner. By the intravenous administration of ^{131}I-labeled hippuric acid and the use of a scintillation probe

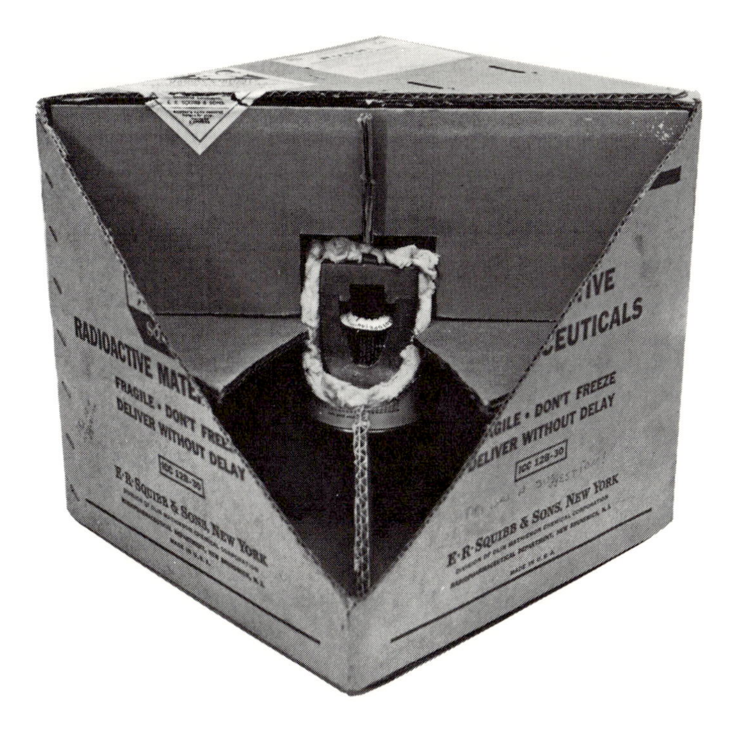

Figure 15.7—CUT-AWAY VIEW
OF A SHIPPING CONTAINER.

positioned directly over each kidney, we may record the radioactivity of each kidney by means of a dual ratemeter and a dual recorder.

Licensure and reports—Until recently all users of radioisotopes in the United States were required to obtain a license for their use from the Atomic Energy Commission. This is still true in most states. Application for license may be made on forms 313 and 313A which are available from the Division of Licensing and Regulation, United States Atomic Energy Commission, Washington 25, D.C.

Effective March 26, 1962, the Commonwealth of Kentucky by mutual agreement assumed certain of the Atomic Energy Commission's regulatory authority in that state. Individuals or organizations desiring to use radioisotopes in Kentucky should contact the State Commissioner of Health, Frankfort, Kentucky. Similar agreements have been made with the State of Mississippi, effective July 1, 1962, and the State of California, effective September 1, 1962. Recently the State of New York submitted a proposal to the AEC for a similar program.

While the above state programs replace AEC regulations, most states have radiation control programs which supplement those of the AEC. State regulations frequently require registration of radiation sources. Similar registration is desirable with the local municipality, especially with fire and police officials, even though not always required.

PROCEDURE 15.1 Blood Volume (Radio-Iodinated Serum Albumin Method)

RATIONALE:

Plasma volume is determined by means of an isotope dilution technique using radio-iodinated serum albumin. If the hematocrit (HCT) is measured, both the blood volume and the red cell volume can be calculated.

*I*odinated ^{131}I *H*uman *S*erum *A*lbumin (IHSA) is prepared by mild iodination of human serum albumin with iodine-131, the iodine adds to the tyrosine in a position ortho to the hydroxyl group. IHSA should be stored at about 5° C.

Within about 10 minutes after an intravenous injection of IHSA it has become thoroughly mixed with the circulating blood. The ^{131}I tag of the IHSA does not transfer to other protein molecules as the injected iodinated albumin is diluted by the normal protein pool of the plasma. After about an hour approximately 10 per cent of the iodinated albumin is lost from the plasma. During the first week approximately 25 to 50 per cent of the radioactivity is eliminated (52-60).

PROCEDURE:

Preparation of Diluted IHSA Solution

1. From the concentration (μCi/ml) of the commercial stock IHSA solution, calculate the volume (A) required to provide 10 microcuries of activity.

2. By means of sterile technique, dilute a volume A of stock IHSA solution to 12 ml with isotonic saline solution. Mix well. Exactly 10 ml (B) of this solution will be injected; the remaining 2 ml (C) will be used for preparation of the standard.

Determination of Hematocrit and Radioactivity Blank

3. Using a heparinized syringe, 8-10 ml of blood is withdrawn from a vein.

4. Part of this blood is used to measure the hematocrit (HCT); the rest of the blood is centrifuged.

5. Exactly 2 ml of plasma from the centrifuged sample is transferred to a counting tube and the activity is measured in a well counter. This is the background count or plasma blank R_{pb} which must be subtracted from all other measurements of radioactivity on plasma. This blank also serves as a check for residual activity from previous tests.

Administration of Diluted IHSA Solution

6. Exactly 10 ml of diluted IHSA solution (B from step 2) is injected into the antecubital vein. To assure complete delivery of the radioactive solution, wash the needle and syringe thoroughly by drawing a few ml of the patient's blood into the syringe and reinjecting without removal of the needle from the vein.

Withdrawal of Blood Samples

7. Ten minutes after injection, a 5-7 ml blood sample is taken with a heparinized syringe.

8. A hematocrit is determined and the balance of the blood is centrifuged.

9. An accurately measured 2 ml aliquot of plasma is transferred to a counting tube and the radioactivity of the sample is measured in a well counter. The plasma blank R_{pb} is subtracted to give the net sample activity R_{10}.

10. Twenty minutes after injection, steps 7 through 9 are repeated to provide the twenty minute sample activity R_{20}.

11. One ml of diluted IHSA solution (from C of step 2) is diluted to 100 ml with water. Use a volumetric flask for the dilution and mix well. This is solution D.

12. A 2 ml aliquot of solution D is transferred to a counting tube and the radioactivity R_d is measured in the well counter.

13. Into another counting tube is placed 2 ml of water (the same supply of water used in the preparation of solution D should be used) and the activity R_w is measured. The net activity of the standard R_s is the difference of R_d and R_w, that is, $R_s = R_d - R_w$.

CALCULATIONS:

As a solution is diluted, the concentration will vary inversely with the volume. If changes in plasma volume due to the injection of IHSA solution and the removal of blood for testing are neglected, then, for the ten minute test,

$$\text{Plasma volume} = \frac{B\,(100\,R_s)}{R_{10}} = \frac{1000\,R_s}{R_{10}}$$

Plasma volume is also calculated by use of the values for the twenty minute blood sample. Both the ten and twenty minute results should be similar. For accuracy, extrapolation to zero time may be desirable so that correction is made for losses of IHSA from the circulation.

$$\text{Total blood volume} = \frac{\text{Plasma volume}}{1 - (\text{HCT} \times 0.915)^*}$$

$$\text{Red cell volume} = \text{Total blood volume} - \text{Plasma volume}$$

*0.915 is the correction to be applied for trapped plasma in the red cells and for the difference between the venous HCT and total body HCT.

NORMAL VALUES:

Plasma volume	35-45 ml/Kg;	1300-1700 ml/m^2
Total blood volume	65-85 ml/Kg;	2500-3200 ml/m^2
Packed red cell volume	30-40 ml/Kg;	1150-1550 ml/m^2

PROCEDURE 15.2 Blood Volume (Sodium Radiochromate Method)

RATIONALE:

Blood volume is determined by an isotope dilution technique which uses the patient's own red blood cells (RBCs) tagged with Sodium Radiochromate ^{51}Cr Injection U.S.P.

When $Na_2{}^{51}CrO_4$ is incubated with RBCs, approximately 70-95% of the ^{51}Cr is bound to the globin of the erythrocytes. Only the RBCs are labeled; the plasma is not. Exchange of the chromium occurs only slowly or not at all and, as the RBC's are destroyed, the liberated ^{51}Cr apparently is not reutilized. It appears that only hexavalent chromium enters the cells where it is reduced and becomes fixed in the trivalent state. If one begins with trivalent chromium, the cells will not be tagged (54-60).

PROCEDURE:

Preparation of ^{51}Cr-Tagged RBCs

1. Approximately 20 ml of blood, withdrawn from the patient's antecubital vein, is injected aseptically into a pyrogen-free vial containing 10 ml of ACD solution (Acid Citrate Dextrose, an anticoagulant). The solution is mixed gently.

2. About 75μCi of Sodium Radiochromate ^{51}Cr Injection USP is added to the solution containing the blood and the mixture agitated by gentle swirling.

3. The container and its contents are incubated at 37°C for 20 minutes or at room temperature for 40 minutes.

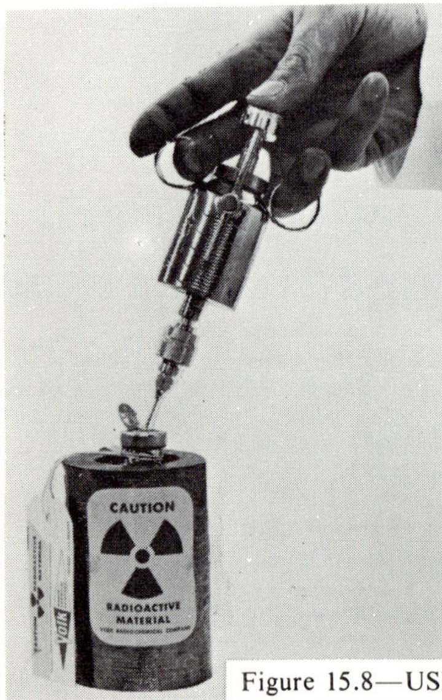

Figure 15.8—USE OF A SHIELDED SYRINGE FOR MEASUREMENT AND ADMINISTRATION OF RADIOACTIVE MATERIALS.

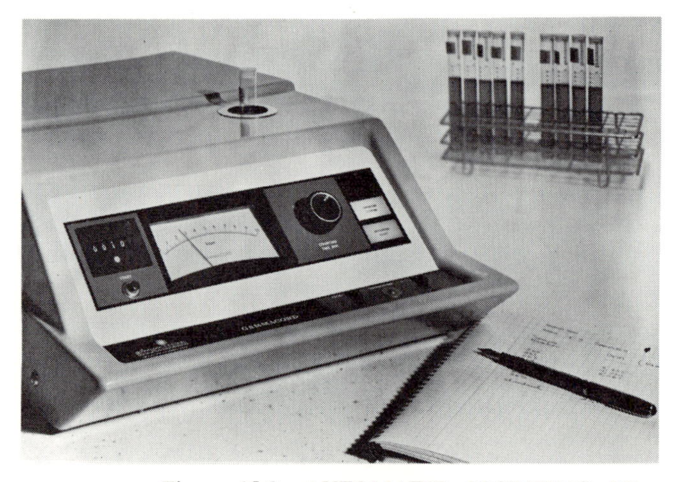

Figure 15.9—AUTOMATIC COUNTING DEVICE FOR MEASUREMENT OF BLOOD VOLUMES (*Courtesy Atomium.*)

4. About 50-100 mg of Sodium ascorbate (as a sterile aqueous solution) is added to the incubated mixture to reduce the uncombined chromate ions to chromic ions. This procedure "tags" the hemoglobin in the RBCs; the reduced chromium is removed rapidly from the blood by the kidneys.

Administration of ^{51}Cr-Tagged Blood

5. Exactly 10 ml of tagged blood (the mixture from step 4) is injected into the antecubital vein. To assure complete delivery of the radioactive solution, a few ml of the patient's blood are drawn into the syringe and reinjected without removal of the needle from the vein.

6. A 10-ml sample of blood is then withdrawn with a heparinized syringe, 15-20 minutes post injection. This sample should be taken from the opposite arm.

Procedure with Remaining Chromated Blood (from step 4)

7. A hematocrit is determined. This is HCT_1.

8. One ml of whole blood from step 4 is diluted to 100 ml with distilled water in a volumetric flask. A 2-ml aliquot of this solution is transferred to a counting tube and the activity measured in a well counter. This is activity R_a.

9. A portion of the blood from step 4 is centrifuged and the plasma removed. A 1 ml aliquot of the plasma is diluted to 50 ml in a volumetric flask with distilled water. A 2-ml aliquot of this dilution is transferred to a counting tube and the activity measured in a well counter. This is activity R_b.

Measurements on the Post-Injection Blood Sample (from step 6)

10. A hematocrit is determined. This is HCT_2.

11. A 2-ml aliquot of the whole blood is transferred to a counting tube and the activity measured in a well counter. This is activity R_c.

12. A portion of the remaining blood is centrifuged. A 2-ml aliquot of the plasma is transferred to a counting tube and the activity measured in a well counter, as above. This activity is R_d.

CALCULATIONS:

Calculations on Chromated Blood (Note: HCT is expressed as a decimal)

A = cpm/ml of blood = $(100\ R_a)/2$
B = cpm/ml of plasma = $(50\ R_b)/2$
C = cpm in plasma/ml of blood = $B\,(1 - HCT_1)$
D = cpm on cells/ml of blood = A − C
E = Activity injected on the cells = D (volume injected) = 10 D
F = % Uptake by red cells = (100 D)/A

Calculations on Post-Injection Samples

G = cpm in plasma/ml of blood = $R_d(1 - HCT_2)/2$
H = cpm on cells/ml of blood = $(R_c/2) - G$
I = cpm/ml of red cells in sample = $H/(HCT_2)$
J = Red Cell Volume = E/I
K = Total Blood Volume = $J/(0.91\ HCT_2)$
L = Plasma Volume = K − J

PROCEDURE 15.3 Red Cell Survival Time (Sodium Radiochromate Method)

RATIONALE:

We know that blood cells live but a limited time and perish constantly in great numbers, even in normal individuals. This destruction of cells, known as phagocytosis, has been noted in the liver, spleen and bone marrow. We also know that cells undergo hemolysis and fragmentation in the blood. By tagging the blood cells radioactively we can measure their rate of destruction.

The patient's own cells are withdrawn, tagged with ^{51}Cr and reinjected. The unbound ^{51}Cr is present as $^{51}CrCl_3$ in the plasma and is essentially cleared from the plasma by excretion within 24 hours.

The radioactivity of the patient's blood is measured for a period of several weeks in a series of blood samples. The time required for the radioactivity of the blood to be reduced to one-half the initial value (after total excretion of the plasma $^{51}CrCl_3$) is the *half-life value* or the *half-survival time* of the blood cells. If all blood samples are saved and counted at the same time, no correction is required for the radiological half-life of ^{51}Cr.

Although the ^{51}Cr is tightly bound to the red cells there is a constant elution loss of about 1% per day. A correction for this loss must be applied for more accurate results (61-65).

PROCEDURE:

Preparation of ^{51}Cr-Tagged RBCs

1. to 4. Follow steps 1 to 4 as outlined in Procedure 2—Blood Volume.

Administration of ^{51}Cr-Tagged Blood

5. Twenty (20) ml of tagged blood (the mixture from step 4) is withdrawn from the vial with a sterile needle and syringe and injected directly into the antecubital vein. One may rinse the syringe by drawing a few ml of the patient's blood into the syringe and reinjecting the blood without removal of the needle from the vein, although a quantitative delivery of the radioactivity is not required.

6. Heparinized blood samples (5 ml) are withdrawn 24 hours post injection and every 48 hours thereafter, until sufficient samples have been accumulated to establish the trend.

Measurements on the Withdrawn Blood Samples

7. A hematocrit (HCT) is determined on each blood sample withdrawn. (HCT is expressed as a decimal).

8. The radioactivity of each sample is measured on a 2 ml aliquot of whole blood in a plastic test tube. Measurements should be made in a well-type scintillation detector with the pulse-height analyzer adjusted to the 0.32 MeV photopeak. If all samples are counted on the same day (after the final blood collection), corrections for radiological decay need not be made.

CALCULATIONS:

The measure of survival of the red blood cells is expressed as the per cent of tagged RBCs remaining in the circulation. The 24-hour blood sample is used as the 100% level.

24-Hour Post-injection Sample

A = net cpm = gross cpm − background cpm

B = net cpm/ml blood = A/2 ml

C = 100% level = cpm/ml of red cells in the sample = B/HCT

Post-injection Samples

(Unless all activities, including the 100% reference activity, are measured on the same day, a correction must be applied for the radiological decay of ^{51}Cr.)

D = net cpm = gross cpm − background cpm

E = net cpm/ml blood = D/2 ml

F = cpm/ml of red cells in sample = E/HCT

Per-cent Survival

% Survival = 100E/C

Half-Survival Time

Data are plotted on semi-logarithmic paper. From the slope of the curve the time required for one-half of the ^{51}Cr activity to disappear from the blood is calculated. This is the half-survival time.

Normal range from 22-23 days, with an average of about 28 days.

PROCEDURE 15.4 Sequestration of Red Blood Cells (Chromium-51 Scanning Technique)

RATIONALE:

Erythrophagocytosis has been observed in the liver, spleen and bone marrow. In certain pathologic conditions in which the destruction of red cells is accelerated, erythrophagocytosis in the spleen has greatly increased. This condition can be detected through the use of radioactively tagged red cells. The cells are tagged with ^{51}Cr and the principal site of sequestration or destruction of red blood cells is determined by external counting over the spleen and the liver (66-68).

PROCEDURE:

Tagging the RBCs with ^{51}Cr

If we are measuring blood cell survival time (Procedure 3), we can determine the sequestration site of the blood cells at the same time simply by making the necessary measurements of activity of the spleen and the liver. Otherwise, it is necessary that we first tag the blood cells as outlined in steps 1 to 5 of Procedure 3.

Measurement of Liver and Spleen Activities

At least 24 hours should elapse before measurements are made of the activities of the liver and spleen in order to allow for the accumulation of phagocytic products in these organs. Activity measurements are made externally by means of a crystal scintillation counter. The detector should be collimated so that it may define the area of activity. Discrimination should be provided by a pulse-height analyzer adjusted to the 0.32 MeV photopeak of ^{51}Cr. A constant source-to-detector distance must be maintained for an accurate comparison of liver and spleen activities. We can then measure the relative activities of the liver and spleen after careful positioning of the detector over each organ in turn.

INTERPRETATION:

We can calculate the ratio of activities, after correcting for background, by dividing the corrected activity of the spleen by the corrected activity of the liver. Representative values for the ratio, spleen:liver, are given below:

Ratio of Activities (*spleen:liver*)	Indication
1:1	Normal
2:1	Splenomegaly
3:1 and 4:1	Hemolytic anemia

PROCEDURE 15.5 Gastro-Intestinal Bleeding

RATIONALE:

Our purpose in this study is to detect and measure any loss of blood by way of the gastro-intestinal tract. This procedure may be performed concurrently with the red cell survival study (Procedure 3).

The red cells are tagged with ^{51}Cr. Gastro-intestinal bleeding will be manifest by the appearance of radioactivity in the stools and, in severe cases, by a decrease in the blood cell survival time. From quantitative measurements of blood and stool activities, the volume of blood in the stool can be calculated. A loss of blood up to 2 ml per day is considered normal (69-73).

PROCEDURE:

Tagging the RBCs and Collecting the Samples

1. From 25 to 30 ml of whole blood is tagged with ^{51}Cr by use of the technique described for the red blood cell survival study (Procedure 3). This blood is re-injected into the patient.

2. Heparinized blood samples are collected, 24, 48 and 72 hours after administration of the tagged blood.

3. All stool samples are collected for 72 hours.

Treatment of Samples

Stool samples are compared with blood samples from the same day. For each day's samples, calculations are made as follows:

4. A-2 ml aliquot of blood is transferred to a plastic test tube and counted in a well counter.

5. Stools for a given day are weighed and then a weighed aliquot representing a uniform sampling is transferred to a Waring blendor with water in the following proportion:

Stool Weight		*Water*
50— 75 g	add	100 ml (or 100 g)
80—150 g	add	200 ml (or 200 g)

A 2-ml aliquot of homogenate is transferred to a tared test tube for radio assay in a well counter. The weight of the 2-ml aliquot is recorded. (Weight is used in the calculation rather than volume so that we may eliminate errors due to incorporated air.)

CALCULATIONS:

6. All measured activities are corrected for background and for decay. (If corresponding blood and stool samples are counted at the same time a correction for decay is not necessary.)

7. Activity per ml of blood is calculated (= observed activity/2).

8. The ml of blood excreted in the stool is computed by use of the following formula:

$$\text{ml blood excreted in stool} = \frac{AB(C+D)}{CEF}$$

where: A = Activity of 2-ml aliquot of homogenized stool (cpm)
B = Total stool weight for day (g)
C = Weight of stool homogenized (g)
D = Weight of water added to homogenize (g)
E = Weight of 2-ml aliquot of homogenized stool (g)
F = Activity per ml of blood (cpm/ml)

PROCEDURE 15.6 Thyroid Uptake

RATIONALE:

The thyroid gland concentrates inorganic iodides from the blood and converts them to organic-bound iodine compounds (iodo-tyrosines and thyronines) through the action of peroxidase enzymes. When a thyroid gland is relatively "iodine starved," i.e., is receiving no more iodine than is found in the normal "inland" diet, the administration of a small dose of radioactive iodine results in a portion of the dose being retained by the thyroid while the remainder of the radioactive isotope is excreted in the urine. The amount of radioactive iodine retained by the thyroid is an index of the thyroid function (74-81).

In the following test, the patient is given an oral dose of carrier-free Sodium Radio-iodide ^{131}I. An identical sample is set aside as a standard. After the elapse of, say, 24 hours, the radioactivity of the thyroid gland is determined with a gamma-sensitive detector and the activity compared to that of the standard when measured under *identical conditions of geometry*.

The radiation emitted from the patient consists of at least three components. These are:

1. Direct radiation from the thyroid—0.36 MeV primaries of ^{131}I plus a small amount of radiation of lower energy.

2. Scattered radiation, resulting especially from Compton interactions of the primary radiation with neck tissue.

3. Radiation from parts of the body other than the thyroid. A "knee count" is taken by some laboratories as a body background.

The error from the third source is normally small after 24 hours and accounts for only 2 to 3% of the observed radiation. Compton scattering, however, may represent 20 to 30% of the observed radiation. If the *surroundings* of the standard ^{131}I and the thyroid gland are different, the measurements will be in error as a consequence of the difference in the extent of Compton scattering. A "phantom neck" is sometimes employed with the standard to produce an amount of Compton scattering equivalent to that produced by the patient's neck tissues. It has been shown that a sheet of lead, 1⁄32″ thick, placed in front of the detector, will remove a large percentage of this low energy, scattered radiation, while at the same time removing only a small amount of the high energy direct radiation. A spectrometer may also be used to eliminate scattered radiation through pulse-height analysis.

While the problems of measurement mentioned above result in an increase in the amount of observed radiation, absorption of radiation by the tissues of the neck tends to decrease the observed activity. Absorption errors are difficult to estimate, but their effect may generally be cancelled or reduced in magnitude through the use of a phantom neck.

The normal range for iodine uptake (for a tracer dose of iodide) is about 10 to 40% in 24 hours. Uptakes from 10 to 15% and from 35 to 45% may be considered borderline. An uptake exceeding 50% is highly suggestive of hyperthyroidism, while an uptake of less than 15% may usually be interpreted as indicative of myxedema.

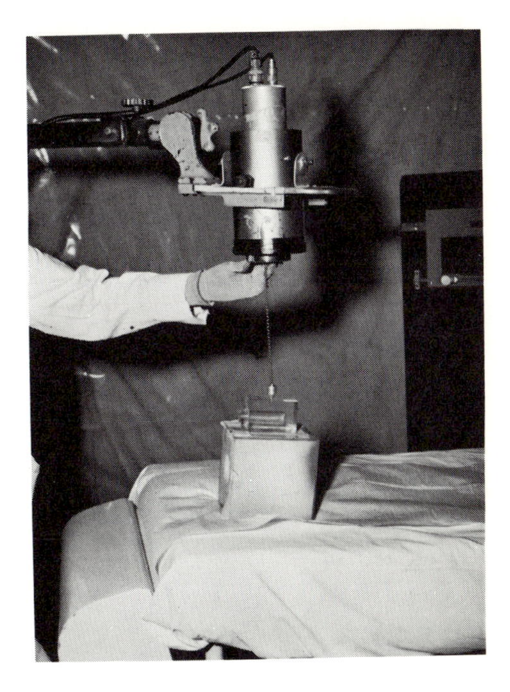

Figure 15.10—CALIBRATION FOR THYROID UPTAKE. The instrument is calibrated against a standard source in a phantom neck.

The use of radio-iodine is contraindicated during the second and third trimesters of pregnancy. The fetal thyroid is sufficiently developed at 12 to 14 weeks to pick up iodine from the maternal circulation. Even a tracer dose given to the mother may be sufficient to inhibit or injure the fetal thyroid.

A number of substances inhibit the uptake of radio-iodine. These substances as well as several interfering procedures, should be avoided prior to the tracer studies. The following table lists some of these substances and procedures and indicates the length of time during which they should be avoided before the tracer studies are conducted.

Substance or procedure	Time
Foods (Fish, cabbage, brussels sprouts, cauliflower, soya beans, rutabaga and turnips)	3 days
Iodine, inorganic, enteral (Lugol's solution, hydriodic acid)	7 days
Iodine, topical (Iodine, tincture or ointment)	14 days
Iodine, topical (Iodoform or other organic compound)	60 days
Iodine, organic, enteral (Enterovioform, Diodoquin or Enterosept)	30 days
Cholecystography (Cholegrafin, Priodax, Telepaque)	6 months
Pyelography (Diodrast, Urokon, Neo-iopax)	30 days
Salpingography (Salpix, Skiodan Acacia)	30 days
Myelography (Pantopaque)	2-10 years
Bronchography (Lipiodol, Iodochlorol)	2-10 years
Desiccated thyroid, and thyroxin	3-5 weeks
Potassium thiocyanate	10 days
Thiouracil and its derivatives, cortisone, ACTH, estrogen derivatives	8 days

PROCEDURE:

1. The patient's history should be reviewed to determine if he has been exposed to substances which inhibit iodine uptake.

2. The patient is given an oral dose of from 10 to 50 microcuries of Sodium Radioiodide ^{131}I Solution or Capsules. A similar dose is set aside as a reference standard. All activities are measured at a standard distance.

3. After 24 hours (or other specified time), the activity of the reference solution or capsule is measured. The use of a phantom neck is recommended.

4. The standard is removed from the phantom and placed behind lead shielding. Background is measured on the empty phantom and subtracted from the activity measured in step 3. The corrected activity of the standard is R_s.

5. The activity of the patient's thyroid is measured. The counter should be positioned at the same standard distance from the neck as was used for measuring the standard activity.

6. The counter is now positioned at the standard distance from the patient's thigh. This activity represents the background for the measurement of the thyroid activity and is subtracted from the result of step 5. The corrected thyroid activity is R_t.

CALCULATIONS:

$$\% \text{ Uptake} = \frac{\text{Radioactivity in patient's gland}}{\text{Radioactivity in dose (Phantom Neck)}} \times 100$$

$$\% \text{ Uptake} = 100\, R_t/R_s$$

PROCEDURE 15.7 Urinary Excretion of ^{131}I

RATIONALE:

Two mechanisms compete for the iodine circulating in the body: (1) uptake by the thyroid, and (2) urinary excretion. The amount of iodine excreted by the kidneys is inversely related to the amount fixed by the thyroid. Urinary excretion may therefore be used as an indirect measure of thyroid function.

An advantage in the use of urinary ^{131}I output over the measure of thyroid uptake rests in the ability to reproduce the counting geometry more accurately. A second advantage is that the patient need not be present when radioactivity measurements are performed. This method has two disadvantages, however, which may introduce serious errors. First, it is necessary to collect a reliable urine specimen. Loss of urine will yield low results. Secondly, accuracy of the test is contingent upon normal kidney function.

With a hyperfunctioning thyroid, generally less than 30% of the dose will appear in the urine in 24 hours; with a hypofunctioning gland, over 80% will usually appear.

The normal range is approximately 40 to 70%. Theoretically, the iodine uptake plus the urinary excretion should equal about 90% of the dose (82-87).

PROCEDURE:

1. A tracer dose of ^{131}I, similar to that given for the uptake study (Procedure 6), is administered orally to the patient. The patient is instructed to save his entire 24-hour urine sample.

2. A standard solution is made by dilution to 2000 ml of the same volume or quantity of tracer dose given the patient. A small amount of non-radioactive sodium iodide should be added to the solution as a carrier.

3. A measured amount of the standard solution is counted, about 50 ml if a Marinelli beaker is used, a smaller volume if a well-type scintillation counter is used. The measured activity, corrected for background, and expressed in cpm, is R_s.

4. The patient's 24-hour urine sample is diluted to 2000 ml, and a volume of this diluted sample, equal to the volume of standard solution used in step 3, is counted in a similar manner. The measured activity of the aliquot of diluted urine, corrected for background, and expressed in cpm, is R_u.

CALCULATIONS:

$$\%\ \text{Urinary excretion} = \frac{\text{Activity of patient's urine}}{\text{Activity of standard}} \times 100$$

$$\%\ \text{Urinary excretion} = 100\ R_u/R_s$$

PROCEDURE 15.8 PBI Conversion Ratio

RATIONALE:

Assimilated inorganic iodine is concentrated in the thyroid gland where it is converted, in part, to thyroxine. Upon release from the thyroid, the thyroxine is found to be reversibly bound with the serum protein. Treatment of the serum or plasma with trichloroacetic acid (TCA), for example, causes the protein to precipitate. Protein-bound iodine will be found in this precipitate. If, on the other hand, the serum or plasma is passed through a suitable ion exchange column, inorganic iodine will be retained on the column while the protein-bound fraction passes on through (88-96).

The conversion ratio is an expression of the fraction (usually indicated as per cent) of the protein-bound radioiodine in the blood compared to the total radioiodine present in the serum or plasma at a given time, i.e., 2, 4, 6, 12, 24 or 72 hours after administration of the dose. In many hospitals a single blood sample is taken at 24 hours.

In humans, an oral dose of 50 μCi of $Na^{131}I$ has been found satisfactory for this test in the case of euthyroid patients, while only 25 μCi need be used if exophthalmic goiter is indicated and as much as 100 μCi may be required in cases of myxedema.

Current procedures for the determination of the PBI conversion ratio are all essentially the same, except for the step involving the separation of the protein-bound iodine from the inorganic iodides. The procedure may be outlined as follows:

1. Administration of the dose.
2. Waiting for a predetermined period of time.
3. Collection of blood sample.
4. Determination of the activity of an aliquot of the serum or plasma.
5. Separation of the protein fraction from the inorganic fraction.
6. Determination of the activity of the protein fraction.
7. Calculation of the ratio-organic $PB^{131}I$/total plasma ^{131}I.

PROCEDURE:

1. Administer an oral dose of from 25 to 100 μCi of Sodium Radio-iodide ^{131}I. The size of the dose will be determined by the condition of the patient as explained above and by the sensitivity of the detector.

2. After a predetermined time—2 to 72 hours—collect 15 ml of heparinized or oxalated blood.

3. Centrifuge the blood and separate the plasma.

4. *Trichloroacetic Acid Method*

a. Count 2 ml of plasma in a well-type scintillation counter. (R_t)

b. To the 2 ml of plasma add 4 ml of 10% TCA. Mix and Centrifuge. Decant and wash the precipitate twice with 4 ml of 10% TCA. Then, measure the activity of the precipitate. (R_p)

c. Calculate the PBI conversion ratio.

Note: Volume correction for well counter—When a well-type scintillation counter is used, a correction must be applied for the differences in the volumes of the solutions being measured. (See experiment 9.1). It is necessary to normalize all readings to a standard volume with the aid of a calibration chart or curve similar to that shown in figure 9.5.

5. *PBI REZIKIT Method (Squibb)*

a. Measure the activity of 2 ml of plasma in a test tube in a well counter. This is the sum of the activities of the thyroxine and the inorganic iodide. (R_t)

b. The sample is then passed through the ion exchange column of the PBI Rezikit for removal of inorganic iodide ions. Withdraw the polyethylene tube containing the resin column from the neck of the test tube and open it by cutting off the sealed portion at the top and removing the cap from the bottom.

c. Apply air pressure to expel the excess water from the resin column by folding the upper part of the polyethylene tube back on itself several times. Return the polyethylene tube to the neck of the test tube.

d. Pour the plasma sample over the resin column and allow it to drain through. Expel the excess of plasma by applying air pressure in the same manner as before.

e. Rinse out the test tube which contained the plasma sample with 2 ml of isotonic saline solution and use the same rinse solution to wash the resin column.

f. Record the volume of the effluent (plasma plus rinse solution) and discard the resin column.

g. Place the test tube containing the effluent in the well scintillation counter for measurement of the protein-bound ^{131}I (i.e., the thyroxine activity). Correct all activities for background and solution volume. (R_p).

6. *IORESIN Method (Abbott)*

a. A small circle of precut filter paper is placed in the bottom of a polyethylene cylinder supplied with the Ioresin kit. Sufficient "Ioresin", suspended in water, is then transferred gently to the cylinder to make a column 1½-2 cm high and the water is allowed to drain through. The Ioresin column is washed once with physiological saline solution and allowed to drain. The excess liquid is expressed by gentle air pressure applied by means of a small rubber bulb held against the cylinder. In removing this rubber bulb, one should take care to break the seal between the bulb and the cylinder by gentle turning and lifting before releasing the pressure in order that the resin column not be disturbed. Should the column be disturbed, it may be rewashed and pressure reapplied.

The column thus prepared is arranged so that it drains into one of the plastic tubes provided. The kit includes a plastic rack which can hold six cylinders as well as a companion block for holding the test tubes in which the solutions from the columns are collected.

b. The activity of a 2-ml sample of whole, undiluted plasma in a plastic test tube is measured in a well counter. The background is determined and subtracted from all values. This is the sum of the thyroxine and inorganic iodide activities. (R_t).

c. The same sample is then poured onto a column and allowed to drain through. When the plasma has stopped dripping from the column, air pressure is again applied with the bulb.

d. A 2-ml portion of isotonic saline is used to wash out the test tube and the wash liquid is added to the column. This is followed by a 1-ml saline wash, the excess liquid being expelled with the bulb.

e. The liquid is then counted in a well counter as above. The activity is corrected for sample volume and background. This tube contains the thyroxine activity only since the inorganic iodide has been retained by the resin column. (R_p).

CALCULATIONS:

PBI Conversion Ratio % = $100R_p/R_t$

PROCEDURE 15.9 Triiodothyronine in Vitro Uptake Methods

RATIONALE:

Circulating thyroid hormone substances, thyroxine and triiodothyronine, are bound by plasma proteins. The binding capacity of plasma proteins is greater for thyroxine than for triiodothyronine (98,99). Red cells also show an affinity for these substances, but to a lesser extent than plasma proteins (99,101). These circulating hormone substances are therefore not bound to RBCs to any appreciable extent until the plasma proteins are relatively saturated. The RBCs appear to function as a passive reservoir for triiodothyronine not bound to protein.

The *in vitro* uptake of ^{131}I-labeled triiodothyronine (T_3), based upon the above facts, was first suggested by Hamolsky, Stein and Freedberg (100) as a test of thyroid function. The test is of special interest because it does not require the administration of radioactive substances to the patient. A few ml of whole blood, withdrawn from the patient, is incubated with T_3 for about two hours at 37° C. The activity retained by the RBCs after washing is compared to the total activity added to the blood. The greater the production of thyroid hormone, the greater will be the degree of saturation of binding sites on the protein. Thus, less T_3 will be bound by the protein and a relatively high activity of T_3 will be found in the RBCs. Conversely, in hypothyroidism, the T_3 activity of the blood cells will be low (97-114).

Hamolsky and his associates (100, 102) observed a dependence of red cell uptake (RCU) on the hematocrit so that all their readings were corrected to a hematocrit of 100%. However, the simple correction (see Procedure A, Step 9) is not adequate to correct for wide variations in hematocrit. Christensen (107) and Adams, Specht and Woodward (105) have suggested a mechanism, based on the mass action law, which accounts for the essential role of the plasma proteins, explains the relatively passive role of the RBCs and accurately corrects for hematocrit.

$$P + T \rightleftharpoons PT$$
$$E + T \rightleftharpoons ET$$

where T = triiodothyronine
P = triiodothyronine binding sites on the protein
E = triiodothyronine binding sites on the erythrocytes

then,

$$(T) = k_1 \frac{(PT)}{(P)} = k_2 \frac{(ET)}{(E)}$$

and,

$$\frac{(ET)}{(PT)} = K \frac{(E)}{(P)} = \frac{Q}{1 - Q}$$

where, Q = the ratio of activity in the RBCs to the total activity
= Hamolsky RCU

Adams, Specht and Woodward suggest that the total binding sites on erythrocytes are proportional to the hematocrit, while the protein binding sites are proportional to the plasmacrit. Hence,

$$\frac{Q}{1 - Q} = K' \frac{HCT}{1 - HCT}$$

and

$$K' = \frac{Q\,(1 - HCT)}{HCT\,(1 - Q)}$$

The constant K′ is referred to as the binding coefficient. It, rather than Q, is used as the criterion of thyroid function.

Recognizing the passive nature of the RBCs, Rabinowitz and Shapiro (110,111,114) suggested the use of Sephadex to bind the residual T_3 quantitatively. In this method, the activity of the plasma or serum is measured rather than the RBC activity. Thus the numerical results are roughly complementary to those obtained by use of the RBC-T_3 method of Hamolsky.

PROCEDURE:

A. *RBC*-T_3 *Method*

1. Dilute ^{131}I-labeled triiodothyronine with isotonic saline so that the final solution contains 1.1×10^{-2} micrograms per 0.1 ml. (Prepare fresh solutions at least once a week and store in the refrigerator in the dark.)
2. Withdraw 10 ml of venous blood using a heparinized syringe.
3. Place 3 ml of blood into each of two 10 ml Erlenmeyer flasks and add 0.1 ml of the diluted T_3 solution (from Step 1) to each flask. Seal the flasks with rubber stoppers, agitate and then incubate for two hours at 37° C with intermittent agitation.
4. At the end of the incubation period, pipet two 1 ml aliquote of whole blood from each flask into separate test tubes. Count each of the four samples for 3 minutes in a well counter and correct for background. This is R_t.
5. Add 10 ml of normal saline (at room temperature) to each aliquot, mix gently, and then centrifuge for 5 minutes at 3,000 rpm. The supernatant is removed and the erythrocytes are washed four more times with ten-fold volumes of isotonic saline.
6. After the final wash do not reconstitute with saline. This leaves approximately 1 ml of solution.
7. Count each sample again for 3 minutes and subtract background. This is R_e.
8. Determine the hematocrit.
9. Calculations and results
   a. The per cent red cell uptake, $\% \text{ RCU} = \frac{100\,R_e}{R_t} \times \frac{100}{HCT}$
   b. Determine the average of the four determinations.
   c. Normal values: 11 to 17%.

B. *Sephadex Method*

1. Prepare *M*/15 phosphate buffer, pH 7.5 as follows: 2.3 g NaH_2PO_4; 14.1 g Na_2HPO_4; 1755 ml water.

2. Dilute 0.1 ml labeled T_3 to 50 ml with phosphate buffer.

3. Add 0.1 ml of diluted T_3 to 3 ml of serum. Stopper.

4. Incubate in water bath for 15 minutes at 37° C.

5. Slurry 2 g Sephadex Gel-25 in 10 ml phosphate buffer into chromatographic column.

6. Allow column to settle evenly and place a filter paper disc on top of the Sephadex Gel-25.

7. Pipet 1 ml incubated serum into test tube for counting. Subtract background. This is R_t. It represents the total radioactive iodine present in the serum.

8. Carefully pipet 1 ml incubated serum onto the column.

9. Add 1 ml phosphate buffer to the column *as soon as* the serum passes below the top of the column.

10. Add 12 to 15 ml phosphate buffer to the column after the 1 ml of buffer has passed below the filter paper.

11. Collect ten 1-ml fractions in test tubes as soon as the serum has passed through two-thirds of the column.

12. Count the fractions in a well counter and deduct background from all measurements. Typical results follow:

Fraction #1	Little or no activity
Fractions #2-4	Protein-bound ^{131}I only
Fractions #5-6	Little or no activity
Fractions #7-9	^{131}I as inorganic iodide only

13. Calculations:

a. Calculate the sum of counts in those fractions containing protein-bound iodine ^{131}I only (e.g, fractions #2-4) this is R_p.

b. Calculate the sum of counts in those fractions containing ^{131}I as inorganic iodide only. This is R_I.

c. Calculate the net counts for the protein-bound iodine in the serum, R_s.

$$R_s = R_t - R_I$$

d. Calculate the per cent serum protein uptake (% SPU).

$$\% \text{ SPU} = 100 \, R_p/R_s$$

e. Values for % SPU

Normal	73 to 86%
Hyperthroid	below 73%
Hypothyroid	above 86%

PROCEDURE 15.10 The Schilling Test for Pernicious Anemia

RATIONALE:

In a normal individual, over 50% of an oral dose of vitamin B_{12} is absorbed through the walls of the gastrointestinal tract. This absorption occurs only in the presence of the intrinsic factor of Castle with which the vitamin must presumably combine in order to pass through the intestinal walls. (The biochemical defect in pernicious anemia is the failure of the gastric mucosa to elaborate intrinsic factor). By means of ^{60}Co-labeled vitamin B_{12} it can be shown that over half of an oral dose soon appears in the blood. Normally only a small amount of activity appears in the urine, but if a large "flushing" dose (1,000 μg) of non-radioactive vitamin B_{12} is given parenterally, within an hour after the tagged oral dose, the renal threshold is exceeded and radioactivity is observed in the urine.

In the patient with pernicious anemia there is a deficiency of intrinsic factor which causes poor absorption of the vitamin so that most of the ingested B_{12} is found in the feces. The degree of absorption or of fecal excretion can be measured by the use of labeled vitamin B_{12}.

Other anemias, such as those associated with sprue and idiopathic steatorrhea, are also accompanied by a decrease in vitamin B_{12} absorption. They may be differentiated from pernicious anemia through the oral administration of intrinsic factor. A marked increase in vitamin B_{12} absorption results in the case of pernicious anemia but not in the case of sprue and other malabsorption syndromes (115-123).

In order to reduce the radiation dose received by the patient, and at the same time to improve the sensitivity of the test, some workers have used cobalt-57 and cobalt-58 instead of cobalt-60 as the radioactive tag for vitamin B_{12} (cyanocobalamin).

PROCEDURE:

Administration of tracer dose of radioactive vitamin B_{12}

1. The patient should be in an overnight fasting state.
2. The patient should empty bladder.
3. A tracer dose of about 0.5 μg, possessing an activity of about 0.5 μc, is administered orally to the fasting patient. This test dose may be in the form of a capsule or a solution.

Administration of the "flushing dose"

4. Within two hours after oral administration of the isotopic material, the subject is given an intramuscular injection of 1,000 μg of non-radioactive vitamin B_{12}. (The usual time of injection of the vitamin B_{12} is one hour after the tracer dose).
5. Patient may now have black coffee until lunch.

Collection of the sample

6. The subject is instructed to save all urine voided during the 24-hour period following the tracer dose. A valid 24-hour urine collection is essential for an accurate determination.

Preparation of the standard

7. Prepare a standard either by dissolving a capsule similar to that ingested by the patient, by diluting a volume of the solution given to the subject, or by using a reference standard furnished by the supplier, so that 1,000 ml of the standard solution contains 10% of the ingested dose (i.e., 0.05 μCi per liter).

8. Transfer the 1,000 ml of standard solution to a one-liter polyethylene bottle. Position the bottle on the crystal of a well scintillation counter and make a count. Measure the background by using a similar one-liter bottle filled with distilled water. The activity corrected for background is R_s.

Preparation of the sample

9. Measure the volume of the 24-hour urine sample. This volume, expressed in liters, is V_u.

10. Transfer 1,000 ml of urine to a one-liter polyethylene bottle. If the volume of urine is less than one liter, dilute to this volume with water.

11. Position the bottle on the crystal of a well scintillation counter and make a count. Measure the background by using a similar bottle filled with water. The activity corrected for background is R_u.

Calculations

12. Calculate the % excretion by the formula:

$$\% \text{ excretion} = 10\, V_u R_u / R_s$$

Interpretation of results

13. The following conditions are indicated by the urinary excretion of ^{60}Co-labeled vitamin B_{12} (without administration of intrinsic factor):

0 to 4%	Abnormal
5 to 6%	Questionable
7 to 35%	Normal

Differentiation of pernicious anemia

14. When the urinary excretion of vitamin B_{12} is low, the patient is given a capsule of intrinsic factor and the above test (steps 1 through 12) is repeated.

15. An increase in the urinary excretion of vitamin B_{12} (with the administration of intrinsic factor) is indicative of pernicious anemia. Anticipated ranges of urinary B_{12} excretion with intrinsic factor follow:

0 to 4%	Sprue or other dysfunction
5 to 6%	Questionable
7 to 30%	Pernicious Anemia

PROCEDURE 15.11 Kinetics of Iron Metabolism

RATIONALE:

If the plasma iron is labeled by intravenous injection of Radio-ferrous ^{59}Fe Citrate, it is possible to obtain a comprehensive evaluation of the kinetics of iron metabolism. Among the parameters which can be measured are the following:

1. Plasma iron clearance half time, PIC (Hours)
2. Plasma volume, V_p (Liters)
3. Hematocrit, HCT (As a fraction or as per cent)
4. Blood volume, V_b (Liters)
5. Incorporation of red cell iron (i.e., % utilization)
6. Daily iron clearance, DIC (Plasma iron turnover or plasma iron transport rate (mg/24 hrs)
7. Daily hemoglobin formation, DHF (Grams/day)
8. Per cent daily hemaglobin replacement, % DHR (%/24 hrs)

The iron turnover, or the number of times per day the total iron content of the plasma is replaced by iron obtained through the catabolism of the red cells, can be estimated in the following way:

On the average, the normal individual possesses a blood volume of about 5,000 ml. Since the hemoglobin content of whole blood is about 15 grams per 100 ml, the average person has about 750 grams of hemoglobin in his blood stream. Hemoglobin contains 0.334% of iron. Therefore, the hemoglobin iron content per person is about 2.5 grams. If the nominal life of a red cell is taken as 110 to 125 days, then approximately 20 to 23 milligrams of iron are released per day to the plasma by the red cells. The total content of iron in the plasma normally equals only 2 to 3 mg per person. The plasma has to supply the 20 to 23 mg of iron necessary for replacing the hemoglobin lost daily by catabolism as well as the one mg per day excreted in the urine. Therefore, the plasma iron turnover must be of the order of eight or more times per day. *In vivo* organ monitoring is also frequently employed while determining iron kinetics (124-132).

Variations in the basic procedures for the measurement of the parameters listed above are in current use. The procedures presented here are those used at many hospitals in Philadelphia.

PROCEDURE:

Preparation and injection of the dose

1. About 35 ml of venous blood are withdrawn from a fasting patient and transferred aseptically to a heparinized test tube for transport to the laboratory.

2. In the laboratory the blood is transferred to a prepared vial containing about 5 ml of ACD solution.

3. To the ACD-blood mixture there is added 10 μCi of Radio-ferrous ^{59}Fe Citrate. The blood is mixed well and allowed to incubate at room temperature for 15 to 30 minutes with occasional gentle agitation.

4. Exactly 30 ml of the ACD-blood-^{59}Fe mixture is injected intravenously into the patient. Care should be taken to ensure that sufficient surplus of the injection mixture remains available for the preparation of standards.

Calibration of activity of injected dose—R_d

5. Exactly 2 ml of ACD-blood-^{59}Fe mixture from step 3 is diluted to 500 ml with distilled water in a volumetric flask and mixed well.

6. Exactly 2 ml of this diluted sample is placed in a well counter and the activity measured. The background is also measured. The corrected activity is R_a.

7. The activity R_d of the dose injected into the patient is calculated, according to the formula:

$$R_d = (30/2) \times (500/2) \times R_a = 3{,}750\ R_a$$

8. A second calibration of the dose is performed as a check by use of the plasma fraction. For this procedure, 5 ml of the ACD-Blood-^{59}Fe mixture is centrifuged and the resulting plasma carefully transferred to a 500 ml volumetric flask.

9. To assure complete transfer of the plasma to the flask, the experimenter should wash the cells two or three times with normal saline and transfer the washings to the flask with the plasma. The plasma and washings are diluted to 500 ml with distilled water and the contents of the flask mixed thoroughly.

10. Measure the activity of exactly 2 ml of this diluted plasma in a well counter. Correct for background. The corrected activity is R_c.

11. Calculate the activity R'_d of the dose injected into the patient:

$$R'_d = (30/5) \times (500/2)\ R_c = 1{,}500\ R_c$$

(The two activities R_d and R'_d should be identical since the ^{59}Fe does not move into the red cells to any appreciable extent in this short time. The two calibrations serve as a check on each other).

Collection and treatment of blood samples on day of injection

12. By means of a heparinized syringe, 10-ml samples of blood are withdrawn at 10, 30, 60 and 120 minutes following injection of the dose. The exact time of each withdrawal should be noted.

13. Measure the activity of 2 ml of each sample of whole blood in a well counter. Correct each measurement for background and divide by 2 to obtain the activity per ml. The corrected activities are R_{10}, R_{30}, R_{60} and R_{120} respectively (in cpm/ml).

14. Centrifuge 5 ml of each sample of blood. Transfer a 2-ml aliquot of each plasma sample to a test tube and measure the activity in a well counter. These activities, corrected for background and divided by 2, are R_{p10}, R_{p30}, R_{p60} and R_{p120} respectively (in cpm/ml of plasma).

Collection and treatment of daily blood samples

15. Following the day of injection, blood samples are drawn daily; the day and hour should be recorded. Only 3 to 5 ml of blood need be drawn, a heparinized syringe being used. Take samples for 10 days.

16. Exactly 2 ml of each blood sample is transferred to a test tube and the activity measured in a well counter. Each activity is corrected for background and divided by 2. The corrected activities are R_1, R_2, $R_3 \ldots\ldots R_{10}$, respectively. It is not necessary to measure the plasma activity because essentially all of the activity appears in the cells after the first day.

Supplemental determinations

17. To calculate the per cent incorporation of iron into the hemoglobin, one must first know the blood volume. The blood volume can be measured simultaneously by use of the ^{51}Cr method. Pulse height discrimination is used to measure the ^{51}Cr and the ^{59}Fe independently. As additional information, the red cell survival time can also be measured, if desired, since the cells will already be properly tagged.

18. A standard hematocrit is measured.

19. A standard plasma iron determination is required for the calculation of iron turnover.

20. Determine the blood hemoglobin.

CALCULATIONS:

Plasma iron clearance half-time—PIC

1. Plot the plasma activities, R_{p10}, R_{p30}, R_{p60}, and R_{p120}, as functions of time. If plotted on semi-logarithmic paper, the data should fall on a straight line.

2. Draw the best straight line through the points and extrapolate back to zero time (time of injection). The activity at zero time is R_{po}.

3. From the graph, determine the time required for the activity to decrease to a value of $R_{po}/2$. This time is the plasma iron clearance half-time, PIC.

4. Repeat steps 1, 2 and 3 using the whole blood activities R_{b10}, R_{b30}, R_{b60} and R_{b120}. The value of the whole blood activity at the time of injection is R_{bo}, and is determined by extrapolation of the line back to zero time. The value for the half-time should agree with that obtained in step 3 and will serve as a check on the determination.

Plasma volume—V_p

5. Calculate the plasma volume from the relationship:

$$V_p \text{ (liters)} = \frac{R_d}{1{,}000\, R_{po}}$$

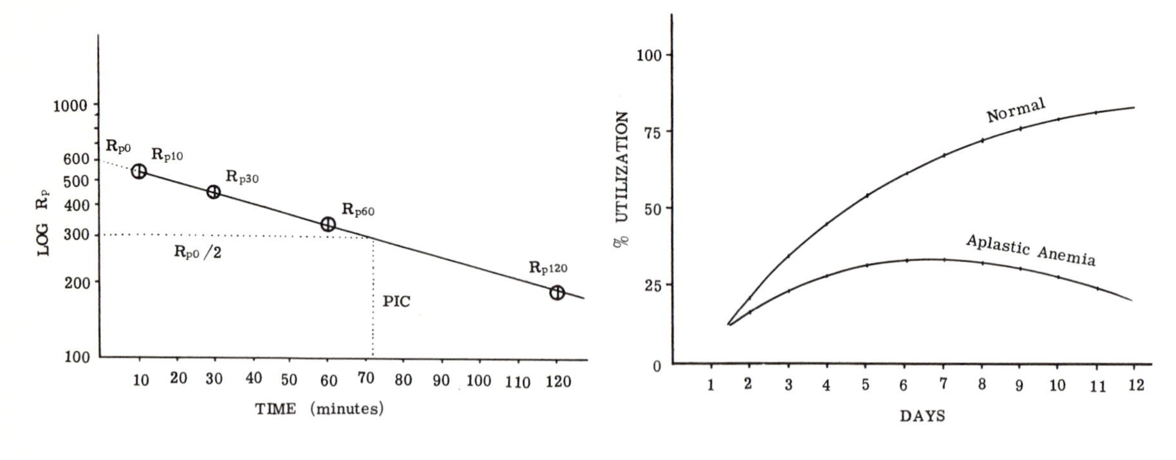

Figure 15.11—CALCULATION OF PLASMA IRON CLEARANCE HALFTIME, PIC, FROM THE RATE OF PLASMA IRON CLEARANCE.

Figure 15.12—NORMAL AND ABNORMAL UTILIZATION OF IRON FOR THE FORMATION OF HEMOGLOBIN.

Hematocrit—HCT

6. Calculate the hematocrit from the relationship:

$$HCT = 1 - \text{Plasmacrit} = 1 - (R_{bo}/R_{po})$$

7. Compare this value of hematocrit with that determined in step 18 of the procedures. The two values should agree with each other and serve as a check on the technique.

Blood volume —V_b

8. Blood volume is calculated by use of the relationship:

$$V_b \text{ (liters)} = \frac{R_d}{1{,}000\ R_{bo}}$$

9. Since plasmacrit is the ratio of plasma volume to blood volume, the calculation in step 8 can be checked through the relationship:

$$\text{Plasmacrit} = R_{bo}/R_{po} = V_p/V_b$$

or,

$$V_b = V_p (R_{po}/R_{bo})$$

Red cell incorporation (% Utilization)

10. Calculate the % utilization of iron (i.e., % incorporated into the red cells) on the first day after injection by use of the relationship:

$$\% \text{ Utilization} = 100\ R_1/R_{bo}$$

11. Calculate the % utilization of iron by the second day by use of the relationship:

$$\% \text{ Utilization} = 100\ R_2/R_{bo}$$

12. In a similar way, calculate the % utilization for the third, fourth, etc. day until the tenth day or the last day a blood sample was withdrawn.

13. On linear graph paper, plot the % utilization as a function of time.

Daily iron clearance—DIC (Plasma iron turnover; Plasma iron transport rate)

14. The relationship between the rate at which iron is removed from the plasma (RPC) and the plasma iron clearance half-time (PIC) is analogous to the relationship between the decay constant and the half-life of a radioisotope.

$$\text{RPC} = 0.693/\text{PIC}$$

If PIC is expressed in hours, then RPC is the fraction of the total iron removed from the plasma per hour. The fraction of iron removed daily (i.e., in 24 hours) is therefore equal to (24) (0.693)/PIC and the daily iron clearance (DIC), expressed as mg of plasma iron per 24 hours is given by:

$$\text{DIC} = \frac{(24)\,(0.693)\,(10)\ Fe_p V_p}{(\text{PIC})} = \frac{166\ Fe_p V_p}{(\text{PIC})}$$

where Fe_p is the plasma or serum iron (mg %) from step 19 of the procedure (page 551) and V_p is the plasma volume (liters).

Daily hemoglobin formation—DHF

15. Hemoglobin contains 0.334% iron. Using this value, calculate the daily production of hemoglobin, DHF (grams/day), from the daily rate of plasma iron clearance, DIC (mg/day). Because not all iron is utilized for the production of hemoglobin, at least directly, a correction must be applied for the % utilization.

$$\text{DHF} = \frac{(\text{DIC})\,(\%\ \text{Utilization})}{3.34}$$

Per Cent daily hemoglobin replacement—% DHR

16. The total body hemoglobin (Total Hgb) in grams, is calculated from the value of the blood hemoglobin (step 20 on page 551) expressed in grams %, and blood volume V_b expressed in liters.

$$\text{Total Hgb} = 10\ (\text{grams}\ \%\ \text{Hgb})\ (V_b)$$

17. The per cent daily hemoglobin replacement is then given by:

$$\%\ \text{DHR} = 100\ (\text{DHF})\ /\ (\text{Total Hgb})$$

PROCEDURE 15.12 Fat Absorption

RATIONALE:

Oils and fats are composed almost entirely of glycerides—esters of glycerol and fatty acids. Before absorption can occur through the intestinal wall these esters must be hydrolysed to fatty acids and glycerol by the action of the fat digesting enzymes in the stomach and intestines. Following absorption, the fatty acids and glycerol recombine to form neutral triglycerides for distribution throughout the body.

Studies of the absorption of orally administered radioactive fats and fatty acids provide useful clinical information in certain disorders of the gastro-intestinal tract. Although the earlier work was done by measurement of fecal excretion of the labeled fat, some recent studies have shown that the determination of blood levels of radioactivity is an easier and more accurate procedure, others prefer counting the entire stool container.

Fats can be labeled with carbon-14 or with iodine, either ^{131}I or ^{125}I. An iodine tag is introduced into the fat molecule by iodination of one of the unsaturated fatty acids, e.g., oleic acid. The iodine tag alters the chemical composition of the fat and causes some minor changes in the absorption rate when compared to a carbon-14-labeled fat, but the small difference does not decrease the value of the test as an empirical measure of fat absorption (133-141).

PROCEDURE:

Administration of the dose

1. For two days prior to the study, the patient is given an oral daily dose of 10 drops of Lugol's solution and instructed to eat nothing on the morning of the test.

2. The fasting subject is given 25 to 50 μCi of Radio-iodinated ^{131}I glycerol trioleate in the form of a capsule.

3. This dose is followed immediately by a "cold" meal emulsified in a blender and consisting of 48% peanut oil, 48% water and 4% Tween 80. The cold meal* is given on the basis of 1.0 ml per Kg of body weight.

Sample collection

4. Blood samples (3 to 5 ml) are taken in heparinized tubes every hour for 6 or 7 hours beginning 2 hours after administration of the dose. Additional samples are taken at about 12 and 24 hours after the dose. Note the exact time of withdrawal of each sample. The patient should have nothing more than black coffee or tea for three hours after administration of the dose. Then he may eat normally.

5. If desired, total stools are collected from 0 to 24 hours and from 24 to 48 hours.

Preparation of counting standards

6. A dose containing the same activity as that ingested by the patient is dissolved in 10% dioxane in water and diluted to 100 ml with the same solvent in a volumetric flask. This is solution S_a.

*LIPOMUL® (*Upjohn*), a palatable emulsion containing 67% vegetable oil can be used for the cold meal.

7. An accurately measured 1-ml aliquot of solution S_a is diluted to 100 ml in a second volumetric flask using 10% dioxane in water. This is solution S_b. It has a concentration equivalent to a dilution of the original dose to 10 liters.

8. Measure the activity of a 2-ml aliquot of solution S_b in a well counter. The measured activity, corrected for background, is R_{bs}.

9. If stools are collected, a dose similar to that ingested by the patient is dissolved in 100 ml of 10% dioxane in water and placed in a container similar to that used for stool collection.

10. The container is placed on top of the crystal of a well counter for measurement of the activity. The activity, corrected for background, is R_{ss}.

Preparation of the samples

11. Exactly 2 ml of each blood sample is transferred to a test tube and the activity measured in a well counter. The corrected activities are R_{b2}, R_{b3}, R_{b4}, ... R_{b12} and R_{b24}.

12. If stool samples are collected, measure the activities by placing the sample container on a well scintillation detector. The corrected activities are R_{s24} and R_{s48}.

Calculations

13. Blood volume V_b may be estimated as 75 ml per Kg of body weight. If a more accurate determination is required, blood volume can be determined as directed in procedure 2, ^{51}Cr being measured in the presence of ^{131}I by pulse-height discrimination. Express V_b in liters for the calculation.

14. Calculate the % of administered dose absorbed and present in the blood for each of the blood samples. The example given is for the 2 hour specimen.

$$\% \text{ of dose} = \frac{R_{b2} V_b}{10 R_{bs}} \times 100 = \frac{10 R_{b2} V_b}{R_{bs}}$$

15. The % of dose excreted in the stool is given by:

$$\% \text{ of dose} = \frac{R_{s24}}{R_{ss}} \times 100$$

(If a more accurate measure of excretion in the stool is desired, the stools must be homogenized with 10% dioxane in water as a diluent and diluted to a specific volume. The activity of a 2-ml aliquot is then measured in a well counter. The standard must be diluted to the same volume and a 2-ml aliquot counted in a similar manner.

INTERPRETATION:

The plasma lipid radioiodine level rises over a six hour period to about 12% of the administered dose. The normal level is above 8%. In the case of sprue, pancreatitis, etc., the plasma level is much lower.

Further studies are made if the blood radioactivity is low and if the following conditions should be found to exist:

1. *Pancreatic function is abnormal*—There is a deficiency of pancreatic lipase and the fat is therefore not being split into fatty acids and glycerol, so that absorption does not take place.

2. *Pancreatic function is normal*—Hydrolysis of the fat occurs but the products of hydrolysis—fatty acids and glycerol—are not being absorbed because of intrinsic absorptive defects in the small intestine. A differentiation between cases 1 and 2 frequently can be made by repeating the test with Radio-iodinated ^{131}I Oleic Acid in place of Radio-iodinated ^{131}I Glyceryl trioleate.

PROCEDURE 15.13 Renal Function

The excretion of certain compounds is almost entirely by way of the kidneys. If both kidneys are functioning properly, each should excrete approximately 50% of these compounds or any other substance with a blood concentration in excess of the renal threshold. The performance of the kidneys can be determined by injection of a radioactive compound which is quickly and exclusively excreted by the kidneys. The radioactive tag is selected from those nuclides which emit gamma radiation to permit external detection of the isotope. The relative concentration of the tagged compound in each kidney can then be measured by means of two identical crystal scintillation detectors, one being positioned over each kidney. Renal malfunction is indicated if the measured activities are unequal (142-149).

The renal function test was introduced by Winter and co-workers (142-143) in 1956. In their original work they used iodopyracet. However, 10 to 15% of this compound is excreted through the liver. One of the best materials now known is sodium o-iodohippurate tagged with either ^{131}I or ^{125}I. Tubis and Nordyke (147-148) report this compound to be excreted rapidly and exclusively by the kidneys. The use of this compound, therefore, increases the accuracy of the test and at the same time reduces the time required to perform it.

PROCEDURE:

Preparation of the dose

1. Prepare a dilute solution of Sodium o-Iodohippurate ^{131}I, using sterile saline. The final concentration should be about 5 to 10 microcuries per ml.

Location of the kidneys

2. The location of the kidneys must be known with accuracy since the counters must be "looking" directly at the kidneys for valid results. The kidneys may be located by use of a roentgenogram. The locations are marked on the patient's back.

3. Alternatively, inject an initial dose of approximately 2 μCi of the diluted sodium iodohippurate. Locate the kidneys with the scintillation detectors and mark the areas

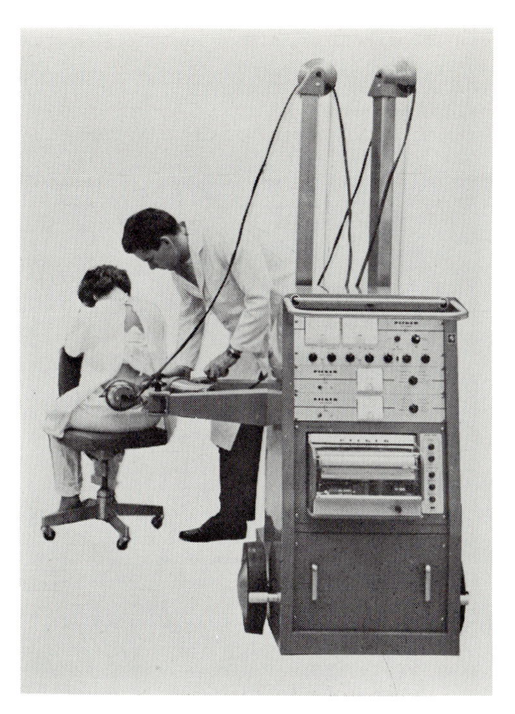

Figure 15.13—KIDNEY FUNCTION TEST.

on the patient's back. The areas giving miximum activity are noted. Allow 15 minutes for the material to be excreted before commencing the test.

Performance of test

4. With the patient in a sitting position, adjust the detectors so they are perpendicular to the body surface directly over the kidneys. Adjust the ratemeters and recorder before beginning the run. Typical settings are:

Recorder speed	—	12 inches per hour
Ratemeter scale	—	3,000 cpm full scale
Time constant	—	1.5 seconds

5. Inject the dose intravenously. The dose should be calculated on the basis of 0.2 to 0.3 μCi per Kg of body weight or a total dose of up to about 35 μCi.

REFERENCES

TEXTBOOKS ON CLINICAL APPLICATIONS

1. Behrens, C. P., "Atomic Medicine", 3rd ed., Williams & Wilkins, 1959.
2. Beierwaltes, W. H., P. C. Johnson, and A. J. Solari, "Clinical Use of Radioisotopes", Philadelphia, W. B. Saunders Co., 1957.
3. Blahd, W. H., F. K. Bauer, and B. Cassen, "Practice of Nuclear Medicine", Chas. C. Thomas, 1958.

4. Fields, T., and L. Seed, "Clinical Use of Radioisotopes", Year Book Publishers, 2nd ed., 1962.
5. Hahn, P. F., "Therapeutic Use of Artificial Radioisotopes", New York, John Wiley & Sons, 2nd ed., 1962.
6. Low-Beer, B. V. A., "Clinical Use of Radioactive Isotopes", Springfield, Ill., Chas. C. Thomas, 1950.
7. Owen, C., "Diagnostic Use of Radioisotopes", Chas. C. Thomas, 1959.
8. Quimby, E. H., and S. Feitelberg, "Radioactive Isotopes in Medicine and Biology"—"Basic Physics and Instrumentation", 2nd ed., Philadelphia, Lea & Febiger, 1963.
9. Silver, S., "Radioactive Isotopes in Medicine and Biology"—"Medicine", 2nd ed., Philadelphia, Lea & Febiger, 1962.
10. Veall, N., and H. Vetter, "Radioisotope Techniques in Clinical Research and Diagnosis", London, Butterworth and Co., 1958.

GENERAL REFERENCES

1. Technical data sheet, Bull. A-7839, E. R. Squibb & Sons, April 1962.
2. Flocks, R. H., et al., "The Treatment of Carcinoma of the Prostate with Radioactive Gold." In *Therapeutic Use of Artificial Radioisotopes*, P. F. Hahn, ed., New York, Wiley, 1956.
3. Sherman, A. I., "Carcinoma of the Cervix Treated with Radioactive Gold Colloids." In *Therapeutic Use of Artificial Radioisotopes*, P. F. Hahn, ed., New York, Wiley, 1956.
4. Nelson, C., "Use of Radioactive Gold in the Treatment of Carcinoma of the Bladder", South. M. J. 48, 245 (March 1955).
5. Technical data sheet, Bull. A-8097, E. R. Squibb & Sons, June 1962.
6. Technical data sheet, Bull. A-8096, E. R. Squibb & Sons, June 1962.
7. Fields, T., E. Kaplan, and M. Terrill, "Simplified Technique for Blood Volume Determination Using I-131 HSA", J. Lab. & Clin. Med. 43, 332 (1954).
8. Erickson, J. R., J. B. McCormick, and L. Seed, "Improved Method for Determination of Blood Volume Using Radioactive Iodinated Human Serum Albumin", Science 118, 595 (1953).
9. Prentice, T. C., M. Siri, N. I. Berlin, G. M. Hyde, R. J. Parsons, E. E. Joiner, and J. H. Lawrence, "Studies of Total Body Water with Tritium", J. Clin. Investig. 31, 412 (1951).
10. Forbes, G. B., and A. Perley, "Estimation of Total Body Sodium by Isotope Dilution Studies on Young Adults", J. Clin. Investig. 30, 558 (1951).
11. Corsa, L., Jr., J. M. Olney, Jr., R. M. Steenberg, N. R. Ball, and F. D. Moore, "Measurement of Exchangeable Potassium in Man by Isotope Dilution", J. Clin. Investig. 29, 1280 (1950).
12. Robinson, C. V., W. L. Arons, and A. K. Solomon, "Improved Method for Simultaneous Determination of Exchangeable Body Sodium and Potassium", J. Clin. Investig. 34, 134 (1955).
13. Wallace, G. B., and B. B. Brodie, "Distribution of Administered Bromide in Comparison with Chloride and Its Relation to Body Fluids", J. Pharmacol. and Exper. Ther. 65, 214 (1939).
14. Wright, H. P., S. B. Ashborne, and D. G. Edmonds, "Measurement of Rate of Venous Blood Flow in Legs of Women at Term and in Puerperium, Using Radioactive Sodium", J. Obst. & Gyn. Brit. Emp. 56, 36 (1949).
15. Quimby, E. H., "Isotope Studies of Bloof Flow and Blood Cells", Am. J. Roentgenol. 75, 1068 (1956).
16. MacIntyre, W. J., J. P. Storlaasli, H. Krieger, W. Pritchard, and H. L. Friedell, "I-131-Labeled Serum Albumin; Its Use in the Study of Cardiac Output and Peripheral Vascular Flow", Radiology 59, 849 (1952).
17. Huff, R. L., D. D. Feller, and G. Bogardus, "Cardiac Output by Body Surface Counting of I-131 Human Serum Albumin", J. Clin. Investig. 33, 944 (1954).
18. Conway, H., B. Roswit, R. B. Start, and R. Yalow, "Radioactive Sodium Clearance as a Test of Circulatory Efficiency of Tubed Pedicles and Flaps", Proc. Soc. Exper. Biol. & Med. 77, 348 (1951).
19. Kiehn, C. L., J. Benson, D. M. Glober, and M. Berg, "Study of Revascularization of Blood Vessel Grafts by Means of Radioactive Phosphorus", A. M. A. Arch. Surg. 65, 477 (1951).
20. Hamolsky, M. M., A. Golodetz, and A. S. Freedberg, "The Plasma Protein-Thyroid Hormone Complex in Man. Parts I, II and III". J. Clin. Endoc. Metabolism 17, 33 (1957); 19, 103 (1959); 19, 92 (1959).
21. Shapiro, B., and Rabinowitz, J. L., "A Chromatographic Method Utilizing Sephadex for the Separation of Free Iodide, Protein-Bound and Unbound Triiodothyronine in Sera. A) Clinical Correlations with the Hamolsky T-3-RBC Uptake Method (108 cases)", J. Nuclear Med. 3, 417 (1962).
22. Rabinowitz, J. L., B. Shapiro, and P. Johnson, "Sephadex Chromatographic Test in the Evaluation of Thyroid Function", J. Nuclear Med. 4, 139 (1963).
23. Shilling, R. F., D. V. Clatanoff, and D. R. Korst, "Intrinsic Factor Studies. III Further Observations Utilizing the Urinary Radioactivity Test in Subjects with Achlorhydria, Pernicious Anemia or Total Gastrectomy", J. Lab. & Clin. Med. 45, 926 (1955).
24. Belcher, E. N., I. G. F. Gilbert, and L. F. Lamberton, "Experimental Studies with Radioactive Iron", Brit. J. Radiol. 27, 387 (1954).
25. Badenock, J., and S. T. Callender, "Use of Radioactive Iron in Investigation of Anemia", Brit. J. Radiol. 27, 381 (1954).

26. LeRoy, G. V., "Clinical Research Using Compounds Labeled with Radioactive Carbon and Hydrogen as Tracers", Ann. Int. Med. 42, 239 (1955).
27. Rabinowitz, J. L., R. M. Myerson, and G. T. Wohl, "Deposition of C-14-Labelled Cholesterol in the Atheromatous Aorta", Proc. Soc. Exp. Biol. & Med. 105, 241 (1960).
28. Rabinowitz, J. L., J. L. LaFair, H. D. Strauss, and H. C. Allen, "Carbon Isotope Effects in Enzyme Systems II. Studies with Formic Acid Dehydrogenase", Biochim. Biophys. Acta. 27, 544 (1958).
29. Stanley, M. M., and S. J. Thannhauser, "Absorption and Deposition of Orally Administered I-131-Labelled Neutral Fat in Man", J. Lab. & Clin. Med. 34, 1634 (1949).
30. Ruffin, J. N., W. W. Shingleton, G. T. Baylin, J. C. Hymans, J. K. Isley, A. P. Sanders, and M. F. Sohnar, "I-131-Labelled Fat in Study of Intestinal Absorption", N. Eng. J. Med. 255, 595 (1956).
31. Maln, J. R., K. Reentsha, and H. G. Barker, "Comparative Fat and Fatty Acid Intestinal Absorption Test Utilizing Radioactive Labelling—Results in Normal Subjects", Proc. Soc. Exper. Biol. & Med. 92, 471 (1956).
32. Beres, P., J. Wenger, and J. B. Kirsner, "The Use of I-131 Triolein in the Study of Absorptive Disorders in Man", Gastroenterology 32, 1 (1957).
33. Berkowitz, D., D. Sklaroff, A. Woldow, A. G. Jacobs, and W. Likoff, "Blood Absorptive Patterns of Isotopically-Labelled Fat and Fatty Acid", Ann. Int. Med. 50, 247 (1959).
34. Likoff, W., D. Berkowitz, A. Woldow, A. G. Jacobs, and D. M. Sklaroff, "Radioactive Fat Absorption Patterns, Their Significance in Coronary Artery Atherosclerosis", Circulation 18, 1118 (1958).
35. Seller, R. H., J. Brachfeld, H. Sandberg, and S. Bellet, "Use of I-131 Labelled Fat in the Study of Lipid Handling in Patients with Coronary Artery Disease", Am. J. Med. 27, 231 (1959).
36. Taplin, G. V., O. H. Meredith, Jr., and H. Kade, "The Radioactive (I-131-tagged) Rose Bengal Uptake-Excretion Test for Liver Function Using External Gamma-Ray Scintillation Counting Techniques", J. Lab. &Clin. Med. 45, 665 (1955).
37. Taplin, G. V., O. H. Meredith, Jr., H. Kade, and C. C. Winter, "The Radioisotopes Renogram. An External Test for Individual Kidney Function and Upper Urinary Tract Patency", J. Lab. & Clin. Med. 48, 886 (1956).
38. Schloesser, L. L., et al., "Radioactivity over the Spleen and Liver Following the Transfusion of Chromium-51 Labeled Erythrocytes in Hemolytic Anemia", J. Clin. Investigation 36, 1470 (1957).
39. Friedell, H. L., C. I. Thomas, and J. S. Krohmer, "Beta-Ray Application to Eye, with Description of Applicator Utilizing Sr-90 and Its Clinical Use", Am. J. Ophth. 33, 525 (1950).
40. Dunphy, E. B., K. K. Dreisler, J. B. Cadisan, and W. Sweet, "Uptake of Radioactive Phosphorus by Intraocular Neoplasms", Am. J. Ophth. 37, 45 (1954).
41. Haskins, M. E., M. L. Wagner, M. Irker, and B. P. Widmann, "The Early Diagnosis of Prostatic Malignancy by Use of P-32", Am. J. Roent. Radium and Nuclear Med. 85, 99 (1961).
42. Selverstone, B., and J. C. White, "Evaluation of the Radioactive Mapping Technic in Surgery of Brain Tumors", Am. J. Surgery 134, 387 (1951).
43. Wagner, H. N., Jr., J. G. McAfee and J. M. Mozley, "Medical Radioisotope Scanning", J.A.M.A. 174, 162 (1960).
44. Taplin, G. V., E. K. Dove, D. E. Johnson, and H. S. Kaplan, "Suspensions of Radioalbumin Aggregates for Photoscanning the Liver, Spleen, Lung and other Organs", J. Nucl. Med. 5, 259 (1964).
45. "The Medical Use of Radioisotopes. Recommendations and Requirements by the Atomic Energy Commission." Isotopes Extension, Division of Civilian Application, U. S. Atomic Energy Comission, Oak Ridge, Tennessee, February 1957.
46. "Starting an Isotope Unit", Abbott Laboratories, North Chicago, Illinois (1960).
47. "Radioisotopes in Medicine. A General Guide for Physicians and Hospital Personnel", Abbott Laboratories (Rev. July 1961).
48. "Handbook of Rules for Administration of Radioactive Materials to Patients", M. D. Anderson Hospital and Tumor Institute, Texas Medical Center, Houston, Texas. Reproduced and distributed through special permission by E. R. Squibb & Sons (1959).
49. Latiolais, C. J., P. F. Parker, G. Hutchinson, and R. A. Statler, "Radioisotopes in Hospital Pharmacy", Bull. Am. Soc. Hosp. Pharm. 12, 372 (1955).
50. Wagner, H. N., Jr., J. G. McAfee, and J. M. Mozley, "Medical Radioisotope Scanning", J.A.M.A. 174, 162 (1960).
51. Bull. N-5261, Picker X-Ray Corp., White Plains, N. Y.

Blood Volume, RBC Survival Time, Sequestration and G. I. Bleeding

52. Lewis, J. W., "Blood and Plasma Volume Determinations by Means of Radioactive Iodinated Serum Albumin", Rocky Mountain M. J., 54, 813 (1957).
53. Twigg, H., L. Nakayama, and P. Goodwin, "Blood Volume Studies in Humans Using Radioiodinated Human Serum Albumin", Maryland M. J. 7, 150 (1958).
54. Gray, S. J., and H. Frank, "The Simultaneous Determination of Red Cell Mass and Plasma Volume in Man with Radioactive Sodium Chromate and Chromic Chloride", J. Clin. Invest. 32, 1000 (1953).
55. Small, W. J., and M. C. Verloop, "Determination of Blood Volume Using Radioactive Cr^{51}: Modification of the Original Technique", J. Lab. &Clin. Med. 47, 255 (1956).
56. Albert, C. A., H. N. J. Eccleston, A. Raffii, C. H. Hunter, E. H. Henley, and S. N. Albert, "A Rapid Method for Preparing Washed Red Cells Tagged with Cr^{51}", J. Lab. & Clin. Med. 54, 300 (1959).
57. Cooper, M., and C. A. Owen, "Labeling Human Erythrocytes with Radiochromium", J. Lab. &Clin. Med. 47, 65 (1956).

58. Cunningham, T. A., et al., "The Effect of Prior Contact Between Acid Citrate Dextrose and Sodium Radiochromate Solutions on the Efficiency with which Cr^{51} Labels Red Cells", J. Lab. & Clin. Med. 50, 778 (1957).
59. Gray, S. J., and K. Sterlin, "The Tagging of Red Cells and Plasma Proteins with Radioactive Chromium", J. Clin. Invest. 29, 1604 (1950).
60. Sterling, K., and S. J. Gray, "Determination of the Circulating Red Cell Volume in Man by Radioactive Chromium", J. Clin. Invest. 29, 1614 (1950).
61. Sutherland, D. A., et al., "The Survival of Human Erythrocytes Estimated by Means of Cells Tagged with Radioactive Chromium: A Study of the Normal State", J. Lab. & Clin. Med. 43, 717 (1954).
62. Necheles, T. F., et al., "Radioactive Sodium Chromate for the Study of Survival of Red Blood Cells. I. The Effect of Radioactive Sodium Chromate on Red Cells", J. Lab & Clin. Med. 42, 358 (1953).
63. Eadie, G. S., et al., "The Potential Life Span and Ultimate Survival of Fresh Red Blood Cells in Normal Healthy Recipients as Studied by Simultaneous Cr^{51} Tagging and Differential Hemolysis", J. Clin. Invest. 34, 629 (1955).
64. Ebaugh, F. G., Jr., et al., "The Use of Radioactive $Chromium^{51}$ as an Erythrocyte Tagging Agent for the Determination of Red Cell Survival *in Vivo*", J. Clin. Invest. 32, 1260 (1953).
65. Strumia, M. M., et al., "Uses and Limitations of Survival Studies of Erythrocytes Tagged with Cr^{51}", Blood 10, 429 (1955).
66. Jandl, J., et al., "Clinical Determination of the Site of Red Cell Destruction", J. Clin. Invest. 35, 842 (1956).
67. McCurdy, P. R., and C. E. Rath, "Splenectomy in Hemolytic Anemia: Results Predicted by Body Scanning after Injection of Cr^{51}-Tagged Red Cells", New England J. Med. 259, 459 (1958).
68. Schloesser, L. L., et al., "Radioactivity over the Spleen and Liver Following the Transfusion of $Chromium^{51}$-Labeled Erythrocytes in Hemolytic Anemia", J. Clin. Invest. 36, 1470 (1957).
69. Bannerman, R. M., "Measurement of Gastro-intestinal Bleeding Using Radioactive Chromium", Brit. M. J. II, 1032 (1957).
70. Ebaugh, F. G., Jr., et al., "Quantitative Measurement of Gastrointestinal Bloos Loss. I. The Use of Radioactive Cr^{51} in Patients with Gastrointestinal Hemorrhage", Am. J. Med. 25, 169 (1958).
71. Jones, H. C. H., "Measurement of Red Cell Loss from the Gastrointestinal Tract Using Radioactive Chromium", Brit. M. J., I, 493 (1958).
72. Owen, C. S., Jr., et al., "Radiochromium-Labeled Erythrocytes for the Detection of Gastrointestinal Hemorrhage", J. Lab. & Clin. Med. 44, 238 (1954).
73. Roche, M., et al., "Study of Urinary and Fecal Excretion of Radioactive Chromium Cr^{51} in Man. Its Use in the Measurement of Intestinal Blood Loss Associated with Hookworm Infection", J. Clin. Invest. 36, 1183 (1957).

Thyroid Function Methods

74. Hertz, S., A. Roberts, and R. D. Evans, "Radioactive Iodine as an Indicator in the Study of Thyroid Physiology", Proc. Soc. Exper. Biol. & Med. 38, 510 (1938).
75. Hamilton, J. G., and M. H. Soley, "Studies in Iodine Metabolism by Use of a New Radioactive Isotope of Iodine", Am. J. Physiol. 127, 557 (1939); 131, 135 (1940).
76. Oddie, T. H., "Analysis of Radio-Iodine Uptake and Excretion Curves", Brit. J. Radiol. 22, 261 (1949).
77. Luellen, T. J., et al., "Relative Measurement *in Vivo* of Accumulation of Radio-Iodine by the Human Thyroid Gland: Comparison with Radioactivity in Peripheral Tissues", J. Clin. Invest. 28, 207 (1949).
78. Riggs, D. S., "Quantitative Aspects of Iodine Metabolism in Man", Pharmacological Reviews 4, 284 (1952).
79. Hare, E. H., and C. P. Haigh, "Variations in the Iodine Activity of the Normal Human Thyroid as Measured by the 24-Hour I-131 Uptake", Clin. Sci. 14, 441 (1955).
80. Bishopric, G. A., N. H. Garrett, and W. M. Nicholson, "The Thyroid Uptake of Radioactive Iodine as Modified by the Iodine-Restricted Diet", J. Clin. Endocrin. and Metabol. 15, 592 (1955).
81. Rall, J. E., "The Role of Radioactive Iodine in the Diagnosis of Thyroid Disease", Am. J. Med. 20, 719 (1956).
82. Keating, F. R., Jr., M. H. Power, J. Berkson, and S. F. Haines, "The Urinary Excretion of Radioiodine in Various Thyroid States", J. Clin. Invest. 26, 1138 (1947).
83. Childs, D. S., Jr., F. R. Keating, Jr., J. E. Rall, M. M. D. Williams, and M. H. Power, "The Effect of Varying Quantities of Inorganic Iodide (Carrier) on the Urinary Excretion and Thyroidal Accumulation of Radioiodine in Exophthalmic Goiter", J. Clin. Invest. 29, 726 (1950).
84. Perry, W. F., and J. F. S. Hughes, "Urinary Excretion and Thyroid Uptake of Iodine in Renal Disease", J. Clin. Invest. 34, 454 (1952).
85. Berson, S. A., R. S. Yalow, J. Sorrentino, and B. Roswit, "The Determination of Thyroidal and Renal Plasma I-131 Clearance Rates as a Routine Diagnostic Test of Thyroid Dysfunction", J. Clin. Invest. 31, 141 (1952).
86. Fraser, R., Q. J. G. Hobson, D. G. Arnott, and E. W. Emery, "The Urinary Excretion of Radioiodine as a Clinical Test of Thyroid Function", Quart. J. Med. 22, 99 (1953).
87. Hlad, C. J., Jr., and N. S. Bricker, "Renal Function and I-131 Clearance in Hyperthyroidism and Myxedema", J. Clin. Endocrin. and Metabol. 14, 1539 (1954).
88. Clar, D. E., R. H. Moe, and E. E. Adams, "Rate of Conversion of Administered Inorganic Radioactive Iodine into Protein-Bound Iodine of Plasma as an Aid in the Evaluation of Thyroid Function", Surgery 26, 331 (1949).
89. Harsha, W. N., "Evaluation of the Conversion of Radioactive Inorganic Iodine to Protein-Bound Iodine as a Diagnostic Aid in Thyroid Dysfunction", J. Clin. Endocrin. 11, 1524 (1951).

90. Sheline, G. E., M. D. Moore, A. Kappas, and D. E. Clar, "A Correlation Between the Serum Protein-Bound Iodine and the Radioiodine Conversion Ratio in Various Thyroid States", J. Clin. Endocrin. 11, 91, (1951).
91. Scott, K. G., and W. A. Reilly, "Use of Anionic Exchange Resin for the Determination of Protein-Bound I^{131} in Human Plasma", Metabolism 3, 506 (1954).
92. Van Middleworth, L., C. E. Nurnberger, and A. Lipscomb, "Simplified Sensitive Test for Thyroid Function, Using Protein-Bound I-131", J. Clin. Endocrin. 14, 1056 (1954).
93. Paley, K. R., E. S. Sobel, and R. S. Yalow, "A Comparison of Thyroidal Plasma I-131 Clearance and the Plasma Protein-Bound I-131 Tests for the Diagnosis of Hyperthyroidism", J. Clin Endocrin. 15, 995 (1955).
94. Clarke, K. H., and E. V. Sherriff, "The Diagnostic Value of Protein-Bound Iodine and 48 Hour Protein-Bound I-131 as Indices of Thyroid Disorder", M. J. Australia II, 89 (1955).
95. Fields, T., D. S. Kinnory, E. Kaplan, Y. T. Oester and E. N. Bowser, "The Determination of Protein-Bound Iodine With Anion Exchange Resin Column", J. Lab. & Clin. Med. 47, 333 (1956).
96. Zieve, L., W. D. Vogel, and A. L. Schultz, "Determination of Protein-Bound Radioiodine with an Anion Exchange Resin", J. Lab. & Clin. Med. 47, 663 (1956).
97. Gross, J., and R. Pitt-Rivers, "3:5:3'-Triiodothyronine. 2.Physiological Activity", Biochem. J. 53, 652-657 (1953).
98. Crispell, K. R., and J. Coleman, "A Study of the Relative Binding Capacity of Plasma Proteins, Intact Human Red Cells, and Human Red Cell Stroma for Radioactive I-131 Labeled L-Thyrodine", J. Clin. Invest. 35, 475 (1956).
99. Crispell, K. R., S. Kahana and H. Hyer, "The Effect of Plasma on the *In Vitro* Uptake or Binding by Human Red Cells of Radioactive I-131 Labeled L-Thyroxine and L-Triiodothyronine", J. Clin. Invest. 35, 121 (1956).
100. Hamolsky, M. W., M. Stein, and A. S. Freedberg, "The Thyroid Hormone-Plasma Protein Complex in Man. II. A New *In Vitro* Method for Study of 'Uptake' of Labeled Hormonal Components by Human Erythrocytes", J. Clin. Endocrin. & Metab. 17, 33 (1957).
101. Crispell, K. R., J. Coleman, and H. Hyer, "Factors Affecting the Binding Capacity of Human Erythrocytes for I-131-Labeled L-Thyroxine and L-Triiodothyronine", J. Clin. Endocrin. & Metab. 17, 1305 (1957).
102. Hamolsky, M. W., A. Golodetz, and A. S. Freedberg, "The Plasma Protein-Thyroid Hormone Complex in Man. III. Further Studies on the Use of the *In Vitro* Red Blood Cell Uptake of I-131 L-Triiodothyronine as a Diagnostic Test of Thyroid Function", J. Clin. Endocrin. & Metab. 19, 103 (1959).
103. Robbins, L. R., "Experience with the *In Vitro* Erythrocyte Uptake of I-131-Labeled L-Triiodothyronine in a Routine Clinical Laboratory", J. Clin. Endocrin. & Metab. 19, 1292 (1959).
104. Ureles, A. L., and M. Murray, "The Erythrocyte Uptake of I-131-L-Triiodothyronine as a Measure of Thyroid Function", J. Lab. & Clin. Med. 54, 178 (1959).
105. Adams, R., N. Specht, and J. Woodward, "Labeling of Erythrocytes *In Vitro* With Radioiodine-Tagged L-Triiodothyronine as an Index of Thyroid Function: An Improved Hematocrit Correction", J. Clin. Endocrin. & Metab. 20, 1366 (1960).
106. Barrett, O., A. Berman, and J. G. Maier, "Uptake of I-131 L-Triiodothyronine in Various Erythrocyte Abnormalities", J. Clin. Endocrin. & Metab. 20, 1467 (1960).
107. Christensen, L. K., "Triiodothyronine Uptake by Erythrocytes", Acta Med. Scand. 166, 141 (1960).
108. Hlad, C. J., Jr., M. Erwin, and H. Elrick, "Observations on the Binding Properties of Triiodothyronine", J. Nuclear Med. Spec. 1959 Conv. Issue, p. 38 (1960).
109. Meade, R. C., "Possible Errors in the Determination of Red Blood Cell Uptake of I-131-Triiodothyronine", J. Clin. Endocrin. & Metab. 20, 480 (1960).
110. Rabinowitz, J., and B. Shapiro, "Observations on the Uptake of I-131-Triiodothyronine by Erythrocytes", J. Nuclear Med. 3, 309 (1962).
111. Shapiro, B., and J. Rabinowitz, "A Chromatographic Method Utilizing Sephadex for the Separation of Free Iodide, Protein-Bound and Unbound Triiodothyronine in Sera. A) Clinical Correlations with the Hamolsky T-3-RBC Uptake Method", J. Nuclear Med. 3, 417-421 (1962).
112. Walfish, P. G., A. Britton, R. Volpe, and C. Ezrin, "Quantitative Plasma Binding Capacity Studies of Thyroxine-Binding Globulin by Use of a Modification of the Erythrocyte L-Triiodothyronine-I-131 Uptake Test", J. Clin. Endocrin. & Metab. 22, 178-186 (1962).
113. Parrow, A., "The Plasma Protein Binding of Triiodothyronine and Its Relation to the Uptake of I-131-Labeled Triiodothyronine by Human Erythrocytes *In Vitro*", Scand. J. Clin. & Lab. Invest. 14, 192 (1962).
114. Rabinowitz, J., B. Shapiro, and P. Johnson, "'Sephadex Chromatographic' Test in the Evaluation of Thyroid Function", J. Nuclear Med. 4, 139 (1963).

Pernicious Anemia

115. Schilling, R. F., "Intrinsic Factor Studies. II. The Effect of Gastric Juice on the Urinary Excretion of Radioactivity after the Oral Administration of Radioactive Vitamin B_{12}", J. Lab. & Clin. Med. 42, 860 (1953).
116. Schilling, R. F., D. V. Clatanoff, and D. R. Korst, "Intrinsic Factor Studies. III. Further Observations Utilizing the Urinary Radioactivity Test in Subjects with Achlorhydria, Pernicious Anemia, or Total Gastractomy", J. Lab. & Clin. Med. 45, 926 (1955).
117. Krevans, J. R., C. L. Conley, and M. V. Sachs, "Radioactive Tracer Tests for the Recognition and Identification of Vitamin B_{12} Deficiency States", J. Chronic Disease 3, 234 (1956).

118. Meyer, L. M., et al., "Vitamin B_{12} Distribution Determined by Surface Body Counting Following Parental Administration of Co^{60} B_{12}", Proc. Soc. Exper. Biol. & Med. 91, 129 (1956).
119. Rath, C. E., et al., "Value and Limitations of $Co^{60}B_{12}$ Test", Blood 11, 96 (1956).
120. Reisner, E. H., Jr., et al., "Applications of Urinary Tracer Test (of Schilling) as an Index of Vitamin B_{12} Absorption", Am. J. Clin. Nutrition 4, 134 (1956).
121. Arias, I. M., et al., "Absorption of Radioactive Vitamin B_{12} in Nonanemic Patients with Combined System Disease", New England J. Med. 255, 164 (1956).
122. Goldberg, S. R., et al., "Radioactive Vitamin B_{12} Studies. Experience with the Urinary Excretion Test and the Measurement of Absorbed Plasma Radioactivity", J. Lab. & Clin. Med. 49, 582 (1957).
123. Grasbeck, R., et al., "Biliary and Fecal Vitamin B_{12} Excretion in Man. An Isotope Study", Proc. Soc. Exper. Biol. & Med. 97, 780 (1958).

Iron Kinetics

124. Dubach, R., C. V. Moore, and V. Minnich, "Studies in Iron Utilization and Metabolism. V. Utilization of Intravenously Injected Radioactive Iron for Hb Synthesis and an Evaluation of the Radioactive Iron Method for Studying Iron Absorption", J. Lab. & Clin. Med. 31, 1201 (1946).
125. Finch, C. A., J. G. Gibson II, W. C. Peacock, and R. G. Fluharty, "Iron Metabolism; Utilization of Intravenous Radioactive Iron", Blood 4, 905 (1949).
126. Huff, R. L., T. G. Hennessy, R. E. Austin, J. F. Garcia, B. M. Roberts, and J. H. Lawrence, "Plasma and Red Cell Iron Turnover in Normal Subjects and in Patients Having Various Hematopoietic Disorders", J. Clin. Invest. 29, 1041 (1950).
127. Loeffler, R. K., D. A. Rappaport, and V. P. Collins, "Radio-iron Citrate as Tracer to Determine Disappearance Rate of Plasma Iron in Normal Subjects", Proc. Soc. Exper. Biol. & Med. 88, 441 (1955).
128. Spencer, R. P., T. G. Mitchell, and E. R. King, "The Use of Radioisotopes in Diagnositc Hematologic Studies. II. Fe^{59} Erythrokinetic Studies", Amer. J. Clin. Path. 28, 123 (1957).
129. Bothwell, T. H., H. V. Hurtado, D. M. Donohue, and C. A. Finch, "Erythrokinetics. IV. The Plasma Iron Turnover as a Measure of Erythropoiesis", Blood 12, 409 (1957).
130. Finch, C. A., D. H. Coleman, A. G. Motulsky, D. M. Donohue, and R. H. Reiff, "Erythrokinetics in Pernicious Anemia", Blood 11, 807 (1956).
131. Teng, C. T., V. P. Collins, and W. D. West, "A Critical Appraisal of the Diagnositc and Prognostic Value of Radioiron Study in Hematologic Disorders", Am. J. Roentgenol. 84, 687 (1960).
132. Pollycove, M., and R. Mortimer, "The Quantitative Determination of Iron Kinetics and Hemoglobin Synthesis in Human Subjects", J. Clin. Invest. 40, 753 (1961).

Fat Absorption

133. Hoffman, M. C., "Radioactive Iodine-labeled Fat", J. Lab. Clin. Med. 41, 521 (1953).
134. Ruffin, J. M., et al., "I-131 Labeled Fat in the Study of Intestinal Absorption", New England J. Med. 255, 594 (1956).
135. Sanders, A. P., et al., "Radioiodine Recovery in the Feces Following an I-131 Labeled Fat Test Meal", Am. J. Roentgenol. 75, 386 (1956).
136. Beres, P., et al., "The Use of I-131 Triolein in the Study of Absorptive Disorders in Man", Gastroenterology 2, 1 (1957).
137. Isley, J. K., et al., "Use of I-131 Labeled Oleic Acid in the Study of Gastrointestinal Function", Proc. Soc. Exper. Biol. & Med. 94, 807 (1957).
138. Grossman, M. I., and P. H. Jordan, Jr., "The Radioiodinated Triolein Test for Steatorrhea", Gastroenterology 34, 892 (1958).
139. Kaplan, E., et al., "Intestinal Absorption of Iodine-131 Labeled Triolein and Oleic Acid in Normal Subjects and in Steatorrhea", Gastroenterology 34, 901 (1958).
140. Rufin, F., W. M. Blahd, R. A. Nordyke, and M. I. Grossman, "Reliability of I-131 Triolein Test in the Detection of Steatorrhea", Gastroenterology 41, 220 (1961).
141. Rothfeld, B., and J. L. Rabinowitz, "A Comparison of Measurements of Fat Absorption by I-131 and C-14 Labeled Fats", Am. J. Dig. Dis. 9, 263 (1964).

Renal Function

142. Winter, C. C., "A Clinical Study of a New Renal Function Test: the Radioactive Diodrast Renogram", J. Urol. 76, 182 (1956).
143. Taplin, G. V., O. M. Meredith, H. Kade, and C. C. Winter, "The Radioisotope Renogram, An External Test for Individual Kidney Function and Upper Urinary Tract Patency", J. Lab. & Clin. Med. 48, 886 (1956).
144. Winter, C. C., and G. V. Taplin, "A Clinical Comparison and Analysis of Radioactive Diodrast, Hypaque, Miokon and Urokon Renograms as Tests of Kidney Function", J. Urol. 79, 573 (1958).
145. Winter, C. C., "Further Experiences with the Radioisotope Renogram", Am. J. Roentgenol. 82, 862 (1959).
146. Serratto, M., J. T. Grayhack, and D. P. Earle, "A Clinical Evaluation of the Iodopyracet (Diodrast) Renogram", A.M.A. Arch. Int. Med. 103, 851 (1959).
147. Nordyke, R. A., M. Tubis, and W. H. Blahd, "Simultaneous Comparison of Individual Kidney Function Using Radioiodinated Hippuran", Clin. Res. 8, 116 (1960).
148. Tubis, M., E. Posnick and R. A. Nordyke, "Preparation and Use of I-131 Labeled Sodium Iodohippurate in Kidney Function Tests", Proc. Soc. Exptl. Biol & Med. 103, 497 (1960).
149. Mulrow, P. J., A. Huvos, and D. L. Buchanan, ' Measurement of Residual Urine", J. Lab. & Clin.Med. 57, 109 (1961).

APPENDIX

APPENDIX A The MKS and CGS Systems of Units

On January 1, 1940 the meter-kilogram-second system of units (mks system) replaced the centimeter-gram-second system (cgs system) which had been official since 1881. While the cgs system is somewhat more convenient to use for certain applications, the mks system has several distinct advantages:

1. the mks system conforms with the international standards of length (the meter) and mass (the kilogram).

2. definitions of the "practical" or common electrical units, e.g., the ampere, volt, ohm, coulomb, farad and henry, all of long standing and in common use, are identical to the mks definitions. These practical units thus become official.

3. there is no longer any need for two distinct electrical systems since the electrostatic cgs units (esu) and the electromagnetic cgs units (emu) become unnecessary.

Mechanical Units

Entity	Symbol	Dimension	MKS System		CGS System
Length	l	l	1 meter (m)	=	100 centimeters (cm)
Mass	m	m	1 kilogram (kg)	=	1000 grams (g)
Time	t	t	1 second (sec)	=	1 second (sec)
Wavelength	λ	l	1 m	=	100 cm
Area	A	l^2	1 m^2	=	10^4 cm^2
Volume	V	l^3	1 m^3	=	10^6 cm^3
Angular Velocity	ω	t^{-1}	1 radian/sec	=	1 radian/sec
Frequency	f, ν	t^{-1}	1 cycle/sec	=	1 cycle/sec
Density	ρ	ml^{-3}	1 kg/m^3	=	10^{-3} g/cm^3
Velocity	v	lt^{-1}	1 m/sec	=	100 cm/sec
Acceleration	a	lt^{-2}	1 m/sec^2	=	100 cm/sec^2
Flow	—	$l^3\ t^{-1}$	1 m^3/sec	=	10^6 cm^3/sec
Rotational Inertia	I	ml^2	1 kg-m^2	=	10^7 g-cm^2
Angular Momentum	p_ϕ	ml^2t^{-1}	1 kg-m^2 /sec	=	10^7 g-cm^2/sec
Force	F	mlt^{-2}	1 newton (nt)	=	10^5 dynes (dyn)
Momentum	p	mlt^{-1}	1 kg-m/sec (= 1 nt-sec)	=	10^5 g-cm/sec
Pressure	P	$ml\ ^{1}t^{-2}$	1 nt/m^2	=	10 dyn/cm^2
Work	W				
Energy, total	E	ml^2t^{-2}	1 joule (= 1 nt-m) (= 1 watt-sec)	=	10^7 erg (= 10^7 dyn-cm)
potential	U				
kinetic	T				
Power	P	ml^2t^{-3}	1 watt (= 1 joule/sec)	=	10^7 erg/sec

Electrical Units

Entity	Symbol	MKS System		CGS System esu		emu	Ratio esu/emu
Quantity	Q	1 coulomb (coul)	=	3×10^9 statcoulomb (escoulomb)	=	0.1 abcoulomb	3×10^{10}
Resistance	R	1 ohm (Ω)	=	$\frac{1}{9} \times 10^{-11}$ statohm	=	10^9 abohm	$1/(3 \times 10^{10})^2$
Potential	E	1 volt (V)	=	$\frac{1}{3} \times 10^{-2}$ statvolt (esvolt)	=	10^8 abvolt	$1/(3 \times 10^{10})$
Current	I	1 ampere (amp)	=	3×10^9 statampere (esampere)	=	0.1 abampere	3×10^{10}
Capacitance	C	1 farad (f)	=	9×10^{11} statfarad	=	10^{-9} abfarad	$(3 \times 10^{10})^2$
Inductance	L	1 henry (h)	=	$\frac{1}{9} \times 10^{-11}$ stathenry	=	10^9 abhenry	$1/(3 \times 10^{10})^2$
Magnetic flux	Φ	1 weber			=	10^8 maxwell	
Magnetic flux density	B	1 weber/m^2			=	10^4 gauss	
Magnetic field strength	H	1 amp turn/m			=	$4\pi/10^3$ oersted	

Selected Electrical and Energy Relationships								
E	=	IR	1 volt	=	1 amp	×	1 ohm
P	=	IE	1 watt	=	1 amp	×	1 volt
Q	=	CE	1 coulomb	=	1 farad	×	1 volt
Q	=	It	1 coulomb	=	1 ampere	×	1 sec
W	=	QE	1 joule	=	1 coulomb	×	1 volt
W	=	Pt	1 joule	=	1 watt	×	1 sec
F	=	m*a*	1 newton	=	1 kilogram	×	1 m/sec^2

Selected Physical Constants in MKS Units	
Electronic charge, e	= 1.60 × 10^{-19} coulombs
	(1 coulomb = 6.28 × 10^{18} electronic charges)
	(1 abcoulomb = 6.28 × 10^{19} electronic charges)
	(1 statcoulomb = 2.09 × 10^{9} electronic charges)
1 electron volt, eV	= 1.60 × 10^{-19} joules (volt-coulombs or watt-sec)
Electronic charge to mass ratio, e/m	= 1.76 × 10^{11} coul/kg
Electronic mass, m	= 9.103 × 10^{-31} kg
1 atomic mass unit, amu	= 1.659 × 10^{-27} kg
Planck's constant, h	= 6.610 × 10^{-34} joule-sec
Speed of light, c	= 2.998 × 10^{8} m/sec

APPENDIX B The Greek Alphabet

Alpha	Α	α
Beta	Β	β
Gamma	Γ	γ
Delta	Δ	δ
Epsilon	Ε	ϵ
Zeta	Ζ	ζ
Eta	Η	η
Theta	Θ	θ
Iota	Ι	ι
Kappa	Κ	κ
Lambda	Λ	λ
Mu	Μ	μ
Nu	Ν	ν
Xi	Ξ	ξ
Omicron	Ο	o
Pi	Π	π
Rho	Ρ	ρ
Sigma	Σ	σ
Tau	Τ	τ
Upsilon	Υ	υ
Phi	Φ	ϕ
Chi	Χ	χ
Psi	Ψ	ψ
Omega	Ω	ω

APPENDIX C Units, Constants and Conversion Factors

ATOMIC

Mass of ^{12}C atom	=	Exactly 12 amu
1 atomic mass unit (amu)	=	1.661×10^{-24} g
Proton charge	=	4.803×10^{-10} esu
Proton rest mass (M_p)	=	1.007277 amu
Proton rest mass (M_p)	=	1.6725×10^{-24} g
Neutron rest mass (M_n)	=	1.008665 amu
Neutron rest mass (M_n)	=	1.6748×10^{-24} g
Mass of hydrogen atom (M_H)	=	1.007825 amu
Mass of hydrogen atom (M_H)	=	1.6734×10^{-24} g
Avogadro constant (N_0)	=	6.0225×10^{23}/g-mole
Boltzmann constant (k)	=	1.3805×10^{-16} erg/deg
Faraday constant ($\mathfrak{F}$)	=	9.6487×10^{4} coulomb/g-equiv

ENERGY

1 electron volt (eV)	=	1.6020×10^{-12} erg
	=	1.07×10^{-9} amu
1 million electron volts (MeV)	=	10^{6} eV
1 billion electron volts (BeV)	=	10^{9} eV
1 ion pair in air	≈	32.5 eV
1 watt	=	1 joule/sec
	=	10^{7} ergs/sec
1 horsepower	=	746 watts
1 British thermal unit (BTU)	=	1055.18 joules
	=	252 calories
1 electron volt/molecule	=	23,052 calories/mole
1 atomic mass unit (amu)	=	9.31×10^{2} MeV
	=	1.49×10^{-3} ergs
	=	3.56×10^{-11} calories
	=	4.15×10^{-17} kilowatt-hours
1 MeV	=	1.07×10^{-3} amu
	=	1.60×10^{-6} erg
	=	3.83×10^{-14} calories
	=	4.45×10^{-20} kilowatt-hours
1 Erg	=	6.71×10^{2} amu
	=	6.24×10^{5} MeV
	=	2.39×10^{-8} calories
	=	2.78×10^{-14} kilowatt-hours
1 Calorie	=	2.81×10^{10} amu
	=	2.62×10^{13} MeV
	=	4.18×10^{7} ergs
	=	1.16×10^{-6} kilowatt-hours
1 Kilowatt-hour	=	2.41×10^{16} amu
	=	2.25×10^{19} MeV
	=	3.60×10^{13} ergs
	=	8.60×10^{5} calories

ELECTRONIC

Electronic charge (mks system)	=	1.6021×10^{-19} coulomb
(cgs system)	=	1.6021×10^{-20} emu (abcoulomb)
(cgs system)	=	4.8030×10^{-10} esu (statcoulomb)
Electronic rest mass (m_0)	=	9.1091×10^{-28} g
	=	5.486×10^{-4} amu
Electron rest energy (mc^2)	=	0.51101 MeV
1 Coulomb	=	6.28×10^{18} electrons

RADIATION

Planck's constant (h)	=	6.6256×10^{-27} erg-s
	=	6.6256×10^{-34} J-s
($\hbar = h/2\pi$)	=	1.0545×10^{-27} erg-s
Speed of light in vacuum (c)	=	2.9979×10^{10} cm/sec
Photon wavelength at 1 eV	=	$12{,}395 \times 10^{-8}$ cm
Rydberg constant for hydrogen (R_H)	=	1.0967758×10^{5} cm^{-1}
Rydberg constant for infinite mass (R_∞)	=	1.0973731×10^{5} cm^{-1}

RADIOACTIVITY

1 curie (Ci)	=	3.7×10^{10} dps = 2.22×10^{12} dpm
1 millicurie (mCi)	=	10^{-3} curies = 3.7×10^{7} dps
1 microcurie (μCi)	=	10^{-6} curies = 3.7×10^{4} dps
1 rutherford (rd)	=	1×10^{6} dps = 6.00×10^{7} dpm
	=	1/37 millicurie
1 millirutherford (mrd)	=	10^{-3} rutherfords = 10^{3} dps
1 microrutherford (μrd)	=	10^{-6} rutherfords = 1 dps

RADIATION DOSE

1 roentgen (r)	=	1 esu/cc of standard air
	=	2.083×10^{9} ion pairs/cc of standard air
	=	1.61×10^{12} ion pairs/gram standard air
	=	6.77×10^{4} MeV/cc of standard air
	=	5.24×10^{7} MeV/gram of standard air
	=	83.8 erg/gram air
1 milliroentgen	=	10^{-3} roentgens
1 rad	=	100 ergs/g
1 rep	=	84 ergs/g in air
	=	93 ergs/g in water or tissue

NATURAL LOGARITHMS

ln 10	=	2.30258
e	=	2.71828

MASS

1 Kilogram (kg)	=	2.205 pounds
1 Pound (lb)	=	453.592 g

LENGTH

1 centimeter (cm)	=	0.3937 inches
1 micron (μ)	=	10^{-4} cm
1 millimicron (mμ)	=	10^{-7} cm
1 Angstrom unit (A)	=	10^{-8} cm
1 Fermi	=	10^{-13} cm

AREA

1 barn (b)	=	10^{-24} cm^2
1 millibarn (mb)	=	10^{-27} cm^2
1 microbarn (μb)	=	10^{-30} cm^2

TIME

Number of seconds in 1 day	=	86,400
Number of seconds in 1 week	=	6.048×10^5
Number of seconds in 365 days	=	3.156×10^7

DENSITY

Aluminum	2.70	Gold	19.32	Lead	11.35
Cadmium	8.65	Indium	7.28	Silver	10.50
Copper	8.94	Iron	7.86	Uranium	18.68

APPENDIX D Glossary of Symbols and Abbreviations

SYMBOLS

A	Disintegration rate ($= -dN/dt$)
A	Mass number
a	Radius of Bohr orbit
a	Acceleration
B	Binding energy
B	Buildup factor
C	Concentration
C	Capacitance
C	Millicurie strength of a source
c	Velocity of light
D_β, D_γ	Total beta dose; total gamma dose
d	Distance, cm
d	Deuteron
E	Total energy
E_α, E_β	Kinetic energy of an alpha particle; kinetic energy of a beta particle
E_m	Maximum energy (e.g., of a beta ray spectrum)
$\bar{E}_\beta$	Average beta energy
E	Single electrode potential
E_c	Cell potential
E^0	Standard electrode potential
e	Base of Naperian (natural) logarithms (= 2.71828)
e	Elementary charge; electronic charge
e^+, e^-	Positron; negatron
f_A, f_B, f_H, f_S, f_W	Counting geometry factors or coefficients for air scattering, backscattering, side-scattering, self absorption, and window and air absorption respectively
F	Force
F	Packing fraction
$\mathfrak{F}$	Faraday
f	Frequency
G	Physical geometry factor
G_n	Gaussian probability of obtaining a count of n
G	Gravitational constant (6.673×10^{-11} nt-m^2/kg^2)
h	Planck's constant

$\hbar$	Unit of angular momentum ($= h/2\pi$)
I	Intensity of radiation
I	Current
I	Rotational inertia
K	Linear pair production attenuation coefficient
K_{sp}	Solubility product constant
k	Reaction rate constant
k	Boltzmann constant ($= R/N_0$)
L	Inductance
l	Length
l	Quantum number for orbital angular momentum
M	Isotopic mass of neutral atom, amu
M_H, M_{He}	Isotopic mass of neutral hydrogen atom and neutral helium atom respectively, amu
M_n	Neutron mass, amu
M_p	Proton mass, amu
M′	Nuclear mass, amu
M	Molarity
m	Relativistic mass of a particle with rest mass of m_0, g
m	Mass of a substance, g or mg
m	Magnetic quantum number
N	Neutron number
N	Number of atoms per cm^3 in a target
N	Number of atoms
N	Number of observations (statistics)
N	Normality
N_0	Avogadro constant
n	Neutron
n	Observed number of random events (statistics)
$\bar{n}$	Mean value of n for a number of observations (statistics)
n	Principle quantum number
P	Probable error (statistics)
P_n	Poisson probability of obtaining a count of n
P	Pressure; power
p	Proton
p	Momentum
p	Probability of occurrence of a random event
p_ϕ	Angular momentum ($= I\omega$)
Q	Energy change in a nuclear reaction
Q	Charge of electricity

R	Corrected counting rate
R	Nuclear radius
R_b, R_s, R_t	Corrected count rate for background, sample and total activities.
R_∞	Rydberg constant
R	Range of a particle
R	Molar gas constant
R_i	Internal resistance of a cell
R_m	Maximum range of a particle
RC	Time constant of a circuit with resistance R and capacitance C
r	Observed count rate (= n/t)
r_b, r_s, r_t	Observed background, sample and total count rates
S	Area
S	Separation energy
S	Stopping power
S	Specific activity
S_a, S_n	Apparent specific activity; normalized specific activity
s	Spin quantum number
s	Sample standard deviation (statistics)
s_D	Sample standard deviation of a difference (statistics)
s^2	Sample variance (statistics)
T	Kinetic energy
T	Resolving time of a nuclear particle detector
t	Time
$t_{1/2}$	Half-life; reaction half-time
t_b, t_r, t_e	Biological, radiological and effective half-lives
U	Potential energy
u	Apparent absorption coefficient for a beta particle, cm^2/mg
u	Wave velocity
V	Volts
V	Volume
v	Velocity
W	Work
X	Absorber thickness, cm
$X_{1/2}$	Half-value thickness or half-value layer, cm
x	Absorber thickness, mg/cm^2
Z	Atomic number
Ze	Nuclear charge

Symbol	Meaning
α	Alpha particle
β	Velocity relative to velocity of light (= v/c)
β, β^-, β^+	Beta particle, negatron and positron
Γ	Specific gamma-ray emission or specific gamma-ray output
γ	Gamma ray or other photon
ϵ	Efficiency
ϵ	Dielectric constant (cgs); Permittivity (mks)
λ	Wavelength
λ	Decay constant
λ_c	Compton wavelength (= h/m_0c)
μ	Actual or true mean count
μ	Total linear absorption coefficient, cm^{-1}.
μ	Reduced mass
ν	Frequency, sec^{-1}.
ν	Neutrino
$\bar{\nu}$	Antineutrino
$\bar{\nu}$	Wave number (= $1/\lambda$)
π	Ratio of circumference of a circle to its diameter (= 3.14159)
ρ	Density, g/cm^3
Σ	Macroscopic cross section, cm^{-1}.
σ	Cross section of a target, cm^2 per target particle
σ	True standard deviation (statistics)
σ^2	Variance
$\sigma, \sigma_s, \sigma_a$	Linear coefficients for Compton collision, Compton scattering and Compton absorption.
σ_D	Standard deviation of the difference
τ	Mean life (= $1/\lambda$)
τ_e	Effective mean life
τ	Photoelectric linear absorption coefficient
τ	Relative error (statistics)
Φ	Particle fluence (e.g., neutrons per second passing through a surface)
ϕ	Particle flux density (e.g., neutrons/sec/cm^2)
φ	Work function or electron binding energy
X^2	Pearson's chi square function
Ψ	Amplitude factor (wave mechanics)
ψ	Wave function or eigenfunction
Ω	Ohms (electrical resistance)
ω	Overvoltage
ω	Angular velocity

ABBREVIATIONS

A	Angstrom (= 10^{-8} cm)
abs	Absolute
A.C.	Alternating current
amu	Atomic mass unit
b	Barn
b.p.	Boiling point
coul	Coulomb
cgs	Centimeter-gram-second
cm	Centimeter
cpm	Counts per minute
cps	Counts per second
Ci	Curie
CRM	Counting rate meter
D.C.	Direct current
DE	Dose equivalent
dpm	Disintegrations per minute
dps	Disintegrations per second
dyn	Dyne
EC	Electron capture
emu	Electromagnetic unit
esu	Electrostatic unit
eV	Electron volt
f	Farad
g	Gram
h	Hour
HCT	Hematocrit
HVL	Half-value layer
IHSA	Iodinated human serum albumin
IP	Ion product
I. P.	Ion pairs
IT	Internal transition
keV	Kilo-electron volt
LD_{50}	Leathal dose 50%
ln	Naperian logarithm
log	Base-10 logarithm
m	Meter
meq	Milli-equivalent
MeV	Million-electron volt
mb	Millibarn
mCi	Millicurie
Min	Minute
mg	Milligram
mks	Meter-kilogram-second
ml	Milliliter
m.p.	Melting point
mr	Milliroentgen
MCi	Megacurie
nt	Newton
PBI	Protein-bound iodine
r	Roentgen
rad	Radiation dose (an absorbed dose of 100 ergs/g)
RBC	Red blood cell
RBE	Relative biological effectiveness
rem	Roentgen equivalent man
rep	Roentgen equivalent physical
sec	Second
SPU	Serum protein uptake
TCA	Trichloroacetic Acid
T-3	Tri-iodothyronine
μCi	Microcurie
μg	Microgram
μl	Microliter
μsec	Microsecond

PREFIXES

T	tera	10^{12}	c	centi	10^{-2}
G	giga	10^{9}	m	milli	10^{-3}
M	mega	10^{6}	μ	micro	10^{-6}
k	kilo	10^{3}	n	nano	10^{-9}
h	hecto	10^{2}	p	pico	10^{-12}
dk	deka	10	B	billion*	10^{6} (U.S.)
d	deci	10^{-1}			10^{12} (U.K.)

* giga or tera are preferred to billion to eliminate ambiguity

APPENDIX E Glossary of Nuclear Terms

A—Symbol for mass number. The sum of Z (atomic number) and N (neutron number).

ABSORBER—Any material used to absorb radiation for aa specific purpose. Absorbers are used (a) for the determination of the characteristics of a specific radiation; (b) as shields for reducing the intensity of radiation for safe handling and storage; and (c) for removing a particular component from a complex source of radiation. Calibrated absorbers are usually made in the form of metal discs.

ABSORPTION—A process in which all or part of the energy of incident radiation is transferred to the matter through which it passes by various interactions with the basic particles (electrons, nuclei, etc.) of which the matter consists. This process results in a reduction of the radiation intensity or a reduction in the number of particles emerging from the absorbing material relative to the number of incident particles. Also, a nuclear reaction or process in which a particle is absorbed by a nucleus, frequently resulting in the release of a different particle by the nucleus.

ABSORPTION COEFFICIENT—The rate of change in the intensity of a beam of radiation as it passes through matter. The *linear absorption coefficient* is the fractional decrease in beam intensity per unit of distance. The *mass absorption coefficient* is the fractional decrease in beam intensity per unit of surface density (cm^2/g).

ACCELERATOR—Commonly referred to as an "atom smasher." A device used to impart a high kinetic energy to a charged particle to cause it to undergo nuclear or particle reactions. From the standpoint of the associated "temperature" in the light of the kinetic theory, the accelerator occupies the same position with respect to nuclear reactions that the Bunsen burner occupies in the field of chemical reactions. Common accelerators are the cyclotron, synchrotron, Van de Graaff accelerator and betatron.

ACD SOLUTION—Anticoagulant Acid Citrate Dextrose Solution, a sterile solution of citric acid (7.3 g), sodium citrate (22 g), and dextrose (24.5 g) in water for injection (q.s., 1000 ml) used to preserve whole blood.

ACTINIDES—The series of elements having atomic numbers 89 through 103. The actinide series occupies the last row of the periodic table. Elements in this series result from filling the 5f shell and are tripositive.

ACTION—The product of work and time. Units of action are the erg-second and the joule-second. Planck's constant is a constant of action.

ACTIVATION—The process of causing a substance to become artificially radioactive by subjecting it to bombardment by neutrons or other particles.

ACTIVATION ANALYSIS—An analytical procedure permitting the detection and measurement of trace quantities of elements following their exposure to a flux of neutrons.

ACTIVATION ENERGY—The energy required for a reaction to occur. From the kinetic point of view, the minimum collision velocity (energy) required for interaction between the colliding particles.

ACTIVITY—The strength of a radioactive source. In absolute units, it relates to the number of radioactive atoms decaying per unit of time; in relative terms, it is expressed in terms of the number of recorded counts per unit of time. Also a synonym for radioactivity. Absolute activity is usually expressed in curies or millicuries.

ACUTE—Having a short and relatively severe course; not chronic. (Clinic)

AIR MONITOR—A detecting device, used for control and warning purposes, to measure the amount of radioactivity present in the air.

ALPHA CHAMBER—A sensing device or counting tube for the detection and measurement of alpha particles possessing some degree of discrimination against radiation of other types.

ALPHA DECAY—The radioactive transmutation of one element into another by the emission of an alpha particle.

ALPHA PARTICLE—A particle which is identical to the helium nucleus, consisting of two protons and two neutrons. It carries a positive charge of 2.

ALPHA RADIATION—The radiation consisting of alpha particles emitted by certain radioactive atoms.

ALPHA RAYS—A stream of helium nuclei. The helium nucleus has a mass number of 4 and an atomic number of 2. It consists of two protons and two neutrons.

ANGSTROM—A unit of length equal to 10^{-8} centimeters.

ANGULAR ACCELERATION—The time rate of change of angular velocity. $\alpha = d\omega/dt$. Dimensions—t^{-2}.

ANGULAR MOMENTUM—Also called the moment of momentum. It is the product of the angular velocity and the moment of inertia. Dimensions—ml^2t^{-1}.

ANGULAR VELOCITY—Time rate of motion about an axis. $\omega = \theta/t$. Dimensions—t^{-1}.

ANION—A negatively charged ion.

ANNIHILATION—The reaction between a pair of anti-particles resulting in the disappearance of the particles and the production of an equivalent amount of energy in the form of photons; e.g., the interaction between an electron and a positron. These reactions result in the conversion of mass into electromagnetic radiation.

ANNIHILATION RADIATION—The radiation produced by the reaction between a particle and an anti-particle resulting in the annihilation of both. The radiation is electromagnetic possessing a total energy equivalent to the masses of the particles annihilated.

ANODE—The electrode by which electrons leave an electrolyte.

ANTICOINCIDENCE CIRCUIT—An electronic circuit with two inputs but only one output, so designed that an output pulse will be produced if a pulse is received at only one of the inputs. If a pulse is received by each input within a given time no output results.

ANTIMATTER—Matter in which the ordinary nuclear particles (neutrons, protons, electrons, etc.) are conceived to be replaced by their corresponding antiparticles (antineutrons, antiprotons, positrons, etc.). Normal matter and antimatter would mutually annihilate each other upon contact and be converted into gamma rays.

ANTINEUTRINO—The particle emitted simultaneously with a negatron during radioactive decay. The anti-neutrino is the anti-particle of the neutrino. It has no charge and possibly no rest mass. It is sometimes considered a hole in the Dirac lattice of neutrinos theoretically permeating all space and matter.

ANTI-PROTON—A particle possessing exactly the same properties of a proton except that its charge is negative.

ATOMIC ENERGY—A misnomer for nuclear energy but accepted because of common usage to denote the energy released in nuclear reactions.

ATOMIC MASS—The mass of a neutral atom expressed in atomic mass units. The *atomic mass unit* is defined as exactly one-twelfth the mass of the carbon isotope ^{12}C.

ATOMIC NUMBER—The positive charge of the nucleus expressed as multiples of the electronic charge e. The symbol for atomic number is Z.

ATOMIC WEIGHT—The average weight of the neutral atoms of an element existing as a mixture of isotopes in the same ratio as found in nature.

ATOM SMASHER—(See accelerator). Colloquial term for accelerator.

ATTENUATION—The decrease in the intensity of radiation caused by the absorption and scattering of the radiation as it passes through matter.

AUGER EFFECT—A process involving the transition of an orbital electron from an excited state to a lower energy level resulting in the production of an x-ray. The x-ray so produced may not escape from the sphere of influence of the atom before colliding with a second orbital electron to which it imparts all its energy. This electron, called an *Auger electron*, is ejected from the atom with a kinetic energy equal to that of the x-ray less the binding energy of the electron.

AUTORADIOGRAPH (Radioautograph)—A photographic record showing the location of radioactivity in an object prepared by placing the radioactive object in contact with unexposed film or photographic emulsion followed by the usual developing process.

AVALANCHE—A process in which a strong electrostatic field accelerates a single charged particle, imparting to the particle sufficient energy to produce additional charged particles through ionization. Repetition of this process results in the production of an extremely large number of charged particles, often called a Townsend avalanche.

AVERAGE LIFE—Synonym for mean life.

BACKGROUND—In measurements of radioactivity, the observed count, in the absence of a sample, caused by cosmic radiation, instrument noise, power line fluctuations, etc. Also called background radiation.

BACKSCATTERING—The deflection of radiation by the sample support or shielding, amounting to more than a 90° change in direction, such that radiation emitted in a direction other than toward the sensitive volume of a counter is deflected toward the counter with a resultant increase in the observed counting rate.

BARN—A unit of area used for expressing the area of nuclear cross sections. 1 barn $= 10^{-24}$ cm^2.

BETA DECAY—The radioactive disintegration of a nucleus resulting in the emission of an electron (beta particle). If the beta is negative, the process results in an increase of atomic number by one unit, but the atomic mass remains the same. Positron decay results in a decrease in atomic number by one unit.

BETA EMITTER—Any radioactive nuclide that decays by beta decay with the emission of a beta particle.

BETA PARTICLE—An electron, either positive or negative. Positive beta particles are called positrons β^+. Negative beta particles are sometimes called negatrons β^-. The term beta particle and the symbol β are reserved for electrons originating in a nucleus.

BETA RAY—Synonym for a stream of beta particles.

BETA RAY SPECTROMETER—An instrument used to measure the energy distribution of beta particles.

BETATRON—A device for accelerating beta particles by induction by a process analogous to the induction of a flow of current in the secondary winding of a transformer.

BEV—Symbol for billion electron volts. 1 BeV $= 10^9$ eV. It is a unit of energy.

BEVATRON—A synchrotron designed to accelerate protons.

BINARY SCALER—A counting device utilizing the binary system of numbers.

BINDING ENERGY—The energy with which a particle is held to an atom or nucleus. *Electron binding energy* is a synonym for ionization potential. *Nuclear binding energy* for a neutron, proton or alpha particle is equal to the difference in mass between the original nucleus and the sum of the product particles.

BIOLOGICAL HALF LIFE—The time required for one-half of an administered substance to be excreted from the body or from an organ or section of living tissue.

BLOOD DYSCRASIA—An persistent change from normal of one or more of the blood components.

BODY BURDEN—The amount of radioactive material present in the body of man or animals.

BOHR RADIUS—The orbital radius (cm) calculated from the Bohr equation,

$$a = n^2h^2/4\pi^2mZe^2$$
$$= (0.529 \times 10^{-8})\ n^2$$

where n is the principal quantum number.

BOMBARDMENT—The act of subjecting a substance to a flux of neutrons or other high energy particles.

BONE SEEKER—A radioisotope that tends to lodge in the bones when it is introduced into the body. Example: strontium-90, which behaves chemically like calcium.

BORON COUNTER—An ionization chamber containing a boron compound for the detection of neutrons.

BRAGG CURVE—A curve showing the specific ionization of an ionizing particle as a function of distance or energy.

BRANCHING—The phenomenon involving the radioactive decay of certain species of nuclei by a choice of modes, a given statistical fraction of the nuclei of a species undergoing decay by one process, the remainder by a different process. The *branching fraction* is the fraction decaying by a particular mode and the *branching ratio* is the ratio of the two branching fractions.

BREEDER REACTOR—A nuclear reactor so designed that it produces more fuel than it consumes. This is accomplished by producing in the nuclear reaction more atoms of fissionable plutonium than the number of uranium atoms consumed.

BREMSSTRAHLUNG—Electromagnetic radiation emitted when a charged particle undergoes acceleration, especially by inelastic collision with another charged particle. X-rays are a typical example being produced by a similar process.

BY-PRODUCT MATERIAL—In atomic energy law, any radioactive material (except source or fissionable material) obtained in the process of producing or using source or fissionable material. Includes fission products and many other radioisotopes produced in nuclear reactors.

CALORIE—The quantity of heat required to raise the temperature of 1 gram of water 1 degree centigrade.

CALUTRON—A device used to separate isotopes. It is based on the principle of the mass spectrometer.

CAPACITANCE—Ratio of the charge on a condenser to the voltage across the plates:

$$C = Q/E$$

where C is the capacitance in farads, Q the charge in coulombs, and E the voltage.

CAPTURE—A process involving a collision between a particle and a nucleus resulting in the retention of the particle within the nucleus and the emission of electromagnetic radiation called *Capture gamma rays*.

CARBON CYCLE—A series of thermonuclear reactions proposed by Hans Bethe to account for the tremendous release of energy from the sun.

CARRIER—Stable atoms which are mixed with radioactive atoms of the same element (i.e., same atomic number) in the same chemical form for the purpose of carrying out a chemical process.

CARRIER-FREE—The adjective applied to a nuclide which is essentially free of its stable isotopes.

CATAPHORESIS—See Electrophoresis.

CATHODE—The electrode by which electrons enter an electrolyte or gaseous conductor.

CATHODE RAYS—Electrons, emitted from the cathode of any gas or vacuum tube.

CERENKOV RADIATION—Visible light emitted by charged particles as they pass from a transparent medium of low refractive index to a transparent medium of high refractive index when their velocity in the first medium exceeds the velocity of light in the second.

CHAIN REACTION—Any reaction in which one of the products formed is capable of initiating further reactions of the same type.

CHARGE—The fuel (fissionable material) used in reactor to sustain a chain reaction.

CLOUD CHAMBER—A chamber in which a supersaturated vapor can be produced in which the paths of ionizing radiation can be observed.

COFFIN—A thick walled container used for transporting radioactive substances.

COHERENT SCATTERING—A type of scattering in which a definite phase relationship exists between the incident and scattered photons or particles. Bragg scattering is an example of coherent scattering. The Bragg equation $n\lambda \doteq 2d \sin \theta$ defines the conditions for scattering. For particles, λ is the deBroglie wavelength.

COINCIDENCE—Occurrence of two or more events simultaneously. It usually refers to ionizing events.

COINCIDENCE CIRCUIT—An electronic circuit having two inputs but only one output. An output pulse is produced only when an input pulse is received at both inputs simultaneously or within a given interval known as the concidence time.

COINCIDENCE COUNTING—A method of counting employing a coincidence circuit so that an event is recorded only if events are detected in two sensing devices simultaneously. Such counting methods are used to reduce background counts.

COINCIDENCE LOSS—The loss of register of events caused by their occurring within a span of time too short to be resolved by the electronic circuit. Also referred to as *counting loss*, or *resolving time loss*. The correction applied is termed the coincidence correction.

COLLIMATOR—An apparatus, often consisting of a pair of slits, used to confine radiation to a narrow beam.

COLLISION—Any interaction between photons or particles involving an interchange of momentum, kinetic energy or charge.

COMPTON EFFECT—The collision of a photon and an electron in which the direction of the photon is changed and its energy reduced. The electron is thus set in motion by the collision and is called the *Compton recoil electron.*

COMPTON WAVELENGTH—For the electron the compton wavelength λ_0 is defined as

$$\lambda_c = h/m_ec = 2.43 \times 10^{-10} \text{cm}$$

(See also deBroglie wavelength, with which it is sometimes confused.)

CONTAMINATION—The presence of unwanted radioactive matter, or the "soiling" of objects or materials with "radioactive dirt."

CONTROL ROD—A rod composed of a neutron-absorbing element such as cadmium used to control the rate of fission in a reactor.

CONVERTER—A reactor which uses one kind of fuel and produces another. For example a converter charged with uranium isotopes might consume Uranium-235 and produce plutonium from Uranium-238.

COPRECIPITATION—The precipitation of a trace substance along with a weighable quantity of another precipitating substance.

CORE—The region in a nuclear reactor where the fission process occurs.

COSMIC RAYS—Radiation originating outside the earth's atmosphere, consisting of particles capable of producing ionizing events, primarily protons and nuclei, some of which have energies as high as 10^{15} eV. *Secondary cosmic rays* are produced by the interaction of *primary cosmic rays* with the earth's atmosphere.

COSMOTRON—A proton-synchrotron. The magnets of the Brookhaven machine weigh 2,200 tons.

COULOMB—A unit quantity of electricity equal to 6.28×10^{18} electrons. It is that quantity of electricity required to deposit 0.0011180 grams of silver.

COULOMB FORCE—The electrostatic force of attraction or repulsion exhibited between charged particles.

COUNTER—A device for indication of, and often for recording, nuclear radiation.

COUNTING LOSS—See coincidence loss.

COUNTING-RATE METER—An instrument which indicates the average time rate of occurrence of events. Connected to a Geiger tube, it will register counts per minute.

CRITICAL MASS—The mass of fissionable material required to sustain a nuclear reaction.

CROSS SECTION—The apparent cross-sectional area of a nucleus as calculated on the basis of the probability of occurrence of a reaction by collision with a particle. It may or may not coincide with value for the geometrical cross-sectional area, πR^2.

CURIE—A unit of radioactivity defined as 3.70×10^{10} disintegrations per second. It also refers to that quantity of a nuclide containing 3.70×10^{10} disintegrating atoms per second.

CUTIE-PIE—A "Gun-type," portable instrument for measuring the radiation level.

CYCLOTRON—A circular magnetic resonant accelerator.

CYTOPLASM—The protoplasm of a cell exclusive of that of the nucleus.

DAUGHTER—The decay product produced by a radioactive nuclide. When the *parent* nuclide undergoes decay a *daughter* nuclide is produced.

DAUGHTER (DAUGHTER ELEMENT)—The nuclide formed by the radioactive decay of another nuclide which in this context is called the "parent."

DEAD TIME—The length of time immediately following an impulse that the instrument remains insensitive and unable to record another ionizing event. (See coincidence loss).

DECAY—The radioactive disintegration of a nucleus.

DECAY SERIES—Synonym for decay chain. A series of radioactive decompositions in which radioactive daughters are produced, these radioactive daughters then becoming parents by producing radioactive daughters of their own.

de BOROGLIE WAVELENGTH—The wavelength associated with a moving particle of mass m, momentum p, and velocity v, derived by equating the expressions $E = mc^2$ and $E = h\nu$.

DECADE SCALER—A recording instrument having a scaling factor of ten.

DECAY CONSTANT—Synonym for disintegration constant.

DEE—An electrode of a cyclotron so named because it is shaped like the letter "D".

DEUTERIUM—An isotope of hydrogen having one proton and one neutron in the nucleus. It is often called heavy hydrogen.

DEUTERON—A particle consisting of one proton and one neutron and having a positive charge of 1. It is also the deuterium nucleus.

DISCRIMINATOR—An electronic device capable of accepting or rejecting a pulse according to the pulse height or voltage.

DISINTEGRATION—A spontaneous, radioactive transformation of one species of nucleus into a nucleus of a different type, usually accompanied by the emission of radiation.

DOERNER-HOSKINS DISTRIBUTION LAW—A law explaining the logarithmic distribution of activity occurring in certain cases of coprecipitation.

DOPPLER EFFECT—A shift in the measured frequency of a wave pattern caused by movement of the receiving device or wave source. The moving receiver will intercept more or fewer waver per unit time depending on whether it is moving toward or away from the source of the waves. By analogy, in a reactor, since fission cross sections depend on the relative velocity of the neutrons and the uranium atoms (neutron movement can be considered wave motion), vibration of the uranium atoms in a fuel element due to the increased operating temperature leads to the Doppler effect. This Doppler effect can vary the reactivity of the reactor.

DOSE—the amount of ionizing radiation energy absorbed per unit mass of irradiated material at a specific location, such as a part of the human body. Measured in reps, rems, and rads.

DOSE, AIR—The dose of radiation in roentgens delivered to a point in free air.

DOSE, INTEGRAL—The total, accumulative dose of radiation received by an individual, usually expressed in terms of gram-roentgens.

DOSE METER—Synonym for dosimeter. Any instrument which measures radiation dosage.

DOSE, PERMISSIBLE—The dose of radiation which can be received by the body without resulting in the production of any harmful effects.

DOSE RATE—The dose of radiation received per unit of time.

DOSIMETER—A device (dose meter) used to measure the radiation dose to which an individual has been exposed.

DYNAMIC-CONDENSER ELECTROMETER—A vibrating-reed electrometer used as a direct current voltage amplifier by converting D.C. to A.C., amplifying the A.C. voltage and rectifying the amplified voltage back to D.C.

DYNE—The unit of force which, when acting upon a mass of 1 gram, will cause the mass to accelerate at the rate of 1 cm/sec^2.

DYNODE—One of the intermediate electrodes in a photomultiplier tube.

EFFECTIVE HALF-LIFE—The half-life of a radioisotope in a biological system as a result of the combined effects of the biological half-life and the radiological half-life.

ELASTIC COLLISION—A collision between particles in which there is no change in total kinetic energy of the particles and/or quanta.

ELECTROMAGNETIC RADIATION—Radiation consisting of electric and magnetic waves that travel at the speed of light. Examples: light, radio waves, gamma rays, x-rays. All can be transmitted through a vacuum.

ELECTROMETER—An instrument designed to measure electrical potential. Such instruments have an extremely high input impedance so that a minimum load is placed across the measured potential.

ELECTRON—An elementary particle of nature having a charge of one and a mass of 9.1×10^{-28} grams. If the charge is positive, it is called positron; if negative, it is sometimes called a negatron, but usually the term electron implies a negative charge.

ELECTRON CAPTURE—A type of radioactive transformation in which an orbital electron collides with the nucleus and also unites with the nucleus to form a new nuclide having the same mass number but an atomic number diminished by one.

ELECTRON MULTIPLIER TUBE—A phototube of high sensitivity as a result of electron amplification within the tube. Such amplification may reach a factor of 10^6.

ELECTRON VOLT—The kinetic energy gained by an electron after passing through a potential difference of 1 volt. Symbol, eV.

ELECTROPHORESIS—The migration of amphoteric molecules or colloidal particles in an electrolysis cell as the result of an applied potential difference.

ELECTROSCOPE—An instrument for measuring an electrical charge by means of the forces exerted between charged bodies.

ELUTION—Desorption of ions from an ion exchange column by means of an eluant.

EMULSION, NUCLEAR—A photographic emulsion specially prepared for the detection of nuclear particles.

ENDOERGIC REACTION—A nuclear reaction which absorbs energy.

ENDERGONIC REACTION—See endoergic reaction.

ENZYME—A protein that accelerates (catalyzes) specific transformations of material as in the digestion of foods.

ERG—The unit of work done by a force of 1 dyne acting through a distance of 1 cm.

ERYTHROCYTES (RED CORPUSCLES)—Circular, biconcave, disc shaped, blood cells containing no nucleus. They contain the oxygen carrying pigment, hemoglobin.

EV—Abbreviation for electron volt.

EXCHANGE REACTION—The transfer of atoms, radioactive or stable, of the same element, between molecules, or from one site to another on the same molecule.

EXOERGIC REACTION— A nuclear reaction which liberates energy.

EXERGONIC REACTION—See exoergic reaction.

EXPONENTIAL DECAY—A decrease in the amount of radioactive substances by a first order rate reaction.

FALLOUT—Debris (radioactive material) that resettles to earth after a nuclear explosion. Fallout takes two forms. The first, called "local fallout," consists of the denser particles injected into the atmosphere by the explosion. They descend to earth within 24 hours near the site of the detonation and in an area extending downwind for some distance (often hundreds of miles), depending on meteorological conditions and the yield of the detonation. The other form, called "worldwide fallout," consists of lighter particles which ascend into the upper troposphere and stratosphere and are distributed over a wide area of the earth by atmospheric circulation. They then are brought to earth, mainly by rain and snow, over periods ranging from months to years.

FARAD—A condenser has a capacitance of 1 farad when a potential difference of 1 volt will charge it with 1 coulomb of electricity.

FAST NEUTRONS—Neutrons having energies in excess of about 0.1 MeV.

FEATHER ANALYSIS—A technique first described by N. Feather in 1938 for measuring the range of beta particles by comparison with beta particles of known energy distribution.

FEMUR—The thigh bone.

FILM BADGE—A piece of photographic film contained in a lightproof holder and worn by an individual in order to measure the amount of radiation to which he is exposed.

FISSION—The splitting of a heavey nucleus into two approximately equal pieces.

FLUORESCENCE—The extranuclear emission of photons caused by excitation of an atom.

FLUX—The number of particles or photons passing through a unit of area per unit of time.

FUSION—The joining of light nuclei to form a heavier nucleus. Particle velocities corresponding to millions of degrees are required for this process.

G—A factor used in radiation chemistry to indicate the number of chemical processes (broken chemical bonds) produced per 100 eV when a substance is irradiated.
GAMMA-RAY SPECTROMETER—An instrument designed to measure the distribution of gamma ray energies.
GAMMA RAYS—Electromagnetic radiation having its origin in an atomic nucleus.
GAS AMPLIFICATION—The ratio of the charge collected at the electrodes of an ionization chamber to the charge produced by the incident radiation.
GAUSS—The cgs electromagnetic unit of flux density. An emf of 1 abvolt will be induced in a conductor 1 cm long moving perpendicular to a flux of 1 gauss at a velocity of 1 cm/sec.
GEIGER COUNTER—An ionization chamber operating in the Geiger region.
GEIGER REGION—That region of the characteristic curve for counting tubes where the charge produced in the tube per count is independent of the degree of ionization produced by the initial radiation.
GENE—An hereditary germinal factor or unit in the chromosome which carries an hereditarily transmissible character.
GENETICS—The science which deals with the origin of the characteristics of the individual; a study of heredity.
GEOMETRY FACTOR—The solid angle at the radioactive source subtended by the sensitive volume of the detector divided by the total solid angle (4π).
GRAM ATOM—Synonym for gram atomic weight. The mass of an element, in grams, numerically equal to the atomic weight.
GRAM EQUIVALENT—The mass of a substance which will react with or is otherwise equivalent to one gram atom of hydrogen.
GRAM MOLE—Mass of a substance, in grams, numerically equal to the molecular weight.
GRAM MOLECULAR WEIGHT—Synonym for gram mole.
GRAVITATIONAL CONSTANT—The proportionality constant G relating the force F exerted between two bodies to their masses m and m′ and the distance d between them.

$$F = G \frac{mm'}{d^2}$$

where $G = 6.673 \times 10^{-11}$ nt-m²/kg² $= 6.673 \times 10^{-8}$ dyne-cm²/g²
GROUND STATE—The lowest energy level of a nucleus, atom or molecule.
HALF-LIFE—The time required for one-half of a given number of radioactive atoms to undergo decay. Symbol, $t_{1/2}$.
HALF-LIFE, BIOLOGICAL—The time required for a biological system, such as a man or an animal, to eliminate, by natural processes, half the amount of a substance which has entered it.
HALF-THICKNESS—The thickness of any absorber required to reduce the intensity of a beam of radiation by one-half. Synonym for Half-value layer.
HALF-TIME OF EXCHANGE—In a chemical reaction involving the exchange of atoms, the time required for one-half of the atoms to be measurably exchanged.
HALF-VALUE LAYER—The thickness of an absorbing material required to reduce the intensity of a beam of x- or gamma radiation to one-half its original intensity. Synonym for Half-thickness.
HARDNESS (X-RAYS)—A term referring to the penetrating power of x-rays. Soft x-rays, of lower frequency and hence lower energy, are less penetrating. Hard x-rays, having a higher frequency and greater energy, are more penetrating.
HEAVY HYDROGEN—Synonym for deuterium.
HEAVY WATER—Water in which the hydrogen has been replaced by deuterium. Thus, its formula is D_2O. Its density is greater than that of H_2O. Density = 1.076 g/ml at 20° C.
HEMATOCRIT—An expression of the volume of the red blood cells per unit volume of circulating blood. Abbreviation—HCT.
HEMATOPOIETIC SYSTEM—The blood making system; the bone marrow, the spleen, and lymph nodes.
HEPARIN—A substance which prevents the clotting of blood.
HOLD-BACK CARRIER—Synonym: Hold-back agent. Inactive isotopes of a particular nuclide employed to retard the coprecipitation or absorption of a radionuclide.
HOT—A colloquial term meaning highly radioactive.
HYPERON—A heavy particle. All hyperons have a mass greater than that of the neutron. Hyperons include the Λ, Σ^+, Σ^-, Ξ^-, Ξ° and Ω^- particles.
INDUCED RADIOACTIVITY—Artificial radioactivity or that produced by a nuclear reaction.
INELASTIC COLLISION—A collision between particles or photons in which there is a change in the total kinetic energy of the particles and/or quanta.
INERTIA—Resistance offered by mass to a change in its state of rest or motion. Dimension—m.
INTENSITY OF RADIATION—Amount of radiant energy emitted in a specified direction per unit time and per unit surface area.
ION—A charged atom or radical. The term is occasionally used for a free electron or other charged nuclear particle. e.g., see ion pair.
ION EXCHANGE—A chemical process involving the reversible exchange of ions between ions in solution and ions bound to an insoluble substance or resin.
ION PAIR—A positive ion and an electron or negative ion produced by the action of ionizing radiation.
IONIZATION—The process of adding electrons to, or knocking electrons from, atoms or molecules, thereby creating ions. High temperatures, electrical discharges, and nuclear radiation can cause ionization.
IONIZATION CHAMBER—An instrument for measuring radiation. It operates by measuring the ions produced by the radiation in a given volume and consists of two electrodes between which an electric field is maintained to collect the charge.

IONIZATION POTENTIAL—The potential necessary to separate one electron from an atom with the formation of a positive ion with one elementary charge. The average value for air is about 32.5 eV.

IONIZING EVENT—Event in which an ion is produced.

IONIZING PARTICLE—Any charged particle possessing sufficient energy to produce ion pairs.

ISOBAR—One of a group of nuclides having the same total number of particles (neutrons + protons) in the nucleus but with these particles so proportioned as to result in different values of Z; e.g., ^{3}H and ^{3}He.

ISOMER, NUCLEAR—One of two or more nuclides having the same number of neutrons and protons in the nucleus, (same Z and same A) but existing in different energy states.

ISOTONE—Any one of several nuclides having the same number of neutrons in the nucleus but differing in the number of protons.

ISOTOPE—One of a group of nuclides of the same element (same Z), having the same number of protons in the nucleus but differing in the number of neutrons, resulting in different values of A. Sometimes used as a general synonym for nuclide, but this use is not recommended.

ISOTOPE EFFECT—The effects caused by the difference in isotopic mass on reaction rates and equilibria.

K-CAPTURE—A term used for "K-electron capture" or "electron capture."

KEV—(KeV) Symbol for kilo-electron-volt or 1000 eV.

KILOCURIE—One thousand curies.

KINETIC ENERGY—The energy which a body possesses by virtue of its mass and velocity. The Newtonian kinetic energy is

$$T = \frac{1}{2}mv^2$$

The relativistic kinetic energy is given by $T = (m-m_0)c^2$

LABEL (isotopic)—A labeled compound or molecule is one containing one or more radioactive atoms (or stable atoms of different mass) as a part of its structure. Such radioactive or stable atoms are called labels or tags.

LABILE—Synonym for unstable.

LAMBDA—A unit of volume. Synonym for microliter.

LANTHANIDES—Rare earth elements having atomic numbers 57 to 71 inclusive.

LATENT PERIOD—The period or state of seeming inactivity between the time of exposure to an agent and the beginning of the response.

LD_{50}—Lethal dose-50%. Synonym for median lethal dose.

LEPTON—Term derived from word for smallest Greek coin which refers to the group of small particles, the electron, positron and neutrino.

LETHAL DOSE—A dose of ionizing radiation sufficient to cause death. Median lethal dose (MLD, LD_{50}) is the dose required to kill half of the individuals in a large group similarly exposed within a specified period of time. The MLD for man is about 400 roentgens.

LEUKEMIA—A malignant disease characterized by excessive leukocytes in the blood.

LEUKOCYTES (LEUCOCYTE)—White blood cell; a group of nucleated, amoeboid cells found in blood which engulf foreign particles and micro-organisms.

LINEAR ABSORPTION COEFFICIENT—(See absorption coefficient).

LINEAR ACCELERATOR—A device for accelerating charged particles. It consists of electrodes, arranged in a straight line, so proportioned that as the potential of the electrodes is varied the particles are accelerated.

LINEAR AMPLIFIER—An electronic amplifier of special design for amplifying pulses so that the size of the output pulse will be proportioned to the size of the input pulse.

LYMPHOCYTE—A type of leukocyte (white blood cell) which arises in the lymph glands.

MAGIC NUMBERS—The integers corresponding to the numbers of protons or number of neutrons which have been observed to contribute to a stable nucleus. Observed numbers are 2, 8, 20, 28, 50, 82 and 126.

MASS ABSORPTION COEFFICIENT—(See absorption coefficient).

MASS DEFECT—The difference between the mass of an atom and the sum of the free masses of its constituents.

MASS-ENERGY EQUIVALENCE—The equivalence between mass and energy as expressed by the Einstein equation, $E = mc^2$.

MASS NUMBER—The number, indicated by the symbol A, which represents the total number of nucleons in a nucleus.

MASS SPECTROMETER—A device for measuring the mass of individual particles by passing them through electrostatic and magnetic fields.

MAXIMUM PERMISSIBLE DOSE—That dose of ionizing radiation which competent authorities have established as the maximum that can be absorbed without undue risk to human health.

MAXWELLIAN DISTRIBUTION—The velocity distribution of molecules in thermal equilibrium as computed on the basis of the kinetic theory.

MEAN FREE PATH—Average distance traveled by a particle between collisions.

MEAN LIFE—The average life of a radioactive atom. Symbol = τ. It is equal to the reciprocal of the decay constant. $\tau = 1/\lambda$.

MECHANICAL REGISTER—An electrically actuated mechanical device for recording counts or impulses.

MEDIAN LETHAL DOSE (MLD)—The dose of radiation required to produce death in 50% of the individuals, animals or organisms, of an irradiated population.

MESON—A particle with a mass greater than that of the electron but less than that of the proton.

METABOLISM—The sum of all the physical and chemical processes by which living organized substance is produced and maintained and by which energy is made available for the uses of the organism.

METABOLITE—Any substance produced by metabolism.

METASTABLE STATE—An excited energy state of a nucleus which returns to the ground state by the emission of a gamma ray over a measurable half-life.

MEV—(MeV) Symbol for one million electron volts, or 10^6 eV.

MICROCURIE—One millionth of a curie. Symbol μCi.

MILLICURIE—One thousandth of a curie. Symbol mCi.

MILLIROENTGEN—One thousandth of a roentgen. Symbol mr.

MODERATOR—A material used to slow neutrons in a reactor. These slow neutrons are particularly effective in causing fission. Neutrons are slowed down when they collide with atoms of light elements such as hydrogen and carbon, two common moderators.

MOLE—Synonym for Gram Molecular Weight.

MOMENT OF INERTIA—A measure of the effective mass in rotation. Moment of inertia bears the same relationship to rotational motion that mass bears to translational motion. Symbol *I*. Dimension—ml^2.

$$I = \Sigma (mr^2)$$

MOMENTUM—Product of mass and volocity. Symbol p. Dimensions—mlt^{-1}.

MONITOR—A radiation detector used to determine the safety of a working area.

MONOCHROMATIC RADIATION—Electromagnetic radiation of a single wavelength. That is, all photons have the same energy.

MU-MESON—An elementary particle with 207 times the mass of an electron. It may have a single positive or negative charge. Not really a meson. (See Chapter 1).

MUON—(See Mu-meson).

MUTATION—A permanent transmissible change in the chacters or traits of an offspring from those of its parents.

N—Symbol for neutron number, the number of neutrons in a given nucleus.

NEGATIVE ELECTRON—A synonym for electron. Also called a negatron.

NEGATRON—A negative electron or ordinary electron. Synonym for electron.

NEUTRINO—A neutral particle having possibly zero mass. Dirac proposed the neutrino to account for that part of beta decay energy not associated with the emitted beta particle.

NEUTRON—A neutral elementary particle having a mass number of one. In the free state (outside the nucleus), it is unstable, having a half-life of about 12 minutes. It decays by the process $n = p + e^- + \nu$.

NEUTRON NUMBER—The number of neutrons in a nucleus. Symbol N, where $N = A - Z$.

NORMALIZED PLATEAU SLOPE—The figure of merit for a counter tube calculated as the percentage change in counting rate divided by the percentage change in voltage, using the threshold value as a base.

NUCLEAR EMULSION—A photographic emulsion specially prepared for the detection of nuclear particles. Such emulsions are generally thicker and contain a higher concentration of silver halide than those used for standard photography.

NUCLEAR ENERGY—The energy released by a nuclear reaction. Measured in terms of the Q-value.

NUCLEAR FISSION—(See fission).

NUCLEAR ISOMERS—Isotopes of elements (i.e., nuclides having the same Z, N and A) in different quantum or energy states. Isomers of an isotope will exhibit different properties such as mode of decay.

NUCLEAR REACTOR—A device for supporting a self-sustained nuclear chain reaction under controlled conditions.

NUCLEON—Any particle found as a constituent of the nucleus. According to current theory this would include the proton and neutron.

NUCLEONICS—The application of nuclear science in the fields of chemistry, bacteriology, physics, etc. including industrial applications.

NUCLEON NUMBER—Synonym for mass number.

NUCLEUS—The denser, spherical structure within a cell which controls the activity of that cell and carries the hereditary material (chromatin).

NUCLIDE—Any one of the more than one thousand species of atoms characterized by the number of protons and number of neutrons in the nucleus.

OPERATING VOLTAGE—The voltage across a counting tube or radiation detector in the quiescent state.

ORBITAL-ELECTRON CAPTURE—(See electron capture).

OVERVOLTAGE—For a Geiger-Müller tube, the difference between the operating voltage and the threshold voltage.

p—Symbol for the proton.

PACKING FRACTION—The ratio (M-A)/A. This is the difference between the atomic mass M and the mass number A divided by the mass number. It is positive for most nuclides of $A < 6$ and $A > 175$, and negative for the others. The packing fraction is related to the binding energy and indicates nuclear stability. The smaller the packing fraction the more stable the nucleus. The value of the packing fraction is frequently multiplied by 10,000 so that it can be expressed by a small number.

PAIR PRODUCTION—A photon, having an energy in excess of 1.02 MeV, is capable of undergoing an energy to mass transition in a strong magnetic field (such as that in the vicinity of a nucleus) in which the energy $h\nu$ of the photon is converted to mass. The quantity of mass created is related to the energy by $E = mc^2$. A positron and an electron are produced in the process. Excess energy of the photon $h\nu - 2\,mc^2$ appears as the kinetic energy of the newly formed particles.

PARENT—A radionuclide which upon decay yields a specific daughter. Thus, ^{32}P is the parent of ^{32}S.

PARTITION COEFFICIENT—Synonym for distribution ratio. The ratio of the equilibrium concentrations of a given solute between two immiscible solvents.

PERIOD—Currently used as a synonym for half-life. Formerly used to designate mean life also.

PHOSPHOR—A material, such as zinc sulfide, which gives off visible light when struck by nuclear radiation. The inside face of a television picture tube is coated with a phosphor.

PHOTODISINTEGRATION—A nuclear reaction involving the interaction of a nucleus with a gamma ray.

PHOTOELECTRIC EFFECT—The inelastic collision of a photon with an orbital electron in which the electron is ejected from the atom.

PHOTOELECTRON— The electron ejected in the photoelectric effect.

PHOTOMULTIPLIER TUBE—A phototube of exceptionally high sensitivity, the electron or electrons released at the photocathode initiating a cascade from one dynode to another with a resultant electron amplification of as high as 10^9.

PHOTON—A quantum of electromagnetic radiation. The energy of a photon of frequency ν is equal to $h\nu$.

PHOTONEUTRON—A neutron ejected from a nucleus by a (γ, n) reaction, i.e., by a photonuclear reaction.

PHOTONUCLEAR REACTION—Synonym for photodisintegration.

PIG—A container used to store or ship radioactive substances.

PILE—A nuclear reactor.

PI-MESON—The mass of a charged pion is about 273 times that of an electron. An electrically neutral pion has a mass 264 times that of an electron.

PION—(See Pi-meson). Contraction of pi-meson.

PITCHBLENDE—An ore of uranium and radium.

PLANCK CONSTANT—A universal constant h relating the energy of a photon to its frequency, $E = h\nu$. It has the value 6.6256×10^{-27} erg sec.

PLASMACRIT—An expression of the volume of plasma per unit volume of whole circulating blood. Plasmacrit = 1 − HCT.

PLATEAU—That portion of a characteristic counter curve which is almost independent of voltage.

PLATEAU SLOPE—(See normalized plateau slope.)

POCKET CHAMBER—A small ionization chamber, having a physical appearance similar to a fountain pen, used to monitor the amount of radiation to which an individual is exposed.

POLYCYTHEMIA VERA—A disease characterized by persistent overabundance of red blood cells due to excessive formation in the bone marrow.

POSITIVE RAYS—A stream of positive ions or particles produced in a potential gradient by ionizing agents.

POSITRON—A positive electron.

POSITRONIUM—An "atom" consisting of a positron and an electron, one revolving about the other. It may be considered the counterpart of the hydrogen atom in which the positron replaces the proton. It has a mean life of about 10^{-7} sec. It is destroyed by electron-positron annihilation.

POTENTIAL DIFFERENCE—The difference in electrical potential between any two points in a circuit.

POTENTIAL GRADIENT—The rate of change of potential with distance. Unit = volts/cm.

POWER—The time rate of doing work. Dimensions—ml^2t^{-3}.

PRIMARY ELECTRON—The electron ejected from an atom by an initial ionizing event, as caused by a photon or beta particle.

PROPORTIONAL COUNTER—A radiation detector designed to provide linear gas amplification of the primary ionization, thus providing an output pulse whose height is proportional to the energy of the incident radiation.

PROTON—A positively charged elementary particle having a mass number of 1. The nucleus of a hydrogen atom of mass 1.

PULSE AMPLIFIER—An amplifier specifically designed for the amplification of pulses generated by radiation sensing devices.

Q (Q value)—The disintegration energy or energy released per atom for a particular decay process or other nuclear reaction.

QUANTUM—At the atomic level it becomes evident that observable quantities (energy, momentum, etc.) are not continuously variable but are variable only in discrete steps. The possible change is a multiple of a unit called the quantum.

QUANTUM LEVEL—An energy level of an electron, distinct from any other of its energy levels by discrete quantities dependent upon Planck's constant.

QUANTUM NUMBER—One of a set of integral or half-integral numbers, one for each degree of freedom, which determine the state of an atomic system in terms of constants of nature.

QUANTUM THEORY—The concept that energy is radiated in discrete units of energy called quanta.

QUENCHING—The process of limiting the discharge of an ionization detector, either externally by a momentary reduction in applied potential to the tube through suitable electronic circuitry, or internally by the introduction of a quenching agent, such as butane or chlorine.

RABBIT—A capsule in which a substance is placed for irradiation in a reactor.

RAD—Radiation absorbed dose. The basic unit of absorbed dose of ionizing radiation. One rad is equal to the absorption of 100 ergs of radiation energy per gram of matter.

RADIATION—A term or originally signifying the propagation of electromagnetic radiation (visible light, infrared, x-rays, etc.) through space, but extended to include corpuscular radiation, such as alpha particles, beta particles, neutrons and electrons.

RADIATION CHEMISTRY—That branch of chemistry dealing with the effects of high energy radiation on chemical substances and on chemical reactions.

RADIOACTIVE SERIES—A succession of radioactive nuclides, one decaying to the next until a stable nucleus is formed.

RADIOCHEMISTRY—Those phases of chemistry involving the application of radioisotopes in the study of chemical problems as well as the study of the chemical behavior of small quantities of radioactive substances.

RADIO COLLOID—Aggregates produced by the clumping of molecules containing radioactive atoms.

RADIOGRAPHY—A process analogous to the making of an x-ray picture of an object by the selective absorption of radiation by the object, a radioactive source being used in place of an x-ray tube.

RADIOISOTOPE—Synonym for radioactive isotope. Any isotope which is unstable, thus undergoing decay with the emission of a characteristic radiation.

RADIONUCLIDE—Synonym for radioactive nuclide.
RANGE—The thickness of an absorbing material required to remove or absorb all detectable radiation of a particular type.
RATE METER—A counting rate-meter used to determine continuously the time rate of events such as ionization.
REACTOR—(See nuclear reactor.)
RECOVERY TIME—The time, following detection of a pulse, which must elapse before a second pulse can be detected. Also called resolving time or coincidence time.
RELATIVE BIOLOGICAL EFFECTIVENESS—The relative effectiveness of a given kind of ionizing radiation in producing a biological response as compared with 250,000 electron volt gamma rays.
RELATIVE PLATEAU SLOPE—The relative increase in the number of counts as a function of voltage expressed in percentage per 100 volts increase above the Geiger threshold.
RELATIVISTIC MASS—The mass m of a particle or body in motion. It is related to the rest mass m_0, the velocity of the particle v, and the velocity of light c, by the equation

$$m = \frac{m_0}{\sqrt{1-(v^2/c^2)}}$$

REP—Abbreviation for roentgen equivalent physical.
RESOLUTION—Ability to separate counts which occur very close together in time.
RESOLVING TIME—The minimum time which must elapse between two events to permit their being recorded as individual events. Applied to counting tubes it becomes synonymous with recovery time. See also coincidence loss.
RESONANCE CAPTURE—A nuclear reaction involving the capture of a particle of specific energy by a resonance level of a nucleus. The nucleus presents a very large cross section to particles possessing this particular energy.
REST MASS—The mass m_0 of a particle at rest. It represents the Newtonian mass and does not include the additional mass acquired by a particle in motion by the relativistic effect.
R-METER—A meter calibrated to indicate roentgens.
ROENTGEN—The quantity of x- or gamma radiation such that the associated corpuscular emission per 0.001293 grams of air (i.e., 1 ml at 0° C and 760 mm) produces, in air, ions carrying 1 electrostatic unit of quantity of electricity of either sign.
ROENTGEN EQUIVALENT MAN (REM)—That quantity of radiation which when absorbed by man produces an effect equivalent to the absorption of one roentgen of x— or gamma radiation.
ROENTGEN EQUIVALENT PHYSICAL (REP) — The amount of ionizing radiation which is capable of producing 1.615×10^{12} ion pairs per gram of tissue or that amount which will be absorbed by tissue to the extent of 93 ergs per gram. This unit is used in particular to measure beta radiation.
ROENTGEN RAYS—Synonym for x-rays.
RUTHERFORD—A unit of decay rate defined as the amount of substance undergoing 10^6 disintegrations per second.
SATURATION CURRENT—The resulting current when the applied potential to an ionization chamber is sufficient to collect all ions. The current is dependent only on the amount of ionizing radiation and is independent of the applied voltage.
SCALER—An electronic device which produces an output pulse for a given number of input pulses. Term generally applied to a device for indicating the total number of observed pulses or events.
SCATTERING—The change in direction produced by the collision of a photon or particle with another particle or field.
SCINTILLATION—The flash of light produced in a phosphor by radiation.
SCINTILLATION COUNTER—A counter employing a phosphor, photomultiplier tube and associated circuits for the detection of radiation.
SCINTILLATION SPECTROMETER—A scintillation counter designed to permit the measurement of the energy distribution of radiation.
SECONDARY ELECTRON—An electron ejected from an atom or molecule upon collision with a charged particle or photon.
SECULAR EQUILIBRIUM—Equilibrium of decay, established only after a relatively long time, between a long lived parent and a short lived daughter.
SELF-ABSORPTION—The absorption of radiation by the radioactive substance itself. This phenomenon presents a particular problem in the measurement of weak radiation in a bulky sample. This process tends to decrease the observed activity of a sample.
SELF-QUENCHED COUNTER TUBE—A Geiger tube which is quenched internally by incorporation of a quenching gas (isobutane, chlorine, etc.) with the counting gas.
SELF-SCATTERING—The scattering of radiation by the material emitting the radiation. This process tends to increase the observed activity of a sample.
SENSITIVE VOLUME—The volume of a radiation detector through which the radiation must pass in order to be detected.
SHIELD—A wall to protect personnel from harmful radiation.
SLOW NEUTRONS—Neutrons possessing a kinetic energy equal to or less than about 100 eV.
SLUG—A fuel element for a nuclear reactor.
SOMATIC EFFECTS OF RADIATION—Effects limited to the exposed individual, as distinguished from genetic effects. Large radiation doses can be fatal. Smaller doses may make the individual noticeably ill or may merely produce temporary changes in blood-cell levels detectable only in the laboratory.
SOURCE—Any material which emits radiation.
SPALLATION—A nuclear reaction in which a small fragment of a bombarded nucleus splits off, the small fragment usually breaking further into several smaller fragments or particles.

SPARK CHAMBER—An instrument for detecting and measuring nuclear radiation. Analogous to the cloud chamber. It consists of numerous electrically charged metal plates mounted in a parallel array, the spaces between the plates being filled with inert gas. Ionizing radiation causes sparks to jump between the plates along its path through the chamber.

SPECIFIC ACTIVITY—(1) The activity or decay rate of a radioisotope per unit of mass of the sample (e.g., microcuries per millimole, disintegrations per second per milligram). (2) The relative activity per unit of mass (counts per minute per milligram).

SPECIFIC IONIZATION—The number of ion pairs produced per unit of distance along the track of an ionizing particle.

SPECTRUM—A plot of the intensity of radiation as a function of wavelength or frequency.

SPIN—A term used in nuclear physics to describe the angular momentum of particles and nuclei.

SPURIOUS COUNTS—Counts caused by malfunction of the apparatus, improper use of the apparatus or from outside influence (motors, etc.).

STABLE ISOTOPE—An isotope which is not radioactive.

STARTING VOLTAGE—The minimum potential which must be applied to a counting tube for its operation in conjunction with its associated circuitry.

STATISTICAL ERROR—Errors in counting caused by the random decay of atoms.

STRAGGLING—The variation in the range of particles which were initially all of the same energy.

SYNCHROTRON—An accelerator used to achieve higher velocities for atomic particles than is possible in a conventional cyclotron.

SYNDROME—A set of symptoms and signs occurring together in a single disease.

TAGGED ATOM—An isotopic tracer or a radioactive atom in a molecule.

TARGET—The material subjected to bombardment by radiation, high energy particles or high energy nuclei for the purpose of producing a nuclear reaction.

THERAPY—The treatment of disease.

THERMAL NEUTRONS—Neutrons having energies corresponding to room temperature; that is, approximately 0.025 eV which is the kinetic energy of a molecule at about 300° K.

THERMONUCLEAR REACTION—A nuclear reaction, the activation energy for which is provided by the thermal agitation of the reacting nuclei.

THYROID GLAND—One of the endocrine glands, located in the neck, which produces the hormone thyroxin.

THYROXIN—The iodine containing hormone of the thyroid gland which strongly influences growth and the metabolic rate. Symbol—T-4.

TIME-CONSTANT—Measure of the time required for a capacitor to charge or discharge in a circuit. It is numerically equal in seconds to the product of the resistance in megohms and capacity in microfarads.

TRACER—An isotopic tracer is an isotope used to tag or follow a chemical reaction or process such that its location and concentration can later be determined.

TRANSMUTATION—A process in which one nuclide is transformed into a nuclide of a different element.

TRITIUM—A hydrogen isotope of mass three. Its nucleus contains one proton and two neutrons.

TRITON—Tritium nucleus consisting of one proton and two neutrons.

TUMOR—A mass of new tissue which persists and grows independently of its surrounding structures. It may be malignant or benign.

UNCERTAINTY PRINCIPLE—The indeterminancy principle of Heisenberg which postulates that any physical measurement disturbs the system being measured, thus limiting the accuracy of measurements, especially at the atomic, nuclear and particulate levels of magnitude.

VALENCE—The combining power of an element measured by the number of atoms of hydrogen, or the equivalent, with which one atom of the element will combine or react.

VALENCE ELECTRONS—The orbital electrons which participate in chemical reactions through their loss, gain or sharing.

VAN de GRAAFF GENERATOR—An electrostatic generator. An insulated electrode is charged to a high voltage by a moving conveyor belt.

VELOCITY—Time rate of motion in a specific direction. Velocity is a vector quantity. Dimensions—lt^{-1}.

VELOCITY OF A WAVE—Velocity of propagation in terms of wave length and frequency υ. The equation is $v = \upsilon\lambda$.

VIBRATING-REED ELECTROMETER—An extremely sensitive instrument used to measure very small potentials. It converts the small D.C. potential to A.C. by means of a small vibrating capacitor, the A.C. is then amplified and reconverted, if desired, back to D.C.

WAVE LENGTH—Distance between any two similar points of two consecutive waves.

WAVE MOTION—Progressive disturbance propagated in a medium by periodic vibration of the particles of the medium. *Transverse* wave motion is that in which the vibration of the particles is perpendicular to the direction of propagation. *Longitudinal* wave motion is that in which the vibration of the particles is parallel to the direction of propagation.

WEIGHT—Force with which a body is attracted toward the earth; cgs unit: gm-cm/sec^2.

WILSON CHAMBER—A cloud chamber.

WORK—The transfer of energy by the application of a force through a distance. Product of a force and the distance through which it moves; cgs unit: gm-cm^2/sec^2.

X-RAYS—Electromagnetic radiation in the region below 100 angstroms.

X-UNIT—A unit used to express the wavelengths of x-rays. It is known as the Siegbahn x-unit. The conversion factor from x-units to angstroms = $\Lambda = 1.002063 \times 10^{-3}$.

ZEEMAN EFFECT—The splitting of spectral line by a strong magnetic field.

ZWITTERION—An amphoteric dipolar ion possessing a positive charge at one position of its structure and a negative charge at another.

APPENDIX F Exponential Functions

To determine the extent of radioactive decay by use of the relation $A = A^0 e^{-\lambda t}$, let $\lambda t = x$. The corresponding value of e^{-x} obtained from the table is the decay factor $e^{-\lambda t}$ and is equal to the fraction of activity remaining at time t.

x	e^{-x}	x	e^{-x}	x	e^{-x}	x	e^{-x}	x	e^{-x}
0.00	1.000	0.40	0.670	0.80	0.449	1.20	0.301	1.60	0.201
0.01	0.990	0.41	.663	0.81	.448	1.21	.298	1.61	.199
0.02	.980	0.42	.657	0.82	.440	1.22	.295	1.62	.197
0.03	.970	0.43	.650	0.83	.436	1.23	.292	1.63	.195
0.04	.960	0.44	.644	0.84	.431	1.24	.289	1.64	.193
0.05	0.951	0.45	0.637	0.85	0.427	1.25	0.286	1.65	0.192
0.06	.941	0.46	.631	0.86	.423	1.26	.283	1.66	.190
0.07	.932	0.47	.625	0.87	.418	1.27	.280	1.67	.188
0.08	.923	0.48	.618	0.88	.414	1.28	.278	1.68	.186
0.09	.913	0.49	.612	0.89	.410	1.29	.275	1.69	.184
0.10	0.904	0.50	0.606	0.90	0.406	1.30	0.272	1.70	0.182
0.11	.895	0.51	.600	0.91	.402	1.31	.269	1.71	.180
0.12	.886	0.52	.594	0.92	.398	1.32	.267	1.72	.179
0.13	.878	0.53	.588	0.93	.394	1.33	.264	1.73	.177
0.14	.869	0.54	.582	0.94	.390	1.34	.261	1.74	.175
0.15	0.860	0.55	0.576	0.95	0.386	1.35	0.259	1.75	0.173
0.16	.852	0.56	.571	0.96	.382	1.36	.256	1.76	.172
0.17	.843	0.57	.565	0.97	.379	1.37	.254	1.77	.170
0.18	.835	0.58	.559	0.98	.375	1.38	.251	1.78	.168
0.19	.826	0.59	.554	0.99	.371	1.39	.249	1.79	.166
0.20	0.818	0.60	0.548	1.00	0.367	1.40	0.246	1.80	0.165
0.21	.810	0.61	.543	1.01	.364	1.41	.244	1.81	.163
0.22	.802	0.62	.537	1.02	.360	1.42	.241	1.82	.162
0.23	.794	0.63	.532	1.03	.357	1.43	.239	1.83	.160
0.24	.786	0.64	.527	1.04	.353	1.44	.236	1.84	.158
0.25	0.778	0.65	0.522	1.05	0.349	1.45	0.234	1.85	0.157
0.26	.771	0.66	.516	1.06	.346	1.46	.232	1.86	.155
0.27	.763	0.67	.511	1.07	.343	1.47	.229	1.87	.154
0.28	.755	0.68	.506	1.08	.339	1.48	.227	1.88	.152
0.29	.748	0.69	.501	1.09	.336	1.49	.225	1.89	.151
0.30	0.740	0.70	0.496	1.10	0.332	1.50	0.223	1.90	0.149
0.31	.733	0.71	.491	1.11	.329	1.51	.220	1.91	.148
0.32	.726	0.72	.486	1.12	.326	1.52	.218	1.92	.146
0.33	.718	0.73	.481	1.13	.323	1.53	.216	1.93	.145
0.34	.711	0.74	.477	1.14	.319	1.54	.214	1.94	.143
0.35	0.704	0.75	0.472	1.15	0.316	1.55	0.212	1.95	0.142
0.36	.697	0.76	.467	1.16	.313	1.56	.210	1.96	.140
0.37	.690	0.77	.463	1.17	.310	1.57	.208	1.97	.139
0.38	.683	0.78	.458	1.18	.307	1.58	.205	1.98	.138
0.39	.677	0.79	.453	1.19	.304	1.59	.203	1.99	.136

x	e^{-x}	x	e^{-x}	x	e^{-x}	x	e^{-x}	x	e^{-x}
2.00	0.1353	2.45	0.0862	2.90	0.0550	3.35	0.0350	3.80	0.0224
2.01	.1339	2.46	.0854	2.91	.0544	3.36	.0347	3.81	.0221
2.02	.1326	2.47	.0845	2.92	.0539	3.37	.0343	3.82	.0219
2.03	.1313	2.48	.0837	2.93	.0533	3.38	.0340	3.83	.0217
2.04	.1300	2.49	.0829	2.94	.0528	3.39	.0337	3.84	.0215
2.05	0.1287	2.50	0.0820	2.95	0.0523	3.40	0.0334	3.85	0.02123
2.06	.1274	2.51	.0812	2.96	.0518	3.41	.0330	3.86	.02024
2.07	.1261	2.52	.0804	2.97	.0513	3.42	.0327	3.87	.01925
2.08	.1249	2.53	.0796	2.98	.0507	3.43	.0323	3.88	.01831
2.09	.1236	2.54	.0788	2.99	.0502	3.44	.0321	3.89	.01742
2.10	0.1224	2.55	0.0780	3.00	0.0497	3.45	0.0317	4.10	0.01657
2.11	.1212	2.56	.0773	3.01	.0492	3.46	.0314	4.15	.01576
2.12	.1200	2.57	.0765	3.02	.0488	3.47	.0311	4.20	.01499
2.13	.1188	2.58	.0757	3.03	.0483	3.48	.0308	4.25	.01426
2.14	.1176	2.59	.0750	3.04	.0478	3.49	.0305	4.30	.01356
2.15	0.1164	2.60	0.0742	3.05	0.0473	3.50	0.0302	4.35	0.01290
2.16	.1153	2.61	.0735	3.06	.0468	3.51	.0299	4.40	.01227
2.17	.1141	2.62	.0728	3.07	.0464	3.52	.0296	4.45	.01167
2.18	.1130	2.63	.0720	3.08	.0459	3.53	.0293	4.50	.01110
2.19	.1119	2.64	.0713	3.09	.0455	3.54	.0290	4.55	.01056
2.20	0.1108	2.65	0.0706	3.10	0.0450	3.55	0.0287	4.60	0.01005
2.21	.1097	2.66	.0699	3.11	.0446	3.56	.0284	4.65	.00956
2.22	.1086	2.67	.0692	3.12	.0441	3.57	.0282	4.70	.00909
2.23	.1075	2.68	.0685	3.13	.0437	3.58	.0279	4.75	.00865
2.24	.1064	2.69	.0678	3.14	.0432	3.59	.0276	4.80	.00823
2.25	0.1053	2.70	0.0672	3.15	0.0428	3.60	0.0273	4.85	0.00782
2.26	.1043	2.71	.0665	3.16	.0424	3.61	.0271	4.90	.00744
2.27	.1033	2.72	.0658	3.17	.0420	3.62	.0268	4.95	.00708
2.28	.1022	2.73	.0652	3.18	.0415	3.63	.0265	5.00	.00673
2.29	.1012	2.74	.0645	3.19	.0411	3.64	.0263	5.10	.00609
2.30	0.1002	2.75	0.0639	3.20	0.0407	3.65	0.0260	5.20	0.00552
2.31	.0992	2.76	.0632	3.21	.0403	3.66	.0257	5.30	.00499
2.32	.0982	2.77	.0626	3.22	.0399	3.67	.0255	5.40	.00452
2.33	.0972	2.78	.0620	3.23	.0395	3.68	.0252	5.50	.00409
2.34	.0963	2.79	.0614	3.24	.0391	3.69	.0250	5.60	.00370
2.35	0.0953	2.80	0.0608	3.25	0.0387	3.70	0.0247	5.70	0.00335
2.36	.0944	2.81	.0602	3.26	.0383	3.71	.0245	5.80	.00303
2.37	.0934	2.82	.0596	3.27	.0380	3.72	.0242	5.90	.00274
2.38	.0925	2.83	.0590	3.28	.0376	3.73	.0240	6.00	.00248
2.39	.0916	2.84	.0584	3.29	.0372	3.74	.0238	6.25	.00193
2.40	0.0907	2.85	0.0578	3.30	0.0368	3.75	0.0235	6.50	0.00150
2.41	.0898	2.86	.0572	3.31	.0365	3.76	.0233	6.75	.00117
2.42	.0889	2.87	.0566	3.32	.0361	3.77	.0231	7.00	.00091
2.43	.0880	2.88	.0561	3.33	.0357	3.78	.0228	7.25	.00071
2.44	.0871	2.89	.0555	3.34	.0354	3.79	.0226	7.50	.00055

APPENDIX G Problems

CHAPTER 1—*ATOMIC AND MOLECULAR STRUCTURE*

1.1. Calculate the energy equivalent to the rest mass of an electron. Express the answer in (a) ergs and (b) MeV.

Answers: (a) 8.18×10^{-7} ergs; (b) 0.510 MeV.

1.2. (a) Because an antiproton can be created only in a pair with a proton, to produce an antiproton we need an amount of energy equivalent to the total mass of the two particles. Calculate the energy required for the antiproton-proton pair in BeV.

Answer: 1.878 BeV.

(b) In practice, the reaction considered in part (a) is accomplished in a bevatron by hurling, for example, a proton against another proton. After the collision there are four particles: the original protons plus the antiproton-proton pair. Thus the generation of an antiproton by this method requires energy for the creation of the proton-antiproton pair, as calculated in part (a), plus the kinetic energy of the four emerging particles. If the average kinetic energy of each of these particles is 1 BeV, calculate their average velocity and the total energy required for the reaction.

Answers: 2.64×10^{10} cm/sec; 5.878 BeV.

1.3. Calculate the fraction of alpha particles, impinging upon a gold foil 0.0004 mm thick, scattered through an angle equal to or greater than (a) 90°, (b) 45°.

Answers: (a) 4.76×10^{-5}

1.4. For the first three orbits of the Bohr hydrogen atom, calculate the: (a) radius, (b) linear velocity, (c) frequency of revolution and (d) the frequency of transition from n=2 to n=1 and from n=3 to n=1 (i.e., the Lyman series).

Answers: (a), (b) and (c) See table 1.1. (d) $\nu_1 = 2.4 \times 10^{15}$, $\nu_2 = 2.9 \times 10^{15}$, $\nu_3 = 3.1 \times 10^{15}$

1.5. Repeat question 1.4 for Fermium. Discuss the limitations of the Bohr theory when applied to Fermium.

1.6. Derive the expression $p = \sqrt{ze^2\mu\, a}$ for the momentum of an orbital electron.

1.7. At what velocity is the relativistic mass double the rest mass?

1.8. An electron is traveling at 99% of the velocity of light. Calculate its (a) mass, (b) total energy, (c) kinetic energy, and (d) associated de Broglie wavelength.

1.9. Show that the number of de Broglie wavelengths in an orbit is equal to the principal quantum number.

1.10. For an electron in the ground state of the Bohr helium atom, calculate its (a) orbital radius, (b) velocity, and (c) the number of revolutions per second about the nucleus.

1.11. What is the de Broglie wavelength of the electron in problem 1.10 and how long is the circumference of the orbit in de Broglie wavelengths?

1.12. Calculate (a) the radius of the second Bohr orbit (n = 2) for helium and (b) the velocity of an electron in this orbit.

Answers: (a) 1.05×10^{-8} cm (b) 2.18×10^{8} cm/sec

1.13. Calculate the ratio of the relativistic mass to the rest mass for a particle traveling at a velocity 0.99 c.

Answer: $m = 7.07m_0$

1.14. For the particle in problem 1.13 calculate the ratio of (a) the total energy and (b) the kinetic energy to the potential (or rest) energy.

Answers: (a) $E = 7.07m_0c^2$ (b) $T = 6.07m_0c^2$

CHAPTER 2—*THE ATOMIC NUCLEUS*

2.1. For the reaction

$$C + O_2 \longrightarrow CO_2 \qquad -\Delta H = 97 \text{ Kcal}$$

the mass of CO_2 produced is not equal to the sum of the masses of C and O_2 consumed. Calculate the loss in mass per mole. Why is this loss difficult to prove in the chemical laboratory?

Answer: 4.51×10^{-9} g/mole.

2.2. Calculate the rate of energy liberation (in calories per hour) for 1.00 g of pure radium, free of its decay products.

Answer: 24.1 calories/hour.

2.3. Compute the Q values for the following reactions.

(a) ^{24}Mg (d,p) ^{25}Mg

(b) $^{27}Al_{13} + {}^{4}He_2 \longrightarrow {}^{30}P_{15} + {}^{1}n_0$

Answer: (b) —0.003139 amu or —2.94 MeV.

2.4. Balance the following nuclear equations

$^{27}Al_{13} + {}^{4}He_2 \longrightarrow {}^{30}P_{15} +$

$^{106}Pd_{46} + {}^{1}n_0 \longrightarrow {}^{106}Rh_{45} +$

^{24}Mg (d, p)________

^{27}Al (n, γ)________

^{9}Be (α, n)________

2.5. Show that a mass of 1 amu is equivalent to 931.48 MeV of energy.

2.6. Calculate the binding energy of the deuteron.

2.7. From a consideration of the binding energies of ^{12}C and ^{13}C predict which nucleus should be the more stable. Does the natural abundance of these isotopes support your prediction? How is the situation influenced by the even-odd rules and other considerations.

In the same manner explain the natural abundances of ^{14}N and ^{15}N, ^{10}B and ^{11}B, ^{6}Li and ^{7}Li and ^{16}O, ^{17}O and ^{18}O.

2.8. By means of nuclear energetics predict if alpha decay of ^{212}Po is spontaneous.

2.9. Calculate the packing fractions for the isotopes considered in problem 2.7. Is there any apparent relationship between packing fraction and isotopic abundance?

CHAPTER 3—*INTRODUCTION TO RADIATION MEASUREMENT*

3.1. The following data were recorded for a characteristic counting rate curve for a Geiger-Müller tube.

Voltage	cpm	Voltage	cpm
1000	0	1400	2083
1025	85	1450	2072
1050	1050	1500	2095
1100	2010	1550	2069
1150	2023	1600	2071
1200	2045	1650	2110
1250	2031	1700	2123
1300	2042	1750	2493
1350	2078		

Plot the data and determine the best values for (a) starting potential, (b) threshold voltage, (c) operating voltage, (d) length of the plateau and (e) the slope of the plateau curve.

3.2. Sketch a family of Geiger-Müller characteristic curves indicating how the tube characteristics would be expected to change with aging of the tube. Assume the tube is organic quenched.

3.3. A solution of ^{32}P has an activity of 20 mCi per ml. Calculate the dilution factor required so that 100 microliters of the diluted solution will give 5,000 cpm at an assumed counting efficiency of 7%.

Answer: 6.21×10^4.

CHAPTER 4—*INDETERMINATE ERRORS IN MEASUREMENT (STATISTICS)*

4.1. Calculate the probability of obtaining "heads" with (a) one flip of a coin, (b) on each of two flips, (c) on all of n flips.

Answers: (a) 0.5, (b) 0.25, (c) 2^{-n}.

4.2. If the true count is 12, calculate (a) the Poisson probability and (b) the Gaussian probability of obtaining a count of 9. (c) What is the error $(n-\mu)$? (d) What is the Gaussian probability of obtaining an error greater than this value?

Answers: (a) 0.0874, (b) 0.0794, (c) 3, (d) 0.43.

4.3. Duplicate samples, each for one minute, gave activities of 5079 cpm and 4841 cpm, respectively. Calculate (a) the probable per cent error in the average activity and (b) the odds against 1 that the observed deviation from the mean was due to a statistical variation rather than to experimental error.

Answers: (a) 0.96%, (b) 0.017

4.4. Refer to the data of table 4.1. For a single observation what is the probability of obtaining an error greater than (a) 3, (b) 6, (c) 12? What error provides (d) a 90% confidence? (e) a 99% confidence?

Answers: (a) 0.74, (b) 0.509, (c) 0.226, (d) 16.3, (e) 25.4.

4.5. For the following cases calculate how long the sample should be counted for the most efficient utilization of time. Background is counted for 1 minute in each case.

	Sample Activity (cpm)	Background (cpm)
(a)	50	50
(b)	100	50
(c)	50	25

Answers: (a) 1.0 min. (b) 1.5 min. (c) 1.5 min.

4.6. A series of one-minute counts was made using the same sample and geometry to determine whether or not the instrument was functioning properly. (a) Draw your own conclusion from the following data by applying the chi-square test.

cpm	cpm	cpm	cpm
2111	2217	2119	2153
2120	2247	2120	2122
2126	2159	2264	2159
2215	2175	2135	2122
2180	2148	2186	2185

(b) Apply Chauvenet's criterion to the above data and determine whether or not any should be discarded.

4.7. In the measurement of the activity of a sample, a total of 6,400 counts was recorded in one minute.

(a) What is the standard deviation?
(b) What is the probability of obtaining a counting error greater than 100 cpm?
(c) What is the probable error?
(d) There is a 50-50 chance that the error is greater than ________ cpm.
(e) There is a 90% confidence that the error is less than ________ cpm.

Answers: (a) 80 cpm, (b) 0.2 or 20%,
(c) 54 cpm, (d) 54 cpm, (e) 130 cpm.

4.8. If precision is to be increased ten fold, by what factor must the time for measurement of activity be increased?

4.9. The following activities were observed for samples from the two populations A and B. One-minute measurements were made.

Set A — 114, 105, 101, 95, 86
Set B — 99, 150, 84, 80, 86

For each population:

a. Calculate the standard deviation σ. Criticize the method used to calculate σ.
b. Calculate the sample standard deviation S. Do the data appear to fit a normal distribution? Give reason for your answer.
c. What criterion may be used to estimate the acceptance or rejection of a suspect piece of data? Which, if any of these data should be rejected? Why?
d. What is the probability of observing a count of 80 or less? Does the exclusion of suspect data alter the conclusions?

e. 90% of one-minute observations should lie within what limits of activity?
f. What is the two-sigma error?
g. What is the probable per cent error?

Comparing the two populations:

h. What is the standard error of the difference?
i. What is the probability that the two populations are different?

CHAPTER 5—*DETERMINATE ERRORS IN RADIOACTIVITY MEASUREMENT*

5.1. It is desired to determine the best detector, shielding, and counting arrangement to use for measuring the activity of a large number of a particular type of sample. The activity of a radioactive source, representative of the samples to be measured, is determined with a variety of counters, shields, etc. From the results determine which arrangement is best. List in order of preference.

Counting arrangement used	Sample Activity plus background cpm	Background cpm
1	500	70
2	300	40
3	620	150
4	75	5
5	150	20

5.2. If the resolving time of a detector is 300 microseconds, calculate the activity, corrected for resolving time, for an observed activity of (a) 1,000 cpm, (b) 10,000 cpm and (c) 100,000 cpm.

5.3. The "dead time" of a Geiger-Müller counter was determined by the method of paired sources. Source #1 gave an activity of 2,388 cpm, source #2 an activity of 2,214 cpm, and the two sources together 4,504 cpm. Calculate (a) the resolving time in minutes, (b) the resolving time in microseconds (c), the corrected activity of source #1, (d) of source #2 and (e) of the two sources combined.

5.4. The following data were recorded using identical samples:

"Weightless" polystyrene mount	1360 cpm
Silver backing	2176 cpm

Calculate (a) the per cent backscatter and (b) the backscattering factor f_B.

5.5. It is desired to radioassay ^{14}C by solid counting as $Ba^{14}CO_3$. To permit correction for self absorption a set of $BaCO_3$ samples was prepared from a solution of sodium carbonate by precipitation with $BaCl_2$ and NH_4Cl, collection on sintered stainless steel disc filters and drying to constant weight.

Sample No.	Wt. Sample	cpm (corrected for background)	cpm/mg
1	2.4 mg	1020	________
2	9.4 mg	2715	________
3	19.2 mg	4276	________
4	35.1 mg	4848	________
5	68.8 mg	4814	________
6	92.0 mg	4960	________

Since all samples were made from a common carbonate solution the true specific activity of the $BaCO_3$ is the same for all six samples.

A. From the above data make plots of:
   (1) Activity (cpm) vs. sample weight.
   (2) "Apparent Specific Activity" (cpm/mg) vs Sample Mass.

B. In the course of a distribution study, two ^{14}C samples are radioassayed as $BaCO_3$.

	Mass	cpm corrected for background
Sample 1	19 mg	1500
Sample 2	38 mg	2500

From these data calculate:
   (1) The ratio of the total activities (corrected for self absorption).
   (2) The ratio of the specific activities (corrected for self absorption).

5.6. A gamma source registers 2000 counts per minute at a distance of 8 cm from the detector. (a) What will the activity be at 12 cm from the source? (b) At what distance would the activity be 4000 cpm? and (c) where would it be reduced to 1000 cpm?

5.7. (a) The observed activity of a sample is 3,500 cpm. Calculate the true activity if the resolving time is 300 microseconds.

Answer: 3,562 cpm.

(b) The observed activity of a sample is 3,500 cpm. Calculate the true activity using the approximate coincidence correction of 0.5% per 1000 cpm.

Answer: 3,561 cpm.

CHAPTER 6—*RADIOACTIVE DECAY*

6.1. The half-life of sodium-24 is 14.9 hours. Calculate
   (a) The decay constant, λ
   (b) The average life, τ
   (c) The number of atoms required to produce 1 microcurie of activity

(d) The number of atoms required to produce 1000 disintegrations per minute

Answers: (a) $4.65 \times 10^{-2}\ h^{-1}$ (b) 21.5 h (c) 2.86×10^9 (d) 1.29×10^6

6.2. A certain active substance (which has no radioactive parent) has a half-life of 8.0 days. What fraction of the initial amount will be left after (a) 4 days, (b) 16 days, (c) 32 days, (d) 57 days?

Answers: (a) 0.706, (b) ¼, (c) 1/16, (d) 1/140.

6.3. How long would a sample of radium-D (^{210}Pb) have to be observed before the decay amounted to 3%?

Answer: 359 days or 0.985 years.

6.4. Find the number of disintegrations of Uranium I atoms occurring per minute in 1 mg of ordinary uranium.

Answer: 741 dpm.

6.5. 1 g of radium is separated from its decay products and then placed in a sealed vessel. How much helium will accumulate in the vessel in 60 days? Express answer in ml at S.T.P.

Answer: 0.028 ml.

6.6. Compute (a) the weight of 1 curie of radium, (b) the weight of 1 curie of ^{32}P, and (c) the weight of 1 curie of ^{14}C.

6.7. The activity of a certain beta-emitting sample was measured over a period of time, as follows:

Time (days)	Activity (cpm)	Time (days)	Activity (cpm)
0.0	9900	4.0	652
0.5	6350	5.0	503
1.0	4050	6.0	430
1.5	2682	7.0	379
2.0	1820	8.0	345
2.5	1310	10.0	290
3.0	989	12.0	245
3.5	786	14.0	207

Plot the decay curve on semi-log paper and analyze it into components. Calculate the half-lives and initial activities of the component activities.

6.8. Plot decay, growth and total activity curves for RaE (^{210}Bi) and RaF (^{210}Po). Assume an activity of 1000 dps for RaE at t_0.

Note: 1. Calculate the number of RaE atoms at t_0.

2. Calculate the number of RaF atoms at various times (about 15 points up to 40 weeks).

3. Calculate the RaF activity at various times from the number of RaF atoms.
4. Add RaE and RaF activities to obtain total activity.

6.9. A solution contained 1.0 millicurie of gold-198 per ml. Calculate the activity of 25 lambdas of this solution exactly 10 days later. Express results in (a) microcuries and (b) disintegrations per second.

6.10. Refer to a chart of the nuclides. Compile a list of 15 radioactive nuclides having a high neutron-proton ratio and a list of 15 having a low neutron-proton ratio. For each, indicate the major mode of decay. Explain your findings in terms of the stability diagonal.

6.11. For ^{76}As, calculate (a) the decay constant λ expressed in days $^{-1}$, (b) the average life τ expressed in hours and (c) the number of atoms required to produce one millicurie of activity.

Answers: (a) 0.602 days^{-1} (b) 39.7 hrs (c) 5.16×10^{12} atoms

6.12. For each of the following statements, list *all* types of radioactive decay processes which apply.

a. n/p ratio too high.
b. Z decreases by 1.
c. A does not change.
d. N increases by 1.
e. A photon but no particle is emitted by the nucleus.
f. Monoenergetic particles are emitted by the nucleus.
g. Annihilation radiation is produced by an interaction of the emitted particle.
h. The nucleus experiences an appreciable recoil.
i. Neutrinos are emitted.
j. The mass of the daughter nucleus is greater than that of the parent.

6.13. The half life of gold-198 is 64.8 hours. Calculate

a. The decay constant λ.
b. The average life τ.
c. The number of atoms required to produce 1 microcurie of activity.
d. The number of atoms required to produce 1000.
e. Disintegrations per minute.
f. The weight of 1 microcurie of Au-198.

6.14. A radioactive substance decays through 5 half-lives. The number of atoms which have undergone decay is N_x. What fraction N_x/N_0 of the original atoms N_0 have disintegrated? (Express as an exponential).

6.15. A solution contains 10.0 millicuries of Mn-52 per ml of solution. Half-life = 6.2 days. Calculate the activity in 50 lambdas of this sample exactly 20 days later. Express answer in (a) microcuries and (b) disintegrations per second.

6.16. Compile a list of 15 alpha emitters. Within what range of atomic numbers are most alpha emitters found?

6.17. Complete the following chart.

Type of Decay or Process	Change in			Examples
	Z	N	A	
Alpha emission				
Beta emission				
Positron emission				
Gamma emission				
Electron Capture				
Isomeric Transition				
Internal conversion				

CHAPTER 7—*PROPERTIES OF RADIATION*

7.1. The absorption curve for a sample emitting both beta and gamma rays was measured with the use of aluminum absorbers. The following data were obtained:

Absorber thickness (g/cm^2)	Activity (cpm)	Absorber thickness (g/cm^2)	Activity (cpm)
0.000	8800	0.800	146
0.050	6300	1.000	135
0.100	4600	2.000	114
0.150	3300	3.000	103
0.200	2400	5.000	82
0.300	1250	7.000	67
0.400	645	9.000	54
0.500	335	11.000	44
0.600	178	14.000	32

From these data calculate (a) the approximate maximum energy of the beta spectrum, (b) the energy of the gamma ray, (c) the mass absorption coefficient and (d) the linear absorption ceofficient for the gamma ray in lead.

7.2. Calculate the number of ion pairs produced by a 5.3 MeV alpha particle if all of its energy is expended in air.

Answer: 1.63×10^5

7.3. If the range of an alpha particle is 5.0 cm in air, estimate its range in (a) water, (b) muscle tissue, (c) aluminum, and (d) lead.

Answers: (a) 6.5×10^{-3}cm or 65μ, (b) 65μ, (c) 24μ, (d) 5.8μ

7.4. For an electron traveling at 95% of the velocity of light, calculate (a) its mass in grams, (b) the ratio m/m_0, (c) its kinetic energy in ergs, and (d) its kinetic energy in MeV.

Answers: (b) 3.2 (d) 1.12 MeV

7.5. Calculate the energy in electron volts of (a) a photon of red light at 7,000 A, (b) a photon of ultraviolet light at 320 millimicrons and (c) the wavelength of a 50 keV x-ray and (d) of a 1.0 MeV gamma ray.

Answers: (a) 1.77 eV, (b) 3.88 eV, (c) 2.5×10^{-9}cm, (d) 1.24×10^{-10}cm

7.6. (a) For an incident 1.0 MeV photon, calculate the energy of a Compton photon scattered at an angle of 90° from the incident photon.

(b) Similarly, calculate the energy of the Compton photon, also at 90°, resulting from an 0.50 MeV incident photon.

Answers: (a) 0.338 MeV, (b) 0.252 MeV

7.7. If the energy of a 2.30 MeV photon, remaining after pair production, is evenly divided between the electron and the positron, calculate the kinetic energy of each.

Answer: 0.64 MeV

7.8. Calculate the fraction of incident radiation passing through 10 half value layers of absorber. What is the half value layer of ^{60}Co in lead? Express your answer in g/cm^2, in cm and in inches.

7.9. (a) Calculate the energy (in MeV) of a Compton scattered photon emerging from the collision at 45° from the path of an incident photon of 1.3 MeV.

Answer: 0.745 MeV

(b) At what velocity is the mass of a particle equal to twice its rest mass?

Answer: 2.60×10^{10} cm/sec

7.10. Calculate the angle at which the energy of Compton-scattered (a) photons and (b) electrons, is maximum.

7.11. The half value layer for Co-60 in lead is 0.51 inches.

(a) Calculate the mass absorption coefficient in cm^2/g.

(b) How many inches of lead shielding is required to reduce the dose rate received from 1 curie of cobalt-60 to a value of 6.25 mr/hr at a distance of 100 cm from the source.

Answers: (a) 0.047 cm^2/g (b) 3.9 inches

CHAPTER 8 — *RADIATION DETECTION BASED ON ION COLLECTION*

8.1. Calculate the intensity of radiation in mr/hr at a distance of (a) 10 cm from 1.0 millicuries of radium, (b) at 100 cm distance. Assume the radium to be in equilibrium with its decay products (0.5 mm filtration).

Answers: (a) 84 mr/hr, (b) 0.84 mr/hr

8.2. How much cobalt-60 is required to give the same radiation intensity (at the same distance from the source) as 1.0 millicurie of radium in equilibrium with its decay products?

Answer: 0.656 mCi

8.3. What would be the effect on the gamma spectrum of a pure gamma emitter if the following instrumental changes were made in the scintillation spectrometer:—

(a) the spectrum was scanned slowly from 2 MeV to essentially 0 MeV.

(b) Repeat step (a) with a narrower window.

(c) Repeat step (a) with the upper discriminator out (window wide open)

(d) Repeat step (a) with amplifier gain doubled. (How would this affect the range of energies scanned?

CHAPTER 9—*SCINTILLATION TECHNIQUES AND NUCLEAR EMULSIONS*

9.1. (a) Why is it useful to have two channels for liquid scintillation counting?

(b) What would be the advantage of having three channels?

(c) What are the disadvantages of liquid scintillation counters compared to crystal scintillation counters?

9.2. As the high voltage applied to the photomultipliers of a liquid scintillation counter is increased, the count rate observed in the upper channel is observed to first increase and then to decrease. How can this behavior be explained?

9.3. Two samples containing carbon-14 are compared in a liquid scintillation counter. The ratio of activities observed in the two windows is observed to change indicating quenching in one or both of the samples. Consequently 100 microliters of an internal standard (specific activity 62,320 dpm per ml) are added to each sample and to a blank.

	Sample Activity (cpm) (lower window/upper window)	Sample plus Standard (cpm) (lower window/upper window)
Blank	24/13	1863/2484
Sample #1	1305/1737	3157/4211
Sample #2	1632/1310	3185/2552

(a) In which sample(s) has quenching occurred? (b) What is the ^{14}C efficiency of each window? (c) What is the overall ^{14}C efficiency? (d) Calculate the relative activities of the two samples.

9.4. What is the best choice of (a) detector, (b) absorber, if required, (c) recording or indicating device, and (d) any accessory equipment required for each of the following radiological assays. (The underscored nuclide is to be measured). The choice need not be restricted to instrumentation discussed in this chapter.

1. Thin sample of $\underline{^{32}P}$ in presence of ^{35}S.
2. $\underline{^{14}C}$ as CO_2 gas.
3. $\underline{^{131}I}$ in aqueous solution with ^{32}P.
4. $\underline{^{210}Po}$ in presence of ^{210}Bi, prepared as a thin sample on a platinum foil.
5. A 10 mc source of $\underline{^{60}Co}$.
6. A weak source of $\underline{^{226}Ra}$ in equilibrium with its decay products, prepared as an aqueous solution in a sealed ampule.
7. Tritium in the presence of ^{14}C.
8. $\underline{^{131}I}$ in vivo.

9. ^{15}N in an amino acid.
10. ^{59}Fe in a suspension of red blood cells.
11. A mixture of ^{131}I and ^{60}Co.

CHAPTER 10—*RADIOCHEMICAL SEPARATION TECHNIQUES*

10.1. Is it possible to precipitate the iodide of carrier-free NaI solution as AgI by the addition of silver nitrate? Consider the iodide ^{131}I solution to have an activity of 1 millicurie per ml.

10.2. Calculate the activity, in millicuries per ml, of a carrier-free $^{59}FeCl_3$ solution which is just saturated with respect to $Fe(OH)_3$ at a pH of 7.0.

10.3. Compute the potential of a zinc electrode immersed in (a) 0.1 M $ZnSO_4$ solution, (b) 1.0 M $ZnSO_4$ solution, (c) 2.0 M $ZnSO_4$ solution. Assume unit activity coefficients.

Answers: (a) 0.793 volts, (b) 0.763 volts, (c) 0.754 volts

10.4. A solution is 1 m with respect to each of the following ions: Ag^+, Na^+, Zn^{++}, Ni^{++}, Cu^{++}. Is it possible to separate these ions by electrodeposition? Indicate the order in which they will deposit.

10.5. Suggest a procedure for separating a mixture of ^{60}Co, ^{59}Fe, ^{64}Cu, ^{115}Cd, and ^{51}Cr, all elements being present in solution as a mixture of their chlorides. Carriers may be used.

CHAPTER 11—*NUCLEAR ANALYSIS*

11.1. (a) What is the natural radioactivity of ^{40}K in disintegrations per minute per gram of ^{40}K.

(b) From the % abundance of ^{40}K, calculate the disintegrations per minute per gram of ordinary KCl.

Answers: (a) 1.53×10^7 dpm (b) 947 dpm

11.2. A sample of ^{32}P (10 lambdas of solution evaporated to dryness) was assayed in a windowless flow-counter having exactly 2π geometry. The sample was mounted on an "infinitely" thick silver planchet. At the time of assay (1400 on March 20) the counting rate was 5,000 cpm (corrected for background and dead time). Estimate the absolute activity of the sample at 1000 on March 10. Give answer in dps/ml. Assume $f_B = 1.6$.

Answer: 1.71×10^4 dps/ml

11.3. A nuclide was found to have a half-life of about 39 seconds. It emits alpha particles with a measured energy of 6.4 MeV. No beta radiation was detected. What is the nuclide?

11.4. A shipment of ^{131}I is to be calibrated against a Ra-DEF standard. 50 lambdas of the original solution are diluted to 10 ml. Of this solution, 1 ml is further diluted to 100 ml. 50 lambdas of the final dilution are then transferred to a standard silver disc and dried. The activity of this sample when compared with that of the Ra-DEF standard similarly mounted, and extrapolated to zero total absorber, are 8,360 cpm and 9,120 cpm, respectively. The activity of the standard is 1.93×10^3 disintegrations per second. Calculate the activity of the sample in mCi/ml.

Answer: 19.03 mCi/ml

11.5. A shipment of $Na_2H^{32}PO_4$ solution is calibrated against a ^{204}Tl standard as follows: A 100μl aliquot is diluted to 5 ml. Of this dilution, a 1 ml aliquot is further diluted to 10 ml and 25 μl of this final dilution is transferred to a planchet and dried. The activity of this sample and that of the ^{204}Tl standard, similarly mounted and both extrapolated to zero total absorber, are observed to be 8,620 cpm and 6328 cpm, respectively. The activity of the ^{204}Tl standard is 1.70×10^3 dps. Calculate the activity of the original $Na_2H^{32}PO_4$ solution in mCi/ml.

CHAPTER 12—*RADIOISOTOPES IN CHEMISTRY*

12.1. A protein hydrolysate is to be assayed for glycine. You add exactly 5.0 mg of glycine having a specific of 0.36 microcuries per milligram. From the hydrolysate you are able to isolate 0.26 milligrams of highly purified gylcine having a specific activity of 0.0093 microcuries per milligram. How much glycine was in the original sample?

12.2. (a) Calculate the millicuries of ^{24}Na produced during an irradiation time of 24 hours by the reaction

$$^{23}Na\ (n,\gamma)\ ^{24}Na$$

from the following data: ^{23}Na is 100% abundant; cross section = 0.6 barns; half-life of ^{24}Na = 15.06 hours; target material = 1.2 grams of Na_2CO_3; flux density = 5×10^{11} n $cm^{-2}sec^{-1}$. (b) Calculate the specific activity.

12.3. ^{32}P is produced by neutron bombardment of ^{32}S by the reaction

$$^{32}S_{16} + {}_0n^1 \longrightarrow {}^{32}P + {}^1p_1$$

^{32}S is 95.1% abundant; cross section = 0.49 barns; target material = 1.0 Kg natural sulfur; flux density = 5×10^{11} n $cm^{-2}sec^{-1}$; irradiation time = 14 days. Calculate (a) the curies of ^{32}P produced and (b) the number of atoms of ^{32}P produced.

12.4. Natural sulfur contains 0.74% of ^{33}S which produces ^{33}P according to the reaction

$$^{33}S\ (n,\ p)\ ^{33}P$$

For this reaction the neutron cross section is 0.0023 barns. Calculate the quantity of ^{33}P produced as a contaminant of the ^{32}P in problem 12.3.

12.5. A 5 kilogram batch of crude penicillin is assayed by isotope dilution: To a 1 gram sample of the batch was added 10 mg pure penicillin having an activity of 10,600 cpm; only 1.40 mg of pure penicillin was recovered having an activity of 280 cpm. What is the penicillin content of the batch?

Answer: 215 g

CHAPTER 13—*ISOTOPIC TRACERS IN BIOLOGY*

13.1. 10 microcuries of ^{24}Na was injected into a 180 g rat. If we assume an even distribution of the ^{24}N throughout the body of the rat, what activity would be observed (in counts per minute) on 0.1 g of the tissue as a sample, measured in an instrument providing 5% counting efficiency?

Answer: 618 cpm

13.2. If the measured effective half-life of ^{24}Na is 0.61 days, calculate the estimated biological half-life.

Answer: 25 hours

13.3. 100 μCi of ^{32}P were injected into a rat. Urine and feces were collected. At the end of the fifth day the pooled excreta were assayed. The total ^{32}P content of the excreta at time of assay was 45 μCi. Calculate (a) the per cent of ^{32}P excreted and (b) the per cent retained by the animal by the fifth day.

Answers: (a) 57.4% (b) 42.6%

13.4. Calculate, in millicuries, the amount of ^{32}P required for an experimental run in which a phosphorus compound is synthesized with a 30% yield, purified with a 12% yield, and administered to a 200 g animal which retains 60% of the administered dose. A 1 gram sample of animal bone is assayed one month after the compound was synthesized with a 7% counting efficiency. A count rate of 300 cpm is desired for satisfactory precision.

CHAPTER 14—*RADIATION BIOLOGY*

14.1. Calculate the dose rate 10 feet from a two-curie, unshielded gamma source. Assume one gamma ray is emitted per disintegration and that the gamma energy is (a) 0.5 MeV, (b) 1.0 MeV, (c) 2.0 MeV and (d) 4.0 MeV. Note the relationship between gamma energy and calculated dose rate.

14.2. List the types of free radicals one would expect to be produced in tissue upon exposure to radiation and indicate for each whether an oxidizing or reducing agent or possibly both depending upon the other reactants.

14.3. List several radiomimetic agents and discuss, if possible, theories to explain how they produce effects similar to those caused by radiation.

CHAPTER 15—*NUCLEAR MEDICINE*

15.1. The IHSA method was used to measure the blood volume of a patient. A dilution of IHSA was prepared and 10 ml of the diluted solution was injected into the patient. To calibrate the IHSA solution, 1 ml of the above dilution was further diluted to 100 ml; the activity of 2 ml of this second dilution was 13,602 cpm. After 10 minutes a 10-ml sample of blood was removed with a heparinized syringe and centrifuged. The activity of 2 ml of plasma was 4,035 cpm. The hematocrit was 0.52. From the patient's height (5′2″) and weight (97 Kg) his surface area was estimated to be 2.23 square meters. For (a) plasma, (b) whole blood and (c) red cells, calculate total volume, volume per kilogram and volume per square meter. Do the values fall within the normal range?

Answers: (a) plasma— 3,371 ml; 35 ml/kg; 1,512 ml/sq. m
(b) whole blood—6,433 ml; 66 ml/kg; 2,885 ml/sq. m.
(c) red cells— 3,062 ml; 32 ml/kg; 1.373 ml/sq. m.

15.2. The sodium radiochromate method was used to measure blood volume. Blood cells were tagged by incubating 20 ml of the patient's blood with $Na_2{}^{51}Cr_2O_7$. The hematocrit of the tagged blood was 0.31. Radioactivity of the whole, tagged blood was measured by diluting 1 ml to 100 ml. The activity of 2 ml of this dilution was 6,880 cpm. Another portion of the tagged blood was centrifuged and 1 ml of the plasma was diluted to 50 ml. The activity of 2 ml of diluted plasma was 984 cpm.

Exactly 10 ml of the calibrated tagged blood was injected intravenously. After 15 minutes a 10 ml blood sample was withdrawn using a heparinized syringe. This blood had a hematocrit of 0.43. The activity of 2 ml of undiluted blood was 1910 cpm while the activity of 2 ml of plasma obtained by centrifugation was 34 cpm.

From the patient's height (5′ 4″) and weight (62.7 kg) his surface area was estimated to be 1.65 square meters. For (a) red cells, (b) whole blood and (c) plasma, calculate total volume, volume per kilogram and volume per square meter. Determine if the values fall within the normal ranges.

Answers: (a) red cells— 1,485 ml; 24 ml/kg; 900 ml/sq. m.
(b) whole blood—3,808 ml; 61 ml/kg; 2,308 ml/sq. m.
(c) plasma— 2,323 ml; 37 ml/kg; 1,408 ml/sq. m.

15.3. Red cell survival time was determined by the sodium radiochromate method. The patient's cells were tagged by incubation with $Na_2{}^{51}Cr_2O_7$ and reinjected. Blood was withdrawn at intervals as indicated for measurement of whole blood activity and hematocrit. Calculate the half-survival time.

Day	cpm/ml of Whole Blood	Hematocrit	Day	cpm/ml of Whole Blood	Hematocrit
1	341	40.5	7	254	36.5
2	298	41.0	9	235	35.5
3	304	41.5	11	226	34.9
4	289	41.5	15	192	35.5
5	267	40.0	17	181	33.0
6	270	37.2	21	153	33.5

Answer: 23 days

15.4. To measure thyroid uptake, patients are given a capsule of $Na^{131}I$ which is taken orally and a second identical capsule is retained as a standard. Results on three patients are tabulated below. The recorded activities are in cpm.

	Patient #1	Patient #2	Patient #3
Standard capsule	4,644	8,219	1,957
Background	138	142	172
Neck activity	1,549	1,905	1,177
Thigh activity	360	596	159

Answers: Patient #1 26.4%; #2 16.2%; #3 57.0%

15.5. A patient received 10 μCi of carrier-free $Na^{131}I$, the gross activity of which was 685 cpm. The room background was 35 cpm. After 24 hours radioassay of the thyroid gland, with similar geometry and using the same equipment, gave a neck activity of 280 cpm. Room background was now 38 cpm. The patient's body background (thigh) was 85 cpm. A 24-hour urine specimen was also collected; it assayed 428 cpm. Calculate the per cent uptake of ^{131}I, the per cent of ^{131}I excreted and the per cent unaccounted for.

Answers: 30%, 60% and 10%.

15.6. Protein-bound iodine was determined on three patients by the trichloroacetic acid method. The following activities, expressed in cpm, were observed:

Patient	Protein Fraction	Total Plasma
#1	41 cpm	179 cpm
#2	771 cpm	1,372 cpm
#3	17 cpm	591 cpm

Express the PBI conversion ratio as a per cent for each patient. Which are normal?

Answers: #1 23%; #2 56%; #3 3%. Only #1 is normal.

15.7. Triiodothyronine *in-vitro* uptake studies were performed on three patients by the Homolsky method. The following results were obtained.

Patient	#1	#2	#3
Net activity of washed RBCs	1,162 cpm	522 cpm	1,650 cpm
Net activity of unwashed RBCs	27,908 cpm	11,075 cpm	13,872 cpm
Hematocrit	45.0%	33.5%	49.5%

Calculate the per cent red cell uptake for each patient. Which are normal?

Answers: #1 9.2%; #2 14.1%; #3 24.0%; #2 is normal.

15.8. Triiodothyronine *in-vitro* uptake studies were performed on three patients by the sephadex method. The following data were obtained.

Fraction	Patient A	Patient B	Patient C
1	0 cpm/ml	0 cpm/ml	13 cpm/ml
2	938	578	1,371
3	2,678	3,661	3,991
4	257	457	591
5	24	21	55
6	25	12	53
7	57	32	123
8	58	59	208
9	45	56	76
10	10	18	59

Calculate the serum protein uptake for each patient. Which are normal?

Answers: A 83.9%; B 90.2%; C 64.2%; A is normal.

15.9. In a study of the kinetics of iron matabolism and plasma iron clearance half-time was determined. Calculate the value from the following data:

Time post injection Sample Withdrawn	Blood Activity cpm/ml	Plasma Activity cpm/ml
10 min.	202	287
30 "	147	218
45 "	128	179
60 "	98	142
90 "	59	85
120 "	34	60

Answers: From blood activity, 50 min.; From plasma activity, 44 min.

APPENDIX H Reference Books

1905—1935

1. DUNCAN, R. K., *The New Knowledge."* A. S. Barnes (1905).
2. SODDY, F., *"The Chemistry of the Radioelements."* Longmans Green (1911).
3. FAJANS, K., *"Radioactivity."* Methuen (1923).
4. LODGE, SIR OLIVER J., *"Atoms and Rays: and introduction to modern views on atomic structure and radiation."* Doran (1924).
5. LIND, S. C., *"The Chemical Effects of Alpha Particles and Electrons."* New York, Chemical Catalog (ACS Monograph No. 2) (1928).
6. RUTHERFORD, E., J. CHADWICK and C. D. ELLIS, *"Radiations from Radioactive Substances."* Cambridge Univ. Press (1930).
7. WHITE, H. E., *"Introduction to Atomic Spectra."* McGraw-Hill (1934) $3.95.
8. COMPTON, A. H. and S. K. ALLISON, *"X-Rays in Theory and Experiment."* Van Nostradd (1935).

1936

9. DUGGAR, B. M., *"Biological Effects of Radiation."* McGraw Hill.
10. FEATHER, N., *"An Introduction to Nuclear Physics."* Cambridge Univ. Press.
11. HAHN, O., *"Applied Radiochemistry."* Cornell Univ. Press; Oxford Univ. Press.
12. RASETTI, F., *"Elements of Nuclear Physics."* Prentice-Hall.

1938

13. HEVESY, G. VON. and F. A. PANETH, *"A Manual of Radioactivity."* Oxford Univ. Press.
14. SPEAKMAN, *"Modern Atomic Theory. An Elementary Introduction."* St. Martins. $2.50.
15. STRONG, J., *"Procedures in Experimental Physics."* Prentice-Hall (Prentice-Hall Physics Series).

1940

16. GAMOW, G. W., *"Mr. Tompkins in Wonderland."* Cambridge.

1941

17. ELLINGER, F., *"The Biologic Fundamentals of Radiation Therapy."* Van Nostrand. $4.75.

1942

18. LEWIS, W. B., *"Electrical Counting with Special Reference to Counting Alpha and Beta Particles."* Cambridge bridge Univ. Press.
19. STRANATHAN, J. D., *"The Particles of Modern Physics."* Blakiston.

1944

20. GAMOW, G. W., *"Mr. Tompkins Explores the Atom."* Cambridge (Macmillan)
21. HEITLER, W., *"The Quantum Theory of Radiation."* Oxford Univ. Press (International Series on Physics).
22. HERZBERG, G., *"Atomic Spectra and Atomic Structure."* Dover.

1945

23. SMYTH, H. de W., *"Atomic Energy for Military Purposes: The Official Report on the Development of the Atomic Bomb under the Auspices of the U. S. Government, 1940-1945."* Princeton Univ. Press.

1946

24. BORN, M., *"Atomic Physics."* Blackie and Son. $1.95.
25. GAMOW, G., *"Atomic Energy in Cosmic and Human Life: Fifty Years of Radioactivity."* Cambridge Univ. Press; Macmillan.
26. GLASSTONE, S., *"Textbook of Physical Chemistry."* Van Nostrand.
27. MATTAUCH, J., *Nuclear Physics Tables"*, and FLUEGGE, S., *"An Introduction to Nuclear Physics."* Interscience.

1947

28. ANDRADE, E. N. daC., *"The Atom and its Energy."* London, G. Bell and Sons Ltd.
29. FRISCH, O. R., *"Meet the Atoms."* London, Sigma Books.
30. MILLIKEN, R. A., *"Electrons (+ and −), Protons, Neutrons, Mesotrons, and Cosmic Rays."* University of Chicago Press (Rev. ed.). $9.00.
31. WILSON, D. W., A. O. C. WIER and S. P. REIMANN (eds.) *"Preparation and Measurement of Isotopic Tracers."* A Symposium. Edward Bros.

1948

32. BLOOM, W., (ed.), *"Histopathology of Irradiation from External and Internal Sources."* McGraw-Hill (National Nuclear Energy Series, Div. 4, Vol. 22). $10.75.
33. GOODMAN, C. D., *"The Science and Engineering of Nuclear Power."* Cambridge; Addison-Wesley (1947-48). 2 Vols.
34. HEVESY, G. VON, *"Radioactive Indicators; Their Application in Biochemistry, Animal Physiology and Pathology."* Interscience.
35. HOAG, J. B. and S. A. KORFF, *"Electron and Nuclear Physics."* Van Nostrand. $7.00.
36. NEWMAN, R. W., and B. S. MILLER, *"The Control of Atomic Energy."* McGraw-Hill.
37. PAULI, W., *"Meson Theory of Nuclear Forces."* Interscience.

38. ROBERTSON, J. K., *"Radiology Physics; An Introductory Course for Medical or Premedical Students and for all Radiologists."* Van Nostrand.
39. ROSENFELD, L., *"Nuclear Forces."* Interscience, Part III.
40. STEPHENS, W. E. (ed.), *"Nuclear Fission and Atomic Energy."* Science Press.
41. TOLANSKY, S., *"Introduction to Atomic Physics."* Longmans Green.

1949

42. BEYER, R. T. (ed.), *"Foundations of Nuclear Physics."* Dover. $1.75.
43. CALVIN, M., C. HEIDELBERGER, J. C. REID, B. M. TOLBERT and P. E. YANKWICH, *"Isotopic Carbon; Techniques in its Measurement and Chemical Manipulation."* Wiley. $9.25.
44. CURIE, E., *"Madame Curie"*. Doubleday. $5.00.
45. CURRAN, S. C. and J. D. CRAGGS, *"Counting Tubes; Theory and Applications."* London, Butterworths Scientific Publications.
46. ELMORE, W. C., and M. SANDS (eds.), *"Electronics; Experimental Techniques."* McGraw-Hill. $3.75.
47. KAPLAN, I. I., *"Clinical Radiation Therapy."* Hoeber.
48. OLDENBERG, O., *"Introduction to Atomic Physics."* McGraw-Hill. $5.00.
49. RICE, F. and E. TELLER, *"The Structure of Matter."* Wiley.
50. ROSSI, B. B. and H. H. STAUB, *"Ionization Chambers and Counters; Experimental Techniques."* McGraw-Hill (National Nuclear Energy Series). $3.25.
51. SCHWEITZER, G. K. and I. B. WHITNEY, *"Radioactive Tracer Techniques."* Van Nostrand.
52. SEABORG, G. T., J. J. KATZ, and W. M. MANNING (eds.), *"The Transuranium Elements."* McGraw-Hill. $23.75 set.
53. SIRI, W. E., *"Isotopic Tracers and Nuclear Radiations; with Applications to Biology and Medicine."* McGraw-Hill.
54. YAGODA, H. J., *"Radioactive Measurements with Nuclear Emulsions."* Wiley.

1950

55. BRODA, E., *"Advances in Radiochemistry."* Elsevier. $5.00.
56. COCKCROFT, SIR JOHN D., *"The Development and Future of Nuclear Energy."* Oxford, Clarendon Press.
57. DIXON, SIR ARTHUR L., *"Atomic Energy for the Layman."* London, Chantry Publications Ltd.
58. FERMI, E., *"Nuclear Physics."* Univ. Chicago Press. $3.50.
59. HAHN, O., *"New Atoms."* Elsevier.
60. LOW-BEER, B. V. A., *"The Clinical Use of Radioactive Isotopes."* Thomas.
61. OREAR, J., A. H. ROSENFELD, and R. A. SCHLUTER, *"Nuclear Physics."* Univ. of Chicago Press. $3.50.
62. SOODAK, H. and E. C. CAMPBELL, *"Elementary Pile Theory."* Wiley. $3.90.
63. WILLIAMS, R. R., *"Principles of Nuclear Chemistry."* Van Nostrand.

1951

64. CORYELL, C. D., *"Radiochemical Studies: The Fission Products."* McGraw-Hill. $27.75.
65. DUSHMAN, S., *"Fundamentals of Atomic Physics."* McGraw-Hill.
66. FERMI, E., *"Elementary Particles."* Yale. $3.00.
67. GLASCOCK, R., *"Labelled Atoms."* Interscience.
68. GRAY, D. E., and J. H. MARTENS, *"Radiation Monitoring in Atomic Defense."* Van Nostrand. $3.50.
69. GROBMAN, A. B., *"Our Atomic Heritage."* Univ. of Florida. $2.95.
70. GUEST, G. D., *"Radioisotopes, Industrial Applications."* Pitman.
71. HAHN, P. F., *"A Manual of Artificial Radioisotope Therapy."* Academic Press.
72. OVEREND, W. G., *"The Use of Tracer Elements in Biology."* London, Heinemann.
73. POLLARD, E. C. and W. L. DAVIDSON, *"Applied Nuclear Physics."* 2nd Ed. Wiley. $6.00.
74. STONE, R. S., *"Industrial Medicine on the Plutonium Project; Survey and Collected Papers."* McGraw-Hill (National Nuclear Energy Series).
75. TANNENBAUM, A. (ed.), *"Toxicology of Uranium; Survey and Collected Papers."* McGraw. $4.75.
76. TAYLOR, D., *"Measurement of Radioisotopes."* Wiley.
77. WAHL, A. C. and N. A. BONNER (eds.), *"Radioactivity Applied to Chemistry."* Wiley.
78. ZIRKLE, R. E. (ed.), *Effects of External Beta Radiation."* McGraw-Hill. $3.25.

1952

79. BLATT, J. M., and V. F. WEISSKOPF, *"Theoretical Nuclear Physics."* Wiley. $13.50.
80. BLEULER, E. and G. J. GOLDSMITH, *"Experimental Nucleonics."* Rinehard. $6.50.
81. COOK, G. B. and J. F. DUNCAN, *"Modern Radiochemical Practice."* Oxford Press.
82. FEATHER, N., *"Nuclear Stability Rules."* Cambridge Univ. Press.
83. GLASSER, O., E. H. QUIMBY, L. S. TAYLOR and J. L. WEATHERWAX, *"Physical Foundations of Radiology."* Harper & Bros. $6.50.
84. GLASSTONE, S. and M. C. EDLUND, *"The Elements of Nuclear Reactor Theory."* Van Nostrand. $5.50.
85. ISARD, W. and V. WHITNEY, *"Atomic Powers; An Economic and Social Analysis."* Blakiston.
86. KENNEDY, J. W., *"Radioactive Atoms and Isotopic Tracers."* Penna. State College.
87. KESSLER, J., *"You and the Atom."* Kessler, Glen Oaks, L. I., New York.
88. NICKSON, J. J. (ed.), *"Symposium of Radiobiology."* Wiley. $11.25.
89. ROSSI, B. B., *"High-Energy Particles."* Prentice Hall. $12.50.

90. YATES, R. F., *"Atomic Experiments for Boys."* Harper and Bros. $2.50.
91. THORNDIKE, A. M., *"Mesons; A Summary of Experimental Facts."* McGraw-Hill. $6.00.

1953

92. BIRKS, J. B., *"Scintillation Counters."* McGraw-Hill and Pergamon Press.
93. BRADFORD, J. R., *"Radioisotopes in Industry."* Reinhold. $8.00.
94. CHADWICK, SIR JAMES, *"Radioactivity and Radioactive Substances."* 4th Ed. London, Sir Isaac Pitman and Sons, Ltd.
95. CURRAN, *"Luminescence and the Scintillation Counter."* Academic Press.
96. DELARIO, A. J., *"Roentgen Radium and Radioisotope Therapy."* Lea and Febiger. $7.50.
97. GILLISPIE, A., *"Signal, Noise and Resolution in Nuclear Counter Amplifiers."* McGraw-Hill.
98. HUGHES, D. J., *"Neutron Optics."* Interscience. $3.85.
99. HEISENBERG, WERNER, *"Nuclear Physics."* Philosophical Library. $4.75.
100. HUGHES, D. J., *"Pile Neutron Research."* Addison. $10.00.
101. *"Handbook of Nuclear Techniques."* Nucleonics.
102. MORSE, P. M. and H. FESHBACH, *"Methods of Theoretical Physics"*, Parts 1 and 2. McGraw-Hill. $30.00.
103. National Research Council Conference on Nuclear Glossary, *"A Glossary of Terms in Nuclear Science and Technology."* Amer. Soc. Mechanical Engineers, N.Y. $7.00.
104. SACHS, R. G., *"Nuclear Theory."* Addison-Wesley. $8.50.
105. SACKS, J. *"Isotopic Tracers in Biochemistry and Physiology."* McGraw-Hill. $8.50.
106. SEGRE, E. (ed.), *"Experimental Nuclear Physics."* Wiley. (Vol. I, $16.50); (Vol. II, $13.00).
107. SNEED, M. C., J. L. MAYNARD and R. C. BRASTED (eds.), *"Comprehensive Inorganic Chemistry"*, (The Actinide Elements). Van Nostrand. $5.50.
108. SPEAR, F. G., *"Radiations and Living Cells."* Wiley. $3.50.
109. WHITEHOUSE, W. J. and J. L. PUTMAN, *"Radioactive Isotopes."* Clarendon Press.

1954

110. BEELER, N. F. and F. M. BRANLEY, *"Experiments with Atomics."* Crowell. $2.50.
111. BLAIR, H. A., *"Biological Effects of External Radiation."* McGraw-Hill. $7.00.
112. CHERONIS, N. D., *"Technique of Organic Chemistry, Vol. VI, Micro and Semimicro Methods."* Chapter XIII —Microsyntheses with Tracer Elements, by A. R. RONZIO. Interscience Pub.
113. DEBYE, P. J. W., *"Collected Papers of Peter J. W, Debye."* Interscience. $9.50.
114. FERMI, L., *"Atoms in the Family."* Univ. of Chicago Press. $4.00.
115. GLASCOCK, R. F., *"Isotopic Gas Analysis for Biochemists."* Academic Press. $6.80.
116. HECHT, S., *"Explaining the Atom."* Viking Press. $3.75.
117. HOLLANDER, A., *"Radiation Biology,"* Vol. I—Ionizing Radiations. McGraw-Hill. $17.50.
118. HUNTLEY, H. E., *"Nuclear Species."* St. Martins Press. $4.50.
119. NEVENZEL, J. C., R. F. RILEY, D. R. HOWTON, and G. STEENBERG, *"Bibliography of Synthesis with Carbon Isotopes."* Office of Technical Services, Dept. of Commerce, Washington, D.C. $1.05.
120. SEABORG, G. T. and J. J. KATZ, *"The Actinide Elements."* (National Nuclear Energy Series). McGraw-Hill.
121. WEYL, C., S. R. WARREN, Jr. and D. B. O'NEILL, *"Radiologic Physics."* Baltimore, C. C. Thomas.
122. ZIRKLE, R. E., *"Biological Effects of External X and Gamma Radiation."* Part I). McGraw-Hill. $7.25.
123. WAKEFIELD, E. H., *"Nuclear Reactors for Industry and Universities."* Instruments Publishing Co. $2.00.

1955

124. BACQ, Z. M., *"Fundamentals of Radiobiology."* Academic Press.
125. BLACKWOOD, O., RUARK, ET AL., *"An Outline of Atomic Physics,"* 3rd ed. Wiley. $7.50.
126. BOYD, G. A., *"Autoradiography in Biology and Medicine."* Academic Press. $12.00.
127. CLARK, G. L., *"Applied X-Rays"* (4th ed.) International Series in Pure and Applied Physics. McGraw-Hill. $12.50.
128. COMAR, C. L., *"Radioisotopes in Biology and Agriculture."* McGraw-Hill. $9.50.
129. EINSTEIN, A., *"The Meaning of Relativity"* (5th ed., including the Relativistic Theory of the Non-Symmetric Field). Princeton Univ. Press. $3.75.
130. EVANS, R. D., *"The Atomic Nucleus."* McGraw-Hill.
131. FEENBERG, E., *"Shell Theory of the Nucleus."* Princeton.
132. GLASSTONE, S., *"Principles of Nuclear Reactor Engineering."* Van Nostrand. $7.95.
133. GREEN, A. E. S., *"Nuclear Physics."* McGraw-Hill. $9.00.
134. HALLIDAY, D., *"Introductory Nuclear Physics"*, 2nd Ed. Wiley. $7.50.
135. HARNWELL, G. P. and W. E. STEPHENS, *"Atomic Physics."*
136. HAUSNER, H. H. and S. B. ROBOFF, *"Materials for Nuclear Power Reactors."* Reinhold. $3.50.
137. HOLLAENDER, *"Radiation Biology."* Vol. II. Ultraviolet Radiations. McGraw-Hill. $8.00.
138. HUGHES, D. J. and J. A. HARVEY, *"Neutron Cross Sections."* McGraw-Hill. $12.00.
139. HYDE, M. O., *"Atoms Today and Tomorrow."* McGraw-Hill. $8.00.
140. JOUCH, J. M. and F. ROHRLICH, *"The Theory of Photons and Electrons."* Addison. $10.00.
141. KAPLAN, I., *"Nuclear Physics."* Addison-Wesley. $8.50.
142. KORFF, S. A., *"Electron and Nuclear Counters; Theory and Use."* Van Nostrand.

143. LEA, D. E., *"Actions of Radiations on Living Cells,"* 2nd Ed. Cambridge Univ. Press. $6.50.
144. LIBBY, W. F., *"Radiocarbon Dating,"* 2nd Ed. Univ. of Chicago Press. $4.50.
145. MAYER, M. G. and J. H. D. JENSEN, *"Elementary Theory of Nuclear Shell Structure."* Wiley. $7.50.
146. PEASLEE, *"Elements of Atomic Physics."* Prentice-Hall. $7.50.
147. Physics Staff of the Univ. of Pittsburgh, *"Outline of Atomic Physics."* 3rd Ed. Wiley. $7.50.
148. RICHTMYER, F. K., E. H. KENNARD and T. LAURITSEN, *"Introduction to Modern Physics."* McGraw-Hill. $8.50.
149. SEMAT, H., *"Introduction to Atomic Nuclear Physics."* 3rd Ed. Rinehart. $6.50.
150. SHANKLAND, R. S., *"Atomic and Nuclear Physics."* Macmillan. $7.75.
151. SIEGBAHN, *"Beta-and Gamma-Ray Spectroscopy."* Interscience. $20.00.
152. WAY, KING, McGINNIS and VAN LIESHOUT, *"Nuclear Level Schemes."* Supt. of Documents, U. S. Govt. Printing Office, Washington, D. C. $1.75.
153. WOODBURY, D. O.,*"Atoms for Peace."* Dodd, Mead & Co. $3.50.

1956

154. ARONOFF, S., *"Techniques of Radiobiochemistry."* Iowa State College Press. $5.95.
155. AUERBACH, C., *"Genetics in the Atomic Age."* Essential. $2.00.
156. BETHE, H. A. and P. MORRISON, *"Elementary Nuclear Theory."* Wiley. $6.25.
157. BRONOWSKI, J., *"The Atom."* London, Newman Neame Take Home Books Ltd.
158. COMPTON, A. H., *"Atomic Quest, A Personal Narrative."* Oxford. $5.00.
159. CROWTHER, J. G., *"Nuclear Energy in Industry."* Pitman. $3.95.
160. CULLITY, B. D., *"Elements of X-Ray Diffraction."* Addison Wesley.
161. HAHN, P. F. (Ed.), *"Therapeutic Use of Artificial Radioisotopes."* Wiley. $10.00.
162. HANNAN, R. S., *"Research on the Science and Technology of Food Preservation by Ionizing Radiations."* Chemical Pub. $4.50.
163. HINE, G. J. and BROWNELL, G. L. *"Radiation Dosimetry."* Academic Press.
164. JAY, K., *"Calder Hall: The Story of Britain's First Atomic Power Station."* Harcourt, Brace. $3.00.
165. LAPP, R., *"Atoms and People."* Harper Bros. $4.00.
166. LUNTZ, J. D., *"Handbook of Radioisotope Applications."* Nucleonics, New York.
167. MASSEY, H. S. W., *"Atoms and Energy."* Philosophical Library, Inc. $4.75.
168. McCUE, J. J. G. and K. W. SHERK, *"The World of Atoms: An Introduction to Physical Science."* The Ronald Press. $6.50.
169. OPPENHEIMER, J. R., *"The Constitution of Matter."* (Condon Lectures) Oregon State Institution of Higher Education. $1.00.
170. ROCKWELL, T. (ed.), *"Reactor Shielding Design Manual."* McGraw-Hill. $6.00.
171. SCHULTZ, M. A., *"Control of Nuclear Reactors and Power Plants."* McGraw-Hill. $7.50.
172. SPROULL, R. L., *"Modern Physics."* Wiley. $7.75.
173. THOMAS, M., *"Atomic Energy and Congress."* Univ. of Michigan Press. $4.75.
174. THOMPSON, SIR GEORGE, *"The Atom."* 5th Ed. London, New York, Toronto, Oxford University Press.
175. WENDT, G., *"You and The Atom."* William Morrow & Co. $1.95.

1957

176. ALEXANDER, P., *"Atomic Radiation and Life."* Penguin Books.
177. American Society of Mechanical Engineers, *"Glossary of Terms in Nuclear Science and Technology."* A.S.M.E., 29 W. 39th St., New York 18, N. Y.
178. BAGNALL, K. W., *"Chemistry of the Rare Radioelements."* Academic Press. $5.00.
179. BEIERWALTES, W. H., P. C. JOHNSON and A. J. SOLARI, *"Clinical Use of Radioisotopes."* Saunders. $11.50.
180. BURR, J. G., Jr., *"Tracer Applications for the Study of Organic Reactions."* Interscience. $7.50.
181. COMAR, C. L., *"Atomic Energy and Agriculture."* American Assoc. for the Advancement of Science. $9.50.
182. CORK, J. M., *"Radioactivity and Nuclear Physics."* Van Nostrand. $7.75.
183. CURTIS, H. J. and H. QUASTLER, *"Mammalian Aspects of Basic Mechanisms in Radiobiology."* Nuclear Science Series Report No. 21, National Academy of Sciences—National Research Council (Pub. 513). $2.00
184. DAVISDON, H. O., *"Biological Effects of Whole-Body Gamma Radiation on Human Beings."* Johns Hopkins. $3.00.
185. DEAN, G., *"Report on the Atom."* 2nd Ed. Alfred A. Knopf. $5.00.
186. DICK, W. E., *"Atomic Energy in Agriculture."* Philosophical Library. $6.00.
187. DIENES, G. J. and G. H. VINEYARD, *"Radiation Effects in Solids."* Interscience. $6.50.
188. DUNLAP, H. A. and H. TUCH, *"Atoms at your Service."* Harper. $3.50.
189. ELLINGER, F. P., *"Medical Radiation Biology."* Thomas. $20.00.
190. FERMI, L., *"Atoms for the World."* Univ. of Chicago Press. $3.75.
191. FIELD, F. H. and J. L. FRANKLIN, *"Electron Impact Phenomena and the Properties of Gaseous Ions."* Academic Press. $10.00.
192. GOWGILL, R. W., and A. B. PARDEE, *"Experiments in Biochemical Research Techniques."* Wiley. $3.50.
193. HALNAN, K. E., *"Atomic Energy in Medicine."* New York Philosophical Library; London, Butterworths. $6.00.
194. HARTREE, D. R., *"The Calculation of Atomic Structures."* Wiley.
195. HINTENBERGER, H. (ed.), *"Nuclear Masses and Their Determination."* Pergamon Press. $14.00.
196. HUGHES, D. J., *"On Nuclear Energy: Its Potential for Peacetime Uses."* Harvard Univ. Press. $4.75.
197. HUGHES, D. J. and J. A. HARVEY, *"Neutron Cross Sections."* (Supplement to 1955 edition). U. S. Govt. Printing Office. $1.75.

198. KAMEN, M. D., *"Isotopic Tracers in Biology."* 3rd Ed. Academic Press. $12.00.
199. KATZ, J. J. and G. T. SEABORG, *"Chemistry of the Actinide Elements."* Wiley. $14.00.
200. KOCH, H. W. and R. W. JOHNSON, *"Multichannel Pulse Height Analyzers;* Proceedings of an Informal Conference." Nuclear Science Series Report No. 20. National Academy of Sciences—National Research Council (Pub. 467) $2.00.
201. MANDER, J., *"Atoms at Work."* London, George Newnes Ltd.
202. MANN, M., *"Peacetime Uses of Atomic Energy."* Crowell. $4.50.
203. NEVENZEL, J. C., D. R. HOWTON, R. F. RILEY and G. STEINBERG. *"A Bibliography of Syntheses with Carbon Isotopes, 1953-4."* California University Report UCLA 395.
204. PRICE, B. T., *"Radiation Shielding."* Pergamon Press. $10.00.
205. ROWLAND, J., *"Ernest Rutherford: Atom Pioneer."* Philosophical Library. $4.75.
206. SCHUBERT and LAPP, *"Radiation: What it is and How it Affects You."* Viking Press. $3.95.
207. SELMAN, J., *"Fundamentals of X-Ray and Radium Physics."* Thomas. $8.50.
208. SMULLEN, W., *"Basic Foundation of Isotope Technique for Technicians."* Thomas. $4.75.
209. SORENSEN, POUL, *"Isotope Dilution Analysis."* Den Polytekniske Laereanstalt, Danmarks Tekniske Hoskole, Kobenhavn K.
210. STENSTRON, K. W., *"Manual of Radiation Therapy."* Thomas. $4.50.
211. STOKLEY, J., *"The New World of the Atom."* Ives Washburn, Inc. $5.50.
212. TAYLOR, D., *"The Measurement of Radioisotopes."* 2nd Ed. Methuen, London.
213. WENDT, G., *"The Prospects of Nuclear Power and Technology."* Van Nostrand. $4.95.
214. YOE AND KOCHS (eds.), *"Trace Analysis."* Wiley. $12.00.

1958

215. ALLEN, J. S., *"The Neutrino."* Princeton Univ. Press. $4.50.
216. ARNOTT, D. G., *"Our Nuclear Adventure."* Philosophical Library. $6.00.
217. BELCHEM, R. F. K., *"A Guide to Nuclear Energy."* London, Sir Isaac Pitman and Sons, Ltd.
218. BELL, C. G. and F. N. HAYES, *"Liquid Scintillation Counting."* Pergamon Press. $10.00.
219. BOHR, N., *"Atomic Physics and Human Knowledge."* Wiley. $3.95.
220. BOURSNELL, J. C., *"Safety Techniques for Radioactive Tracers."* Cambridge Univ. Press.
221. BOVEY, F. A., *"The Effects of Radiation on Natural and Synthetic High Polymers."* Interscience. $8.00.
222. BRAESTRUP, C. B., *"Radiation Protection."* Charles C. Thomas. $10.50.
223. BRUCER, M., *"Radioisotope Scanning."* Superintendent of Documents, U. S. Government Printing Office, Wash. 25, D. C. $1.00.
224. CLAUSE, W. D., *"Radiation Biology and Medicine."* Addison Wesley.
225. EISENBUD, L. and E. P. WIGNER, *"Nuclear Structure."* Princeton Univ. Press. $4.00.
226. ENGSTRÖM, A., R. BJÖRNERSTEDT, C. J. CLEMEDSON and A. NELSON, *"Bone and Radiostrontium."* Wiley & Sons. $8.75.
227. ETHERINGTON, H., *"Nuclear Engineering Handbook."* McGraw-Hill. $25.00.
228. EXTERMANN, R. C., *"Radioisotopes in Scientific Research."* Vol. I. Research with Radioisotopes in Physics and Industry. Pergamon Press. $22.50.
229. EXTERMANN, R. C. *"The Peaceful Uses of Atomic Energy."* (12 vols.) Pergamon Press. $15.00 ea.
230. FAIRES, R. A., and B. H. PARKS, *"Radioisotope Laboratory Techniques."* Pitman. $5.75.
231. FREEMAN, M. B., *"The Story of Albert Einstein:* A Biography for Young Readers." Random House. $2.95.
232. FRENCH, A. P., *"Principles of Modern Physics."* Wiley. $6.75.
233. FRISCH, O. R., *"The Nuclear Handbook."* Van Nostrand. $8.50.
234. GAMOW, G., *"Matter, Earth and Sky."* Prentice-Hall. $10.00.
235. GIBBS, R. C. and K. WAY, *"A Directory to Nuclear Data Tabulations."* U. S. Atomic Energy Commission, Washington, D. C.
236. GLASSER, O., *"Dr. W. C. Röntgen."* (2nd ed.) Thomas. $4.50.
237. GLASSTONE, S., *"Source Book on Atomic Energy."* (2nd ed.) Van Nostrand.
238. GOLDSTEIN, H., *"The Attenuation of Gamma Rays and Neutrons in Reactor Shields (U. S. Atomic Energy Commission)."* U. S. Government Printing Office. $2.00.
239. HARWOOD, J. J., H. H. HAUSNER, J. G. MORSE, and W. G. BAUCH (eds.), *"Effects of Radiation on Materials."* Reinhold.
240. HOAG, J. B., *"Nuclear Reactor Equipments."* Van Nostrand. $6.75.
241. HOOPER, J. E., and M. SCHARFF, *"The Cosmic Radiation."* Wiley & Sons. $2.75.
242. HUGES, and SCHWARTZ, *"Neutron Cross Sections."* Superintendent of Documents. $4.50.
243. JACKSON, J. D., *"The Physics of Elementary Particles."* Princeton Univ. Press. $4.50.
244. KOPFERMANN, HANS, *"Nuclear Moments."* Academic Press. $14.50.
245. LANDAU, C., and YA. SMORODINSKY, *"Lectures on Nuclear Theory."* Consultants Bureau, Inc. $15.00.
246. LANSDELL, NORMAN, *"The Atom and the Energy Revolution."* Penguin Books, Inc. $0.85.
247. McCORMICK, J. A., *Isotopes in Biochemistry and Biosynthesis of Labeled Compounds."* U. S. Atomic Energy Commission Report TID-3513.
248 McCORMICK, J. A., *"Isotopes: A Bibliography of U. S. Research and Application 1955-1957."* U. S. Atomic Energy Commission, Washington, D. C. $2.25.
249. MATHER, K. B. and P. SWAN, *"Nuclear Scattering."* Cambridge Univ. Press. $14.50.

250. MANN, W. B., and H. H. SELIGER, *"Preparation, Maintenance and Application of Standards of Radioactivity."* National Bureau of Standards Circular 594. U. S. Govt. Printing Office. $0.35.
251. MURRAY, A., and D. L. WILLIAMS, *"Organic Synthesis with Isotopes."* Part. I. Compounds of Isotopic Carbon. Interscience.
252. MURRAY, A., and D. L. WILLIAMS, *"Organic Synthesis with Isotopes."* Part II. Interscience.
253. NOKES, M. E., *"Radioactivity Measuring Instruments: Guide to Their Construction and Use."* Philosophical Library.
254. PRICE, W. J., *"Nuclear Radiation Detection."* McGraw-Hill.
255. QUIMBY, FEITELBERG, and SILVER, *"Radioactive Isotopes in Clinical Practice."* Lea & Febiger. $10.00.
256. ROSE, M. E., *"Internal Conversion Coefficients."* Interscience.
257. RUECHARDT, E., *"Light: Visible and Invisible."* University of Michigan Press. $4.50.
258. SCHENBERG, S., *"Laboratory Experiments with Radioisotopes for High School Science Demonstrations."* Catalog No. Y3. At 7:2 R 11/18/958. Superintendent of Documents, Washington, D. C. $0.25.
259. SCOTT, T. W., *"Controlled Thermonuclear Processes."* U. S. Atomic Energy Commission, Washington 25, D. C. $1.00.
260. SEABORG, G. T. *"Transuranium Elements."* Yale Univ. Press.
261. SEABORG, G. T., and E. G. VALENS, *"Elements of the Universe."* Dutton. $3.95.
262. SELIGER, H. H. and W. B. MANN, *"Preparation, Maintenance, and Application of Standards of Radioactivity."* U. S. Dept. of Commerce, National Bureau of Standards, Washington 25, D. C. $0.35.
263. SMORODINSKY, YA and C. LANDAU, *"Lectures on Nuclear Theory."* Consultants Bureau, Inc. $15.00.
264. SPARROW, A. H., J. P. BINNINGTON, and V. POND, *"Bibliography on the Effects of Ionizing Radiations on Plants."* Office of Technical Services, Dept. of Commerce, Washington 25, D. C. $2.25.
265. STANG, L. G., *"Hot Laboratory Equipment."* Supt. of Documents, Washington, D. C. $2.50.
266. STEPHENSON, R., *"Introduction to Nuclear Engineering."* 2nd Ed. McGraw-Hill. $9.50.
267. SWAN, M., *"Nuclear Scattering."* Cambridge Univ. Press. $14.50.
268. TAVARES, C. A., *"Cancer and the Atomic Age."* Vantage. $3.50.
269. TAYLOR, E. O., *"Nuclear Reactors for Power Generation."* Philosophical Library, Inc. $7.50.
270. TELLER, E., and A. L. LATTER, *"Our Nuclear Future: Facts, Dangers and Opportunities." Criterion Books, Inc. $3.50.*
271. United Nations, *"Atomic Energy: Glossary of Technical Terms."* Columbia University Press. $4.00.
272. United States Department of Commerce, National Bureau of Standards, *"Radioactivity."* Handbook 86. National Bureau of Standards, Washington, D. C.
273. VEALL, N., and H. VETTER, *"Radioisotope Techniques in Clinical Research and Diagnosis."* Butterworth.
274. WASHTELL, C. C. H., *"An Introduction to Radiation Counters and Detectors."* Newnes, London.
275. WEINBERG, A. M., and E. P. WIGNER, *"The Physical Theory of Neutron Chain Reactors."* University of Chicago Press. $15.00.

1959

276. BALDIN, A., V. GOLDANSKII, and I. ROZENTAL, *"Kinematics of Nuclear Reactions."* Oxford. $6.10.
277. BEHRENS, C. P., *"Atomic Medicine."* 3rd Ed. Williams & Wilkins.
278. BIRKS, L. S., *"X-Ray Spectrochemical Analysis."* (Chemical Analysis Series-Vol. XI) Wiley & Sons. $7.50.
279. BISHOP, A. S., *"Project Sherwood—The U. S. Program in Controlled Fusion."* Addison-Wesley. $5.75.
280. BLAHD, W. H., F. K. BAVER and B. CASSEN, *"Practice of Nuclear Medicine."* Thomas.
281. BLATZ, H., *"Radiation Hygiene Handbook."* McGraw-Hill. $27.50.
282. BRAGG, SIR WM., *"The Universe of Light."* Dover Publications. $1.85.
283. BAUNBEK, W., *"The Pursuit of the Atom."* Emerson Books. $3.95.
284. BROWN, S. C., *"Basic Data of Plasma Physics."* Wiley. $6.50.
285. BRUES, A. (ed.), *"Low Level Irradiation."* American Assoc. for Advancement of Science. $3.75.
286. CLAUS, WALTER D., *"Radiation Biology and Medicine."* Addison-Wesley. $11.50.
287. CURTISS, L. F., *"Introduction to Neutron Physics."* Van Nostrand. $9.75.
288. DANFORTH, J. P., and R. P. STAPP, *"Radioisotopes in Industry: Training Program."* General Motors Corp. Research Labs., Warren, Mich. $10.00 set.
289. DUNHAM, C. L., *"Radioactive Fallout."* Office of Technical Services, Dept. of Commerce, Washington, D. C. $1.25.
290. ELTON, L. R. B., *"Introductory Nuclear Theory."* Interscience. $6.40.
291. ENDT, P. M., and M. DEMEUR, *"Nuclear Reactions."* Interscience. $12.50.
292. FANO, U. and L. FANO, *"Basic Physics of Atoms and Molecules."* Wiley. $10.00.
293. FRANCIS, G. E., W. MULLIGAN, and A. WORWALL, *"Isotopic Tracers."* 2nd Ed. Oxford Press. $8.40.
294. FRISCH, O. R., F. A. PANETH, F. LAVES and P. ROSBAUD, *"Trends in Atomic Physics."* Interscience. $7.50.
295. GOLDSTEIN, H., *"Fundamental Aspects of Reactor Shielding."* Addison-Wesley.
296. HAMILTON, J., *"The Theory of Elementary Particles."* Oxford Univ. Press. $12.00.
297. HANDLOSER, J. S., *"Health Physics Instrumentation."* Pergamon Press.
298. HENNESSY, T. G., *"Radiobiology at the Intra-Cellular Level."* Pergamon Press. $8.50.
299. HETMAN, J., *"Trace Techniques."* Vol. I. Standard Scientific Supply Corp., New York. $3.50.
300. HOFFMAN, B., *"The Strange Story of the Quantum."* 2nd Ed. Dover. $1.45.
301. HUGHES, D. J., *"The Neutron Story."* Doubleday Anchor Books. $0.95.
302. JELLEY, J. V., *"Čerenkov Radiation and Its Applications."* Pergamon Press. $10.00.

303. LETTENMEYER, L., *"Dictionary of Atomic Terminology."* Philosophical Library. $6.00.
304. MARTIN, C. N., *"The Thirteen Steps to the Atom (A Photographic Exploration)."* Watts. $4.95
305. McMAHON, J. J. and A. BERMAN, *"Radioisotopes in Industry."* National Industrial Conference Board. $2.75.
306. MURRAY, A. and D. L. WILLIAMS, *"Organic Synthesis with Isotopes."* Part I, Compounds of Isotopic Carbon; Part II, Organic Compounds Labeled with (Other Isotopes). Interscience. $25.00.
307. OWEN, C., *"Diagnostic Use of Radioisotopes."* Thomas.
308. PAULI, W., *"Theory of Relativity."* Pergamon Press. $60.00.
309. ROCHLIN, R. S., and W. W. SCHULTZ, *"Radioisotopes for Industry."* Reinhold. $4.75.
310. ROCHMAN, A., *"Introduction to Nuclear Engineering."* Simmons-Boardman. $16.00.
311. SAMUEL, D., and P. F. STECKEL, *"Bibliography of the Stable Isotopes of Oxygen."* Pergamon Press. $7.50.
312. SARBACHER, R. I., *"Encyclopedic Dictionary of Electronics and Nuclear Engineering."* Prentice-Hall. $35.00.
313. SEABORG, G. T., *"The New Elements, A Symposium."* Chemical Education. $1.00.
314. SEGRE, E., *"Experimental Nuclear Physics."* Vol. III. Wiley. $23.00.
315. SEMAT, H., and H. WHITE, *"Introduction to Atomics and Nuclear Physics."* 3rd Ed. Holt, Rinehart and Winston. $2.00.
316. SMITH, D. E., *"Proceedings of the International Congress of Radiation Research."* Academic Press. $10.50.
317. STERNBERG, J., *"The Use of Radioactive Isotopes in the Study of Experimental Tuberculosis."* L'Institut de Microbiologie et d'Hydiène de l'Université de Montréal.
318. WALLACE, B. and Th. DOBZHANSKY, *"Radiation, Genes and Man."* Holt.
319. WHYTE, G. N., *"Principles of Radiation Dosimetry."* Wiley. $7.00.
320. WIGNER, E. P., *"Group Theory and its Application to the Quantum Mechanics of Atomic Spectra."* Academic Press. $8.00.

1960

321. AJZENBERG-SELOVE, F. (ed.), *"Nuclear Spectroscopy."* Academic Press. Part A and B, $16.00 each.
322. ALLEN, W. D., *"Neutron Detection."* Philosophical Library. $10.00.
323. ALLIS, W. P., *"Nuclear Fusion."* Van Nostrand. $12.50.
324. BEARD, H. C., *"The Radiochemistry of Arsenic."* Dept. of Commerce, Washington 25, D. C. $0.50.
325. BENDER, A., *"Let's Explore with the Electron."* Sentinel Books. $1.00.
326. BISHOP, A. S., *"Project Sherwood—The U. S. Program in Controlled Fusion."* Doubleday. $1.25.
327. BRAGG, SIR WM., *"The Universe of Light."* Dover. $1.85.
328. BRODA, E., *"Radioactive Isotopes in Biochemistry."* Van Nostrand. $11.50.
329. BRODA, E., and T. SCHONFELD, *"Technical Applications of Radioactivity."* Van Nostrand.
330. BROWNING, E., *"Harmful Effects of Ionizing Radiations."* Van Nostrand. $3.00.
331. BURTON, M., et al., *"Comparative Effects of Radiation."* Wiley & Sons. $8.50.
332. CALDECOTT, R. S. and L. A. SNYDER (eds.), *"Radioisotopes in the Biosphere."* University of Minnesota Press.
333. CAMERON, J. F. and R. J. PICKETT, *"Geiger Gas Counting Methods of Assaying Tritiated Hydrogen and Tritiated Water."* Atomic Energy Research Establishment, Harwell Report AERE-R 3092.
334. CHANDRASEKHAR, S., *"Plasma Physics."* University of Chicago Press. $1.75.
335. CHANDRASEKHAR, S., *"Radiative Transfer."* Dover. $2.25.
336. CHARLESBY, A., *"Atomic Radiation and Polymers."* Vol. I. Pergamon Press $17.50.
337. CRONKITE, E., and V. BOWD, *"Radiation Injury in Man."* C. C. Thomas. $6.50.
338. CROUTHAMEL, C. E., *"Applied Gamma Ray Spectrometry."* Pergamon Press. $6.50.
339. CURRY, D., III, and B. R. NEWMAN, *"The Challenge of Fusion."* Van Nostrand. $5.50.
340. DE BROGLIE, L., *"Non-Linear Wave Mechanics."* Elsevier. $11.00.
341. DESROSIER, N. W., and H. M. ROSESTOCK, *"Radiation Technology in Food, Agriculture and Biology."* Avi. $12.50.
342. DEVOE, J. R., *"The Radiochemistry of Cadmium."* Dept. of Commerce, Washington 25, D.C. $0.75.
343. DUCKWORTH, H. E., *"Proceedings of the International Conference on Nuclidic Masses."* Toronto University Press. $10.00.
344. DUQUESNE, M., *"Matter and Antimatter."* Harper and Row. $1.75.
345. EDELMANN, A., *"Radioactivity for Pharmaceutical and Allied Research Laboratories."* Academic Press. $6.00.
346. ERRERA, M., and A. FORSSBERG, *"Mechanisms in Radiobiology."* Vol. II. Multicellular Organisms. Academic Press. $13.00.
347. FAIRES, R. A. and B. H. PARKS, *"Radioisotope Laboratory Techniques."* 2nd Ed. London, Newnes.
348. FOWLER, J. M., *"Fallout: A Study of Superbombs, Strontium 90, and Survival."* Basic Books. $5.50.
349. GLASSTONE, S., and R. H. LOVBERG, *"Controlled Thermonuclear Reactions, An Introduction to Theory and Experiment."* Van Nostrand. $5.60.
350. Gordon, G., *"The Chemical Elements and Their Isotopes."* Museum of Science & Industry. $1.05.
351. HAHN, W. F., and J. C. NEFF, *"American Strategy for the Nuclear Age."* Doubleday. $1.45.
352. HARRIS, W. H., *"Dive: The Story of an Atomic Submarine."* Harper. $2.95.
353. HECHT, SELIG, *"Explaining the Atom."* Viking Press. $1.25.
354. HALLAENDER, A., *"Radiation Protection and Recovery."* Pergamon Press. $10.00.
355. HUDIS, J., *"The Radiochemistry of Carbon, Nitrogen and Oxygen."* National Academy of Sciences, National Research Council Report NAS-NS 3019.
356. HUTCHINSON, F. W., *"Nuclear Radiation Engineering: An Introduction."* Ronald Press Co. $6.00.
357. HYDE, E. K., *"The Radiochemistry of Francium."* Dept. of Commerce, Washington 25, D.C. $0.50.

358. International Atomic Energy Agency, *"The Application of Radioisotopes in Biology."* Vienna, International Atomic Energy Agency.
359. International Atomic Energy Agency, *"Metrology of Radionuclides."* Vienna, International Atomic Energy Agency.
360. International Atomic Energy Agency, *"Tritium: Dosage, Preparation de Molecules Marquees et Applications Biologiques."* (Review series No. 2) Vienna, International Atomic Energy Agency.
361. JACOBS, A. M., D. E. KLINE and F. J. REMICK, *"Basic Principles of Nuclear Science and Reactors."* Van Nostrand. $6.50.
362. KOCH, R. C., *"Activation Analysis Handbook."* Academic Press. $9.00.
363. LEYMONIE, C., *"Radioactive Tracers in Physical Metallurgy."* Wiley. $8.50.
364. LIVERHANT, S. E., *"Elementary Introduction to Nuclear Reactor Physics."* Wiley. $9.75.
365. LOCK, W. O., *"High Energy Nuclear Physics."* Wiley. $3.25.
366. MELANDER, L., *"Isotopic Effects on Reaction Rates."* Ronald Press. $6.00.
367. MILLS, M. M., et al., *"Modern Nuclear Technology.' '* (A Survey for Industry and Business). McGraw-Hill. $9.50.
368. NEFF, J. C., and W. F. HAHN, *"American Strategy for the Nuclear Age."* Doubleday. $1.45.
369. OVERMAN, R. T., and H. M. CLARK, *"Radioisotope Techniques."* McGraw-Hill. $10.00.
370. PAYNE, B. R., *"Radioactiviation Analysis."* Butterworth. $5.75.
371. PEARSON, F. J., and OSBORNE, R. R., *"Practical Nucleonics."* A Course of Experiments in Nuclear Physics. Spon, London.
372. PEPINSKY, R., *"Computing Methods and the Phase Problem in X-ray Crystal Analysis."* Pergamon Press. $9.00.
373. PUTNAM, J. E., *"Isotopes."* Houghton Mifflin. $1.95.
374. ROMAN, P., *"Theory of Elementary Particles."* Interscience. $12.00.
375. ROMER, R., *"The Restless Atom."* Doubleday. $0.95.
376. RUSSELL, R. D., and R. M. FARQUHAR, *"Lead Isotopes in Geology."* Interscience. $9.00.
377. SAUNDERS, B. C., and R. E. D. CLARK, *"Order and Chaos in the World of Atoms."* Dover. $2.75.
378. SHANKLAND, R. S., *"Atomic and Nuclear Physics."* Macmillan. $8.75.
379. SHARPE, J., *"Nuclear Radiation."* Simmons-Boardman. $2.75.
380. SHILLING, C. W., *"Radiation: Use and Control in Industrial Applications."* Grune and Stratton, New York. $6.75.
381. SLATER, J. C., *"Quantum Theory of Atomic Structure."* McGraw-Hill. $13.00.
382. SOLOMON, A. K., *"Why Smash Atoms?"* Penguin Books. $0.95.
383. SWALLOW, A. J., *"Radiation Chemistry of Organic Compounds."* Pergamon Press. $15.00.
384. VAN MELSEN, A. G., *"From Atomos to Atom."* Harper Torchbooks. $1.45.
385. WALLACE, B., and TH. DOBZHANSKY, *"Radiation, Genes, and Man."* Holt.
386. WASHTELL, C. C. H., *"An Introduction to Radiation Counters and Detectors."* Philosophical Library. $7.50.
387. WEHR, M. R., and J. A. RICHARDS, Jr., *"Physics of the Atom."* Addison-Wesley.
388. WILSON, R. R. and R. LITTAUER, *"Accelerators; Machines of Modern Physics."* Doubleday. $0.95.
389. ALLEN, A. O., *"Radiation Chemistry of Water and Aqueous Solutions."* Van Nostrand. $6.00.
390. AMPHLETT, C. B., *"Treatment and Disposal of Radioactive Wastes."* Pergamon Press. $12.00.
391. ANDREWS, H. L., *"Radiation Biophysics."* Prentice-Hall.
392. BACQ, Z. M. and P. ALEXANDER, *"Fundamentals of Radiobiology."* 2nd Ed. Rev. Pergamon Press. $12.00.
393. BALDIN, A., *"Kinematics of Nuclear Reactions."* Oxford Univ. Press. $6.10.
394. BILLINGTON, D. S., and J. H. CRAWFORD, *"Radiation Damage in Solids."* Princeton Univ. Press. $12.50.
395. BRINKMAN, G. A., *"Standardization of Radioisotopes by* $4\pi(a,\beta)$ *and* $4\pi(a,\beta)\equiv\gamma$ *Coincidence Counting Techniques with Liquid and Plastic Scintillators."* Ultgeverij Excelsion, Rijswijk.
396. BUSH, G. L., and A. A. SILVIDI, *"The Atom: A Simplified Description."* Barnes & Noble. $1.25.
397. CATCH, J. R., *"Carbon-14 Compounds."* Butterworth. $5.50.
398. CHOPPIN, G. R., *"Experimental Nuclear Chemistry."* Prentice-Hall. $6.95.
399. COFFINBERRY, A. S. and W. N. MINER, *"The Metal Plutonium."* Univ. of Chicago Press. $9.50.
400. COLLINS, J. C., *"Radioactive Wastes: Their Treatment and Disposal."* Wiley.
401. COOK, C. S., *"Modern Atomic and Nuclear Physics."* Van Nostrand.
402. CURIE, MARIE, *"Radioactive Substances."* Philosophical Library. $0.95.
403. DEVOE, J. R., *"Radioactive Contamination of Materials Used in Scientific Research."* National Academy of Sciences-National Research Council, Washington 25, D. C. $2.00.
404. DRELL, S. D., and F. ZACHARIASEN, *"Electromagnetic Structure of Nucleons."* Oxford University Press. $2.00.
405. ELTON, L. R. B., *"Nuclear Sizes."* Oxford University Press.
406. EMELEUS, H. J., and A. G. SHARPE, *"Advances in Inorganic Chemistry and Radiochemistry."* Vol. 3. Academic Press. $12.00.
407. ERRERA, M., and A. FORSSBERG, *"Mechanisms in Radiobiology."* Vol. I, General Principles. Academic Press. $16.00.
408. FIELDS, T. and L. SEED (ed), *"Clinical Use of Radioisotopes."* The Year Book Publishers. $9.50.
409. FLAGG, J. R., *"Chemical Processing of Reactor Fuels."* Academic Press. $17.50.
410. FRISCH, O. R., *"Atomic Physics Today."* Basic Books.
411. GAMOW, G., *"The Atom and Its Nucleus."* Prentice-Hall.
412. GAMOW, G., *"Biography of Physics."* Harper & Row. $5.95.
413. GLASSNER, A., *"Introduction to Nuclear Science."* Van Nostrand. $3.75.
414. GLUECKAUF, E. (ed), *"Atomic Energy Waste: Its Nature, Use, and Disposal."* Interscience. $14.00.

415. GRIFFITH, T. C., and E. A. POWERS, *"Nuclear Forces and the Few-Nucleon Problem."* Vols. 1 and 2. Pergamon Press.
416. HAISSINSKY, M., *"The Chemical & Biological Action of Radiations."* Vol. 5. Academic Press. $9.50.
417. HARPER, W. R., *"Basic Principles of Fission Reactors."* Interscience. $7.50.
418. HARRIS, R. J. C., *"The Initial Effects of Ionizing Radiations on Cells."* Academic Press. $12.00.
419. HEFTMANN, E., *"Chromatography."* Reinhold. $17.50.
420. ILBERY, P. L. T., *"Radiobiology."* Butterworth. $11.00.
421. Illinois University, *"Mechanism of the Radiation-Induced Addition of Tritium to Carbon-Carbon Double Bonds."* Illinois Univ. Report TID-14787.
422. JACOBS, A. M., D. E. KLINE, and F. J. REMICK, *"Basic Principles of Nuclear Science and Reactors."* Van Nostrand.
423. KEPPLE, R. E., *"Sodium Iodide and Sodium Iodide Crystals: Their Use in Scintillation Counting and Spectrometry."* Argonne National Laboratory Report ANL-6446.
424. KLEMPARSKAYA, N. N., and O. G. ALEKSLYEVA, et als., *"Problems of Infection Immunity and Allergy in Acute Radiation Diseases."* Pergamon Press.
425. KOHL, J., R. D. ZENTNER and N. R. LUKENS, *"Radioisotope Application Engineering."* Van Nostrand. $12.50.
426. LAJTHA, L. G., *"The Use of Isotopes in Haematology."* Oxford, Blackwells.
427. LIND, S. C., *"Radiation Chemistry of Gases."* Reinhold. $12.50.
428. MANN, M., *"Peacetime Uses of Atomic Energy."* 3rd Ed. Viking Press. Revised. $1.65.
429. MITCHNER, M., *"Radiation and Waves in Plasmas."* Stanford Univ. Press. $4.50.
430. MURPHY, G., *"Elements of Nuclear Engineering."* Wiley. $7.50.
431. MURRAY, R. L., *"Introduction to Nuclear Engineering."* 2nd Ed. Prentice-Hall.
432. PARRATT, L. G., *"Probability and Experimental Errors in Science."* Wiley. $6.50.
433. ROSE, M. E., *"Relativistic Election Theory."* Wiley.
434. SANDERS, J. H., *"The Fundamental Atomic Constants."* University Press. $1.60.
435. SCHULTZ, V., and A. W. KLEMENT, Jr., *"Radioecology."* Reinhold. $16.50.
436. SNELL, A. H., *"Nuclear Instrumentation and Methods."* Wiley. $12.00.
437. STEINBERG, B. P., *"Counting Methods for the Assay of Radioactive Samples."* Argonne National Laboratory Report ANL-6361.
438. United States Department of Commerce, National Bureau of Standards, *"A Manual of Radioactivity Procedures." Handbook 80. National Bureau of Standards, Washington, D. C.*
439. United States Department of Commerce, National Bureau of Standards, *"Report of the International Commission on Radiological Units and Measurements (ICRU) 1959."* Handbook 78. National Bureau of Standards, Washington, D. C.
440. WEISMAN, M. L., B. H. EGGLESTON and F. E. ARMSTRONG, *"Reproducibility of Tritium Analysis of Organic Compounds Using a Liquid Scintillation Spectrometer."* United States Department of the Interior, Bureau of Mines Report BM-RI 5801.
441. WENDT, G. (ed.), *"Atoms for Industry World Survey."* International Publications.
442. WHIEMAN, M. L., F. G. SCHWARTZ, and B. H. McCLESTON, *"Susceptibility of Organic Compounds to Tritium Exchange Labeling."* U. S. Department of the Interior, Bureau of Mines, Report BM-RI 5717.
443. WILLIAMS, W. S. C., *"An Introduction to Elementary Particles."* Academic Press. $11.00.
444. ZIMMER, K. G., *"Studies on Quantitative Radiation Biology."* Oliver & Boyd, London.

1962

445. Atomic Energy Commission, *"Radioactive Fallout. A Bibliography of the World's Literature."* Dept. of Commerce, Washington 25, D. C. $3.00.
446. Australian Atomic Energy Commission, *"Technological Use of Radiation."* Cambridge Univ. Press. $5.50.
447. BARNES, D. E. et al., *"Newnes Concise Encyclopaedia of Nuclear Energy."* George Newnes, Ltd., London.
448. Basic Systems, Inc., *"Chemistry I: Atomic Structure and Bonding."* Appleton-Century-Crofts, Inc. $3.24. Response book $.64. Teacher's Manual free.
449. BATES, D. R. (ed.), *"Atomic and Molecular Processes."* Academic Press. $19.50.
450. BLEICH, A. R., *"The Story of X-Rays from Röntgen to Isotope."* Dover. $1.35.
451. BLIZARD, E. P., and L. S. ABBOTT, *"Reactor Handbook."* Vol. III, Part B. Shielding. 2nd Ed. Wiley. $9.00.
452. BOHR, NIELS, *"Atomic Theory and the Description of Nature."* Cambridge Univ. Press. $2.75.
453. BORN, MAX, *"Einstein's Theory of Relativity."* Dover. $2.00.
454. BRADLEY, W. H., and R. H. HUEBNER, *"Nuclear Spectrometer Applications."* Nuclear Measurements Corp., Indianapolis. $10.00.
455. BUSH, G. L., and A. A. SILVIDI, *"The Atom."* Barnes and Noble. $1.25.
456. CALDER, R., *"Living with the Atom."* University of Chicago Press. $5.95.
457. CARO, McDONELL and SPICER, *"Introduction to Atomic and Nuclear Physics."* Aldine Publishing Co., Chicago. $6.00.
458. CHADWICK, SIR JAMES, *"The Collected Papers of Lord Rutherford of Nelson."* Wiley. $19.50.
459. CHAPIRO, A., *"Radiation Chemistry of Polymeric Systems."* Wiley. $21.00.
460. COMPTON, D. M., and A. H. SCHOEN, *"The Mossbauer Effect."* Wiley. $16.50.
461. ENDT, P. M., *"Nuclear Reactions."* Vol. 2. Interscience.
462. FRAUENFELDER, H., *"The Mössbauer Effect."* W. A. Benjamin. $4.95.
463. GENUNCHE, A., *"Bibliography of Carbon-14 Measurements."* Institutul de Fizicã Atomicã, Bucharest Report IFA/CO/21.

464. HALEY, T. J., R. S. SNIDER, and S. M. LINDE, *"Response of the Nervous System to Ionizing Radiation."* Academic Press. $18.00.
465. HARVEY, B. J., *"Introduction to Nuclear Physics and Chemistry."* Prentice-Hall. $12.00.
466. HENLEY, E. J., and H. KOUTS, *"Advances in Nuclear Science and Technology."* Vol. I. Academic Press. $12.00.
467. HERBER, R. H., *"Inorganic Isotopic Synthesis."* W. A. Benjamin. $7.50.
468. HILL, J. F., *"Textbook of Reactor Physics."* George Allen and Unwin, London.
469. International Atomic Energy Agency, *"Radioisotopes in Soil-Plant Nutrition Studies."* National Agency for International Publications, New York. $9.00.
470. International Atomic Energy Agency, *"Radioactive Dating."* National Agency for International Publications, New York. $8.50.
471. International Atomic Energy Agency, *"Use of Radioisotopes in Animal Biology and the Medical Sciences."* Academic Press. Vol. 1, $16.00; Vol. 2, $9.50.
472. International Atomic Energy Agency, *"Thermodynamics of Nuclear Materials."* National Agency for International Publications, New York. $11.00.
473. International Atomic Energy Agency, *"International Directory of Radioisotopes."* 2nd Ed. Nucleonics. $9.00.
474. International Atomic Energy Agency, *"Diagnosis and Treatment of Acute Radiation Injury."* Columbia University Press. $8.00.
475. International Atomic Energy Agency, *"Radiation Damage in Solids."* International Publications, Inc., New York. Vol. I, $7.00; Vol. II, $6.00.
476. International Atomic Energy Agency, *"Whole-Body Counting."* International Publications, Inc., New York. $10.00.
477. International Atomic Energy Agency, *"Application of Isotope Techniques in Hydrology."* National Agency for International Publications, Inc., New York. $1.00.
478. International Atomic Energy Agency, *"Nuclear Electronics, Vol. I, II and III."* Vienna, International Atomic Energy Agency.
479. International Atomic Energy Agency, *"Tritium in the Physical and Biological Sciences. Vol. I and II."* Vienna, International Atomic Energy Agency.
480. International Atomic Energy Agency and United Nations Educational, Scientific and Cultural Organization, *"Radioisotopes in the Physical Sciences and Industry."* Vol. I, II and III. Vienna, International Atomic Energy Agency.
481. ISRAEL, H., and A. KREBS, *"Nuclear Radiation in Geophysics."* Academic Press. $18.00.
482. KAUFMANN, A. R., *"Nuclear Reactor Fuel Elements."* Interscience.
483. KUHN, H. G., *"Atomic Spectra."* Academic Press. $13.00.
484. LETAVET, A. A., and E. B. KURLYANDSKAYA, *"The Toxicology of Radioactive Substances."* Vol. I. Pergamon Press. $12.50.
485. LIPKIN, H. J., *"Beta Decay for Pedestrians."* Interscience.
486. LIVINGSTON, M. S., and J. P. BLEWETT, *"Particle Accelerators."* McGraw-Hill. $17.50.
487. MITCHELL, A. C. G., and M. W. ZEMANSKY, *"Resonance Radiation and Excited Atoms."* Cambridge. $6.00.
488. NIGHTINGALE, R. E., *"Nuclear Graphite."* Academic Press. $15.80.
489. O'KELLEY, G. D. (ed.), *"Applications of Computers to Nuclear & Radiochemistry."* Department of Commerce, Washington 25, D. C. $2.50.
490. O'KELLEY, G. D., *"Detection and Measurement of Nuclear Radiation."* National Academy of Sciences, National Research Council Report NAS-NS 3105.
491. PRESTON, M. A., *"Physics of the Nucleus."* Addison-Wesley. $15.00.
492. Radio-chemical Centre, *"The Radiochemical Manual. Part I. Physical Data."* Amhevshire, England.
493. RAMAKRISHNAN, A., *"Elementary Particles and Cosmic Rays."* Pergamon Press. $15.00.
494. RUSSELL, C. R., *"Reactor Safeguards."* Pergamon Press. $15.00.
495. SCOTT, E. C. and F. A. KANDA, *"The Nature of Atoms and Molecules: A General Chemistry."* Harper & Row. $8.00.
496. SEMAT, H., *"Introduction to Atomic and Nuclear Physics."* 4th Ed. Holt, Rinehart and Winston. $7.50.
497. SHEPPARD, C. W., *"Basic Principles of the Tracer Method."* Wiley. $8.00.
498. SILVER, S., *"Radioactive Isotopes in Medicine and Biology."* Lea and Febiger. $8.00.
499. SLATER, D. N., *"Gamma-Rays of Radionuclides in Order of Increasing Energy."* Butterworth. $9.00.
500. SNELL, ARTHUR H., *"Nuclear Instruments and Their Uses."* Vol. I. Interscience. $7.50.
501. TALMI, I., and A. DE-SHABIT, *"Nuclear Shell Theory."* Academic Press.
502. THOMSON, J. F., *"Radiation Protection in Mammals."* Reinhold. $8.50.
503. TIEVSKY, G., *"Ionizing Radiation."* C. C. Thomas. $8.00.
504. U. S. Atomic Energy Commission, *"Reactor Handbook."* 2nd Ed. Vol. III, Part A. John Wiley & Sons. $10.75. Part B. $9.00.
505. WACHSMANN, F., and G. BARTH, *"Moving Field Radiation Therapy.'* University of Chicago Press. $10.95.
506. WEHR, M. R., and J. A. RICHARDS, Jr., *"Introductory Atomic Physics."* Addison-Wesley. $8.75.
507. WENZEL, M., and P. E. SCHULZE, *"Tritium-markierung, Darstellung, Messung und Anwendung nach Wilzback ^{3}H-markierter Verbindungen."* Berlin, Walter de Gruyter.
508. WILLIAMS, KATHARINE, et al., *"Radiation and Health."* Little Brown & Co. $7.00.
509. WILLIAMS, I. R., and M. W. WILLIAMS, *"Basic Nuclear Physics."* George Newnes, Ltd., London.
510. YEATER, M. L., *"Neutron Physics."* Vol. 2. Academic Press. $12.00.

1963

511. ADLER, I., *"Inside the Nucleus."* John Day Co. $4.95.
512. BARKAS, W. H., *"Nuclear Research Emulsions."* 1. Techniques and Theory. Academic Press. $18.00.
513. BARNES, D. E., R. BATCHELOR, A. G. MADDOCK, J. A. SMEDLEY and D. TAYLOR, Concise Encyclo- *paedia of Nuclear Energy."* Interscience. $25.00.
514. BARNES, D. E. and D. TAYLOR, *"Radiation Hazards and Protection."* 2nd Ed. Pitman. $8.50.
515. BIRKS, J. B., *"Rutherford at Manchester."* Benjamin. $12.50.
516. BOHR, N., *"Essays 1958-1962 on Atomic Physics and Human Knowledge."* John Wiley & Sons. $5.00.
517. BOLT, R. O., and J. G. CARROLL, *"Radiation Effects on Organic Materials."* Academic Press. $13.50.
518. BROWN, B., *"Experimental Nucleonics."* Prentice-Hall. $8.95.
519. BURCHAM, W. E., *"Nuclear Physics, An Introduction."* McGraw-Hill. $12.00.
520. BUSH, H. D., *"Atomic and Nuclear Physics."* (Theoretical Principles). Prentice-Hall. $8.95.
521. CASIMIR, H. B. G., *"On the Interaction Between Atomic Nuclei and Electrons."* Freeman. $2.00.
522. CLARK, G. L. (ed.), *"The Encyclopedia of X-Rays and Gamma Rays."* Reinhold. $35.00.
523. CRAIG, H. (ed.), et al., *"Isotopic & Cosmic Chemistry."* North-Holland. $15.00.
524. DEARNALEY, G., *"Semiconductor Counters for Nuclear Radiations."* Wiley $8.75.
525. DEBROGLIE, L., *"Introduction to the Vigier Theory of Elementary Particles."* Elsevier. $11.00.
526. DE-SHALIT, A., and I. TALMI, *"Nuclear Shell Theory."* Academic Press. $14.50.
527. DEVOE, J. R., *"Application of Distillation Techniques to Radiochemical Separations."* Dept. of Commerce, Washington 25, D. C. $0.50.
528. EISENBUD, M., *"Environmental Radioactivity."* McGraw-Hill.
529. ELSON, L. A., *"Radiation and Radiomimetic Chemicals."* Butterworth. $5.00.
530. ENDT, P. M., and P. B. SMITH, *"Nuclear Reactions."* Interscience. $18.50.
531. FEINSTEIN, R. N., *"Implications of Organic Peroxides in Radiobiology."* Academic Press. $9.00.
532. FORD, K. W., *"The World of Elementary Particles."* Blaisdell. $2.95.
533. FULTON, T., *"Elementary Particle Physics and Field Theory."* (1962 Brandeis Lectures, Vol. 1). Benjamin. $5.95.
534. GEL'MAN, A. D., A. I. MOSKVIN, L. M. ZAITSEV, and M. P. MEFOD'EVA, *"Complex Compounds of Transuranium Elements."* Consultants Bureau. $12.50.
535. GOLDBERGER, M. L. and K. M. WATSON, *"Collision Theory."* Wiley & Sons. $19.95.
536. GREEN, T. S., *"Thermonuclear Power."* George Newnes, Ltd.
537. GRODZINS, M. and E. RABINOWITCH (eds.), *The Atomic Age."* Basic Books, New York. $10.00.
538. HANSON, N. R., *"The Concept of the Positron."* Cambridge University Press. $5.95.
539. HARRIS, R. J. C., *"Cellular Basis & Aetaology of Late Somatic Effects of Ionizing Radiation."* Academic Press. $12.00.
540. HARRIS, L., and A. L. LOEB, *"Introduction to Wave Mechanics."* McGraw-Hill. $8.95.
541. HASTED, J. B., *"Physics of Atomic Collisions."* Butterworth. $26.00.
542. HECKMAN, H., and P. W. STARRING, *"Nuclear Physics and Fundamental Particles."* Holt, Rinehart and Winston.
543. HERMAN, F., and S. SKILLMAN, *"Atomic Structure Calculations."* Prentice-Hall. $13.00.
544. HILL, R. D., *"Tracking Down Particles."* W. A. Benjamin. $2.95.
545. HOCHSTRASSER, R. M., *"Behavior of Electrons in Atoms."* W. A. Benjamin. $.395 cloth.
546. HOFSTADTER, R., *"Nuclear and Nucleon Structure."* W. A. Benjamin. $6.95.
547. HOGERTON, J. F. (ed.), *"The Atomic Energy Deskbook."* Reinhold. $11.00.
548. International Atomic Energy Agency, *"Neutron Dosimetry."* Vols. I & II. International Publications. $13.00 ea.
549. International Atomic Energy Agency, *"Radiation Damage in Solids."* Vol. 3. International Publications. $4.00.
550. International Atomic Energy Agency, *"Chemistry Research and Chemical Techniques Based on Research Reactors."* National Agency for International Publications. $6.00.
551. International Atomic Energy Agency, *"Radioisotopes in Hydrology."* National Agency for International Publications, Inc., New York. $9.00.
552. International Commission on Radiological Protection, *"Radiation Protection."* Report of Committee IV (1953-1959) on Protection Against Electromagnetic Radiation Above 3MeV and Electrons, Neutrons & Protons. The Macmillan Co., Pergamon Press. $3.25.
553. ISBIN, H. S., *"Introductory Nuclear Reactor Theory."* Reinhold. $22.50.
554. JEFFRIES, C. D., *"Dynamic Nuclear Orientation."* Wiley. $5.95.
555. JOHNSON, N. R., E. EICHLER, and G. D. O'KELLEY, *"Nuclear Chemistry."* Interscience. $8.00.
556. JUNGE, C. E., *"Air Chemistry and Radioactivity."* Academic Press. $13.50.
557. KAPLAN, IRVING, *"Nuclear Physics, 2nd Ed."* Addison-Wesley.
558. KING, C. D. G., *"Nuclear Power Systems, An Introductory Text."* Harper & Row. $13.00.
559. KOFOEDHANSEN, O., *"The Negotiators, the Challenge of the Atomic Age."* Ejnar Munksgaard, Copenhagen.
560. KURTSIN, I. T., *"Effects of Ionizing on the Digestive System."* Elsevier.
561. LAPP, R. E., and H. L. ANDREWS, *"Nuclear Radiation Physics."* 3rd Ed. Prentice-Hall. $12.00.
562. LARSON, D. B., *"The Case Against the Nuclear Atom."* North Pacific Publishers. $4.50.
563. LEBEDINSKIY, A. V., and Z. N. NAKHIL'NITSKAYA, *"Effects of Ionizing Radiation on the Nervous System."* Elsevier.
564. LETAVET, A. A., *"Toxicology of Radioactive Substances."* In two volumes. Vol. II. $7.50. Pergamon Press.
565. LEVI SETTI, R., *"Elementary Particles."* Univ. of Chicago Press. $2.00.

566. LEYMONIE, C., *"Radioactive Tracers in Physical Metallurgy."* Wiley. $8.50.
567. LITTLEFIELD, T. A., and N. THORLEY, *"Atomic and Nuclear Physics."* Van Nostrand. $9.75.
568. LONGMIRE, C. L., *"Elementary Plasma Physics."* Wiley. $9.75.
569. LOTHIAN, G. F., *"Electrons in Atoms."* Butterworth. $6.95.
570. MARION, J. B. and J. L. FOWLER, *"Fast Neutron Physics, Part II: Experiments and Theory."* Wiley.
571. MILLIKAN, R. A., *"The Electron."* University of Chicago Press. $2.45.
572. MUELLER, W. M. and M. FAY (eds.), *"Advances in X-Ray Analysis."* Vol. VI — 11th Annual Conference. Plenum Press. $17.50.
573. NEEL, J., *"Changing Perspectus on Genetic Effects of Radiation."* Thomas.
574. OVERMAN, R. T., *"Basic Concepts of Nuclear Chemistry."* Reinhold. $1.95.
575. PARRISH, W., and M. MACK, *"Data for X-Ray Analysis."* 2nd Ed. Philips' Technical Library, Eindhoven, Holland. Vol. I, $4.50; Vol. II, $5.50; Vol. III, $5.50.
576. PAULING, L., and S. GOUDSMIT, *"The Structure of Line Spectra."* McGraw-Hill. $2.75.
577. PEARSON, F. J., *"Nuclear Power Technology."* Oxford Univ. Press. $6.75.
578. PETERSON, S. and R. G. WYMER, *"Chemistry in Nuclear Technology."* Addison-Wesley.
579. PHILLIPS, G. C., et al., *"Progress in Fast Neutron Physics."* University of Chicago Press. $8.50.
580. PURDOM, C. E., *"Genetic Effects of Radiations."* Academic Press. $7.00.
581. QUIMBY, E. H., and S. FEITELBERG, *"Radioactive Isotopes in Medicine and Biology."* $8.00.
582. Radiochemical Centre, *"The Radiochemical Manual."* Part 2. Amersham Radiochemical Centre.
583. RANDERATH, K., *"Thin-Layer Chromatography."* Academic Press. $8.00.
584. RANKAMA, K., *"Progress in Isotope Geology."* Wiley & Sons. $20.00.
585. ROSECRANCE, R. N. (ed.), *"The Dispersion of Nuclear Weapons."* Columbia University Press. $7.50.
586. SCHMEISER, K., *"Radionuclide."* Springer-Verlag, Berlin, West Germany.
587. SCHRAM, E., *"Organic Scintillation Detectors."* Elsevier. $7.50.
588. SCHULTZ, V., and A. W. KLEMENT, Jr., *"Radioecology."* Reinhold. $16.50.
589. SEABORG, G. T., *"Man-Made Transuranium Elements."* Prentice-Hall. $1.50.
590. SIMNAD, M. T., and L. R. ZUMWALT (eds.), *"Materials & Fuels for High-Temperature Nuclear Energy Applications."* The M.I.T. Press. $12.00.
591. United States Atomic Energy Commission, *"Plasma Physics and Controlled Nuclear Fusion Research."* U. S. Atomic Energy Commission. $10.50.
592. United States Department of Commerce, National Bureau of Standards, *"Clinical Dosimetry."* Handbook. 87. National Bureau of Standards, Washington, D. C.
593. United States Department of Commerce, National Bureau of Standards, *"Radio-biological Dosimetry."* Handbook 88. National Bureau of Standards, Washington, D. C.
594. VALENTE, F. A., *"A Manual of Experiments in Reactor Physics."* Macmillan. $7.50.
595. VASIL'EV, I. M., *"Effect of Ionizing Radiations on Plants."* Publishing House of the Academy of Sciences, USSR, Moscow. Office of Technical Services, Washington, D. C. $3.50.
596. WEISSKOPF, V. F., *"Nuclear Physics."* Academic Press. $7.50.
597. WHITE, H. E., *"Introduction to Atomic & Nuclear Physics."* Van Nostrand. $9.75.
598. WILSON, R., *"The Nucleon-Nucleon Interaction, Experimental and Phenomenological Aspects."* Wiley. $6.00.
599. WINTER, C. C., *"Radioisotope Renography."* Williams and Wilkins, Baltimore.
600. WOLFENDALE, A., *"Cosmic Rays."* George Newnes, Ltd.
601. YUAN, L. C. L., and C-S WU., *"Nuclear Physics."* Vol. V, Part B. Academic Press. $22.50.
602. ZYRYANOVA, L. N., *"Once-Forbidden Beta-Transitions."* Macmillan. $5.00.

1964

603. ADAMS, J. A. S., *"The Natural Radiation Environment."* University of Chicago Press. $15.00.
604. AMELINCKY, S., R. STRUMANE, J. NICHOL, and R. GEVERS, *"The Interaction of Radiation with Solids."* Wiley. $40.00.
605. AMPHLETT, C. B., *"Inorganic Ion Exchanger."* Elsevier. $6.50.
606. ANDERSON, D. L., *"The Discovery of the Electron."* Van Nostrand. $1.50.
607. ANDRADE, E. N. da C., *"Rutherford and the Nature of the Atom."* Doubleday. $1.25.
608. Anonymous, *"Radioisotope Applications in Industry, A Survey."* International Publications.
609. AUGENSTEIN, L. et als., *"International Symposium on Physical Processes in Radiation Biology."* Academic Press.
610. AUGENSTEIN, L. et. als., *"Advances in Radiation Biology."* Vol. 1. Academic Press.
611. BAK, T. A., *"Photons and Photon Interactions."* W. A. Benjamin. $12.50.
612. BALABUKHA, V. S., *"Chemical Protection of the Body Against Ionizing Radiation."* Macmillan. $8.50.
613. BEHRENS, C. F., and E. R. KING, *"Atomic Medicine."* 4th Ed. Williams & Wilkins. $18.00.
614. BLATZ, H., *"Introduction to Radiation Health."* McGraw-Hill.
615. BONDARENKO, I. I. (ed.), *"Group Constants for Nuclear Reactor Calculations."* Consultants Bureau Enterprises, including Plenum Press. $17.50.
616. BOOTH, V. H., *"The Structure of Atoms."* Macmillan. $2.95.
617. BROWN, G. E., *"Unified Theory of Nuclear Models."* Wiley & Sons. $7.25.
618. BURNAZYAN, A. E., et als., *"Radiation Medicine."* Pergamon Press.
619. CARLSON, W. D. and F. X. GASSNER (eds.), *"Effects of Ionizing Radiation on the Reproductive Systems."* Macmillan Co. $14.00.
620. CHARLESBY, A., *"Radiation Sources."* Macmillan Co., Pergamon Press. $12.00.

621. CHASE, G. D., S RITUPER, and J. W. SULCOSKI, *"Experiments in Nuclear Science."* Burgess. $3.50. (Teacher's Guide, $2.45).
622. CHOPPIN, GREGORY R., *"Nuclei and Radioactivity."* Benjamin. $1.95.
623, COMMISSARIAT A L'ENERGIE ATOMIQUE, *"Dictionnaire Des Science et Techniques Nucleaires."* Presses Universitaires de France, Paris.
624. COOK, S. C., *"Structure of Atomic Nuclei."* D. Van Nostrand. $1.50.
625. CORLISS-HARVEY, *"Radioisotopic Power Generation."* Prentice-Hall. $14.75.
626. CUNINGHAME, J. G., *"Introduction to the Atomic Nucleus."* Elsevier. $9.00.
627. DA C. ANDRADE, E. N., *"Rutherford and the Nature of the Atom."* Doubleday. $1.25.
628. DANER, M., *"Radiation Hygiene."* William & Wilkins.
629. DANIELS, J. M., *"Oriented Nuclei: Polarized Targets and Beams."* Academic Press.
630. DAUVILLIER, A., *"Cosmic Dust."* Philosophical Library. $15.00.
631. DE BENEDETTI, SERGIO, *"Nuclear Interactions."* Wiley. $16.00.
632. DEL VECCHIO, A. (ed.), *"Concise Dictionary of Atomics."* Philosophical Library. $6.00.
633. DEVOE, J. R. (ed.), *"Radiochemical Analysis: Activation Analysis, Instrumentation, Radiation Techniques and Radioisotope Techniques, July 1963 to June 1964."* Supt. of Documents, U. S. Government Printing Office, Washington, D. C. $0.50.
634. EAVES, G., *"Principles of Radiation Protection."* Gordon & Breach, New York. $8.25.
635. EGELSTAFF, P. A. (ed.), *"The Inelastic Scattering of Thermal Neutrons in Solid and Liquid State Research."* Academic Press.
636. EMELÉUS, H. J., and A. G. SHARPE, *"Advances in Inorganic Chemistry and Radiochemistry."* Academic Press. $16.00.
637. FREIDLANDER, G., and J. KENNEDY, and J. M. MILLER, *"Nuclear & Radiochemistry."* 2nd Ed. Wiley.
638. FRISCH, D. H., and A. M. THORNDIKE, *"Elementary Particles."* Van Nostrand. $1.75.
639. FRISCH, O. R., *"Progress in Nuclear Physics."* Macmillan. Vol. IX. $15.00.
640. GOL'DANSKII, V. I., *"The Mossbauer Effect and Its Applications in Chemistry."* Consultants Bureau Enterprises, Inc. $12.50.
641. GOL'DENBLAT, I. I., and N. A. NIKOLAENKO, *"Calculation of Thermal Stresses in Nuclear Reactors."* Consultants Bureau, New York. $15.00.
642. GOLDSCHMIDT, B., *"The Atomic Adventure."* Pergamon Press. $4.50.
643. GOWING, M., *"Britain & Atomic Energy 1939-1945."* St. Martin's Press. $12.50.
644. HAISSINSKY, M., *"Nuclear Chemistry and Its Applications."* Addison-Wesley. $22.50.
645. HAISSINSKY, M., and J. P. ADLOFF, *"Radiochemical Survey of the Elements."* Elsevier.
646. HALEY, T. J., and R. S. SNIDER, *"Response of the Nervous System to Ionizing Radiation."* Little Brown & Co. $18.50.
647. HANSEN, N. J., *"Analytical Chemistry, Vol. IV, Part I, Solid State Charged Particle."* Macmillan, Pergamon Press. $4.25 paper.
648. HEATH, R. L., *"Scintillation Spectrometry Gamma-Ray Spectrum Catalogue."* (IDO-16880) Office of Technical Services, U. S. Department of Commerce, Washington, D. C. $6.00.
649. HENLEY, E. J., (ed.), *"Advances in Nuclear Science and Technology."* Academic Press. $14.00.
650. Hine, G. J., and G. L. BROWNELL, *"Radiation Dosimetry."* Vol. 1 & 2. Academic Press.
651. HOCHSTRASSER, R. M., *"Behavior of Electrons in Atoms: Structure, Spectra, and Photochemistry of Atoms."* Benjamin. $3.95 cloth.
652. HOFSTADDTER, R. and L. I. SCHIFF, *"Nucleon Structure."* Stanford University Press. $12.50.
653. HYDE, E. K., *"The Nuclear Properties of the Heavy Elements."* Vols. I, II and III. Prentice Hall. Vol. I, $15.00; Vol. II, $25.00; Vol. III, $18.00.
654. International Atomic Energy Agency, *"Photonuclear Reactions."* International Publications, Inc. $8.00.
655. International Atomic Energy Agency, *"Directory of Nuclear Reactors."* Vol. V. Research, Test and Experimental Reactors. International Publications, Inc. $7.00.
656. International Atomic Energy Agency, *"Directory of Whole-Body Radioactivity Monitors."* International Publications, Inc. $14.00.
657. International Atomic Energy Agency, *"Technology of Radioactive Waste Management Avoiding Environmental Disposal."* International Publications. $3.00.
658. International Atomic Energy Agency, *"Isotope Techniques for Hydrology."* International Publications. $1.00.
659. International Atomic Energy Agency, *"Operating Experience with Power Reactors, Vol. I and II."* International Publications. Vol. I, $10.00; Vol. II, $8.50.
660. International Atomic Energy Agency, *"Radiation Damage in Reactor Materials."* International Publications. $15.00.
661. International Atomic Energy Agency, *"Radioisotope Applications In Industry. A Survey."* International Publications. $2.50.
662. International Atomic Energy Agency, *"Radiological Health & Safety In Mining and Milling of Nuclear Materials, Vol. II."* International Publications. $11.00.
663. International Union of Pure and Applied Chemistry and International Atomic Energy Agency, *"Isotope Mass Effects in Chemistry and Biology."* Butterworth. $14.95.
664. JACCHIA, E., *"Radiation Hazards in Atomic Age."* International Publications, Inc.
665. JACOB, M., and G. F. CHEW, *"Strong Interaction Physics."* Benjamin. $10.00 Cloth.
666. JENKINS, E. N., *"An Introduction to Radioactivity."* Butterworth. $5.95.
667. JOHNS, H. E., *"The Physics of Radiology."* (Revised second printing). Thomas. $8.50.

668. KIADO, AKADEMIAI, *"Radiation Chemistry."* Alkotmany U., Budapest V., Hungary. $14.00.
669. KING, C. D. GREGG, *"Nuclear Power Systems."* Macmillan. $13.00.
670. KIRCHER, J. F., and R. E. BOWMAN (eds.), *"Effects of Radiation on Materials and Components."* Reinhold. $22.50.
671. KNISELY, R. M., W. N. TAUXE, and E. B. ANDERSON (eds.), *"Dynamic Clinical Studies with Radioisotopes."* U. S. Atomic Energy Commission, Washington, D. C. $4.50.
672. KUZIN, A. M., *"Radiation Biochemistry."* Davey & Co. $15.25.
673. LAMBIE, D. A., *"Techniques in the Use of Radioisotopes in Analysis."* Van Nostrand.
674. LANE, A. M., *"Nuclear Theory."* Benjamin. $8.00.
675. LEHNERT, B., *"Dynamics of Charged Particles."* Wiley & Sons. $11.50.
676. LINK, L. E. (ed.), *"Reactor Technology."* U. S. Atomic Energy Commission, Oak Ridge, Tenn. $6.50.
677. MCCABE, C. L., and C. L. BAUER, *"Metals, Atoms & Alloys."* McGraw-Hill. $2.50.
678. MCLAIN, S. and J. H. MARTENS, (eds.), *"Reactor Handbook: Engineering."* Vol. 4. Interscience. $25.40.
679. MCLEAN, F. C., and A. M. BUDY, *"Radiation, Isotopes, and Bone."* Academic Press. $5.95.
680. MARCHUK, G. I., *"Theory and Methods of Nuclear Reactor Calculations."* Consultants Bureau. $40.00.
681. MARTIN, J. (ed.), *"Nuclear Engineering."* Part X. American Institute of Chemical Engineers, New York. $15; $3.00 to AIChE members.
682. MOSES, A. J., *"Nuclear Techniques in Analytical Chemistry."* Macmillan. $6.00.
683. National Academy of Sciences—National Research Council, *"Studies in Penetration of Charged Particles in Matter."* Nuclear Science Series Report No. 29; Publication No. 1133 from the Academy, Washington, D. C. $7.00.
684. National Academy of Sciences, *"Instrumentation Techniques in Nuclear Pulse Analysis."* Printing & Publishing Office, National Academy of Sciences, Washington, D. C. $5.00.
685. National Bureau of Standards, *"Physical Aspects of Irradiation."* Handbook 85. Report 10b. Supt. of Documents, U. S. Government Printing Office, Washington, D. C. $0.70.
686. National Bureau of Standards, *"Safety Standard for Non-Medical X-Ray and Sealed Gamma-Ray Sources."* Handbook 93. Supt. of Documents, U. S. Government Printing Office, Washington, D. C. $0.30.
687. O'BRIEN, R. D., and L. S. WOLFE, *"Radiation, Radioactivity, and Insects."* Academic Press. $5.95.
688. OLFE, D. B. and V. ZAKKAY, *"Supersonic Flow, Chemical Processes and Radioactive Transfer."* Pergamon Press, Macmillan Co. $25.00.
689. OSGOOD, T. H., A. E. RUARK and E. HUTCHISON, *"Atoms, Radiation and Nuclei."* Wiley & Sons. $2.25.
690. PATTEE, H. H., V. E. COSSLETT and A. ENGSTROM (eds.), *"X-Ray Optics and X-Ray Microanalysis."* Academic Press. $22.00.
691. PAVAN, C. (ed.) et al., *"Mammalian Cytogenetics & Related Problems in Radiobiology."* Macmillan Co., Pergamon Press. $15.00.
692. PEDERSEN, E. S., *"Nuclear Energy in Space."* Prentice-Hall. $19.95.
693. PIJCK, J., *"Radiochemistry of Chromium."* Clearinghouse for Federal Scientific and Technical Information, National Bureau of Standards, U. S. Dept. of Commerce, Springfield, Va. $0.75.
694. PRICE, J. W., *"Nuclear Radiation Detection."* McGraw-Hill.
695. QUINN, J. L., *"Scintillation Scanning in Clinical Medicine."* Saunders Co. $11.50.
696. Radiochemical Centre, *"Radioactive Isotope Dilution Analysis."* Radiochemical Centre, Amersham, Buckinghamshire, England. Free of charge.
697. RATNER, B. S., *"Accelerators of Charged Particles."* Macmillan Co. $3.50.
698. RING, F., JR. (ed.), *"Hot Laboratory Operations and Equipment."* The American Society of Mechanical Engineers. $17.50.
699. ROBERTS, J. T., *"Nuclear Magnetic Resonance: Applications to Organic Chemistry."* McGraw-Hill. $6.50.
700. ROMER, A., *"The Discovery of Radioactivity and Transmutation."* Dover. $1.65.
701. RUSSELL, L. E., (ed.) et al., *"Carbides in Nuclear Energy, Vol. I: Physical and Chemical Properties; Phase Diagrams; Vol. II: Preparation, Febrication, Irradiation Behavior."* St. Martin's Press. $40.00.
702. SEABORG, G. T., *"Education and the Atom."* McGraw-Hill. $20.00.
703. SEGRE, E. (ed.), *"Annual Review of Nuclear Science, Vol. 14."* Annual Reviews, Inc., Palo Alto, Calif. $8.50.
704. SHARPE, J., *"Nuclear Radiation Detectors."* Wiley & Sons. $4.95.
705. SHILLING, C. W., *"Atomic Energy Encyclopedia in the Life Sciences."* Saunders.
706. SHUMILOVSKII, N. N., and L. V. MEL'TTSER, *"Radioactive Isotopes in Instrumentation and Control."* Macmillan. $10.00.
707. SIEGBAHN, KAI (ed.), *"Alpha-, Beta-, and Gamma-Ray Spectroscopy."* 2 Vols. North Holland Publishing Co., Amsterdam, The Netherlands. $50.00.
708. SPINKS, J. W. T. and R. J. WOODS, *"An Introduction to Radiation Chemistry."* Wiley. $12.75.
709. TALIAFERRO, W. H., L. G. TALIAFERRO, and B. N. JAROSLOW, *"Radiation and Immune Mechanisms."* Academic Press. $5.95.
710. TAYLOR, D., *"Neutron Irradiation and Activation Analysis."* Newnes, London. $7.00.
711. TOPCHIEV, A. V., *"Radiolysis of Hydrocarbons."* Elsevier. $11.00.
712. UHRIG, R. E. (ed.), *"Noise Analysis in Nuclear Systems."* U. S. Atomic Energy Commission, Oak Ridge, Tenn. $3.75.
713. U. S. Atomic Energy Commission, *"Reactor Handbook."* 2nd Ed. (Vol. IV—Engineering). Wiley. $28.00.
714. UREY, H. C., *"Isotopic and Cosmic Chemistry."* North Holland Publishing Co., Amsterdam, The Netherlands. $15.00.
715. VECCHIO, Alfred Del, *"Concise Dictionary of Atomics."* Philosophical Library. $6.00.

716. VERHAAR, B. J. (ed.), *"Selected Topics in Nuclear Spectroscopy."* Wiley. $12.50.
717. WATT, D. E. and D. RAMSDEN, *"High Sensitivity Counting Techniques."* Pergamon Press.
718. WEINSTEIN, R., A. BOLTAX, and G. LANZA, *"Nuclear Engineering Fundamentals."* McGraw-Hill. $20.00.
719. WERTHEIM, G. T., *"Mössbauer Effect: Principles and Applications."* Academic Press. $2.45.
720. WHITE, H. E., *"Introduction to Atomic Spectra."* McGraw-Hill. $3.95.
721. WOLF, G., *"Isotopes in Biology."* Academic Press. $2.45.
722. ZICHICHI, A. (ed.), *"Elementary Particle Physics: Strong, Electromagnetic and Weak Interactions."* Benjamin, Inc. $4.95 paper, $9.00 cloth.
723. ZINN, W. H., F. K. PITTMAN, and J. F. HOGERTON, *"Nuclear Power, USA."* McGraw-Hill. $20.00.

1965

724. Anonymous, *"Entymology of Radiation Disinfectation of Grain: A Collection of Original Research Papers."* Pergamon Press. $5.00.
725. APANASEVICH, P. A. and V. S. AIZENSHTADT, *"Tables for the Energy & Photon Distribution in Equilibrium Radiation Spectra."* Macmillan Co., Pergamon Press. $12.75.
726. Atomic Industrial Forum, *"Atomforum '64."* International Publications, Inc., UNESCO Publications Center, New York. $15.00 (3 volumes) Purchased Ind. $6.00 per volume.
727. CHADDERTON, L. T., *"Radiation Damage in Crystals."* Wiley. $6.75.
728. DOSTOVISKY, S., *"The Isotopes of Oxygen."* Pergamon Press.
729. EBERT, M. and A. HOWARD (eds.), *"Current Topics in Radiation Research."* North Hollard Publiching Co., Amsterdam. $8.40. (Vol. I).
730. FARLEY, F. J. M., *"Progress in Nuclear Techniques and Instrumentation."* Wiley. $17.50.
731. GOLDWASSER, E. L., *"Optics, Waves, Atoms, and Nuclei: An Introduction."* W. A. Benjamin. $6.00.
732. GREEN, A. E. S. and P. J. WYATT, *"Intorduction to Atomic Physics and Space Science."* Addison-Wesley.
733. HARVEY, B. G., *"Nuclear Chemistry."* Prentice-Hall. $3.95.
734. HODGETTS, J. E., *"Administering the Atom for Peace."* Atherton Press. $6.95.
735. HOPPER, V. D., *"Cosmic Radiation & High Energy Interactions."* Prentice-Hall, Logos Press. $12.95.
736. International Atomic Energy Agency, *"Use of Radioisotopes in Animal Nutrition and Physiology."* International Publications, Inc. $11.50.
737. International Atomic Energy Agency, *"Management of Radioactive Wastes Produced by Radioisotope Users."* International Publications, Inc. $1.00.
738. International Atomic Energy Agency, *"Radiochemical Methods of Analysis."* Vol. 1. International Publications, Inc. $7.00.
739. International Atomic Energy Agency, *"Atomic Energy Review."* Vol. 2, No. 4. International Publications, Inc. $10.00 (Annual subscription).
740. International Atomic Energy Agency, *"Civil Liability for Nuclear Damage."* International Publications, Inc. $10.00.
741. International Atomic Energy Agency, *"Physics and Material Problems of Reactor Control Rods."* International Publications, Inc. $15.00.
742. International Atomic Energy Agency, *"Low-Background High-Efficiency Geiger-Mueller Counter."* Tech. Rept. #33. International Publications, Inc. $1.00.
743. International Atomic Energy Agency, *"Manual for the Operation of Research Reactors."* International Publications, Inc. $4.50.
744. International Atomic Energy Agency, *"Medical Uses of* Ca^{47}." Tech. Rept. #32. International Publications, Inc. $4.00.
745. International Atomic Energy Agency, *"Reactor Shielding."* Tech. Rept. #34. International Publications, Inc. $3.50.
746. International Atomic Energy Agency, *"Training in Radiological Protection: Curricula and Programming."* Tech. Rept. #31. International Publications, Inc. $2.50.
747. International Atomic Energy Agency, *"Medical Radioisotope Scanning."* International Publications, Inc. $11.50—Vol. I; $12.00—Vol. II.
748. International Atomic Energy Agency, *"International Directory of Isotopes."* 3rd Ed. International Publications, Inc. $9.00.
749. KEEPIN, G. R., *"Elements of Nuclear Kinetics."* Addison-Wesley. $12.50.
750. KEEPIN, G. R., *"Physics of Nuclear Kinetics."* Addison-Wesley.
751. KLINEBERG, O. (ed.), *"Social Implications of the Peaceful Uses of Nuclear Energy."* International Publications, Inc. $2.00.
752. KRAYEVSKII, N. A., *"Studies in the Pathology of Radiation Disease."* Pergamon. $15.00.
753. KUBA, J., et al., *"Coincidence Tables for Atomic Spectroscopy."* American Elsevier. $23.50.
754. LANGHAAR, J. W., and L. B. SHAPPERT, (eds.), *"Nuclear Engineering."* Part XIV. American Institute of Chemical Engineers, New York. $3.00 to members; $15.00 to nonmembers.
755. MARON, S. H., and C. F. PRUTTON, *"Principles of Physical Chemistry."* 4th Ed. Macmillan Co. $9.95.
756. MAWSON, C. A., *"Management of Radioactive Wastes."* Van Nostrand. $6.95.
757. MOSES, A. J., *"Nuclear Techniques in Analytical Chemistry."* Macmillan. $6.50.
758. NESMEYANOV, A. N., *"Handbook of Radiochemical Exercises."* Pergamon. $12.00.
759. PARSEGIAN, V. L., *"Industrial Management in the Atomic Age."* Addison-Wesley.
760. RAU, HANS (ed.), *"Dictionary of Nuclear Physics and Nuclear Chemistry."* Reinhold. $8.75.
761. ROGINSKII, S. Z., and S. E. SHNOL, *"Isotopes in Biochemistry."* Davey & Co. $15.25.

762. ROHRLICH, F., *"Classical Charged Particles."* Addison-Wesley. $12.50.
763. ROTHCHILD, S., *"Advances in Tracer Methodology."* Plenum Press. Vol. I, (1962); Vol. II, (1965). $12.00.
764. SPIERS, F. W., and G. W. REED, *"Radiation Dosimetry."* Academic Press. $14.00.
765. STEWART, A. T., *"Perpetual Motion: Electrons & Atoms in Crystals."* Doubleday & Co. $1.25 paper.
766. United Nations, *"Report of the United Nations Scientific Committee on the Effects of Atomic Radiation."* International Publications, Inc. $1.50.
767. VAVILOV, V. S., *"Effects of Radiation on Semiconductors."* Consultants Bureau, including Plenum Press. $15.00.
768. VERESHCHINSKII, I. V., and A. K. PIKAEV, *"Introduction to Radiation Chemistry."* Daniel Davey & Co. $15.25.
769. WANG, C. H., and D. L. WILLIS, *"Radiotracer Methodology in Biological Science."* Prentice-Hall. $16.00.
770. World Health Organization, *"Protection Against Ionizing Radiations."* International Publications, Inc. $2.00.

APPENDIX I Periodicals

ADVANCES IN INORGANIC CHEMISTRY AND RADIOCHEMISTRY—Academic Press, New York
ADVANCES IN NUCLEAR SCIENCE AND TECHNOLOGY—Academic Press, New York
ADVANCES IN RADIATION BIOLOGY—Academic Press, New York
AMERICAN JOURNAL OF PHYSICS—American Institute of Physics, New York
AMERICAN JOURNAL OF ROENTGENOLOGY, RADIUM AND NUCLEAR MEDICINE—C. Thomas, Springfield, Illinois
ANNUAL REVIEW OF NUCLEAR SCIENCE—Annual Reviews, Palo Alto, California
ATOMENERGIE—Frankfurt, Germany
ATOMES—1100 fr. 4 Place de l'Odeon, Paris, France
ATOMICS—Technical Publishing Co., Barrington, Illinois
ATOMICS AND NUCLEAR ENERGY—Leonard Hill Technical Group, London, England
ATOMLIGHT—New England Nuclear Corporation, Boston, Massachusetts
ATOMPRAXIS—G. Braun, Karlsruhe, Germany
BULLETIN OF THE ATOMIC SCIENTISTS—Educational Foundation for Nuclear Science, Inc., Chicago, Illinois
CANADIAN NUCLEAR TECHNOLOGY—MacLean & Hunter, Toronto, Canada
INTERNATIONAL BIBLIOGRAPHY ON ATOMIC ENERGY—United Nations, Department of Security, Council Affairs, Atomic Energy Commission Group, New York
INTERNATIONAL JOURNAL OF APPLIED RADIATION AND ISOTOPES—Pergamon Press, New York, Oxford, London, Paris
INTERNATIONAL JOURNAL OF RADIATION BIOLOGY AND RELATED STUDIES IN PHYSICS, CHEMISTRY AND MEDICINE—Taylor & Francis, Ltd., London, England
ISOTOPE INDEX, THE—Scientific Equipment Co., Indianapolis, Indiana
ISOTOPES AND RADIATION TECHNOLOGY—Superintendent of Documents, Washington, D. C.
JOURNAL OF APPLIED PHYSICS—American Institute of Physics, New York
JOURNAL OF THE BRITISH NUCLEAR ENERGY SOCIETY, THE—London, England
JOURNAL OF INORGANIC AND NUCLEAR CHEMISTRY—Pergamon Press, Ltd., New York
JOURNAL OF LABELLED COMPOUNDS—Presses Académiques Européennes—Brussels, Belgium
JOURNAL OF NUCLEAR MEDICINE—S. N. Turiel & Associates, Inc., 333 North Michigan Ave., Chicago, Illinois
NUCLEAR APPLICATIONS—American Nuclear Society, Hinsdale, Illinois
NUCLEAR ENGINEERING—Temple Press, Ltd., London, England
NUCLEAR NEWS—American Nuclear Society, Hinsdale, Illinois
NUCLEAR SAFETY—Superintendent of Documents, Washington, D. C.
NUCLEAR SCIENCE ABSTRACTS—Superintendent of Documents, Washington, D. C.
NUCLEAR SCIENCE AND ENGINEERING—American Nuclear Society, Hinsdale, Illinois
NUCLEONICS—McGraw-Hill, New York (Publication discontinued June 1967)
OAK RIDGE INSTITUTE OF NUCLEAR STUDIES NEWSLETTER—ORINS, Oak Ridge, Tennessee
PROGRESS IN NUCLEAR ENERGY—Pergamon Press, New York

SERIES
I PHYSICS AND MATHEMATICS
II REACTORS
III PROCESS CHEMISTRY
IV TECHNOLOGY ENGINEERING AND SAFETY
V METALLURGY AND FUELS
VI BIOLOGICAL SCIENCES
VII MEDICAL SCIENCES
VIII ECONOMICS
IX ANALYTICAL CHEMISTRY
X LAW AND ADMINISTRATION
XI PLASMA PHYSICS AND THERMONUCLEAR RESEARCH
XII HEALTH PHYSICS

PROGRESS IN NUCLEAR TECHNIQUES AND INSTRUMENTATION—North-Holland Publishing Company, Amsterdam, The Netherlands
RADIATION BOTANY—Pergamon Press, New York
RADIATION RESEARCH—Academic Press, New York
RADIOCARBON (ANNUAL)—American Journal of Science, Yale University, New Haven, Connecticut
REVIEW OF SCIENTIFIC INSTRUMENTS—American Institute of Physics, New York
REVIEWS OF MODERN PHYSICS—American Institute of Physics, New York
SOVIET RADIOCHEMISTRY (RADIOKHIMIYA)—Translated—Consultants Bureau Enterprises, Inc., New York

APPENDIX J Handbooks of the National Bureau of Standards

Handbooks are available from the Superintendent of Documents, Government Printing Office, Washington 25, D. C.

No.		
48	Control and Removal of Radioactive Contamination in Laboratories	$0.15
49	Recommendations for Waste Disposal of Phosphorus-32 and Iodine-131 for Medical Users	.15
51	Radiological Monitoring Methods and Instruments	.20
53	Recommendations for the Disposal of Carbon-14 Wastes	.15
55	Protection Against Betatron-Synchrotron Radiations up to 100 Million Electron Volts	.25
57	Photographic Dosimetry of X-and Gamma Rays	.15
58	Radioactive-Waste Disposal in the Ocean	.20
59	Permissible Dose from External Sources of Ionizing Radiation	.35
63	Protection Against Neutron Radiation up to 30 Million Electron Volts	.40
64	Design of Free-Air Ionization Chambers	.20
65	Safe Handling of Bodies Containing Radioactive Isotopes	.15
66	Safe Design and Use of Industrial Beta-Ray Sources	.20
69	Maximum Permissible Body Burdens and Maximum Permissible Concentrations of Radionuclides in Air and in Water for Occupational Exposure	.35
72	Measurement of Neutron Flux and Spectra for Physical and Biological Applications	.35
73	Protection Against Radiations from Sealed Gamma Sources	.30
75	Measurement of Absorbed Dose of Neutrons and of Mixtures of Neutrons and Gamma Rays	.35
76	Medical X-Ray Protection up to Three Million Volts	.25
78	Report of the International Commission of Radiological Units and Measurements, 1959	.65
79	Stopping Powers for Use with Cavity Chambers	.35
80	A Manual of Radioactivity Procedures	.50
84	Radiation Quantities and Units (ICRU Report 10a)	.20
85	Physical Aspects of Irradiation (ICRU Report 10b)*	
86	Radioactivity (ICRU Report 10c)*	
87	Clinical Dosimetry (ICRU Report 10d)	.40
88	Radiobiological Dosimetry (ICRU Report 10e)	.25
89	Methods of Evaluating Radiological Equipment and Materials (ICRU Report 10f)	.35
91	Experimental Statistics	4.25
93	Safety Standard for Non-Medical X-Ray and Sealed Gamma-Ray Sources	.30

*In preparation.

APPENDIX K The Wave Equation

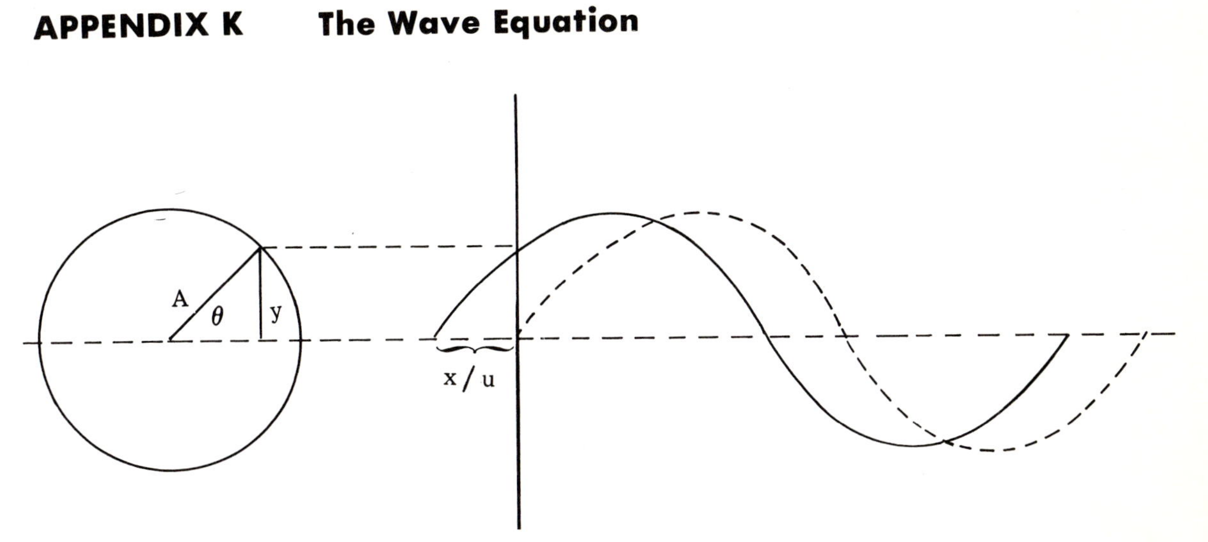

Simple harmonic motion can be represented by a sine function of the type

$$y = A \sin \theta = A \sin \omega t \tag{K-1}$$

where y is the vertical displacement and A is the amplitude. If the period (i.e., time for a complete revolution) be represented by τ, then angular velocity $\omega = 2\pi/\tau$, and the frequency $\nu = 1/\tau$, and

$$y = A \sin \frac{2\pi}{\tau} t = A \sin 2\pi\nu t \tag{K-2}$$

The time interval between two waves, measured as the quotient of the distance x and the velocity u, is x/u. Then the relation

$$y = A \sin 2\pi\nu \left(t - \frac{x}{u}\right) \tag{K-3}$$

accounts for all waves since zero time is arbitrary, i.e., the wave has no beginning or end. Equation K-3 is then modified, by use of the wave number $\tau = 1/\lambda$, to

$$y = A \sin 2\pi(\nu t - \nu x) \tag{K-4}$$

Differentiation of equation K-4 with respect to x gives

$$\frac{\partial y}{\partial x} = A \cos (2\pi\nu t - 2\pi\bar{\nu}x)(-2\pi\bar{\nu})$$

$$\frac{\partial^2 y}{\partial x^2} = -A \sin (2\pi\nu t - 2\pi\bar{\nu}x)(-2\pi\bar{\nu})(-2\pi\bar{\nu})$$

$$= -4\pi^2\nu^2 A \sin 2\pi(\nu t - \bar{\nu}x) \tag{K-5}$$

Substitution of equation K-4 into K-5 gives

$$\frac{\partial^2 y}{\partial x^2} = -4\pi^2\nu^2 y \tag{K-6}$$

Differentiation of equation K-4 with respect to t gives

$$\frac{\partial y}{\partial t} = A\cos(2\pi\nu t - 2\pi\bar{\nu}x)(-2\pi\nu)$$

$$\frac{\partial^2 y}{\partial t^2} = -A\sin(2\pi\nu t - 2\pi\bar{\nu}x)(-2\pi\nu)(-2\pi\nu)$$

$$= -4\pi^2\nu^2 A\sin 2\pi(\nu t - \bar{\nu}x) \tag{K-7}$$

Substitution of equation K-4 into K-7 gives

$$\frac{\partial^2 y}{\partial t^2} = -4\pi^2\nu^2 y \tag{K-8}$$

Division of equation K-6 by K-8 yields the wave function for one dimension:

$$\frac{\partial^2 y}{\partial x^2} - \frac{1}{u^2}\frac{\partial^2 y}{\partial t^2} = 0 \tag{K-9}$$

It will be noted that equations K-9 and 1.61 are identical.

INDEX